Lecture Notes in Physics

Volume 936

The Lecture Notes in Physics

The series Lecture Notes in Physics (LNP), founded in 1969, reports new developments in physics research and teaching-quickly and informally, but with a high quality and the explicit aim to summarize and communicate current knowledge in an accessible way. Books published in this series are conceived as bridging material between advanced graduate textbooks and the forefront of research and to serve three purposes:

- to be a compact and modern up-to-date source of reference on a well-defined topic
- to serve as an accessible introduction to the field to postgraduate students and nonspecialist researchers from related areas
- to be a source of advanced teaching material for specialized seminars, courses and schools

Both monographs and multi-author volumes will be considered for publication. Edited volumes should, however, consist of a very limited number of contributions only. Proceedings will not be considered for LNP.

Volumes published in LNP are disseminated both in print and in electronic formats, the electronic archive being available at springerlink.com. The series content is indexed, abstracted and referenced by many abstracting and information services, bibliographic networks, subscription agencies, library networks, and consortia.

Proposals should be sent to a member of the Editorial Board, or directly to the managing editor at Springer:

Christian Caron
Springer Heidelberg
Physics Editorial Department I
Tiergartenstrasse 17
69121 Heidelberg/Germany
christian.caron@springer.com

More information about this series at http://www.springer.com/series/5304

Morten Hjorth-Jensen • Maria Paola Lombardo •
Ubirajara van Kolck
Editors

An Advanced Course in Computational Nuclear Physics

Bridging the Scales from Quarks to Neutron Stars

Springer

Editors
Morten Hjorth-Jensen
National Superconducting Cyclotron Laboratory and Department of Physics and Astronomy
Michigan State University
East Lansing, Michigan
USA

Department of Physics
University of Oslo
Oslo, Norway

Maria Paola Lombardo
INFN, Laboratori Nazionali di Frascati
Frascati Roma, Italy

Ubirajara van Kolck
Department of Physics
University of Arizona
Tucson, AZ
USA

ISSN 0075-8450 ISSN 1616-6361 (electronic)
Lecture Notes in Physics
ISBN 978-3-319-53335-3 ISBN 978-3-319-53336-0 (eBook)
DOI 10.1007/978-3-319-53336-0

Library of Congress Control Number: 2017937055

Printed on acid-free paper

This Springer imprint is published by Springer Nature
The registered company is Springer International Publishing AG
The registered company address is: Gewerbestrasse 11, 6330 Cham, Switzerland

Preface

This graduate-level text collects and synthesizes ten series of lectures on the nuclear quantum many-body problem—starting from our present understanding of the underlying forces with a presentation of recent advances within the field of lattice quantum chromodynamics via effective field theories to central many-body methods like Monte Carlo methods, coupled cluster theories, the similarity renormalization group approach, Green's function methods, and large-scale diagonalization approaches.

In particular algorithmic and computational advances show promise for breakthroughs in predictive power including proper error estimates and a better understanding of the underlying effective degrees of freedom and of the respective forces at play.

Enabled by recent advances in theoretical, experimental, and numerical techniques, the modern and state-of-the art applications considered in this volume span the entire range from our smallest components, quarks and gluons, as the mediators of the strong force to the computation of the equation of state for neutron star matter.

The present lectures provide a proper exposition of the underlying theoretical and algorithmic approaches as well as strong ties to the numerical implementation of the exposed methods. Several of the lectures provide links to actual numerical software and benchmark calculations, allowing eventual readers, based upon the available material, to develop their own programs for tackling challenging nuclear many-body problems.

East Lansing, MI, USA and Oslo, Norway — Morten Hjorth-Jensen
Frascati Roma, Italy — Maria Paola Lombardo
Tucson, AZ, USA and Paris, France — Ubirajara van Kolck

Contents

Contributors

Carlo Barbieri Department of Physics, University of Surrey, Guildford, UK

Scott K. Bogner Department of Physics and Astronomy and National Superconducting Cyclotron Laboratory, Michigan State University, East Lansing, MI, USA

Arianna Carbone Institut für Kernphysik, Technische Universität Darmstadt, Darmstadt, Germany

ExtreMe Matter Institute EMMI, GSI Helmholtzzentrum für Schwerionenforschung GmbH, Darmstadt, Germany

Gaute Hagen Physics Division, Oak Ridge National Laboratory, Oak Ridge, TN, USA

Department of Physics and Astronomy, University of Tennessee, Knoxville, TN, USA

Hans-Werner Hammer Institut für Kernphysik, Technische Universität Darmstadt, Darmstadt, Germany

Tetsuo Hatsuda Nishina Center, RIKEN, Saitama, Japan

Heiko Hergert Department of Physics and Astronomy and National Superconducting Cyclotron Laboratory, Michigan State University, East Lansing, MI, USA

Morten Hjorth-Jensen National Superconducting Cyclotron Laboratory and Department of Physics and Astronomy, Michigan State University, East Lansing, Michigan, USA

Department of Physics, University of Oslo, Oslo, Norway

Gustav R. Jansen National Center for Computational Sciences and Physics Division, Oak Ridge National Laboratory, Oak Ridge, TN, USA

Sebastian König Department of Physics, The Ohio State University, Columbus, OH, USA

Dean Lee Department of Physics, North Carolina State University, Raleigh, NC, USA

Justin G. Lietz Department of Physics and Astronomy and National Superconducting Cyclotron Laboratory, Michigan State University, East Lansing, MI, USA

Maria Paola Lombardo INFN, Laboratori Nazionali di Frascati, Frascati, Italy

Titus D. Morris Department of Physics and Astronomy, University of Tennessee, Knoxville, TN, USA

Physics Division, Oak Ridge National Laboratory, Oak Ridge, TN, USA

Amy Nicholson Department of Physics, University of California, Berkeley, Berkeley, CA, USA

Samuel J. Novario Department of Physics and Astronomy and National Superconducting Cyclotron Laboratory, Michigan State University, East Lansing, MI, USA

Giuseppina Orlandini Dipartimento di Fisica, Università di Trento, Trento, Italy

Trento Institute for Fundamental Physics and Applications - I.N.F.N, Trento, Italy

Nathan M. Parzuchowski Department of Physics and Astronomy and National Superconducting Cyclotron Laboratory, Michigan State University, East Lansing, MI, USA

Francesco Pederiva Physics Department, University of Trento, and INFN-TIFPA, Trento, Italy

Alessandro Roggero Institute for Nuclear Theory, University of Washington, Seattle, WA, USA

Thomas Schäfer Department of Physics, North Carolina State University, Raleigh, NC, USA

Kevin E. Schmidt Department of Physics and Astronomy, Arizona State University, Tempe, AZ, USA

Ubirajara van Kolck Institut de Physique Nucleaire, Orsay, France

Department of Physics, University of Arizona, Tucson, AZ, USA

Fei Yuan Department of Physics and Astronomy and National Superconducting Cyclotron Laboratory, Michigan State University, East Lansing, MI, USA

Chapter 1
Motivation and Overarching Aims

Morten Hjorth-Jensen, Maria Paola Lombardo, and Ubirajara van Kolck

Nuclear physics has recently experienced several discoveries and technological advances that address the fundamental questions of the field, in particular how nuclei emerge from the strong dynamics of quantum chromodynamics (QCD). Many of these advances have been made possible by significant investments in frontier research facilities worldwide over the last two decades. Some of these discoveries are the detection of perhaps the most exotic state of matter, the quark-gluon plasma, which is believed to have existed in the very first moments of the Universe. Recent experiments have validated the standard solar model and established that neutrinos have mass. High-precision measurements of the quark structure of the nucleon are challenging existing theoretical understanding. Nuclear physicists have started to explore a completely unknown landscape of nuclei with extreme neutron-to-proton ratios using radioactive and short-lived ions, including rare and very neutron-rich isotopes. These experiments push us towards the extremes of nuclear stability. Moreover, these rare nuclei lie at the heart of nucleosynthesis processes in the Universe and are therefore an important component in the puzzle of matter generation in the Universe.

M. Hjorth-Jensen (✉)
National Superconducting Cyclotron Laboratory and Department of Physics and Astronomy, Michigan State University, East Lansing, Michigan, USA

Department of Physics, University of Oslo, Oslo, Norway
e-mail: hjensen@msu.edu

M.P. Lombardo
INFN, Laboratori Nazionali di Frascati, Frascati, Italy
e-mail: mariapaola.lombardo@lnf.infn.it

U. van Kolck
Institut de Physique Nucleaire, Orsay, France

Department of Physics, University of Arizona, Tucson, AZ, USA
e-mail: vankolck@ipno.in2p3.fr

M. Hjorth-Jensen et al. (eds.), *An Advanced Course in Computational Nuclear Physics*, Lecture Notes in Physics 936, DOI 10.1007/978-3-319-53336-0_1

A firm experimental and theoretical understanding of nuclear stability in terms of the basic constituents is a huge intellectual endeavor. Experiments indicate that developing a comprehensive description of all nuclei and their reactions requires theoretical and experimental investigations of rare isotopes with unusual neutron-to-proton ratios that are very different from their stable counterparts. These rare nuclei are difficult to produce and study experimentally since they can have extremely short lifetimes. To study theoretically these nuclear systems entails being able to solve a complicated quantum-mechanical many-body problem in order to address important issues such as whether we can explain from first-principle methods the existence of magic numbers and their eventual vanishing with increasing neutron numbers, how the binding energy of neutron-rich nuclei behaves, or the radii, neutron skins, and many other probes that extract information about many-body correlations as nuclei evolve towards their limits of stability. These are all fundamental questions which, combined with recent experimental and theoretical advances, will allow us to advance our basic knowledge about the limits of stability of matter, and, hopefully, help us in gaining a better understanding of visible matter.

Accompanying the experimental developments, a qualitative change has swept the nuclear theory landscape thanks to a combination of techniques that are allowing, for the first time, to construct links between QCD and the nuclear many-body problem. This transformation has been brought by a dramatic improvement in the capability of numerical calculations both in QCD, via lattice simulations, and in the nuclear many-body problem via first principle or ab initio many-body methods that employ non-relativistic Hamiltonians. Simultaneously, effective field theories attempt at building a bridge between the two numerical approaches, allowing to convert the results of lattice QCD into input Hamiltonians that can be used in ab initio methods. Furthermore, algorithmic and computational advances hold promise for breakthroughs in predictive power including proper error estimates, enhancing the already strong ties between theory and experiment. These advances include better ab initio many-body methods as well as a better understanding of the underlying effective degrees of freedom and the respective forces at play. Similarly, we have recently witnessed a significant improvement in numerical algorithms and high-performance computing. This provides us with important new insights about the stability of nuclear matter and allows us to relate these novel understandings to the underlying laws of motion, the corresponding forces and the pertinent fundamental building blocks of nuclear matter.

It is within this framework the present set of lectures finds its rationale. This text collects and synthesizes ten series of lectures on the nuclear many-body problem, starting from our present understanding of the underlying forces with a presentation of recent advances within the field of lattice QCD, via effective field theories to central many-body methods like various Monte Carlo approaches, coupled-cluster theory, the similarity renormalization group approach, Green's function methods and large-scale diagonalization methods. The applications span from our smallest components, quarks and gluons as the mediators of the strong force to the computation of the equation of state for infinite nuclear matter and neutron star matter. The lectures provide a proper exposition of the underlying theoretical and algorithmic approaches as well as strong ties to the numerical implementation of the exposed methods.

The next chapter, by Thomas Schäfer, aims at a brief introduction to quantum chromodynamics (QCD), the QCD phase diagram, and non-equilibrium phenomena in QCD. This chapter emphasizes the aspects of the theory that can be addressed using computational methods. In Chap. 3, Tetsuo Hatsuda presents several basic concepts and applications of lattice quantum chromodynamics (LQCD), ending the chapter by presenting recent LQCD results on baryon-baryon interactions. These results are extremely promising since they allow for a better understanding of the links between QCD and effective field theories with say nucleons and pions only. The latter provide the necessary degrees of freedom and inputs for defining the nuclear Hamiltonians that enter the solution of the various nuclear many-body methods discussed in Chaps. 7 through 11. Chapter 4 by Hans-Werner Hammer and Sebastian König presents the general theoretical aspects of nuclear effective field theories. In the two subsequent chapters by Amy Nicholson and Dean Lee, the authors apply lattice techniques to nuclear effective field theories involving nucleons and pions as the basic degrees of freedom. These authors give a detailed exposure, with exercises and numerical codes, of lattice techniques applied to effective field theory, explaining the theory and algorithms relevant to lattice simulations of nuclear few- and many-body systems. Chapter 7 by Giuseppina Orlandini gives an overview of several ab initio approaches currently used to study nuclear structure properties and reactions. Chapter 8 by Justin Lietz, Sam Novario et al. introduces a computational approach to infinite nuclear matter employing Hartree-Fock theory, many-body perturbation theory and coupled cluster theory, with an extensive discussion of computational topics. Many of these basic ingredients are used in the next three chapters. Chapter 9 by Francesco Pederiva, Alessandro Roggero and Kevin Schmidt reviews Quantum Monte Carlo methods for solving the many-body Schrödinger equation for an arbitrary Hamiltonian ending the discussion with the newly developed Configuration Interaction Monte Carlo algorithm. Comparisons are made with coupled cluster theory for the equation of state of infinite neutron star matter from Chap. 8. Chapter 10 by Heiko Hergert et al. presents applications of the In-Medium Similarity Renormalization Group method to studies of infinite nuclear matter. Finally Chap. 11, by Carlo Barbieri and Arianna Carbone, presents the fundamental techniques and working equations of many-body Green's function theory for calculating ground state properties and the spectral strength, with applications to infinite neutron star matter and comparisons with the results from Chaps. 8–10.

The first five chapters are thus meant to expose the reader to the most recent developments in our understanding of the strong interaction, linking QCD with effective field theories. With the appropriate and pertinent effective degrees of freedom we can in turn define various effective non-relativistic Hamiltonians and embark on our studies of widely used methods for solving the non-relativistic Schrödinger equation. Spanning from Monte Carlo methods to various wave function based methods like full configuration interaction theory, coupled cluster theory, similarity renormalization group approaches and Green's function theory,

Chaps. 7 through 11 aim at presenting these methods to the reader, with applications to infinite nuclear matter in Chaps. 8 through 11. Studies of infinite matter play a central role in nuclear physics. The determination of for example the equation of state (EoS), which is intimately linked with our capability to solve the nuclear many-body problem, has important consequences for neutron star properties like the mass range, the mass-radius relationship, the thickness of the crust and the rate by which a neutron star cools down over time. The EoS is also an important ingredient in studies of the energy release in supernova explosions. Infinite matter offers also several technical simplifications to the many-body problem compared with finite nuclei, as discussed in Chaps. 8 through 11. In these chapters we provide benchmark calculations and compare different many-body methods using a simplified model for the nuclear forces. However, the formalism and codes we present can easily be extended to include interaction models based on effective field theories, as well as other systems, spanning from the homogeneous electron gas in two and three dimensions to finite systems like nuclei. The various chapters propose exercises meant to deepen the theoretical concepts that are discussed. Actual numerical software allows the reader to build upon the theoretical concepts and develop her/his own insights about these methods. These codes can serve as a starting point for developing own programs for tackling complicated many-body problems. Proper benchmarks for the various programs are also provided, allowing thereby potential readers and users to check the correctness, installation and compilation of the various programs. All codes are properly linked and available via the github link https://github.com/ManyBodyPhysics/LectureNotesPhysics/tree/master/Programs.

Acknowledgements The different chapters are based on lectures given at the Doctoral Training program *Computational Nuclear Physics - Hadrons, Nuclei and Dense Matter* held at the European Center for Theoretical Nuclear Physics and Related Areas (ECT*) in Trento, Italy, from April 13 to May 22, 2015, the Nuclear Talent course *Many-body methods for nuclear physics*, held at Grand Accelerateur National d'Ions Lourds (GANIL), Caen, France, from July 5 to July 25, 2015, and the Nuclear Talent course *High-performance computing and computational tools for nuclear physics* held at North Carolina State University from July 11 to July 29, 2016. For more information about the Nuclear Talent courses see http://www.nucleartalent.org. For the Doctoral Training program of the ECT*, see http://www.ectstar.eu/node/799. The support for organizing these series of lectures from the ECT*, GANIL and the University of Basse Normadie at Caen, North Carolina State University, Los Alamos National Laboratory and Michigan State University is greatly acknowledged. Two of the lectures (Chaps. 8 and 10) are co-authored by graduate students (from Michigan State University) who attended the abovementioned Nuclear Talent courses.

The work of Morten Hjorth-Jensen is supported by NSF Grant No. PHY-1404159 (Michigan State University).

The work of Ubirajara van Kolck is supported in part by the European Union Research and Innovation program Horizon 2020 under grant number 654002 and by the DOE, Office of Science, Office of Nuclear Physics, under award number DE-FG02-04ER41338.

Chapter 2
Quantum Chromodynamics

Thomas Schäfer

2.1 Introduction

The goal of this chapter is to provide a brief summary of Quantum Chromodynamics (QCD) and the QCD phase diagram, and to give an introduction to computational methods that are being used to study different aspects of QCD. Quantum Chromodynamics is a remarkable theory in many respects. QCD is an almost parameter free theory. Indeed, in the context of nuclear physics QCD is completely characterized by the masses of the up, down, and strange quark, and a reasonable caricature of nuclear physics emerges in the even simpler case in which the up and down quark are taken to be massless, and the strange quark is infinitely heavy. QCD nevertheless accounts for the incredible richness of the phase diagram of strongly interacting matter. QCD describes finite nuclei, normal and superfluid states of nuclear matter, color superconductors, hadronic gases, quark gluon plasma, and many other states. This rich variety of states is reflected in the large number of computational methods that have been brought to bear on problems in QCD. This includes a large number of methods for the structure and excitations of finite Fermi systems, quantum Monte Carlo methods, and a variety of tools for equilibrium and non-equilibrium statistical mechanics.

The bulk of this book is devoted to the study of few and many nucleon systems. Summarizing everything else in one brief chapter is obviously out of the question, both because of limitations of space and because of my limited expertise. I will therefore be very selective, and focus on a number of very simple yet powerful ideas. This reflects, in part, my background, which is not primarily in computational physics. It also reflects my conviction that progress in computational physics is

---

T. Schäfer (✉)
Department of Physics, North Carolina State University, Raleigh, NC 27695, USA
e-mail: tmschaef@ncsu.edu

M. Hjorth-Jensen et al. (eds.), *An Advanced Course in Computational Nuclear Physics*, Lecture Notes in Physics 936, DOI 10.1007/978-3-319-53336-0_2

unfortunately often reflected in increasingly complicated codes that obscure the simplicity of the underlying methods.

2.2 Path Integrals and the Metropolis Algorithm

Consider a simple quantum mechanical problem, the motion of a particle in a one-dimensional potential. In order to be specific I will focus on the double well potential $V(x) = \lambda(x^2 - \eta^2)^2$, where η and λ are parameters. The Hamiltonian is

$$H = \frac{p^2}{2m} + \lambda(x^2 - \eta^2)^2 . \tag{2.1}$$

Using a change of variables I can set $2m = \lambda = 1$. This implies that there is only one physical parameter in this problem, the barrier separation η. The regime $\eta \gg 1$ corresponds to the limit in which the system has two almost degenerate minima that are split by semi-classical tunneling events. The energy eigenstates and wave functions are solutions of the eigenvalue problem $H|n\rangle = |n\rangle E_n$. Once the eigenstates are known I can compute all possible correlation functions

$$\Pi_n(t_1, t_2, \ldots, t_n) = \langle 0|x(t_1)x(t_2)\ldots x(t_n)|0\rangle , \tag{2.2}$$

by inserting complete sets of states. An alternative to the Hamiltonian formulation of the problem is the Feynman path integral [1]. The path integral for the anharmonic oscillator is given by

$$\langle x_1|e^{-iHt_f}|x_0\rangle = \int_{x(0)=x_0}^{x(t_f)=x_1} \mathscr{D}x\, e^{iS}, \qquad S = \int_0^{t_f} dt \left(\frac{1}{4}\dot{x}^4 - (x^2 - \eta^2)^2\right). \tag{2.3}$$

This expression contains a rapidly oscillating phase factor e^{iS}, which prohibits any direct numerical attempt at computing the path integral. The standard approach is based on analytic continuation to imaginary time $\tau = it$. This is also referred to as Euclidean time, because the Minkowski interval $dx^2 - dt^2$ turns into the Euclidean expression $dx^2 + d\tau^2$. In the following I will consider the euclidean partition function

$$Z(T) = \int \mathscr{D}x\, e^{-S_E}, \qquad S_E = \int_0^{\beta} d\tau \left(\frac{1}{4}\dot{x}^4 + (x^2 - \eta^2)^2\right), \tag{2.4}$$

where $\beta = 1/T$ is the inverse temperature and we assume periodic boundary conditions $x(0) = x(\beta)$. To see that Eq. (2.4) is indeed the partition function we can use Eq. (2.3) to express the path integral in terms of the eigenvalues of the Hamiltonian, $Z(T) = \sum_n \exp(-E_n/T)$. In the following I will describe numerical simulations using a discretized version of the euclidean action. For this purpose I discretize the euclidean time coordinate $\tau_j = ja$, $i = 1, \ldots n$ where $a = \beta/n$ is the length of time interval. The discretized action is given by

$$S = \sum_{i=1}^{n} \left\{ \frac{1}{4a}(x_i - x_{i-1})^2 + a(x_i^2 - \eta^2)^2 \right\}, \tag{2.5}$$

where $x_i = x(\tau_i)$. I consider periodic boundary conditions $x_0 = x_n$. The discretized euclidean path integral is formally equivalent to the partition function of a statistical system of (continuous) "spins" x_i arranged on a one-dimensional lattice. This statistical system can be studied using standard Monte-Carlo sampling methods. In the following I will use the Metropolis algorithm [2]. Detailed numerical studies of the Euclidean path integral can be found in [3–6].

The Metropolis method generates an ensemble of configurations $\{x_i\}^{(k)}$ where $i = 1, \ldots, n$ labels the lattice points and $k = 1, \ldots, N_{conf}$ labels the configurations. Quantum mechanical averages are computed by averaging observables over many configurations,

$$\langle \mathcal{O} \rangle = \lim_{N_{conf} \to \infty} \frac{1}{N_{conf}} \sum_{k=1}^{N_{conf}} \mathcal{O}^{(k)} \tag{2.6}$$

where $\mathcal{O}^{(k)}$ is the value of the classical observable $\mathcal{O}$ in the configuration $\{x_i\}^{(k)}$. The configurations are generated using Metropolis updates $\{x_i\}^{(k)} \to \{x_i\}^{(k+1)}$. The update consists of a sweep through the lattice during which a trial update $x_i^{(k+1)} = x_i^{(k)} + \delta x$ is performed for every lattice site. Here, δx is a random number. The trial update is accepted with probability

$$P\left(x_i^{(k)} \to x_i^{(k+1)}\right) = \min\left\{\exp(-\Delta S), 1\right\}, \tag{2.7}$$

where ΔS is the change in the action Eq. (2.5). This ensures that the configurations $\{x_i\}^{(k)}$ are distributed according the "Boltzmann" distribution $\exp(-S)$. The distribution of δx is arbitrary as long as the trial update is micro-reversible, i. e. is equally likely to change $x_i^{(k)}$ to $x_i^{(k+1)}$ and back. The initial configuration is arbitrary. In order to study equilibration it is useful to compare an ordered (cold) start with $\{x_i\}^{(0)} = \{\eta\}$ to a disordered (hot) start $\{x_i\}^{(0)} = \{r_i\}$, where r_i is a random variable.

The advantage of the Metropolis algorithm is its simplicity and robustness. The only parameter to adjust is the distribution of δx. A simple choice is to take δx to be a Gaussian random number, and choose the width of the distribution so that the average acceptance rate for the trial updates is around 50%. Fluctuations of $\mathcal{O}$ provide an estimate in the error of $\langle \mathcal{O} \rangle$. The uncertainty is given by

$$\Delta\langle \mathcal{O} \rangle = \sqrt{\frac{\langle \mathcal{O}^2 \rangle - \langle \mathcal{O} \rangle^2}{N_{conf}}}. \tag{2.8}$$

This requires some care, because the error estimate is based on the assumption that the configurations are statistically independent. In practice this can be monitored by computing the auto-correlation "time" in successive measurements $\mathcal{O}(\{x_i\}^{(k)})$.

I have written a simple fortran code that implements the Metropolis algorithm for Euclidean path integrals [6]. The most important part of that code is a sweep through the lattice with a Metropolis update on every site τ_j:

```
do j=1,n-1

        nhit = nhit+1

        xpm = (x(j)-x(j-1))/a
        xpp = (x(j+1)-x(j))/a
        t = 1.0/4.0*(xpm**2+xpp**2)
        v = (x(j)**2-f**2)**2
        sold = a*(t+v)

        xnew = x(j) + delx*(2.0*ran2(iseed)-1.0)

        xpm = (xnew-x(j-1))/a
        xpp = (x(j+1)-xnew)/a
        t = 1.0/4.0*(xpm**2+xpp**2)
        v = (xnew**2-f**2)**2
        snew = a*(t+v)
        dels = snew-sold

        p = ran2(iseed)
        if (exp(-dels) .gt. p) then
           x(j) = xnew
           nacc = nacc + 1
        endif

enddo
```

Here, `sold` is the local action corresponding to the initial value of `x(j)`, and `snew` is the action after the trial update. The trial update is accepted if `exp(-dels)` is greater that the random variable `p`. The function `ran2()` generates a random number between 0 and 1, and `nacc/nhit` measures the acceptance rate. A typical path is shown in Fig. 2.1. An important feature of the paths in the double well potential is the presence of tunneling events. Indeed, in the semi-classical regime $\eta \gg 1$, a typical path can be understood as Gaussian fluctuations superimposed on a series of tunneling events (instantons).

The path integral method does not provide direct access to the eigenvalues of the Hamiltonian, but it can be used to compute imaginary time correlation functions

$$\Pi_n^E(\tau_1,\ldots,\tau_n) = \langle x(\tau_1)\ldots x(\tau_n)\rangle. \tag{2.9}$$

Note that the average is carried out with respect to the partition function in Eq. (2.4). In the limit $\beta \to \infty$ this corresponds to the ground state expectation value. A very

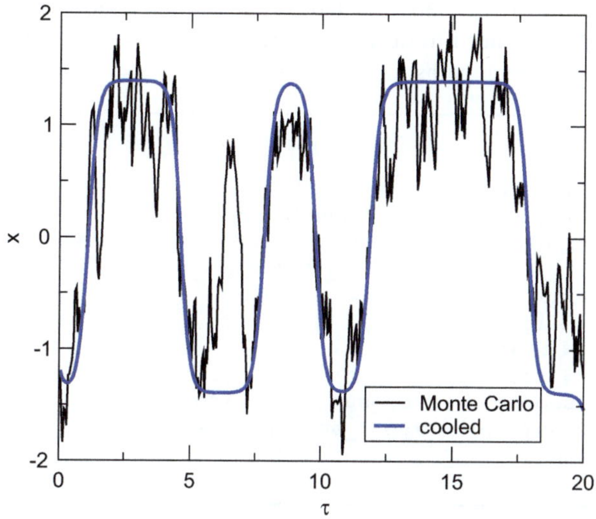

Fig. 2.1 Typical euclidean path obtained in a Monte Carlo simulation of the discretized euclidean action of the double well potential for $\eta = 1.4$. The lattice spacing in the euclidean time direction is $a = 0.05$ and the total number of lattice points is $N_\tau = 800$. The *blue curve* shows the corresponding smooth path obtained by running 100 cooling sweeps on the original path

important observable is the two-point function $\Pi^E(\tau) \equiv \Pi_2^E(0,\tau)$. The euclidean correlation functions is related to the eigenstates of the Hamiltonian via a spectral representations. This representation is obtained by inserting a complete set of states into Eq. (2.9). The result is

$$\Pi^E(\tau) = \sum_n |\langle 0|x|n\rangle|^2 \exp(-(E_n - E_0)\tau), \tag{2.10}$$

where E_n is the energy of the state $|n\rangle$. This can be written as

$$\Pi^E(\tau) = \int dE\, \rho(E) \exp(-(E - E_0)\tau), \tag{2.11}$$

where $\rho(E)$ is the spectral function. In the case of the double well potential there are only bound states and the spectral function is a sum of delta-functions. Equation (2.10) shows that the euclidean correlation function is easy to construct once the energy eigenvalues and eigenfunctions are known. The inverse problem is well defined in principle, but numerically much more difficult. The excitation energy of the first excited state $\Delta E_1 = E_1 - E_0$ is easy to extract from the exponential decay of the two-point functions, but higher states are more difficult to compute. A technique for determining the spectral function from Euclidean correlation functions is the maximum entropy image reconstruction method, see [7, 8].

The calculation of correlation functions in a Monte Carlo simulation is very straightforward. All I need to do is multiply the values of $x(\tau_i)$ for a given path, and then average over all paths:

```
do ic=1,nc

     ncor = ncor + 1
     ip0  = int( (n-np)*ran2(iseed) )
     x0  = x(ip0)

     do ip=1,np
        x1  = x(ip0+ip)
        xcor  = x0*x1
        x2cor= xcor**2
        xcor_sum(ip)  = xcor_sum(ip)  + xcor
        xcor2_sum(ip)  = xcor2_sum(ip)  + xcor**2
   enddo
enddo
```

The advantages of this method are that it is extremely robust, that it requires no knowledge (or preconceived notion) of what the wave function looks like, and that it can explore a very complicated configuration space. On the other hand, in the case of one-dimensional quantum mechanics, the Metropolis method is very inefficient. Using direct diagonalization in a finite basis it is not difficult to compute the energies of the first several states in the potential in Eq. (2.1) with very high accuracy, $\Delta E/E_0 \sim O(10^{-6})$ or better. On the other hand, using the Monte Carlo method, it is quite difficult to achieve an accuracy of $O(10^{-2})$ for observable other than $(E_1 - E_0)/E_0$. The advantage of the Monte Carlo method is that the computational cost scales much more favorably in high dimensional systems, such as quantum mechanics of many particles, or quantum field theory.

The Monte Carlo method also does not directly provide the ground state energy, or the partition function and free energy at finite temperature. In quantum mechanics we can compute the ground state energy from the expectation value of the Hamiltonian $\langle H\rangle = \langle T + V\rangle$ in the limit $\beta \to \infty$. The expectation value of the kinetic energy is singular as $a \to 0$, but this problem can be overcome by using the Virial theorem

$$\langle H\rangle = \left\langle \frac{x}{2}V' + V\right\rangle . \tag{2.12}$$

There is no simple analog of this method in quantum field theory. A method for computing the free energy which does generalize to quantum field theory is the adiabatic switching technique. The idea is to start from a reference system for which the free energy is known and calculate the free energy difference to the real system using Monte Carlo methods. For this purpose I write the action as

$$S_\alpha = S_0 + \alpha \Delta S\,, \tag{2.13}$$

where S_0 is the action of the reference system, ΔS is defined by $\Delta S = S - S_0$ where S is the full action, and α can be viewed as a coupling constant. The action S_α interpolates between the physical system for $\alpha = 1$ and the reference system for $\alpha = 0$. Integrating the relation $\partial \log Z(\alpha)/(\partial\alpha) = -\langle \Delta S \rangle_\alpha$ I find

$$\log(Z(\alpha = 1)) = \log(Z(\alpha = 0)) - \int_0^1 d\alpha' \, \langle \Delta S \rangle_{\alpha'} \,, \tag{2.14}$$

where $\langle . \rangle_\alpha$ is computed using the action S_α. In the case of the anharmonic oscillator it is natural to use the harmonic oscillator as a reference system. In that case the reference partition function is

$$Z(\alpha = 0) = \sum_n \exp(-\beta E_n^0) = \frac{\exp(-\beta\omega_0/2)}{1 - \exp(-\beta\omega_0)}, \tag{2.15}$$

where ω_0 is the oscillator constant. Note that the free energy $F = T \log(Z)$ of the anharmonic oscillator should be independent of the reference frequency ω_0. The integral over the coupling constant α can be calculated in a Monte Carlo simulation by slowly changing α from 0 to 1 during the simulation. Free energy calculations of this type play an important role in quantum chemistry, and more efficient methods for determining ΔF have been developed [9].

2.3 Quantum Chromodynamics

2.3.1 *QCD at Zero Temperature and Density*

The rich phenomenology of strong interacting matter is encoded in a deceptively simple Lagrangian. The fundamental fields in the Lagrangian are quark fields $q^c_{\alpha f}$ and gluon fields A^a_μ. Here, $\alpha = 1, \ldots, 4$ is a Dirac spinor index, $c = 1, \ldots, N_c$ with $N_c = 3$ is a color index, and $f = up, down, strange, charm, bottom, top$ is a flavor index. Interactions in QCD are governed by the color degrees of freedom. The gluon field A^a_μ is a vector field labeled by an index $a = 1, \ldots, N_c^2 - 1$ in the adjoint representation. The N_c^2- multiplet of gluon fields can be used to construct a matrix valued field $A_\mu = A^a_\mu \frac{\lambda^a}{2}$, where λ^a is a set of traceless, Hermitian, $N_c \times N_c$ matrices. The QCD Lagrangian is

$$\mathscr{L} = -\frac{1}{4} G^a_{\mu\nu} G^a_{\mu\nu} + \sum_f^{N_f} \bar{q}_f (i\gamma^\mu D_\mu - m_f) q_f \,, \tag{2.16}$$

where $G^a_{\mu\nu}$ is the QCD field strength tensor defined by

$$G^a_{\mu\nu} = \partial_\mu A^a_\nu - \partial_\nu A^a_\mu + g f^{abc} A^b_\mu A^c_\nu \,, \tag{2.17}$$

and $f^{abc} = 4i\,\mathrm{Tr}([\lambda^a, \lambda^b]\lambda^c)$ are the $SU(N_c)$ structure constants. The action of the covariant derivative on the quark fields is

$$iD_\mu q = \left(i\partial_\mu + gA^a_\mu \frac{\lambda^a}{2} \right) q \,, \tag{2.18}$$

and m_f is the mass of the quarks. The terms in Eq. (2.16) describe the interaction between quarks and gluons, as well as nonlinear three and four-gluon interactions. Note that, except for the number of flavors and their masses, the structure of the QCD Lagrangian is completely fixed by the local $SU(N_c)$ color symmetry.

A natural starting point for studying the phase diagram of hadronic matter is to consider the light flavors (up, down, and strange) as approximately massless, and the heavy flavors (charm, bottom, top) as infinitely massive. In this limit the QCD Lagrangian is completely characterized by two integer valued parameters, the number of colors $N_c = 3$ and flavors $N_f = 3$, and a single dimensionless coupling constant g. Quantum fluctuations cause the coupling constant to become scale dependent [10, 11]. At one-loop order the running coupling constant is

$$g^2(q^2) = \frac{16\pi^2}{b_0 \log(q^2/\Lambda^2_{QCD})} \,, \qquad b_0 = \frac{11}{3} N_c - \frac{2}{3} N_f \,, \tag{2.19}$$

where q is a characteristic momentum and N_f is the number of active flavors. The scale dependence of the coupling implies that, as a quantum theory, QCD is not governed by a dimensionless coupling but by a dimensionful scale, the QCD scale parameter Λ_{QCD}. This phenomenon is known as dimensional transmutation [12].

A crucial aspect of the scale dependence of the coupling in QCD is that the effective interaction decreases as the energy or momentum scale is increased. This feature of QCD is called asymptotic freedom [10, 11]. It implies that high energy interactions can be analyzed using perturbative QCD. The flip side of asymptotic freedom is anti-screening, or confinement: The effective interaction between quarks increases with distance, and quarks are permanently confined into hadrons. The absence of colored states in the spectrum implies that the use of perturbation theory is subtle, even at high energy. Quantities that can be computed perturbatively either involve a sum over many hadronic states, or allow for a factorization of perturbative interactions and non-perturbative matrix elements.

If quarks are massless then QCD observables are dimensionless ratios like m_p/Λ_{QCD}, where m_p is the mass of the proton. This implies that the QCD scale is not a parameter of the theory, but reflects a choice of units. In the real world QCD is part of the standard model, quarks acquire masses by electroweak symmetry breaking, and the QCD scale is fixed by value of the coupling constant at the weak scale.

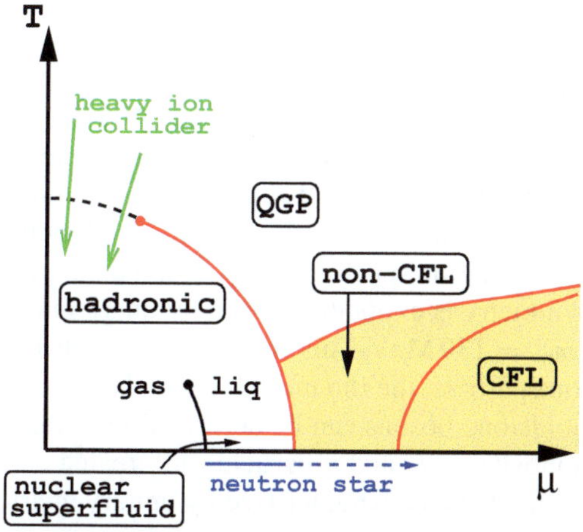

Fig. 2.2 Schematic phase diagram of QCD as a function of temperature T and baryon chemical potential μ. The quark gluon plasma phase is labeled QGP, and CFL refers to the color superconducting phase that is predicted to occur at asymptotically large chemical potential. The critical endpoints of the chiral and nuclear liquid-gas phase transitions, are denoted by *red* and *black points*, respectively. The chiral pseudo-critical line associated with the crossover transition at low temperature is shown as a *dashed line*. The *green arrows* indicate the regions of the phase diagram that can be studied by the experimental heavy ion programs at RHIC and the LHC

Experiments determine the value of the QCD fine structure constant $\alpha_s = g^2/(4\pi)$ at the position of the Z boson pole, $\alpha_s(m_z) = 0.1184 \pm 0.0007$ [13]. The numerical value of Λ_{QCD} depends on the renormalization scheme used in computing quantum corrections to the coupling constant. Physical observables, as well as the value of b_0, are independent of this choice. In the modified minimal subtraction ($\overline{MS}$) scheme the scale parameter is $\Lambda_{QCD} \simeq 200\,\text{MeV}$ [13].

A schematic phase diagram of QCD is shown in Fig. 2.2. In this figure I show the phases of strongly interacting matter as a function of the temperature T and the baryon chemical potential μ. The chemical potential μ controls the baryon density ρ, defined as 1/3 times the number density of quarks minus the number density of anti-quarks. In the following I will explain that the basic structure of the phase diagram is determined by asymptotic freedom and the symmetries of QCD. For more detailed reviews see [14–16].

At small temperature and chemical potential the interaction between quarks is dominated by large distances and the effective coupling is strong. This implies that quarks and gluons are permanently confined in color singlet hadrons, with masses of order Λ_{QCD}. The proton, for example, has a mass of $m_p = 935\,\text{MeV}$. A simplistic view of the structure of the proton is that it is a bound state of three constituent quarks with effective masses $m_Q \simeq m_p/3 \simeq \Lambda_{QCD}$. These masses should be compared to the bare up and down quark masses which are of the order 10 MeV.

As a consequence of strong interactions between virtual quarks and anti-quarks in the QCD ground state a vacuum condensate of $\bar{q}q$ pairs is generated, $\langle\bar{q}q\rangle \simeq -\Lambda_{QCD}^3$ [17–19]. This vacuum expectation value spontaneously breaks the approximate chiral $SU(3)_L \times SU(3)_R$ flavor symmetry of the QCD Lagrangian down to its diagonal subgroup, the flavor symmetry $SU(3)_V$. Spontaneous chiral symmetry breaking implies the existence of Goldstone bosons, massless modes with the quantum numbers of the generators of the broken axial symmetry $SU(3)_A$. The corresponding excitations in the spectrum of QCD are the π, K and η mesons. The $SU(3)_L \times SU(3)_R$ symmetry is explicitly broken by quark masses, and the mass of the charged pion is $m_\pi = 139\,\mathrm{MeV}$. This scale can be compared to the mass of the lightest non-Goldstone particle, the rho meson, which has a mass $m_\rho = 770\,\mathrm{MeV}$.

At low energy Goldstone bosons can be described in terms of an effective field theory in which composite π, K and η particles are treated as fundamental fields. The Goldstone boson field can be parametrized by unitary matrices

$$\Sigma = \exp(i\lambda^a\phi^a/f_\pi)\,, \tag{2.20}$$

where λ^a are the Gell-Mann matrices for $SU(3)$ flavor and $f_\pi = 93\,\mathrm{MeV}$ is the pion decay constant. For example, $\pi^0 = \phi^3$ and $\pi^\pm = (\phi_1 \pm i\phi_2)/2$ describe the neutral and charged pion. Other components of ϕ^a describe the neutral and charged kaons, as well as the eta. The eta prime, which is the $SU(3)_F$ singlet meson, acquires a large mass because of the axial anomaly, and is not a Goldstone boson. The axial anomaly refers to the fact that the flavor singlet axial current, which is conserved in massless QCD at the classical level, is not conserved if quantum effects are taken into account. The divergence of the axial current $A_\mu = \bar{q}\gamma_\mu\gamma_5 q$ is

$$\partial_\mu A^\mu = \frac{g^2 N_f}{32\pi^2}\epsilon^{\mu\nu\alpha\beta}G^a_{\mu\nu}G^a_{\alpha\beta}\,. \tag{2.21}$$

The right hand side is the topological charge density, which I will discuss in more detail in Sect. 2.4.3.

At low energy the effective Lagrangian for the chiral field can be organized as a derivative expansion in gradients of Σ. Higher derivative terms describe interactions that scale as either the momentum or the energy of the Goldstone boson. Since Goldstone bosons are approximately massless, the energy is of the same order of magnitude as the momentum. We will see that the expansion parameter is $p/(4\pi f_\pi)$. At leading order in (∂/f_π) there is only one possible term which is consistent with chiral symmetry, Lorentz invariance and the discrete symmetries C, P, T. This is the Lagrangian of the non-linear sigma model

$$\mathscr{L} = \frac{f_\pi^2}{4}\mathrm{Tr}\left[\partial_\mu\Sigma\partial^\mu\Sigma^\dagger\right] + \left[B\mathrm{Tr}(M\Sigma^\dagger) + h.c.\right] + \ldots\,, \tag{2.22}$$

where the term proportional to B takes into account explicit symmetry breaking. Here, $M = \text{diag}(m_u, m_d, m_s)$ is the quark mass matrix and B is a low energy constant that I will fix below.

First, I will show that the parameter f_π controls the pion decay amplitude. For this purpose I have to gauge the weak $SU(2)_L$ symmetry of the non-linear sigma model. As usual, this is achieved by promoting the derivative to a gauge covariant operator $\nabla_\mu \Sigma = \partial_\mu \Sigma + ig_w W_\mu \Sigma$ where W_μ is the charged weak gauge boson and g_w is the weak coupling constant. The gauged non-linear sigma model gives a pion-W boson interaction

$$\mathscr{L} = g_w f_\pi W_\mu^\pm \partial^\mu \pi^\mp . \tag{2.23}$$

This term contributes to the amplitude $\mathscr{A}$ for the decay $\pi^\pm \to W^\pm \to e^\pm \nu_e$. I get $\mathscr{A} = g_w f_\pi q_\mu$, where q_μ is the momentum of the pion. This result can be compared to the standard definition of f_π in terms of the weak axial current matrix element of the pion, $\langle 0|A_\mu^a|\pi^b\rangle = f_\pi q_\mu \delta^{ab}$. This comparison shows that the coefficient of the kinetic term in the non-linear sigma model is indeed the weak decay constant of the pion.

In the ground state $\Sigma = 1$ and the ground state energy is $E_{vac} = -2B\text{Tr}[M]$. Using the relation $\langle \bar{q}q\rangle = \partial E_{vac}/(\partial m)$ we find $\langle \bar{q}q\rangle = -2B$. Fluctuations around $\Sigma = 1$ determine the masses of the Goldstone bosons. The pion mass satisfies the Gell-Mann-Oaks-Renner relation (GMOR) [17]

$$m_\pi^2 f_\pi^2 = -(m_u + m_d)\langle \bar{q}q\rangle \tag{2.24}$$

and analogous relations exist for the kaon and eta masses. This result shows the characteristic non-analytic dependence of the pion mass on the quark masses, $m_\pi \sim \sqrt{m_q}$.

2.3.2 QCD at Finite Temperature

The structured of QCD at high temperature can be analyzed using the assumption that quarks and gluons are approximately free. We will see that this assumption is internally consistent, and that it is confirmed by lattice calculations. If the temperature is large then quarks and gluons have thermal momenta $p \sim T \gg \Lambda_{QCD}$. Asymptotic freedom implies that these particles are weakly interacting, and that they form a plasma of mobile color charges, the quark gluon plasma (QGP) [20, 21]. The pressure of a gas of quarks and gluons is

$$P = \frac{\pi^2 T^4}{90}\left(2\left(N_c^2 - 1\right) + 4N_c N_f \frac{7}{8}\right) . \tag{2.25}$$

This is the Stefan-Boltzmann law, where $2(N_c^2 - 1)$ is the number of bosonic degrees of freedom, and $4N_cN_F$ is the number of fermions. The factor 7/8 takes into account the difference between Bose and Fermi statistics. The pressure of a QGP is parametrically much bigger than the pressure of a pion gas, indicating that the QGP at high temperature is thermodynamically stable.

The argument that the QGP at asymptotically high temperature is weakly coupled is somewhat more subtle than it might appear at first glance. If two quarks or gluons in the plasma interact via large angle scattering then the momentum transfer is large, and asymptotic freedom implies that the effective coupling is weak. However, the color Coulomb interaction is dominated by small angle scattering, and it is not immediately clear why the effective interaction that governs small angle scattering is weak. The basic observation is that in a high temperature plasma there is a large thermal population ($n \sim T^3$) of mobile color charges that screen the interaction at distances beyond the Debye length $r_D \sim 1/(gT)$. We also note that even in the limit $T \gg \Lambda_{QCD}$ the QGP contains a non-perturbative sector of static magnetic color fields [22]. This sector of the theory, corresponding to energies below the magnetic screening scale $m_M \lesssim g^2T$, is strongly coupled, but it does not contribute to thermodynamic or transport properties of the plasma in the limit $T \to \infty$.

The quark gluon plasma exhibits neither color confinement nor chiral symmetry breaking. This implies that the high temperature phase must be separated from the low temperature hadronic phase by a phase transition. The order of this transition is very sensitive to the values of the quark masses. In QCD with massless u, d and infinitely massive s, c, b, t quarks the transition is second order [23]. In the case of massless (or sufficiently light) u, d, s quarks the transition is first order. Lattice simulations show that for realistic quark masses, $m_u \simeq m_d \simeq 10\,\text{MeV}$ and $m_s \simeq 120\,\text{MeV}$, the phase transition is a rapid crossover [24, 25]. The transition temperature, defined in terms of the chiral susceptibility, is $T_c \simeq 151 \pm 3 \pm 3\,\text{MeV}$ [26, 27], which is consistent with the result $154 \pm 9\,\text{MeV}$ reported in [25, 28].

The phase transition is expected to strengthen as a function of chemical potential, so that there is a critical baryon chemical potential μ at which the crossover turns into a first order phase transition [29]. This critical point is the endpoint of the chiral phase transition. Because of the fermion sign problem, which I will discuss in Sect. 2.4.4, it is very difficult to locate the critical endpoint using simulations on the lattice. Model calculations typically predict the existence of a critical point, but do not constrain its location. A number of exploratory lattice calculations have been performed [30–35], but at the time I am writing these notes it has not been demonstrated conclusively that the transition strengthens with increasing baryon chemical potential [36]. The critical endpoint is important because, with the exception of the endpoint of the nuclear liquid-gas transition, it is the only thermodynamically stable point in the QCD phase diagram at which the correlation length diverges. This means that the critical endpoint may manifest itself in heavy ion collisions in terms of enhanced fluctuation observables [37].

2.3.3 High Baryon Density QCD

The origin of the phase diagram, $T = \mu = 0$, corresponds to the vacuum state of QCD. If we stay on the $T = 0$ line and increase the chemical potential μ then there is no change initially. At zero temperature the chemical potential μ is the energy required to add a baryon to the system, and QCD has a large mass gap for baryonic states. The first non-vacuum state we encounter along the $T = 0$ axis of the phase diagram is nuclear matter, a strongly correlated superfluid composed of approximately non-relativistic neutrons and protons. Nuclear matter is self-bound, and the baryon density changes discontinuously at the onset transition, from $\rho = 0$ to nuclear matter saturation density $\rho = \rho_0 \simeq 0.15\,\mathrm{fm}^{-3}$. The discontinuity decreases as nuclear matter is heated, and the nuclear-liquid gas phase transition ends in a critical point at $T \simeq 18\,\mathrm{MeV}$ and $\rho \simeq \rho_0/3$ [38–40]. Hot hadronic matter can be described quite accurately as a weakly interacting gas of hadronic resonances. Empirically, the density of states for both mesons and baryons grows exponentially. A system of this type is called a Hagedorn gas, and it is known that a Hagedorn gas has a limiting temperature. It is also known that an exponential density of states can be realized using the string model of hadronic resonances.

In the regime $\mu \gg \Lambda_{QCD}$ we can use arguments similar to those in the limit $T \gg \Lambda_{QCD}$ to establish that quarks and gluons are weakly coupled. At low temperature non-interacting quarks form a Fermi surface, where all states below the Fermi energy $E_F \simeq \mu/3$ are filled, and all states above the Fermi energy are empty. Interactions take place near the Fermi surface, and the corresponding interaction is weak. The main difference between cold quark matter and the hot QGP is that the large density of states near the quark Fermi surface implies that even weak interactions can cause qualitative changes in the ground state of dense matter. In particular, attractive interactions between pairs of quarks $(\mathbf{p}_F, -\mathbf{p}_F)$ on opposite sides of the Fermi surface leads to color superconductivity and the formation of a $\langle qq \rangle$ diquark condensate.

Since quarks carry many different quantum numbers, color, flavor, and spin, a variety of superconducting phases are possible. The most symmetric of these, known as the color-flavor locked (CFL) phase, is predicted to exist at asymptotically high density [41, 42]. In the CFL phase the diquark order parameter is

$$\langle q^A_{\alpha f} q^B_{\beta g} \rangle = (C\gamma_5)_{\alpha\beta} \epsilon^{ABC} \epsilon_{fgh} \delta^h_C \Phi \,, \tag{2.26}$$

where $C\gamma_5$ is an anti-symmetric (spin zero) Dirac matrix, and Φ determines the magnitude of the gap near the Fermi surface. This order parameter has a number of interesting properties. It breaks the $U(1)$ symmetry associated with baryon number, leading to superfluidity, and it breaks the chiral $SU(3)_L \times SU(3)_R$ symmetry. Except for Goldstone modes the spectrum is fully gapped. Fermions acquire a BCS-pairing gap, and gauge fields are screened by the color Meissner effect. This implies that the CFL phase, even though it is predicted to occur in a very dense liquid of quarks, exhibits many properties of superfluid nuclear matter.

The CFL order parameter describes equal pair-condensates $\langle ud \rangle = \langle us \rangle = \langle ds \rangle$ of all three light quark flavors. As the density is lowered effects of the non-zero strange quark mass become important, and less symmetric phases are predicted to appear [14]. Phases that have been theoretically explored include Bose condensates of pions and kaons, hyperon matter, states with inhomogeneous quark-anti-quark or diquark condensates, and less symmetric color superconducting phases. The regimes of moderate baryon chemical potential in the phase diagram shown in Fig. 2.2 is largely conjecture. Empirical evidence shows that at low μ there is a nuclear matter phase with broken chiral symmetry and zero strangeness, and weak coupling calculations indicate that at high μ we find the CFL phase with broken chiral symmetry but non-zero strangeness. In principle the two phases could be separated by a single onset transition for strangeness [43, 44], but model calculation support a richer picture in which one or more first order transitions intervene, as indicated in Fig. 2.2.

2.4 Lattice QCD

2.4.1 The Wilson Action

Symmetry arguments and perturbative calculations can be used to establish general features of the QCD phase diagram, but quantitative results can only be obtained using numerical calculations based on lattice QCD. The same is true for the masses of hadrons, their properties, and interactions. Lattice QCD is based on the euclidean path integral representation of the partition function, see the contribution by Hatsuda and [45–49] for introductions. More detailed reviews of the lattice field theory approach to hot and dense QCD can be found in [50, 51].

The euclidean partition function for QCD is

$$Z(T, \mu, V) = \int \mathscr{D}A_\mu \, \mathscr{D}q_f \, \mathscr{D}\bar{q}_f \, \exp(-S_E) , \tag{2.27}$$

where S_E is the euclidean action

$$S_E = -\int_0^\beta d\tau \int_V d^3x \, \mathscr{L}^E , \tag{2.28}$$

$\beta = T^{-1}$ is the inverse temperature and $\mathscr{L}^E$ is the Euclidean Lagrangian, which is obtained by analytically continuing Eq. (2.16) to imaginary time $\tau = it$. As in the quantum mechanical example in Eq. (2.4) the temperature enters via the boundary condition on the fields in the imaginary time direction. Gauge fields and fermions obey periodic and anti-periodic boundary conditions, respectively. The chemical potential enters through its coupling to the conserved baryon density

$$\mathscr{L}^E(\mu) = \mathscr{L}^E(0) + \mu \bar{q}_f \gamma_0 q_f\,. \tag{2.29}$$

In his pioneering work Wilson proposed to discretize the action on a $N_\tau \times N_\sigma^3$ space-time lattice with lattice spacings a_τ and a_σ [52]. In many cases $a_\sigma = a_\tau = a$, but we will encounter an exception in Sect. 2.5.4. when we discuss the Hamiltonian formulation of the theory.

At finite temperature we have to ensure that the spatial volume is larger than the inverse temperature, $L > \beta$. Here, $\beta = N_\tau a_\tau$, $L = N_\sigma a_\sigma$, and $V = L^3$ is the volume. Thermodynamic quantities are determined by taking derivatives of the partition function. The energy and baryon density are given by

$$\mathscr{E} = -\frac{1}{V}\left.\frac{\partial \log Z}{\partial \beta}\right|_{\beta\mu}, \tag{2.30}$$

$$\rho = \frac{1}{\beta V}\left.\frac{\partial \log Z}{\partial \mu}\right|_{\beta}. \tag{2.31}$$

The discretized action for the gauge fields originally suggested by Wilson is given by

$$S_W = -\frac{2}{g^2}\sum_n \sum_{\mu<\nu} \mathrm{Re}\,\mathrm{Tr}\left[W_{\mu\nu}(n) - 1\right] \tag{2.32}$$

where $W_{\mu\nu}(n)$ is the plaquette, the product of gauge links around an elementary loop on the lattice,

$$W_{\mu\nu}(n) = U_\mu(n)U_\nu(n+\hat{\mu})U_{-\mu}(n+\hat{\mu}+\hat{\nu})U_{-\nu}(n+\hat{\nu})\,. \tag{2.33}$$

Here, $n = (n_\tau, n_i)$ labels lattice sites and $\hat{\mu}$ is a unit vector in the μ-direction. The gauge links $U_\mu(n)$ are $SU(N_c)$ matrices. We can think of the gauge links as line integrals

$$U_\mu(n) = \exp(iaA_\mu(n))\,, \tag{2.34}$$

and of the plaquettes as fluxes

$$W_{\mu\nu}(n)) = \exp(ia^2 G_{\mu\nu}(n))\,, \tag{2.35}$$

but the fundamental variables in the path integral are the (compact) group variables U_μ, not the (non-compact) gauge potentials A_μ. In particular, the path integral in pure gauge QCD takes the form

$$Z = \int \prod_{n,\mu} dU_\mu(n) \, \exp(-S_W) , \tag{2.36}$$

where dU is the Haar measure on $SU(N_c)$. The Haar measure describes the correct integration measure for the gauge group. Some group integrals are discussed by Hatsuda, but part of the beauty of the Metropolis method is that we never have to explicitly construct $dU_\mu(n)$.

Using Eq. (2.34) we can check that the Wilson action reduces to continuum pure gauge theory in the limit $a \to 0$. We note that the gauge invariance of QCD is maintained exactly, even on a finite lattice, but that Lorentz invariance is only restored in the continuum limit. We also observe that classical scale invariance implies that the massless QCD action is independent of a. The continuum limit is taken by adjusting the bare coupling at the scale of the lattice spacing according to asymptotic freedom, see Eq. (2.19). In practice the lattice spacing is not small enough to ensure the accuracy of this method, and more sophisticated scale setting procedures are used [50, 51].

Monte Carlo simulations of the path integral Eq. (2.36) can be performed using the Metropolis algorithm explained in Sect. 2.2:

- Initialize the link variables with random $SU(N_c)$ matrices. A simple algorithm is based on writing U in terms of N_c complex row vectors $\mathbf{u}_i$. Take each vector to be random unit vector and then use the Gram-Schmidt method to orthogonalize the different vectors, $\mathbf{u}_i \cdot \mathbf{u}_j^* = \delta_{ij}$. This ensures that U is unitary and distributed according to the $SU(N_c)$ Haar measure [53].
- Sweep through the lattice and update individual link variables. For this purpose multiply the link variable by a random $SU(N_c)$ matrix, $U_\mu \to RU_\mu$. Compute the change in the Wilson action and accept the update with probability $\exp(-\Delta S_W)$.
- Compute physical observables. The simplest observable is the average plaquette $\langle W_{\mu\nu} \rangle$, which can be related to the equation of state, see Eq. (2.30). More complicated observables include the correlation function between plaquettes, and the Wilson loop

$$W(\mathscr{C}) = Tr\,[L(\mathscr{C})] , \qquad L(\mathscr{C}) = \prod_{(n,\mu)\in\mathscr{C}} U_\mu(n) , \tag{2.37}$$

where $L(\mathscr{C})$ is the product of link variables around a closed loop. The average Wilson loop is related to the potential between two static charges in the fundamental representation

$$V(R) = -\lim_{T\to\infty} \frac{1}{T} \log\left[\langle W(\mathscr{C})\rangle\right] , \tag{2.38}$$

where $R \times T$ is the area of a rectangular loop $\mathscr{C}$.

- Tune to the continuum limit $a \to 0$ by adjusting the coupling constant according to the asymptotic freedom formula Eq. (2.19). Note that the Lambda parameter for the lattice regulator is quite small, $\Lambda_{lat} = \Lambda_{\bar{MS}}/28.8$ [54]. Also observe that we have to increase N_σ, N_τ to keep the physical volume constant as $a \to 0$. Indeed, once the continuum limit $a \to 0$ is reached we have to study the infinite volume (thermodynamic) limit $V \to \infty$. This is more difficult than it appears, because $a \to 0$, corresponding to $g \to 0$, is a critical point of the partition function (2.36), and simulations exhibit critical slowing down.

Metropolis simulations with the pure gauge Wilson action are very simple and robust. As an illustration I provide a simple Z_2 lattice gauge theory code written by M. Creutz in the Appendix. Reasonable results for the heavy quark potential can be obtained on fairly coarse lattices, for example an 8^4 lattice with a spacing $a \simeq 0.25\,\text{fm}$ [55]. However, accurate results with controlled error bars require significant computational resources. In practice the perturbative relation between a and g^2 is only valid on very fine lattices, and the scale setting has to be done non-perturbatively. Also, determining the spectrum of pure gauge theory is difficult. Purely gluonic states, glueballs, are quite heavy, with masses in the range $m \simeq 1.6\,\text{GeV}$ and higher. This implies that gluonic correlation functions are short range, requiring a resolution $a \simeq 0.1$ fm or better. Finally, simulations on fine lattices are affected by critical slowing down. Indeed, finding an efficient method for updating gauge fields on very fine lattices, analogous to the cluster algorithms for spin models [56], is an important unsolved problem.

2.4.2 Fermions on the Lattice

The main difficulty in lattice QCD is related to the presence of light fermions. The fermion action is of the form

$$S_F = a^4 \sum_{m,n} \bar{q}(m) D_{mn} q(n)\,. \tag{2.39}$$

Formally, the integration over the fermion fields can be performed exactly, resulting in the determinant of the Dirac operator $\det(D(A_\mu, \mu))$. Several methods exist for discretizing the Dirac operator D, and for sampling the determinant. Different discretization schemes differ in the degree to which chiral symmetry is maintained on a finite lattice. The original formulation due to Wilson [52] preserves no chiral symmetry, the staggered Fermion scheme [57] maintains a subset of the full chiral symmetry, while the domain wall [58] and overlap methods [59] aim to preserve the full chiral symmetry on a discrete lattice.

The central difficulty in implementing these methods is that the fermion determinant is a very non-local object. While updating a single gauge link only requires recalculating a small number of plaquettes (6 in $d = 4$ dimensions) in the Wilson action, recalculating the fermion action requires computing the determinant of a (very sparse) matrix of size $(N_\tau N_\sigma^3) \times (N_\tau N_\sigma^3)$ or larger. This is clearly impractical. Fermion algorithms rely on a number of tricks. The first is the observation that the Dirac operator has a property called γ_5-hermiticity, $\gamma_5 D \gamma_5 = D^\dagger$, which implies that $\det(D)$ is real. The determinant of a two-flavor theory is then real and positive. This allows us to rewrite the fermion determinant as a path integral over a bosonic field with a non-local but positive action

$$\det(D_u)\det(D_d) = \det(DD^\dagger) = \int \mathscr{D}\phi\mathscr{D}\phi^\dagger \, \exp(-\phi^\dagger(DD^\dagger)^{-1}\phi) \, . \tag{2.40}$$

The path integral over the pseudofermion field ϕ can be sampled using a combination of deterministic methods like molecular dynamics and stochastic methods such as the Metropolis algorithm. These combined algorithms are known as Hybrid Monte Carlo (HMC) methods. Codes that implement the HMC algorithm for pseudofermions are significantly more complicated than the Metropolis algorithm for the pure gauge Wilson action discussed above, and I refer the interested reader to the more specialized literature [60]. I also note that since these algorithms involve the calculation of D^{-1} the computational cost increases as the quark masses are lowered.

The calculation of correlation functions also differs from the bosonic case. Consider, for example, an operator with the quantum numbers of a charged pion, $J_\pi(x) = \bar{u}^a(x)\gamma_5 d^a(x)$. Since the fermion action is quadratic the correlation function in a given gauge configuration can be computed exactly in terms of the fermion propagator. The full correlation function is

$$\Pi_\pi(x) = \langle J_\pi(x) J_\pi(0)\rangle = \langle \mathrm{Tr}\,[S(x,0)\gamma_5 S(0,x)\gamma_5]\rangle \, , \tag{2.41}$$

where $S(x,y) = \langle x|D^{-1}|y\rangle$ is the fermion propagator, and we have assumed exact isospin symmetry so that the propagator of the up quark is equal to the propagator of the down quark. Note that the interaction between quarks is encoded in the average over all gauge fields. The one-gluon exchange interaction, for example, corresponds to a perturbative fluctuation in the gauge field that modifies the two quark propagators. An operator with the quantum number of the proton is $\eta_\alpha(x) = \epsilon_{abc}(u^a(x)C\gamma_\mu u^b(x))(\gamma^\mu\gamma_5 d^c(x))_\alpha$. The correlation function is

$$\Pi_{\alpha\beta}(x) = 2\epsilon_{abc}\epsilon_{a'b'c'}\Big\langle \Big(\gamma_\mu\gamma_5 S^{cc'}(0,x)\gamma_\nu\gamma_5\Big)_{\alpha\beta} \mathrm{Tr}\Big[\gamma_\mu S^{aa'}(0,x)\gamma_\nu C(S^{bb'}(0,x))^T C\Big]\Big\rangle . \tag{2.42}$$

Note that meson correlation function involves one forward and one backward going propagator, whereas the propagators in the baryon correlation function are all forward going. A difficulty arises when we consider flavor singlet $\bar{q}q$ currents such as $J_{\eta'} = (\bar{u}^a(x)\gamma_5 u^a(x) + \bar{d}^a(x)\gamma_5 d^a(x))/\sqrt{2}$, which has the quantum numbers

Fig. 2.3 Topological objects in lattice QCD (figure courtesy of S. Sharma, see [62]). This picture shows a slice through a low lying eigenstate of the Dirac operator in lattice QCD

of the η' meson. We find

$$\Pi_{\eta'}(x) = \langle J_{\eta'}(x) J_{\eta'}(0)\rangle = \langle \mathrm{Tr}\,[S(x,0)\gamma_5 S(0,x)\gamma_5] - 2\mathrm{Tr}\,[S(x,x)\gamma_5]\,\mathrm{Tr}\,[S(0,0)\gamma_5]\rangle\ , \tag{2.43}$$

which involve propagators $S(x,x)$ that loop back to the same point. These contributions are known as quark-line disconnected diagrams, and difficult to treat numerically, see [61] for a recent discussion.

2.4.3 *The QCD Vacuum*

It is natural to hope that lattice QCD can provide us with an intuitive picture of what the QCD vacuum looks like, similar to the picture of the quantum mechanical ground state shown in Fig. 2.3. This turns out to be more complicated, for a number of reasons. The first is that the field in QCD is a $SU(3)$ matrix, which is hard to visualize. The second, more important, problem is related to quantum fluctuations. In QCD there is no obvious separation of scales that would allow us to clearly separate perturbative fluctuations from large semi-classical fluctuations.

This has led to the idea to eliminate short range fluctuations by some kind of filtering or smoothing algorithm. The simplest of these is known as cooling [63]. In the cooling method we modify the Metropolis algorithm so that only updates that reduce the action are accepted. Since the update algorithm is local, this will tend to eliminate small structures but preserve larger objects. A modern version of cooling is gradient flow [64]. In the gradient flow method we continue the gauge fields to a

5th "time" dimension. In this direction the fields satisfy a differential equation

$$\partial_\tau A_\mu = D^\nu G_{\mu\nu} \,, \tag{2.44}$$

where $A_\mu(\tau = 0)$ is the four-dimensional gauge field and the rhs is computed from the gauge potentials evaluated at the flow time τ. The Lorentz indices remain four-dimensional. The rhs of the flow equations is the classical equation of motion, so that the gradient flow tends to drive gauge fields towards the closest classical solution. The only finite action solutions of the euclidean field equations on R^4 are instantons [65, 66]. Instantons and anti-instantons are characterized by integer values $Q_{top} = \pm 1$ of the topological charge

$$Q_{top} = \int d^4x\, q(x) \,, \qquad q(x) = \frac{g^2}{64\pi^2}\epsilon^{\mu\nu\alpha\beta} G^a_{\mu\nu} G^a_{\alpha\beta} \,. \tag{2.45}$$

Exact higher charge solutions exist, but the QCD vacuum is dominated by configurations with both instantons and anti-instantons. These gauge field configurations are only approximate solutions of the equations of motion [66]. Under cooling or gradient flow instantons and anti-instantons will eventually annihilate and evolve to an exact multi-instantons solution with $Q_{top} = N_I - N_A$, where $N_{I,A}$ are the numbers of (anti)instantons. However, the $N_I + N_A$ topological objects are preserved for flow times that are much longer than the decay time of ordinary quantum fluctuations, and the total number of well separated instantons and anti-instantons can be determined.

The average topological charge is zero, but the pure gauge vacuum is characterized by a non-zero topological susceptibility

$$\chi_{top} = \frac{1}{V}\langle Q^2_{top}\rangle \,, \tag{2.46}$$

where V is the euclidean four-volume. The topological charge can be determined using the naive lattice discretization of Eq. (2.45), but this operator is very noisy, and in general not an integer. This problem can be addressed using the cooling or gradient flow algorithms discussed above. Recent lattice calculations based on these methods give $\chi_{top} = (190 \pm 5\,\text{MeV})^4$ [67, 68]. A simple picture of the QCD vacuum which is consistent with this value is the dilute instanton liquid model, which assumes that the topological susceptibility is determined by Poisson fluctuations in an ensemble of instantons and anti-instantons with an average density $(N_I + N_A)/V \simeq 1\,\text{fm}^{-4}$ [66]. This is an approximate picture, and more complicated configurations involving monopoles and fractional charges are needed to understand the large N_c limit and the role of confinement [69].

Another important development is the use of fermionic methods to analyze the vacuum structure of QCD. In a given gauge configuration the quark propagator can written as

$$S(x,y) = \sum_{\lambda} \frac{\psi_\lambda(x)\psi_\lambda^\dagger(y)}{\lambda + im}, \tag{2.47}$$

where ψ_λ is an eigenvector of the Dirac operator with eigenvalue λ: $D\psi_\lambda = (\lambda + im)\psi_\lambda$. Note that this is not how propagators are typically determined in lattice QCD, because the calculation of the complete spectrum is numerically very expensive. Gamma five hermiticity implies that eigenvalues come in pairs $\pm\lambda$. The quark condensate is given by

$$\langle \bar{q}q \rangle = -i \int d^4x \, \langle \mathrm{Tr}\,[S(x,x)] \rangle = -\left\langle \sum_{\lambda>0} \frac{2m}{\lambda^2 + m^2} \right\rangle. \tag{2.48}$$

Here, I have ignored the contribution from exact zero modes because the density of zero modes is suppressed by m^{N_f}. This factor comes from the determinant in the measure. If we were to ignore the determinant (this is called the quenched approximation), then the quark condensate would diverge as $1/m$. We observe that a finite value of the quark condensate in the chiral limit $m \to 0$ requires an accumulation of eigenvalues near zero. This can be made more explicit by introducing the density of states

$$\rho(\nu) = \left\langle \sum_{\lambda \geq 0} \delta(\lambda - \nu) \right\rangle. \tag{2.49}$$

The chiral condensate in the thermodynamic and chiral limits is given by

$$\langle \bar{q}q \rangle = -\pi\rho(0). \tag{2.50}$$

This is known as the Banks-Casher relation [70]. Note that it is essential to take the thermodynamic $V \to \infty$ limit before the chiral limit $m \to 0$.

Exact zero modes of the Dirac operator are related to topology. The Dirac operator has one left handed zero mode in the field of an instanton, and a right handed zero mode in the field of an anti-instanton. This is consistent with the Atiyah-Singer index theorem, which states that the topological charge is equal to the index of the Dirac operator, the difference between the number of left and right handed zero modes, $Q_{top} = N_f(n_L - n_R)$. These results suggest that it is possible to give a purely fermionic definition of the topological charge density.

On the lattice, this can be achieved for a class of Dirac operators that satisfy the Ginsparg-Wilson relation [71]

$$D\gamma_5 + \gamma_5 D = aD\gamma_5 D, \tag{2.51}$$

where a is the lattice spacing. In the continuum limit we recover the expected relation $D\gamma_5 + \gamma_5 D = 0$ for the massless Dirac operator. The important observation is that the fermionic topological density

$$q_f(n) = \frac{1}{2a^3} \mathrm{tr}_{CD} \left[\gamma_5 D(n,n)\right] , \tag{2.52}$$

where tr_{CD} is a color-Dirac trace, satisfies the index theorem

$$Q_{top} = a^4 \sum_n q_f(n) \tag{2.53}$$

on a discrete lattice. Figure 2.3 shows the absolute square of $q_f(x)$ constructed from lying eigenmodes of the Dirac operator. We observe that fermionic operators can indeed be used to probe the topological content of the QCD vacuum directly, without the need for filtering or smoothing.

The existence of zero mode implies that the topological susceptibility is zero if at least one quark flavor is massless. This is because the path integral measure contains the fermion determinant, which is vanishes if $m = 0$ and $Q_{top} \neq 0$. We can be more precise using the chiral lagrangian Eq. (2.22). In order to keep track of topology we add to the QCD action a topological term $S_\theta = i\theta Q_{top}$. Then the topological susceptibility is given by the second derivative of the free energy with respect to θ. Since every zero mode in the Dirac operator contributes a factor $\det(M)$ to the partition function we know that θ enters the effective lagrangian in the combination $\theta + \arg(\det(M))$. The vacuum energy is determined by

$$V = -B\mathrm{Tr}\left[Me^{i\theta/N_f}\Sigma^\dagger\right] + h.c. , \tag{2.54}$$

and we observe that the topological susceptibility in QCD with degenerate quark masses is proportional to $m\langle\bar{q}q\rangle$. Note that Eq. (2.54) is consistent with the vanishing of χ_{top} for $m_u = 0$. If $m_u = 0$ and $m_d \neq 0$ then Eq. (2.54) is minimized by $\Sigma = \exp(i\phi\tau_3)$ with $\phi = \theta/2$, and the vacuum energy is independent of θ.

It is tempting to think that exact zero modes, governed by topology, and approximate zero modes, connected to chiral symmetry breaking, are related. This is the basis of the instanton liquid model [66]. In the instanton liquid model we consider an ensemble of instantons and anti-instantons with no (or small) net topology. The exact zero modes of individual instantons are lifted, and form a zero mode zone. The density of eigenvalues in the zero mode zone determines the chiral condensate via the Banks-Casher relation. This model predicts the correct order of magnitude of $\langle\bar{q}q\rangle$, but the calculation cannot be systematically improved because chiral symmetry breaking requires strong coupling. Recently, we showed that the connection of chiral symmetry breaking, instantons and monopoles can be made precise in a certain limit of QCD. The idea is to compactify QCD on $R^3 \times S_1$, where the size of the circle is much smaller than Λ_{QCD}^{-1}, and the fermions satisfy non-thermal (twisted) boundary conditions [72].

2.4.4 Lattice QCD at Finite Baryon Density

In Sect. 2.4.2 I discussed some of the difficulties that appear when we discretize the Dirac operator. A separate, more serious, issue with fermions is that for $\mu \neq 0$ the Dirac operator does not satisfy γ_5-hermiticity. This implies that the fermion determinant is no longer real, and that standard importance sampling methods fail. This is the "sign" problem already mentioned in Sect. 2.3.2. To understand the severity of the problem consider a generic expectation value

$$\langle \mathcal{O} \rangle = \frac{\int dU \, \det(D) \, \mathcal{O} \, e^{-S}}{\int dU \, \det(D) \, e^{-S}} . \tag{2.55}$$

If the determinant is complex I can write this as

$$\langle \mathcal{O} \rangle = \frac{\int dU \, |\det(D)| \, \mathcal{O} e^{i\varphi} \, e^{-S}}{\int dU \, |\det(D)| \, e^{i\varphi} e^{-S}} \equiv \frac{\langle \mathcal{O} e^{i\varphi} \rangle_{pq}}{\langle e^{i\varphi} \rangle_{pq}} , \tag{2.56}$$

where $\langle . \rangle_{pq}$ refers to a phase quenched average. This average can be computed using the Metropolis (or HMC) algorithm. The problem is that the average phase $\langle e^{i\varphi} \rangle_{pq}$ is very small. This follows from the fact that the average phase can be expressed as the ratio of two partition functions

$$\langle e^{i\varphi} \rangle_{pq} = \frac{\int dU \, \det(D) \, e^{-S}}{\int dU \, |\det(D)| \, e^{-S}} = \frac{Z}{Z_{pq}} = e^{-V\Delta F} , \tag{2.57}$$

where ΔF is the free energy density difference, and V is the volume of the system. This shows that the phase is exponentially small, and that the ratio Eq. (2.56) is very difficult to compute.

As a specific example consider QCD with two degenerate flavors, up and down, and a baryon chemical potential $\mu_u = \mu_d = \mu_B/3$. Then $\det(D) = \det(D_u)\det(D_d)$ and $|\det(D)| = \det(D_u)\det(D_d)^*$. The phase quenched partition function Z_{pq} can be interpreted as the partition function of QCD with a non-zero isospin chemical potential $\mu_u = -\mu_d = \mu_I/2$. The small μ behavior of both the isospin and baryon number theories at $T = 0$ is easily understood. The isospin theory has a second order phase transition at $\mu_I = m_\pi$ which corresponds to the onset of pion condensation. The baryon theory has a first order transition at $\mu_B = m_p - B$, where $B \simeq 15\,\text{MeV}$ is the binding energy of infinite nuclear matter. This implies that for $\mu > m_\pi$ the partition functions Z and Z_{pq} describe very different physical systems, and the sign problem is severe.

The sign problem may manifest itself in different ways. Consider, for example, an attempt to study the correlation function of A nucleons in a QCD ensemble generated at $\mu_B = 0$. For large A this correlation function determines the binding energy of nuclear matter. There are two difficulties with this approach. The first is that the operator contains $3A$ quark fields, so that the correlator has up to $(3A)!$

contractions. This is not prohibitive, because the number of contractions can be reduced using symmetries and iterative algorithms. Indeed, correlators for medium mass nuclei have been computed [49]. The second, more serious, problem is the signal-to-noise ratio. The variance of the correlator C is

$$\mathrm{var}(C) = \langle CC^{\dagger} \rangle - \langle C \rangle^2 . \tag{2.58}$$

The A nucleon correlator C contains $3A$ forward going quark propagators, and $CC^{\dagger}$ consists of $3A$ forward and $3A$ backward propagators. This implies that $CC^{\dagger}$ couples to a state of $3A$ mesons. Since the lightest meson is the pion and the lightest baryon is the proton the signal-to-noise of an A nucleon correlation function is

$$\frac{\mathcal{S}}{\mathcal{N}} \sim \exp(-A(m_p - 3m_{\pi}/2)\tau) . \tag{2.59}$$

In order to resolve the ground state with a given A we have to make τ sufficiently large so that excited states with the same A are suppressed. For $A = 1$ there is a πN continuum starting an excitation energy $\Delta E = m_{\pi}$, and the first resonance at $\Delta E = m_{\Delta} - m_N \simeq 300\,\mathrm{MeV}$. This means that we have to consider $\tau \gtrsim 1\,\mathrm{fm}$. For multi-nucleon states the situation is more complicated, because there are many closely spaced multi-nucleon states in a finite volume. The problem is studied, for example, in [73]. The conclusion is that different bound and scattering states are separated by 10 s of MeV, requiring $\tau \gtrsim 4\,\mathrm{fm}$. It may be possible to improve on this estimate by using variationally improved sources, but even for $\tau \simeq 2\,\mathrm{fm}$ the signal to noise is extremely poor for $A \gtrsim 4$. This shows that in simulations with fixed A the sign problem manifests itself as a noise problem. This is not surprising. One way to think about the sign problem is to view it as an overlap problem. The configurations that contribute to Z_{pq} have poor overlap with those that contribute to Z. The same phenomenon is at work here. Configurations generated at $\mu_B = 0$ reflect vacuum physics, and the lightest fermionic fluctuation is a pion. Large cancellations are required to explore the physics of multi-baryon states.

There are many attempts to find direct solutions to the sign problem, but at this time the only regime in which controlled calculations are feasible is the regime of small μ and high T. In this region the partition function can be expanded in a Taylor series in μ/T. The corresponding expansion coefficients are generalized susceptibilities that can be determined from lattice simulations at zero chemical potential. The susceptibilities not only determine the equation of state at finite baryon density, but also control fluctuations of conserved charges.

In addition to methods that are restricted to the regime $\mu \lesssim \pi T$, a number of proposals to explore QCD at high baryon density are being pursued. This includes new approaches, like integration over Lefshetz thimbles [74, 75], as well as novel variants of old approaches, like the complex Langevin method [76, 77], or the use of dual variables [78]. The ultimate promise of these methods is still unclear, but the central importance of the sign problem to computational physics continues to attract new ideas.

2.4.5 Real Time Properties

The basic trick in lattice QCD is the continuation of the path integral to imaginary time. This makes it possible to calculate the path integral by importance sampling, but it implies that we only have direct access to imaginary time correlation functions. For many observables this is not a serious problem. Thermodynamic observables, for example, are static quantities and no analytic continuation is necessary. The ground state contribution to a hadron correlation function is $\Pi(\tau) \sim e^{-m_H\tau}$ which is trivially continued to $\Pi(t) \sim e^{-im_H t}$. However, difficulties arise if one studies excited states, in particular resonances, the interaction between hadrons, or the real time response of many body systems at finite temperature and density.

Significant progress has been made in studying scattering processes, at least in the elastic regime. This is discussed in some of the later chapters of this book. Here, I will concentrate on the calculation of real time response functions. The prototypical example is the calculation of the shear viscosity of a QCD plasma using the retarded correlation function of the stress tensor T^{xy},

$$G_R^{xy,xy}(\omega,\mathbf{k}) = -i\int dt \int d^3x\, e^{i(\omega t-\mathbf{k}\cdot\mathbf{x})}\Theta(t)\langle[T^{xy}(\mathbf{x},t),T^{xy}(0,0)]\rangle\,, \tag{2.60}$$

The associated spectral function is defined by $\rho(\omega,\mathbf{k}) = -\operatorname{Im} G_R(\omega,\mathbf{k})$. The imaginary part of the retarded correlator is a measure of dissipation. This relation can be made more precise using fluid dynamics, which is an effective theory of the response function in the low energy, small momentum limit [79, 80].

Linearized fluid dynamics shows that the static response function is determined by the pressure of the fluid, and that the leading energy and momentum dependence is governed by transport coefficients. These relations can be used to derive Kubo formulas, expressions for the transport coefficients in terms of retarded correlation functions. The Kubo relation for the shear viscosity is

$$\eta = \lim_{\omega\to 0}\lim_{k\to 0}\frac{\rho^{xy,xy}(\omega,\mathbf{k})}{\omega}\,, \tag{2.61}$$

and similar results hold for the bulk viscosity, the thermal conductivity, and heavy quark diffusion constants.

The spectral function contains information about the physical excitations that carry the response. The euclidean path integral does not provide direct access to the retarded correlator or the spectral function. Lattice calculations are based on the relation between the spectral function and the imaginary energy (Matsubara) correlation function

$$G_E(i\omega_n) = \int\frac{d\omega}{2\pi}\frac{\rho(\omega)}{\omega - i\omega_n}\,, \tag{2.62}$$

where $\omega_n = 2\pi nT$ is the Matsubara frequency. The imaginary time correlation function is

$$G_E(\tau) = \int \frac{d\omega}{2\pi} K(\omega, \tau)\rho(\omega) , \tag{2.63}$$

where the kernel $K(\omega, \tau)$ is given by

$$K(\omega, \tau) = \frac{\cosh[\omega(\tau - 1/(2T))]}{\sinh[\omega/(2T)]} = [1 + n_B(\omega)]\, e^{-\omega\tau} + n_B(\omega)e^{\omega\tau} , \tag{2.64}$$

and $n_B(\omega)$ is the Bose distribution function. The imaginary time correlation function Eq. (2.63) was studied on the lattice in [81–84]. The basic idea for calculating transport coefficients is to numerically compute $G_E(\tau)$, invert the integral transform in Eq. (2.63) to obtain the spectral functions $\rho(\omega)$, and then study the limit $\omega \to 0$.

The problem is that $G_E(\tau)$ is computed on a small number of discrete lattice sites, and that the imaginary time correlator at distances on the order of $\beta/2$ is not very sensitive to the slope of the spectral function at low energy. Recent attempts to to address these problems and to obtain numerically stable spectral functions and reliable error estimates are based on Bayesian methods such as the maximum entropy method mentioned in Sect. 2.2, see [85, 86]. It is also possible to optimize the contribution from the transport peak by measuring the correlation functions of conserved charges, such as energy and momentum density, at non-zero spatial momentum [87, 88]. A possible issue with lattice calculations is that effects of poor resolution tend to bias the result towards small values of η/s, where s is the entropy density. The finite temperature spectral function satisfies the sum rule [89]

$$\frac{2}{\pi} \int d\omega \; [\eta(\omega) - \eta_{T=0}(\omega)] = \frac{3}{10} sT , \tag{2.65}$$

where $\eta(\omega) = \rho(\omega)/\omega$. On the lattice it is difficult to resolve sharp features in the spectral function. Roughly, the resolution is limited by the lowest Matsubara frequency πT. I will therefore assume that the $T \neq 0$ spectral function is a Lorentzian with width πT

$$\eta(\omega) - \eta_{T=0}(\omega) \simeq \frac{\eta(0)(\pi T)^2}{\omega^2 + (\pi T)^2} . \tag{2.66}$$

Then the integral on the lhs is equal to $\eta(0)\pi T$, and the sum rule predicts $\eta/s \sim 3/(10\pi)$, quite close to $\eta/s = 1/(4\pi)$. The lesson is that it is easy to obtain small values of η/s, and much more difficult to obtain large values of η/s, predicted by perturbative QCD [90].

The first calculation of the shear viscosity on the lattice was performed by Karsch and Wyld [81]. More recently, the problem of computing the shear and bulk viscosity in a pure gauge plasma near T_c was revisited by Meyer [82, 88].

He obtains $\eta/s = 0.102(56)$ and $\zeta/s = 0.065(17)$ at $T = 1.24T_c$. Shear viscosity is only weakly dependent on temperature, but bulk viscosity is strongly peaked near T_c. The value of η/s is consistent with experimental results, and with the prediction from holographic duality, $\eta/s = 1/(4\pi)$ [91].

2.5 Nonequilibrium QCD

In the remainder of this chapter I will discuss a number of coarse grained approaches to the non-equilibrium dynamics of QCD. These method are relevant to the study of nuclear collisions, in particular in the ultra-relativistic regime. This regime is explored experimentally at the Relativistic Heavy Ion Collider (RHIC) at Brookhaven National Laboratory and the Large Hadron Collider (LHC) at CERN. A rough time line of a heavy ion collision is shown in Fig. 2.4. Initial nucleon-nucleon collisions release a large number of quarks and gluons. This process is described by the full non-equilibrium quantum field theory, but there are a number of approximate descriptions that may be useful in certain regimes. The first is a classical field theory description in terms of highly occupied classical gluon fields. The second is a kinetic theory in terms of quark and gluon quasi-particles. Finally, there is a new approach, which is a description in terms of a dual gravitational theory.

Theories of the initial state demonstrate that there is a tendency towards local equilibration. If local equilibrium is achieved then a simpler theory, fluid dynamics is applicable. Fluid dynamics is very efficient in the sense that it deals with a small number of variables, the conserved densities of particle number, energy and momentum, and that it has very few parameters, an equation of state and a set of transport coefficients. The fluid dynamic stage of a heavy ion collision has a finite duration. Eventually the density becomes too low and local equilibrium can no longer be maintained. At this point kinetic theory is again relevant, now

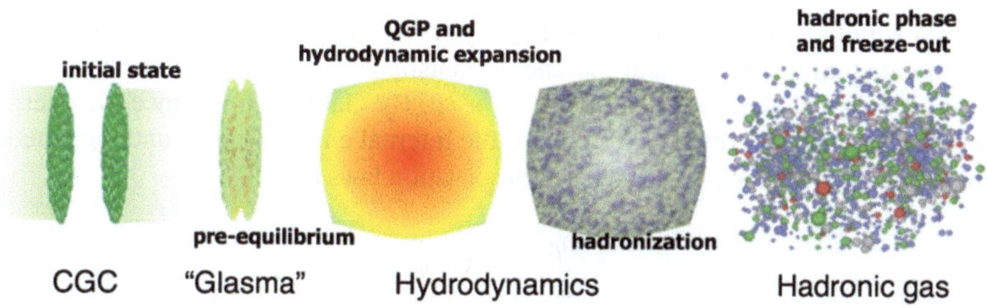

Fig. 2.4 Schematic time evolution of a heavy ion collision. Figure courtesy of S. Bass. CGC refers to the color glass condensate, a semi-classical model of the overpopulated gluon configuration in the initial state of a heavy ion collision. Glasma refers to the non-equilibrium evolution of this state into a locally equilibrated plasma. Hydrodynamics is the theory of the time evolution of a locally equilibrated fireball, and hadronic phase refers to the late time kinetic stage of the collision

formulated in terms of hadronic quasi-particles. All the theories we have mentioned, fluid dynamics, kinetic theory, classical field theory, and holography, have reached a high degree of sophistication and I will point to text books and review for detailed introductions. Nevertheless, the basic ideas are quite simple, and I will provide some examples in the following sections.

2.5.1 Fluid Dynamics

I begin with fluid dynamics, because it is the most general and in some ways the simplest non-equilibrium theory. It is important to remember, however, that fluid dynamics is a very rich framework, both mathematically and in terms of the range of phenomena that one may encounter. In the following I will focus on the non-relativistic theory. There is no fundamental difference between the relativistic and non-relativistic theories, but some simplifications appear in the non-relativistic regime. Non-relativistic fluid dynamics is used in many areas of physics, including the physics of cold atomic Fermi gases and neutron stars. The relativistic theory is relevant to high energy heavy ion collisions and supernova explosions. Introductions to relativistic fluid dynamics can be found in [92–94].

Fluid dynamics reduces the complicated non-equilibrium many-body problem to equations of motion for the conserved charges. The reason that this is possible is the separation of scales between the microscopic collision time τ_{micro}, and the relaxation time τ_{macro} of hydrodynamic variables. A generic perturbation of the system decays on a time scale on the order of τ_{micro}, irrespective of the typical length scale involved. Here, τ_{micro} is determined by microscopic time scales, such as the typical collision time between quasi-particles. A fluctuation of a conserved charge, on the other hand, cannot decay locally and has to relax by diffusion or propagation. The relevant time scale τ_{macro} increases with the length scale of the perturbation. As a consequence, when we focus on sufficiently large scales we can assume $\tau_{macro} \gg \tau_{micro}$, and focus on the evolution of conserved charges.

In a simple non-relativistic fluid the conserved charges are the mass density ρ, the momentum density $\boldsymbol{\pi}$, and the energy density $\mathcal{E}$. The momentum density can be used to define the fluid velocity, $\mathbf{u} = \boldsymbol{\pi}/\rho$. By Galilean invariance the energy density can then be written as the sum of the internal energy density and the kinetic energy density, $\mathcal{E} = \mathcal{E}_0 + \frac{1}{2}\rho u^2$. The conservation laws are

$$\frac{\partial \rho}{\partial t} = -\boldsymbol{\nabla} \cdot \boldsymbol{\pi}, \tag{2.67}$$

$$\frac{\partial \pi_i}{\partial t} = -\nabla_j \Pi_{ij}, \tag{2.68}$$

$$\frac{\partial \mathcal{E}}{\partial t} = -\boldsymbol{\nabla} \cdot \mathbf{j}^{\epsilon}. \tag{2.69}$$

In order for these equations to close we have to specify constitutive relations for the stress tensor Π_{ij} and the energy current $\mathbf{j}^\epsilon$. Since fluid dynamics is an effective long wavelength theory we expect that the currents can be systematically expanded in gradients of the hydrodynamic variables ρ, $\mathbf{u}$ and $\mathcal{E}_0$. At leading order the stress tensor contains no derivatives and the structure is completely fixed by rotational symmetry and Galilean invariance. We have

$$\Pi_{ij} = \rho u_i u_j + P\delta_{ij} + \delta\Pi_{ij}\,, \tag{2.70}$$

where $P = P(\rho, \mathcal{E}_0)$ is the equation of state and $\delta\Pi_{ij}$ contains gradient terms. The approximation $\delta\Pi_{ij} = 0$ is called ideal fluid dynamics, and the equation of motion for $\boldsymbol{\pi}$ is known as the Euler equation. Ideal fluid dynamics is time reversal invariant and the entropy is conserved. If gradient terms are included then time reversal invariance is broken and the entropy increases. We will refer to $\delta\Pi_{ij}$ as the dissipative stresses. At first order in the gradient expansion $\delta\Pi_{ij}$ can be written as $\delta\Pi_{ij} = -\eta\sigma_{ij} - \zeta\delta_{ij}\langle\sigma\rangle$ with

$$\sigma_{ij} = \nabla_i u_j + \nabla_j u_i - \frac{2}{3}\delta_{ij}\langle\sigma\rangle\,, \qquad \langle\sigma\rangle = \boldsymbol{\nabla}\cdot\mathbf{u}\,. \tag{2.71}$$

The dissipative stresses are determined by two transport coefficients, the shear viscosity η and the bulk viscosity ζ. The energy current is given by

$$\mathbf{j}^{\,\epsilon} = \mathbf{u}w + \delta\mathbf{j}^\epsilon\,, \tag{2.72}$$

where $w = P + \mathcal{E}$ is the enthalpy. At leading order in the gradient expansion

$$\delta j_i^\epsilon = u_j\delta\Pi_{ij} - \kappa\nabla_i T\,, \tag{2.73}$$

where κ is the thermal conductivity. The second law of thermodynamics implies that η, ζ and κ must be positive. The equation of motion for $\boldsymbol{\pi}$ at first order in gradients is known as the Navier-Stokes equation, and Eq. (2.73) is Fourier's law of heat conduction.

It is sometimes useful to rewrite the fluid dynamic equations using the comoving derivatives $D_t = \partial_t + \mathbf{u}\cdot\boldsymbol{\nabla}$. The equations are

$$D_t\rho = -\rho\boldsymbol{\nabla}\cdot\mathbf{u}\,, \tag{2.74}$$

$$D_t u_i = -\frac{1}{\rho}\nabla_j\left(\delta_{ij}P + \delta\Pi_{ij}\right)\,, \tag{2.75}$$

$$D_t\epsilon = -\frac{1}{\rho}\nabla_i\left(u_i P + \delta j_i^{\mathcal{E}}\right)\,, \tag{2.76}$$

where $\epsilon = \mathscr{E}/\rho$ is the energy per mass. This is called the Lagrangian form of the equations, in contrast to the Eulerian form given above. The Eulerian form is more naturally implemented on a fixed space-time lattice, whereas the Lagrangian form lends itself to a discretization where the computational cell is dragged along with the fluid.

2.5.2 Computational Fluid Dynamics

The fluid dynamic equations form a set of partial differential equations (PDEs) that can be solved in a variety of ways. I will focus here on grid based methods. The main difficulties that a numerical method needs to address are: (1) The existence of surfaces of discontinuity (shocks), (2) the need to implement global conservation laws exactly, even on a coarse lattice, (3) the existence of instabilities (turbulence), and the need to deal with solutions that involve many different length scales.

In the following I will discuss a numerical scheme that addresses these issues in a fairly efficient way, the PPM algorithm of Collela and Woodward [95], as implemented in the VH1 code by Blondin and Lufkin [96] and extended to viscous fluids in [97]. The first observation is that it is sufficient to construct a 1-d algorithm. Higher dimensional methods can be constructed by combining updates in different directions. Note that the coordinate system can be curvilinear, for example 3-d spherical or cylindrical coordinates, or the Milne coordinate system that is used for longitudinally expanding quark gluon plasmas.

The basic 1-d algorithm consists of a Lagrangian time step followed by a remap onto an Eulerian grid. I will denote the 1-d velocity by u and write the equation of mass conservation in terms of a mass variable m

$$\frac{\partial \tau}{\partial t} - \frac{\partial u}{\partial m} = 0\,, \tag{2.77}$$

where $\tau = \rho^{-1}$ and

$$m(r) = \int_{r_0}^{r} dr\, \rho(r)\,. \tag{2.78}$$

Here, I restrict myself to flat coordinate systems. In curvilinear coordinates Eqs. (2.77) and (2.78) contain suitable volume factors. Equation (2.77) is solved by

$$\frac{dr}{dt} = u\,(m(r), t)\ , \tag{2.79}$$

which is the equation for the Lagrange coordinate. In terms of the mass coordinate $m(r)$ the momentum and energy equations are

$$\frac{\partial u}{\partial t} + \frac{\partial P}{\partial m} = 0\,, \tag{2.80}$$

$$\frac{\partial \epsilon}{\partial t} + \frac{\partial (uP)}{\partial m} = 0\,, \tag{2.81}$$

where I have only written down the ideal contributions to the stress tensor and energy current. To put these equations on a grid I focus on the mass integrated quantities

$$U_j^n = \frac{1}{\Delta m_j} \int_{m_{j-1/2}}^{m_{j+1/2}} U(m, t^n) dm \tag{2.82}$$

where U is any of the hydrodynamic variables (τ, u, ϵ), Δm_j is the mass contained in the cell j, and $m_{j+1/2} = \sum_k^j \Delta m_k$. We can now integrate the conservation laws ((2.80), (2.77)). The result is

$$u_j^{n+1} = u_j^n + \frac{\Delta t}{\Delta m_j} \left(\bar{P}_{j-1/2} - \bar{P}_{j+1/2}\right)\,, \tag{2.83}$$

$$\epsilon_j^{n+1} = \epsilon_j^n + \frac{\Delta t}{\Delta m_j} \left(\bar{u}_{j-1/2}\bar{P}_{j-1/2} - \bar{u}_{j+1/2}\bar{P}_{j+1/2}\right)\,, \tag{2.84}$$

where I have introduced the cell face averages $\bar{u}_{j\pm 1/2}$ and $\bar{P}_{j\pm 1/2}$. These quantities can be obtained by parabolic interpolation from the cell integrated values. The PPM scheme introduced in [95] uses a method for constructing cell face averages which conserves the cell integrated variables.

This scheme addresses the second issue mentioned above. The first issue, the existence of shocks, can be taken into account by refining the method for calculating the cell face averages. The observation is that one can make use of exact solution of the equations of fluid dynamics in the case of piecewise constant one-dimensional flows, known as the Riemann problem. We can view $\bar{u}_{j+1/2}$ and $\bar{P}_{j+1/2}$ as the solution of a Riemann problem with left state u_j, P_j and right state u_{j+1}, P_{j+1}. The PPM code contains a simple iterative Riemann solver described in [95]. Using $\bar{u}_{j\pm 1/2}$ and $\bar{P}_{j\pm 1/2}$ the Lagrange step is given by:

```
do n = nmin-3, nmax+3

! density evolution. lagrangian code, so all we have to do is
    watch the
! change in the geometry.

  r(n)  = r(n) * ( dvol1(n) / dvol(n) )
  r(n)  = max(r(n),smallr)
```

```
! velocity evolution due to pressure acceleration and forces.

  uold (n) = u(n)
  u(n) = u(n) - dtbdm(n)*(pmid(n+1)-pmid(n))*0.5*(amid(n+1)+amid
       (n)) &
       + 0.5*dt*(fict0(n)+fict1(n))

! total energy evolution

  e(n) = e(n) - dtbdm(n)*(amid(n+1)*upmid(n+1) - amid(n)*upmid(n
       ))
  q(n) = e(n) - 0.5*(u(n)**2+v(n)**2+w(n)**2)
  p(n) = max(r(n)*q(n)*gamm,smallp)

enddo
```

Here, `r(n)` is the density, `u(n)` is the velocity, and `e(n)` is the energy per mass. The transverse components of the velocity are `v(n)`, `w(n)`. In cartesian coordinates the volume and area factors `dvol(n)`, `amid(n)` are equal to unity, and the fictitious forces `fict(n)` vanish.

After the Lagrange step the hydrodynamic variables have to be remapped onto a fixed Eulerian grid. This can be done using the parabolic interpolation mentioned above. The advantage of the remap step is that it is simple to change the grid resolution in the process. Finally, we have to specify the time step and grid resolution. The grid resolution is determined by the requirement that $(\Delta x)\nabla_x U \ll U$, where Δx is the cell size, and U is any of the hydrodynamic variables. Note that there is no need to worry about discontinuities, because shocks are captured by the Riemann solver. Also note that the PPM scheme has at least second order accuracy, so that relatively coarse grids can be used. The time step is determined by the Courant criterion $c\Delta x \leq \Delta t$, where c is the maximum of the speed of sound and the local flow velocity. This criterion ensures that the domain of dependence of any physical variable does not exceed the cell size.

In general, modern hydro codes are very fast and efficient. The main difficulty is that $3 + 1$ dimensional simulations may require a lot of memory, and that some physical phenomena, such as turbulent convection and shock instabilities in supernovae, require very high resolution. One of the frontiers of numerical hydrodynamics is the problem of dealing with systems that transition from fluid dynamics to ballistic behavior at either early or late times, or systems in which the density varies by a very large factor. Problems of this type arise in the early and late time dynamics of heavy ion collisions, the dilute corona of cold atomic gases, and the transition from hydrodynamics to free streaming in the neutrino transport in a supernova explosions. Recent progress in this direction includes the development of the anisotropic hydrodynamics method [98–101], and applications of the lattice Boltzmann method to problems in nuclear and atomic physics [102, 103].

In the relativistic regime recent progress includes the development of stable and causal viscous fluid dynamics codes [92, 94]. The problem with a naive implementation of the relativistic Navier-Stokes equation derived by Landau is that viscous stresses are determined by the instantaneous value of the shear strain $\nabla_i u_j$. This leads to acausal propagation of shear waves and possible instabilities. This is not a fundamental problem with fluid dynamics. Acausal behavior occurs in the regime of high wave numbers in which fluid dynamics is not expected to be reliable. However, high wave number instabilities prohibit numerical implementations. The solution is to go to next order in the gradient expansion, which includes the finite relaxation time of viscous stresses. In practice, second order fluid dynamics codes are usually based on the idea of transient fluid dynamics. In this method, the shear stresses $\delta\Pi_{ij}$ are promoted to fluid dynamic variables, which satisfy separate fluid dynamic equations, see [92, 94].

2.5.3 Kinetic Theory

Fluid dynamics is based on the assumption of local thermal equilibrium and requires the mean free path to be small compared to the characteristic scales of the problem. When this condition is not satisfied a more microscopic approach to the non-equilibrium problem is required. The simplest method of this type is kinetic theory, which is based on the existence of well defined quasi-particles. This implies, in particular, that the width of a quasi-particle has to be small compared to its energy. In this case we can define the phase space density $f(\mathbf{x}, \mathbf{p}, t)$ of quasi-particles. In general, there can be many different kinds of quasi-particles, labeled by their spin, charge, and other quantum numbers. The phase space distribution determines the conserved densities that enter the hydrodynamic description. For example, the mass density is given by

$$\rho(\mathbf{x}, t) = \int d\Gamma \; mf(\mathbf{x}, \mathbf{p}, t) \,, \tag{2.85}$$

where $d\Gamma = d^3p/(2\pi)^3$. The momentum density is

$$\pi(\mathbf{x}, t) = \int d\Gamma \; mv_p f(\mathbf{x}, \mathbf{p}, t) \,, \tag{2.86}$$

where $v_p = \nabla_p E_p$ is the quasi-particle velocity and E_p is the quasi-particle energy. In general, the quasi-particle energy can be a functional of the phase distribution $f(\mathbf{x}, \mathbf{p}, t)$. This takes into account possible in-medium modifications of particle properties. If E_p is a functional of $f(\mathbf{x}, \mathbf{p}, t)$ then the total energy of the system is not just given by the integral of $E_p f(\mathbf{x}, \mathbf{p}, t)$. Instead, we must construct an energy density functional $\mathscr{E}[f]$ that satisfies [104]

$$E_p = \frac{\delta \mathscr{E}}{\delta f_p} . \tag{2.87}$$

The equation of motion for the distribution function is the Boltzmann equation

$$\left(\partial_t + \mathbf{v} \cdot \nabla_x - \mathbf{F} \cdot \nabla_p\right) f(\mathbf{x}, \mathbf{p}, t) = C[f] , \tag{2.88}$$

where $\mathbf{F} = -\nabla_x E_p$ is a force, and $C[f_p]$ is the collision term. For dilute systems the collision term is dominated by binary scattering and

$$C[f_p] = - \prod_{i=2,3,4} \left(\int d\Gamma_i \right) w(1,2;3,4) \left(f_1 f_2 - f_3 f_4 \right) , \tag{2.89}$$

where $f_i = f(\mathbf{x}, \mathbf{p}_i, t)$. The transition rate is given by

$$w(1,2;3,4) = (2\pi)^4 \delta\Big(\sum_i E_i \Big) \delta\Big(\sum_i \mathbf{p}_i \Big) \, |\mathscr{A}|^2 , \tag{2.90}$$

where $\mathscr{A}$ is the scattering amplitude. For non-relativistic s-wave scattering $\mathscr{A} = 4\pi a/m$, where a is the scattering length.

The Boltzmann equation is a 6 + 1 dimensional partial integro-differential equation, and direct methods of integration, similar to those used in computational fluid dynamics, are impractical. Standard methods for solving the Boltzmann equation rely on sampling phase space using Monte Carlo methods. In nuclear physics the test particle method for solving the Boltzmann equation was popularized by Bertsch and Das Gupta [105]. Below, I will present a simple non-relativistic algorithm described by Lepers et al. [106].

The main idea is to represent the distribution as a sum of delta functions

$$f(\mathbf{x}, \mathbf{p}, t) = \frac{N}{N_t} \sum_{i=1}^{N_t} (2\pi)^3 \delta(\mathbf{p} - \mathbf{p}_i(t)) \delta(\mathbf{x} - \mathbf{x}_i(t)) , \tag{2.91}$$

where N is the number of particles, the integral of $f(\mathbf{x}, \mathbf{p}, t)$ over phase space, and N_t is the number of test particles. In typical applications $N_t \gg N$, but if N is already very large it is possible to run simulations with $N_t < N$. Phase space averages can be computed as averages over test particles

$$\bar{F} = \frac{1}{N} \int d^3x \int d\Gamma \, f(\mathbf{x}, \mathbf{p}, t) F(\mathbf{x}, \mathbf{p}) = \frac{1}{N_t} \sum_{i=1}^{N_t} F(\mathbf{x}_i, \mathbf{p}_i) . \tag{2.92}$$

In practice this requires some smoothing, and the delta functions are replaced by Gaussian distributions

$$\delta(\mathbf{p}-\mathbf{p}_i)\delta(\mathbf{x}-\mathbf{x}_i) \rightarrow g_{w_p}(\mathbf{p}-\mathbf{p}_i)g_{w_x}(\mathbf{x}-\mathbf{x}_i)\,, \tag{2.93}$$

where $g_w(\mathbf{x})$ is a normalized Gaussian with width w. The widths w_x and w_p are chosen such that the delta function singularities are smoothed out, but physical structures of the distribution function $f(\mathbf{x}, \mathbf{p}, t)$ are preserved.

If there is no collision term the equation of motion for the distribution function is Hamilton's equation for the test particle positions and momenta

$$\frac{d\mathbf{x}_i}{dt} = \frac{\mathbf{p}_i}{m}\,, \qquad \frac{d\mathbf{p}_i}{dt} = \mathbf{F}_i\,. \tag{2.94}$$

These equations can be solved with high accuracy using a staggered leapfrog algorithm

$$\mathbf{v}_i(t_{n+1/2}) = \mathbf{v}_i(t_n) + \mathbf{a}_i(t_n)\Delta t/2\,, \tag{2.95}$$

$$\mathbf{r}_i(t_{n+1}) = \mathbf{r}_i(t_n) + \mathbf{v}_i(t_{n+1/2})\Delta t\,, \tag{2.96}$$

$$\mathbf{v}_i(t_{n+1}) = \mathbf{v}_i(t_{n+1/2}) + \mathbf{a}_i(t_{n+1})\Delta t/2\,, \tag{2.97}$$

where $\mathbf{a}_i = \mathbf{F}_i/m$ is the acceleration of particle i, and $\Delta t = t_{n+1} - t_n$ is the time step of the algorithm. The size of the time step depends on the specific problem, but a good check is provided by monitoring conservation of energy.

The collision term is treated stochastically, by allowing the test particles to collide with the scaled cross section $\sigma_t = (N/N_t)\sigma$. To determine whether a collision occurs we go through all pairs of particles and compute the relative distance $\mathbf{r}_{ij} = \mathbf{r}_i - \mathbf{r}_j$ and velocity $\mathbf{v}_{ij} = \mathbf{v}_i - \mathbf{v}_j$. We then determine whether on the current trajectory the time of closest approach will be reached during the next time step. This happens if $t_{min} = t_n - \mathbf{r}_{ij} \cdot \mathbf{v}_{ij}/\mathbf{v}_{ij}^2$ satisfies $|t_{min} - t_n| \leq \Delta t/2$. In that case we compute

$$r_{min}^2 = \mathbf{r}_{ij}^2 - \frac{(\mathbf{r}_{ij} \cdot \mathbf{v}_{ij})^2}{\mathbf{v}_{ij}^2} \tag{2.98}$$

and check if $\pi r_{min}^2 < \sigma_t$. If this condition is satisfied then the collision is allowed to take place. For an s-wave elastic collision we propagate the particles to t_{min}, randomize their relative velocity $\mathbf{v}_{ij}$, and then propagate them back to t_n. Higher partial wave amplitudes are easy to implement by randomizing $\mathbf{v}_{ij}$ with suitable probability distributions. After all pairs have been checked we perform the velocity and position update in Eqs. (2.95)–(2.97).

There are a number of refinements that can be included. At low temperature Pauli-blocking has to be taken into account. This can be done by computing the phase space densities $f(\mathbf{r}_i, \mathbf{p}_i, t)$ for the collision products, and accepting the collision with probability $(1 - f_i)(1 - f_j)$. At higher energies relativistic effects are important. Relativistic effects in the particle propagation are easy to incorporate, but the treatment of the collision term is more subtle. The problem is that a finite collision cross section, treated geometrically, will lead to instantaneous interactions at a distance. Additional difficulties arise from the treatment of resonances, pair production and annihilation, n-body processes, etc. There are a number of codes on the market that address these issues, and that have been tuned against existing data on pp, pA and AA interactions in the relativistic regime. Examples include UrQMD [107], GiBUU [108], HSD [109], and others.

At high energies the initial pp collisions are very inelastic, and one has to rely on Monte Carlo generators developed in the high energy physics community. A possible alternative is to use a purely partonic kinetic theory that involves scattering between quark and gluon quasi-particles. There are some subtleties with this approach, having to do with the problem of how to include screening and damping of the exchanged gluons, soft gluon radiation, etc. I will not attempt to discuss these issues here, and I refer the reader to the original literature [110, 111].

2.5.4 Classical Field Theory

An interesting simplification occurs if the occupation numbers are large, $f \gg 1$. This is argued to happen for the gluons in the initial state of a heavy ion collision [112]. In this limit the classical kinetic theory is equivalent to a classical field theory [113]. Indeed, if the occupations numbers are non-perturbative, $f \gtrsim 1/g$, the kinetic theory no longer applies, and we have to rely on classical field theory. In general the classical action is not known, but in the weak coupling limit the bare QCD action can be used.

Classical QCD simulation have been used to study a number of issues, such as particle production from an overpopulated gluon field, and the possible approach to thermal equilibrium. Instabilities in the classical field evolution may play an important role in speeding up the equilibration process. Here, I will briefly describe a method for solving classical evolution equations on a space-time lattice, following the recent review [114].

In order to construct a Hamiltonian approach to lattice QCD I start from the Wilson action in Minkowski space with separate coupling constants β_0 and β_s in the temporal and spatial direction

$$S[U] = -\frac{\beta_0}{2N_c}\sum_x\sum_{i=1}^{3}\mathrm{Tr}\left(W_{0i}(x) + W_{0i}^\dagger(x) - 2\right)$$
$$+\frac{\beta_s}{2N_c}\sum_x\sum_{i<j}\mathrm{Tr}\left(W_{ij}(x) + W_{ij}^\dagger(x) - 2\right), \tag{2.99}$$

In the continuum limit, we expect

$$\beta_0 = \frac{2N_c a}{g^2\Delta t}, \qquad \beta_s = \frac{2N_c\Delta t}{g^2 a}. \tag{2.100}$$

where a and Δt are spatial and temporal lattice spacings. In order to construct a Hamiltonian we have to fix the gauge freedom of the theory. Here, I will use the temporal axial gauge, $A_0 = 0$. In this case the canonical variables are the spatial gauge potentials and the conjugate momenta are the electric fields. On the lattice the gauge $A_0 = 0$ corresponds to setting all temporal gauge links to the identity, $U_0(x) = 1$. The canonical variables are given by the spatial gauge links $U_j(x)$, and the conjugate momenta are the temporal plaquettes $W_{0j}(x)$. In the continuum limit

$$A_j^a(x) = \frac{2i}{ag}\mathrm{Tr}\big[\lambda^a U_j(x)\big], \tag{2.101}$$

$$E_j^a(x) = \frac{2i}{ag\Delta t}\mathrm{Tr}\big[\lambda^a W_{0j}(x)\big]. \tag{2.102}$$

Varying the action Eq. (2.99) with respect to $U_j(x)$ gives an equation of motion for E_j

$$E_j^a(t+\Delta t,\mathbf{x}) = E_j^a(t,\mathbf{x}) + \frac{i\Delta t}{ga^3}\sum_k\Big\{\mathrm{Tr}\left[\lambda^a U_j(x)U_k(x+\hat{j})U_j^\dagger(x+\hat{k})U_k^\dagger(x)\right]$$
$$+\,\mathrm{Tr}\left[\lambda^a U_j(x)U_k^\dagger(x+\hat{j}-\hat{k})U_j^\dagger(x-\hat{k})U_k(x-\hat{k})\right]\Big\}. \tag{2.103}$$

We note that $E_j^a(t+\Delta t,\mathbf{x})$ is determined by the electric fields and the spatial gauge links at time t. Using Eq. (2.102) and the electric field E_j^a at time $t+\Delta t$ we can compute the temporal plaquette $W_{0j}(x)$ at $t+\Delta t$. This result can be used to evolve the spacelike gauge links

$$U_j(t+\Delta t,\mathbf{x}) = W_{0j}(x)U_j(x). \tag{2.104}$$

Together, Eqs. (2.103) and (2.104) describe a staggered leapfrog algorithm, similar to Eqs. (2.95)–(2.97) above. An important constraint on the numerical evolution is provided by Gauss law. Varying the lattice action with respect to U_0 before imposing temporal axial gauge gives

$$\sum_j \left[E_j^a(x) - U_j^\dagger(x-\hat{j}) E_j^a(x-\hat{j}) U_j(x-\hat{j}) \right] = 0\,. \tag{2.105}$$

This constraint is preserved by the evolution equations.

The classical field equations are exactly scale invariant and there is no dependence on the coupling constant g. Physical quantities, like the energy momentum tensor, explicitly depend on g. In practice classical field simulations require a model for the initial conditions and the corresponding coupling. The initial conditions are typically an ensemble of gauge fields distributed according to some distribution, for example an anisotropic Gaussian in momentum space. The anisotropy is assumed to be a consequence of the strong longitudinal expansion of the initial state of a heavy ion collision. Physical observables are determined by averages the evolved fields over the initial ensemble.

Note that a purely classical field evolution does not thermalize. A thermal ensemble of classical fields would satisfy the equipartition law, and the total energy would be dominated by modes near the lattice cutoff. This is the Rayleigh-Jeans UV catastrophe. However, classical field evolution has interesting non-thermal fixed points [115], which may play a role in thermalization.

The classical field framework has been extended in a variety of ways. One direction is the inclusion of quantum fluctuations on top of the classical field [116]. Another problem is the inclusion of modes that are not highly populated. In the hard thermal loop approximation one can show that hard modes can be described as colored particles interacting with the classical field corresponding to the soft modes [117]. The equations of motion for the colored particles are known as Wong's equations [118]. Numerical studies can be found in [119].

2.5.5 Nonequilibrium QCD: Holography

A new approach to quantum fields in and out of equilibrium is provided by the AdS/CFT correspondence [120–124]. The AdS/CFT correspondence is a holographic duality. It asserts that the dynamics of a quantum field theory defined on the boundary of a higher dimensional space is encoded in boundary correlation functions of a gravitational theory in the bulk. The correspondence is simplest if the boundary theory is strongly coupled and contains a large number N of degrees of freedom. In this case the bulk theory is simply classical Einstein gravity. The partition function of the boundary quantum field theory (QFT) is

$$Z_{QFT}[J_i] = \exp\left(-S\left[\phi_i|_{\partial M} = J_i\right]\right)\,, \tag{2.106}$$

where J_i is a set of sources in the field theory, S is the gravitational action, ϕ_i is a dual set of fields in the gravitational theory, and ∂M is the boundary of AdS_5. The fields ϕ_i satisfy classical equations of motions subject to boundary conditions on ∂M.

The original construction involves a black hole in AdS_5 and is dual to a relativistic fluid governed by a generalization of QCD known as $\mathcal{N} = 4$ super Yang-Mills theory. This theory is considered in the limit of a large number of colors N_c. The gravitational theory is Einstein gravity with additional matter fields that are not relevant here. The AdS_5 black hole metric is

$$ds^2 = \frac{(\pi T R_a)^2}{u}\left(-f(u)dt^2 + d\mathbf{x}^2\right) + \frac{R_a^2}{4u^2 f(u)}du^2 , \tag{2.107}$$

where $\mathbf{x}, t$ are Minkowski space coordinates, and u is a "radial" coordinate where $u = 1$ is the location of the black hole horizon and $u = 0$ is the boundary. T is the temperature, R_a is the AdS radius, and $f(u) = 1 - u^2$.

It is instructive to check that this metric does indeed provide a solution to the Einstein equations with a negative cosmological constant. This can be done using a simple Mathematica script. I begin by defining the metric and its inverse:

```
(* metric *)
(* ------ *)
n = 5;
coord = {t, x, y, z, u};
f[u_] := 1 - u^2
metric = DiagonalMatrix[{-f[u]/u*(Pi*T*Ra)^2, (Pi*T*Ra)^2/u, (Pi
    *T*Ra)^2/
   u, (Pi*T*Ra)^2/u, Ra^2/(4*u^2*f[u])}]
inversemetric = Simplify[Inverse[metric]]
```

From the metric I compute the Christoffel symbols

$$\Gamma^{\mu}_{\alpha\beta} = \frac{1}{2} g^{\mu\nu}\left(\partial_\alpha g_{\nu\beta} + \partial_\beta g_{\nu\alpha} - \partial_\mu g_{\alpha\beta}\right) , \tag{2.108}$$

the Riemann tensor

$$R^{\mu}{}_{\nu\alpha\beta} = \partial_\alpha \Gamma^{\mu}_{\nu\beta} - \partial_\beta \Gamma^{\mu}_{\nu\alpha} + \Gamma^{\rho}_{\nu\beta}\Gamma^{\mu}_{\rho\alpha} - \Gamma^{\rho}_{\nu\alpha}\Gamma^{\mu}_{\rho\beta} , \tag{2.109}$$

the Ricci tensor $R_{\alpha\beta} = R^{\mu}{}_{\alpha\mu\beta}$, and the scalar curvature $R = R^{\mu}{}_{\mu}$. Finally, I compute the Einstein tensor $G_{\mu\nu} = R_{\mu\nu} - \frac{1}{2} g_{\mu\nu} R$.

```
(* Christoffel Symbols *)
(* ------------------- *)
affine := affine = Simplify[
  Table[(1/2)*
    Sum[(inversemetric[[i, s]])*(D[metric[[s, j]], coord[[k]]] +
       D[metric[[s, k]], coord[[j]]] -
       D[metric[[j, k]], coord[[s]]]), {s, 1, n}], {i, 1, n}, {j
          , 1,
     n}, {k, 1, n}]]
```

```
(* Riemann Tensor *)
(* -------------- *)
riemann := riemann = Simplify[Table[
   D[affine[[i, j, l]], coord[[k]]] -
    D[affine[[i, j, k]], coord[[l]]] +
    Sum[affine[[s, j, l]]*affine[[i, k, s]] -
      affine[[s, j, k]]*affine[[i, l, s]], {s, 1, n}], {i, 1, n}
        , {j,
     1, n}, {k, 1, n}, {l, 1, n}]]

(* Ricci Tensor *)
(* ------------ *)
ricci := ricci = Simplify[
   Table[Sum[riemann[[i, j, i, l]], {i, 1, n}], {j, 1, n}, {l,
        1, n}]]

(* scalar curvature *)
(* ---------------- *)
scalar = Simplify[
  Sum[inversemetric[[i, j]]*ricci[[i, j]], {i, 1, n}, {j, 1, n}
      ]]

(* Einstein tensor *)
(* --------------- *)
einstein = Simplify[ricci - (1/2)*scalar*metric]
```

Now I can check the equation of motion, $G_{\mu\nu} = \frac{\Lambda}{2} g_{\mu\nu}$, where the cosmological constant is determined by the AdS radius R.

```
(* Field equation with cosmological constant *)
(* ----------------------------------------- *)
lam = 12/Ra^2;
Simplify[einstein - lam/2*metric]
```

In the boundary theory the metric couples to the stress tensor $\Pi_{\mu\nu}$. Correlation functions of the stress tensor can be found by linearizing the bulk action around the AdS_5 solution, $g_{\mu\nu} = g^0_{\mu\nu} + \delta g_{\mu\nu}$. Small oscillations of the off-diagonal strain δg^y_x are particularly simple, because the equation of motion for $\phi \equiv g^y_x$ is that of a minimally coupled scalar

$$\frac{1}{\sqrt{-g}} \partial_\mu \left(\sqrt{-g} g^{\mu\nu} \partial_\nu \phi \right) = 0 . \tag{2.110}$$

The wave equation can be obtained using the metric coefficients defined above.

```
(* \sqrt{-g} g^{\mu\nu} \partial_{nu} \Phi(t,z,u) *)
(* ------------------------------------------- *)
SqrtG = Simplify[Sqrt[-Det[metric]], {Ra > 0, T > 0, u > 0}]
dnuPhi = Table[D[Phi[t, z, u], coord[[i]]], {i, 1, n}];
DnuPhi = SqrtG*inversemetric.dnuPhi;
```

```
(* Laplacian, up to factor \sqrt{-g} *)
(* -------------------------------- *)
DPhi = FullSimplify[Sum[D[DnuPhi[[nu]], coord[[nu]]], {nu, 1, n}
    ]]

(* harmonic space and time dependence *)
(* --------------------------------- *)
DPhiS =
 DPhi /. { D[Phi[t, z, u], {z, 2}] -> -k^2*fp,
  D[Phi[t, z, u], {t, 2}] -> -w^2*fp,
  D[Phi[t, z, u], {u, 2}] -> fpPP, D[Phi[t, z, u], {u, 1}] ->
     fpP}
```

In the case of harmonic dependence on the Minkowski coordinates $\delta g_x^y = \phi_k(u)e^{ikx-i\omega t}$ the fluctuations are governed by the wave equation

$$\phi_k''(u) - \frac{1+u^2}{uf(u)}\phi_k'(u) + \frac{\omega^2 - k^2 f(u)}{(2\pi T)^2 uf(u)^2}\phi_k(u) = 0\,. \tag{2.111}$$

This differential equation has two linearly independent solutions. The retarded correlation function corresponds to picking a solution that is purely infalling at the horizon [121]. The retarded correlation function $G_R(\omega, k)$ defined in Eq. (2.60) is determined by inserting the solution into the Einstein-Hilbert action, and then computing the variation with respect to the boundary value of δg_x^y.

The infalling solution can be expressed as

$$\phi_k(u) = (1-u)^{-i\mathfrak{w}/2} F_k(u) \tag{2.112}$$

where $\mathfrak{w} = \omega/(2\pi T)$ and the first factor describes the near horizon behavior. The function $F_k(u)$ can be obtained as an expansion in $\mathfrak{w}$ and $\mathfrak{k} = k/(2\pi T)$. At second order in $O(\mathfrak{w}$ and $\mathfrak{k}$ the solution is [125]

$$F_k(u) = 1 - \frac{i\mathfrak{w}}{2}\log\left(\frac{1+u}{2}\right) + \frac{\mathfrak{w}^2}{8}\left\{\left[8 - \frac{8\mathfrak{k}^2}{\mathfrak{w}^2} + \log\left(\frac{1+u}{2}\right)\right]\log\left(\frac{1+u}{2}\right) - 4Li_2\left(\frac{1-u}{2}\right)\right\}\,. \tag{2.113}$$

In the opposite limit, $\mathfrak{w} \gg 1$, the wave equation can be solved using a WKB approximation [126]. For $\mathfrak{k} = 0$ the result is

$$\phi_k(u) = \pi\mathfrak{w}^2 \frac{u}{\sqrt{1-u^2}}\left[iJ_2\left(2\mathfrak{w}\sqrt{u}\right) - Y_2\left(2\mathfrak{w}\sqrt{u}\right)\right]\,. \tag{2.114}$$

In the intermediate regime the wave equation can be solved numerically. A standard method is to start from the near horizon result given in Eq. (2.112) and integrate outwards towards the boundary. The retarded correlation function is given by the variation of the boundary action with respect to the field. For this purpose we consider the quadratic part of the Einstein-Hilbert action and use the AdS/CFT correspondence to express Newton's constant in terms of gauge theory parameters. We find

$$S = -\frac{\pi^2 N^2 T^4}{8} \int du \int d^4x \frac{f(u)}{u} (\partial_u \phi)^2 + \dots . \tag{2.115}$$

The boundary action follows after an integration by parts. The retarded Green function is determined by the second variational derivative with respect to the boundary value of the field [125, 127],

$$G_R(\mathfrak{w}, \mathfrak{k}) = -\frac{\pi^2 N^2 T^4}{4} \left[\frac{f(u)\partial_u \phi_k(u)}{u\phi_k(u)} \right]_{u\to 0} . \tag{2.116}$$

Finally, the spectral function is given by $\eta(\omega, k) = -\omega^{-1} Im\, G_R(\omega, k)$. Below is a short Mathematica script that determines the spectral function numerically.

```
(* equation of motion for minimally coupled scalar *)
(* with harmonic space and time dependence *)
(* ------------------------------------------- *)
f[u_] := 1 - u^2
EomPhi = phi''[u] - (1 + u^2)/(u f[u]) phi'[u]
  + (w^2 - q^2 f[u])/(u f[u]^2) phi[u]

(* boundary solution *)
(* ----------------- *)
phiHorizon[u_] := (1-u)^(-I*w/2)

(* numerically integrate from Horizon to boundary *)
(* ------------------------------------------- *)
SolPhi[omega_, qq_] := Block[{w = omega, q = qq},
  NDSolve[
   {0 == EomPhi,
    phi[epsH] == phiHorizon[epsH],
    phi'[epsH] == phiHorizon'[epsH]},
   phi[u],
   {u, epsB, epsH}]][[1, 1, 2]]

(* retarded correlator from boundary action *)
(* --------------------------------------- *)
Gret[omega_, qq_] := (f[u]/u D[solPhi[omega, qq], u]/solPhi[
    omega, qq] )
 /. {u -> epsB}
```

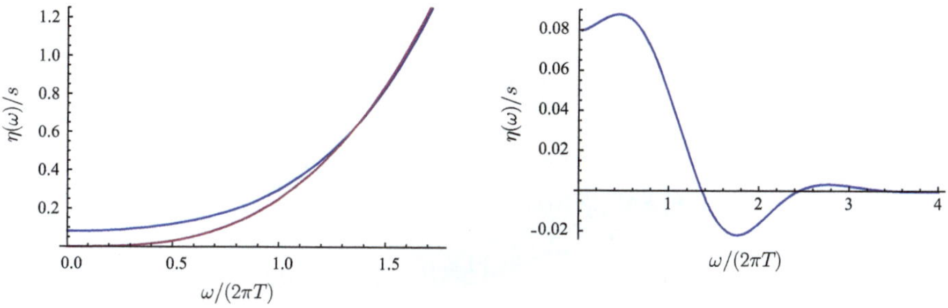

Fig. 2.5 Viscosity spectral function in a $\mathcal{N} = 4$ SUSY Yang Mills plasma. The spectral function is computed in the large N_c limit of a strongly coupled plasma using the AdS/CFT correspondence. The figure in the *left panel* shows $\eta(\omega)/s$ (*blue*) and the zero temperature counterpart $\eta_{T=0}(\omega)/s$ (*red*) as a function of ω. The figure in the *right panel* shows the finite temperature part $[\eta(\omega) - \eta_{T=0}(\omega)]/s$. The figures were generated using the script described below Eq. (2.116)

The spectral function for $k = 0$ is shown in Fig. 2.5. This is an interesting result because it represent a systematic calculation of a real time observable in the strong coupling limit of a quantum field theory. As explained in Sect. 2.4.5 the corresponding lattice calculation is very difficult, and existing results are difficult to improve upon. We also note that the result is quite different from expectations at weak coupling. At weak coupling we expect the spectral function to show a narrow transport peak at zero energy [80].

So far we have only considered calculations very close to equilibrium, corresponding to small perturbations of the AdS_5 Schwarzschild solution. In order to address the problem of initial state dynamics and thermalization we have to consider initial conditions that mimic colliding nuclei. Recent work focuses on colliding shock waves in asymptotically AdS_5 spaces. In the strong coupling limit the evolution of the shock waves is a problem in numerical relativity. Special methods have been developed to deal with problems in AdS space [129]. These methods are quite different from the techniques employed in connection with black hole or neutron star mergers in asymptotically flat Minkowski space time. A typical result is shown in Fig. 2.6. The calculations demonstrate fast "hydrodynamization", that means a rapid decay of non-hydrodynamic modes. At somewhat longer time scales thermal equilibration is achieved. This corresponds to the formation of an event horizon in the bulk. In general, it was realized that there is a fluid-gravity correspondence, an equivalence between dynamic space times containing a horizon and solutions of the Navier-Stokes equation [130]. This correspondence can be used to study, both analytically and numerically, difficult problems in fluid dynamics.

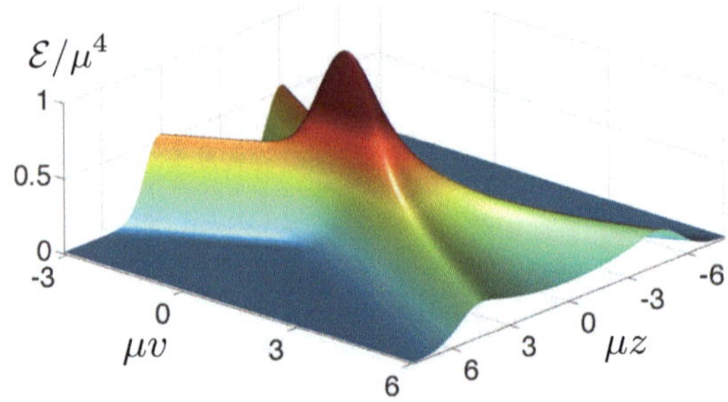

Fig. 2.6 Energy density of colliding shock waves in AdS_5 space [128]. The figure shows the energy density $\mathcal{E}/\mu^4$ on the boundary of AdS_5 as a function of the time coordinate v and the longitudinal direction z. The shocks are infinitely extended in the transverse direction. The parameter μ sets the overall scale

2.6 Outlook and Acknowledgments

I hope this brief review provides a flavor of the breadth of computational problems that are related QCD. This includes many issues that are at the forefront of computational physics, like the sign problem in euclidean QCD at finite baryon density, and the challenge to extract real time correlation functions from the euclidean path integral. It also includes many problems that are of great interest to mathematicians. Both the Yang-Mills existence and mass gap as well as the Navier-Stokes existence and smoothness problems are among the Clay Millenium Prize problems [131, 132]. Interesting work on the Boltzmann equation was recently recognized with a Fields medal [133], and gradient flow plays an important role in the proof of the Poincare conjecture [134].

Acknowledgements The euclidean path integral simulation in quantum mechanics is described in [6], and the programs are available at https://www.physics.ncsu.edu/schaefer/\penalty\z@physics/. A simple Z_2 lattice gauge code can be found in the Appendix. You should be able to extend this code to $SU(2)$ and $SU(3)$. Modern lattice QCD tools can be found on the chroma website http://github.com/JeffersonLab/chroma. The VH1 hydro code is described in [96] and can be downloaded at http://wonka.\penalty\z@physics.ncsu.edu/pub/VH-1/. Dissipative and anisotropic versions are available on request. There are a number of relativistic hydro codes on the web. An example is the VISHNU code [135] which is available at https://u.osu.edu/vishnu/. Both UrQMD http://urqmd.org/ and GiBUU https://gibuu.\penalty\z@hepforge.org/ are also available online.

The mathematica notebooks in Sect. 2.5.5 are adapted from notebooks available on Jim Hartle's website http://web.\penalty\z@physics.ucsb.edu/~gravitybook/. Much more sophisticated tensor packages are easily found on the web. The simple script for solving the wave equation in AdS_5 is adapted from a notebook written by Matthias Kaminski. A set of lecture notes and mathematica notebooks for solving the Einstein equations numerically on asymptotically AdS spaces can be found on Wilke van der Schee's website https://sites.google.com/site/\penalty\z@wilkevanderschee/ads-numerics. Thomas Schäfer work is supported by the US Department of Energy grant DE-FG02-03ER41260.

Appendix: Z_2 Gauge Theory

This is a simple Monte Carlo program for Z_2 gauge theory written by M. Creutz [136].

```
/* Z_2 lattice gauge simulation */
/* Michael Creutz <creutz@bnl.gov> */
/* http://thy.phy.bnl.gov/~creutz/z2.c */

#include <stdio.h>
#include <stdlib.h>
#include <math.h>

/* the lattice is of dimensions SIZE**4 */
#define SIZE 6
int link[SIZE][SIZE][SIZE][SIZE][4]; /* last index gives link
    direction */

/* utility functions */
void moveup(int x[],int d) {
  x[d]+=1;
  if (x[d]>=SIZE) x[d]-=SIZE;
  return;
}
void movedown(int x[],int d) {
  x[d]-=1;
  if (x[d]<0) x[d]+=SIZE;
  return;
}
void coldstart(){ /* set all links to unity */
  int x[4],d;
  for (x[0]=0;x[0]<SIZE;x[0]++)
    for (x[1]=0;x[1]<SIZE;x[1]++)
      for (x[2]=0;x[2]<SIZE;x[2]++)
        for (x[3]=0;x[3]<SIZE;x[3]++)
          for (d=0;d<4;d++)
      link[x[0]][x[1]][x[2]][x[3]][d]=1;
  return;
}
/* for a random start: call coldstart() and then update once at
    beta=0 */

/* do a Monte Carlo sweep; return energy */
double update(double beta){
  int x[4],d,dperp,staple,staplesum;
  double bplus,bminus,action=0.0;
  for (x[0]=0; x[0]<SIZE; x[0]++)
    for (x[1]=0; x[1]<SIZE; x[1]++)
      for (x[2]=0; x[2]<SIZE; x[2]++)
        for (x[3]=0; x[3]<SIZE; x[3]++)
          for (d=0; d<4; d++) {
            staplesum=0;
```

```
        for (dperp=0;dperp<4;dperp++){
          if (dperp!=d){
            /* move around thusly:
               dperp      6--5
               ^          | |
               |          1--4
               |          | |
               -----> d   2--3 */
            /* plaquette 1234 */
            movedown(x,dperp);
            staple=link[x[0]][x[1]][x[2]][x[3]][dperp]
              *link[x[0]][x[1]][x[2]][x[3]][d];
            moveup(x,d);
            staple*=link[x[0]][x[1]][x[2]][x[3]][dperp];
            moveup(x,dperp);
            staplesum+=staple;
            /* plaquette 1456 */
            staple=link[x[0]][x[1]][x[2]][x[3]][dperp];
            moveup(x,dperp);
            movedown(x,d);
            staple*=link[x[0]][x[1]][x[2]][x[3]][d];
            movedown(x,dperp);
            staple*=link[x[0]][x[1]][x[2]][x[3]][dperp];
            staplesum+=staple;
          }
    }
        /* calculate the Boltzmann weight */
        bplus=exp(beta*staplesum);
        bminus=1/bplus;
        bplus=bplus/(bplus+bminus);
        /* the heatbath algorithm */
        if ( drand48() < bplus ){
          link[x[0]][x[1]][x[2]][x[3]][d]=1;
          action+=staplesum;
        }
        else{
          link[x[0]][x[1]][x[2]][x[3]][d]=-1;
          action-=staplesum;
        }
      }
  action /= (SIZE*SIZE*SIZE*SIZE*4*6);
  return 1.-action;
}

/*******************************/
int main(){
  double beta, dbeta, action;
  srand48(1234L); /* initialize random number generator */
  /* do your experiment here; this example is a thermal cycle */
  dbeta=.01;
  coldstart();
  /* heat it up */
  for (beta=1; beta>0.0; beta-=dbeta){
    action=update(beta);
```

```
    printf("%g\t%g\n",beta,action);
  }
  printf("\n\n");
  /* cool it down */
  for (beta=0; beta<1.0; beta+=dbeta){
    action=update(beta);
    printf("%g\t%g\n",beta,action);
  }
  printf("\n\n");
  exit(0);
}
```

References

1. R.P. Feynman, A.R. Hibbs, *Quantum Mechanics and Path Integrals* (McGraw-Hill, New York, 1965)
2. N. Metropolis, A.W. Rosenbluth, M.N. Rosenbluth, A.H. Teller, E. Teller, J. Chem. Phys. **21**, 1087 (1953)
3. M. Creutz, B. Freedman, Ann. Phys. **132**, 427 (1981)
4. E.V. Shuryak, O.V. Zhirov, Nucl. Phys. B **242**, 393 (1984)
5. E.V. Shuryak, Nucl. Phys. B **302**, 621 (1988)
6. T. Schäfer, Instantons and Monte Carlo methods in quantum mechanics (2004). arXiv:hep-lat/0411010
7. M. Jarrell, J.E. Gubernatis, Phys. Rep. **269**, 133 (1996)
8. M. Asakawa, T. Hatsuda, Y. Nakahara, Prog. Part. Nucl. Phys. **46**, 459 (2001)
9. C. Jarzynski, Phys. Rev. Lett. **78**, 2690 (1997)
10. D.J. Gross, F. Wilczek, Phys. Rev. Lett. **30**, 1343 (1973)
11. H.D. Politzer, Phys. Rev. Lett. **30**, 1346 (1973)
12. S.R. Coleman, E.J. Weinberg, Phys. Rev. D **7**, 1888 (1973)
13. K. Nakamura et al., J. Phys. G **37**, 075021 (2010)
14. M.G. Alford, A. Schmitt, K. Rajagopal, T. Schäfer, Rev. Mod. Phys. **80**, 1455 (2008)
15. A. Adams, L.D. Carr, T. Schäfer, P. Steinberg, J.E. Thomas, New J. Phys. **14**, 115009 (2012)
16. P. Braun-Munzinger, V. Koch, T. Schäfer, J. Stachel, Phys. Rep. **621**, 76 (2016)
17. M. Gell-Mann, R.J. Oakes, B. Renner, Phys. Rev. **175**, 2195 (1968)
18. S.R. Coleman, E. Witten, Phys. Rev. Lett. **45**, 100 (1980)
19. G. 't Hooft, NATO Sci. B **59**, 135 (1980)
20. E.V. Shuryak, Sov. Phys. JETP **47**, 212 (1978) [Zh. Eksp. Teor. Fiz. **74**, 408 (1978)]
21. E.V. Shuryak, Phys. Lett. B **78**, 150 (1978) [Yad. Fiz. **28**, 796 (1978)]
22. A.D. Linde, Phys. Lett. B **96**, 289 (1980)
23. R.D. Pisarski, F. Wilczek, Phys. Rev. D **29**, 338 (1984)
24. Y. Aoki, G. Endrodi, Z. Fodor, S.D. Katz, K.K. Szabo, Nature **443**, 675 (2006)
25. A. Bazavov et al., Phys. Rev. D **85**, 054503 (2012)
26. Y. Aoki, Z. Fodor, S.D. Katz, K.K. Szabo, Phys. Lett. B **643**, 46 (2006)
27. Y. Aoki, S. Borsanyi, S. Durr, Z. Fodor, S.D. Katz, S. Krieg, K.K. Szabo, J. High Energy Phys. **06**, 088 (2009)
28. A. Bazavov et al., Phys. Rev. D **90**(9), 094503 (2014)
29. M.A. Stephanov, Prog. Theor. Phys. Suppl. **153**, 139 (2004)
30. Z. Fodor, S.D. Katz, J. High Energy Phys. **03**, 014 (2002)
31. C. Allton, S. Ejiri, S. Hands, O. Kaczmarek, F. Karsch et al., Phys. Rev. D **66**, 074507 (2002)

32. F. Karsch, C.R. Allton, S. Ejiri, S.J. Hands, O. Kaczmarek, E. Laermann, C. Schmidt, Nucl. Phys. Proc. Suppl. **129**, 614 (2003/2004)
33. Z. Fodor, S. Katz, J. High Energy Phys. **0404**, 050 (2004)
34. R.V. Gavai, S. Gupta, Phys. Rev. D **78**, 114503 (2008)
35. S. Datta, R.V. Gavai, S. Gupta, Nucl. Phys. A **904–905**, 883c (2013)
36. P. de Forcrand, O. Philipsen, Phys. Rev. Lett. **105**, 152001 (2010)
37. M.A. Stephanov, K. Rajagopal, E.V. Shuryak, Phys. Rev. Lett. **81**, 4816 (1998)
38. G. Sauer, H. Chandra, U. Mosel, Nucl. Phys. A **264**, 221 (1976)
39. J. Pochodzalla et al., Phys. Rev. Lett. **75**, 1040 (1995)
40. J.B. Elliott, P.T. Lake, L.G. Moretto, L. Phair, Phys. Rev. C **87**(5), 054622 (2013)
41. M.G. Alford, K. Rajagopal, F. Wilczek, Nucl. Phys. B **537**, 443 (1999)
42. T. Schäfer, Nucl. Phys. B **575**, 269 (2000)
43. T. Schäfer, F. Wilczek, Phys. Rev. Lett. **82**, 3956 (1999)
44. T. Hatsuda, M. Tachibana, N. Yamamoto, G. Baym, Phys. Rev. Lett. **97**, 122001 (2006)
45. M. Creutz, *Quarks, Gluons, and Lattices* (Cambridge University Press, Cambridge, 1983)
46. I. Montvay, G. Münster, *Quantum Fields on a Lattice* (Cambridge University Press, Cambridge, 1994)
47. J. Smit, *Introduction to Quantum Fields on a Lattice* (Cambridge University Press, Cambridge, 2002)
48. C. Gattringer, C.B. Lang, *Quantum Chromodynamics on the Lattice* (Springer, Berlin, 2009)
49. H.W. Lin, H.B. Meyer, *Lattice QCD for Nuclear Physics* (Springer, Berlin, 2014)
50. Z. Fodor, S.D. Katz, The Phase diagram of quantum chromodynamics (2009). arXiv:0908.3341
51. H.T. Ding, F. Karsch, S. Mukherjee, Thermodynamics of strong-interaction matter from Lattice QCD (2015). arXiv:1504.05274
52. K.G. Wilson, Phys. Rev. D **10**, 2445 (1974) [45 (1974)]
53. F. Mezzadri, How to generate random matrices from the classical compact groups (2006). arXiv:math-ph/0609050
54. A. Hasenfratz, P. Hasenfratz, Phys. Lett. B **93**, 165 (1980). [241(1980)]
55. G.P. Lepage, Strong interactions at low and intermediate energies, in *Proceedings of the 13th Annual Hampton University Graduate Studies (HUGS), Newport News, USA, May 26-June 12, 1998* (1998), p. 49
56. U. Wolff, Phys. Rev. Lett. **62**, 361 (1989)
57. J.B. Kogut, L. Susskind, Phys. Rev. D **11**, 395 (1975)
58. D.B. Kaplan, Phys. Lett. B **288**, 342 (1992)
59. H. Neuberger, Phys. Lett. B **417**, 141 (1998)
60. M. Luscher, Modern perspectives in lattice QCD: Quantum field theory and high performance computing, in *Proceedings of the International School, 93rd Session, Les Houches, France, August 3–28, 2009* (2010), p. 331
61. E. Endress, C. Pena, K. Sivalingam, Comput. Phys. Commun. **195**, 35 (2015)
62. V. Dick, F. Karsch, E. Laermann, S. Mukherjee, S. Sharma, Phys. Rev. D **91**(9), 094504 (2015)
63. M. Teper, Phys. Lett. B **171**, 86 (1986)
64. M. Lüscher, J. High Energy Phys. **08**, 071 (2010) [Erratum: J. High Energy Phys. **03**, 092 (2014)]
65. A.A. Belavin, A.M. Polyakov, A.S. Schwartz, Yu.S. Tyupkin, Phys. Lett. B **59**, 85 (1975)
66. T. Schäfer, E.V. Shuryak, Rev. Mod. Phys. **70**, 323 (1998)
67. L. Del Debbio, L. Giusti, C. Pica, Phys. Rev. Lett. **94**, 032003 (2005)
68. M. Ce, C. Consonni, G.P. Engel, L. Giusti, PoS LATTICE **2014**, 353 (2014)
69. E. Poppitz, T. Schäfer, M. Ünsal, J. High Energy Phys. **03**, 087 (2013)
70. T. Banks, A. Casher, Nucl. Phys. B **169**, 103 (1980)
71. P.H. Ginsparg, K.G. Wilson, Phys. Rev. D **25**, 2649 (1982)

72. A. Cherman, T. Schäfer, M. Unsal, Chiral lagrangian from duality and monopole operators in compactified QCD (2016). arXiv:1604.06108
73. S.R. Beane, P.F. Bedaque, A. Parreno, M.J. Savage, Phys. Lett. B **585**, 106 (2004)
74. M. Cristoforetti, F. Di Renzo, L. Scorzato, Phys. Rev. D **86**, 074506 (2012)
75. G. Aarts, L. Bongiovanni, E. Seiler, D. Sexty, J. High Energy Phys. **10**, 159 (2014)
76. G. Aarts, E. Seiler, I.O. Stamatescu, Phys. Rev. D **81**, 054508 (2010)
77. D. Sexty, Phys. Lett. B **729**, 108 (2014)
78. T. Kloiber, C. Gattringer, PoS LATTICE **2013**, 206 (2014)
79. T. Schäfer, D. Teaney, Rep. Prog. Phys. **72**, 126001 (2009)
80. T. Schäfer, Ann. Rev. Nucl. Part. Sci. **64**, 125 (2014)
81. F. Karsch, H.W. Wyld, Phys. Rev. D **35**, 2518 (1987)
82. H.B. Meyer, Phys. Rev. D **76**, 101701 (2007)
83. H.B. Meyer, Phys. Rev. Lett. **100**, 162001 (2008)
84. S. Sakai, A. Nakamura, PoS LATTICE **2007**, 221 (2007) [AIP Conf. Proc. **893**, 5 (2007)]
85. G. Aarts, C. Allton, J. Foley, S. Hands, S. Kim, Phys. Rev. Lett. **99**, 022002 (2007)
86. G. Aarts, PoS LATTICE **2007**, 001 (2007)
87. G. Aarts, C. Allton, J. Foley, S. Hands, S. Kim, Nucl. Phys. A **785**, 202 (2007)
88. H.B. Meyer, J. High Energy Phys. **08**, 031 (2008)
89. P. Romatschke, D.T. Son, Phys. Rev. D **80**, 065021 (2009)
90. P.B. Arnold, G.D. Moore, L.G. Yaffe, J. High Energy Phys. **11**, 001 (2000)
91. P. Kovtun, D.T. Son, A.O. Starinets, Phys. Rev. Lett. **94**, 111601 (2005)
92. P. Romatschke, Int. J. Mod. Phys. E **19**, 1 (2010)
93. L. Rezzolla, O. Zanotti, *Relativistic Hydrodynamics* (Oxford University Press, Oxford, 2013)
94. S. Jeon, U. Heinz, Int. J. Mod. Phys. E **24**(10), 1530010 (2015)
95. P. Colella, P.R. Woodward, J. Comput. Phys. **54**, 174 (1984)
96. J.M. Blondin, E.A. Lufkin, Astrophys. J. Suppl. **88**, 589 (1993)
97. T. Schäfer, Phys. Rev. A **82**, 063629 (2010)
98. W. Florkowski, R. Ryblewski, Phys. Rev. C **83**, 034907 (2011)
99. M. Martinez, M. Strickland, Nucl. Phys. A **848**, 183 (2010)
100. M. Bluhm, T. Schäfer, Phys. Rev. A **92**(4), 043602 (2015)
101. M. Bluhm, T. Schäfer, Phys. Rev. Lett. **116**(11), 115301 (2016)
102. P. Romatschke, M. Mendoza, S. Succi, Phys. Rev. C **84**, 034903 (2011)
103. J. Brewer, M. Mendoza, R.E. Young, P. Romatschke, Phys. Rev. A **93**(1), 013618 (2016)
104. L.P. Kadanoff, G. Baym, *Quantum Statistical Mechanics* (W. A. Benjamin, New York, 1962)
105. G.F. Bertsch, S.D. Gupta, Phys. Rep. **160**, 189 (1988)
106. T. Lepers, D. Davesne, S. Chiacchiera, M. Urban, Phys. Rev. A **82**, 023609 (2010)
107. S.A. Bass et al., Prog. Part. Nucl. Phys. **41**, 255 (1998) [Prog. Part. Nucl. Phys. **41**, 225 (1998)]
108. O. Buss, T. Gaitanos, K. Gallmeister, H. van Hees, M. Kaskulov, O. Lalakulich, A.B. Larionov, T. Leitner, J. Weil, U. Mosel, Phys. Rep. **512**, 1 (2012)
109. W. Ehehalt, W. Cassing, Nucl. Phys. A **602**, 449 (1996)
110. K. Geiger, B. Muller, Nucl. Phys. B **369**, 600 (1992)
111. Z. Xu, C. Greiner, Phys. Rev. C **71**, 064901 (2005)
112. L.D. McLerran, R. Venugopalan, Phys. Rev. D **49**, 2233 (1994)
113. A.H. Mueller, D.T. Son, Phys. Lett. B **582**, 279 (2004)
114. S. Mrowczynski, B. Schenke, M. Strickland, Phys. Rep. (2016, in press). Preprint, arXiv:1603.08946
115. J. Berges, A. Rothkopf, J. Schmidt, Phys. Rev. Lett. **101**, 041603 (2008)
116. K. Dusling, T. Epelbaum, F. Gelis, R. Venugopalan, Nucl. Phys. A **850**, 69 (2011)
117. D.F. Litim, C. Manuel, Phys. Rep. **364**, 451 (2002)
118. S.K. Wong, Nuovo Cim. A **65**, 689 (1970)
119. C.R. Hu, B. Muller, Phys. Lett. B **409**, 377 (1997)
120. J.M. Maldacena, Int. J. Theor. Phys. **38**, 1113 (1999) [Adv. Theor. Math. Phys. **2**, 231 (1998)]
121. D.T. Son, A.O. Starinets, Ann. Rev. Nucl. Part. Sci. **57**, 95 (2007)

122. S.S. Gubser, A. Karch, Ann. Rev. Nucl. Part. Sci. **59**, 145 (2009)
123. J. Casalderrey-Solana, H. Liu, D. Mateos, K. Rajagopal, U.A. Wiedemann, (2011)
124. O. DeWolfe, S.S. Gubser, C. Rosen, D. Teaney, Prog. Part. Nucl. Phys. **75**, 86 (2014)
125. G. Policastro, D.T. Son, A.O. Starinets, J. High Energy Phys. **09**, 043 (2002)
126. D. Teaney, Phys. Rev. D **74**, 045025 (2006)
127. D.T. Son, A.O. Starinets, J. High Energy Phys. **03**, 052 (2006)
128. P.M. Chesler, L.G. Yaffe, Phys. Rev. Lett. **106**, 021601 (2011)
129. P.M. Chesler, L.G. Yaffe, J. High Energy Phys. **07**, 086 (2014)
130. M. Rangamani, Class. Quant. Grav. **26**, 224003 (2009)
131. A. Jaffe, E. Witten, Quantum Yang Mills theory, official description of millennium prize problems (2000). www.claymath.org
132. C. Fefferman, Existence and smoothness of the Navier-Stokes Equation, official description of millennium prize problems (2000). www.claymath.org
133. C. Mouhot, C. Villani, Acta Math. **207**, 29 (2011)
134. G. Perelman, The entropy formula for the Ricci flow and its geometric applications (2002). arXiv:math/0211159 [math.DG]
135. C. Shen, Z. Qiu, H. Song, J. Bernhard, S. Bass, U. Heinz, Comput. Phys. Commun. **199**, 61 (2016)
136. M. Creutz, Comput. Sci. Eng. **6**, 80 (2004)

Chapter 3
Lattice Quantum Chromodynamics

Tetsuo Hatsuda

3.1 Introduction

In this chapter, we introduce lattice quantum chromodynamics (LQCD) originally proposed by Wilson in 1974 [1]. What makes the LQCD unique and powerful is that it can allow first-principle, gauge invariant and non-perturbative calculations of strongly interacting quarks and gluons. After the first numerical attempts by Creutz [2], LQCD simulations have been extensively applied to study heavy quark potentials, hadron masses, hadronic matrix elements, QCD phase transition at finite temperature, and so on. In recent years, there are also progresses in deriving the baryon-baryon interactions, which are particularly relevant to nuclear physics and astrophysics. Throughout this chapter, we will focus on the theoretical concepts, numerical techniques and some applications to hadron masses and nuclear forces. LQCD at finite temperature and/or baryon density will not be covered. The interested readers may want to consult the following review articles for further details, or may even want to run the open source codes or to use the open LQCD configurations.

Review Articles Here is a list a few articles which are useful to learn more about LQCD.

- Comprehensive review on QCD can be seen in [3].
- The origin of LQCD is discussed in [4].
- The basic concepts of LQCD are summarized in the monographs [5, 6].
- Recent progresses of LQCD can be seen in the reviews [7, 8] and references therein.

T. Hatsuda (✉)
iTHEMS (interdisciplinary Theoretical and Mathematical Sciences) Program, RIKEN, Saitama 351-0198, Japan
e-mail: thatsuda@riken.jp

M. Hjorth-Jensen et al. (eds.), *An Advanced Course in Computational Nuclear Physics*, Lecture Notes in Physics 936, DOI 10.1007/978-3-319-53336-0_3

Open Source Codes It takes a lot of time and energy to develop the LQCD code from scratch. To lower the bar, several source codes for LQCD simulations have been released for public use.

- Bridge++: http://bridge.kek.jp/Lattice-code/index_e.html
- Lattice ToolKit: http://nio-mon.riise.hiroshima-u.ac.jp/LTK/
- OpenQCD: http://luscher.web.cern.ch/luscher/openQCD/index.html
- USQCD: http://usqcd-software.github.io/

LQCD Configurations Outputs of large scale LQCD simulations is a *big data* called "LQCD configurations". Physicists can study various aspects of QCD by using these configurations. The International Lattice Data Grid (ILDG) is a project to share the configurations around the world.

- ILDG http://plone.jldg.org/wiki/index.php/Main_Page

3.1.1 Euclidean QCD Action

LQCD is formulated on the Euclidean spacetime lattice. Observables in the Minkowski spacetime are obtained by the analytic continuation of the imaginary-time τ to the real-time t. In terms of the time evolution operator in quantum mechanics, it corresponds to the continuation from the imaginary time evolution, $e^{-H\tau}$, to the real time evolution, e^{-iHt}, where H being the Hamiltonian of the system. The functional integral representation of the the Euclidean QCD partition function $\mathscr{Z}$ on a finite spatial box L^3 and the temperature T is given by

$$\mathscr{Z}(T, V, J) = \int [dAd\bar{q}dq] e^{-\int_0^{1/T} d\tau \int_{L^3} d^3x \left(\mathscr{L}_{\rm QCD}^{\rm E} + J\Xi\right)} \tag{3.1}$$

where the Euclidean QCD Lagrangian in terms of quarks $q^{\alpha=1,2,3}$ and gluons $A_{\mu=1,2,3,4}^{b=1,\cdots,8}$ is given by

$$\mathscr{L}_{\rm QCD}^{\rm E} = \bar{q}^{\alpha}(\Gamma_{\mu} D_{\mu}^{\alpha\beta} + m\delta^{\alpha\beta}) q^{\beta} + \frac{1}{4} G_{\mu\nu}^{b} G_{\mu\nu}^{b}, \tag{3.2}$$

Here the Euclidean version of the γ-matrices, Γ_{μ}, is defined in the Appendix [Four vectors and Dirac matrices]. The quark mass matrix in the flavor space $(u, d, s, \cdots)$ is denoted by m with the flavor indices suppressed. The color covariant derivative is defined by

$$D_{\mu}^{\alpha\beta} = \partial_{\mu}\delta^{\alpha\beta} + igA_{\mu}^{\alpha\beta}, \tag{3.3}$$

with $x_{\mu} = (\tau, \mathbf{x})$, $\partial_{\mu} = (\partial_{\tau}, \nabla)$ and the 3×3 matrix field, $A_{\mu} = A_{\mu}^{b} t^{b}$. The explicit form of the color SU(3) generators $t^{b=1,\cdots,8}$ is given in the Appendix [SU(N) algebra]. The field strength tensor is $G_{\mu\nu} = G_{\mu\nu}^{b} t^{b}$ with $G_{\mu\nu}^{b} = \partial_{\mu} A_{\nu}^{b} - \partial_{\nu} A_{\mu}^{b} - g f_{bcd} A_{\mu}^{c} A_{\nu}^{d}$. The arbitrary external fields (such as the external source of the quarks and gluons, external electroweak fields etc) are denoted by J, while the corresponding dynamical operators are denoted by $\Xi(A, \bar{q}, q)$. The functional integration measure for the c-number gluons and the Grassmann-number quarks is defined by

$$[dAd\bar{q}dq] \equiv \prod_{x,\text{color,spin,flavor}} dA_\mu^b(x) d\bar{q}^\alpha(x) dq^\beta(x). \tag{3.4}$$

See Appendix [Gaussian and Grassmann integrals] for the examples of integration with these measures. The temporal boundary condition of the gluon (quark) field is periodic (anti-periodic) due to its c-number (Grassmann-number) character; $A_\mu^b(\tau = 0, \mathbf{x}) = A_\mu^b(\tau = 1/T, \mathbf{x})$, $\bar{q}^\alpha(\tau = 0, \mathbf{x}) = -\bar{q}^\alpha(\tau = 1/T, \mathbf{x})$, and $q^\beta(\tau = 0, \mathbf{x}) = -q^\beta(\tau = 1/T, \mathbf{x})$. On the other hand, the spatial boundary conditions are not constrained and can be taken to be either periodic or anti-periodic; the difference should disappear in the thermodynamic limit, $L \to \infty$. Throughout this chapter, we take $T = 1/L$ to study hadrons and hadron-hadron interactions at zero temperature in the thermodynamic limit.

Further details of the functional integral formulation of the general many-body systems of fermions and bosons can be seen in the textbook [9].

3.1.2 Quantum Fluctuations

In weak coupling perturbation theory, one assumes that the QCD coupling g is small and expands the partition function $\mathscr{Z}$ in terms of a power series of g. Such a perturbative expansion in QCD is justified, however, only in limited circumstances such as at extreme high temperature/density or at very short distances. This is because the renormalized QCD coupling (or often called the running coupling) becomes small only in the processes with the energy scale much above 1 GeV. This is called the asymptotic freedom. To calculate low-energy hadron properties below 1 GeV, we need to go beyond perturbation theory and to evaluate the functional integral with full quantum fluctuations. The lattice QCD provides a way to carry out this task numerically in a gauge invariant manner.

3.2 Lattice QCD: Theoretical Basis

3.2.1 Wilson Line

Let us first start with the *Wilson line* which is defined on a path P connecting the point y and x in the continuous Euclidean spacetime as shown in Fig. 3.1a. By parametrizing the path in terms of a coordinate $z(s)$ with $z(s = 0) = y$ and

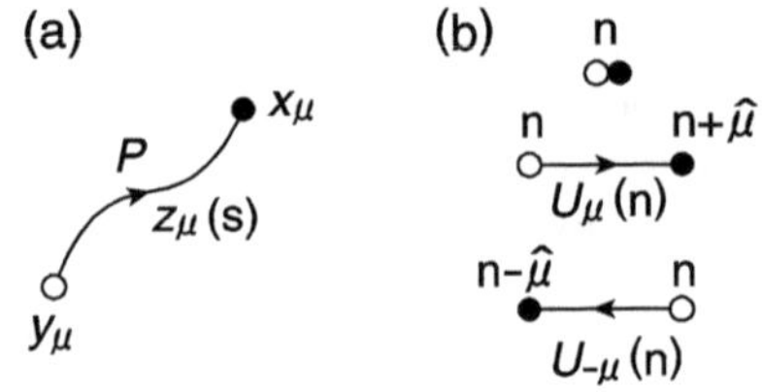

Fig. 3.1 (**a**) The Wilson line in the Euclidean spacetime. (**b**) Basic quark bilinears with gauge invariance

$z(s = 1) = x$, the Wilson line reads

$$U_P(x, y; A) = \mathrm{P}\exp\left(ig \int_P dz_\mu A_\mu\right) = \mathrm{P}\exp\left(ig \int_0^1 ds\, \lambda_\mu A_\mu\right)$$
$$= \sum_{n=0}^{\infty} \frac{(ig)^n}{n!} \int_0^1 ds_1 \int_0^1 ds_2 \cdots \int_0^1 ds_n\, \mathrm{P}[\lambda \cdot A(s_1)\ \cdots\ \lambda \cdot A(s_n)], \tag{3.5}$$

where $\lambda_\mu = dz_\mu/ds$. The path ordered symbol P is necessary because $A_\mu = A_\mu^b t^b$ is a matrix in the color space.

The Wilson line has the following properties which can be proved from the definition of U_P (Exercise 3.1) :

(i) It can be broken into parts at any arbitrary points on the path;

$$U_P(x, y; A) = U_{P_2}(x, z(s); A) U_{P_1}(z(s), y; A). \tag{3.6}$$

(ii) It satisfies a differential equation,

$$\frac{d}{ds} U_P(z(s), y; A) = [ig\lambda(s) \cdot A(z(s))]\ U_P(z(s), y; A). \tag{3.7}$$

(iii) Under the local gauge transformation $A^V_\mu(x) = V(x)[A_\mu(x)V^\dagger(x) - (i/g)V(x)(\partial_\mu V^\dagger(x))$, it transforms covariantly,

$$U_P(x, y; A) \to U_P(x, y; A^V) = V(x) U_P(x, y; A) V^\dagger(y). \tag{3.8}$$

The Wilson line is useful to define the non-local and gauge-invariant objects. In particular, the gauge-invariant quark bilinear $\bar{q}(x)U_P(x, y; A)q(y)$, and the gauge-invariant *Wilson loop* tr $U_P(x, x; A)$ turn out to be important building blocks to define the QCD action on the lattice. Here "tr" implies the trace over color indices.

3.2.2 Lattice Gluons

Let us consider a four dimensional hyper-cubic lattice with a lattice spacing a and the four dimensional volume L^4. Each lattice site is specified by n_μ corresponding to the Euclidean coordinates through $x_\mu = an_\mu$ (see Fig. 3.2). The *link variable* (the shortest Wilson line on the lattice) is an SU(3) matrix connecting the neighboring sites n and $n + \hat{\mu}$,

$$U_\mu(n) = \exp\left(igaA_\mu(n)\right). \tag{3.9}$$

Here $\hat{\mu}$ implies a vector pointing the direction of μ with a length a. Since it is the minimal Wilson line, we do not need the path ordering symbol P. Also, any non-minimal Wilson line on the lattice is represented by a product of link-variables. For later purpose, we introduce the link variable pointing the opposite direction as $U_{-\mu}(n + \hat{\mu}) = [U_\mu(n)]^\dagger$.

Let us now define the smallest closed loop;

$$U_{\mu\nu}(n) = U_\nu^\dagger(n)U_\mu^\dagger(n + \hat{\nu})U_\nu(n + \hat{\mu})U_\mu(n). \tag{3.10}$$

Under local gauge transformation (rotation under arbitrary SU(3) matrix $V(n)$), we have

$$U_\mu(n) \to V(n)U_\mu(n)V^\dagger(n + \hat{\mu}), \quad U_{\mu\nu}(n) \to U_{\mu\nu}^V(n) = V(n)U_{\mu\nu}(n)V^\dagger(n), \tag{3.11}$$

which are the direct consequence of Eq. (3.8).

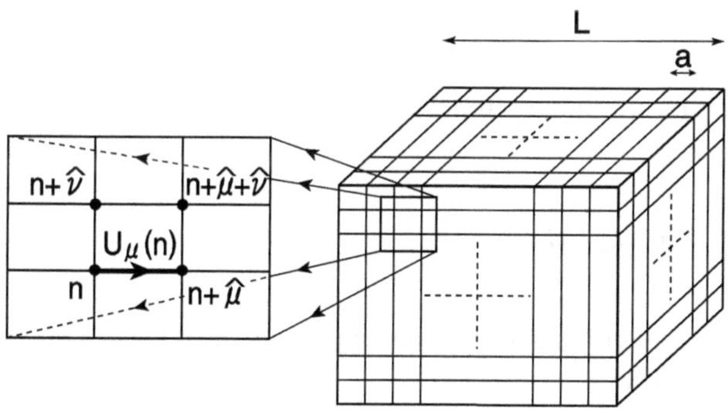

Fig. 3.2 A hypercubic lattice in Euclidean spacetime with a lattice constant a and the lattice size L. Quarks $q(n)$ (gluons $U_\mu(n)$) are defined on the sites (links)

In the naive continuum limit where $a \to 0$, we have

$$U_\mu(n) - 1 = igaA_\mu(n) + O(a^2), \tag{3.12}$$

$$U_{\mu\nu}(n) - 1 = \exp\left(iga^2t^b(G^b_{\mu\nu}(n) + O(a^3))\right) - 1 = iga^2G_{\mu\nu}(n) + O(a^4), \tag{3.13}$$

$$\begin{aligned}\mathrm{tr}\left(U_{\mu\nu}(n) - 1\right) &= \mathrm{tr}\left[iga^2t^b(G^b_{\mu\nu}(n) + O(a^3)) - \frac{1}{2}g^2a^4G_{\mu\nu}(n)^2 + O(a^5)\right] \\ &= -\frac{1}{4}g^2a^4(G^b_{\mu\nu}(n))^2 + O(a^5).\end{aligned} \tag{3.14}$$

Here Eq. (3.13) is obtained by using the Baker-Campbell-Hausdorff formula, $\exp X \cdot \exp Y = \exp(X + Y + [X, Y]/2 + \cdots)$ (Exercise 3.2).

Finally, a gluon action on the lattice, which reduces to the Yang-Mills action in the naive continuum limit ($a \to 0$), reads

$$\begin{aligned}S_{\mathrm{G}} &= \beta \sum_{\mathrm{Pl}} \left[1 - \frac{1}{N_c}\mathrm{Re}\;\mathrm{tr}\; U_{\mu\nu}(n)\right] \\ &= \beta a^4 \sum_n \sum_{\mu<\nu} \left[1 - \frac{1}{2N_c}\mathrm{tr}\left(U_{\mu\nu}(n) + U^\dagger_{\mu\nu}(n)\right)\right] \\ &= \frac{1}{g^2}\sum_n \sum_{\mu\neq\nu} \mathrm{tr}\left[1 - U_{\mu\nu}(n)\right] \quad \xrightarrow{a\to 0} \quad \frac{1}{4}\int d^4x\, G^b_{\mu\nu}(x)^2,\end{aligned} \tag{3.15}$$

where $\sum_{\mathrm{Pl}}$ is a sum over all non-oriented *plaquettes* (minimum square tile on the lattice with the area a^2). Note that $\beta \equiv \frac{2N_c}{g^2}$ with N_c being the number of colors ($N_c = 3$ for QCD) should not be confused with the inverse temperature. The lattice gluon action is not unique in the sense that one may add arbitrary non-minimal terms which vanish in the continuum limit ($a \to 0$).

3.2.3 Lattice Fermions

There exist three types of gauge invariant objects made of nearest neighbor fermions as shown in Fig. 3.1b;

$$\bar{q}(n)q(n), \;\; \bar{q}(n + \hat{\mu})U_\mu(n)q(n), \;\; \bar{q}(n - \hat{\mu})U_{-\mu}(n)q(n). \tag{3.16}$$

Here one may put any γ-matrices between $\bar{q}$ and q without spoiling the color gauge invariance. A special combination of the above terms is called the Wilson's fermion action

$$S_{\rm F} = a^4 \sum_n \left[m\bar{q}(n)q(n) - \frac{1}{2a}\sum_{\pm\mu} \bar{q}(n+\hat{\mu})\Gamma_\mu U_\mu(n)q(n) \right.$$
$$\left. - \frac{r}{2a}\sum_{\pm\mu}\left(\bar{q}(n+\hat{\mu})U_\mu(n)q(n) - \bar{q}(n)q(n)\right)\right]$$
$$\equiv a^4 \sum_{n',n} \bar{q}(n')\left(m\delta_{n',n} + D_{\rm W}(n',n;U)\right)q(n) \tag{3.17}$$
$$\xrightarrow[a\to 0]{} \int d^4x\, \bar{q}(x)\left(\Gamma_\mu D_\mu + m - \frac{ar}{2}D_\mu^2\right)q(x), \tag{3.18}$$

where the Wilson's Dirac operator in Eq. (3.17) with the Wilson's parameter r reads

$$D_{\rm W}(n',n;U) = -\frac{1}{2a}\sum_{\pm\mu}\left[\delta_{n',n+\hat{\mu}}(r+\Gamma_\mu)U_\mu(n) - r\delta_{n',n}\right]. \tag{3.19}$$

To take the continuum limit in Eq. (3.18), we use the midpoint prescription, $(f(x+a) - f(x-a))/2a = f'(x) + O(a^2)$ and $(f(x+a)+f(x-a)-2f(x))/a^2 = f''(x) + O(a^2)$. One of the important properties of $D_{\rm W}(n',n;U)$ is its Γ_5 Hermiticity (Exercise 3.3),

$$\Gamma_5 D_{\rm W} \Gamma_5 = D_{\rm W}^\dagger, \tag{3.20}$$

where Γ_5 is given in Appendix [Four vectors and Dirac matrices]. Note that the Hermitian conjugate is taken for color, spin and spacetime.

The dispersion relation (relation between the energy and momentum) for free fermion can be obtained from Eq. (3.17) by taking $U_\mu = 1$ (or equivalently $g = 0$) and substituting the Fourier transform, $q(n) = \int_{-\pi/a}^{\pi/a} \frac{d^4p}{(2\pi)^4} e^{ip_\mu n_\mu} q(p)$. This leads to $S_{\rm F}^{\rm (free)} = \int \frac{d^4p}{(2\pi)^4}\bar{q}(-p)\mathscr{G}_{\rm F}^{-1}q(p)$ with the free fermion propagator (Exercise 3.4),

$$\mathscr{G}_{\rm F}(p) = \frac{-i\sum_\mu \bar{p}_\mu \Gamma_\mu + m(p)}{\sum_\mu \bar{p}_\mu^2 + m^2(p)}, \tag{3.21}$$

$$\bar{p}_\mu = \frac{1}{a}\sin(p_\mu a),\ \ m(p) = m(0) + \frac{r}{a}\sum_\mu\left(1-\cos(p_\mu a)\right). \tag{3.22}$$

Since $\sin(p_\mu a)$ becomes zero for $p_\mu a$=$(0,0,0,0)$, $(\pi,0,0,0)$,$\cdot$ $(0,\pi,\pi,\pi)$, (π,π,π,π), there arise $2^4 = 16$ degenerate fermions if we take $r = 0$. This is called the fermion doubling problem on the lattice. In fact, there is a no-go theorem by Nielsen and Ninomiya: The fermion doubling always exists, if the free fermion action on the lattice has (i) bilinearity in quark field, (ii) translational invariance, (iii) hermiticity (in the Minkowski spacetime), (iv) locality in spacetime, and (v) exact chiral symmetry. Indeed, (i)–(v) are all satisfied for $r = 0$.

The doubling in the dispersion relation in the Minkowski spacetime is easily seen by Wick rotating $p_4 \to iE$ in Eq. (3.21). As an illustration, let us take the case with massless fermion ($m(0) = 0$) in (1+1)-dimension. Then, the zero of the denominator after the Wick rotation for small a gives,

$$E^2(p) \simeq \left(\frac{1}{a}\sin(pa)\right)^2 + \left(\frac{r}{a}(1-\cos(pa))\right)^2, \tag{3.23}$$

whose positive energy solution is plotted in Fig. 3.3 for several values of r. One finds that the unphysical massless pole at $pa = \pi$ is lifted up as r increases.

In general, $r \neq 0$ leads to a mass splitting of 16 fermions: $m(p) \simeq m(0)$ ($^\forall p_\mu \to 0$) and $m(p) = m(0) + \frac{2r}{a}N_\pi$ ($^\exists p_\mu \to \pi/a$), where $N_\pi (= 1, 2, 3, 4)$ being the number of π's in $p_\mu a$. This implies that we can select only one light fermion by choosing $m(0) \simeq 0$ and all the other 15 fermions have masses of $O(1/a)$ for positive r. A price to pay is that the non-vanishing r breaks chiral symmetry explicitly for finite a, i.e. $\{\gamma_5, D_\mathrm{W}\} \neq 0$ even for $m(0) = 0$. Namely, the Nielsen-Ninomiya's no-go theorem is evaded by breaking the condition (v).

Better way to evade the no-go theorem is to break the condition (v) in a way that the definition of chiral symmetry is modified. Suppose we consider a modified chiral rotation in the flavor space,

$$q \to \mathrm{e}^{-i\theta_\mathrm{A}\hat{\Gamma}_5} q, \quad \bar{q} \to \bar{q}\mathrm{e}^{-i\theta_\mathrm{A}\Gamma_5} \ \text{with} \ \hat{\Gamma}_5 = \Gamma_5(1 - 2aD_\mathrm{GW}), \tag{3.24}$$

which reduces to the standard axial rotation for $a \to 0$. Here D_GW is a generalized Dirac operator which is constructed so that $\bar{q}D_\mathrm{GW}q$ is invariant under Eq. (3.24) even for finite a;

$$\Gamma_5 D_\mathrm{GW} + D_\mathrm{GW}\hat{\Gamma}_5 = 0, \tag{3.25}$$

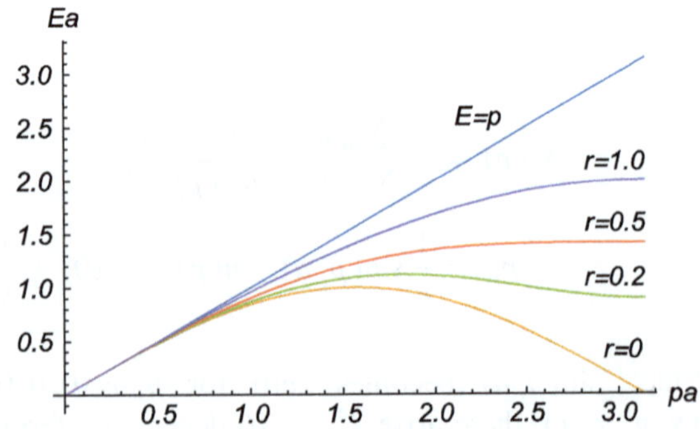

Fig. 3.3 The dispersion relation for massless fermion in (1+1)-dimension on the lattice for different values of the Wilson's parameter r. The linear dispersion in the continuum ($E = p$) is also shown for comparison

or equivalently $\{\Gamma_5, D_{\rm GW}\} = 2aD_{\rm GW}\Gamma_5 D_{\rm GW}$. This is called the Ginsparg-Wilson relation. An explicit form of $D_{\rm GW}$ may be constructed as

$$D_{\rm GW} = \frac{1}{2a}\left(1 + \frac{X}{\sqrt{X^\dagger X}}\right) \quad \text{with } X \equiv D_{\rm W}^{(r=1)} - m_0, \tag{3.26}$$

where $m_0 a$ being a dimensionless parameter of $O(1)$. Unlike the case of m in the Wilson fermion, m_0 is not directly related to the physical fermion mass. Nevertheless, if we choose the region $0 < m_0 a < 2$, there exists an exact massless mode for $N_\pi = 0$ for finite a, and other 15 modes have a large mass $(2/a)(2N_\pi - m_0 a) > 0$ (Exercise 3.5).

Going back to the Wilson's fermion action, Eq. (3.17). it can be conveniently rewritten as

$$S_{\rm F} = \sum_{n',n} \bar{\psi}(n')F(n',n;U)\psi(n), \tag{3.27}$$

$$F(n',n;U) = \delta_{n'n} - \kappa \sum_{\pm\mu} \delta_{n',n+\hat{\mu}}(r + \Gamma_\mu)U_\mu(n), \tag{3.28}$$

where we have redefined the quark field as $\psi = a^{3/2}q/\sqrt{2\kappa}$ with $\kappa = [2(ma + 4r)]^{-1}$ being the *hopping parameter*. If the quark mass m is large, κ is small and the "hopping" to the neighboring lattice site is suppressed.

3.2.4 Partition Function on the Lattice

The functional integration over quarks and gluons in continuum QCD in Eq. (3.1) is now transformed to the integration over quarks on each site and gluons on each link in lattice QCD. With Eqs. (3.15) and (3.27), the partition function without the external field ($J = 0$) reads

$$\mathscr{Z} = \int [dUd\bar{\psi}d\psi]{\rm e}^{-S_{\rm G}(U)-S_{\rm F}(\bar{\psi},\psi,U)} = \int [dU]\,{\rm Det}\,F(U)\,{\rm e}^{-S_{\rm G}(U)} = \int [dU]\ {\rm e}^{-S_{\rm eff}(U)} \tag{3.29}$$

To obtain the second equality, we have explicitly carried out the integration over the Grassmann variables, $\bar{\psi}$ and ψ, by using the formula in Appendix [Gaussian and Grassmann integrals]. Here Det implies the determinant in spacetime, color, flavor and spin degrees of freedom. In the third equality, the exponent is defined as

$$S_{\rm eff}(U) \equiv S_{\rm G}(U) - {\rm lnDet}F(U). \tag{3.30}$$

The integration over the group element $[dU] = \prod_{\mu,n} dU_\mu(n)$ can be defined through the *Haar measure* dU which has the property, $d(V_{\rm L}UV_{\rm R}^\dagger) = dU$ with $V_{\rm L,R}$

being arbitrary group elements. Such a measure is unique for compact groups such as SU(N). If we parametrize the group element as $U = \exp(i\theta_a t^a)$, one can define the distance in the group space as $ds^2 = g_{ab}(\theta)d\theta_a d\theta_b$, where the metric is given by $g_{ab} = \mathrm{tr}(L_a L_b) = \mathrm{tr}(R_a R_b)$ with

$$L_a = -iU^{-1}(\partial U/\partial\theta_a), \quad R_a = -i(\partial U/\partial\theta_a)U^{-1}. \tag{3.31}$$

Then the Haar measure can be explicitly written as

$$dU = \mathscr{N}\sqrt{\det g}\prod_a d\theta_a, \tag{3.32}$$

with an overall normalization factor $\mathscr{N}$.

The followings are some examples of the SU(N) group integration, which can be proved by the invariant property of the Haar measure (except for the first one which determines the normalization of the measure) (Exercise 3.6):

$$\int dU\ 1 = 1, \tag{3.33}$$

$$\int dU\ U_{ij} = 0, \tag{3.34}$$

$$\int dU\ U_{ij}U^{\dagger}_{k\ell} = \frac{1}{N}\delta_{i\ell}\delta_{jk}, \tag{3.35}$$

$$\int dU\ U_{ij}U_{k\ell}U^{\dagger}_{i'j'}U^{\dagger}_{k'\ell'}$$
$$= \frac{\delta_{ij'}\delta_{ji'}\delta_{k\ell'}\delta_{\ell k'} + (j' \leftrightarrow \ell', i' \leftrightarrow k')}{N^2-1} - \frac{\delta_{ij'}\delta_{jk'}\delta_{k\ell'}\delta_{\ell i'} + (j' \leftrightarrow \ell', i' \leftrightarrow k')}{N}, \tag{3.36}$$

$$\int dU\ U_{i_1 j_1}\cdots U_{i_N j_N} = \frac{1}{N!}\epsilon_{i_1\cdots i_N}\epsilon_{j_1\cdots j_N}. \tag{3.37}$$

Similar to the statistical systems such as the Ising model, observables are obtained by averaging over the statistical weight as

$$\langle\mathscr{O}\rangle = \frac{1}{\mathscr{Z}}\int[dU]\ \mathscr{O}(U)\ \mathrm{e}^{-S_{\mathrm{eff}}(U)}. \tag{3.38}$$

Due to the gauge invariance of the Haar measure, gauge non-invariant quantities have vanishing expectation values (Elitzer's theorem). For example, consider the expectation value of the link variable,

$$\langle U_\mu(n)\rangle = \frac{1}{\mathscr{Z}}\int[dU]U_\mu(n)\ \mathrm{e}^{-S_{\mathrm{eff}}(U)} = V(n)\langle U_\mu(n)\rangle, \tag{3.39}$$

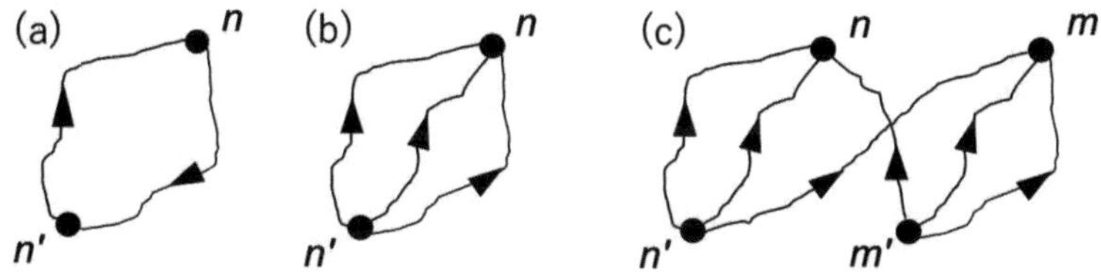

Fig. 3.4 (**a**) Single meson correlation representing the propagation of a meson created at point n' and absorbed at point n. (**b**) Single baryonic correlation representing the propagation of a baryon created at point n' and absorbed at point n. (**c**) Two baryon correlation which contains the information on baryon-baryon interaction

where we have made a change of variable, $U_\mu(n) \to V(n)U_\mu(n)$ with $V(n)$ being the $SU(N)$ matrix, and used the gauge invariance of the Haar measure as well as Det $F(U)$ and $S_G(U)$. Since Eq. (3.39) must be true for arbitrary $V(n)$, we have $\langle U_\mu(n)\rangle = 0$.

Some examples of the non-vanishing observables are shown in Fig. 3.4: (a) and (b) correspond to the mesic and baryonic correlations, respectively, while (c) is a correlation related to the baryon-baryon interactions. The filled circles are the spacetime points where the quarks and anti-quarks are created or absorbed. Each line with arrow indicates the quark propagator $F^{-1}(n, n'; U)$ connecting two spacetime points n and n'. Thus, the explicit forms of the mesic and baryonic correlations are

$$C_M(n, n') = \frac{1}{\mathcal{Z}} \int [dU]\, F^{-1}_{\alpha\beta}(n, n'; U) F^{-1}_{\beta\alpha}(n', n; U)\, e^{-S_{\rm eff}(U)}, \tag{3.40}$$

$$C_B(n, n') = \frac{1}{\mathcal{Z}} \int [dU]\, \epsilon_{\alpha\beta\gamma}\epsilon_{\alpha'\beta'\gamma'} F^{-1}_{\alpha\alpha'}(n, n'; U) F^{-1}_{\beta\beta'}(n, n'; U) F^{-1}_{\gamma\gamma'}(n, n'; U)\, e^{-S_{\rm eff}(U)}, \tag{3.41}$$

where all the color indices are contracted so that C_M and C_B are gauge invariant. Other quantum numbers such as spin and flavor associated with F^{-1} are not shown explicitly. Spacetime, spin and flavor dependences of $C_M(n, n')$ in (a) and $C_B(n, n')$ in (b) have all the information on the hadronic states in various different channels, while $C_{BB}(n, m, n', m')$ in (c) has the information on baryon-baryon interactions.

3.2.5 *Strong Coupling Expansion and Quark Confinement*

One of the remarkable properties of QCD is the confinement of quarks inside hadrons. Simplest setup to see this phenomena is to consider the potential $V(R)$ between an infinitely heavy quark Q and an anti-quark $\bar{Q}$ with a fixed separation R.

Fig. 3.5 A rectangular Wilson loop with the temporal (spatial) size $\mathscr{T}$ (R)

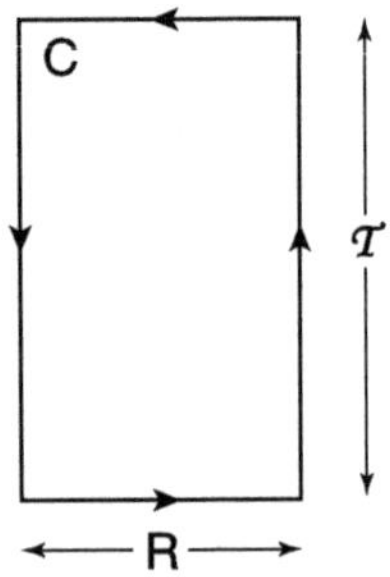

It corresponds to Fig. 3.5 and can be written as

$$\langle W(C)\rangle = \langle \mathrm{tr} \prod_{\mathrm{link}\in C} U_\mu(n)\rangle \tag{3.42}$$

$$\propto \mathrm{e}^{-V(R)\mathscr{T}} \simeq \exp\left[-\left(KR + b + \frac{c}{R} + \cdots\right)\mathscr{T}\right], \tag{3.43}$$

where we have taken a limit $\mathscr{T} \gg R \to \infty$ in Eq. (3.43). Remembering the fact that the real time t and the imaginary time τ are related as $\tau = it$, the exponential falloff of $\langle W(C)\rangle$ in τ implies the temporal oscillation in t, and its R-dependent coefficient is nothing but the interaction energy between Q and $\bar{Q}$.

In Eq. (3.43), $K > 0$ implies the existence of a string-like linear confining potential. It also implies the area law of the Wilson loop $\langle W(C)\rangle \sim \exp(-K\mathscr{A})$ where $\mathscr{A} = R \times \mathscr{T}$ is the area inside the path C. In full QCD where pair creation of light quarks are allowed, the linear rising potential becomes eventually flat at long distances due to the breaking of the string, $Q\bar{Q} \to (Q\bar{q})(q\bar{Q})$.

To make the analysis simple, let us now consider the SU(N_c) Yang-Mills theory without light quarks: This is called the quenched approximation and corresponds to take $F(U) = 1$. In this case, the Wilson loop can be evaluated analytically in the strong coupling regime ($g \to \infty$). First of all, S_{G} is proportional to $1/g^2$, so that one can make an expansion, $\exp(-S_{\mathrm{G}}) = 1 - S_{\mathrm{G}} + S_{\mathrm{G}}^2/2 + \cdots$ and finds (Exercise 3.7)

$$\langle W(C)\rangle = \frac{1}{\mathscr{Z}} \int [dU]\, \mathrm{tr} \prod_{\mathrm{link}\in C} U_\mu(n) \sum_{\ell=0}^{\infty} \frac{1}{\ell!}(-S_{\mathrm{G}})^\ell. \tag{3.44}$$

Only the first three integrals, Eqs. (3.33)–(3.35), are necessary to extract the leading contribution to $\langle W(C)\rangle$ in the strong coupling. Key observation is that all the U's from the Wilson loop and $U^\dagger$'s from $(-S_{\mathrm{G}})^\ell$ should be paired in the leading order of $1/g^2$ in Eq. (3.44). This means that the area inside the Wilson loop is tiled up with minimum number of plaquettes as shown in Fig. 3.6. All the structures other than the minimal surface are higher orders in $1/g^2$.

In the evaluation of the numerator of Eq. (3.44), each plaquette has a contribution $1/g^2$. Also each integration on the link gives a factor $1/N_c$ and the contraction of the

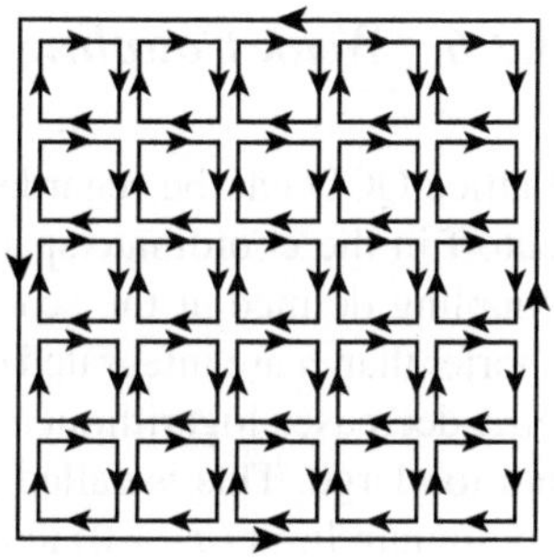

Fig. 3.6 A minimum surface in which the Wilson loop is tiled up by the fundamental plaquettes in the strong coupling limit

color indices gives a factor N_c on each site. On the other hand, $\mathscr{Z}$ (the denominator of Eq. (3.44) is unity in the leading order. Thus, one arrives at the formula in the lowest order of the strong coupling expansion,

$$\frac{1}{N_c}\langle W(C)\rangle \xrightarrow[g^2\to\infty]{} \frac{1}{N_c}\cdot\left(\frac{1}{g^2}\right)^{N_{\rm plaq}}\cdot\left(\frac{1}{N_c}\right)^{N_{\rm link}}\cdot N_c^{N_{\rm site}}$$

$$= \left(\frac{1}{N_c g^2}\right)^{\frac{R\mathscr{T}}{a^2}} = \exp\left(-\frac{\ln N_c g^2}{a^2}R\mathscr{T}\right), \tag{3.45}$$

where we have used a relation, $N_{\rm link} - N_{\rm site} + 1 = N_{\rm plaq}$ and $N_{\rm plaq}a^2 = R\mathscr{T} = \mathscr{A}$. Since it shows the area law, the confinement is naturally obtained in the strong coupling with the linear rising potential,

$$V(R) = KR \text{ with } K = \frac{1}{a^2}\ln(N_c g^2). \tag{3.46}$$

If we consider higher orders of the strong coupling expansion, "rough" surfaces should be taken into account. Nevertheless, the confining feature is stable against small perturbations in $1/g^2$. In fact, there exits a theorem that, for sufficiently large g, the strong coupling expansion converges and shows confinement for all compact gauge groups in all spacetime dimensions.

A question here is that whether the real world corresponds to the strong coupling region discussed above. The answer is no; the real world corresponds to the weak coupling regime. For compact QED (quantum electrodynamics formulated in terms of the U(1) link variable), the confinement $K > 0$ in the strong coupling regime changes to $K = 0$ in the weak coupling regime. On the other hand, in QCD in four spacetime dimensions with $N_c = 3$, the confinement feature is expected to persist even in the weak coupling regime. Indeed, there are strong evidences for this statement from LQCD simulations. Its analytic proof, however, is still missing and is being one of the most challenging problems in mathematical physics.

3.2.6 Weak Coupling Expansion and Continuum Limit

Lattice QCD can be regarded as an effective field theory with an ultraviolet (UV) cutoff in the coordinate space. The gauge coupling g is then interpreted as a bare coupling defined at the scale a where quantum fluctuations with the wave length shorter than a are integrated out. In non-Abelian gauge theories, it can be shown that $g(a)$ decreases logarithmically as a decreases unless the number of matter fields is not too large. This is called the *asymptotic freedom*, and is essential for taking the continuum limit ($a \to 0$) to remove the lattice artifact.

For simplicity, let us consider the case with massless fermions, where observables $\mathcal{O}$ such as the string tension and the hadron masses depend only on the coupling g and the regularization scale a. Then, from the dimensional ground, one can write

$$\mathcal{O}(g(a), a) = a^{-d} X(g(a)), \tag{3.47}$$

where d is the mass-dimension of $\mathcal{O}$ and X is a dimensionless function of g. The a-independence of the observable implies

$$a\frac{d\mathcal{O}}{da} = \left(a\frac{\partial}{\partial a} - \beta_{\mathrm{LAT}}\frac{\partial}{\partial g}\right)\mathcal{O}(g(a), a) = 0, \quad \beta_{\mathrm{LAT}}(g) = -a\frac{dg(a)}{da}. \tag{3.48}$$

By integrating the first equation in Eq. (3.48), we find

$$X(g) = \exp\left(-d\int^{g} \frac{dg'}{\beta_{\mathrm{LAT}}(g')}\right). \tag{3.49}$$

Suppose that the beta-function can be expanded in terms of g for small a: $\beta_{\mathrm{LAT}}(g) = -\beta_0 g^3 - \beta_1 g^5 + \cdots$. Here β_0 and β_1 can be shown to be independent of the regularization scheme and are known to be

$$\beta_0 = \frac{1}{(4\pi)^2}\left(\frac{11N_c}{3} - \frac{2N_f}{3}\right), \quad \beta_1 = \frac{1}{(4\pi)^4}\left(\frac{34N_c^2}{3} - \frac{38N_f}{3}\right), \tag{3.50}$$

for QCD with N_c colors and N_f fermions.

By integrating the second equation in Eq. (3.48) with the above expansion of $\beta_{\mathrm{LAT}}(g)$, one finds

$$a = \Lambda_{\mathrm{LAT}}^{-1} \cdot \exp\left(-\frac{1}{2\beta_0 g^2}\right) \cdot (\beta_0 g^2)^{-\frac{\beta_1}{2\beta_0^2}} \cdot (1 + O(g^2)). \tag{3.51}$$

Here Λ_{LAT} is called the scale parameter on the lattice, and $g(a)$ can be expressed in terms of a and Λ_{LAT} (Exercise 3.8),

$$\frac{1}{g^2(a)} = \beta_0 \ln\left(\frac{1}{a^2\Lambda_{\mathrm{LAT}}^2}\right) + \frac{\beta_1}{\beta_0}\ln\ln\left(\frac{1}{a^2\Lambda_{\mathrm{LAT}}^2}\right) + \cdots. \tag{3.52}$$

This is the asymptotic freedom in which $g(a)$ decreases as a decreases. This also justified the assumption that the beta-function can be expanded by $g(a)$ for small a.

Direct way to extract the actual value of $\Lambda_{\rm LAT}$ is to carry out numerical simulations of a certain physical quantity (such as the string tension) and compare the result with the experimental value. For example, the string tension, which has mass-dimension two ($d = 2$) should behave as

$$Ka^2 = C_K \exp\left(-\frac{1}{\beta_0 g^2}\right) (\beta_0 g^2)^{-\beta_1/\beta_0^2} = C_K \Lambda_{\rm LAT}^2, \tag{3.53}$$

with C_K being a dimensionless numerical constant independent of g. As can be seen from this example, the functional form of the physical quantities for $g \sim 0$ is severely constrained. This is called the *asymptotic scaling* which is used to check whether the system is close enough to the continuum limit. Shown in Fig. 3.7 is a

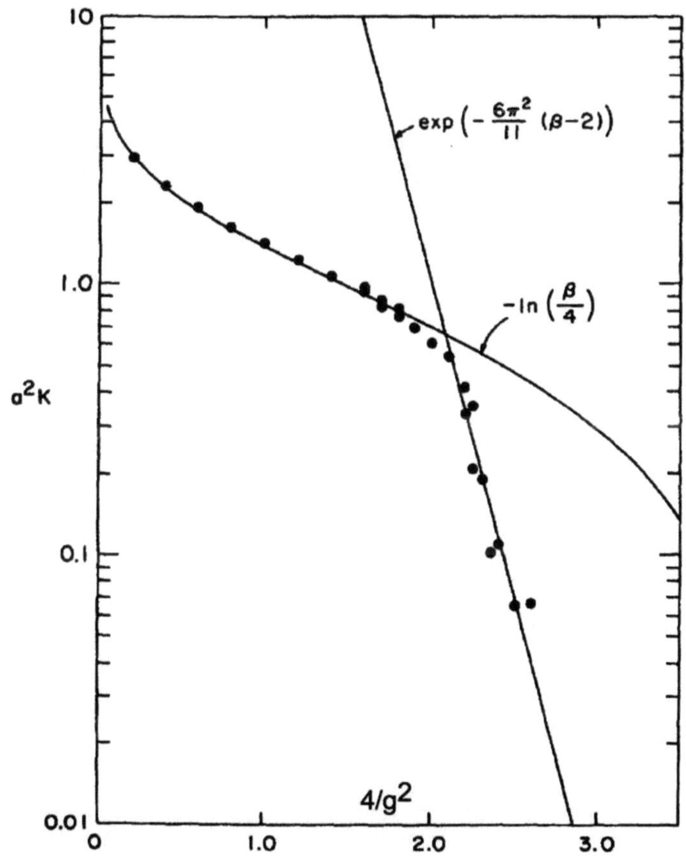

Fig. 3.7 Crossover behavior of the dimensionless string tension Ka^2 from the strong coupling regime $\beta = 2N_c/g^2 \to 0$ to the weak coupling (asymptotic scaling) regime $\beta = 2N_c/g^2 \to \infty$ for SU($N_c = 2$) Yang-Mills theory. The figure is adapted from [2]

historic numerical study, which shows a crossover of Ka^2 from the strong coupling regime to the weak coupling regime in SU(2) Yangs-Mills theory.

3.2.7 Running Coupling

Let us now consider an observable $\mathcal{O}$ which depends not only on $g(a)$ and a but also on some external dimensionful parameter. For concreteness, we consider the heavy quark potential $V(R; g(a), a)$ in the quenched approximation. Since it has the dimension of energy, one may write

$$V(R, g(a), a) = R^{-1}\tilde{V}(R/a, g(a)). \tag{3.54}$$

Then the cutoff independence of the observable, $a\frac{d}{da}V(R, g(a), a) = 0$, leads to

$$\left(\lambda\frac{\partial}{\partial\lambda} + \beta_{\mathrm{LAT}}\frac{\partial}{\partial g}\right)\tilde{V}(\lambda, g) = 0, \tag{3.55}$$

where we have introduced a dimensionless scaling parameter λ through $R = \lambda a$.

The solution of the *renormalization group equation*, Eq. (3.55), reads

$$\tilde{V}(\lambda, g(a)) = \tilde{V}(1, \bar{g}(\lambda)). \tag{3.56}$$

Here $\bar{g}(\lambda)$ is called the *running coupling* which is a solution of

$$\lambda\frac{d\bar{g}}{d\lambda} = -\beta_{\mathrm{LAT}}(\bar{g}(\lambda)), \tag{3.57}$$

with the boundary condition, $\bar{g}(\lambda = 1) = g(a)$. One can show that Eq. (3.56) satisfies Eq. (3.55) explicitly by applying the partial derivatives or more generally by the method of characteristics in Appendix [Method of characteristics]. Then, we eventually arrive at the formula

$$V(R, g(a), a) = \frac{a}{R}V(a, \bar{g}(R/a), a). \tag{3.58}$$

If R is in the interval, $a < R \ll \Lambda_{\mathrm{LAT}}^{-1}$, the running coupling $\bar{g}$ is small enough, so that one may use the perturbative expansion of β_{LAT} to obtain

$$\bar{g}^2(R/a) \simeq \frac{g^2(a)}{1 - 2\beta_0 g^2(a)\ln(R/a)} = \frac{1}{2\beta_0\ln(1/(R\Lambda_{\mathrm{LAT}}))}. \tag{3.59}$$

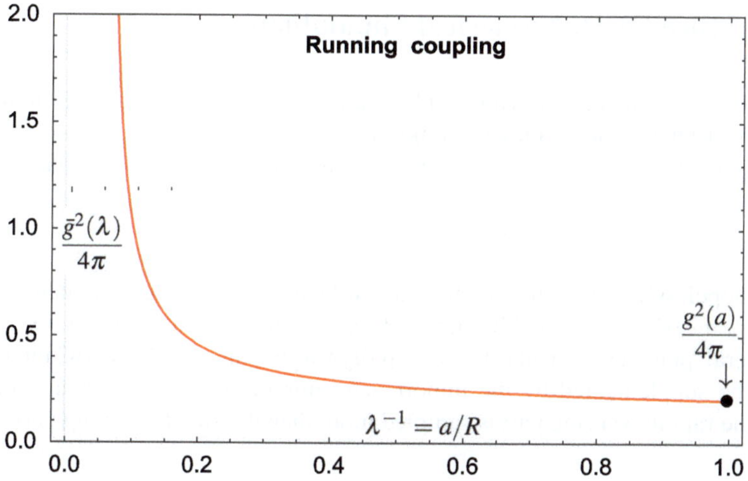

Fig. 3.8 Perturbative running coupling $\frac{1}{4\pi}\bar{g}^2(\lambda)$ as a function of λ^{-1}. In the short distance limit ($\lambda = 1$), the running coupling coincides with the bare coupling $\bar{g}(\lambda = 1) = g(a)$, while, in the long distance regime ($\lambda \ll 1$), the running coupling grows

In Fig. 3.8, the behavior of $\frac{1}{4\pi}\bar{g}^2(\lambda)$ as a function of λ^{-1} is shown. The bare coupling $\frac{1}{4\pi}g^2(a)$ appears as the boundary condition at shortest distance $\lambda = R/a = 1$, while $\frac{1}{4\pi}\bar{g}^2(\lambda)$ grows as λ increases. The latter implies that the strong interaction has anti-screening feature.

For R sufficiently close to a, one may evaluate the potential by using perturbation theory as $V(a, \bar{g}(R/a), a) = -C_F\frac{\bar{g}(R/a)^2}{4\pi a}$, so that we finally obtain

$$V(R, g(a), a) \simeq -C_F\frac{\bar{g}^2(R/a)}{4\pi R} \quad (a < R \ll \Lambda_{\text{LAT}}^{-1}), \tag{3.60}$$

where $C_F = 4/3$ for $N_c = 3$ is given in Appendix [SU(N) algebra]. Equation (3.60) is nothing but the Coulomb potential with the running coupling constant. Note that the left hand side is a-independent, while the right hand side has logarithmic a-dependence through $\bar{g}$. This is due to the use of perturbation theory; such a logarithmic a-dependence is cancelled successively by higher order terms. Note also that the similar analysis can be done for any other observables. The QCD thermal pressure at finite temperature $P(T, g(a), a)$ is a typical example, in which the weak coupling analysis (i.e. the description by quark-gluon plasma picture) is justified under the condition $\Lambda_{\text{LAT}} \ll T < a^{-1}$.

3.3 Lattice QCD: Numerical Simulations

Suppose we have a lattice having N_s (N_τ) number of sites in each spatial (temporal) direction. Then the total number of links is $N_s^3 \times N_\tau \times 4$. Therefore the total number of gluon integrations $\int [dU]$ for a moderate lattice size $N_s = N_\tau = 32$ reads

$$(N_s^3 \times N_\tau \times 4)_{\text{links}} \times 8_{\text{color}} \sim 3 \times 10^7. \tag{3.61}$$

This is hopelessly a large number for standard methods of numerical integration. In this case, the Monte Carlo (MC) integration, which is a statistical way to evaluate the integral, plays a powerful role. For rapidly varying integrand, the MC integration should be supplemented by the importance sampling to have better accuracy, in which the rapidly varying part is sampled more than the slowly varying part.

3.3.1 Importance Sampling

Let us consider the general partition function $\mathscr{Z} = \int [d\phi] \exp(-S(\phi))$, with some c-number field ϕ and try to calculate an observable $\mathscr{O}$ by

$$\langle \mathscr{O} \rangle = \frac{1}{\mathscr{Z}} \int [d\phi]\ \mathscr{O}(\phi) e^{-S(\phi)}. \tag{3.62}$$

The basic procedure of the MC integration with the *importance sampling* consists of two steps:

(I) Generate a set of field configurations, $\{\phi^{(1)}, \phi^{(2)}, \cdots, \phi^{(N)}\}$, with $\phi^{(n)}$ being arranged to appear with a probability in "equilibrium", $W_{\text{eq}}[\phi] = \mathscr{Z}^{-1} \exp(-S(\phi))$.

(II) The field configurations thus generated are used to calculate the expectation value,

$$\langle \mathscr{O} \rangle = \frac{1}{N} \sum_{n=1}^{N} \mathscr{O}^{(n)} \pm \sqrt{\frac{\sigma^2}{N}}, \quad \sigma^2 = \frac{1}{N-1} \sum_{n=1}^{N} \langle \mathscr{O}^{(n)} - \langle \mathscr{O} \rangle \rangle^2, \tag{3.63}$$

with $\mathscr{O}^{(n)} = \mathscr{O}(\phi^{(n)})$.

3.3.2 Markov Chain Monte Carlo (MCMC)

For large number of integration variables such as Eq. (3.61), it is essential to develop an appropriate scheme to carry out Step (I) in Sect. 3.3.1. The *Markov chain Monte Carlo (MCMC)* method is one of such schemes.

Let us consider a chain of configurations generated successively starting from an initial configuration,

$$\phi_0 \to \phi_1 \to \phi_2 \to \cdots \to \phi_i \to \phi_{i+1} \to \cdots, \tag{3.64}$$

where the "*update*" of the i-th configuration (ϕ) to the $(i+1)$-th configuration (ϕ') is governed by the conditional probability P or equivalently the transition matrix $\mathbf{T}$,

$$P(\phi \to \phi') = (\mathbf{T})_{\phi\phi'} \tag{3.65}$$

which has the property, $\sum_{\phi'} P(\phi \to \phi') = 1$. Equation (3.64) generated by Eq. (3.65) is called the *Markov chain* since it is governed by the Markov process where the conditional probability depends only on the neighbouring pair. The probability distribution $W[\phi]$ (with the properties, $W[\phi] \geq 0$ and $\sum_\phi W[\phi] = 1$) is updated successively by Eq. (3.65),

$$W'[\phi'] = \sum_\phi W[\phi]P(\phi \to \phi'). \tag{3.66}$$

If the Markov chain is *irreducible* (any ϕ and ϕ' are connected with each other) and *aperiodic* (absence of ϕ which appears periodically),[1] there exists a theorem that the Markov chain has a unique equilibrium distribution W_{eq} satisfying

$$W_{\text{eq}}[\phi'] = \sum_\phi W_{\text{eq}}[\phi]P(\phi \to \phi'), \tag{3.67}$$

and it can be reached by $\mathbf{T}^\infty$ starting from arbitrary initial distribution. For a heuristic proof of this theorem, see Exercise 3.9. For mathematical proof, see [10].

As is easily seen, a sufficient but not necessary condition for P to lead W_{eq} is the *detailed balance*:

$$W_{\text{eq}}[\phi]P(\phi \to \phi') = W_{\text{eq}}[\phi']P(\phi' \to \phi). \tag{3.68}$$

There also exits specific algorithm without the detailed balance in MCMC [11].

The Markov chain Eq. (3.64) takes certain *thermalization time* to reach equilibration. Also, the nearby field configurations are strongly correlated during the *autocorrelation time*. To calculate the actual average in Eq. (3.63), we then need to discard non-themalized configurations and also thin out the configurations to avoid the autocorrelations. This is schematically shown for an observable $\mathcal{O}$ in Fig. 3.9. The thermalization time can be estimated by monitoring the behavior of $\mathcal{O}$ under

[1]Rigorous definitions are as follows. (1) The Markov chain is said to be irreducible if one can find a finite positive integer $n(< \infty)$ such that $(\mathbf{T}^n)_{\phi\phi'} > 0$ for all ϕ and ϕ'. (2) The period, $d(\phi)$, is defined by the greatest common divisor of the set of positive integers $n(\geq 1)$ such that $(\mathbf{T}^n)_{\phi\phi} > 0$ is satisfied. If $d(\phi) = 1$ for all ϕ, the Markov chain is said to be aperiodic [10].

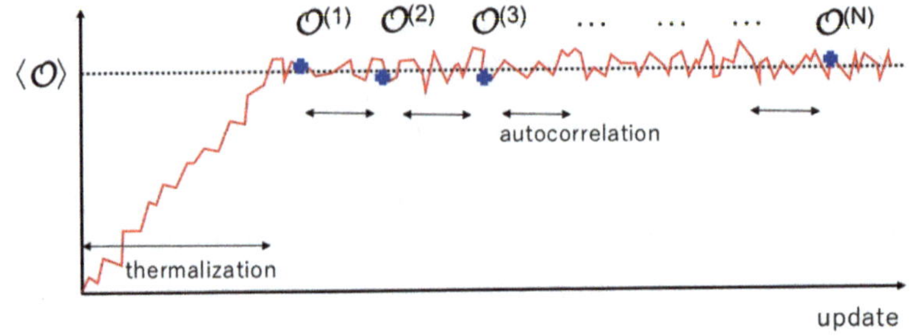

Fig. 3.9 Schematic illustration on the behavior of $\mathcal{O}$ under successive update starting from certain initial configuration. Blue crosses correspond to $\mathcal{O}^{(n)}$ to be used for the actual average in Eq. (3.63)

successive update, while the autocorrelation time can be estimated by calculating the correlations of $\mathcal{O}$ for different configurations.

3.3.3 Hybrid Monte Carlo (HMC)

Most widely used method for generating configurations in LQCD is the hybrid Monte Carlo (HMC) method [12] and its variations. The basic procedure of the HMC can be summarized as follows: First, we rewrite the partition function by introducing a conjugate momentum field π, so that $\mathcal{Z}$ is transformed to a phase space functional integral,

$$\mathcal{Z} = \int [d\phi]\, e^{-S(\phi)} = \int [d\Phi]\, e^{-H(\Phi)}, \quad H(\Phi) = \frac{1}{2}\pi^2 + S(\phi), \tag{3.69}$$

where $\Phi \equiv (\phi, \pi)$ and $[d\Phi] \equiv [d\phi d\pi]$. Then we follow the steps below:

1. Start with arbitrary chosen initial configuration, ϕ.
2. Generate π with the Gaussian distribution,

$$P_{\mathrm{G}}(\pi) \propto \exp(-\pi^2/2). \tag{3.70}$$

3. Evolve Φ under transition probability P_{H} with the *reversibility condition*,

$$P_{\mathrm{H}}(\Phi \to \Phi') = P_{\mathrm{H}}(\Phi'_r \to \Phi_r), \quad \Phi_r \equiv (\phi, -\pi). \tag{3.71}$$

4. Accept the configuration Φ' with the probability,

$$P_{\mathrm{A}}(\Phi \to \Phi') = \min.\{1, e^{-\Delta H}\}, \tag{3.72}$$

   where $\Delta H = H(\Phi') - H(\Phi)$. This is called the *Metropolis test* [13].
5. If the new configuration Φ' is accepted, go to Step 2 with ϕ'. Otherwise, keep the original ϕ and go to Step 2.

The above procedure satisfies the detailed balance Eq. (3.68) with $W_{\rm eq}[\phi] = \exp(-S(\phi))$. In fact, the Step 4 satisfies the detail balance in phase space (Exercise 3.10),

$$e^{-H(\Phi)}P_{\rm A}(\Phi \to \Phi') = e^{-H(\Phi')}P_{\rm A}(\Phi' \to \Phi). \tag{3.73}$$

Then, we have

$$\begin{aligned} e^{-S(\phi)}P(\phi \to \phi') &= e^{-S(\phi)} \int [d\pi d\pi'] P_{\rm G}(\pi) P_{\rm H}(\Phi \to \Phi') P_{\rm A}(\Phi \to \Phi'), \\ &= \int [d\pi d\pi']\, e^{-H(\Phi)} P_{\rm H}(\Phi \to \Phi') P_{\rm A}(\Phi \to \Phi'), \\ &= \int [d\pi d\pi']\, e^{-H(\Phi')} P_{\rm H}(\Phi \to \Phi') P_{\rm A}(\Phi' \to \Phi), \\ &= \int [d\pi d\pi']\, e^{-H(\Phi')} P_{\rm H}(\Phi'_r \to \Phi_r) P_{\rm A}(\Phi' \to \Phi), \\ &= \int [d\pi d\pi']\, e^{-H(\Phi')} P_{\rm H}(\Phi' \to \Phi) P_{\rm A}(\Phi' \to \Phi), \\ &= e^{-S(\phi')} \int [d\pi d\pi'] P_{\rm G}(\pi') P_{\rm H}(\Phi' \to \Phi) P_{\rm A}(\Phi' \to \Phi) \\ &= e^{-S(\phi')} P(\phi' \to \phi), \end{aligned} \tag{3.74}$$

where we have used Eq. (3.71) to obtain the 4th line, and also used $H(\Phi) = H(\Phi_r)$ to obtain the 5th line.

Note that $P_{\rm H}$ can be chosen to be any transition probability as long as it satisfies Eq. (3.71). In practice, the deterministic procedure based on the Molecular Dynamics (MD) evolution along the "computer" time s is useful:

$$\frac{d}{ds}\begin{pmatrix}\phi \\ \pi\end{pmatrix} = \begin{pmatrix}0 & 1 \\ -1 & 0\end{pmatrix}\begin{pmatrix}\delta H(\phi,\pi)/\delta\phi \\ \delta H(\phi,\pi)/\delta\pi\end{pmatrix} = \begin{pmatrix}\pi \\ -\delta S(\phi)/\delta\phi\end{pmatrix}, \tag{3.75}$$

which leads to

$$P_{\rm H}(\Phi \to \Phi') = \delta(\Phi' - \Phi(s)), \tag{3.76}$$

on the phase space trajectories, $\Phi = \Phi(0) \to \Phi(s)$.

If we do not introduce the MD before the Metropolis test $P_{\rm A}$, the procedure is essentially the MCMC with the Metropolis test. It becomes, however, very slow for non-local action such as Eq. (3.30) where $S_{\rm G}(U)$ is local in spacetime while $\mathrm{LnDet}F(U)$ is non-local. The MD is a nice way to evolve the whole variables on the lattice at once. The computer time s needs to be discretized with a step size ε, which brings inevitable numerical error in MD. However, the Metropolis test in Step 4 eliminates such error so that no extrapolation to $\epsilon = 0$ is required in HMC.

There are numerical algorithms in MD to satisfy the reversibility and preserve the phase space area exactly for finite ε. The *leapfrog integrator* is one of such algorithms widely used in LQCD (see Appendix [Leapfrog integrator in molecular dynamics]). Since this conserves the Hamiltonian with $0(\varepsilon^2)$ accuracy, the acceptance rate in Step 4 can be kept high.

In LQCD simulations, we need to treat the unitary matrices $U_\mu(n)$ as dynamical variables, i.e. the MD should be performed on the SU(N_c) group manifold. The appropriate choice of the conjugate momentum would be the element of the Lie algebra, $P_l = R_l^a t^a = -i(dU_l/ds)U_l^{-1}$ (see Eq. (3.31)) where we have abbreviated the link index n and site index μ as l for simplicity. This leads to the equation of motion for U_l,

$$\frac{dU_l}{ds} = iP_l U_l. \tag{3.77}$$

The effective Hamiltonian is naturally written as

$$H = \mathrm{tr} \sum_l P_l^2 + S_{\mathrm{eff}}(U), \tag{3.78}$$

where tr is over color indices with the normalization given in Eq. (3.105). Then the time-parameter independence $\frac{dH}{ds} = 0$ leads to the equation of motion for P_l (Exercise 3.11),

$$\frac{dP_l}{ds} = -i \sum_{i,j} t^a \, (t^a U_l)_{ij} \, \frac{\partial S_{\mathrm{eff}}(U)}{\partial (U_l)_{ij}}. \tag{3.79}$$

In the actual simulations, the lnDet$F(U)$ part of the effective action is treated by introducing a set of bosonic variables (pseudofermions) through the identity,

$$\mathrm{Det}F = (\mathrm{Det}F^{-1})^{-1} = \int [d\chi^* d\chi] \, \exp\left(-\sum_{IJ} \chi_I^* F_{IJ}^{-1} \chi_J \right), \tag{3.80}$$

where I and J stand for all possible internal and spacetime indices carried by F. For further details of HMC (and its variations) with pseudofermions, consult the recent review [14] and references therein.

3.3.4 Error Estimate

There are two kinds of errors in the data obtained from LQCD simulations.

Systematic Errors
They are related to the lattice spacing a, the lattice volume L^3, and the quark masses (m). During the continuum extrapolation ($a \to 0$) and the thermodynamic

extrapolation ($L \to \infty$) under the guidance of the asymptotic scaling for small a and the finite size scaling for large L, some systematic errors are brought in. Also, one often needs to make extrapolation to the physical quark mass by using lattice data with heavier quark masses. This also brings some systematic errors.

Statistical Error

It originates from the importance sampling. A very useful procedure to estimate such error commonly used in LQCD is the *jackknife resampling method*. (The name originates from the "jackknife" which is an easy and portable tool for general purposes). Let us consider the mean and the unbiased variance of a certain quantity $\mathcal{O}$,

$$\langle \mathcal{O} \rangle = \frac{1}{N} \sum_{n=1}^{N} \mathcal{O}^{(n)} \pm \sqrt{\frac{\sigma^2(\mathcal{O})}{N}}, \quad \sigma^2(\mathcal{O}) = \left(\frac{N}{N-1} \right) \frac{1}{N} \sum_{n=1}^{N} (\mathcal{O}^{(n)} - \langle \mathcal{O} \rangle)^2, \tag{3.81}$$

where the factor $\frac{N}{N-1}$ is called the Bessel's correction. The jackknife samples are obtained by

$$\mathcal{O}_J^{(n)} = \frac{1}{N-1} \sum_{n' \neq n} \mathcal{O}^{(n')} \quad (n = 1, \cdots, N). \tag{3.82}$$

If we need to make a quick estimate of the mean and the variance of a function $f(\mathcal{O})$, we have

$$\langle f(\mathcal{O}_J) \rangle = \frac{1}{N} \sum_{n=1}^{N} f(\mathcal{O}_J^{(n)}) \pm \sqrt{\frac{\sigma_J^2(f)}{N}}, \quad \sigma_J^2(f) = (N-1) \sum_{n=1}^{N} (f(\mathcal{O}_J^{(n)}) - \langle f(\mathcal{O}_J) \rangle)^2. \tag{3.83}$$

For $f(\mathcal{O}) = \mathcal{O}$, we recover the original mean and variance; $\langle \mathcal{O}_J \rangle = \langle \mathcal{O} \rangle$ and $\sigma_J^2(\mathcal{O}) = \sigma^2(\mathcal{O})$ (Exercise 3.12). One can generalize this procedure by dividing N into $N_b = N/n_b$ with the bin-size n_b and create the N_b jackknife samples. Equation (3.83) corresponds to the case with $n_b = 1$.

3.3.5 Heavy Quark Potential

As one of the examples of the accurate inter-quark potential obtained from LQCD simulations, we show in Fig. 3.10 the dimensionless $Q\bar{Q}$ potential $[V(R) - V(R_0)] \times R_0$ as a function of R/R_0 extracted from the calculation of the Wilson loop in the quenched approximation. R is a distance between the heavy quarks and R_0 is called the *Sommer scale* defined by $R^2 \frac{dV(R)}{dR}\Big|_{R=R_0} = 1.65$.

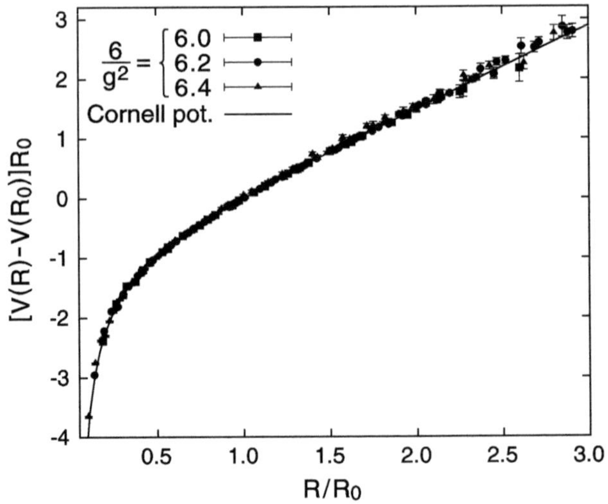

Fig. 3.10 A dimensionless $Q\bar{Q}$ potential as a function of a dimensionless quark–anti-quark separation R/R_0 with R_0 being the Sommer scale. Different symbols correspond to different lattice couplings $g(a)$ and hence different lattice spacings. The solid line shows an empirical Cornell potential. The figure is adapted from [15]

Simulations with different lattice couplings $6/g^2$ correspond to different lattice spacings a. The latter can be fixed, e.g., by taking a phenomenological value $R_0 \simeq 0.5$ fm. The lattice spacings in the physical unit in the figure are $a = 0.094$ fm (squares: $\beta = 6/g^2 = 6.0$), $a = 0.069$ fm (circles: $\beta = 6.2$) and $a = 0.051$ fm (triangles: $\beta = 6.4$). Since there exists no appreciable a dependence of the potential, the system is already close enough to the continuum limit.

Figure 3.10 clearly shows that the heavy quark potential has a linear confining part at long distance and an attractive Coulombic part at short distance. The LQCD results agree not only qualitatively but also quantitatively with an empirical linear + Coulomb potential (the Cornell potential) shown by the solid line, $V(r) = Kr - b/r + \text{const}$ with $b = 0.295$.

3.3.6 Masses of Light Hadrons

Meson masses and baryon masses can be calculated with high accuracy by LQCD simulations with dynamical quarks. The starting point is the hadronic correlation functions $C_{\mathrm{H=M,B}}(n, n')$ in Eqs. (3.40), (3.41) integrated over the spatial coordinates, $\mathbf{n}$ and $\mathbf{n}'$,

$$C_{\mathrm{H}}(\tau) = \sum_{\mathbf{n},\mathbf{n}'} C_{\mathrm{H}}(n, n') \xrightarrow[\tau\to\infty]{} |Z_{\mathrm{H}}|^2 \mathrm{e}^{-M_{\mathrm{H}}\tau}, \tag{3.84}$$

where $\tau = (n_4 - n'_4)a$ is the temporal distance between the source at n' and the sink at n, and $M_{\rm H}$ ($Z_{\rm H}$) corresponds to the mass (the pole residue) of a lightest bound state in each channel. If the temporal extent of the lattice is infinite, one can extract the hadron mass from the formula, $M_{\rm H} = -(1/\tau)\ln C_{\rm H}(\tau)|_{\tau\to\infty}$. In practice, the *effective mass* defined below is more useful,

$$aM_{\rm H}^{\rm eff}(\tau) = \ln\left(\frac{C_{\rm H}(\tau)}{C_{\rm H}(\tau+a)}\right). \tag{3.85}$$

The asymptotic plateau of the effective mass at large τ corresponds to the hadron mass. In actual simulations, the temporal extent is limited ($0 \le \tau/a \le N_\tau$), so that the exponential damping of Eq. (3.84) is replaced by $C_{\rm H} \to \exp[-M_{\rm H}\tau] \pm \exp[-M_{\rm H}(N_\tau a - \tau)]$ where $+(-)$ for the periodic (anti-periodic) boundary condition.

Shown in the left panel of Fig. 3.11 is the dimensionless effective masses ($aM_{\rm H}^{\rm eff}$) against $\tau/a = (n_4 - n'_4)$ [16]. Data points are the effective masses for the pion (π), the kaon (K), the nucleon (N), the cascade baryon (Ξ) and the omega baryon (Ω) calculated on the lattice with $a \simeq 0.085$ fm and the pion mass $M_\pi \simeq 190$ MeV. Reasonable plateau above $\tau/a > 9$ can be seen within the error bars.

Shown in the right panel of Fig. 3.11 is the M_π^2-dependence of the N and Ω masses for three different values of a [16]. The crosses are the values extrapolated to the continuum limit and to the physical pion mass. The N and Ω masses predicted from LQCD and corresponding experimental numbers are

$$M_N^{\rm LQCD} = 0.936(25)(22)\ {\rm GeV}, \quad M_\Omega^{\rm LQCD} = 1.676(20)(15)\ {\rm GeV}, \tag{3.86}$$

$$M_N^{\rm exp.} = 0.939\ {\rm GeV}, \quad M_\Omega^{\rm exp.} = 1.672\ {\rm GeV}. \tag{3.87}$$

Note that the numbers in the first (second) parenthesis in Eq. (3.86) represent the statistical (systematic) errors on the last digits.

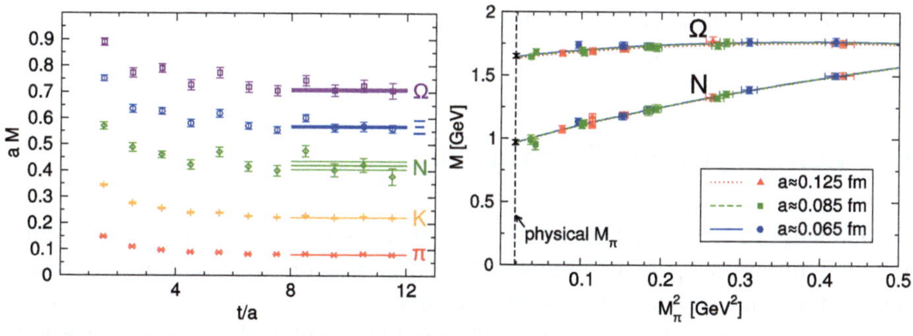

Fig. 3.11 (*Left*) The effective masses of hadrons against the temporal separation between the source and the sink. (*Right*) Hadron masses under the changes of the (pion mass)^2 as well as the lattice spacing a. The figures are taken from [16]

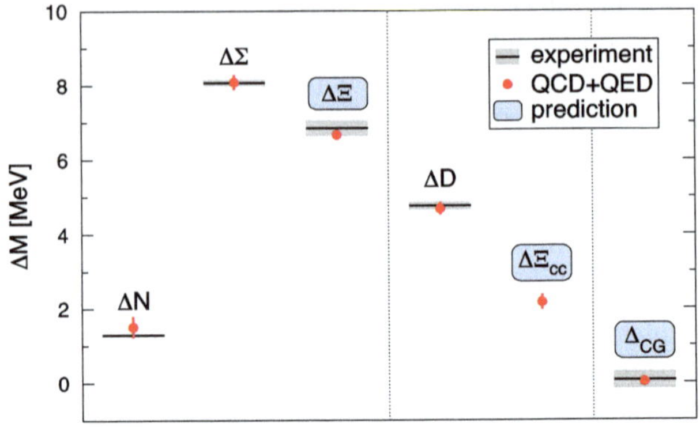

Fig. 3.12 Mass splittings in channels that are stable under the strong and electromagnetic interactions. $\Delta N = M_n - M_p$, $\Delta\Sigma = M_{\Sigma^-} - M_{\Sigma^+}$, $\Delta\Xi = M_{\Xi^-} - M_{\Xi^+}$, $\Delta D = M_{D^\pm} - M_{D^0}$, $\Delta\Xi_{cc} = \Delta\Xi_{cc}^{++} - \Delta\Xi_{cc}^{+}$, $\Delta CG = \Delta N - \Delta\Sigma + \Delta\Xi$. The figure is taken from [17]

Shown in Fig. 3.12 are the high precision numerical results of the hadron mass splittings obtained by the QCD+QED lattice simulations with dynamical u, d, s, c quarks [17]. The horizontal lines are the experimental values and the grey shaded regions represent the experimental errors. Red dots are the lattice results with their uncertainties denoted by the vertical error bars. The neutron-proton mass differences from numerical simulations and the corresponding experimental numbers are

$$(M_n - M_p)^{\text{LQCD+QED}} = \Delta N = 1.51(16)(23)\,\text{MeV}, \tag{3.88}$$

$$(M_n - M_p)^{\text{exp.}} = 1.29\,\text{MeV}. \tag{3.89}$$

Since all hadrons are composite particles of quarks and gluons, there are numerous excited states [18]. To extract the properties of the excited hadrons from LQCD, only looking at the asymptotic form as shown in Eq. (3.84) is not sufficient, and more sophisticated methods such as the maximal entropy method (MEM) and the variational method are required. Interested readers should consult the reviews [19, 20] and references there in.

3.4 Lattice QCD and Nuclear Force

Understanding of the nuclear force from QCD is one of the most challenging problems in nuclear physics. Below the pion production threshold, the notion of the *NN* potential (either in the coordinate space or in the momentum space) has been known to be useful, since it can be used not only to describe the two-body system but also to study nuclear many-body problems through ab-initio calculations (Consult other chapters of this volume).

Several high precision phenomenological *NN* forces have been constructed to reproduce the neutron-proton and proton-proton scattering data (about 4500 data points) with a $\chi^2/\text{dof} \sim 1$. However, they have typically 20–40 fitting parameters: e.g. the CD Bonn potential, AV18 potential and N^3LO chiral effective field theory have 38, 40, and 24 parameters, respectively [21]. If one tries to extend these to hyperon-nucleon and hyperon-hyperon interactions, the task becomes extremely tough since the number of parameters increases and the scattering data are scarce.

Under this situation, it is highly desirable to study baryon-baryon interactions from first principle LQCD simulations, where all the hadronic interactions are controlled only by the QCD coupling g and the quark mass m whose values are pretty well determined at present by the precision QCD simulations [22].

The finite volume method (FVM), a theoretical framework to study hadron-hadron interactions from LQCD, was first proposed by Lüscher [23]: For two hadrons in a finite box with a spatial size L^3, an exact relation between the energy spectra in the box and the elastic scattering phase shift can be derived. If the range of the hadronic interaction R_{QCD} is sufficiently smaller than the size of the box $R_{\text{QCD}} < L/2$, the behavior of the equal-time Bethe-Salpeter amplitude (or more precisely the Nambu-Bethe-Salpeter (NBS) amplitude) $\psi(\mathbf{r})$ in the interval $R_{\text{QCD}} < |\mathbf{r}| < L/2$ has sufficient information to relate the phase shift and the energy shift $\Delta E = M_{\text{HH}} - 2M_{\text{H}}$.

The HAL QCD method was proposed as another theoretical framework to study the hadron-hadron interactions from LQCD by Ishii et al. [24] and was further developed by HAL QCD Collaboration [25]. The starting point is the same equal-time NBS amplitude $\psi(\mathbf{r})$: Instead of looking at the amplitude outside the range of the interaction, the internal region $|\mathbf{r}| < R_{\text{QCD}}$ is considered and an energy-independent non-local potential $U(\mathbf{r}, \mathbf{r}')$ is deduced from $\psi(\mathbf{r})$. Since $U(\mathbf{r}, \mathbf{r}')$ in QCD is spatially localized due to the confinement of quarks and gluons, it is affected only weakly by the finite lattice volume. Physical quantities such as the scattering phase shifts, bound state spectra, and resonance energies can be calculated by solving the integro-differential equation satisfied by $\psi(\mathbf{r})$ with $U(\mathbf{r}, \mathbf{r}')$.

Recently, a detailed comparison between the FVM and the HAL QCD method has been carried out: Although they agree with each other quite accurately for non-resonant pion-pion scattering, large signal to noise ratio inherent in the effective mass $\Delta E^{\text{eff}}(\tau)$ for baryon-baryon scatterings prevents FVM to extract scattering observables [26]. Therefore, we will focus on the HAL QCD method below.

3.4.1 Master Equation for Baryon-Baryon Interaction

Let us consider the baryon-baryon correlation in Fig. 3.4c and define the equal-time NBS amplitude $\psi_\ell(\mathbf{r}, \tau)$ from its large τ behavior:

$$C_{\text{BB}}(\mathbf{r}, \tau) = \sum_{\mathbf{n}', \mathbf{m}'} C_{\text{BB}}(n, m, n', m')\big|_{n_4 = m_4, n_4' = m_4'} \to \sum_{\ell} a_\ell \psi_\ell(\mathbf{r}, \tau) e^{-E_\ell \tau}, \tag{3.90}$$

where $\mathbf{r} = (\mathbf{n} - \mathbf{m})a$, $\tau = (n_4 - n_4')a$, and $\psi_\ell(\mathbf{r}, \tau)$ being the NBS wave function for ℓ-th scattering state on the lattice. For large lattice size, E_ℓ is very dense, so that it is impossible to identify each level. This causes a fatal problem in FVM as mentioned above. On the other hand, if we define $C_{BB}(\mathbf{r};\tau) = \mathscr{R}(\mathbf{r},\tau)e^{-2M_B\tau}$, the following integro-differential equation can be derived below the inelastic threshold ($\tau > M_\pi^{-1}$),

$$\left\{\frac{1}{4M_B}\frac{\partial^2}{\partial\tau^2} - \frac{\partial}{\partial\tau} - H_0\right\}\mathscr{R}(\mathbf{r},\tau) = \int d^3r' U(\mathbf{r},\mathbf{r}')\mathscr{R}(\mathbf{r}',\tau), \qquad (3.91)$$

with $H_0 = -\nabla^2/M_B$. This is the master equation which has the correct information of the S-matrix and hence the scattering phase shift for elastic BB scatterings [25].

If we further focus on the energies much below the inelastic threshold, the velocity expansion of $U(\mathbf{r},\mathbf{r}')$ in terms of its non-locality can be adopted. In fact, the potential with hermiticity, rotational invariance, parity symmetry, and time-reversal invariance may be expanded as [27]

$$U(\mathbf{r},\mathbf{r}') = V(\mathbf{r},\mathbf{v})\delta(\mathbf{r}-\mathbf{r}'), \qquad (3.92)$$

$$V(\mathbf{r},\mathbf{v}) = \underbrace{V_C(r) + V_T(r)S_{12}}_{\text{LO}} + \underbrace{V_{LS}(r)\mathbf{L}\cdot\mathbf{S}}_{\text{NLO}} + \underbrace{O(\mathbf{v}^2)}_{\text{N}^2\text{LO}} + \cdots, \qquad (3.93)$$

where $\mathbf{v} = \mathbf{p}/(M_B/2)$, $\mathbf{L} = \mathbf{r}\times\mathbf{p}$, $\mathbf{p} = -i\nabla$ and $S_{12} = 3(\boldsymbol{\sigma}_1\cdot\mathbf{r})(\boldsymbol{\sigma}_2\cdot\mathbf{r})/r^2 - \boldsymbol{\sigma}_1\cdot\boldsymbol{\sigma}_2$. The central potential V_C and the tensor potential V_T are classified as the leading order (LO) potentials since they are of $O(\mathbf{v}^0)$. The next-to-leading (NLO) potential of $O(\mathbf{v})$ is the spin-orbit potential $V_{LS}(r)$.

3.4.2 Baryon-Baryon Interaction in Flavor SU(3) Limit

To obtain qualitative understandings of the nuclear force from QCD, the S-wave interaction between octet baryons in the flavour SU(3) limit would be a good starting point. In this case, two baryon states with a given angular momentum are labelled by the irreducible flavour multiplets,

$$\mathbf{8}\otimes\mathbf{8} = \underbrace{\mathbf{27}\oplus\mathbf{8}_s\oplus\mathbf{1}}_{\text{symmetric}} \oplus \underbrace{\mathbf{10}^*\oplus\mathbf{10}\oplus\mathbf{8}_a}_{\text{anti-symmetric}}\ . \qquad (3.94)$$

Here "symmetric" and "anti-symmetric" stand for the symmetry under the flavour exchange of two baryons. For the system in the orbital S-wave, the Pauli principle between two baryons imposes **27**, $\mathbf{8}_s$ and **1** to be spin singlet ($^1S_0$) while $\mathbf{10}^*$, **10** and $\mathbf{8}_a$ to be spin triplet ($^3S_1$). Since there are no mixings among different multiplets in the SU(3) limit, one may define the corresponding potentials as

$$^1S_0 \ :\ V^{(\mathbf{27})}(r),\ V^{(\mathbf{8}_s)}(r),\ V^{(\mathbf{1})}(r), \qquad (3.95)$$

$$^3S_1 \ :\ V^{(\mathbf{10}^*)}(r),\ V^{(\mathbf{10})}(r),\ V^{(\mathbf{8}_a)}(r)\ . \qquad (3.96)$$

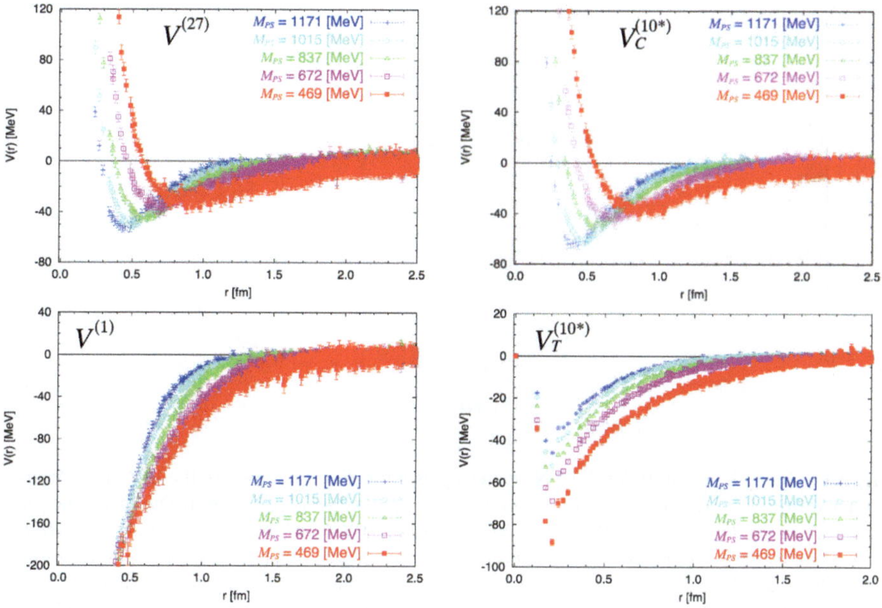

Fig. 3.13 The baryon-baryon potentials from LQCD simulations in the flavour SU(3) limit with several different masses of pseudo-scalar meson. The figures are taken from [29]

The diagonal potential ($B_1B_2 \rightarrow B_1B_2$) and the off-diagonal potentials ($B_1B_2 \rightarrow B_3B_4$) in the particle basis, are obtained by suitable combinations of $V^{(\alpha)}(r)$ with $\alpha = \mathbf{27}, \mathbf{8}_s, \mathbf{1}, \mathbf{10}^*, \mathbf{10}, \mathbf{8}_a$.

Shown in Fig. 3.13 are the results of the exploratory study of the potentials obtained by LQCD simulations on the lattice with $a \simeq 0.12$ fm, $L \simeq 4$ fm and 3 degenerate flavours [29]. Corresponding pion mass ranges from 469 to 1171 MeV.

- The upper left panel is the central potential $V^{(\mathbf{27})}(r)$ to which the ${}^1S_0$ nucleon-nucleon potential belongs. It has a repulsive core at short distance and an attractive pocket at intermediate distance. As the pion mass decreases, the repulsive core gets stronger and the attractive tail gets longer.
- As shown in the lower left panel, the structure of the potential is quite different for $V^{(\mathbf{1})}(r)$ to which the flavour singlet H dibaryon (composed of *uuddss*) belongs. There is no repulsive core and the attraction increases as the pion mass decreases. Such a feature is consistent with the notion of the quark Pauli principle previous discussed in phenomenological quark models [28].
- The upper and lower right panels of Fig. 3.13 are the central potential and the tensor potential of $V^{(\mathbf{10}^*)}(r)$, respectively. The ${}^3S_1$ nucleon-nucleon potential belongs to this channel. The central part has a similar structure as the ${}^1S_0$ channel, while the tensor part has strong attraction and grows rapidly as the pin mass decreases. The latter aspect is qualitatively consistent with the one-pion-exchange picture at long distances.

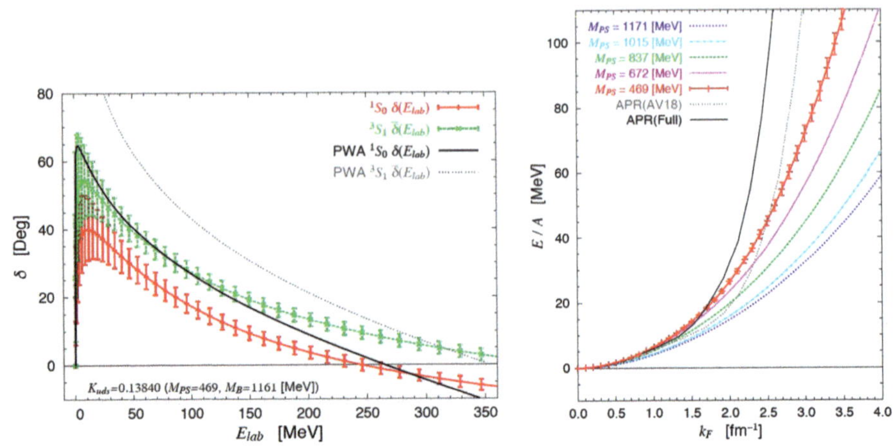

Fig. 3.14 (*Left*) Phase shifts of the NN scattering as a function of energy in the laboratory frame, extracted from LQCD data at the pion mass 469 MeV in the flavor-SU(3) limit. The *black* and *gray dashed lines* are the results of the partial wave analysis (PWA) of the experimental data. (*Right*) Ground state energy per neutron for the pure neutron matter as a function of the Fermi momentum. The APR with *black dotted line* (*black solid line*) corresponds to the empirical equation of state without (with) the phenomenological three nucleon force [31]. The figures are taken from [30]

Shown in Fig. 3.14 (left) is the nucleon-nucleon scattering phase shifts in the $^1S_0$ channel and $^3S_1$ channel obtained by using the potentials, $V^{(\mathbf{27})}(r)$ and $V^{(\mathbf{10^*})}(r)$, for the pion mass 469 MeV [30]. The qualitative feature of the phase shift in the $^1S_0$ channel is similar to the experimental one denoted by the black solid line, despite the fact that pion mass in the simulation is more than 3 times heavier than the physical value. In the $^3S_1$ channel, the deuteron bound state is not formed yet due to heavy pion mass, so that the phase shift starts from 0 at zero energy in contrast the experimental one denoted by the black dashed line. There is however a tendency that the attraction in the $^3S_1$ channel is larger than the $^1S_0$ channel even for the heavy pion mass. Shown in Fig. 3.14 (right) is the energy per particle E/A as a function of the fermi momentum k_F for pure neutron matter calculated by using the Brueckner-Hartree-Fock method with the neutron-neutron potential in the $^1S_0$ channel in Fig. 3.14 (left). As the pion mass decreases, the equation of state becomes stiffer due to the growth of the repulsive core. The APR with black dotted line (black solid line) corresponds to the empirical equation of state without (with) the phenomenological three nucleon force [31].

We note that calculations of the baryon-baryon interactions with (2+1)-flavour LQCD on a large volume ($L \simeq 8.2$ fm, $a \simeq 0.085$ fm) at nearly the physical quark mass ($m_\pi \simeq 146$ MeV, $m_K \simeq 525$ MeV) are underway [32].

3.5 Exercises

3.1 Prove the properties of the Wilson line, Eqs. (3.6)–(3.8).

3.2 Derive the expression on $U_{\mu\nu}(n)-1$ in Eq. (3.13) by using the Baker-Campbell-Hausdorff formula.

3.3 Show the Γ_5 Hermiticity of the Dirac operator in Eq. (3.20).

3.4 Derive the free fermion propagator on the lattice in the momentum representation, Eqs. (3.21) and (3.22).

3.5 Analyze the dispersion relation of the free fermion associated with the Dirac operator, $D_{\rm GW}$ in Eq. (3.26).

3.6 Derive the group integration formulas, Eqs. (3.34)–(3.37), by taking appropriate contractions of the color indices.

3.7 Derive the formula for the Wilson loop in the strong coupling limit Eq. (3.44) by using the group integration formulas Eqs. (3.33)–(3.35).

3.8 Derive the lattice coupling $g(a)$ as a function of the lattice spacing a, Eq. (3.52).

3.9 Show the convergence of $W[\phi]$ to the equilibrium distribution $W_{\rm eq}[\phi]$ under the Markov process by introducing the distance $D = \sum_\phi \left|W[\phi] - W_{\rm eq}[\phi]\right|$ and by studying its behavior under update.

3.10 Prove that the Metropolis test in Eq. (3.72) satisfies the detailed balance Eq. (3.73).

3.11 Derive the equation of motion for P_l in Eq. (3.79) from Eqs. (3.77) and (3.78).

3.12 Prove that the jackknife average and variance for $f(\mathscr{O}) = \mathscr{O}$ reduce to the standard mean and unbiased variance, respectively.

3.13 Prove that the leapfrog integrator satisfies the reversibility Eq. (3.71) exactly. Also prove that the leapfrog integrator preserves the phase space area exactly by evaluate the Jacobian, $d\phi' d\pi' = J d\phi d\pi$.

Acknowledgements The author thanks Takumi Doi and Atsushi Nakamura for useful comments and information on various aspects of LQCD simulations. He also thank the members of HAL QCD Collaboration for fruitful discussions on the hadron-hadron interactions on the lattice. This work was supported in part by MEXT SPIRE and JICFuS and also by RIKEN iTHES Project.

Appendix

Four Vectors and Dirac Matrices

In the (3+1)-dimensional Minkowski spacetime, coordinates, derivatives and four vectors with $\mu = 0, 1, 2, 3$ are

$$x^\mu = (t, \mathbf{x}), \;\; \partial^\mu = (\partial_t, -\nabla), \;\; A^\mu = (A^0, \mathbf{A}). \tag{3.97}$$

In the 4-dimensional Euclidean space, we define the corresponding vectors for $\mu = 4, 1, 2, 3$ as

$$(x_\mu)^{\rm E} = (\tau = it, \mathbf{x}), \;\; (\partial_\mu)^{\rm E} = (\partial_\tau = -i\partial_t, \nabla), \;\; (A_\mu)^{\rm E} = (A_4 = iA^0, \mathbf{A}). \tag{3.98}$$

In the (3+1)-dimensional Minkowski spacetime with the metric $g^{\mu\nu} = \mathrm{diag}(1, -1, -1, -1)$, the Dirac matrices satisfy the following relations for $\mu = 0, 1, 2, 3$,

$$\{\gamma^\mu, \gamma^\nu\} = 2g^{\mu\nu}, \;\; (\gamma^\mu)^\dagger = \gamma^0\gamma^\mu\gamma^0, \;\; \gamma^5 = i\gamma^0\gamma^1\gamma^2\gamma^3 = \gamma_5 = (\gamma_5)^\dagger\,. \tag{3.99}$$

In the standard Dirac representation, we have

$$\gamma^0 = \begin{pmatrix} 1 & 0 \\ 0 & -1 \end{pmatrix}, \;\; \gamma^j = \begin{pmatrix} 0 & \sigma_j \\ -\sigma_j & 0 \end{pmatrix}, \;\; \gamma^5 = \begin{pmatrix} 0 & 1 \\ 1 & 0 \end{pmatrix}, \tag{3.100}$$

where σ_j are the Pauli matrices; $\sigma_1 = \begin{pmatrix} 0 & 1 \\ 1 & 0 \end{pmatrix}, \sigma_2 = \begin{pmatrix} 0 & -i \\ i & 0 \end{pmatrix}, \sigma_3 = \begin{pmatrix} 1 & 0 \\ 0 & -1 \end{pmatrix}$.

In the 4-dimensional Euclidean space with the metric $\delta_{\mu\nu} = \mathrm{diag}(1, 1, 1, 1)$, we define the Euclidean Dirac matrices as

$$\Gamma_\mu \equiv \left(\gamma_4 = \gamma^0, -i\boldsymbol{\gamma}\right), \;\; \Gamma_{-\mu} \equiv -\Gamma_\mu, \;\; \text{and} \;\; \Gamma_5 \equiv \gamma^5, \tag{3.101}$$

which satisfy the relations,

$$\{\Gamma_\mu, \Gamma_\nu\} = 2\delta_{\mu\nu}, \;\; \Gamma_\mu^\dagger = \Gamma_\mu, \quad (\text{for } \mu = 1, 2, 3, 4, 5) \tag{3.102}$$

SU(N) Algebra

Let $\mathcal{T}^a$ $(a = 1, \cdots, N^2 - 1)$ are the Hermitian generators of the SU(N) group. They satisfy the Lie algebra

$$\left[\mathcal{T}^a, \mathcal{T}^b\right] = if_{abc}\mathcal{T}^c, \tag{3.103}$$

where f_{abc} is the structure constant being totally anti-symmetric in its indices. $(\mathscr{T}^b)^2$ commutes with every generator $\mathscr{T}^a$ and is called the quadratic Casimir operator.

For $N = 2$, f_{abc} reduces to the anti-symmetric tensor ϵ_{ijk} with $\epsilon_{123} = 1$. For $N = 3$, the non-vanishing components of f_{abc} read $f_{123} = 1, f_{147} = -f_{156} = f_{246} = f_{257} = f_{345} = -f_{367} = 1/2, f_{458} = f_{678} = \sqrt{3}/2$.

In the fundamental representation, $\mathscr{T}^a$ is written by the $N \times N$ matrices t^a as

$$t^a = \frac{1}{2}\lambda_a, \tag{3.104}$$

where λ_a for $N = 2$ reduce to the Pauli matrices σ_i, while those for $N = 3$ reduce to the Gell-Mann matrices.

Some useful relations of t^a for general N are

$$\mathrm{tr}(t^a t^b) = \frac{1}{2}\delta_{ab}, \quad t^a_{ij} t^b_{kl} = \frac{1}{2}(\delta_{il}\delta_{jk} - \frac{1}{N}\delta_{ij}\delta_{kl}), \quad (t^a t^a)_{ij} = C_{\mathrm{F}}\delta_{ij}, \quad \text{with } C_{\mathrm{F}} = \frac{N^2-1}{2N}. \tag{3.105}$$

In the adjoint representation, $\mathscr{T}^a$ is written by $(N^2-1) \times (N^2-1)$ matrices T^a as

$$(T^a)_{bc} = -if_{abc}, \tag{3.106}$$

which satisfy the relations

$$\mathrm{tr}(T^a T^b) = N\delta_{ab}, \quad (T^a T^a)_{bc} = C_{\mathrm{A}}\delta_{bc}, \quad \text{with } C_{\mathrm{A}} = N. \tag{3.107}$$

Gaussian and Grassmann Integrals

Basic Gaussian and Grassmann integrals are

$$\int_{-\infty}^{+\infty} \frac{dx}{\sqrt{2\pi}}\, \mathrm{e}^{-ax^2/2} = \frac{1}{\sqrt{a}}, \tag{3.108}$$

$$\int \frac{dz^* dz}{2\pi i}\, \mathrm{e}^{-b|z|^2} = \frac{1}{b}, \tag{3.109}$$

$$\int d\bar{\xi} d\xi\, \mathrm{e}^{-c\bar{\xi}\xi} = c. \tag{3.110}$$

Here x (z) is a real (complex) number, while $\bar{\xi}$ and ξ are anti-commuting Grassmann numbers ($\{\xi, \bar{\xi}\} = 0$, and $\xi^2 = \bar{\xi}^2 = 0$). a and b are assumed to be real and positive numbers, while c is an arbitrary complex number. Equation (3.109) can be shown by rewriting the integral in terms of the real and imaginary parts of z or in terms of the polar coordinates of z. Equation (3.110) can be shown by noting that $e^{-c\bar{\xi}\xi} = 1 - c\bar{\xi}\xi$ and $\int d\xi = \partial/\partial\xi$ (integral = derivative) for Grassmann variables.

Generalization of the above results to the case of multiple variables is straightforward. For $x = (x_1, \cdots, x_n)$, $z = (z_1, \cdots, z_n)$, $\xi = (\xi_1, \cdots, \xi_n)$, and $\bar{\xi} = (\bar{\xi}_1, \cdots, \bar{\xi}_n)$ with $\{\xi_k, \xi_l\} = \{\bar{\xi}_k, \bar{\xi}_l\} = \{\xi_k, \bar{\xi}_l\} = 0$, we have

$$\int \prod_{l=1}^{n} \frac{dx_l}{\sqrt{2\pi}} \, e^{-\frac{1}{2}xAx} = \frac{1}{\sqrt{\text{Det}\, A}}, \tag{3.111}$$

$$\int \prod_{l=1}^{n} \frac{dz_l^* dz_l}{2\pi i} \, e^{-z^* Bz} = \frac{1}{\text{Det}\, B}, \tag{3.112}$$

$$\int \prod_{l=1}^{n} d\bar{\xi}_l d\xi_l \, e^{-\bar{\xi}C\xi} = \text{Det}\, C. \tag{3.113}$$

Here A is a non-singular and real-symmetric matrix whose eigenvalues a_l satisfy $a_l > 0$ for all l. B is a non-singular complex matrix whose complex eigenvalues b_l obtained by the biunitary transformation ($UBV^\dagger$) satisfy Re $b_l > 0$ for all l. C is an arbitrary complex matrix with no conditions. Note that B and C do not have to be Hermitian matrices. In field theories, the label "l" summarizes all possible indices including spin, flavor, color, spacetime points etc. and "Det" denotes the determinant for all these indices.

Method of Characteristics

We need to construct a general solution of the following partial differential equation,

$$\left(\lambda \frac{\partial}{\partial \lambda} + \beta(g) \frac{\partial}{\partial g}\right) f(\lambda, g) = 0. \tag{3.114}$$

For this purpose, we introduce the running coupling $\bar{g}(\lambda)$ through $\lambda d\bar{g}/d\lambda = -\beta(\bar{g})$ whose formal solution reads

$$\lambda = \exp\left(-\int_{g}^{\bar{g}(\lambda)} \frac{dg'}{\beta(g')}\right). \tag{3.115}$$

Then the solution of Eq. (3.114) can be written as

$$f(\lambda, g) = f(1, \bar{g}(\lambda)). \tag{3.116}$$

This can be explicitly checked by applying the partial derivative on both sides,

$$\begin{aligned}\lambda\partial_\lambda f(\lambda,g) &= \lambda(\partial_\lambda \bar{g})(\partial_{\bar{g}} f) = -\beta(\bar{g})(\partial_{\bar{g}} f),\\ \beta\partial_g f(\lambda,g) &= \beta(g)\,(\partial\bar{g}/\partial g)|_\lambda\,(\partial_{\bar{g}} f) = \beta(\bar{g})(\partial_{\bar{g}} f).\end{aligned} \tag{3.117}$$

where we have used the relation $\partial\bar{g}/\partial g = \beta(\bar{g})/\beta(g)$ obtained from Eq. (3.115).

In general, the first-order partial differential equation (PDE) can be transformed to a set of ordinary differential equations (ODEs) and can be solved by *the method of characteristics*. As an illustration, let us consider the following PDE,

$$a(t,x)\partial_t u(t,x) + b(t,x)\partial_x f(t,x) = c(t,x). \tag{3.118}$$

This is equivalent to the coupled ODEs,

$$\frac{d\bar{t}}{ds} = a(\bar{t},\bar{x}), \quad \frac{d\bar{x}}{ds} = b(\bar{t},\bar{x}), \quad \frac{df(\bar{t},\bar{x})}{ds} = a(\bar{t},\bar{x}), \tag{3.119}$$

where s parametrizes the "flow" of the coordinates. $(\bar{t}(s),\bar{x}(s))$. This is called the *characteristic curve* as shown in Fig. 3.15.

The function f can be obtained by integrating the last equation of Eq. (3.119) on the characteristic curve from the initial point $(t_\mathrm{i},x_\mathrm{i})$ to the final point $(t_\mathrm{f},x_\mathrm{f}) \equiv (t,x)$,

$$f(t,x) = f(t_\mathrm{i},x_\mathrm{i}) + h(t,x,t_\mathrm{i},x_\mathrm{i}) \tag{3.120}$$

where h stands for an integration of the known function $c(\bar{t},\bar{x})$ on the characteristic curve. Equation (3.120) implies that the desired function at (t,x) is obtained essential by a "pullback" of the point to $(t_\mathrm{i},x_\mathrm{i})$ along the characteristic curve. It is a straightforward exercise to generalize the above derivation to the system with more coordinates, $(t,\mathbf{x})$.

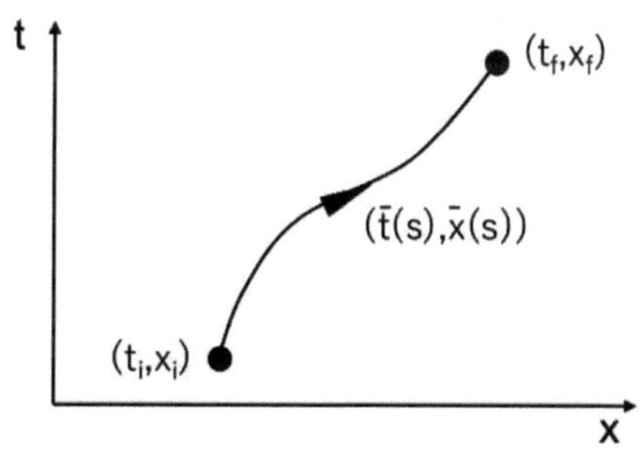

Fig. 3.15 Schematic illustration of the characteristic curve

Leapfrog Integrator in Molecular Dynamics

Let us start with a Tayler expansion of the field ϕ:

$$\begin{aligned}\phi(s+\varepsilon) &= \phi(s) + \varepsilon\dot{\phi}(s) + \frac{\varepsilon^2}{2}\ddot{\phi}(s) + O(\varepsilon^3), \\ &= \phi(s) + \varepsilon\pi(s) + \frac{\varepsilon^2}{2}\dot{\pi}(s) + O(\varepsilon^3), \\ &= \phi(s) + \varepsilon\pi(s+\varepsilon/2) + O(\varepsilon^3), \end{aligned} \tag{3.121}$$

where we have used the equation of motion, $\dot{\phi}(s) \equiv d\phi(s)/ds = \pi(s)$. To evaluate $\pi(s+\varepsilon/2)$, we take the midpoint prescription which does not have $O(\varepsilon^2)$ error,

$$\begin{aligned}\pi(s+\varepsilon/2) &= \pi(s-\varepsilon/2) + \varepsilon\dot{\pi}(s) + O(\varepsilon^3) \\ &= \pi(s-\varepsilon/2) - \varepsilon\frac{\delta S(\phi)}{\delta\phi(s)} + O(\varepsilon^3). \end{aligned} \tag{3.122}$$

Equations (3.121) and (3.122) give a procedure to move the molecular dynamics one-step forward, $(\phi(s), \pi(s-\varepsilon/2)) \to (\phi(s+\varepsilon), \pi(s+\varepsilon/2))$. The initial and final steps need to receive special care,

$$\pi(\varepsilon/2) = \pi(0) - \frac{1}{2}\varepsilon\frac{\delta S(\phi)}{\delta\phi(s)} + O(\varepsilon^2), \quad \pi(s_f) = \pi(s_f - \varepsilon/2) - \frac{1}{2}\varepsilon\frac{\delta S(\phi)}{\delta\phi(s_f)} + O(\varepsilon^2), \tag{3.123}$$

which have only $O(\epsilon^2)$ accuracy. An illustration of this leapfrog integrator is shown in Fig. 3.16. Since the initial and final steps introduce $O(\varepsilon^2)$ error irrespective of the length of the MD trajectory, and the intermediate steps introduce $O(\varepsilon^3) \times \varepsilon^{-1} = O(\varepsilon^2)$ error as a whole, one finds $\Delta H = O(\varepsilon^2)$ after one MD trajectory before the Metropolis test.

The leapfrog integrator satisfies the reversibility and symplectic property, which can be checked explicitly by using the above definitions (Exercise 3.13).

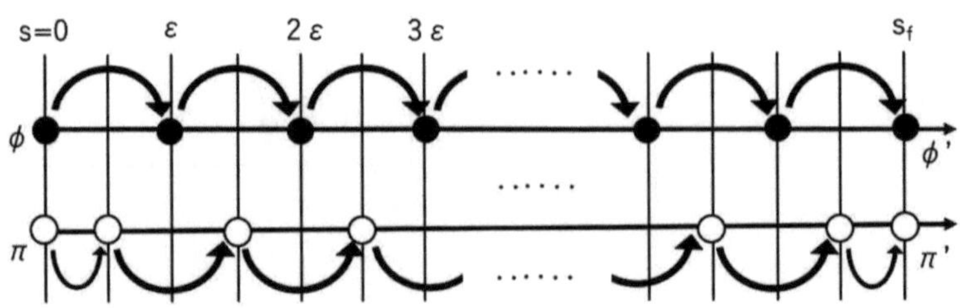

Fig. 3.16 The leapfrog integrator

References

1. K.G. Wilson, Phys. Rev. D **10**, 2445 (1974)
2. M. Creutz, Phys. Rev. D **21**, 2308 (1980)
3. N. Brambilla et al., Eur. Phys. J. C **74**, 2981 (2014)
4. K.G. Wilson, Nucl. Phys. Proc. Suppl. **140**, 3 (2005)
5. M. Creutz, *Quarks, Gluons and Lattices* (Cambridge University Press, Cambridge, 1985)
6. H.J. Rothe, World Sci. Lect. Notes Phys. **82**, 1 (2012)
7. C. Hoelbling, Acta Phys. Polon. B **45**, 2143 (2014)
8. A. Ukawa, J. Stat. Phys. **160**, 1081 (2015)
9. J.W. Negele, H. Orland, *Quantum Many Particle Systems* (Addison-Wesley, Redwood, 1988)
10. O. Häggström, *Finite Markov Chains and Algorithmic Applications* (Cambridge University Press, Cambridge, 2002)
11. H. Suwa, S. Todo, Phys. Rev. Lett. **105**, 120603 (2010)
12. S. Duane, A.D. Kennedy, B.J. Pendleton, D. Roweth, Phys. Lett. B **195**, 216 (1987)
13. N. Metropolis, A.W. Rosenbluth, M.N. Rosenbluth, A.H. Teller, E. Teller, J. Chem. Phys. **21**, 1087 (1953)
14. S. Schaefer, PoS LATTICE **2012**, 001 (2012)
15. G.S. Bali, Phys. Rep. **343**, 1 (2001)
16. S. Durr et al., Science **322**, 1224 (2008)
17. S. Borsanyi et al., Science **347**, 1452 (2015)
18. The Review of Particle Physics (2015). http://pdg.lbl.gov/
19. M. Asakawa, T. Hatsuda, Y. Nakahara, Prog. Part. Nucl. Phys. **46**, 459 (2001)
20. Z. Fodor, C. Hoelbling, Rev. Mod. Phys. **84**, 449 (2012)
21. R. Machleidt, Int. J. Mod. Phys. **E26**, 1740018 (2017)
22. S. Aoki et al., Eur. Phys. J. C **74**, 2890 (2014)
23. M. Lüscher, Nucl. Phys. B **354**, 531 (1991)
24. N. Ishii, S. Aoki, T. Hatsuda, Phys. Rev. Lett. **99**, 022001 (2007)
25. N. Ishii et al., HAL QCD collaboration. Phys. Lett. B **712**, 437 (2012)
26. T. Iritani, HAL QCD collaboration. JHEP **1610**, 101 (2016)
27. S. Okubo, R.E. Marshak, Ann. Phys. **4**, 166 (1958)
28. M. Oka, K. Shimizu, K. Yazaki, Prog. Theor. Phys. Suppl. **137**, 1 (2000)
29. T. Inoue et al., HAL QCD collaboration. Nucl. Phys. A **881**, 28 (2012)
30. T. Inoue et al., HAL QCD collaboration. Phys. Rev. Lett. **111**, 112503 (2013)
31. A. Akmal, V.R. Pandharipande, D.G. Ravenhall, Phys. Rev. C **58**, 1804 (1998)
32. T. Doi et al., HAL QCD collaboration. PoS **LATTICE2015**, 086 (2016)

Chapter 4
General Aspects of Effective Field Theories and Few-Body Applications

Hans-Werner Hammer and Sebastian König

4.1 Introduction: Dimensional Analysis and the Separation of Scales

Effective field theory (EFT) provides a general approach to calculate low-energy observables by exploiting scale separation. The origin of the EFT approach can be traced to the development of the renormalization group [1] and the intuitive understanding of ultraviolet divergences in quantum field theory [2]. A concise formulation of the underlying principle was given by Weinberg [3]: If one starts from the most general Lagrangian consistent with the symmetries of the underlying theory, one will get the most general S-matrix consistent with these symmetries. As a rule, such a most general Lagrangian will contain infinitely many terms. Only together with a power counting scheme that orders these terms according to their importance at low energies one obtains a predictive paradigm for a low-energy theory.

The Lagrangian and physical observables are typically expanded in powers of a low-momentum scale M_{lo}, which can be a characteristic external momentum or an internal infrared scale, over a high-momentum scale $M_{\mathrm{hi}} \gg M_{\mathrm{lo}}$.[1] This expansion

[1]Note there are often more than two scales, which complicates the power counting. Here we focus on the simplest case to introduce the general principle.

H.-W. Hammer (✉)
Institut für Kernphysik, Technische Universität Darmstadt, 64289 Darmstadt, Germany
e-mail: Hans-Werner.Hammer@physik.tu-darmstadt.de

S. König
Department of Physics, The Ohio State University, Columbus, OH 43210, USA
e-mail: koenig.389@osu.edu

M. Hjorth-Jensen et al. (eds.), *An Advanced Course in Computational Nuclear Physics*, Lecture Notes in Physics 936, DOI 10.1007/978-3-319-53336-0_4

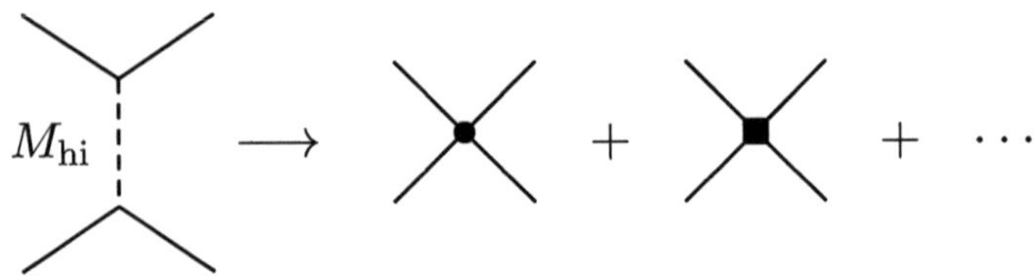

Fig. 4.1 Expansion of heavy-particle exchange between light particles in terms of contact interactions between light particles. The *solid and dashed lines* denote light and heavy particles, respectively. The *circle* and *square* denote contact interactions with zero and two derivatives, in order

provides the basis for the power counting scheme. It depends on the system to which the physical scales $M_{\rm hi}$ and $M_{\rm lo}$ correspond to.

As an example, we take a theory that is made of two particle species, light bosons with mass $M_{\rm lo}$ and heavy bosons with mass $M_{\rm hi} \gg M_{\rm lo}$.[2] We consider now soft processes in which the energies and momenta are of the order of the light particle mass (the so-called soft scale). Under such conditions, the short-distance physics related to the heavy particles can never be resolved explicitly. However, it can be represented by light-particle contact interactions with increasing dimension (number of derivatives). To illustrate this, we consider the scattering of the light particles mediated by heavy-particle exchange, with g the heavy-light coupling constant. The corresponding interaction Lagrangian is given by

$$\mathscr{L}_{\rm int} = g\left(\chi^\dagger\phi\phi + \phi^\dagger\phi^\dagger\chi\right) , \tag{4.1}$$

where ϕ denotes the light boson field and χ is the heavy boson field. As depicted in Fig. 4.1, one can represent such exchange diagrams by a sum of local operators of the light fields with increasing number of derivatives. In a symbolic notation, the leading order scattering amplitude can be written as

$$T \sim \frac{g^2}{{M_{\rm hi}}^2 - q^2} = \frac{g^2}{{M_{\rm hi}}^2} + \frac{g^2 q^2}{{M_{\rm hi}}^4} + \cdots , \tag{4.2}$$

with q^2 the squared 4-momentum transfer. We will come back to this example in more detail in Sect. 4.2.

In many cases, the corresponding high-energy theory is either not known or can not easily be solved. Still, EFT offers a predictive and systematic framework for performing calculations in the light-particle sector. We denote by Q a typical energy or momentum of the order of $M_{\rm lo}$ and by $M_{\rm hi}$ the hard scale where the EFT will break down. In many cases, this scale is set by the masses of the heavy particles not considered explicitly and thus replaced by contact interactions as in the example

[2]For further examples, see the lectures by Kaplan [4, 5].

above. In such a setting, any matrix element or Green's function admits an expansion in the small parameter $Q/M_{\rm hi}$ [3]

$$\mathscr{M} = \sum_{\nu} \left(\frac{Q}{M_{\rm hi}}\right)^{\nu} \mathscr{F}\left(\frac{Q}{\Lambda}, g_i\right) \tag{4.3}$$

where $\mathscr{F}$ is a function of order one (this is the naturalness assumption), Λ a regularization scale (related to the UV divergences appearing in the loop graphs) and the g_i denotes a collection of coupling constants, often called low-energy constants (LECs). These parameterize (encode) the unknown high-energy (short-distance) physics and must be determined by a fit to data (or can be directly calculated if the corresponding high-energy theory is known/can be solved). The counting index ν in general depends on the fields in the effective theory, the number of derivatives and the number of loops. This defines the so-called power counting which allows to categorize all contributions to any matrix element at a given order. It is important to stress that ν must be bounded from below to define a sensible EFT. In QCD, e.g., this is a consequence of the spontaneous breaking of chiral symmetry.

The contributions with the lowest possible value of ν define the so-called leading order (LO) contribution, the first corrections with the second smallest allowed value of ν the next-to-leading order (NLO) terms and so on. In contrast to more conventional perturbation theory, the small parameter is not a dimensionless coupling constant (like, e.g., in Quantum Electrodynamics) but rather a ratio of two scales. Typically, one expands in the ratio of a small energy or momentum and the hard scale $M_{\rm hi}$. A prototype of such a perturbative EFT is chiral perturbation theory, which exploits the strictures of the spontaneous and explicit chiral symmetry breaking in QCD [6, 7]. Here, the light degrees of freedom are the pions, generated through the symmetry violation. Heavier particles like e.g. vector mesons only appear indirectly as they generate local four-pion interactions with four, six, etc. derivatives.

In these lectures, we also consider EFTs with bound states, where certain contributions need to resummed nonperturbatively. In Sect. 4.2, we start with some general considerations. This is followed by the explicit discussion of an EFT for non-relativistic bosons with short-range interactions and large scattering length in Sect. 4.3. The extension of this framework to low-energy nucleons is presented in Sect. 4.4. Finally, we will discuss the inclusion of long-range interactions mediated by photon and pion exchange in Sect. 4.5.

4.2 Theoretical Foundations of Effective Field Theory

As mentioned in the introduction, effective field theories are described by writing down Lagrangians with an infinite number of terms, restricted only by symmetry considerations, and ordered by a scheme referred to as "power counting." In this section, we discuss the meaning and importance of all these ingredients.

4.2.1 *Top-Down vs. Bottom-Up Approaches*

Generally, there are two different motivations for working with an EFT. Given a known quantum field theory, which can be solved to compute a given quantity of interest, it can be beneficial to switch to an effective description valid only in a limited energy regime simply because carrying out the calculation is more efficient with the effective theory. With such a solvable underlying theory, the parameters ("low-energy constants") of the effective theory can be computed directly by considering some number of (simple) processes, i.e., one does not need experimental input beyond what was needed to fix the parameters of the underlying theory. This approach, based on a reduction of expressions from the underlying to the effective picture is called a "top-down" approach.

An alternative procedure, somewhat closer to what we described at the outset, is to start "bottom up," i.e., by simply writing down the effective Lagrangian directly—or more precisely only those terms of the infinitely many which are needed to achieve a given desired accuracy. Being able to do that of course requires that as a first step one has already figured out which terms are allowed and how they should be ordered.

Our approach here is to work top down in the pedagogical sense, i.e., postpone the discussion of the bottom-up approach and its ingredients until later in this section, and instead dive into the matter starting with examples that show how effective low-energy theories can arise from more fundamental ones. We assume that the reader is familiar the material from a standard (relativistic) quantum field theory course.

4.2.1.1 Integrating Out Exchange Particles: Part I

As was also mentioned in the introduction, the very first step in the construction of an EFT is to identify the relevant degrees of freedom to work with, as well as those which are irrelevant and thus do not need to be kept explicitly (with emphasis on the last word, because *implicitly* the physics of left-out degrees of freedom should and does enter in the effective description).

Let us illustrate this by showing how integrating out a "heavy" particle gives rise to contact interactions between the remaining degrees of freedom (see the example in Sect. 4.1). We stress that the particles which are integrated out can in fact be lighter than what is left (like it is the case in pionless EFT)—what really matters for the procedure is which particles are assumed to appear in *asymptotic* states, and what is the typical energy/momentum scale between those. In that spirit, we are not making explicit assumptions about the mass hierarchy of the particles in the following. For the illustration here, we consider two scalar fields (complex and relativistic) with Yukawa interactions and start with a Lagrangian for two species:

$$\mathscr{L} = -\phi^\dagger \left(\Box + m_\phi^2\right)\phi - \chi^\dagger \left(\Box + m_\chi^2\right)\chi + g\left(\phi^\dagger\phi^\dagger\chi + \text{h.c.}\right) . \tag{4.4}$$

Suppose now we are only interested in interactions between ϕ particles at energy scales much smaller than m_χ, so that the explicit χ exchange generated by the interaction term in Eq. (4.4) cannot be resolved. In that case, we can derive a new effective Lagrangian that only contains ϕ degrees of freedom, a process referred to as "integrating out" the field χ (stemming from its implementation in the path-integral formalism). In effect that amounts to using the equations of motion, which we do here. From the Euler-Lagrange equation for $\chi^\dagger$, we directly get

$$\chi = \left(\Box + m_\chi^2\right)^{-1} g\phi\phi\,. \tag{4.5}$$

Defining the Klein-Gordon propagator

$$D_\chi(x-y) = \int \frac{\mathrm{d}^4p}{(2\pi)^4} \mathrm{e}^{-\mathrm{i}p(x-y)} \frac{\mathrm{i}}{p^2 - m_\chi^2 + \mathrm{i}\varepsilon}\,, \tag{4.6}$$

satisfying

$$\left(\Box + m_\chi^2\right) D_\chi(x-y) = -\mathrm{i}\delta^{(4)}(x-y)\,, \tag{4.7}$$

we can write out Eq. (4.5) in configuration space as

$$\chi(x) = \mathrm{i}g \int \mathrm{d}^4y\, D_\chi(x-y)\phi(y)\phi(y)\,. \tag{4.8}$$

Inserting this back into the Lagrangian (4.4), we obtain

$$\mathscr{L}(x) = -\phi^\dagger(x)\left(\Box + m_\phi^2\right)\phi(x) - \mathrm{i}g^2\phi^\dagger(x)\phi^\dagger(x) \int \mathrm{d}^4y\, D_\chi(x-y)\phi(y)\phi(y)\,, \tag{4.9}$$

where we have written out the spacetime dependence of all fields and used Eq. (4.7) to cancel the terms involving $\chi^\dagger(x)$. So far, we have made only exact manipulations, but the resulting Lagrangian (4.9) is *non-local*, i.e., it depends on fields evaluated at different spacetime points. To simplify it further, we want to exploit the fact that χ is considered "heavy" compared to the scales we want to describe. Mathematically, this means that $D_\chi(x-y)$ is peaked at distances that are small compared to $1/m_\chi^2$. There are several ways to implement this knowledge. A particularly intuitive version

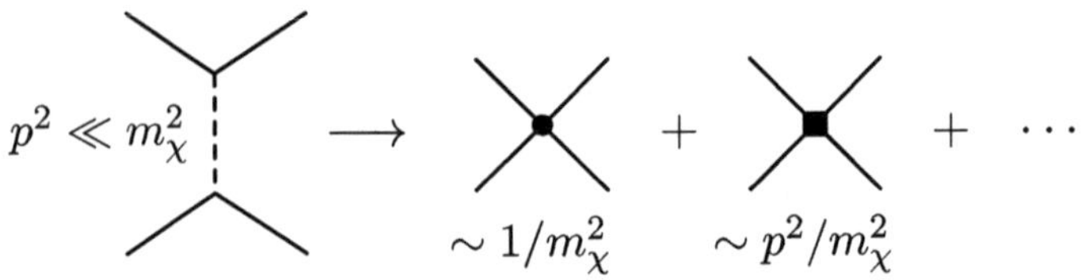

Fig. 4.2 Chain of contact interactions obtained by integrating out an exchange particle

is to expand the propagator (4.6) in momentum space,

$$\frac{\mathrm{i}}{p^2 - m_\chi^2 + \mathrm{i}\varepsilon} = \frac{-\mathrm{i}}{m_\chi^2}\left(1 + \frac{p^2}{m_\chi^2} + \cdots\right), \tag{4.10}$$

and then Fourier-transform back to configuration space. The first term gives a simple delta function, and terms with powers of p^2 induce operators with derivatives acting on $\delta(x-y)$. Inserting the leading term into Eq. (4.9), we arrive at the effective *local* Lagrangian

$$\mathscr{L}_{\text{eff}}(x) = -\phi^\dagger(x)\left(\Box + m_\phi^2\right)\phi(x) - \frac{g^2}{m_\chi^2}\phi^\dagger(x)\phi^\dagger(x)\,\phi(x)\,\phi(x) + \cdots. \tag{4.11}$$

The ellipses contain operators with derivatives acting on $\phi(x)$, obtained from those acting on the delta functions from the propagator after integrating by parts. A diagrammatic illustration of the procedure is shown in Fig. 4.2.

We note that an alternative derivation of the above result, discussed for example in [8], is given by Taylor-expanding the field product $\phi(y)\phi(y)$ about $y = x$ under the integral and then using the properties of the propagator. This directly gives terms with an increasing number of derivatives acting on $\phi^\dagger(x)\phi^\dagger(x)$, and those with an odd number of derivatives are found to vanish, in agreement with Eq. (4.10) featuring only even powers of p^2.

4.2.1.2 Emergence of Many-Body Forces

We now add a third field Φ to the Lagrangian:

$$\begin{aligned}\mathscr{L} = &-\phi^\dagger\left(\Box + m_\phi^2\right)\phi - \chi^\dagger\left(\Box + m_\chi^2\right)\chi - \Phi^\dagger\left(\Box + m_\Phi^2\right)\Phi \\ &+ g\left(\phi^\dagger\phi^\dagger\chi + \text{h.c.}\right) + g'\left(\Phi^\dagger\phi\chi + \text{h.c.}\right).\end{aligned} \tag{4.12}$$

The new interaction term is chosen such that Φ can "decay" into a ϕ and a χ, thus acting like a heavier version of the ϕ. In spite of the simplicity of this bosonic toy model, it is useful to think about ϕ and Φ as the nucleon and its Δ excitation, respectively, and about χ as a pion field. If we first integrate out the Φ field following the procedure described in the previous section, we find

$$\Phi(x) = \mathrm{i}g' \int \mathrm{d}^4y\, D_\Phi(x-y)\phi(y)\chi(y)\,, \tag{4.13}$$

and thus

$$\mathscr{L}_{\text{eff}} = -\phi^\dagger \left(\Box + m_\phi^2\right)\phi - \chi^\dagger \left(\Box + m_\chi^2\right)\chi + g\left(\phi^\dagger\phi^\dagger\chi + \text{h.c.}\right) + \frac{g'^2}{m_\Phi^2}\phi^\dagger\chi^\dagger\phi\chi + \cdots\,, \tag{4.14}$$

where we have only kept the leading (no derivatives) induced contact interaction. Proceeding as before for the χ field, we now get

$$\left(\Box + m_\chi^2\right)\chi = g\phi\phi + \frac{g'^2}{m_\Phi^2}\phi^\dagger\phi\,\chi + \cdots\,. \tag{4.15}$$

This can no longer be solved exactly because we now have a χ on the right-hand side. However, using the general operator identify

$$\left(\hat{A} - \hat{B}\right)^{-1} = \hat{A}^{-1} + \hat{A}^{-1}\hat{B}\hat{A}^{-1} + \cdots\,, \tag{4.16}$$

we can write down a formal iterative solution:

$$\chi = \left(\Box + m_\chi^2\right)^{-1} g\phi\phi + \left(\Box + m_\chi^2\right)^{-1}\frac{g'^2}{m_\Phi^2}\phi^\dagger\phi\left(\Box + m_\chi^2\right)^{-1} g\phi\phi + \cdots\,, \tag{4.17}$$

with each of the inverse differential operators giving a propagator when written out. Those, in turn, each give factors of $-\mathrm{i}/m_\Phi^2$ times a delta function, plus additional terms with derivatives.

4.1 *E*xercise: Derive Eq. (4.16).

Inserting the above result back into Eq. (4.14), we see that in addition to the two-body contact operator $(\phi^\dagger\phi)^2$ obtained previously, we now also get all kinds of higher-body interactions. For example, we get a three-body force through

$$\frac{g'^2}{m_\Phi^2}\phi^\dagger\chi^\dagger\phi\chi \to \frac{g'^2g^2}{m_\Phi^2 m_\chi^4}(\phi^\dagger\phi)^3\,. \tag{4.18}$$

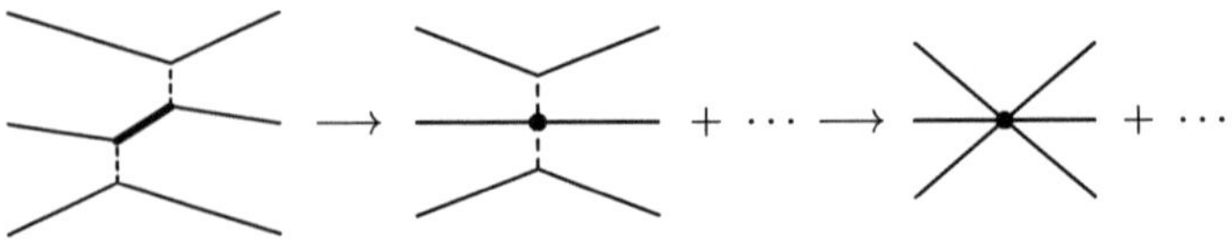

Fig. 4.3 Emergence of a three-body contact interaction

In Fig. 4.3 it is illustrated diagrammatically how such a term arises subsequently, starting from a diagram derived from the original Lagrangian (4.12) with three fields.

4.2.2 Nonrelativistic Field Theory

Relativistic effects and exact Lorentz invariance are not crucial to describe systems at low energies, where "low" means "much smaller than the particles' rest mass." Based on that, one typically starts with a nonrelativistic framework and writes down effective Lagrangians of so-called Schrödinger fields, e.g.,

$$\mathscr{L}_{\phi,\text{free}} = \phi^\dagger \left(\mathrm{i}\partial_t + \frac{\nabla^2}{2m} \right) \phi \tag{4.19}$$

for a free scalar particle, where $\phi^\dagger(t, \mathbf{x})$ is the field operator that creates a particle at time t and position $\mathbf{x}$, and $\phi(t, \mathbf{x})$) correspondingly destroys it. Written in terms of momentum-space ladder operators $\hat{a}_\mathbf{p}$,$\hat{a}_\mathbf{p}^\dagger$ (as they appear in standard many-body quantum mechanics), we have

$$\phi(t, \mathbf{x}) = \int \frac{\mathrm{d}^3 p}{(2\pi)^3} \hat{a}_\mathbf{p}\, \mathrm{e}^{-\mathrm{i}E_\mathbf{p} t} \mathrm{e}^{\mathrm{i}\mathbf{p}\cdot\mathbf{x}} , \tag{4.20}$$

and analogously for $\phi^\dagger(t, \mathbf{x})$. Note that here $E_\mathbf{p} = \mathbf{p}^2/(2m)$ is the kinetic energy alone, and that creation and destruction operators are completely separated. Intuitively, this makes perfect sense: At low energies, virtual particle-antiparticle pairs would be highly off-shell, thus giving rise to very short-range effects that we can simply describe as contact interactions. Other effects, such as self-energy corrections to the particle mass, are automatically accounted for by using the physical value for m in Eq. (4.19). With this in mind, one can proceed in the bottom-up approach and construct an interacting theory by supplementing the free Lagrangian with all allowed contact operators. In particular, like Eq. (4.19) they should all be invariant under Galilei transformations, the low-energy remnant of the Poincaré group. Before we come back to this, however, we find it instructive to explicitly consider the low-energy limit of a relativistic theory.

4.2.2.1 Nonrelativistic Limit of a Bosonic Field

Let us make the connection of Eq. (4.19) to a relativistic complex Klein–Gordon field Φ, the Lagrangian for which can be written as

$$\mathscr{L}_{\varphi,\mathrm{free}} = -\varphi^\dagger \left(\partial_t^2 - \nabla^2 + m^2\right) \varphi \,. \tag{4.21}$$

Using integration by parts, this can be shown to be equivalent to the more common form written with $(\partial_\mu \varphi^\dagger)(\partial^\mu \varphi)$. This implies the Klein–Gordon equation for the field operator,

$$\left(\partial_t^2 - \nabla^2 + m^2\right) \varphi = 0 \,, \tag{4.22}$$

the most general solution of which is typically written as (with a four-vectors $x = (t, \mathbf{x})$, $p = (p_0, \mathbf{p})$, and a Lorentz-invariant integration measure)

$$\varphi(x) = \int \frac{\mathrm{d}^3 p}{(2\pi)^3} \frac{1}{\sqrt{2\omega_\mathbf{p}}} \left(\hat{a}_\mathbf{p}\, \mathrm{e}^{-\mathrm{i}p\cdot x} + \hat{b}_\mathbf{p}^\dagger\, \mathrm{e}^{\mathrm{i}p\cdot x}\right)\Bigg|_{p_0=\omega_\mathbf{p}} \,, \tag{4.23}$$

where $\omega_\mathbf{p} = \sqrt{\mathbf{p}^2 + m^2}$. With this convention where p_0 is chosen positive, modes created by $\hat{a}_\mathbf{p}^\dagger$ correspond to particles (propagating forward in time), whereas $\hat{b}_\mathbf{p}^\dagger$ creates an antiparticle (positive-energy state propagating backwards in time). That we have both stems from the fact that the complex scalar field corresponds to two real ones (completely decoupled in the absence of interactions), each of which comes with its own pair of creation and annihilation operators. To take the nonrelativistic limit, we have to consider the particle and antiparticles separately. Defining

$$\varphi_a(x) = \int \frac{\mathrm{d}^3 p}{(2\pi)^3} \frac{1}{\sqrt{2\omega_\mathbf{p}}} \hat{a}_\mathbf{p}\, \mathrm{e}^{-\mathrm{i}p\cdot x}\Bigg|_{p_0=\omega_\mathbf{p}} \equiv \mathrm{e}^{-\mathrm{i}mt} \phi_a(x) \,, \tag{4.24}$$

and plugging this into the Klein–Gordon equation, we get

$$\mathrm{e}^{-\mathrm{i}mt} \left[\partial_t^2 - 2\mathrm{i}m\, \partial_t - \nabla^2\right] \phi_a(x) = 0 \,, \tag{4.25}$$

where the quadratic mass term has canceled. Since $\phi_a(x) = \mathrm{e}^{\mathrm{i}mt} \varphi_a(x)$, we see from Eq. (4.24) that in the Fourier transform each time derivative acting on $\phi_a(x)$ brings down a factor

$$\omega_p - m = \sqrt{\mathbf{p}^2 + m^2} - m \approx \mathbf{p}^2/(2m) \,, \tag{4.26}$$

i.e., just the kinetic energy $E_\mathbf{p}$ up to corrections of higher order in $1/m$. In the nonrelativistic limit, $E_\mathbf{p} \ll m$, so we see that we can neglect the quadratic time derivative in Eq. (4.25) compared to the other two terms in Eq. (4.25), and then recover the Schrödinger equation for ϕ_a:

$$\left(\mathrm{i}\partial_t + \frac{\boldsymbol{\nabla}^2}{2m}\right)\phi_a(x) = 0\,. \tag{4.27}$$

This establishes the connection to our $\phi(t, \mathbf{x})$ in Eq. (4.19) when we insert an additional factor $\sqrt{2m}$ in the field redefinition to account for the otherwise different normalizations. For the antiparticles, we can carry out an analogous procedure, except that we have to choose the opposite sign for the mass-dependent phase in the field redefinition analogous to Eq. (4.24) because the antiparticle part of $\varphi(x)$ comes with a factor $\mathrm{e}^{+\mathrm{i}p\cdot x}$.

4.2.2.2 Nonrelativistic Fermions

For relativistic Dirac fermions, the nonrelativistic reduction can be carried out with the help of a so-called Foldy-Wouthuysen transformation.[3] The idea behind the approach is to decouple the particle and antiparticle modes contained together in a four-spinor ψ through a sequence of unitary transformations. In the following, we demonstrate this procedure, using an *interacting* model theory to also illustrate what happens to interaction terms in the nonrelativistic limit. Since it will be useful to motivate the pionless EFT discussed in Sect. 4.4, we start with a Lagrangian of the form

$$\begin{aligned}\mathscr{L} = \bar{\psi}\left(\mathrm{i}\not{\partial} - M_N\right)\psi + \frac{1}{2}(\partial^\mu\vec{\pi})\cdot(\partial_\mu\vec{\pi}) - \frac{1}{2}m_\pi^2\vec{\pi}^2 + \frac{1}{2}(\partial^\mu\sigma)\cdot(\partial_\mu\sigma) - \frac{1}{2}m_\sigma^2\sigma^2 \\ - g\bar{\psi}(\sigma - \mathrm{i}\gamma^5\vec{\tau}\cdot\vec{\pi})\psi\,,\end{aligned} \tag{4.28}$$

where the nucleon field ψ is an isospin doublet of Dirac spinors, $\boldsymbol{\pi}$ is an isospin triplet, and σ is an isoscalar. A Lagrangian of this form (plus additional interaction terms among σ and $\boldsymbol{\pi}$), can be obtained from a linear sigma model after spontaneous symmetry breaking (see, for example, [8, Chap. I]) and augmented by an explicit mass term for $\boldsymbol{\pi}$.[4] We denote the Pauli matrices in spin and isospin space as $\boldsymbol{\sigma} = (\sigma^i)$ and $\boldsymbol{\tau} = (\tau^\lambda)$, respectively. For the gamma matrices we use the standard (Dirac)

[3] An alternative way to perform the nonrelativistic reduction is to introduce a "heavy fermion" field [9]. A comparison of this formalism and the Foldy-Wouthuysen transformation can be found in [10].

[4] We stress, however, that this really is a model and not a proper EFT describing QCD.

representation:

$$\gamma^0 = \begin{pmatrix} \mathbf{1} & 0 \\ 0 & -\mathbf{1} \end{pmatrix} , \; \gamma^i = \begin{pmatrix} 0 & \sigma^i \\ -\sigma^i & 0 \end{pmatrix} , \; \gamma^5 = \begin{pmatrix} 0 & \mathbf{1} \\ \mathbf{1} & 0 \end{pmatrix} . \tag{4.29}$$

To perform the nonrelativistic reduction, we start by separating odd and even operators, which are two-by-two block matrices in Dirac space. The result is

$$\mathscr{L}_\psi = \psi^\dagger \left(\hat{E} + \hat{O} - \gamma^0 M_N \right) \psi \, , \tag{4.30}$$

where

$$\hat{E} = \begin{pmatrix} \mathrm{i}\partial_t - g\sigma & 0 \\ 0 & \mathrm{i}\partial_t + g\sigma \end{pmatrix} \text{ and } \hat{O} = \begin{pmatrix} 0 & -\mathrm{i}\vec{\sigma}\cdot\nabla + \mathrm{i}g\vec{\tau}\cdot\vec{\pi} \\ -\mathrm{i}\vec{\sigma}\cdot\nabla - \mathrm{i}g\vec{\tau}\cdot\vec{\pi} & 0 \end{pmatrix} . \tag{4.31}$$

Rotating the phase of the fermion field,

$$\psi \to \tilde{\psi} = \mathrm{e}^{-\mathrm{i}M_N t}\psi \, , \tag{4.32}$$

just like we did for the bosonic field in Eq. (4.24), we can remove the mass term for the upper components:

$$\mathscr{L}_\psi = \tilde{\psi}^\dagger \left(\hat{E} + \hat{O} - (\gamma^0 - \mathbf{1}) M_N \right) \tilde{\psi} \, . \tag{4.33}$$

The Foldy-Wouthuysen transformation is now constructed to (approximately) decouple the upper from the lower components, i.e., nucleons from their antiparticles. To achieve this, we use a sequence of further unitary redefinitions of the fermion field. The first of these is

$$\tilde{\psi} \to \tilde{\psi}' = \mathrm{e}^{-\mathrm{i}\hat{S}}\tilde{\psi} \text{ with } \hat{S} = -\frac{\mathrm{i}\gamma^0\hat{O}}{2M_N} \, . \tag{4.34}$$

Let us consider this transformation up to quadratic order in $1/M_N$. Expanding the exponential, we have

$$\tilde{\psi} = \mathrm{e}^{\mathrm{i}\hat{S}}\tilde{\psi}' = \left(1 + \frac{\gamma^0\hat{O}}{2M_N} + \frac{(\gamma^0\hat{O})^2}{8M_N^2} + \mathscr{O}(1/M_N^3) \right) \tilde{\psi}' \tag{4.35}$$

and likewise

$$\tilde{\psi}^\dagger = \tilde{\psi}'^\dagger \mathrm{e}^{-\mathrm{i}\hat{S}} = \tilde{\psi}'^\dagger \left(1 - \frac{\gamma^0\hat{O}}{2M_N} + \frac{(\gamma^0\hat{O})^2}{8M_N^2} + \mathscr{O}(1/M_N^3) \right) . \tag{4.36}$$

Inserting this into Eq. (4.33) and collecting contributions up to corrections which are $\mathcal{O}(1/M_N^2)$, we get a number of terms:

$$-\frac{\gamma^0\hat{O}\hat{E}}{2M_N}+\frac{\hat{E}\gamma^0\hat{O}}{2M_N}=\frac{\gamma^0[\hat{O},\hat{E}]}{2M_N}, \tag{4.37a}$$

$$\frac{\hat{O}\gamma^0\hat{O}}{2M_N}-(\gamma^0-\mathbf{1})\frac{(\gamma^0\hat{O})^2}{8M_N}-\frac{\gamma^0\hat{O}^2}{2M_N}+\frac{\gamma^0\hat{O}}{2M_N}(\gamma^0-\mathbf{1})\frac{\gamma^0\hat{O}}{2M_N}-\frac{(\gamma^0\hat{O})^2}{8M_N}(\gamma^0-\mathbf{1})=-\frac{\gamma^0\hat{O}^2}{2M_N}, \tag{4.37b}$$

$$\frac{1}{2}\gamma^0\hat{O}(\gamma^0-\mathbf{1})-\frac{1}{2}(\gamma^0-\mathbf{1})\gamma^0\hat{O}=-\hat{O}. \tag{4.37c}$$

Above we have used that

$$\left[\gamma_0,\hat{E}\right]=0\ ,\ \{\gamma_0,\hat{O}\}=0\,, \tag{4.38}$$

and $(\gamma^0)^2=\mathbf{1}$. Collecting everything, we get

$$\mathcal{L}_\psi=\tilde{\psi}'^\dagger\left(\hat{E}-\frac{\gamma^0\hat{O}^2}{2M_N}+\frac{\gamma^0[\hat{O},\hat{E}]}{2M_N}-(\gamma^0-\mathbf{1})M_N\right)\tilde{\psi}'+\mathcal{O}(1/M_N^2)\,. \tag{4.39}$$

The $\hat{O}^2$ term is even and we see that the original odd operator is canceled, but we have generated a new term $\sim\left[\hat{O},\hat{E}\right]$. If we neglect the interaction and consider

$$\hat{E}=\hat{E}_{\text{free}}=\begin{pmatrix}\mathrm{i}\partial_t & 0\\ 0 & \mathrm{i}\partial_t\end{pmatrix}\ \text{and}\ \hat{O}=\hat{O}_{\text{free}}=\begin{pmatrix}0 & -\mathrm{i}\boldsymbol{\sigma}\cdot\boldsymbol{\nabla}\\ -\mathrm{i}\boldsymbol{\sigma}\cdot\boldsymbol{\nabla} & 0\end{pmatrix}, \tag{4.40}$$

we find that $\left[\hat{O}_{\text{free}},\hat{E}_{\text{free}}\right]=0$ (partial derivatives commute) and thus the desired decoupling up to $\mathcal{O}(1/M_N^2)$. For the interacting case, on the other hand, the commutator does not vanish. We see, however, that the new odd contribution is suppressed by a factor $1/M_N$. To push it to the next higher order, we need another rotation:

$$\tilde{\psi}'\to\tilde{\psi}''=\mathrm{e}^{-\mathrm{i}\hat{S}'}\tilde{\psi}\,, \tag{4.41a}$$

with

$$\hat{S}'=-\frac{\mathrm{i}\gamma^0\hat{O}'}{2M_N},\ \hat{O}'=\frac{\gamma^0[\hat{O},\hat{E}]}{2M_N}. \tag{4.41b}$$

After a couple of steps, we arrive at

$$\mathscr{L}_\psi = \tilde{\psi}''^\dagger \left(\hat{E} - \frac{\gamma^0 \hat{O}^2}{2M_N} - (\gamma^0 - \mathbf{1})M_N\right) \tilde{\psi}'' + \mathscr{O}(1/M_N^2)\,, \tag{4.42}$$

i.e., up to $\mathscr{O}(1/M_N^2)$ there are now no odd terms left and the upper and lower components of $\tilde{\psi}''$ are decoupled at this order.

4.2 *Exercise:* Carry out the steps that lead from the transformation (4.41) to Eq. (4.42). Note that it suffices to expand the exponentials up to first order.

In this Lagrangian, we can now write

$$\tilde{\psi}'' = \begin{pmatrix} N \\ n \end{pmatrix} \tag{4.43}$$

and identify the upper ("large") component N—a doublet in both spin and isospin space—with the particle and the lower ("small") component with the antiparticle states. The term $(\gamma^0 - \mathbf{1})M_N$ in Eq. (4.42) ensures that there is no explicit mass term for the field N, whereas that for n comes with a factor two, corresponding to the Dirac mass gap between particles and antiparticles. Let us now write down the Lagrangian obtained for N, omitting the decoupled small components:

$$\mathscr{L}_\psi = N^\dagger \left(\mathrm{i}\partial_t - g\sigma - \frac{1}{2M_N}\left[-\mathrm{i}\vec{\sigma}\cdot\nabla + \mathrm{i}g\vec{\tau}\cdot\vec{\pi}\right]\left[-\mathrm{i}\vec{\sigma}\cdot\nabla - \mathrm{i}g\vec{\tau}\cdot\vec{\pi}\right]\right) N + \cdots\,. \tag{4.44}$$

To simplify this further, we use that[5]

$$(\vec{\sigma}\cdot\nabla)(\vec{\sigma}\cdot\nabla) = \nabla^2 \tag{4.45}$$

and, from the product rule,

$$(\vec{\sigma}\cdot\nabla)(\vec{\tau}\cdot\vec{\pi}) = \vec{\sigma}\cdot(\vec{\tau}\cdot\nabla\vec{\pi}) + (\vec{\tau}\cdot\vec{\pi})(\vec{\sigma}\cdot\nabla) = \sigma^i\tau^a(\partial_i\pi^a) + \tau^a\pi^a\,\sigma^i\partial_i\,. \tag{4.46}$$

In the last step we have written out all indices to clarify the meaning of the two dot products. Collecting everything, we find that the $(\vec{\tau}\cdot\vec{\pi})(\vec{\sigma}\cdot\nabla)$ terms cancel and

[5]Note that Eq. (4.45) is very simple because we have not included a coupling of ψ to the electromagnetic field. If we had done that, the ∇ would be a covariant derivative, $\mathbf{D} = \nabla + \mathrm{i}e\mathbf{A}$, and Eq. (4.45) would generate, among other terms, the magnetic spin coupling $\vec{\sigma}\cdot\mathbf{B}$.

arrive at

$$\mathscr{L} = N^\dagger \left(\mathrm{i}\partial_t + \frac{\boldsymbol{\nabla}^2}{2M_N} \right) N - g\,\sigma N^\dagger N + N^\dagger \left(\frac{g}{2M_N} \vec{\sigma} \cdot (\vec{\tau} \cdot \boldsymbol{\nabla}\vec{\pi}) + \frac{g^2}{2M_N} (\vec{\tau} \cdot \vec{\pi})^2 \right) N$$
$$+ \frac{1}{2}(\partial^\mu \vec{\pi}) \cdot (\partial_\mu \vec{\pi}) - \frac{1}{2} m_\pi^2 \vec{\pi}^2 + \frac{1}{2}(\partial^\mu \sigma) \cdot (\partial_\mu \sigma) - \frac{1}{2} m_\sigma^2 \sigma^2 + \cdots . \tag{4.47}$$

This includes the expected nonrelativistic kinetic term for the fermion field, as well as various interactions with σ and $\vec{\pi}$. Note that the latter two particles are still relativistic and unchanged by the Foldy-Wouthuysen transformation, so that we could simply reinstate their kinetic terms as in Eq. (4.28).

4.2.2.3 Integrating Out Exchange Particles: Part II

With Eq. (4.28) we are now also in a convenient situation to illustrate how we end up with only contact interactions between the nonrelativistic fermions if we integrate out the σ and $\vec{\pi}$ fields. Their equations of motion are

$$\left(\Box + m_\sigma^2\right) \sigma = g\, N^\dagger N \tag{4.48}$$

and

$$\left(\Box + m_\pi^2\right) \pi^\lambda = -\frac{g}{2M_N} \boldsymbol{\nabla} \cdot \left[N^\dagger \vec{\sigma} \tau^\lambda N\right] - \frac{g^2}{M_N} N^\dagger (\vec{\tau} \cdot \vec{\pi}) \tau^\lambda N\,. \tag{4.49}$$

The σ part can be handled exactly as in Sect. 4.2.1.1, giving a leading four-nucleon contact interaction $\sim g^2/m_\sigma^2$ plus a tower of operators with increasing number of derivatives. The $\vec{\pi}$ part is more interesting, but also more complicated due to the derivative in Eq. (4.49). We thus keep the following discussion rather qualitative and leave it as an exercise to work out the details.

In that spirit, we consider only the first term in Eq. (4.49), corresponding to a one-$\vec{\pi}$-exchange operator when substituted back into the Lagrangian. With the propagator $D_\pi(x-y)$ defined in complete analogy to Eq. (4.6), we can write

$$\pi^\lambda(x) = -\frac{\mathrm{i}g}{2M_N} \int \mathrm{d}^4y\, D_\pi(x-y)\, \partial_j^y \left[N^\dagger(y)\, \sigma^j \tau^\lambda N(y)\right] + \cdots , \tag{4.50}$$

and thus get

$$\mathscr{L}_{\text{int}} \sim \left[N^\dagger(x)\, \sigma^i \tau^\lambda N(x)\right] \partial_i^x \int \mathrm{d}^4y\, D_\pi(x-y)\, \partial_j^y \left[N^\dagger(y)\, \sigma^j \tau^\lambda N(y)\right] + \cdots , \tag{4.51}$$

where for the time being we omit the prefactor $g^2/(4M_N^2)$. We integrate by parts to have ∂_j^y act on $D_\pi(x-y)$. The ∂_i^x does this already, so, with all indices written out for clarity:

$$\mathscr{L}_{\text{int}} \sim \int \mathrm{d}^4y\, N^\dagger_{\alpha a}(x)\,(\sigma^i)^\alpha{}_\beta\,(\tau^\lambda)^a{}_b\,N^{\beta b}(x)\left[\partial^i_x\partial^j_y D_\pi(x-y)\right]N^\dagger_{\gamma c}(y)\,(\sigma^j)^\gamma{}_\delta\,(\tau^\lambda)^c{}_d\,N^{\delta d}(y)+\cdots\,. \tag{4.52}$$

From the definition of the propagator we find that

$$\partial^x_i\partial^y_j D_\pi(x-y) = \int \frac{\mathrm{d}^4p}{(2\pi)^4}(\mathrm{i}p_i)(-\mathrm{i}p_j)\mathrm{e}^{-\mathrm{i}p(x-y)}\frac{\mathrm{i}}{p^2-m_\pi^2+\mathrm{i}\varepsilon} = -\partial^x_i\partial^x_j D_\pi(x-y)\,, \tag{4.53}$$

so the partial derivatives can be written fully symmetric in i and j. The various fermion field operators can be rearranged with the help of

$$\begin{aligned}N^{\beta b}(x)N^\dagger_{\gamma c}(y) = -\frac{1}{4}\Big[&(N^\dagger(y)N(x))\,\delta^\beta_\gamma\delta^b_c + (N^\dagger(y)\tau^\kappa N(x))\,\delta^\beta_\gamma\,(\tau^\kappa)^b{}_c \\ &+ (N^\dagger(y)\sigma^k N(x))\,(\sigma^k)^\beta{}_\gamma\delta^b_c + (N^\dagger(y)\,\tau^\kappa\sigma^k N(x))\,(\sigma^k)^\beta{}_\gamma\,(\tau^\kappa)^b{}_c\Big]\,.\end{aligned} \tag{4.54}$$

Using also

$$\sigma^i\sigma^j = \delta^{ij}\mathbf{1} + \mathrm{i}\epsilon^{ijk}\sigma^k\,, \tag{4.55a}$$

$$\sigma^i\sigma^k\sigma^j = \delta^{kj}\sigma^i + \delta^{ki}\sigma^j - \delta^{ij}\sigma^k + \mathrm{i}\epsilon^{ikj}\mathbf{1}\,, \tag{4.55b}$$

$$\sigma^i\sigma^j\sigma^i = -\sigma^j\,, \tag{4.55c}$$

we get four terms from Eq. (4.52) decomposed into contributions symmetric and antisymmetric in i and j, with the latter all vanishing upon contraction with $\partial^x_i\partial^x_j$. The simplest symmetric term comes with a δ^{ij}, yielding $\boldsymbol{\nabla}^2 D_\pi(x-y)$. To see what this generates, we Taylor-expand the fermion fields that depend on y about x, e.g., $N(y) = N(x) + (y-x)^\mu\partial_\mu N(x) + \cdots$. This gives as the leading piece a combination of four fermion operators all evaluated at x, times

$$\begin{aligned}&\int \mathrm{d}^4y\int\frac{\mathrm{d}^4p}{(2\pi)^4}\mathrm{e}^{-\mathrm{i}p\cdot(x-y)}\frac{\mathrm{i}\mathbf{p}^2}{p^2-m_\pi^2+\mathrm{i}\varepsilon} \\ &= \int \mathrm{d}^4y\int\frac{\mathrm{d}^3p}{(3\pi)^4}\int\frac{\mathrm{d}p_0}{2\pi}\,\mathrm{e}^{-\mathrm{i}p(x-y)}\frac{\mathrm{i}\mathbf{p}^2}{p_0^2-\mathbf{p}^2-m_\pi^2+\mathrm{i}\varepsilon}\,.\end{aligned} \tag{4.56}$$

The integral over p_0 can be solved via contour integration. Defining $\omega_{\mathbf{p}} = \sqrt{\mathbf{p}^2 + m_\pi^2}$, we get

$$\int \frac{\mathrm{d}p_0}{2\pi}\,\mathrm{e}^{-\mathrm{i}p_0(x_0-y_0)}\frac{\mathrm{i}}{p_0^2 - \mathbf{p}^2 - m_\pi^2 + \mathrm{i}\varepsilon} = \frac{\mathrm{e}^{-\mathrm{i}\omega'_{\mathbf{p}}|x_0-y_0|}}{2\omega'_{\mathbf{p}}} \quad \text{with } \omega'_{\mathbf{p}} = \omega_{\mathbf{p}} - \frac{\mathrm{i}\varepsilon}{2\omega_{\mathbf{p}}} \equiv \omega_{\mathbf{p}} - \mathrm{i}\varepsilon'\,. \tag{4.57}$$

It is important here to keep track of the small imaginary part, as it allows us to write

$$\int_{-\infty}^{\infty}\mathrm{d}y_0\,\frac{\mathrm{e}^{-\mathrm{i}\omega'_{\mathbf{p}}|x_0-y_0|}}{2\omega'_{\mathbf{p}}} = \int_0^{\infty}\mathrm{d}y_0\frac{\mathrm{e}^{-\mathrm{i}\omega'_{\mathbf{p}}|y_0|}}{\omega'_{\mathbf{p}}} = -\mathrm{i}/(\omega'_{\mathbf{p}})^2\,. \tag{4.58}$$

Collecting the results up to this point, we arrive at

$$\int \mathrm{d}^4y \int \frac{\mathrm{d}^4p}{(2\pi)^4}\,\mathrm{e}^{-\mathrm{i}p(x-y)}\frac{\mathrm{i}\mathbf{p}^2}{p^2 - m_\pi^2 + \mathrm{i}\varepsilon} = -\int \mathrm{d}^3y \int \frac{\mathrm{d}^3p}{(2\pi)^3}\,\mathrm{e}^{\mathrm{i}\mathbf{p}\cdot(\mathbf{x}-\mathbf{y})}\frac{\mathbf{p}^2}{\mathbf{p}^2 + m_\pi^2}\,. \tag{4.59}$$

We finally obtain the desired contact interaction by expanding

$$\frac{\mathbf{p}^2}{\mathbf{p}^2 + m_\pi^2} = \frac{\mathbf{p}^2}{m_\pi^2}\left(1 - \frac{\mathbf{p}^2}{m_\pi^2} + \cdots\right), \tag{4.60}$$

with a leading term $\sim \mathbf{p}^2$, generating a contact interaction $\sim (N^\dagger N)\boldsymbol{\nabla}^2(N^\dagger N)$. This is of course not surprising: after all, the original interaction term in Eq. (4.47) generating the contact operator had a single derivative $\boldsymbol{\nabla}$. Considering other terms coming from Eq. (4.54), one can also find operators like $(N^\dagger\vec{\sigma}\cdot\boldsymbol{\nabla}N)(N^\dagger\vec{\sigma}\cdot\boldsymbol{\nabla}N)$, and it is a useful exercise to work this out in detail. But already from our qualitative discussion here we can infer that the resulting effective theory is an expansion in $\mathbf{p}^2/m_\pi^2$, i.e., its range of validity is determined by three-momenta—rather than the energies—being small compared to m_π.[6]

4.2.2.4 The Schrödinger Field

We conclude this section by looking at the nonrelativistic field theory from a more general perspective, establishing its close connection to the "second quantized" approach to (many-body) quantum mechanics that is used in several later chapters of this volume.

[6]This is assuming $m_\pi < m_\sigma$.

Recall from the beginning of this section that the Lagrangian (4.19) for the free Schrödinger field ϕ is

$$\mathscr{L}_{\phi,\text{free}} = \phi^\dagger \left(\mathrm{i}\partial_t + \frac{\nabla^2}{2m} \right) \phi \,. \tag{4.61}$$

This trivially gives the equation of motion

$$\left(\mathrm{i}\partial_t + \frac{\nabla^2}{2m} \right) \phi = 0 \,, \tag{4.62}$$

which is formally the same as the free Schrödinger equation. However, recall that ϕ here is a field *operator*, i.e., $\phi(x)$ creates a particle at $x = (t, \mathbf{x})$ from the vacuum, so to really get an ordinary Schrödinger equation, we have to act with both sides of Eq. (4.62) on $|0\rangle$, and define the quantum-mechanical one-body state

$$|\phi(t,\mathbf{x})\rangle = \phi(t,\mathbf{x})\,|0\rangle \;. \tag{4.63}$$

If we add to Eq. (4.61) a term $V(x)\phi^\dagger(x)\phi(x)$, we obtain the Schrödinger equation for a particle in a potential $V(x)$. Exactly as for a relativistic field we can define the propagator

$$D_\phi(x-y) = \int \frac{\mathrm{d}^4 q}{(2\pi)^4}\, \mathrm{e}^{-\mathrm{i}p(x-y)} \frac{\mathrm{i}}{p_0 - \frac{\mathbf{p}^2}{2m} + \mathrm{i}\varepsilon} \,, \tag{4.64}$$

satisfying

$$\left(\mathrm{i}\partial_t + \frac{\nabla^2}{2m} \right) D_\phi(x-y) = -\mathrm{i}\delta^{(4)}(x-y) \,. \tag{4.65}$$

Up to a conventional factor i, this is precisely the (retarded) Green's function[7] familiar, for example, from non-relativistic scattering theory (then typically denoted G_0). This will appear again when the Lippmann–Schwinger equation is derived using the field-theory language in Sect. 4.3.

While it is nice and reassuring that we can go back to simple quantum mechanics from the one-body Schrödinger Lagrangian discussed so far, this feature is not very relevant in practice. We can, however, straightforwardly generalize it to the many-body case. To that end, consider a Lagrangian that includes a two-body interaction,

[7] Note that in the nonrelativistic case there is no "Feynman propagator." Particles and particles are decoupled, and the denominator in Eq. (4.64) only has a single pole at $p_0 = \mathbf{p}^2/(2m) - \mathrm{i}\varepsilon$. Flipping the sign of the $\mathrm{i}\varepsilon$ term gives the advanced Green's function.

written in terms of a general non-local potential[8]:

$$\mathscr{L}_{\phi,\text{2-body}}(x) = \phi^\dagger(x)\left(\mathrm{i}\partial_t + \frac{\boldsymbol{\nabla}^2}{2m}\right)\phi(x) + \int \mathrm{d}^4y\,\phi^\dagger(x)\phi(x)\,V(x,y)\,\phi^\dagger(y)\phi(y)\,. \tag{4.66}$$

Note that this has exactly the structure that we found when we integrated out particles in the preceding sections, before expanding the propagators to get simple contact interactions. Such a Lagrangian (possibly including also higher-body forces) is a convenient starting point for example for many-body perturbation theory used to study quantum systems at finite density.

Coming back to effective field theories, we stress that these are *not* defined by putting a given potential into a Lagrangian; in doing that, one merely gets a model written in a convenient way. The EFT instead makes no assumptions on the interaction (besides symmetry constraints). It is thus much more general and not a model, but to be predictive it requires a number of a priori unknown parameters to be fixed and its various terms to be ordered systematically. It is this that we turn to next.

4.2.3 Symmetries and Power Counting

So far, we have discussed how to obtain effective low-energy Lagrangians by integrating out "heavy" degrees of freedom, leaving only those that we want to describe at low energies or rather—as we showed explicitly with the pseudoscalar pion-nucleon model—low momenta. We found the contact interactions generated this way to come with the integrated-out particle's mass in the denominator, and with an increasing number of derivatives as we keep more and more terms from the expansion. These derivatives will turn into powers of momentum, which is a small scale for external states. We furthermore showed how a nonrelativistic reduction generates a chain of operators with an increasing power of the particle's mass in the denominator, thus also giving a hierarchy of terms that eventually restore the original theory's relativistic structure with coupling between particles and antiparticles.

From these procedures it is clear that the terms in the effective Lagrangian should be ordered in a natural way, with the most important ones being those with the least number of large mass scales in the denominator and the least number of derivatives in the numerator. It is also clear that they are restricted in their structure. For example, if we start with a Lorentz-invariant relativistic

[8]A static (time-independent) potential, as it is more common in quantum mechanics, would be a function only of **x** and **y**, and all fields in the interaction term would be evaluated at the same time t.

theory, after the nonrelativistic reduction we will only get terms that are invariant under "small" Lorentz boosts. More precisely, the nonrelativistic operators should be invariant under Galilean transformations (assuming the original theory had rotational invariance, this simply gets inherited by the effective one), and the form of so-called "relativistic corrections" is determined by the expansion of the dispersion relation for positive-energy solutions:

$$E = \sqrt{m^2 + p^2} = m + \frac{p^2}{2m} - \frac{p^4}{8m^3} + \cdots . \tag{4.67}$$

We now turn to discussing the bottom-up approach guided by these principles. To that end, consider the effective Lagrangian for a nonrelativistic bosonic field with contact interactions:

$$\begin{aligned}\mathscr{L} = \phi^\dagger \left(\mathrm{i}\partial_t + \frac{\nabla^2}{2m} \right) \phi + \phi^\dagger \frac{\nabla^4}{8m^3} \phi + \cdots \\ + g_2^{(0)} (\phi^\dagger \phi)^2 + g_2^{(2s/p)} \left((\phi^\dagger \overleftrightarrow{\nabla} \phi)^2 - (\phi^\dagger \phi)(\phi^\dagger (\overleftrightarrow{\nabla})^2 \phi) \mp 2(\phi^\dagger \phi) \nabla^2 (\phi^\dagger \phi) \right) + \cdots \\ + g_3^{(0)} (\phi^\dagger \phi)^3 + \cdots .\end{aligned} \tag{4.68}$$

Here we have used the definition

$$f \overleftrightarrow{\nabla} g = f(\nabla) g - (\nabla f) g \tag{4.69}$$

and conveniently separated the two-body terms with two derivatives into those which contribute to S-wave ($\sim g_2^{(2s)}$) and P-wave ($\sim g_2^{(2p)}$) interactions, respectively. One can of course choose different linear combinations, but a separation by partial waves is typically a good choice for systems with rotational invariance. It is a useful exercise to work out how the structure for the derivative interactions gives the desired result, working in momentum space and considering contractions with external in and out states that have center-of-mass momenta $\pm\mathbf{k}^2$ and $\pm\mathbf{p}/2$, respectively. The structure of the individual terms is determined by the requirement of Galilean invariance,[9] and the EFT paradigm tells us to write down all possible terms with a given number of derivatives (with odd numbers excluded by parity invariance).

[9]See for example [11, Sect. 2.1.1] for a rigorous discussion of the required transformation properties.

4.2.3.1 The Breakdown Scale

As mentioned in the introduction, the most important requirement to construct an EFT is the identification of—at least two, but possibly more—separated scales, ratios of which are used to extract a small expansion parameter. The better the scale separation, the smaller this parameter becomes, and consequently the more precise (and, provided all contributions have been identified correctly, accurate) the theory becomes at any given order in the expansion. In the simplest case, there is one low scale Q associated with the typical momentum of the physical system that we want to describe, and a single large scale M_{hi}, the "breakdown scale" associated with the physics that our EFT does not take into account—in other words: resolve—explicitly. This is exactly the situation that we constructed when we integrated out exchange particles from a given theory in Sects. 4.2.1.1 and 4.2.2.3. By construction, the EFT is not appropriate to describe processes with momenta of the order of or large than the breakdown scale. To emphasize this meaning, the breakdown scale is sometimes also denoted by the letter Λ (with or possibly without some qualifying subscript).[10]

As already mentioned, integrating out degrees of freedom from a given more fundamental theory will naturally yield a breakdown scale set by that particle's mass. But it can also be something more general. For example, although in the situations discussed here so far the particles we were ultimately interested in were already present as degrees of freedom in the original theory, such a scenario is merely a special case. The first step in writing down an effective field theory is to identify what the appropriate—literally: effective—degrees of freedom are for the processes one wants to describe, and they can be different from those of the fundamental theory. This is exactly the case in nuclear physics: while the degrees of freedom in quantum chromodynamics (QCD) are quarks and gluons, describing the binding of nuclei with these is, although possible with state-of-the-art lattice QCD calculations, largely inefficient to say the least. It is much more economical to work with nucleons directly as degrees of freedom, as done in most chapters of this volume, because a detailed knowledge of the internal structure of protons and neutrons is not necessary to describe their binding into nuclei; it is only resolved at much higher energies, for example in deep inelastic scattering. The reason for this is color confinement: the low-energy degrees of freedom of QCD are not quarks and gluons, but color-neutral hadrons. Chiral effective field theory, which we will come back to in Sect. 4.5.3, is designed to work at momenta of the order of pion mass, breaking down at the scale of chiral-symmetry breaking (estimated to be roughly a GeV, but possibly lower).

Other examples are halo EFT, constructed to describe nuclear systems that have the structure of a few nucleons weakly bound to a tight core, which can then

---

[10]We alert the reader that in the literature this is sometimes referred to as the "cutoff of the EFT." We do not use that language to avoid confusion with an (arbitrary) momentum cutoff introduced to regularize divergent loop integrals (discussed).

effectively be treated as a structureless particle. Clearly, such a theory will break down at momenta large enough to probe the core's internal structure. Similarly, one can construct an effective theory for systems of ultracold atomic gases, the constituents of which can be treated as pointlike degrees of freedom without using QED to describe their individual structure, and much less QCD to describe their atomic nuclei.

Whatever the breakdown scale is, once identified it can be used to systematically order terms in the effective Lagrangian by powers of $Q/M_{\rm hi} \ll 1$, and we now turn to discussing how this ordering can be set up.

4.2.3.2 Naïve Dimensional Analysis

In our units with $\hbar = c = 1$, the action

$$S = \int \mathrm{d}^4x\, \mathscr{L}(x) \tag{4.70}$$

has to be a dimensionless quantity. This, in turn, fixes the dimensions for the individual building blocks in the Lagrangian. In a relativistic theory, mass and energy are equivalent and one would simply express everything in terms of a generic mass dimension. For our nonrelativistic framework, on the other hand, energies are *kinetic* energies because the time dependence associated with the rest mass has been absorbed into the field (*cf.* Sects. 4.2.2.1 and 4.2.2.2). This implies that energy and mass scales—as well as time and space—should be counted separately.[11] In fact, it is more natural to consider powers of momentum. To understand what this means, let us start with the kinetic term in Eq. (4.68): $[\nabla^2/(2m)] = \text{momentum}^2/\text{mass}$. The time derivative has to scale in the same way, implying that for time itself we have $[t] = \text{mass}/\text{momentum}^2$, whereas $[x] = \text{momentum}^{-1}$. Consequently, the integration measure scales like $[\mathrm{d}^4x = \mathrm{d}t\,\mathrm{d}^3x] = \text{mass}/\text{momentum}^5$ (to compare, in the relativistic theory one would simply count $[\mathrm{d}^4x] = \text{mass}^{-4} = \text{energy}^{-4}$).

Since the dimension of $\mathscr{L}$ has to cancel that of the measure to give a dimensionless action,we can now infer that our field has to satisfy $[\phi] = \text{momentum}^{3/2}$, i.e., even though it is a scalar field it scales with a fractional dimension (recall that in the relativistic case a scalar would have dimension energy^1). Knowing the scaling of the field and the measure, we can now proceed and deduce that of the various coupling constants.

The basic idea is very simple: each term (operator) in the Lagrangian (4.68) has $2n$ fields and $2m$ derivatives, giving it a total dimension of momentum^{3n+2m}. For example, the $(\phi^\dagger\phi)^2$ term with $2n = 4$ and $m = 0$ has dimension momentum^6. Hence, to get the correct overall dimension $\text{momentum}^5/\text{mass}$ for $\mathscr{L}$, the coupling

[11]This separation would be quite clear if we had not set $c = 1$, which would in fact be more appropriate for a nonrelativistic system. The reason we still do it that it allows us to still energies and momenta in the same units, e.g., in MeV, following the standard convenient in nuclear physics.

constant $g_2^{(0)}$ has to be $\sim 1/(\text{momentum} \times \text{mass})$. Since it is supposed to describe unresolved short-distance details, the momentum scale in the denominator should be the breakdown scale, whereas the mass scale, which as we mentioned is a feature of the nonrelativistic framework and common to all operators, is simply associated with m_ϕ. Of course, counting a single operator does not tell us much: it is the relative order of terms that matters, so we proceed to the $g_2^{(2)}$ interactions. These all come with two derivatives, which are associated with the external (small) momentum scale Q. Hence, we have $2n = 4$ and $2m = 2$, and we need to compensate the two additional powers of momentum in the numerator with two more powers of M_{hi} in the denominator, finding that the $g_2^{(2)}$ interactions are down compared to the $g_2^{(0)}$ term by a factor $(Q/M_{\text{hi}})^2$. This is exactly in line with our picture of the contact terms gradually building up the an unresolved particle exchange through a derivative expansion. For higher-body interactions, it is the larger number of fields that gives a suppression by inverse powers of M_{hi} compared to operators with fewer fields.

This kind of analysis can be much improved if something is known about which unresolved physics is supposed to be represented by which operator, and it is generally more complex if the theory involves different fields. For example, in the EFT for halo nuclei there are contact interactions associated with unresolved pion exchange, as well as those systematically accounting for the internal structure of the core field. Instead of merely putting generic powers of M_{hi} in every denominator, it can be necessary to keep track of several high scales separately to figure out the ordering of terms. Also, it is possible that the external momentum is not the only relevant low-momentum scale in the problem.

This rather abstract discussion will become clearer when we finally discuss concrete EFTs in the following Sects. 4.3 and 4.4. In that context, we will use the scaling of the various terms in the Lagrangian to power-count diagrams as a whole, i.e. to estimate the size of individual contributions composed of vertices and loops to a given physical amplitude of interest. We will then also discuss how the actual so-called scaling dimension of a field in the Lagrangian can turn out to deviate from what we estimated here based purely on dimensional grounds.

4.2.3.3 Fine Tuning

In connection with the previous comment there is another point worth stressing already here: naïve dimensional analysis resides at the beginning of EFT wisdom, not at the end, and in quite a few cases it turns out to be exactly what the name says: naïve. In other words, the actual scaling of a coupling constant can be quite different from what one would infer by counting dimensions, a scenario that is commonly referred to as "fine tuning." To understand why that is, consider, for example, our bosonic toy model from Sect. 4.2.1.1, but now assume that there already is a four-ϕ contact interaction present prior to integrating out the χ field:

$$\mathscr{L} = -\phi^\dagger\left(\Box + m_\phi^2\right)\phi - \chi^\dagger\left(\Box + m_\chi^2\right)\chi + g\left(\phi^\dagger\phi^\dagger\chi + \text{h.c.}\right) + h(\phi^\dagger\phi)^2\,. \tag{4.71}$$

This could, for example, come from unknown (or integrated out) short-distance physics at a yet higher scale. When we now integrate out the χ, the generated non-derivative contact term will combine with the existing one, giving a single operator in the effective low-energy Lagrangian (recall that on dimensional grounds h has to have dimensions of inverse mass squared):

$$\mathscr{L} = -\phi^\dagger\left(\Box + m_\phi^2\right)\phi + \left(h - \frac{g^2}{m_\chi^2}\right)(\phi^\dagger\phi)^2 + \cdots\,. \tag{4.72}$$

Now suppose we had started in the bottom-up approach and simply written down the four-ϕ contact operator with some coefficient c to be determined. According to NDA, we would assume that its scale is set by two powers of the breakdown scale in the denominator, and assuming we actually know about the more fundamental theory, we might have estimated that breakdown scale to be of the order m_χ. From Eq. (4.72) we see that depending on what values g and h take in the underlying theory, the actual size of c might deviate strongly from the naïve expectation, and it could even be set by a low-energy scale of the effective theory. But for this to happen, there would have to be a delicate cancellation between h and g^2/m_χ^2, which is typically deemed unlikely given the a priori vast range of possible values these parameters could take; thus the term "fine tuning." The fact that coupling constants are in fact not simple fixed numbers but get renormalized by loop effects (i.e., depend on a regularization scale with a behavior determined by the renormalization group) justifies this language even more.

4.2.3.4 Loops and Renormalization

It is indeed high time we talk about loops. Our considerations in this section so far have been limited to tree level, which is always only a first approximation in a quantum field theory. In a perturbative theory, loop contributions from virtual intermediate states are added to improve the accuracy of the result. To treat a nonperturbative system such as a bound nucleus, on the other hand, they are absolutely crucial: recall that any finite sum of diagrams in perturbation can never produce a bound state (for example, think about poles in the S-matrix, which cannot be generated through a finite sum of terms). In the field-theory language, this means that an infinite number of diagrams with increasing number of loops has to be summed to get the amplitude with the desired physical properties.

This situation is in fact familiar already from the Schrödinger equation written in the form

$$|\psi\rangle = \hat{G}_0(E)\hat{V}\,|\psi\rangle \ \ , \ \hat{G}_0 = \hat{G}_0(E) = (E - \hat{H}_0)^{-1} \,, \tag{4.73}$$

which can be iterated to get $|\psi\rangle = \hat{G}_0\hat{V}\hat{G}_0\hat{V}\,|\psi\rangle = \cdots$. When these operators are written out in momentum space, each propagator $\hat{G}_0$ corresponds to a loop. More closely related to the amplitude written down in a nonrelativistic field theory, this exercise can be repeated with the Lippmann-Schwinger equation for the T-matrix and its formal solution, the infinite Born series. Exactly this will be recovered in Sect. 4.3.

Of course, even in a nonperturbative theory we do not expect that *all* loop diagrams should be summed up to infinity. Generally, we want the power counting to tell us how to estimate the contribution from a given diagram, including loop diagrams. To do that, we need to know not only which factors we pick up from vertices, but also need an estimate for the integration measure $\mathrm{d}^4q = \mathrm{d}q_0\,\mathrm{d}^3q$. Any loop diagram contributing to an amplitude with external momenta of the order Q will have this scale running through as a whole. It is thus natural to count the contribution from the three-momentum as $\mathrm{d}^3q \sim Q^3$ and, recalling that in the nonrelativistic theory q_0 is a kinetic energy, $\mathrm{d}q_0 \sim Q^2/m$. For each Schrödinger propagator we get, conversely, a factor m/Q^2, as can be seen from Eq. (4.64). These simple rules combined with those for the vertices give an estimate for any diagram in the theory determined by Eq. (4.68).

What this discussion does not cover is the fact that loops in a quantum field theory can be—and mostly are—divergent. Compared to the loops one gets from integrating the Schrödinger or Lippmann-Schwinger equation in quantum mechanics with a potential $\hat{V}$, which are typically all finite, this is different in the EFT simply because our delta-function (contact) interactions are too singular to make direct sense beyond tree level. Of course, this is no different than in any other quantum field theory, and it just means that divergent loops have to be regularized (for example, by imposing a momentum cutoff or with dimensional regularization), and then suitable renormalization conditions have to be imposed to fix the various coupling constants in the effective Lagrangian. These then become functions of the renormalization scale, with a behavior governed by the renormalization group (RG). In Sect. 4.3 this will be discussed in detail for a bosonic EFT that describes, for example, ultracold atomic systems.

The Cutoff

What the regularization of loop integrals does is most transparent with a momentum cutoff. We denote this by Λ and stress again that it has to be distinguished from the EFT breakdown scale M_{hi}. The latter determines the scale beyond which we know our EFT not to be valid. In other words, short-range dynamics corresponding to momenta larger than M_{hi} are, in general, not correctly described by the EFT.

Yet from loop integrals we get contributions from states up to the UV cutoff Λ. Renormalization means to adjust the coupling constants in such a way that they compensate the wrong high-momentum loop contributions in such a way that the physics the EFT is supposed to describe comes out correctly. For momenta up to M_{hi}, we trust the EFT, so it makes sense to keep such states in loops. Hence, one should typically choose $\Lambda > M_{\text{hi}}$. Choosing it lower than the breakdown scale is possible, but this can induce corrections of the order $Q/\Lambda > Q/M_{\text{hi}}$, which is not desirable for the power counting. In the renormalized EFT, any cutoff in the interval $[M_{\text{hi}}, \infty)$ is thus an equally good choice—it does not have to be "taken to infinity." Instead, that phrase should be understood to mean adjusting the couplings at any given finite cutoff. If this procedure is carried out numerically, it can be desirable to keep the cutoff small, but one has to make sure that in principle in *can* be varied arbitrarily.

4.2.4 Matching

The determination of the couplings ("low-energy constants") in the effective Lagrangian is done by expressing a given physical quantity (e.g., a scattering amplitude or related observable) in terms of the couplings and then adjusting them to reproduce a known result. This can be done using experimental input or, when working top-down, by calculating the same amplitude in the more fundamental theory. Generally, this procedure is referred to as "matching." At tree level, this is again exactly what we did by integrating out particles and found the coefficients of the generated contact interactions in terms of the original coupling and mass denominators. Once loop diagrams are involved, the process becomes somewhat more complicated because (a) one has to make sure to use compatible regularization schemes and renormalization scales and (b) loop diagrams with lower-order vertices typically mix with higher-order tree-level diagrams. The latter is a general feature of combined loop and derivative expansions and is thus also important when matching to experimental input. While these comments may sound a bit cryptic here, they will become much clearer in the next section when we finally work with a concrete EFT.

4.3 Effective Field Theory for Strongly Interacting Bosons

We will now use the insights from the previous sections to construct a local effective field theory for identical, spinless bosons with short-range S-wave interactions.[12] For the treatment of higher partial wave interactions the reader is referred to the literature [12–14]. The most general effective Lagrangian consistent with Galilei

[12]See [15] for a similar discussion with a focus on applications in ultracold atoms.

invariance can be written as

$$\mathscr{L} = \phi^\dagger \left(\mathrm{i}\partial_t + \frac{\nabla^2}{2m} \right) \phi - \frac{C_0}{4} \left(\phi^\dagger \phi\right)^2 - \frac{C_2}{4} \left(\nabla(\phi^\dagger \phi)\right)^2 + \frac{D_0}{36} \left(\phi^\dagger \phi\right)^3 + \cdots . \tag{4.74}$$

where m is the mass of the particles and the ellipses denote higher-derivative and/or higher-body interactions. The leading two- and three-body interactions are explicitly written out. The scaling of the coefficients C_0, C_2, D_0, ... depends on the scales of the considered system. Two explicit examples, corresponding to natural and unnaturally large scattering length, are discussed below.

4.3.1 EFT for Short-Range Interactions

We start by considering natural system where all interactions are characterized by only one mass scale M_{hi} that we identify with the formal breakdown scale of the EFT introduced in the previous section. We will see below that this is indeed justified. Since the nonrelativistic boson fields have dimension 3/2, the coupling constants must scale as

$$C_0 \sim \frac{1}{mM_{\mathrm{hi}}} , \quad C_2 \sim \frac{1}{mM_{\mathrm{hi}}{}^3} , \quad \text{and} \quad D_0 \sim \frac{1}{mM_{\mathrm{hi}}{}^4} , \tag{4.75}$$

such that higher dimension operators are strongly suppressed for small momenta $k \ll M_{\mathrm{hi}}$.[13] We first focus on the two-body system and calculate the contribution of the interaction terms in Eq. (4.74) to the scattering amplitude of two particles in perturbation theory. After renormalization, the result reproduces the low-energy expansion of the scattering amplitude for particles with relative momentum k and total energy $E = k^2/m$:

$$T_2(E) = \frac{8\pi}{m} \frac{1}{k \cot \delta_0(k) - \mathrm{i}k} = -\frac{8\pi a}{m} \left(1 - \mathrm{i}ak + (ar_e/2 - a^2)k^2 + \mathcal{O}(k^3)\right) , \tag{4.76}$$

where the effective range expansion for short-range interactions $k \cot \delta_0(k) = -1/a + r_e k^2/2 + \mathcal{O}(k^4)$ has been used.

Since all coefficients of the effective Lagrangian are natural (scaling with inverse powers of M_{hi}), it is sufficient to count the powers of small momenta Q in

[13]Note that coupling constants scale with the particle mass as $1/m$ in nonrelativistic theories. This can be seen by rescaling all energies as $q_0 \to \tilde{q}_0/m$ and all time coordinates as $t \to \tilde{t}m$, so that dimensionful quantities are measured in units of momentum. Demanding that the action is independent of m, it follows that the coupling constants must scale as $1/m$.

scattering amplitudes to determine the scaling of the amplitudes with M_{lo}. The correct dimensions are made up with appropriate factors of M_{hi} contained in the coupling constants [*cf.* Eq. (4.75)]. For a general two-body amplitude with L loops and V_{2i} interaction vertices with $2i$ derivatives, we thus have $T_2 \sim Q^\nu$ where the power ν is given by

$$\nu = 3L + 2 + \sum_i (2i - 2)V_{2i} \geq 0\,. \tag{4.77}$$

Here we have used that loop integrations contribute a factor k^5 and propagators a factor k^{-2} in nonrelativistic theories. The values of the coupling constants C_0 and C_2 can be determined by matching to Eq. (4.76). In the lowest two orders only C_0 contributes.

4.3 *E*xercise: Derive Eq. (4.77) using the topological identity for Feynman diagrams:

$$L = I - V + 1\,, \tag{4.78}$$

with L, V, and I the total number of loops, vertices, and internal lines respectively.

The contact interactions in Eq. (4.74) are ill-defined unless an ultraviolet cutoff is imposed on the momenta in loop diagrams. This can be seen by writing down the off-shell amplitude for two-body scattering at energy E in the center-of-mass frame at second order in perturbation theory:

$$T_2(E) \approx -C_0 - \frac{\mathrm{i}}{2}C_0^2 \int \frac{\mathrm{d}^3q}{(2\pi)^3} \int \frac{\mathrm{d}q_0}{2\pi} \frac{1}{q_0 - q^2/2m + \mathrm{i}\epsilon} \frac{1}{E - q_0 - q^2/2m + \mathrm{i}\epsilon} + \cdots\,.$$

The two terms correspond to the first two diagrams in Fig. 4.4. The intermediate lines have momenta $\pm\boldsymbol{q}$. The integral over q_0 in Eq. (4.79) is easily evaluated using contour integration:

$$T_2(E) \approx -C_0 - \frac{1}{2}C_0^2 \int \frac{\mathrm{d}^3q}{(2\pi)^3} \frac{1}{E - q^2/m + \mathrm{i}\epsilon} + \cdots\,. \tag{4.79}$$

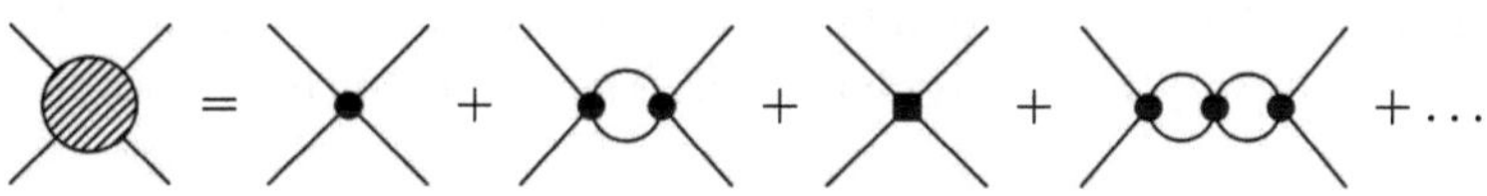

Fig. 4.4 Diagrammatic expression for the two-body scattering amplitude T_2. The *circle* (*square*) denotes a C_0 (C_2) interaction, respectively

The integral over $\boldsymbol{q}$ diverges. It can again be regularized by imposing an ultraviolet cutoff $|\boldsymbol{q}| < \Lambda$. Taking the limit $\Lambda \gg |E|^{1/2}$, the amplitude reduces to[14]

$$T_2(E) \approx -C_0 + \frac{mC_0^2}{4\pi^2}\left(\Lambda - \frac{\pi}{2}\sqrt{-mE - \mathrm{i}\epsilon}\right) + \cdots . \tag{4.80}$$

The dependence on the ultraviolet cutoff Λ can be consistently eliminated by a perturbative renormalization procedure. A simple choice is to eliminate the parameter C_0 in favor of the scattering length a, which is given by Eq. (4.76):

$$a \approx \frac{mC_0}{8\pi}\left(1 - \frac{mC_0\Lambda}{4\pi^2} + \cdots\right) . \tag{4.81}$$

Inverting this expression to obtain C_0 as a function of a we obtain

$$C_0 \approx \frac{8\pi a}{m}\left(1 + \frac{2a\Lambda}{\pi} + \cdots\right) , \tag{4.82}$$

where we have truncated at second order in a. Inserting the expression for C_0 into Eq. (4.80) and expanding to second order in a, we obtain the renormalized expression for the amplitude:

$$T_2(E) \approx -\frac{8\pi a}{m}\left(1 + a\sqrt{-mE - \mathrm{i}\epsilon} + \cdots\right) = -\frac{8\pi a}{m}\left(1 - \mathrm{i}ak + \cdots\right) . \tag{4.83}$$

If we evaluate this at the on-shell point $E = k^2/m$ and insert it into Eq. (4.80), we find that it reproduces the first two terms in the expansion of the universal scattering amplitude in Eq. (4.76) in powers of ka. By calculating $T_2(E)$ to higher order in perturbation theory, we can reproduce the low-momentum expansion of Eq. (4.76) to higher order in ka. At the next order, the C_2 term will contribute at tree level while C_0 will contribute at the two-loop level. Thus a perturbative treatment of the EFT reproduces the low-momentum expansion of the two-body scattering amplitude. The perturbative approximation is valid only if the momentum satisfies $k \ll 1/a$.

A more interesting case occurs when the scattering length is large, but all other effective range coefficients are still determined by the scale M_{hi}: $k \sim 1/|a| \sim M_{\mathrm{lo}} \ll M_{\mathrm{hi}} \sim 1/r_e$. This scenario is able to support shallow bound states with binding momentum of order $1/a$ and is relevant to ultracold atoms close to a Feshbach resonance and to very low-energy nucleons. The scaling of the operators is then

[14] If the calculation was carried out in a frame in which the total momentum of the two scattering particles was nonzero, the simple cutoff $|\boldsymbol{q}| < \Lambda$ would give a result that does not respect Galilean invariance. To obtain a Galilean-invariant result requires either using a more sophisticated cutoff or else imposing the cutoff $|\boldsymbol{q}| < \Lambda$ only after an appropriate shift in the integration variable $\boldsymbol{q}$.

modified to:

$$C_0 \sim \frac{1}{mM_{\mathrm{lo}}}, \quad C_2 \sim \frac{1}{mM_{\mathrm{lo}}{}^2 M_{\mathrm{hi}}}, \quad \text{and} \quad D_0 \sim \frac{1}{mM_{\mathrm{lo}}{}^4}. \tag{4.84}$$

The factors of M_{lo} in amplitudes can now come from small momenta and from the coupling constants. Above we adjusted the scaling of the three-body coupling D_0 as well, foreclosing a result discussed below Eq. (4.98).

With the scaling as in Eq. (4.84), the power counting expression in Eq. (4.77) is therefore modified to

$$\nu = 3L + 2 + \sum_i (i-3)V_{2i} \geq -1\,. \tag{4.85}$$

If we are interested in two-body observables involving energy $E \sim 1/a^2$, such as shallow bound or virtual states, we must resum the diagrams involving only C_0 interactions to all orders [16, 17]. Without this resummation, our EFT would break down not at M_{hi}, but already at the much smaller scale $1/|a| \sim M_{\mathrm{lo}}$. In the scenario assumed here, all higher-derivative two-body interactions (C_2 and beyond) still involve inverse powers of M_{hi} and are thus perturbative.

The resummation of C_0 interactions is most easily accomplished by realizing that the corresponding Feynman diagrams in Fig. 4.4 form a geometric series. Summing the geometric series, the exact expression for the amplitude is

$$T_2(E) = -C_0 \left[1 + \frac{mC_0}{4\pi^2}\left(\Lambda - \frac{\pi}{2}\sqrt{-mE - \mathrm{i}\epsilon}\right)\right]^{-1}. \tag{4.86}$$

Alternatively, we can use the fact that summing the C_0 diagrams in Fig. 4.4 is equivalent to solving the following integral equation:

$$T_2(E) = -C_0 - \frac{\mathrm{i}}{2}C_0 \int \frac{\mathrm{d}^3 q}{(2\pi)^3} \int \frac{\mathrm{d}q_0}{2\pi} \frac{1}{q_0 - q^2/2m + \mathrm{i}\epsilon} \frac{1}{E - q_0 - q^2/2m + \mathrm{i}\epsilon} T_2(E)\,. \tag{4.87}$$

The integral equation is expressed diagrammatically in Fig. 4.5. Since the function $T_2(E)$ is independent of $\boldsymbol{q}$ and q_0, it can be pulled outside of the integral in Eq. (4.87).

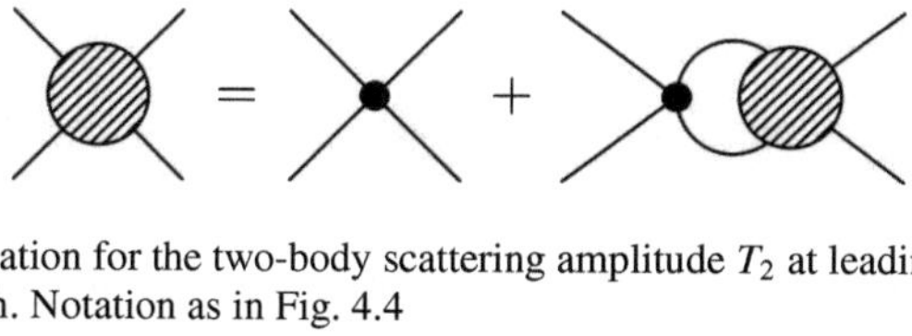

Fig. 4.5 Integral equation for the two-body scattering amplitude T_2 at leading order in the case of large scattering length. Notation as in Fig. 4.4

The integral can again be regularized by imposing an ultraviolet cutoff Λ. The equation is then trivial to solve and the solution is given in Eq. (4.86).

The expression for the resummed two-body amplitude in Eq. (4.86) depends on the parameter C_0 in the Lagrangian and on the ultraviolet cutoff Λ. As in the perturbative case, renormalization can be implemented by eliminating C_0 in favor of a low-energy observable, such as the scattering length a. Matching the resummed expression to the effective range expansion for T_2, we obtain

$$C_0 = \frac{8\pi a}{m}\left(1 - \frac{2a\Lambda}{\pi}\right)^{-1} . \tag{4.88}$$

Given a fixed ultraviolet cutoff Λ, this equation prescribes how the parameter C_0 must be tuned in order to give the scattering length a. Note that for $\Lambda \gg 1/|a|$, the coupling constant C_0 is always negative regardless of the sign of a. Eliminating C_0 in Eq. (4.86) in favor of a, we find that the resummed amplitude reduces to

$$T_2(E) = \frac{8\pi}{m}\frac{1}{-1/a + \sqrt{-mE - \mathrm{i}\epsilon}} , \tag{4.89}$$

which reproduces the effective range expansion of the scattering amplitude by construction. In this simple case, we find that our renormalization prescription eliminates the dependence on Λ completely. In general, we should expect it to only be suppressed by powers of $1/(a\Lambda)$ or mE/Λ^2. A final step of taking the limit $\Lambda \to \infty$ would then be required to obtain results that are completely independent of Λ. The first correction to Eq. (4.89) is given by the C_2 interaction. The corresponding diagrams are shown in Fig. 4.6. After the matching, the final result for the scattering amplitude at next-to-leading order is

$$T_2(E) = \frac{8\pi}{m}\left(\frac{1}{-1/a + \sqrt{-mE - \mathrm{i}\epsilon}} + \frac{r_e mE/2}{(-1/a + \sqrt{-mE - \mathrm{i}\epsilon})^2}\right) , \tag{4.90}$$

where r_e is the effective range. The derivation of this expression will be left as an exercise.

4.4 Exercise: Derive the next-to-leading order correction in Eq. (4.90) by calculating the loop diagrams in Fig. 4.6. Neglect all terms that vanish as $\Lambda \to \infty$. Introduce a next-to-leading order piece of C_0 to cancel the cubic divergence.

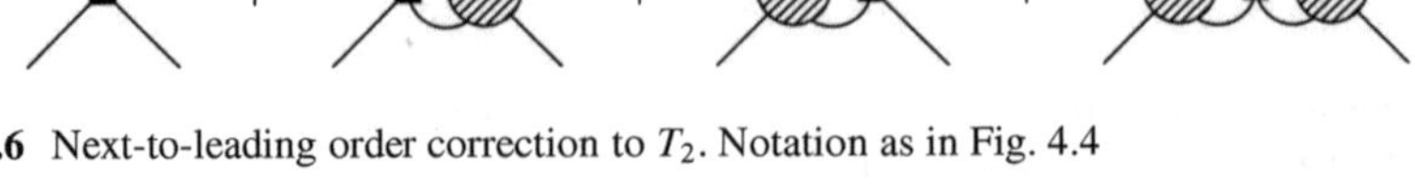

Fig. 4.6 Next-to-leading order correction to T_2. Notation as in Fig. 4.4

4.3.2 Dimer Field Formalism

In applications to systems with more than two particles, it is often useful to rewrite the EFT for short-range interactions specified by the Lagrangian (4.74) using so-called dimer fields d [18]:

$$\begin{aligned}\mathscr{L} = \phi^\dagger \left(\mathrm{i}\partial_t + \frac{\nabla^2}{2m} \right) \phi + g_0 d^\dagger d + g_2 d^\dagger \left(\mathrm{i}\partial_t + \frac{\nabla^2}{4m} \right) d + \cdots \\ - y \left(d^\dagger \phi^2 + \phi^{\dagger 2} d \right) - d_0 d^\dagger d \phi^\dagger \phi + \cdots .\end{aligned} \tag{4.91}$$

One important feature of this Lagrangian is that there is no direct two-body contact interaction term $(\phi^\dagger \phi)^2$. All interactions between ϕ particles are mediated via exchange of a dimer field d, i.e., we have effectively performed a Hubbard–Stratonovich transformation. Eliminating the dimer field d by using its equations of motion, it can be shown that the physics of this EFT is equivalent to the Lagrangian (4.74).

Note that the Lagrangian (4.91) contains one more free parameter than the Lagrangian (4.74). Thus some parameters are redundant. For the leading-order case ($g_n = 0$ for $n \geq 2$), e.g., we find explicitly,

$$C_0 = \frac{4y^2}{g_0}, \tag{4.92}$$

such that y and g_0 are not independent. Higher-order corrections can be obtained by including a kinetic-energy term for the dimer field. The constants g_0 and y then become independent and can be related to combinations of C_0 and C_2 in the theory without dimers. Here, we only discuss the leading-order case.

4.5 *E*xercise: Derive Eq. (4.92) using the classical equation of motion for d.

The bare propagator for the dimer field is simply the constant i/g_0, which corresponds to no propagation in space or time. However, there are corrections to the dimer propagator from the diagrams in Fig. 4.7a which allow the dimer to propagate. This is completely analogous to the geometric series we found we had

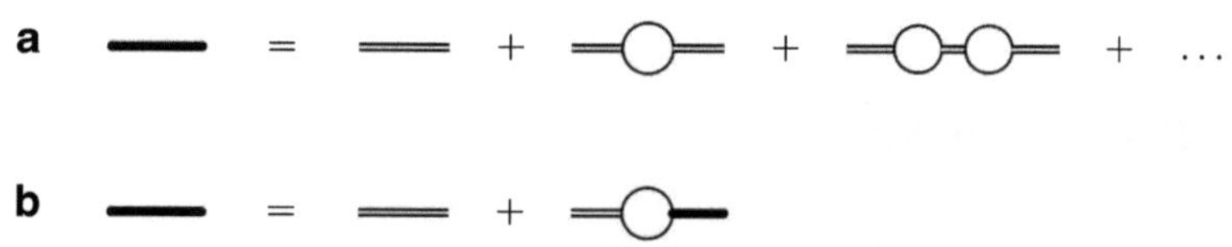

Fig. 4.7 Diagrammatic equations for the full dimer propagator $iD(P_0, P)$. *Thin (thick) solid lines* represent particle (full dimer) propagators. *Double lines* indicate bare dimer propagators. (**a**) perturbative expansion in powers of y, (**b**) integral equation summing the geometric series in (**a**)

to sum to obtain the leading-order scattering amplitude. In Feynman diagrams, we represent the full dimer propagator $\mathrm{i}D(P_0, \boldsymbol{P})$ by a thick solid line. We can calculate the full dimer propagator by solving the simple integral equation shown in Fig. 4.7b. The loop on the right side is just the integral in Eq. (4.79), with E replaced by $P_0 - P^2/(4m)$, where P_0 and $\boldsymbol{P}$ are the energy and momentum of the dimer. The solution for the full dimer propagator is

$$\mathrm{i}D(P_0, P) = \frac{2\pi i}{y^2 m}\left[\frac{2\pi g_0}{y^2 m} + \frac{2}{\pi}\Lambda - \sqrt{-mP_0 + P^2/4 - \mathrm{i}\epsilon}\right]^{-1}, \tag{4.93}$$

where as before Λ is a cutoff on the loop momentum in the bubbles. Using Eq. (4.92) and making the substitution given in Eq. (4.88), the expression for the complete dimer propagator is

$$\mathrm{i}D(P_0, P) = -\frac{2\pi \mathrm{i}}{y^2 m}\left[-1/a + \sqrt{-mP_0 + P^2/4 - \mathrm{i}\epsilon}\right]^{-1}. \tag{4.94}$$

Note that all the dependence on the ultraviolet cutoff is now in the multiplicative factor $1/y^2$. The complete dimer propagator differs from the off-shell two-body amplitude T_2 in Eq. (4.89) only by a multiplicative constant. For $a > 0$, it has a pole at $P_0 = -1/(ma^2) + P^2/4$ corresponding to a dimer of momentum $\boldsymbol{P}$ and binding energy $B_2 = 1/(ma^2)$. As P_0 approaches the dimer pole, the limiting behavior of the propagator is

$$D(P_0, P) \longrightarrow \frac{Z_D}{P_0 - (-1/(ma^2) + P^2/4) + \mathrm{i}\epsilon}, \tag{4.95}$$

where the residue factor is

$$Z_D = \frac{4\pi}{am^2y^2}. \tag{4.96}$$

If we regard the composite operator d as a quantum field that annihilates and creates dimers, then Z_D is the wave function renormalization constant for that field. The renormalized propagator $Z_D^{-1}D(P_0, P)$ is completely independent of the ultraviolet cutoff.

4.3.3 *Three-Body System*

We now study the amplitude for particle-dimer scattering T_3. The simplest diagram we can write down involving only two-body interactions is the exchange of a particle between in- and outgoing dimers. With the scaling of low-energy constants as in

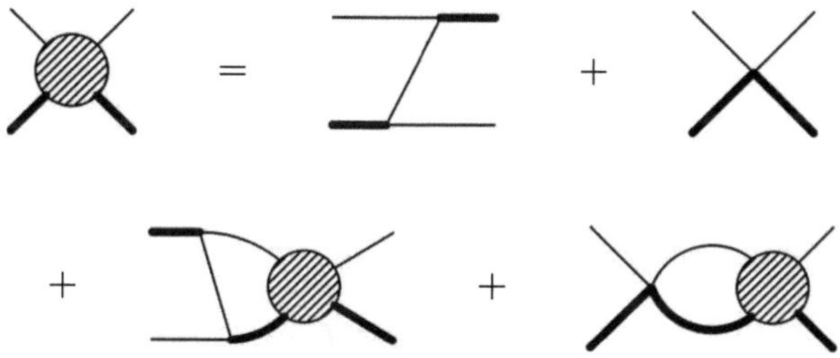

Fig. 4.8 The integral equation for the three-body amplitude T_3. *Thin (thick) solid lines* represent particle (full dimer) propagators. *External lines* are amputated

Eq. (4.84), the power counting implies that all diagrams that are chains of such exchanges are equally important, i.e., they have to be summed up nonperturbatively. Just like in the two-body case, this can be written as an integral equation. Also including the three-body interaction (note $D_0 \to d_0$ in the Lagrangian with dimer fields), we get the result that is shown diagrammatically in Fig. 4.8.[15]

Omitting the three-body interaction, this is exactly the well-known Skorniakov-Ter-Martirosian (STM) integral equation [19], which the EFT with Lagrangian (4.91) reproduces by construction. In addition, EFT provides a clear method to renormalize this equation with a three-body interaction and thus remove its pathologies (discussed below).

In Fig. 4.8, all external lines are understood to be amputated. It simply gives the non-perturbative solution of the three-body problem for the interaction terms proportional to g_0, y, and d_0 in Eq. (4.91).

The two tree diagrams on the right side of Fig. 4.8 constitute the inhomogeneous term in the integral equation. An iterative ansatz for the solution of this equation shows that all diagrams with the g_0, y, and d_0 interactions are generated by the iteration. Note also that the thick black lines in Fig. 4.8 represent the full dimer propagator given in Eq. (4.94).

In the center-of-mass frame, we can take the external momenta of the particle and dimer to be $-\boldsymbol{p}$ and $+\boldsymbol{p}$ for the incoming lines and $-\boldsymbol{k}$ and $+\boldsymbol{k}$ for the outgoing lines. We take their energies to be E_A and $E - E_A$ for the incoming lines and E'_A and $E - E'_A$ for the outgoing lines. The amplitude T_3 is then a function of the momenta $\boldsymbol{p}$ and $\boldsymbol{k}$ and the energies E, E_A and E'_A. The integral equation involves a loop over the momentum $-\boldsymbol{q}$ and energy q_0 of a virtual particle. Using the Feynman rules encoded

[15]Note that this amplitude is well defined even if $a < 0$ and there is no two-body bound state. In this case particle lines must be attached to the external dimer propagators to obtain the 3-particle scattering amplitude.

in the Lagrangian (4.91), we obtain

$$T_3(\boldsymbol{p},\boldsymbol{k};E,E_A,E'_A) = -\left[\frac{4y^2}{E-E_A-E'_A-(\boldsymbol{p}+\boldsymbol{k})^2/(2m)+\mathrm{i}\epsilon}+d_0\right]$$
$$+\frac{2\pi\mathrm{i}}{my^2}\int\frac{\mathrm{d}q_0}{2\pi}\int\frac{\mathrm{d}^3q}{(2\pi)^3}\left[\frac{4y^2}{E-E_A-q_0-(\boldsymbol{p}+\boldsymbol{q})^2/(2m)+\mathrm{i}\epsilon}+d_0\right]$$
$$\times\frac{1}{q_0-q^2/(2m)+\mathrm{i}\epsilon}\,\frac{T_3(\boldsymbol{q},\boldsymbol{k};E,q_0,E'_A)}{1/a-\sqrt{-m(E-q_0)+q^2/4-\mathrm{i}\epsilon}}\,. \tag{4.97}$$

The integral over q_0 can be evaluated by contour integration. This sets $q_0 = q^2/(2m)$, so the amplitude T_3 inside the integral has the incoming particle on-shell.

We obtain a simpler integral equation if we also set the energies of both the initial and final particles in T_3 on-shell: $E_A = p^2/(2m)$, $E'_A = k^2/(2m)$. Thus only the dimer lines have energies that are off-shell. The resulting integral equation is

$$T_3\left(\boldsymbol{p},\boldsymbol{k};E,\frac{p^2}{2m},\frac{k^2}{2m}\right) = -4my^2\left[\frac{1}{mE-(p^2+\boldsymbol{p}\cdot\boldsymbol{k}+k^2)+\mathrm{i}\epsilon}+\frac{d_0}{4my^2}\right]$$
$$-8\pi\int\frac{\mathrm{d}^3q}{(2\pi)^3}\left[\frac{1}{mE-(p^2+\boldsymbol{p}\cdot\boldsymbol{q}+q^2)+\mathrm{i}\epsilon}+\frac{d_0}{4my^2}\right]$$
$$\times\frac{T_3(\boldsymbol{q},\boldsymbol{k};E,q^2/(2m),k^2/(2m))}{-1/a+\sqrt{-mE+3q^2/4-\mathrm{i}\epsilon}}\,. \tag{4.98}$$

This is an integral equation with three integration variables for an amplitude T_3 that depends explicitly on seven independent variables. There is also an additional implicit variable provided by an ultraviolet cutoff $|\boldsymbol{q}| < \Lambda$ on the loop momentum.

If we set $d_0 = 0$ and ignore the ultraviolet cutoff, the integral equation in Eq. (4.98) is equivalent to the Skorniakov-Ter-Martirosian (STM) equation, an integral equation for three particles interacting via zero-range two-body forces derived by Skorniakov and Ter-Martirosian in 1957 [19]. It was shown by Danilov that the STM equation has no unique solution in the case of identical bosons [20]. He also pointed out that a unique solution could be obtained if one three-body binding energy is fixed. Kharchenko was the first to solve the STM equation with a finite ultraviolet cutoff that was tuned to fit observed three-body data. Thus the cutoff was treated as an additional parameter [21]. When we discuss the running of d_0, we will see that this ad hoc procedure is indeed justified and emerges naturally when the three-body equation is renormalized [22].

Here we restrict our attention to the sector of the three-body problem with total orbital angular momentum $L = 0$ where the three-body interaction contributes. For higher L, the original STM equation has a unique solution and can be solved numerically without complication.

The projection onto $L = 0$ can be accomplished by averaging the integral equation over the cosine of the angle between $\boldsymbol{p}$ and $\boldsymbol{k}$: $x = \boldsymbol{p}\cdot\boldsymbol{k}/(pk)$. It is also convenient to multiply the amplitude T_3 by the wave function renormalization factor Z_D given in Eq. (4.96). We will denote the resulting amplitude by T_3^0:

$$T_3^0(p,k;E) \equiv Z_D \int_{-1}^{1} \frac{\mathrm{d}x}{2} T_3\left(\boldsymbol{p},\boldsymbol{k};E,p^2/(2m),k^2/(2m)\right). \tag{4.99}$$

Furthermore, it is convenient to express the three-body coupling constant in the form

$$d_0 = -\frac{4my^2}{\Lambda^2} H(\Lambda). \tag{4.100}$$

Since H is dimensionless, it can only be a function of the dimensionless variables $a\Lambda$ and Λ/Λ_*, where Λ_* is a three-body parameter defined below. We will find that H is a function of Λ/Λ_* only.

The resulting integral equation is:

$$\begin{aligned} T_3^0(p,k;E) &= \frac{16\pi}{ma}\left[\frac{1}{2pk}\ln\left(\frac{p^2+pk+k^2-mE-\mathrm{i}\epsilon}{p^2-pk+k^2-mE-\mathrm{i}\epsilon}\right)+\frac{H(\Lambda)}{\Lambda^2}\right] \\ &+\frac{4}{\pi}\int_0^{\Lambda}\mathrm{d}q\,q^2\left[\frac{1}{2pq}\ln\left(\frac{p^2+pq+q^2-mE-\mathrm{i}\epsilon}{p^2-pq+q^2-mE-\mathrm{i}\epsilon}\right)+\frac{H(\Lambda)}{\Lambda^2}\right] \\ &\times\frac{T_3^0(q,k;E)}{-1/a+\sqrt{3q^2/4-mE-\mathrm{i}\epsilon}}. \end{aligned} \tag{4.101}$$

Note that the ultraviolet cutoff Λ on the integral over q has been made explicit. A change in the endpoint Λ of the loop integral should be compensated by the Λ-dependence of the function H in Eq. (4.101). More specifically, H must be tuned as a function of Λ so that the cutoff dependence of the solution $T_3^0(p,k;E)$ of Eq. (4.101) decreases as a power of Λ. This will guarantee that $T_3^0(p,k;E)$ has a well-behaved limit as $\Lambda \to \infty$. The renormalization group behavior of H will be discussed in detail below. In the next subsection, we show how different three-body observables can be obtained from the solution $T_3^0(p,k;E)$ of Eq. (4.101).

4.6 Fill in the gaps in the above derivation of Eq. (4.101) and generalize the derivation to arbitrary angular momentum L.

4.3.4 Three-Body Observables

The solution $T_3^0(p,k;E)$ to the three-body integral equation (4.101) encodes all information about three-body observables in the sector with total orbital angular momentum quantum number $L = 0$. In particular, it contains information about the binding energies $B_3^{(n)}$ of the three-body bound states [23]. For a given ultraviolet cutoff Λ, the amplitude $T_3^0(p,k;E)$ has a finite number of poles in E corresponding to the bound states whose binding energies are less than about Λ^2. As Λ increases, new poles emerge corresponding to deeper bound states. In the limit $\Lambda \to \infty$, the locations of these poles approach the energies $-B_3^{(n)}$ of the three-body bound states. The residues of the poles of $T_3^0(p,k;E)$ factor into functions of p and functions of k:

$$T_3^0(p,k;E) \longrightarrow \frac{\mathcal{B}^{(n)}(p)\mathcal{B}^{(n)}(k)}{E+B_3^{(n)}}, \qquad \text{as } E \to -B_3^{(n)}. \tag{4.102}$$

Matching the residues of the poles on both sides of Eq. (4.101), we obtain the bound-state equation

$$\begin{aligned}\mathcal{B}^{(n)}(p) = \frac{4}{\pi}\int_0^{\Lambda} \mathrm{d}q\, q^2 \left[\frac{1}{2pq}\ln\frac{p^2+pq+q^2-mE-\mathrm{i}\epsilon}{p^2-pq+q^2-mE-\mathrm{i}\epsilon} + \frac{H(\Lambda)}{\Lambda^2}\right] \\ \times \left[-1/a + \sqrt{3q^2/4-mE-\mathrm{i}\epsilon}\right]^{-1} \mathcal{B}^{(n)}(q)\,.\end{aligned} \tag{4.103}$$

The values of E for which this homogeneous integral equation has solutions are the energies $-B_3^{(n)}$ of the three-body states. For a finite ultraviolet cutoff Λ, the spectrum of $B_3^{(n)}$ is cut off around Λ^2.

The S-wave phase shifts for particle-dimer scattering can be determined from the solution $T_3^0(p,k;E)$ to the integral equation (4.101). The T-matrix element for the elastic scattering of an particle and a dimer with momenta k is given by the amplitude T_3^0 evaluated at the on-shell point $p = k$, $E = -B_2 + 3k^2/(4m)$ and multiplied by a wave function renormalization factor $Z_D^{1/2}$ for each dimer in the initial or final state. It can be represented by the Feynman diagram in Fig. 4.9a. The blob represents the amplitude T_3 or equivalently $Z_D^{-1}T_3^0$. The external double lines are amputated and correspond to asymptotic dimers and are associated with factors $Z_D^{1/2}$

The S-wave contribution to the T-matrix element is

$$T^0_{PD\to PD} = T_3^0(k,k;3k^2/(4m) - 1/(ma^2))\,, \tag{4.104}$$

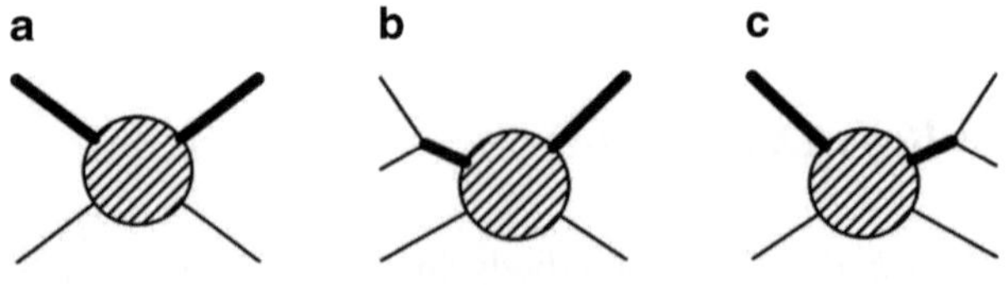

Fig. 4.9 Amplitudes for (**a**) particle-dimer scattering, (**b**) three-body recombination, and (**c**) three-body breakup. Diagrams (**b**) [(**c**)] should be summed over the three pairs of particles that can interact first [last]. Notation as in Fig. 4.8

where $B_2 = 1/(ma^2)$ has been used. Note that the factors of Z_D multiplying T_3^0 cancel. The differential cross section for elastic particle-dimer scattering is

$$d\sigma_{PD\to PD} = \frac{2m}{3k}\left|T_{PD\to PD}(k)\right|^2 \frac{km}{6\pi^2} d\Omega\,. \tag{4.105}$$

The flux factor $2m/(3k)$ is the inverse of the relative velocity of the particle and the dimer. The phase space factor $kmd\Omega/(6\pi^2)$ takes into account energy and momentum conservation and the standard normalization of momentum eigenstates:

$$\int \frac{d^3p_A}{(2\pi)^3}\frac{d^3p_D}{(2\pi)^3}(2\pi)^4\delta^3(\boldsymbol{p}_A+\boldsymbol{p}_D)\delta(p_A^2/(2m)+p_D^2/(4m)-E) = \frac{m}{6\pi^2}(4mE/3)^{1/2}\int d\Omega\,. \tag{4.106}$$

The S-wave phase shift for particle-dimer scattering is related to the T-matrix element via

$$\frac{1}{k\cot\delta_0^{PD}(k) - ik} = \frac{m}{3\pi}T_3^0(k,k;3k^2/(4m)-1/(ma^2))\,. \tag{4.107}$$

In particular, the particle-dimer scattering length is given by

$$a_{PD} = -\frac{m}{3\pi}T_3^0(0,0;-1/(ma^2))\,. \tag{4.108}$$

The threshold rate for three-body recombination can also be obtained from the solution $T_3^0(p,k;E)$ to the three-body integral equation in Eq. (4.101). This is possible only at threshold, because a 3-particle scattering state becomes pure $L=0$ only in the limit that the energies of the particles go to zero. The T-matrix element for the recombination process can be represented by the Feynman diagram in Fig. 4.9b summed over the three pairs of particle lines that can attach to the dimer line. The blob represents the amplitude $Z_D^{-1}T_3^0$ evaluated at the on-shell point $p=0$ $k=2/(\sqrt{3}\,a)$, and $E=0$. The solid line represents the dimer propagator $iD(0,0)$ evaluated at zero energy and momentum $2/(\sqrt{3}\,a)$, which is given by Eq. (4.93). The factor for the particle-dimer vertex is $-i2y$. The wave function renormalization factor $Z_D^{1/2}$ for the final-state dimer is given by Eq. (4.96). In the product of factors multiplying T_3^0, the dependence on y and Λ can be eliminated in favor of the scattering length a. Taking into account a factor of 3 from the three Feynman diagrams, the T-matrix element is

$$T_{PPP\to PD} = 6\sqrt{\pi a^3}\,T_3^0(0,2/(\sqrt{3}a);0)\,. \tag{4.109}$$

The differential rate dR for the recombination of three particles with energies small compared to the dimer binding energy can be expressed as

$$\mathrm{d}R = |T_{PPP \to PD}|^2 \frac{km}{6\pi^2} \,\mathrm{d}\Omega \,, \tag{4.110}$$

where $k = 2/(\sqrt{3}a)$. The threshold rate for three-body breakup can be obtained in a similar way from the Feynman diagram in Fig. 4.9c

The inhomogeneous integral equation for the off-shell particle dimer amplitude, Eq. (4.101), and the homogeneous equation, Eq. (4.103), for the three-body binding energies afford no analytical solution. They are usually solved by discretizing the integrals involved and solving the resulting matrix problems numerically.

4.3.5 Renormalization Group Limit Cycle

The form of the full renormalized dimer propagator in Eq. (4.94) is consistent with the continuous scaling symmetry

$$a \longrightarrow \lambda a \,, \qquad E \longrightarrow \lambda^{-2} E \,, \tag{4.111}$$

for any positive real number λ. In the integral equation (4.101), this scaling symmetry is broken by the ultraviolet cutoff on the integral and by the three-body terms proportional to H/Λ^2. To see that the cutoff and the three-body terms are essential, we set $H = 0$ and take $\Lambda \to \infty$. The resulting integral equation has exact scaling symmetry. We should therefore expect its solution $T_3^0(p, k; E)$ to behave asymptotically as $p \to \infty$ like a pure power of p. Neglecting the inhomogeneous term, neglecting E and $1/a^2$ compared to q^2, and setting $T_3^0 \approx p^{s-1}$, the integral equation reduces to [20]

$$p^{s-1} = \frac{4}{\sqrt{3}\pi p} \int_0^\infty \mathrm{d}q \, q^{s-1} \ln \frac{p^2 + pq + q^2}{p^2 - pq + q^2} \,. \tag{4.112}$$

Making the change of variables $q = xp$, the dependence on p drops out, and we obtain

$$1 = \frac{4}{\sqrt{3}\pi} \int_0^\infty \mathrm{d}x \, x^{s-1} \ln \frac{1 + x + x^2}{1 - x + x^2} \,. \tag{4.113}$$

The integral is a Mellin transform that can be evaluated analytically. The resulting equation for s is

$$1 = \frac{8}{\sqrt{3}s} \frac{\sin(\pi s/6)}{\cos(\pi s/2)} \,. \tag{4.114}$$

The solutions with the lowest values of $|s|$ are purely imaginary: $s = \pm \mathrm{i}s_0$, where $s_0 \approx 1.00624$. The most general asymptotic solution therefore has two arbitrary constants:

$$T_3^0(p,k;E) \longrightarrow A_+ p^{-1+\mathrm{i}s_0} + A_- p^{-1-\mathrm{i}s_0}\,, \qquad \text{as } p \to \infty\,. \tag{4.115}$$

The inhomogeneous term in the integral equation (4.101) will determine one of the constants. The role of the three-body term in the integral equation is to determine the other constant, thus giving the integral equation a unique solution.

By demanding that the solution of the integral equation (4.101) has a well-defined limit as $\Lambda \to \infty$, Bedaque et al. deduced the Λ-dependence of H and therefore of d_0 [22]. The leading dependence on Λ on the right side of the three-body integral equation in Eq. (4.101) as $\Lambda \to \infty$ is a log-periodic term of order Λ^0 that comes from the region $q \sim \Lambda$. There are also contributions of order $1/\Lambda$ from the region $|a|^{-1}, k, |E|^{1/2} \ll q \ll \Lambda$, which have the form

$$\frac{8}{\pi\sqrt{3}} \int^{\Lambda} \mathrm{d}q \left(\frac{1}{q^2} + \frac{H(\Lambda)}{\Lambda^2}\right) \left(A_+ q^{+\mathrm{i}s_0} + A_- q^{-\mathrm{i}s_0}\right). \tag{4.116}$$

The sum of the two terms will decrease even faster as $1/\Lambda^2$ if we choose the function H to have the form

$$H(\Lambda) = \frac{A_+\Lambda^{\mathrm{i}s_0}/(1-\mathrm{i}s_0) + A_-\Lambda^{-\mathrm{i}s_0}/(1+\mathrm{i}s_0)}{A_+\Lambda^{\mathrm{i}s_0}/(1+\mathrm{i}s_0) + A_-\Lambda^{-\mathrm{i}s_0}/(1-\mathrm{i}s_0)}\,. \tag{4.117}$$

The tuning of H that makes the term in Eq. (4.116) decrease like $1/\Lambda^2$ also suppresses the contribution from the region $q \sim \Lambda$ by a power of $1/\Lambda$ so that it goes to 0 in the limit $\Lambda \to \infty$. By choosing $A_\pm = (1+s_0^2)^{1/2}\Lambda_*^{\mp \mathrm{i}s_0}/2$ in Eq. (4.117), we obtain [22]

$$H(\Lambda) \approx \frac{\cos[s_0 \ln(\Lambda/\Lambda_*) + \arctan s_0]}{\cos[s_0 \ln(\Lambda/\Lambda_*) - \arctan s_0]}\,. \tag{4.118}$$

This equation defines a three-body scaling-violation parameter Λ_* with dimensions of momentum. The value of Λ_* can be fixed from a three-body datum. All other three-body observables can then be predicted. If Eq. (4.118) is substituted back into the three-body equation (4.101) for numerical calculations, it must be multiplied by a normalization factor $b \approx 1$ the precise value of which depends on the details of the regularization [24].

Note that H is a π-periodic function of $s_0 \ln(\Lambda/\Lambda_*)$, so Λ_* is defined only up to a multiplicative factor of $(\mathrm{e}^{\pi/s_0})^n$, where n is an integer. Thus the scaling symmetry of Eq. (4.111) is broken to the discrete subgroup of scaling transformations with multiples of the preferred scaling factor $\lambda = \mathrm{e}^{\pi/s_0}$. This discrete scaling symmetry is, e.g., evident in the geometric spectrum of three-body Efimov states [23] in the

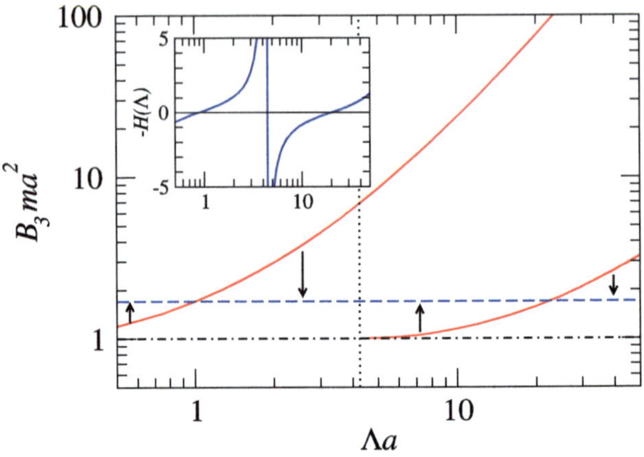

Fig. 4.10 Unrenormalized three-body energies B_3 as a function of the momentum cutoff Λ (*solid lines*). The *dotted line* indicates the cutoff where a new three-body state appears at the particle-dimer threshold (*dash-dotted line*). The *dashed line* shows a hypothetical renormalized energy. The *inset* shows the running of the three-body force $d_0(\Lambda) \sim -H(\Lambda)$ with Λ

unitary limit ($1/a = 0$) that naturally emerge in this EFT:

$$B_3^{(n)} \approx 0.15\lambda^{2(n_*-n)} \frac{\Lambda_*^2}{m}, \tag{4.119}$$

where n_* an integer labeling the state with binding energy $0.15\,\Lambda_*^2/m$.[16] The discrete scaling symmetry becomes also manifest in the log-periodic dependence of three-body observables on the scattering length. This log-periodic behavior is the hallmark signature of a renormalization group limit cycle. It has been observed experimentally in the three-body recombination spectra of ultracold atomic gases close to a Feshbach resonance [25, 26].

The physics of the renormalization procedure is illustrated in Fig. 4.10 where we show the unrenormalized three-body binding energies B_3 in the case of positive scattering length as a function of the cutoff Λ (solid line). As the cutoff Λ is increased, B_3 increases asymptotically as Λ^2. At a certain cutoff (indicated by the dotted line), a new bound state appears at the boson-dimer threshold. This pattern repeats every time the cutoff increases by the discrete scaling factor $\exp(\pi/s_0)$. Now assume that we adopt the renormalization condition that the shallowest state should have a constant energy given by the dashed line. At small values of the cutoff, we need an attractive three-body force to increase the binding energy of the shallowest state as indicated by the arrow. As the cutoff is increased further, the required attractive contribution becomes smaller and around $\Lambda a = 1.1$ a repulsive

[16]For a detailed discussion of the Efimov effect for finite scattering length and applications to ultracold atoms, see [15].

three-body force is required (downward arrow). Around $\Lambda a = 4.25$, a new three-body state appears at threshold and we cannot satisfy the renormalization condition by keeping the first state at the required energy anymore. The number of bound states has changed and there is a new shallow state in the system. At this point the three-body force turns from repulsive to attractive to move the new state to the required energy. The corresponding running of the three-body force with the cutoff Λ is shown in the inset. After renormalization, the first state is still present as a deep state with large binding energy, but for threshold physics its presence can be ignored. This pattern goes on further and further as the cutoff is increased [27].

4.4 Effective Field Theory for Nuclear Few-Body Systems

4.4.1 Overview

Depending on the physics one wishes to describe, there are several effective field theories for low-energy nuclear physics to choose from. They differ in the set of effective degrees of freedom, their expansion point (typical low-energy scale) and range of applicability. Chiral effective field theory includes nucleons and pions and is designed as an expansion about the so-called "chiral limit," i.e., the scenario where the quark masses are exactly zero such that the pions emerge as exactly massless Goldstone bosons from the spontaneous breaking of chiral symmetry. In reality, the quark-masses are nonzero such that the pions become "pseudo-Goldstone" bosons with a small (compared to typical QCD scales like m_ρ or M_N) mass m_π. Chiral EFT takes this as a typical low scale so that its power counting is designed for momenta of the order $Q \sim m_\pi$; we come back to this in Sect. 4.5.3.

For momenta much smaller than m_π, explicit pion exchange cannot be resolved such that these can be regarded as integrated out, much like we did explicitly for the pseudoscalar toy model in Sect. 4.2.2.3. The resulting "pionless" theory has, up to long-range forces that we consider in Sect. 4.5.1) only contact interactions between nucleons left. These contact interactions parameterize not only unresolved pion exchange, but also that of heavier mesons, for which contact terms already exist in Chiral EFT. Pionless EFT is formally very similar to the few-boson EFT discussed in Sect. 4.3, and it despite its simplicity it gives rise to surprisingly rich physics, as we will show in this section.

4.4.2 Pionless Effective Field Theory

The neutron-proton S-wave scattering lengths are experimentally determined to be about 5.4 fm in the ${}^3S_1$ channel, and -23.7 fm in the ${}^1S_0$ channel. What is special about these numbers is that they are large compared to the typical nuclear length scale determined by the pion Compton wavelength, $m_\pi^{-1} \sim 1.4\,\text{fm}$. This estimate comes from the long-range component of the nuclear interaction being determined by one-pion exchange. Naïvely, if we consider the low-energy limit (NN center-of-mass momentum going to zero) we expect that we can integrate out the pions and end up with a contact interaction scaling with the inverse pion mass, and thus a perturbative EFT reproducing natural-sized scattering lengths. The fact that this is not the case is typically interpreted as nature "choosing" the fine-tuned scenario outlined in Sect. 4.2.3.3. In this case, pion exchange[17] combines with shorter range interactions to yield the large S-wave scattering lengths (and the deuteron as an unnaturally shallow bound state), implying that nuclear physics is a strongly coupled—and thus nonperturbative—system at low energies. This is what allows us to write down an EFT that is closely related to the one that describes strongly interacting bosons. What governs the physics of low-energy observables is to a good appropriation just the fact that the scattering lengths are large, so we end up with a short-range EFT much like the one for bosons encountered in Sect. 4.3. Some rather technical new features arise from the fact that nucleons are fermions with spin and isospin.

4.4.3 The Two-Nucleon S-Wave System

The leading-order Lagrangian for pionless EFT can be written as

$$\mathscr{L} = N^\dagger \left(\mathrm{i}\partial_0 + \frac{\boldsymbol{\nabla}^2}{2M_N} + \cdots \right) N - C_{0,s}(N^T \hat{P}_s N)^\dagger (N^T \hat{P}_s N) - C_{0,t}(N^T \hat{P}_t N)^\dagger (N^T \hat{P}_t N) + \cdots , \tag{4.120}$$

with projectors

$$(\hat{P}_t)^i = \sigma^2 \sigma^i \tau^2 / \sqrt{8} \ , \ \ (\hat{P}_s)^\lambda = \sigma^2 \tau^2 \tau^\lambda / \sqrt{8} \tag{4.121}$$

such that $C_{0,s}$ and $C_{0,t}$ refer to the ${}^1S_0$ and ${}^3S_1$ NN channels, respectively. As in previous sections of this chapter, we use σ^i to denote the Pauli matrices in spin space, and write $(\sigma^i)^\alpha{}_\beta$ to refer to their individual entries (with the upper index referring to the row). Conversely, we use the notation $(\tau^\lambda)^a{}_b$ in isospin space.

With the usual cartesian indices $i, \lambda = 1, 2, 3$ the projectors for given i or λ give somewhat unusual combinations of individual states. For example, np configurations are completely contained in the $\lambda = 3$ isospin component, whereas nn and pp are obtained from linear combinations $1 \pm \mathrm{i}2$, In other words, in order to separate the physical states (and likewise to get spin-1 states with $m = 0, \pm 1$

[17] As discussed in Sect. 4.5.3 this actually has to be the exchange of two or more pions, as one-pion exchange does not contribute to S-wave scattering at zero energy.

quantum numbers) one should work instead with a spherical basis. For example, if one wants to include isospin-breaking terms, it is convenient to work with the projectors

$$(\tilde{P}_s)^{-1} = \frac{1}{\sqrt{2}}\left[(\hat{P}_s)^1 - \mathrm{i}(\hat{P}_s)^2\right] \;,\; (\tilde{P}_s)^0 = (\hat{P}_s)^3 \;,\; (\tilde{P}_s)^{+1} = -\frac{1}{\sqrt{2}}\left[(\hat{P}_t)^1 + \mathrm{i}(\hat{P}_s)^2\right] . \tag{4.122}$$

Otherwise, since the difference is a unitary rotation, the choice of basis is arbitrary.

4.4.3.1 Spin and Isospin Projection

To understand why the projectors have been defined as in Eq. (4.121), it is instructive to calculate the tree-level contribution to the amplitude in a given channel. With all spin and isospin vertices written out, the Feynman rule for the four-nucleon vertex in the ${}^3S_1$ channel is

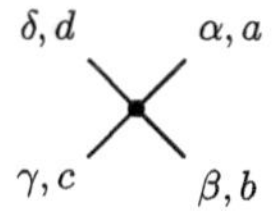

$$\sim \mathrm{i}\frac{C_{0,t}}{8}(\sigma^i\sigma^2)^\alpha{}_\beta(\tau^2)^a{}_b(\sigma^2\sigma^i)^\gamma{}_\delta(\tau^2)^c{}_d\,, \tag{4.123}$$

which is obtained by simply writing out the $(\hat{P}_t)^i$ from Eq. (4.121). Furthermore, this diagram has an associated combinatorial factor 4 because there are two possibilities each to contract the in- and outgoing legs with external nucleon fields.

In order to calculate the T-matrix, we have to write out the Lippmann–Schwinger equation with all indices (and symmetry factors) and then apply the appropriate projectors. For ${}^3S_1$ and isospin 0, the result should have two free spin-1 indices, which we label k and j for the in- and outgoing side, respectively. The inhomogeneous term is just the vertex (4.123) with an additional factor 4. Applying the projectors, we get

$$\begin{aligned}
&\frac{1}{\sqrt{8}}(\sigma^2\sigma^j)^\beta{}_\alpha(\tau^2)^b{}_a \times 4 \times \mathrm{i}\frac{C_{0,t}}{8}(\sigma^i\sigma^2)^\alpha{}_\beta(\tau^2)^a{}_b(\sigma^2\sigma^i)^\gamma{}_\delta(\tau^2)^c{}_d \times \frac{1}{\sqrt{8}}(\tau^2)^d{}_c(\sigma^k\sigma^2)^\delta{}_\gamma \\
&\quad = \mathrm{i}\frac{C_{0,t}}{16}\,\mathrm{Tr}(\sigma^j\sigma^i)\,\mathrm{Tr}(\tau^2\tau^2)\,\mathrm{Tr}(\sigma^i\sigma^k)\,\mathrm{Tr}(\tau^2\tau^2) = \mathrm{i}C_{0,t}\,\delta^{jk}\,,
\end{aligned} \tag{4.124}$$

where we have used the well-known of the Pauli matrices. This is exactly the expected result: the projectors (4.121) have been constructed to give this. The projection of other, more complicated diagram works in the same way. Albeit somewhat tedious, it is technically straightforward. For higher partial waves, one of course has to take into account the coupling of spin and orbital angular momentum.

4.4.3.2 Dibaryon Fields

Just like for bosons, it is convenient to introduce auxiliary dimer—now called dibaryon—field for each of the NN S-wave states. This is done by rewriting the Lagrangian (4.120) as

$$\begin{aligned}\mathscr{L} = N^\dagger \left(\mathrm{i}\partial_0 + \frac{\nabla^2}{2M_N} + \cdots\right) - t^{i\dagger}\left[g_t + \left(\mathrm{i}\partial_0 + \frac{\nabla^2}{4M_N}\right)\right] t^i + y_t\left[t^{i\dagger}\left(N^T P_t^i N\right) + \text{h.c.}\right] \\ - s^{\lambda\dagger}\left[g_s + \left(\mathrm{i}\partial_0 + \frac{\nabla^2}{4M_N}\right)\right] s^\lambda + y_s\left[s^{\lambda\dagger}\left(N^T P_s^\lambda N\right) + \text{h.c.}\right] + \cdots\,, \end{aligned} \tag{4.125}$$

where t (s) denotes a ${}^3S_1$ (${}^1S_0$) dibaryon field and the projection operators $P_{s,t}$ are as defined in Eq. (4.121). The "bare" dibaryon propagators are just $\mathrm{i}/g_{s,t}$, while the full leading-order expressions are obtained by summing up all nucleon bubble insertions. This resummation, which without dibaryon fields gives the NN T-matrix as a bubble chain, reflects the fact we need to generate shallow states (the bound deuteron and the virtual ${}^1S_0$ state) to account for the unnaturally large NN scattering lengths. Pionless EFT is designed to capture this feature.

Omitting spin-isospin factors for simplicity, the resummed propagators are

$$\mathrm{i}\Delta_{s,t}(p_0, \mathbf{p}) = \frac{-\mathrm{i}}{g_{s,t} + {y_{s,t}}^2 I_0(p_0, \mathbf{p})}\,, \tag{4.126}$$

where

$$\begin{aligned} I_0(p_0, \mathbf{p}) &= M_N \int^\Lambda \frac{\mathrm{d}^3 q}{(2\pi)^3} \frac{1}{M_N p_0 - \mathbf{p}^2/4 - \mathbf{q}^2 + \mathrm{i}\varepsilon} \\ &= -\frac{M_N}{4\pi}\left(\frac{2\Lambda}{\pi} - \sqrt{\frac{\mathbf{p}^2}{4} - M_N p_0 - \mathrm{i}\varepsilon}\right) + \mathscr{O}(1/\Lambda) \end{aligned} \tag{4.127}$$

is the familiar bubble integral regularized with a momentum cutoff. The cutoff dependence is absorbed into the parameters $y_{s,t}$ and $g_{s,t}$ to obtain the renormalized propagators. Attaching external nucleon legs gives the NN T-matrix,

$$\mathrm{i}T_{s,t}(k) = (\mathrm{i}y_{s,t})^2\,\mathrm{i}\Delta_{s,t}\big(p_0 = k^2/M_N, \mathbf{p} = \mathbf{0}\big) = \frac{4\pi}{M_N}\frac{\mathrm{i}}{k\cot\delta_{s,t}(k) - \mathrm{i}k}\,, \tag{4.128}$$

so we can match to the effective range expansions for $k\delta_{s,t}(k)$. At leading order, the renormalization condition is to reproduce $k\cot\delta_{s,t}(k) = -1/a_{s,t} + \mathscr{O}(k^2)$, which gives

$$\frac{4\pi g_{s,t}}{M_N {y_{s,t}}^2} = -\frac{1}{a_{s,t}} + \frac{2\Lambda}{\pi}\,. \tag{4.129}$$

Instead of this standard choice of the expansion around the zero-energy threshold, it is convenient to expand the $^3S_1$ channel around the deuteron pole.[18] This is

$$k\cot\delta_t(k) = \gamma_t + \frac{\rho_t}{2}(k^2+\gamma_t^2) + \cdots, \tag{4.130}$$

where $\gamma_t = \sqrt{M_N B_d} \simeq 45.7$ is the deuteron binding momentum and $\rho_t \simeq 1.765$ is the "deuteron effective range." This choice, which sets

$$\frac{4\pi g_t}{M_N y_t{}^2} = -\gamma_t + \frac{2\Lambda}{\pi}, \tag{4.131}$$

gets the exact deuteron pole position right at leading order, but is equivalent to the choice in Eq. (4.129) up to range corrections.

Wavefunction Renormalization

The residue at the deuteron pole gives the deuteron wavefunction renormalization. We find

$$Z_t^{-1} = \mathrm{i}\frac{\partial}{\partial p_0}\frac{1}{\mathrm{i}\Delta_t(p_0,\mathbf{p})}\bigg|_{p_0=-\frac{\gamma_t^2}{M_N},\,\mathbf{p}=0} = \frac{M_N^2 y_t{}^2}{8\pi\gamma_t} \Rightarrow \sqrt{Z_t} = \frac{1}{y_t}\frac{\sqrt{8\pi\gamma_t}}{M_N} \tag{4.132}$$

for the renormalization as in Eq. (4.131). If we directly consider the (off-shell) T-matrix near the pole, we find

$$\begin{aligned} T(E) &= \frac{4\pi}{M_N}\frac{1}{-\gamma_t + \sqrt{-M_N E - \mathrm{i}\varepsilon}} = -\frac{4\pi}{M_N}\frac{\sqrt{-M_N E - \mathrm{i}\varepsilon} + \gamma_t}{-M_N E - \gamma_t^2} \quad \text{for } \varepsilon \to 0 \\ &\sim -\frac{8\pi\gamma_t}{M_N^2}\frac{1}{E + \frac{\gamma_t^2}{M_N}} \quad \text{as } E \to -\gamma_t^2/M_N \ . \end{aligned} \tag{4.133}$$

Comparing to the standard factorization at the pole,[19]

$$T(k,p;E) = -\frac{B^\dagger(k)B(p)}{E+E_B} + \text{terms regular at } E = -E_B\,, \tag{4.134}$$

---

[18]The shallow deuteron bound-state pole is within the radius of convergence of the effective range expansion. The deuteron binding momentum is $\gamma_t = 1a_t + \cdots$, where the ellipses include corrections from the effective range (and higher-order shape parameters).

[19]The minus sign in Eq. (4.134) is a consequence of the convention we use here for the T-matrix.

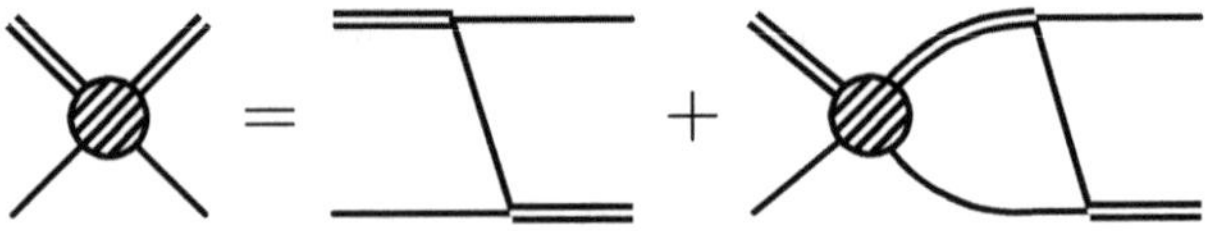

Fig. 4.11 *Nd* quartet-channel integral equation. Nucleons and deuterons are represented as *single* and *double lines*, respectively. The *blob* represents the T-matrix

we can read off from Eq. (4.133) that

$$B(p) = \sqrt{\frac{8\pi\gamma_t}{M_N^2}} = y_t\sqrt{Z_t}\,, \tag{4.135}$$

independent of momentum at this order.

4.4.4 Three Nucleons: Scattering and Bound States

As done in Sect. 4.3.3 for bosons, the dimer/dibaryon formalism allows for a particularly intuitive and simple description of the three-body system. Looking at nucleon-deuteron S-wave scattering,[20] we find that the spin 1 of the deuteron can couple with the spin 1/2 of the nucleon to a total spin of either 3/2 or 1/2. These two cases are referred to as the quartet and doublet channel, respectively.

4.4.4.1 Quartet Channel

In the quartet-channel, the spins of all three nucleons have to be aligned to produce the total spin 3/2. This means that only the deuteron field can appear in intermediate states, and the Pauli principle excludes a three-body contact interaction without derivatives. The resulting integral equation for the *Nd* T-matrix is shown diagrammatically in Fig. 4.11. Compared to Fig. 4.8, we use a different convention here were the T-matrix blob is drawn to the left of the nucleon exchange, and we denote in- and outgoing momenta (in the *Nd* center-of-mass frame) by $\pm\mathbf{k}$ and $\pm\mathbf{p}$, respectively.[21] The energy and momentum dependence are exactly the same as for bosons, but we have to include the additional spin-isospin structure from the

[20]We work in the isospin-symmetric theory here, but in the absence of electromagnetic interactions (discussed in Sect. 4.5.1), the nucleon here should be thought of as a neutron.

[21]Unlike what is done in Sect. 4.3, we also read diagrams from left (incoming particles) to right (outgoing particles). Both conventions can be found in the literature.

vertices. Doing this, the result in its full glory reads:

$$\begin{aligned}\left(\mathrm{i}T_q^{ij}\right)^{\beta b}{}_{\alpha a}(\mathbf{k},\mathbf{p};E) = &-\frac{\mathrm{i}M_N {y_t}^2}{2}\left(\sigma^j\sigma^i\right)^\beta{}_\alpha \delta_a^b \frac{1}{\mathbf{k}^2+\mathbf{k}\cdot\mathbf{p}+\mathbf{p}^2-M_N E-\mathrm{i}\varepsilon} \\ &+\int\frac{\mathrm{d}^3q}{(2\pi)^3}\,\Delta_t\left(E-\frac{\mathbf{q}^2}{2M_N},\mathbf{q}\right)\left(\mathrm{i}T_q^{ik}\right)^{\gamma c}{}_{\alpha a}(E;\mathbf{k},\mathbf{q}) \\ &\times\frac{M_N {y_t}^2}{2}\frac{\left(\sigma^j\sigma^k\right)^\beta{}_\gamma \delta_c^b}{\mathbf{q}^2+\mathbf{q}\cdot\mathbf{p}+\mathbf{p}^2-M_N E-\mathrm{i}\varepsilon}\,. \end{aligned} \tag{4.136}$$

This unprojected amplitude carries spin and isospin indices for the various fields in the initial and final states. To select the overall spin $3/2$ contribution, we take linear combinations as in Eq. (4.122) to select the maximal projections for the in- and outgoing deuterons and $\alpha = \beta = 1$ to get nucleons with spin orientation $+1/2$. We also set $a = b = 2$ to select neutrons. Altogether, this gives

$$\left(\sigma^j\sigma^i\right)^\beta{}_\alpha \delta_a^b \to 2 \tag{4.137}$$

in the inhomogeneous term, and the same factor for the integral part. The fully projected quartet-channel amplitude is

$$T_q = \frac{1}{2}\left(T_q^{11}+\mathrm{i}\left(T_q^{12}-T_q^{21}\right)+T_q^{22}\right)^{12}_{12}\,. \tag{4.138}$$

4.7 *Exercise:* Work out the details leading to Eq. (4.138).

Finally, the S-wave projection of Eq. (4.136) is done by applying the operator $\frac{1}{2}\int_{-1}^{1}\mathrm{d}\cos\theta$, where θ is the angle between $\mathbf{k}$ and $\mathbf{p}$. Introducing a momentum cutoff Λ, the resulting equation can be solved numerically by discretizing the remaining one-dimensional integral. From the result, which we denote by T_q^0, we can calculate observables like scattering phase shifts,

$$\delta_q(k) = \frac{1}{2\mathrm{i}}\ln\left(1+\frac{2\mathrm{i}kM_N}{3\pi}Z_t T_q^0(k,k;E_k)\right)\ ,\ E_k = \frac{3k^2}{4M_N}-\frac{\gamma_t^2}{M_N}\,, \tag{4.139}$$

or the Nd scattering length:

$$a_q = \frac{M_N}{3\pi}\,\mathrm{lim}_{k\to 0}\, Z_t T_q^0(k,k;E_k)\,. \tag{4.140}$$

Note that we have not absorbed the wavefunction renormalization Z_t into T_q^0 but instead chose to keep it explicit in Eqs. (4.139) and (4.140)

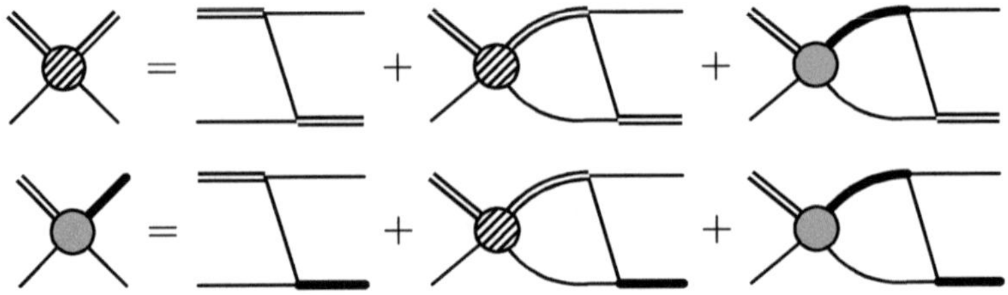

Fig. 4.12 *Nd* doublet-channel integral equation. As in Fig. 4.11, nucleons and deuterons are drawn as *single* and *double lines*, respectively. Additionally, we represent the $^1S_0$ dibaryon as a *thick line*. The *hatched and shaded blobs* are the two components of the doublet-channel $Nd \to Nd$ T-matrix

4.4.4.2 Doublet Channel

The doublet channel (total spin 1/2) has a richer structure, since now also the $^1S_0$ dibaryon can appear as an intermediate state. The result is a coupled channel integral equation, shown diagrammatically in Fig. 4.12. We skip here the technicalities of the spin-isospin projection (for details, see [28] and earlier references therein) and merely quote the result in a compact notation:

$$\begin{pmatrix} T^0_{d,1} \\ T^0_{d,2} \end{pmatrix} = \begin{pmatrix} M_N y_t{}^2/2 \\ -3M_N y_t y_s/2 \end{pmatrix} K + \begin{pmatrix} D_t & 0 \\ 0 & D_s \end{pmatrix} \begin{pmatrix} -M_N y_t{}^2/2 & 3M_N y_t y_s/2 \\ 3M_N y_s y_t/2 & -M_N y_s{}^2/2 \end{pmatrix} K \otimes \begin{pmatrix} T^0_{d,1} \\ T^0_{d,2} \end{pmatrix}, \tag{4.141}$$

where K is the "kernel" function

$$K(k,p;E) = \frac{1}{2kp} \ln\left(\frac{k^2 + p^2 + kp - M_N E - \mathrm{i}\varepsilon}{k^2 + p^2 - kp - M_N E - \mathrm{i}\varepsilon} \right), \tag{4.142}$$

$$EFT - D_{s,t}(q;E) = \Delta_{s,t}\left(E - \frac{q^2}{2M_N}; q \right), \tag{4.143}$$

and the integral operation

$$A \otimes B \equiv \frac{1}{2\pi^2} \int_0^{\Lambda} \mathrm{d}q\, q^2 A(\ldots, q) B(q, \ldots) \tag{4.144}$$

has to be applied within each block. Just like the quartet-channel equation, Eq. (4.141) can be solved numerically by discretizing the integrals, with the additional complication that we now have a 2×2 block matrix. The T-matrix likewise becomes a 2-block vector, the upper part of which is gives the physical $Nd \to Nd$ amplitude.[22]

4.8 *Exercise:* Express the fully projected integral equation for the quartet channel amplitude T_q^0 using the compact notation based on Eqs. (4.142)–(4.144).

Leading-Order Three-Nucleon Force

Studying the doublet-channel solution as a function of increasing UV cutoff Λ, one finds that there is no stable limit as $\Lambda \to \infty$. Instead, the amplitude changes wildly as Λ is varied. This is much unlike the quartet-channel case, which shows a rapid convergence with Λ.

The origin of this behavior was explained by Bedaque et al. [29]. The behavior for large Λ is governed by large momenta, which means that infrared scales like the scattering lengths do not matter. Indeed, one finds that $D_t(E;q)$ and $D_s(E;q)$ have the same leading behavior as $q \to \infty$. To analyze the asymptotic behavior of the amplitude we can thus go to the $SU(4)$ spin-isospin symmetric limit and set $D_t = D_t \equiv D$ as well as $y_t = y_s \equiv y$. In this limit, the two integral equations in Eq. (4.141) can be decoupled by defining $T_{d,\pm} = T_{d,1} \pm T_{d,2}$. For $T_{d,+}$ we find the integral equation

$$T_{d,+}^0(k,p;E) = -M_N y^2 K(k,p;E) + M_N y^2 \int_0^\Lambda \frac{\mathrm{d}q}{2\pi^2} q^2 K(k,q;E) D(q;E) T_{d,+}^0(q,p;E)\,, \tag{4.145}$$

which is formally exactly the same as the three-boson integral equation, Eq. (4.101), in the absence of a three-body force. As discussed in Sect. 4.3.3, this equation does not have a unique solution in the limit $\Lambda \to \infty$. Since $T_{d,1/2}$ are linear combinations of involving $T_{d,+}$, they inherit the same behavior. But the cure is now obvious: a three-nucleon force, which by naïve counting would only enter at higher orders, has to be promoted to leading order in order to make Eq. (4.141) well defined. This three-nucleon force, like the asymptotic amplitudes, is $SU(4)$ symmetric (invariant under arbitrary spin-isospin rotations) and can be written as

$$\mathscr{L}_3 = \frac{h_0}{3} N^\dagger \left[y_t{}^2\, t^{i\dagger} t^j \sigma^i \sigma^j + y_s{}^2\, s^{A\dagger} s^B \tau^A \tau^B - y_t y_s \left(t^{i\dagger} s^A \sigma^i \tau^A + \text{h.c.} \right) \right] N\,, \tag{4.146}$$

[22] Note that this vector is one part of the more general full off-shell amplitude, which is a 2×2 block matrix including the two combinations of dibaryon legs that do not appear in Fig. 4.12.

where the cutoff-running of the three-nucleon coupling is analogous to what we derived for bosons:

$$h_0 = \frac{M_N H(\Lambda)}{\Lambda^2} \,. \tag{4.147}$$

Including this, the coupled doublet-channel integral equation becomes

$$\begin{pmatrix} T^0_{d,1} \\ T^0_{d,2} \end{pmatrix} = \begin{pmatrix} M_N {y_t}^2/2 \\ -3M_N y_t y_s/2 \end{pmatrix} K \\ + \begin{pmatrix} D_t & 0 \\ 0 & D_s \end{pmatrix} \begin{pmatrix} -M_N {y_t}^2/2(K+h_0) & 3M_N y_t y_s/2(K+h_0) \\ 3M_N y_s y_t/2(K+h_0) & -M_N {y_s}^2/2(K+h_0) \end{pmatrix} \otimes \begin{pmatrix} T^0_{d,1} \\ T^0_{d,2} \end{pmatrix} . \tag{4.148}$$

We stress that the requirement to include this three-body force at leading-order is a feature of the nonperturbative physics that can be traced back to the large *NN* scattering lengths. All loop diagrams obtained by iterating the integral equation are individually finite as $\Lambda \to \infty$, yet their infinite sum does not exist in that limit unless the h_0 contact interaction is included.

The Phillips Line

The form of $H(\Lambda)$ is given by Eq. (4.118), but since there is a prefactor that depends on the details of the regularization scheme, in practice one has to determine the appropriate value numerically after choosing a cutoff Λ. This requires a three-body datum as input, which is conveniently chosen to be the triton binding energy—the ^{3}H bound state corresponds to a pole in T_d at $E = -E_B(^3\text{H})$, *cf.* Sect. 4.3.4—or the doublet-channel *nd* scattering length (or, in principle, a phase shift at some fixed energy). Once one of these is fixed, the rest can be predicted. In particular, this means that the existence of a single three-body parameter in pionless EFT at leading order provides a natural explanation of the Phillips line, i.e., the observation that various phenomenological potentials, which were all tuned to produce the same *NN* phase shifts, gave different results for the triton binding energy and doublet S-wave scattering length, which however are strongly correlated. In our framework, we can obtain the line shown in Fig. 4.13 by fitting $H(\Lambda)$ to different values of the scattering length and then calculating $E_B(^3\text{H})$ (or vice versa). Alternatively, one can find the same curve by setting h_0 to zero and varying Λ to move along the line.

With this, we conclude our discussion the three-nucleon system. For more details, like calculations beyond leading order, we encourage the reader to encourage the literature.

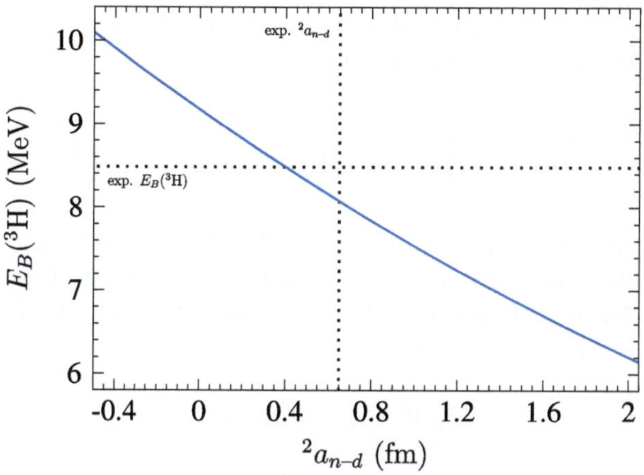

Fig. 4.13 Correlation (Phillips line) between ^{3}H binding energy (in MeV) and doublet *nd* scattering length (in fm) in leading-order pionless EFT

4.5 Beyond Short-Range Interactions: Adding Photons and Pions

4.5.1 Electromagnetic Interactions

So far, we have studied only effective theories where the particles (atoms of nucleons) interact solely via short-range force (regulated contact interactions). While this is, as we have argued, sufficient to describe the strong nuclear interactions at (sufficiently low) energies where the EFT is valid, the electromagnetic interaction does not fit in this scheme. Nevertheless, since almost all systems of interest in low-energy nuclear physics involve more than one proton, the inclusion of such effects is of course important.

Any coupling of photons to charged particles has to be written down in a gauge-invariant way. A natural way to ensure gauge invariance is to replace all derivatives in the effective Lagrangian with covariant ones, i.e.,

$$\partial_\mu \to D_\mu = \partial_\mu + \mathrm{i}eA_\mu\hat{Q}\,, \tag{4.149}$$

where $\hat{Q}$ is the charge operator (for nucleons, for example, $\hat{Q}_N = (\mathbf{1} + \tau_3)/2$). Moreover $e^2 = 4\pi\alpha$ defines the electric unit charge in terms of the fine structure constant $\alpha \approx 1/137$.

Since in the EFT there is an infinite tower of contact interactions with an increasing number of derivatives, we also get an infinite number of photon vertices. Still, merely plugging in the covariant derivative is not enough (after all, this is called *minimal* substitution); it is possible to write down terms which are gauge invariant by themselves, and for the EFT to be complete these have to be included as well. Before we come back to this in Sect. 4.5.1.4, let us first see what we get from gauging the derivatives in the nucleon kinetic term:

$$N^\dagger\left(\mathrm{i}\partial_t + \frac{\boldsymbol{\nabla}^2}{2M_N} + \cdots\right)N \to N^\dagger\left(\mathrm{i}D_t + \frac{\boldsymbol{D}^2}{2M_N} + \cdots\right)N\,. \tag{4.150}$$

In addition to this, we also have to include the photon kinetic and gauge fixing terms to complete the electromagnetic sector. A convenient choice for the our nonrelativistic framework is Coulomb gauge, i.e., demanding that $\boldsymbol{\nabla}\cdot\mathbf{A} = 0$. A covariant way to write is condition is

$$\partial_\mu A^\mu - \eta_\mu\eta_\nu\partial^\nu A^\mu = 0 \tag{4.151}$$

with the timelike unit vector $\eta^\mu = (1, 0, 0, 0)$. Hence, we add to our effective Lagrangian the term

$$\mathscr{L}_{\text{photon}} = -\frac{1}{4}F_{\mu\nu}F^{\mu\nu} - \frac{1}{2\xi}\left(\partial_\mu A^\mu - \eta_\mu\eta_\nu\partial^\nu A^\mu\right)^2 . \tag{4.152}$$

4.5.1.1 The Coulomb Force

From Eq. (4.152) we get the equation of motion for the photon field as

$$\left[\Box g^{\mu\nu} - \partial^\mu\partial^\nu + \frac{1}{\xi}\left(\partial^\mu\partial^\nu - \eta^\nu\eta_\kappa\partial^\mu\partial^\kappa - \eta^\mu\eta_\lambda\partial^\lambda\partial^\nu + \eta^\mu\eta^\nu\eta_\lambda\eta_\kappa\partial^\lambda\partial^\kappa\right)\right]A_\nu(x) = 0\,. \tag{4.153}$$

The photon propagator is defined as the Feynman Green's function for the differential operator acting on $A_\nu(x)$. Writing down the general solution in momentum space and choosing $\xi = 0$ at the end (recall that it is an artificial parameter introduced through enforcing the gauge condition in the path integral [30]), we get

$$D_\gamma^{\mu\nu}(k) = \frac{-\mathrm{i}}{k^2 + \mathrm{i}\varepsilon}\left(g^{\mu\nu} + \frac{k^\mu k^\nu + k^2\eta^\mu\eta^\nu - (k\cdot\eta)(k^\mu\eta^\nu + \eta^\mu k^\nu)}{(k\cdot\eta)^2 - k^2}\right). \tag{4.154}$$

A simple inspection of this expression shows that it vanishes if μ or $\nu = 0$. In other words, A_0 photons do not propagate. Correspondingly, their equation of motion becomes time independent and we can use it to remove A_0 from the effective Lagrangian (the nuclear part plus $\mathscr{L}_{\text{photon}}$). We find

$$\boldsymbol{\nabla}^2 A_0 = -eN^\dagger\hat{Q}N\,, \tag{4.155}$$

which is just the Poisson equation with the nucleon charge density on the right-hand side. Solving this in Fourier space, where $\boldsymbol{\nabla}^2$ turns into the squared three-momentum, we eventually get a term

$$\mathscr{L}_{\text{Coulomb}}(x_0,\mathbf{x}) = -e^2 \int \mathrm{d}^3y\, N^\dagger(x_0,\mathbf{x})N(x_0,\mathbf{x})\,\frac{\mathrm{e}^{-\mathrm{i}\mathbf{q}\cdot(\mathbf{x}-\mathbf{y})}}{\mathbf{q}^2}\left(\frac{\mathbf{1}+\tau^3}{2}\right)N^\dagger(x_0,\mathbf{y})N(x_0,\mathbf{y})\,, \tag{4.156}$$

i.e., static Coulomb potential ($\sim 1/r$ in configuration space) between charged nucleons. This is a non-local interaction that really should be kept in the Lagrangian as a whole. Still, to calculate Feynman diagrams it is convenient to split it up into a vertex

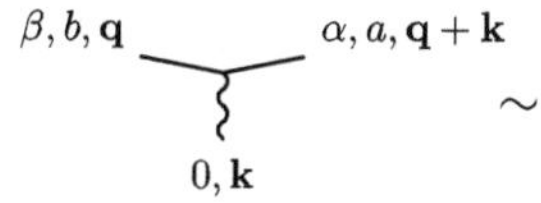

$$\sim -\mathrm{i}e\left(\frac{\mathbf{1}+\tau^3}{2}\right)^a_{\;b}\delta^\alpha{}_\beta\,, \tag{4.157}$$

and a factor $\mathrm{i}/\mathbf{k}^2$ for each "Coulomb-photon propagator," which is really just an expression for the static potential in momentum space. Note that the sign in Eq. (4.157) is arbitrary because these vertices really only come in pairs; we have chosen it here to coincide with what one would naïvely read off from the $N^\dagger \mathrm{i}D_0 N$ term.

Finding that A_0 photons only appear as static internal exchanges goes well in line that physical photons in external states have to be transverse. Still, one might wonder how to treat diagrams with virtual photons coupled to a nuclear system (e.g., electrodisintegration), which should have A_0 contributions. The answer is that the proper way to treat these is to add appropriate external currents to the Lagrangian, which then, in turn, appear on the right-hand side of Eq. (4.155).

4.5.1.2 Coulomb Enhancement and Divergences

The Coulomb force is a long-range interaction: it only falls off like a power law ($\sim 1/r$) in configuration space. In momentum space, it correspondingly has a pole at vanishing momentum transfer ($\mathbf{q}^2 = 0$), i.e., it gives rise to an infrared divergence. There are standard techniques for dealing with this, for example defining so-called Coulomb-modified scattering phase shifts and effective-range expansions well-known from quantum-mechanical scattering theory. These are based on treating Coulomb effects fully nonperturbatively, i.e., resumming to exchange of Coulomb photons to all orders. In the EFT power counting, the need for this resummation is reflected in the fact dressing a given two-body scattering diagram with external momenta of order Q by an additional Coulomb photon gives a factor $\alpha M_N/Q$, i.e., an *enhancement* if $Q \lesssim \alpha M_N$. In Fig. 4.14 we show this for two-photon exchange compared to the single-photon diagram.

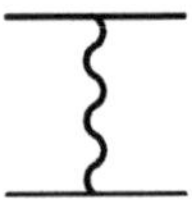

$\sim \frac{\alpha}{Q^2}$

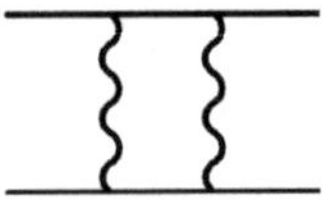

$\sim \frac{Q^5}{M_N}\left(\frac{\alpha}{Q^2}\right)^2\left(\frac{M_N}{Q^2}\right)^2 = \frac{\alpha}{Q^2} \times \frac{\alpha M_N}{Q}$

Fig. 4.14 Infrared enhancement of Coulomb-photon exchange

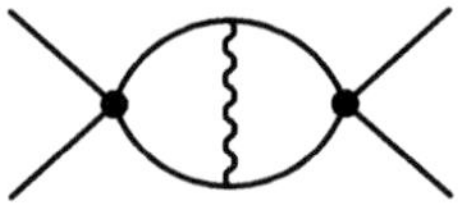

Fig. 4.15 Single photon insertion in a proton-proton bubble diagram

If a problem with Coulomb interactions is solved numerically, the IR divergence has to be regularized in some way to have all quantities well defined. One way to do this "screening" the potential with a photon mass λ, i.e., replacing $\mathbf{q}^2$ with $\mathbf{q}^2 + \lambda^2$ in the Coulomb-photon propagator. If this is done, λ should be kept as small as possible and be extrapolated to zero for all physical observables at the end of the calculation.

In contrast to what one might naïvely expect, Coulomb contributions can also modify the UV behavior of diagrams. Consider, for example the insertion of a single photon within a bubble contributing to proton-proton scattering, as shown in Fig. 4.15. Power counting momenta in this diagram (with C_0 vertices) gives a factor $(Q^5/M_N)^2$ from the two loops, $(M_N/Q^2)^4$ from the nucleon propagators, and an obvious α/Q^2 from the photon, meaning that overall we have Q^0, corresponding to a superficial logarithmic divergence of this diagram (which one can confirm with an explicit calculation). This is a new feature compared to the theory with only short-range interactions, where this particular divergence is absent (all divergences are of power-law type there). Without going into any details, we stress that this divergence has to be accounted for when the theory with Coulomb contributions is renormalized. In particular, this example teaches us that in the EFT the Coulomb force is not merely an "add-on potential" that slightly shifts results, but that it has to be treated consistently along with the short-range interactions.

4.5.1.3 Transverse Photons

Transverse photons come from the quadratic spatial derivative:

$$\mathbf{D}^2 = (\partial^i + \mathrm{i}eA^i\hat{Q})(\partial^i + \mathrm{i}eA^i\hat{Q}) = \nabla^2 + \mathrm{i}e(A^i\hat{Q}\,\partial^i + \partial^i A^i\hat{Q}) - e^2A^2\hat{Q}\,. \tag{4.158}$$

We can rewrite this using

$$\partial^i\left(A^i\hat{Q}N\right) = (\partial^i A^i)\hat{Q}N + A^i\hat{Q}\,\partial^i N\,, \tag{4.159}$$

where the first term vanishes in Coulomb gauge. Hence, we get the vertex

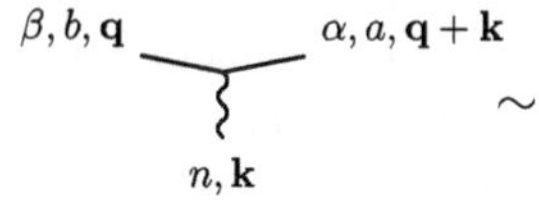

$$\sim -\frac{e}{M_N}\left(\frac{\mathbf{1}+\tau^3}{2}\right)^a_{\ b}\delta^\alpha{}_\beta\,(\mathrm{i}\mathbf{q})^n\,, \tag{4.160}$$

with the momentum dependence coming from the derivative. We leave it as an exercise to write down the Feynman rule for the two-photon term $\sim A^2\hat{Q}$.

Comparing Eq. (4.160) with Eq. (4.157) that a diagram with the exchange of a transverse photon is suppressed by a factor Q^2/M_N^2 compared to the same topology with a Coulomb photon. Transverse photons also have a more complicated propagator than the simple $\mathrm{i}/\mathbf{k}^2$ that we found for Coulomb photons. From Eq. (4.154) we find that

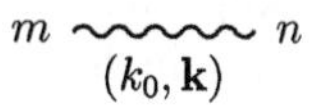

$$\sim \frac{\mathrm{i}}{k_0^2-\mathbf{k}^2+\mathrm{i}\varepsilon}\left(\delta^{mn}-\frac{k^m k^n}{\mathbf{k}^2}\right). \tag{4.161}$$

which now depends on the energy k_0 and thus gives rise to poles in Feynman diagrams. Note that the structure is somewhat different from what one typically sees in QED textbooks that use Lorenz/Feynman gauge.

4.5.1.4 Other External Currents

The covariant derivative alone only gives us photons coupled to the proton's charge. However, as mentioned previously, minimal substitution only gives a subset of all possible electromagnetic terms. For example, the magnetic coupling of photons to the nucleons (both protons and neutrons in this case) is given by

$$\mathscr{L}_{\text{mag}} = \frac{e}{2M_N}N^\dagger(\kappa_0+\kappa_1\tau^3)\,\vec{\sigma}\cdot\mathbf{B}\,N\,. \tag{4.162}$$

where κ_0 and κ_1 are the isoscalar and isovector nucleon magnetic moments, respectively. That is, the low-energy constant of this operator has been fixed directly to the associated single-nucleon observable (similarly to how the nucleon mass is fixed immediately in the nonrelativistic theory). With $\mathbf{B} = \nabla\times\mathbf{A}$ and all indices written out, this is

$$\mathscr{L}_{\text{mag}} = \frac{e}{2M_N}N^\dagger_{\alpha a}\big(\kappa_0\mathbf{1}+\kappa_1\tau^3\big)^a{}_b\big(\sigma^i\big)^\alpha{}_\beta\epsilon_{ijm}\partial^j A^m N^{\beta b}\,. \tag{4.163}$$

and it gives rise to the Feynman rule

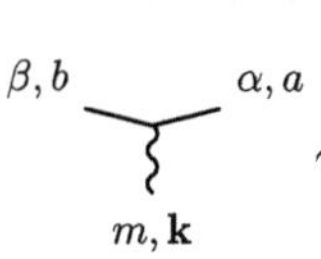

$$\sim \frac{\mathrm{i}e}{2M_N}\big(\kappa_0\mathbf{1}+\kappa_1\tau^3\big)^a{}_b\,\epsilon_{ijm}\big(\sigma^i\big)^\alpha{}_\beta(\mathrm{i}\mathbf{k})^j\,. \tag{4.164}$$

We point out that Eq. (4.163) only gives the leading magnetic coupling. In traditional nuclear physics language, it corresponds to a one-body operator. At higher orders in the EFT, there are additional operators, like a four-nucleon contact interaction with an additional photon. Such many-body terms correspond to what phenomenological approaches model as "meson exchange currents."

4.5.2 Example: Deuteron Breakup

As an application of the things discussed in the previous sections, we now consider the low-energy reaction $d\gamma \leftrightarrow np$. By time-reversal symmetry, the amplitudes for the processes corresponding to the two possible directions of the arrow are the same. For definiteness, we show the simplest diagram for the breakup reaction in Fig. 4.16.

The reaction can be considered in different reference frames. In the lab frame (left panel of Fig. 4.16), the deuteron is initially at rest and then gets hits by a photon with four-momentum $(k_0, \mathbf{k})$. For our theoretical discussion here it is more convenient to take the two outgoing nucleons in their center-of-mass frame, as shown in the right panel of Fig. 4.16. To first order, we can translate between the two frames by boosting all nucleons lines with a momentum $\mathbf{k}/2$. A more careful analysis would keep track of relativistic kinematics (the external photon is never nonrelativistic), but this is not essential for the illustration here, so we can get away with interpreting $(k_0, \mathbf{k})$ to mean different things in the two frames.

In the *NN* center-of-mass frame, the initial and final-state energies are

$$E_i = k_0 + \frac{k^2}{4M_N} - \frac{\gamma_t^2}{M_N}\,, \tag{4.165a}$$

$$E_f = \frac{p^2}{2M_N} + \frac{p^2}{2M_N} = \frac{p^2}{M_N}\,, \tag{4.165b}$$

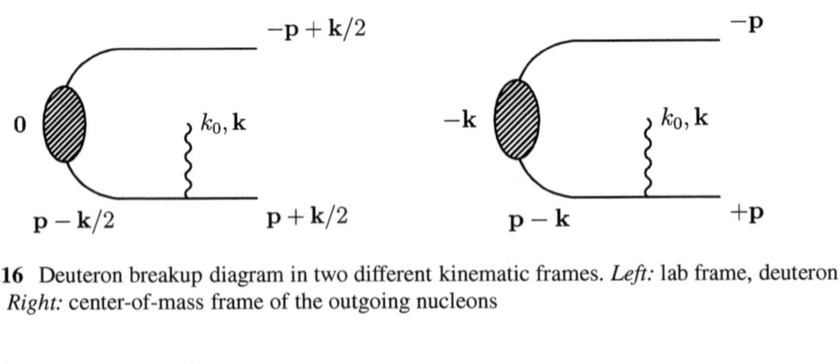

Fig. 4.16 Deuteron breakup diagram in two different kinematic frames. *Left:* lab frame, deuteron at rest. *Right:* center-of-mass frame of the outgoing nucleons

where we have neglected the small deuteron binding energy by setting $M_d = 2M_N$. Conservation of energy implies that

$$p = \sqrt{M_N k_0 - \gamma_t^2 + k^2/4} \Leftrightarrow k_0 = \frac{p^2}{M_N} - \frac{k^2}{4M_N} + \frac{\gamma_t^2}{M_N}\,, \tag{4.166}$$

and the energy assigned to the internal nucleon propagator has to be $-\frac{k_0}{2} - \frac{\gamma_t^2}{2M_N} + \frac{k^2}{8M_N}$.

The blob in Fig. 4.16 represents the deuteron vertex function calculated in Sect. 4.4.3.2. Taking the result found there and adding the spin-isospin structure, we find

i 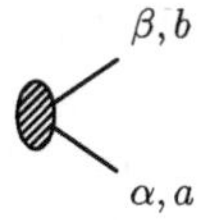

$$\sim \mathrm{i}\frac{1}{\sqrt{8}}\frac{\sqrt{8\pi\gamma_t}}{M_N}(\sigma^i\sigma^2)^\alpha{}_\beta(\tau_2)^a{}_b\,. \tag{4.167}$$

With this, we can finally write down the amplitude. For an E1 transition, the photon couples to the nucleon charge via the Feynman rule given in Eq. (4.160). Combining this with the ingredients above, we get

$$\begin{aligned} &\mathrm{i}\mathscr{M}_{\mathrm{E1}} \\ &= 2 \times \frac{-e}{M_N}\left(\frac{\mathbf{1}+\tau^3}{2}\right)^c{}_a \delta^\gamma_\alpha\, \mathrm{i}(\mathbf{p}-\mathbf{k})^j \frac{\mathrm{i}}{-\frac{k_0}{2} - \frac{\gamma_t^2}{2M_N} + \frac{k^2}{8M_N} - \frac{(\mathbf{p}-\mathbf{k})^2}{2M_N} + \mathrm{i}\varepsilon} \\ &\times \frac{\mathrm{i}}{\sqrt{8}}\frac{\sqrt{8\pi\gamma_t}}{M_N}(\sigma^2\sigma^i)^\beta{}_\gamma(\tau_2)^b{}_c(\boldsymbol{\epsilon}^*_{s_\gamma})^j\,(\boldsymbol{\epsilon}^*_{s_d})^i\,(N_{s_p})^{\alpha a}\,(N_{s_n})_{\beta b}\,. \end{aligned} \tag{4.168}$$

We have included a factor 2 to account for the fact that in drawing the diagram, the photon could be coupled to either nucleon (the isospin projection operator ensures of course that only the proton charge gives a contribution). Moreover, in the second line we have included polarization vectors and spinors for all external particles, the spins of which we denote as s_γ, s_d, s_p, and s_n, respectively. From the amplitude in Eq. (4.168) one can proceed to calculate the corresponding cross section by summing/averaging over the various initial and final states and integrating over the available phase space. We skip these details and close by noting that the isospin part is of course completely fixed ("polarized") by the experimental setup. Hence, the spinor-isospinors are

$$(N_{s_p})^{\alpha a} = N^\alpha_{s_p}\begin{pmatrix}1\\0\end{pmatrix}^a\,, \quad (N_{s_n})_{\beta b} = N^\alpha_{s_p}\begin{pmatrix}1\\0\end{pmatrix}_b\,, \tag{4.169}$$

i.e., we can directly set $a = 1$ (isospin "up", proton) and $b = 2$ (isospin "down", neutron) in the amplitude.

4.9 *Exercise:* Write down the analog of Eq. (4.168) with a magnetically-coupled (M1) photon.

4.5.3 Chiral Effective Field Theory

Pionless EFT provides a simple yet powerful framework to describe few-nucleon systems at very low energies, but its name implies its natural limitation, i.e., the inability to describe physics at energy scales where pion exchange can be resolved. Certainly this becomes important for scattering calculations at momentum scales larger than the pion mass. But nuclear binding also generally increases with increasing number A of bound nucleons, which translates to larger typical momentum scales within the nucleus. There are indications that pionless EFT still converges for $A = 4$, but the question is not fully settled.

The construction of an effective field theory of nucleons and pions was pioneered by Weinberg in the early 1990s [31, 32], proposing a scheme to construct a nuclear potential based on Feynman diagrams from chiral perturbation theory. This theory, which is constructed as an expansion around the so-called "chiral limit," in which the pions are exactly massless Goldstone bosons. The resulting theory, which has been applied with great success in the purely pionic and single-nucleon sector, treats the pion mass as a *small* scale and thus has a power counting designed for typical momenta $Q \sim m_\pi$.

For two or more nucleons, the theory is highly nonperturbative, which motivated Weinberg to develop a scheme where the power counting is applied to the *potential* instead of the amplitude, as we have otherwise done throughout this section. Kaplan et al. [17, 33, 34] proposed a different scheme where pions are included perturbatively on top of a leading order given by pionless EFT. This approach has, however, been found not to converge in channels where pion exchange generates a singular attractive interaction [35]. It is thus understood today that pions in general have to be treated nonperturbatively, in a framework generally referred to as "chiral effective field theory." How exactly this should be implemented, however, is still a matter of debate. Instead of summarizing this here, we refer the reader to the literature (see, e.g., [12, 36–43]).

4.5.3.1 Leading-Order Pion-Nucleon Lagrangian

For a thorough introduction to the field of chiral perturbation theory, we recommend the reader to study the lecture notes of Scherer and Schindler [44] as well as the vast original literature cited therein. It uses an elaborate formalism to construct the most general pion-nucleon Lagrangian that is invariant under chiral transformations (individual rotations of left- and right-handed nucleon fields) plus terms implementing the explicit breaking of chiral symmetry due to finite quark masses. In the conventions of [44], the leading-order pion-nucleon Lagrangian is

$$\mathscr{L}_{\pi N}^{(1)} + \mathscr{L}_{2}^{\pi} = \bar{\psi}\left(\mathrm{i}\not{D} - M_N + \frac{g_A}{2}\gamma^\mu\gamma^5 u_\mu\right)\psi + \frac{f_\pi^2}{4}\mathrm{Tr}\left[(\partial^\mu U)^\dagger(\partial_\mu U) + \chi U^\dagger + U\chi^\dagger\right]. \tag{4.170}$$

Here, ψ is the nucleon Dirac field, and the matrix-valued field

$$U \equiv = \exp\left(\mathrm{i}\frac{\vec{\tau}\cdot\vec{\pi}}{f_\pi}\right) \tag{4.171}$$

collects the pion fields in an exponential representation. D_μ here is the so-called chiral covariant derivative that couples the pion field to the nucleons. The matrix

$$\chi = 2B_0\,\mathrm{diag}(m_q, m_q)\,, \tag{4.172}$$

where m_q is the light quark mass (in the exact isospin limit, $m_u = m_d = m_q$) contains the effect from explicit chiral symmetry breaking. Via the Gell-Mann–Oakes–Renner relation,

$$m_\pi^2 = 2B_0 m_q\,, \tag{4.173}$$

one can show that χ generates a mass term for the pion field.

4.10 *Exercise:* Expand the exponential in Eq. (4.171) to show that $\frac{f_\pi^2}{4}\mathrm{Tr}\left[(\partial^\mu U)^\dagger(\partial_\mu U)\right]$ generates a kinetic term for the pion field $\vec{\pi}$ plus higher-order pion self interactions.

After a couple of steps, which we encourage the reader to follow based on the definitions given in [44], the leading terms in the Lagrangian are found to be

$$\begin{aligned}\mathscr{L}_{\pi N}^{(1)} + \mathscr{L}_{2}^{\pi} &= \bar{\psi}\left(\mathrm{i}\not{\partial} - M_N\right)\psi + \frac{1}{2}(\partial^\mu\vec{\pi})\cdot(\partial_\mu\vec{\pi}) - \frac{1}{2}m_\pi^2\vec{\pi}^2 \\ &\quad - \frac{g_A}{2f_\pi}\bar{\psi}\gamma^\mu\gamma^5(\vec{\tau}\cdot\partial_\mu\vec{\pi})\psi + \frac{\mathrm{i}}{8f_\pi^2}\bar{\psi}\gamma^\mu(\vec{\tau}\cdot\vec{\pi})(\vec{\tau}\cdot\partial_\mu\vec{\pi})\psi + \cdots\,.\end{aligned} \tag{4.174}$$

Comparing this to our pseudoscalar model (4.28) in Sect. 4.2.2.2, we see that the pion-nucleon coupling now comes with explicit derivatives, as required by chiral symmetry. After a Foldy-Wouthuysen transformation, however, one-pion-exchange in the nonrelativistic limit gives the same structure as in Eq. (4.49). There is no explicit σ field in Eq. (4.174); in the chiral theory, this particle only appears as a resonance generated by two-pion exchange.

Acknowledgements We thank Dick Furnstahl and Bira van Kolck and for various stimulating discussions. Moreover, Sebastian König is grateful to Martin Hoferichter for the insights into Coulomb-gauge quantization presented in Sect. 4.5.1.

References

1. K.G. Wilson, Rev. Mod. Phys. **55**, 583 (1983)
2. G.P. Lepage, What is renormalization? (1989). Hep-ph/0506330
3. S. Weinberg, Physica A **96**, 327 (1979)
4. D.B. Kaplan, Effective field theories (1995). Nucl-th/9506035
5. D.B. Kaplan, Five lectures on effective field theory (2005). Nucl-th/0510023
6. J. Gasser, H. Leutwyler, Ann. Phys. **158**, 142 (1984)
7. J. Gasser, H. Leutwyler, Nucl. Phys. B **250**, 465 (1985)
8. J.F. Donoghue, E. Golowich, B.R. Holstein, *Dynamics of the Standard Model* (Cambridge University Press, New York, 1992)
9. E. Jenkins, A.V. Manohar, Phys. Lett. B **255**, 558 (1991). doi:10.1016/0370-2693(91)90266-S
10. A. Gardestig, K. Kubodera, F. Myhrer, Phys. Rev. C **76**, 014005 (2007). doi:10.1103/PhysRevC.76.014005
11. P.R. Hagen, Effective field theory for halo nuclei. Ph.D. thesis, Bonn University (2014)
12. S.R. Beane, P.F. Bedaque, W.C. Haxton, D.R. Phillips, M.J. Savage, From hadrons to nuclei: crossing the border, in *At The Frontier of Particle Physics*. Handbook of QCD, vol. 1 (World Scientific, River Edge, 2001), p. 133
13. C.A. Bertulani, H.W. Hammer, U. van Kolck, Nucl. Phys. A **712**, 37 (2002)
14. P.F. Bedaque, H.W. Hammer, U. van Kolck, Phys. Lett. B **569**, 159 (2003)
15. E. Braaten, H.W. Hammer, Phys. Rep. **428**, 259 (2006)
16. U. van Kolck, Nucl. Phys. A **645**, 273 (1999)
17. D.B. Kaplan, M.J. Savage, M.B. Wise, Nucl. Phys. B **534**, 329 (1998)
18. D.B. Kaplan, Nucl. Phys. B **494**, 471 (1997)
19. G.V. Skorniakov, K.A. Ter-Martirosian, Sov. Phys. JETP **4**, 648 (1957)
20. G. Danilov, Sov. Phys. JETP **13**, 349 (1961)
21. V.F. Kharchenko, Sov. J. Nucl. Phys. **16**, 173 (1973)
22. P.F. Bedaque, H.W. Hammer, U. van Kolck, Phys. Rev. Lett. **82**, 463 (1999)
23. V. Efimov, Phys. Lett. B **33**(8), 563 (1970)
24. E. Braaten, D. Kang, L. Platter, Phys. Rev. Lett. **106**, 153005 (2011)
25. F. Ferlaino, R. Grimm, Physics **3**, 9 (2010)
26. J. Ulmanis, S. Häfner, R. Pires, F. Werner, D.S. Petrov, E.D. Kuhnle, M. Weidemüller, Phys. Rev. A **93**, 022707 (2016)
27. P.F. Bedaque, H.W. Hammer, U. van Kolck, Nucl. Phys. A **646**, 444 (1999)
28. S. König, H.W. Grießhammer, H.W. Hammer, J. Phys. G **42**, 045101 (2015)
29. P.F. Bedaque, H.W. Hammer, U. van Kolck, Nucl. Phys. A **676**, 357 (2000)
30. M.E. Peskin, D.V. Schroeder, *An Introduction to Quantum Field Theory* (Westview Press, Boulder, 1995)
31. S. Weinberg, Phys. Lett. B **251**, 288 (1990)
32. S. Weinberg, Nucl. Phys. B **363**, 3 (1991)
33. D.B. Kaplan, M.J. Savage, M.B. Wise, Nucl. Phys. B **478**, 629 (1996)
34. D.B. Kaplan, M.J. Savage, M.B. Wise, Phys. Lett. B **424**, 390 (1998)
35. S. Fleming, T. Mehen, I.W. Stewart, Nucl. Phys. A **677**, 313 (2000)
36. P.F. Bedaque, U. van Kolck, Annu. Rev. Nucl. Part. Sci. **52**, 339 (2002)
37. A. Nogga, R.G.E. Timmermans, U. van Kolck, Phys. Rev. C **72**, 054006 (2005)
38. M.C. Birse, Phys. Rev. C **74**, 014003 (2006)
39. E. Epelbaum, H.W. Hammer, U.G. Meißner, Rev. Mod. Phys. **81**, 1773 (2009)

40. R. Machleidt, D.R. Entem, Phys. Rep. **503**, 1 (2011)
41. E. Epelbaum, U.G. Meißner, Few Body Syst. **54**, 2175 (2013)
42. H.W. Hammer, A. Nogga, A. Schwenk, Rev. Mod. Phys. **85**, 197 (2013)
43. B. Long, Y. Mei, Phys. Rev. C **93**, 044003 (2016)
44. S. Scherer, M.R. Schindler, Lect. Notes Phys. **830**, 1 (2012)

Chapter 5
Lattice Methods and Effective Field Theory

Amy Nicholson

5.1 Introduction

Quantitative understanding of nuclear physics at low energies from first principles remains one of the most challenging programs in contemporary theoretical physics research. While physicists have for decades used models combined with powerful numerical techniques to successfully reproduce known nuclear structure data and make new predictions, currently the only tools available for tackling this problem that have direct connections to the underlying theory, Quantum Chromodynamics (QCD), as well as quantifiable systematic errors, are Lattice QCD and Effective Field Theory (EFT). In principle, when combined these techniques may be used to not only quantify any bias introduced when altering QCD in order to make it computationally tractable, but also to better understand the connection between QCD and nuclear physics.

The lattice is a tool for discretizing a field theory in order to reduce the path integral, having an infinite number of degrees of freedom, to a finite-dimensional ordinary integral. After rendering the dimension finite (though extremely large), the integral may then be estimated on a computer using Monte Carlo methods. Errors introduced through discretization and truncation of the region of spacetime sampled are controlled through the spatial and temporal lattice spacings, b_s, b_τ, and the number of spatial and temporal points, L, N_τ. Thus, these errors may be quantified through the lattice spacing dependence of the observables, and often may be removed through extrapolation to the continuum and infinite volume limits.

A. Nicholson (✉)
Department of Physics, University of California, Berkeley, Berkeley, CA 94720, USA
e-mail: anicholson@berkeley.edu

M. Hjorth-Jensen et al. (eds.), *An Advanced Course in Computational Nuclear Physics*, Lecture Notes in Physics 936, DOI 10.1007/978-3-319-53336-0_5

LQCD is a powerful and advanced tool for directly calculating low-energy properties of QCD. However, severe computational issues exist when calculating properties of systems with nucleons. Unfortunately, these problems grow rapidly with the number of nucleons in the system.

The first issue is the large number of degrees of freedom involved when using quark fields to create nucleons. In order to calculate a correlation function for a single nucleon in LQCD using quarks (each of which has twelve internal degrees of freedom given by spin and color), one has to perform all possible Wick contractions of the fields in order to build in fermion antisymmetrization. For example, to create a proton using three valence quark operators requires the calculation of two different terms corresponding to interchanging the two up quark sources. The number of contractions involved for a nuclear correlation function grows with atomic number Z and mass number A as $(A+Z)!(2A-Z)!$. For He_4 this corresponds to $\sim 5\times 10^5$ terms[1]!

The second major problem occurs when performing a stochastic estimate of the path integral. A single quark propagator calculated on a given gauge field configuration may be a part of either a light meson or a heavy nucleon. However, the difference cannot be determined until correlations with the other quark fields present are built in by summing over a sufficiently large number of these field configurations.[2] This leads to large fluctuations from configuration to configuration, and a stochastic signal-to-noise ratio, $\mathscr{R}$, which degrades exponentially with the number of nucleons in the system,

$$\mathscr{R} \sim e^{-A(M-3/2m_\pi)\tau} , \tag{5.1}$$

where M is the nucleon mass and m_π is the pion mass [1]. This is currently the major limiting factor for the size of nuclear which can be probed using LQCD. The best calculations we have from LQCD using multiple nucleons to date are in the two-nucleon sector [2–27], while fewer calculations have been performed for three and four nucleon systems [7, 9, 10, 12, 15–19, 28, 29]; however, even for two nucleon systems unphysically large pion masses must be used in order to reduce the noise problem. We will discuss signal-to-noise problems in more detail in Sect. 5.3.1.

Starting from an EFT using nucleons as the fundamental degrees of freedom greatly reduces the consequences from both of these issues. EFTs also enjoy the same benefit as the lattice over traditional model techniques of having quantifiable systematic errors, this time controlled by the cutoff of the EFT compared to the energy regime studied. For chiral EFTs this scale is generally $\Lambda_\chi \sim m_\rho \sim 700$ MeV. Systematic errors can be reduced by going to higher orders in an expansion of p/Λ_χ, where p is the momentum scale probed, with the remaining error given by the size of the first order which is not included. In a potential model there is no controlled expansion, and it is generally unknown how much the results will be

[1] This is a very naïve estimate; far more sophisticated algorithms exist with power-law scaling.

[2] This interpretation of the signal-to-noise problem has been provided by David B. Kaplan.

affected by leaving out any given operator. In addition, field theories provide a rigorous mathematical framework for calculating physical processes, and can be directly translated into a lattice scheme.

In these lecture notes we will explore the use of lattice methods for calculating properties of many-body systems starting from nuclear EFT, rather than QCD. Our discussion will begin with understanding a very basic nuclear EFT, pionless EFT, at leading order. We will then proceed to discretize this theory and set up a framework for performing Monte Carlo calculations of our lattice theory. We will then discuss how to calculate observables using the lattice theory, and how to understand their associated statistical uncertainties. Next we will discuss quantifying and reducing systematic errors. Then we will begin to add terms to our theory going beyond leading order pionless EFT. Finally, we will discuss remaining issues and highlight some successes of the application of these methods by several different groups.

5.2 Basics of Effective Field Theory and Lattice Effective Field Theory

5.2.1 Pionless Effective Field Theory

To develop an EFT we will first write down all possible operators involving the relevant degrees of freedom within some energy range (determined by the cutoff) that are consistent with the symmetries of the underlying theory. Each operator will be multiplied by an unknown low-energy constant which may be fixed by comparing an observable with experiment or lattice QCD. In order to reduce this, in principle, infinite number of operators to a finite number we must also establish a power-counting rule for neglecting operators that do not contribute within some desired accuracy. This is a notoriously difficult problem for nuclear physics, and is in general observable and renormalization scheme dependent. Here, we will only briefly touch upon two common power-counting schemes, the so-called Weinberg and KSW expansions [30–34]. For reviews of these and other power-counting schemes, see [35–37].

The simplest possible nuclear EFT involves non-relativistic nucleon fields interacting via delta functions. This is known as a pionless EFT, and is only relevant for energy scales up to a cutoff $\Lambda \sim m_\pi$. Below this scale, the finite range of pion exchange cannot be resolved, and all interactions appear to be point-like. In this discussion we will closely follow that of [38]. For the moment, let's just consider a theory of two-component (spin up/down) fermion fields, ψ, with the following Lagrangian,

$$\mathscr{L}_{\text{eff}} = \psi^\dagger \left(i\partial_\tau + \frac{\nabla^2}{2M} \right) \psi + g_0 \left(\psi^\dagger \psi \right)^2 + \frac{g_2}{8} \left[(\psi\psi)^\dagger \left(\psi \overleftrightarrow{\nabla}^2 \psi \right) + \text{h.c.} \right] + \cdots , \tag{5.2}$$

where

$$\overleftrightarrow{\nabla}^2 \equiv \overleftarrow{\nabla}^2 - 2\overleftarrow{\nabla} \cdot \overrightarrow{\nabla} + \overrightarrow{\nabla}^2 , \tag{5.3}$$

M represents the nucleon mass, $g_0, g_2, \ldots$ are unknown, low-energy constants (LECs) which may be fixed by comparing to experimental or LQCD results, and all spin indices are suppressed. Because the effective theory involves dynamical degrees of freedom that are only relevant up to a certain scale, we must define a cut-off, Λ, above which the theory breaks down. In general, the LECs scale as $\Lambda^{-\mathrm{dim}(\mathscr{O})}$, where $\mathrm{dim}(\mathscr{O})$ represents the dimension of the operator associated with the LEC. According to naïve power counting, the g_2 term in Eq. (5.2) should be suppressed relative to the g_0 term, because adding a derivative to an operator increases its dimension. One should be careful in practice, however, because naïve power counting does not always hold, as we will see several times throughout these lectures.

5.2.1.1 Two Particle Scattering Amplitude

In order to set the coefficients $g_0, g_2, \ldots$, we may look to experimental scattering data. In particular, if we wish to set the g_0 coefficient we should consider two-particle s-wave scattering because the operator associated with g_0 contains no derivatives. g_2 and other LECs may be set using p- and higher-wave scattering data. Recall that the S-matrix for non-relativistic scattering takes the following form:

$$S = 1 + \frac{iMp}{2\pi} A , \tag{5.4}$$

where p is the scattering momentum and A is the scattering amplitude. For s-wave scattering the amplitude may be written as,

$$A = \frac{4\pi}{M} \frac{1}{p \cot \delta - ip} , \tag{5.5}$$

where δ is the s-wave scattering phase shift. Given a short-range two-body potential, the scattering phase shift has a well-known expansion for low momenta, called the effective range expansion,

$$p \cot \delta = -\frac{1}{a} + \frac{1}{2} r_0 p^2 + r_1 p^4 + \cdots , \tag{5.6}$$

where a is the scattering length, r_0 is the effective range, and r_1 and higher order terms are referred to as shape parameters. The effective range and shape parameters describe the short-range details of the potential, and are generally of order of the appropriate power of the cutoff in a naturally tuned scenario.

The scattering length may be used to describe the asymptotic behavior of the radial wavefunction. In particular, consider two-particles interacting via an

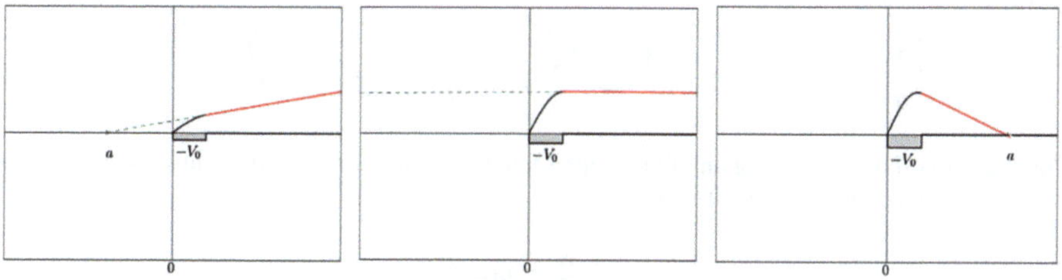

Fig. 5.1 Sketches of two-body radial wavefunctions vs. r corresponding to various scattering lengths. From *left* to *right*: $a < 0$, $a \to \infty$,$a > 0$

attractive square-well potential. If the square-well is sufficiently strongly attractive, the wavefunction turns over and goes to zero at some finite characteristic length. This means the system is bound and the size of the bound state is given by the scattering length, a. On the other hand, if the wavefunction extends over infinite space, then the system is in a scattering state and the scattering length may be determined as the distance from the origin where the asymptote of the wavefunction intersects the horizontal axis (see Fig. 5.1). This implies that the scattering length in the case of a scattering state is negative. If the potential is tuned to give a system which is arbitrarily close to the crossover point from a bound state to a scattering state, corresponding to infinite scattering length, the state is described as being near unitarity, because the unitarity bound on the scattering cross section is saturated at this point. Note that this implies that the scattering length may be any size and is not necessarily associated with the scale set by the cutoff. However, such a scenario requires fine-tuning of the potential. Such fine-tuning is well-known to occur in nuclear physics, with the deuteron and neutron-neutron s-wave scattering being notable examples.

A many-body system composed of two-component fermions with an attractive interaction is known to undergo pairing between the species (higher N-body interactions are prohibited by the Pauli exclusion principle), such as in neutron matter, found in the cores of neutron stars, which is composed of spin up and spin down neutrons. At low temperature, these bosonic pairs condense into a coherent state. If the interaction is only weakly attractive, the system will form a BCS state composed of widely separated Cooper pairs, where the average pair size is much larger than the average interparticle spacing. On the other hand, if the interaction is strongly attractive then the pairs form bosonic bound states which condense into a Bose-Einstein condensate. The crossover between these two states corresponds to the unitary regime, and has been studied extensively in ultracold atom experiments, where the interaction between atoms may be tuned using a Feshbach resonance. In this regime, the average pair size is equal to the interparticle spacing (given by the inverse density), which defines the only scale for the system. Thus, all dimensionful observables one wishes to calculate for this system are determined by the appropriate power of the density times some dimensionless constant. For a review of fermions in the unitary regime, see e.g., [39, 40].

Fig. 5.2 Two-body scattering amplitude represented as a sum of bubble diagrams corresponding to a single contact interaction with coupling g_0

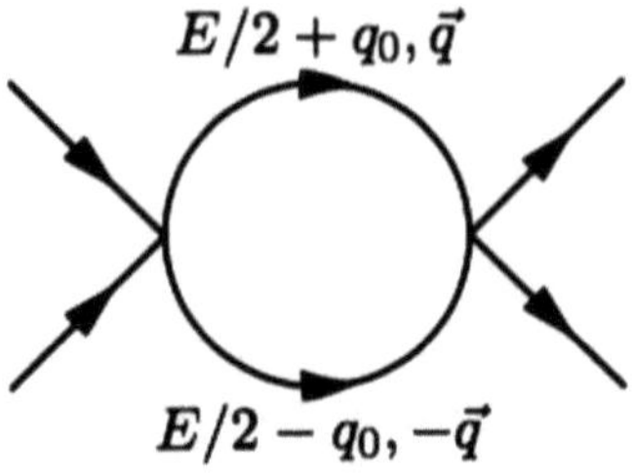

Fig. 5.3 Feynman diagram for a single bubble in Fig. 5.2, giving rise to the loop integral equation (5.7)

5.2.1.2 Two-Body LECs

Returning to our task of setting the couplings using scattering parameters as input, we might consider comparing Eqs. (5.2) and (5.6), to determine the LEC g_0 using the scattering length, g_2 using the effective range, and so forth. To see how this is done in practice we may compute the scattering amplitude A in the effective theory, and match the coefficients to the effective range expansion. Let's begin using only the first interaction term in the effective theory, corresponding to g_0. Diagrammatically, the scattering amplitude may be written as the sum of all possible bubble diagrams (see Fig. 5.2). Because the scattering length may take on any value, as mentioned previously, we cannot assume that the coupling g_0 is small, so we should sum all diagrams non-perturbatively. The first diagram in the sum is given by the tree level result, g_0. If we assume that the system carries energy $E = p^2/M$, then the second diagram may be labeled as in Fig. 5.3, and gives rise to the loop integral,

$$I_0 = i \int \frac{d^4 q}{(2\pi)^4} \frac{1}{\left(E/2 + q_0 - \frac{q^2}{2M} - i\epsilon\right)\left(E/2 - q_0 - \frac{q^2}{2M} + i\epsilon\right)} . \tag{5.7}$$

Performing the integral over q_0 and the solid angle gives

$$I_0 = \frac{1}{2\pi^2} \int^{\pi\Lambda/2} dq \frac{q^2}{\left(E - \frac{q^2}{M}\right)} \tag{5.8}$$

$$= \frac{M}{2\pi^2}\left[\frac{\pi\Lambda}{2} - \sqrt{ME}\tanh^{-1}\left(\frac{\Lambda}{\sqrt{ME}}\right)\right] , \tag{5.9}$$

where I have introduced a hard momentum cutoff, Λ. Removing the cutoff by taking it to infinity results in

$$I_0 \underset{\Lambda\to\infty}{\longrightarrow} \frac{M}{4\pi}\left[\Lambda + ip\right] . \tag{5.10}$$

Because the interaction is separable, the nth bubble diagram is given by n products of this loop function. Thus, the scattering amplitude is factorizable, and may be written

$$A = g_0\left[1 + \sum_n (g_0 I_0)^n\right] \tag{5.11}$$

$$= \frac{g_0}{1 - g_0 I_0} . \tag{5.12}$$

We may now compare Eqs. (5.5), (5.6) and (5.11) to relate the coupling g_0 to the scattering phase shift. This is easiest to do by equating the inverse scattering amplitudes,

$$\frac{1}{A} = \frac{1}{g_0} - \frac{M}{4\pi}\Lambda - \frac{iMp}{4\pi} = -\frac{M}{4\pi a} - \frac{iMp}{4\pi} , \tag{5.13}$$

where I have used Eq. (5.6) cut off at leading order. We now have the relation

$$g_0 = \frac{4\pi}{M}\frac{1}{\Lambda - 1/a} \tag{5.14}$$

between the coupling and the physical scattering length.

Note that the coupling runs with the scale Λ; the particular dependence is determined by the regularization and renormalization scheme chosen. In order to understand the running of the coupling we may examine the beta function. To do so we first define a dimensionless coupling,

$$\hat{g}_0 \equiv -\frac{M\Lambda}{4\pi} g_0 , \tag{5.15}$$

then calculate

$$\beta(\hat{g}_0) \equiv \Lambda \frac{\partial \hat{g}_0}{\partial \Lambda} = -\frac{a\Lambda}{(a\Lambda - 1)^2} = -\hat{g}_0 (\hat{g}_0 - 1) . \tag{5.16}$$

This function is a simple quadratic that is plotted in Fig. 5.4. The beta function has two zeroes, $\hat{g}_0 = 0, 1$, corresponding to fixed points of the theory. At a fixed point, the coupling no longer runs with the scale Λ, and the theory is said to be scale-invariant (or conformal, given some additional conditions). This means that there is no intrinsic scale associated with the theory. The fixed point at $\hat{g}_0 = 0$ is

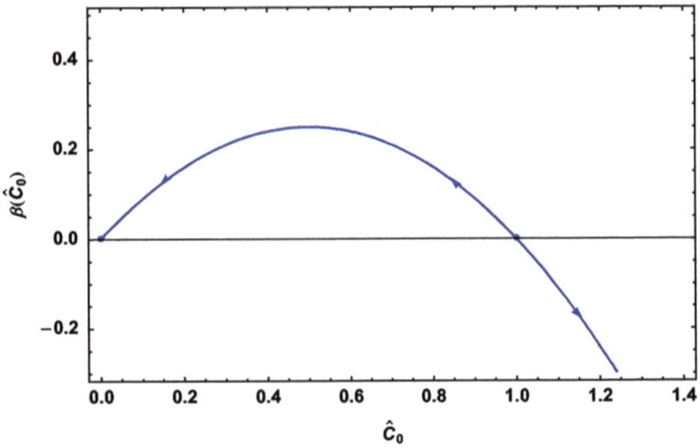

Fig. 5.4 Beta function [Eq. (5.16)] for the two-body contact interaction. *Arrows* represent the direction of flow toward the IR

a trivial fixed point, and corresponds to a non-interacting, free field theory (zero scattering length). The other, non-trivial fixed point at $\hat{g}_0 = 1$ corresponds to a strongly interacting theory with infinite scattering length; this is the unitary regime mentioned previously. Here, not only does the scattering length go to infinity, as does the size of the radial wavefunction, but the energy of the bound state (as approached from $\hat{g}_0 > 1$) goes to zero and all relevant scales have vanished. Note that this is an unstable fixed point; the potential must be finely tuned to this point or else the theory flows away from unitarity as $\Lambda \to 0$ (IR limit).

Generally perturbation theory is an expansion around free field theory, corresponding to a weak coupling expansion. This is the approach used as part of the Weinberg power counting scheme for nuclear EFT [30, 31]. However, in some scattering channels of interest for nuclear theory the scattering length is indeed anomalously large, such as the ${}^1S_0$ and ${}^3S_1$ nucleon-nucleon scattering channels, where

$$a_{{}^1S_0} \sim -24 \text{ fm} , \tag{5.17}$$

$$a_{{}^3S_1} \sim 5 \text{ fm} . \tag{5.18}$$

Such large scattering lengths suggest that an expansion around the strongly coupled fixed-point of unitarity may be a better starting point and lead to better convergence. This approach was taken by Kaplan, Savage, and Wise and led to the KSW power-counting scheme [32–34]. Unfortunately, nuclear physics consists of many scales of different sizes and a consistent power-counting framework with good convergence for all observables has yet to be developed; in general the convergence of a given scheme depends on the scattering channels involved.

Because nuclear physics is not weakly coupled in all channels, non-perturbative methods, such as lattice formulations, will be favorable for studying few- and many-body systems, where two-body pairs may interact through any combination of channels simultaneously. Due to the scale-invariant nature of the unitary regime, it provides a far simpler testbed for numerical calculations of strongly-interacting theories, so we will often use it as our starting point for understanding lattice EFT methods.

5.2.2 Lattice Effective Field Theory

Our starting point for building a lattice EFT will be the path integral formulation of quantum field theory in Euclidean spacetime. The use of Euclidean time allows the exponent of the path integral to be real (in certain cases), a property which will be essential to our later use of stochastic methods for its evaluation. Given a general theory for particles $\psi, \psi^\dagger$ obeying a Lagrangian density

$$\mathscr{L}(\psi^\dagger, \psi) = \psi^\dagger \left(\partial_\tau - \mu\right) \psi + \mathscr{H}\left[\psi^\dagger, \psi\right] , \tag{5.19}$$

where τ is the Euclidean time, μ the chemical potential, and $\mathscr{H}$ is the Hamiltonian density, the Euclidean path integral is given by

$$Z = \int \mathscr{D}\psi^\dagger \mathscr{D}\psi e^{-\int d\tau d^3x\left[\mathscr{L}(\psi^\dagger,\psi)\right]} . \tag{5.20}$$

If the integral over Euclidean time is compact, then the finite time extent β acts as an inverse temperature, and we may draw an analogy with the partition function in statistical mechanics, $Z = tr\left[e^{-H\beta}\right]$. This analogy is often useful when discussing lattice formulations of the path integral. In this work we will generally consider $\mu = 0$ and create non-zero particle density by introducing sources and sinks for particles and calculating correlation functions.

We discretize this theory on a square lattice consisting of $L^3 \times N_\tau$ points, where L is the number of points in all spatial directions, and N_τ is the number of temporal points. We will focus on zero temperature physics, corresponding to large N_τ.[3] We must also define the physical distance between points, the lattice spacings b_s, b_τ, where $b_\tau = b_s^2/M$ by dimensional analysis for non-relativistic theories. The fields are now labeled by discrete points, $\psi(\mathbf{x}, \tau) \to \psi_{\mathbf{n},\tau}$, and continuous integrals are replaced by discrete sums, $\int d^3x \to \sum_{\mathbf{n},\tau}^{L,N_\tau}$.

[3]The explicit condition on N_τ required for extracting zero temperature observables will be discussed in Sect. 5.3.

5.2.2.1 Free Field Theory

To discretize a free field theory, we must discuss discretization of derivatives. The simplest operator which behaves as a single derivative in the continuum limit is a finite difference operator,

$$\partial_{\hat{k}}^{(L)} f_j = \frac{1}{b_s}\left[f_{j+\hat{k}} - f_j\right], \tag{5.21}$$

where $\hat{k}$ is a unit vector in the k-direction. The discretized second derivative operator must involve two hops, and should be a symmetric operator to behave like the Laplacian. A simple possibility is

$$\nabla_L^2 f_j = \sum_k \frac{1}{b_s^2}\left[f_{j+\hat{k}} + f_{j-\hat{k}} - 2f_j\right]. \tag{5.22}$$

We can check the continuum limit by inspecting the corresponding kinetic term in the action,

$$S_{\mathrm{KE}} \propto \sum_j \psi_j^\dagger \nabla_L^2 \psi_j . \tag{5.23}$$

The fields may be expanded in a plane wave basis,

$$\psi_j = \sum_{k=-L/2}^{L/2} \psi_k e^{-\frac{2\pi i}{L} j\cdot k}, \tag{5.24}$$

for spatial indices, j, leading to

$$\sum_j \psi_j^\dagger \nabla_L^2 \psi_j = \frac{1}{b_s^2}\sum_j \sum_{k'} \sum_k \psi_{k'}^\dagger \psi_k \left[e^{\frac{2\pi i}{L} j\cdot k'} e^{\frac{-2\pi i}{L} j\cdot k}\right]\left[e^{\frac{-2\pi i}{L} k} + e^{\frac{2\pi i}{L} k} - 2\right]. \tag{5.25}$$

After performing the sum over j the first piece in brackets gives $\delta_{kk'}$, while the second is proportional to $\sin^2(k\pi/L)$, resulting in,

$$\sum_j \psi_j^\dagger \nabla_L^2 \psi_j = -\frac{4}{b_s^2}\sum_k \psi_k^\dagger \psi_k \sin^2\left(\frac{k\pi}{L}\right). \tag{5.26}$$

Finally, expanding the sine function for small k/L gives,

$$\sum_j \psi_j^\dagger \nabla_L^2 \psi_j = \sum_k \psi_k^\dagger \psi_k \underbrace{\left[-\left(\frac{2\pi k}{b_s L}\right)^2 + \frac{b_s^2}{12}\left(\frac{2\pi k}{b_s L}\right)^4 + \cdots \right]}_{-p^2 + \frac{b_s^2}{12}p^4 + \cdots \underset{b_s \to 0}{\longrightarrow} -p^2}, \tag{5.27}$$

where I've used the finite volume momentum $p = \frac{2\pi k}{b_s L}$ to rewrite the expression in square brackets. Thus, we have the correct continuum limit for the kinetic operator. Note that for larger momenta, approaching the continuum limit requires smaller b_s. However, this is only one possibility for a kinetic term. We can always add higher dimension operators (terms with powers of b_s in front of them), in order to cancel leading order terms in the expansion equation (5.27). This is a form of what's called improvement of the action, and will be discussed in more detail in Sect. 5.4.

Adding a temporal derivative term,

$$\partial_\tau^{(L)} \psi_{\mathbf{x},\tau} = \frac{1}{b_\tau} \left[\psi_{\mathbf{n},\tau} - \psi_{\mathbf{n},\tau-1} \right] , \tag{5.28}$$

we can now write down a simple action for a non-relativistic free-field theory,

$$S_{\text{free}} = \sum_{\tau,\tau'} \frac{1}{b_\tau} \psi_{\tau'}^\dagger \left[K_0 \right]_{\tau,\tau'} \psi_\tau , \tag{5.29}$$

where I've defined a matrix K_0 whose entries are $L^3 \times L^3$ blocks,

$$K_0 \equiv \begin{pmatrix} D & -1 & 0 & 0 & \dots \\ 0 & D & -1 & 0 & \dots \\ 0 & 0 & D & -1 & \dots \\ . & . & . & . & . \\ . & . & . & . & . \\ 1 & . & . & . & . \end{pmatrix} \tag{5.30}$$

where $D \equiv 1 - \frac{b_s^2 \nabla_L^2}{2}$ contains the spatial Laplacian, and therefore connects fields on the same time slice (corresponding to diagonal entries of the matrix K_0), while the temporal derivative contributes the off-diagonal pieces. Note that the choice of "1" in the lower left corner corresponds to anti-periodic boundary conditions, appropriate for fermionic fields. For zero temperature calculations the temporal boundary conditions are irrelevant, and it will often be useful to choose different temporal boundary conditions for computational or theoretical ease.

5.2.2.2 Interactions

Now let's discuss adding interactions to the theory. We'll focus on the first term in a nuclear EFT expansion, the four-fermion interaction:

$$\mathscr{L}_{\text{int}} = \sum_{n} g_0 \psi_{n,\uparrow} \psi_{n,\uparrow} \psi_{n,\downarrow} \psi_{n,\downarrow} \,, \tag{5.31}$$

where $(\uparrow, \downarrow)$ now explicitly label the particles' spins (or alternatively, flavors). Because anti-commuting fields cannot easily be accommodated on a computer, they must be integrated out analytically. The only Grassmann integral we know how to perform analytically is a Gaussian, so the action must be bilinear in the fields. One trick for doing this is called a Hubbard-Stratonovich (HS) transformation, in which auxiliary fields are introduced to mediate the interaction. The key is to use the identity,

$$e^{b_\tau g_0 \psi_\uparrow^\dagger \psi_\uparrow \psi_\downarrow^\dagger \psi_\downarrow} = \frac{1}{\sqrt{2\pi}} \int_{-\infty}^{\infty} d\phi^{-\phi^2/2 - \phi\sqrt{b_\tau g_0}\left(\psi_\uparrow^\dagger \psi_\uparrow + \psi_\downarrow^\dagger \psi_\downarrow\right)} \,, \tag{5.32}$$

where I have dropped the spacetime indices for brevity. This identity may be verified by completing the square in the exponent on the right hand side and performing the Gaussian integral over the auxiliary field ϕ. This form of HS transformation has the auxiliary field acting in what is called the density channel $\left(\psi_\uparrow^\dagger \psi_\uparrow + \psi_\downarrow^\dagger \psi_\downarrow\right)$. It is also possible to choose the so-called BCS channel, $\left(\psi_\uparrow^\dagger \psi_\downarrow^\dagger + \psi_\uparrow \psi_\downarrow\right)$, the usual formulation used in BCS models, however this causes a so-called sign problem when performing Monte Carlo sampling, as will be discussed in detail in Sect. 5.3.1.1. Transformations involving non-Gaussian auxiliary fields may also be used, such as

$$\begin{aligned} Z_2 \text{ field:} \quad & \frac{1}{2} \sum_{\phi = \pm 1} e^{-\phi\sqrt{b_\tau g_0}\left(\psi_\uparrow^\dagger \psi_\uparrow + \psi_\downarrow^\dagger \psi_\downarrow\right)} \\ \text{compact continuous:} \quad & \frac{1}{2\pi} \int_{-\pi}^{\pi} e^{-\sin\phi\sqrt{b_\tau g_0}\left(\psi_\uparrow^\dagger \psi_\uparrow + \psi_\downarrow^\dagger \psi_\downarrow\right)} \,. \end{aligned} \tag{5.33}$$

These formulations may have different pros and cons in terms of computational and theoretical ease for a given problem, and should be chosen accordingly. For example, the Z_2 interaction is conceptually and computationally the simplest interaction, however, it also induces explicit 4- and higher-body interactions in systems involving more than two-components which may not be desired.

5.2.2.3 Importance Sampling

The action may now be written with both kinetic and interaction terms,

$$S = \frac{1}{b_\tau} \sum_{\tau,\tau'} \psi^\dagger_{\tau'} \left[K(\phi)\right]_{\tau'\tau} \psi_\tau \,, \tag{5.34}$$

where the matrix K includes blocks which depend on the auxiliary field ϕ, and also contains non-trivial spin structure that has been suppressed. The partition function can be written

$$Z = \int \mathscr{D}\phi \mathscr{D}\psi^\dagger \mathscr{D}\psi \rho[\phi] e^{-S[\phi,\psi^\dagger\psi]} \,, \tag{5.35}$$

where the integration measure for the ϕ field, $\rho[\phi]$, depends on the formulation chosen,

$$\rho[\phi] = \begin{cases} \prod_n e^{-\phi_n^2/2} & \text{Gaussian} \\ \prod_n \frac{1}{2}\left(\delta_{\phi_{n,1}} + \delta_{\phi_{n,-1}}\right) & Z_2 \\ \prod_n \left(\theta(-\pi + \phi_n)\theta(\pi - \phi_n)\right) & \text{compact continuous} \end{cases} . \tag{5.36}$$

With the action in the bilinear form of Eq. (5.34), the ψ fields can be integrated out analytically, resulting in

$$Z_\phi = \int \mathscr{D}\phi P[\phi] \qquad P[\phi] \equiv \rho[\phi] \det K[\phi] \,. \tag{5.37}$$

Observables take the form

$$\langle \mathscr{O} \rangle = \frac{1}{Z} \int \mathscr{D}\phi P[\phi] \mathscr{O}[\phi] \,. \tag{5.38}$$

Through the use of discretization and a finite volume, the path integral has been converted into a standard integral with finite dimension. However, the dimension is still much too large to imagine calculating it on any conceivable computer, so we must resort to Monte Carlo methods for approximation. The basic idea is to generate a finite set of ϕ field configurations of size N_{cfg} according to the probability measure $P[\phi]$, calculate the observable on each of these configurations, then take the mean as an approximation of the full integral,

$$\langle \mathscr{O} \rangle \approx \frac{1}{N_{\text{cfg}}} \sum_n^{N_{\text{cfg}}} \mathscr{O}(\phi_n) \,. \tag{5.39}$$

Assuming the central limit theorem holds, for $N_{\rm cfg}$ large enough (a non-trivial condition, as will be discussed in Sect. 5.3.2), the distribution of the mean approaches a Gaussian, and the error on the mean falls off with the square root of the sample size.

There are several algorithms on the market for generating field configurations according to a given probability distribution, and I will only briefly mention a few. Lattice calculations are particularly tricky due to the presence of the determinant in Eq. (5.37), which is a highly non-local object and is very costly to compute. One possible algorithm to deal with this is called determinantal Monte Carlo, which implements local changes in ϕ, followed by a simple Metropolis accept/reject step. This process can be rather inefficient due to the local updates. An alternative possibility is Hybrid Monte Carlo, commonly used for lattice QCD calculations, in which global updates of the field are produced using molecular dynamics as a guiding principle. Note that the field ϕ must be continuous in order to use this algorithm due to the use of classical differential equations when generating changes in the field. Also common in lattice QCD calculations is the use of pseudofermion fields as a means for estimating the fermion determinant. Here the determinant is rewritten in terms of a Gaussian integral over bosonic fields, χ,

$$\det K[\phi] \propto \int \mathscr{D}\chi^\dagger \mathscr{D}\chi e^{-\chi^\dagger K^{-1}[\phi]\chi} \; . \tag{5.40}$$

This integral is then evaluated stochastically. These are just a sample of the available algorithms. For more details on these and others in the context of non-relativistic lattice field theory, see [41].

5.2.2.4 Example Formulation

Now that we have developed a general framework for lattice EFT, let's be explicit and make a few choices in order to further our understanding and make calculations simpler. The first choice I'm going to make is to use a $Z_2\phi$ field, so that $\rho[\phi]$ is trivial. The next simplification I'm going to make is to allow the ϕ fields to live only on temporal links,

$$\mathscr{L}_{\rm int} = \sum_{\mathbf{x}} \sqrt{b_\tau g_0}\phi_{\mathbf{x},\tau}\psi^\dagger_{\mathbf{x},\tau}\psi_{\mathbf{x},\tau-1} \; . \tag{5.41}$$

Note that we are free to make this choice, so long as the proper four-fermion interaction is regained in the continuum limit. This choice renders the interaction separable, as it was in our continuum effective theory. This means we may analytically sum two-body bubble chain diagrams as we did previously in order to set the coupling g_0 using some physical observable (see Fig. 5.5).

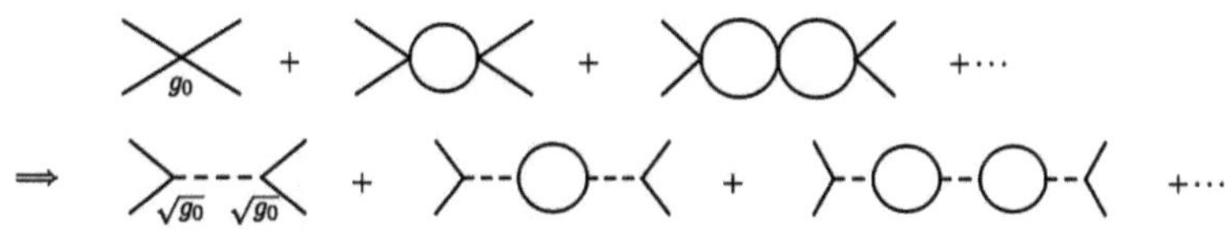

Fig. 5.5 Two-body scattering amplitude of Fig. 5.2, where the contact interaction has been replaced in the second line by exchange of a dimer auxiliary field via a Hubbard-Stratonovich transformation

With this choice we can now write the K-matrix explicitly as

$$K[\phi, N_\tau] \equiv \begin{pmatrix} D & -X(\phi_{N_\tau-1}) & 0 & 0 \; . \; . & . \\ 0 & D & -X(\phi_{N_\tau-2}) & 0 \; . \; . & . \\ . & . & . & . & \\ . & . & . & . & \\ . & . & . & D & X(\phi_0) \\ X(\phi_{N_\tau}) & . & . & 0 & D \end{pmatrix} , \tag{5.42}$$

where $X(\phi_\tau) \equiv 1 - \sqrt{g_0}\phi_\tau$. Now the ϕ-dependence exists only on the upper diagonal, as well as the lower left due to the boundary condition. This block will be eliminated through our final choice: open boundary conditions in time for the ψ fields, $X(\phi_{N_\tau}) = 0$. As mentioned previously, we are free to choose the temporal boundary conditions as we please, so long as we only consider zero temperature (and zero chemical potential) observables.

With this set of choices the matrix K consists purely of diagonal elements, D, and upper diagonal elements, $X(\phi_\tau)$. One property of such a matrix is that the determinant, which is part of the probability distribution, is simply the product of diagonal elements, $\det K = \prod_\tau D$. Note that D is completely independent of the field ϕ. This means that the determinant in this formulation has no impact on the probability distribution $P[\phi]$, and therefore never needs to be explicitly computed, greatly reducing the computational burden. Thus in all of our calculations, performing the path integral over ϕ simply amounts to summing over $\phi = \pm 1$ at each lattice site.

Finally, this form of K also makes the calculation of propagators very simple. The propagator from time 0 to τ may be written,

$$\begin{aligned} K^{-1}(\tau, 0) &= D^{-1}X(\phi_{\tau-1})D^{-1}X(\phi_{\tau-2})D^{-1} \cdots X(\phi_0)D^{-1} \\ &= D^{-1}X(\phi_{\tau-1})K^{-1}(\tau - 1, 0) , \end{aligned} \tag{5.43}$$

where $K^{-1}(0,0) = D^{-1}$, and all entries are $V \times V$ matrices, with $V = L^3$, which may be projected onto the desired state. This form suggests a simple iterative approach to calculating propagators: start with a source (a spatial vector projecting onto some desired quantum numbers and interpolating wavefunction), hit it with the kinetic

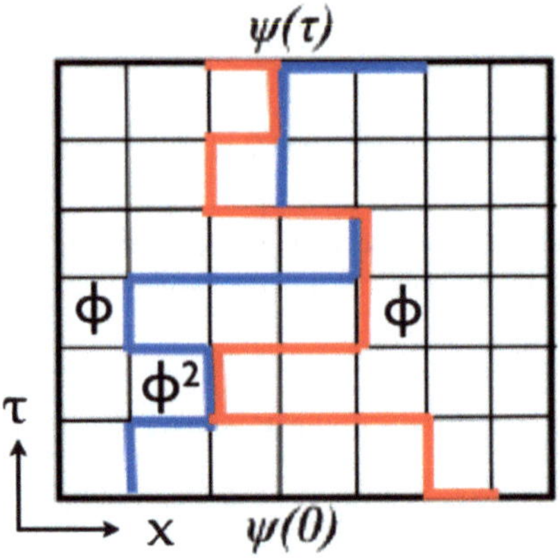

Fig. 5.6 Schematic of a lattice calculation for a two-particle correlation function. The two particles (*red* and *blue lines*) propagate through the lattice between source $\psi(0)$ and sink $\psi(\tau)$, seeing particular values of the auxiliary field, ϕ, on each time link. If two particles occupy the same temporal link, then upon summation over all possible values of ϕ at each link, a non-zero contribution is generated by the interaction term because $\langle\phi^2\rangle \neq 0$

energy operator corresponding to free propagation on the time slice, then hit it with the ϕ field operator on the next time link, then another free kinetic energy operator, and so on, finally projecting onto a chosen sink vector.

As will be discussed further in Sect. 5.4, it is often preferable to calculate the kinetic energy operator in momentum space, while the auxiliary field in $X(\phi)$ must be generated in position space. Thus, Fast Fourier Transforms (FFTs) may be used between each operation to quickly translate between the bases. Example code for generating source vectors, kinetic operators, and interaction operators will be provided in later sections.

A cartoon of this process on the lattice is shown in Fig. 5.6. The choice of Z_2 auxiliary fields also simplifies the understanding of how four-fermion interactions are generated. On every time link, imagine performing the sum over $\phi = \pm 1$. If there is only a single fermion propagator on a given link this gives zero contribution because the term is proportional to $\sum_{\phi=\pm 1} \sqrt{g_0}\phi = 0$. However, on time slices where two propagators overlap, we have instead $\sum_{\phi=\pm 1} g_0\phi^2 = 2g_0$. In sum, anywhere two fermions exist at the same spacetime point a factor of g_0 contributes, corresponding to an interaction.

5.2.2.5 Tuning the Two-Body Interaction

There are several ways to set the two-body coupling. Here we will explore two methods, using different two-body observables. The first involves calculating the two-particle scattering amplitude, and tuning the coupling to reproduce known scattering parameters, to make a connection with our previous calculation for the effective theory. The second method uses instead the energy spectrum of a two-particle system in a box. This powerful method will be useful later when we begin to improve the theory in order to reduce systematic errors.

We have calculated the scattering amplitude previously for our effective theory using a momentum cutoff. For the first method for tuning the coupling, we will calculate it again using our lattice theory with the lattice cutoff as a regulator. First we need the single particle free propagator:

$$G_0(\tau, \mathbf{p}) = \langle \mathbf{p}, \tau | \left(D^{-1}\right)^{\tau+1} | \mathbf{p}, 0 \rangle = \left(1 + \frac{\Delta(p)}{M}\right)^{-(\tau+1)} , \tag{5.44}$$

$$\begin{aligned} \Delta(p) &\equiv -\frac{1}{2} \langle \mathbf{p} | \nabla_L^2 | \mathbf{p} \rangle \\ &= \sum_i \sin^2 \frac{p_i}{2} , \end{aligned}$$

where I've set $b_s = 1$ (we will use this convention from now on until we begin to discuss systematic errors), and have used the previously defined discretized Laplacian operator. I've written the propagator in a mixed $\mathbf{p}, \tau$ representation, as this is often useful in lattice calculations for calculating correlation functions in time when the kinetic operator, D, is diagonal in momentum space.

The diagrammatic two-particle scattering amplitude is shown on the bottom line in Fig. 5.5. Because we have chosen the interaction to be separable, the amplitude can be factorized:

$$A = g_0 \left[1 + \sum_n (g_0 \hat{L})^n \right] = \frac{g_0}{1 - g_0 \hat{L}} , \tag{5.45}$$

where the one loop integral, $\hat{L}$, will be defined below. As before, in order to set a single coupling we need one observable, so we use the effective range expansion for the scattering phase shift to leading order,

$$A = \frac{4\pi}{M} \frac{1}{p \cot \delta - ip} \approx -\frac{4\pi a}{M} . \tag{5.46}$$

Relating Eqs. (5.45), (5.46), we find

$$\frac{1}{g_0} = -\frac{M}{4\pi a} + \hat{L} . \tag{5.47}$$

We will now evaluate the loop integral using the free single particle propagators, Eq. (5.44),

$$
\begin{aligned}
\hat{L} &= \frac{1}{V}\sum_{\mathbf{p}}\sum_{\tau=0}^{\infty}[G_0(\tau,\mathbf{p})]^2 \\
&= \frac{1}{V}\sum_{\mathbf{p}}\sum_{\tau=0}^{\infty}\frac{1}{\left(1+\frac{\Delta(p)}{M}\right)^{2\tau+2}} \\
&= \frac{1}{V}\sum_{\mathbf{p}}\frac{1}{\left(1+\frac{\Delta(p)}{M}\right)^2}\left[1+\sum_{\tau=0}^{\infty}\frac{1}{\left[\left(1+\frac{\Delta(p)}{M}\right)^2\right]^\tau}\right] \\
&= \frac{1}{V}\sum_{\mathbf{p}}\frac{M}{2}\frac{1}{\Delta(p)\left(1+\frac{\Delta(p)}{2M}\right)} .
\end{aligned}
\tag{5.48}
$$

This final sum may be calculated numerically for a given M and L (governing the values of momenta included in the sum), as well as for different possible definitions of the derivative operators contained in Δ, giving the desired coupling, g_0, via Eq. (5.47).

The second method for setting the coupling utilizes the calculation of the ground state energy of two particles. We start with the two-particle correlation function,

$$
C_2(\tau) = \frac{1}{Z}\int \mathcal{D}\phi\mathcal{D}\psi^\dagger\mathcal{D}\psi e^{-S[\psi^\dagger,\psi,\phi]}\Psi^\dagger_{\text{src},2}\Psi_{\text{snk},2} ,
\tag{5.49}
$$

where $\Psi_{\text{src},2(\text{snk},2)}$ is a source (sink) wavefunction involving one spin up and one spin down particle. Integrating out the fermion fields gives,

$$
\begin{aligned}
C_2(\tau) &= \frac{1}{Z_\phi}\int \mathcal{D}\phi P[\phi]\langle\Psi_{\text{snk},2}|K^{-1}(\tau,0)\otimes K^{-1}(\tau,0)|\Psi_{\text{src},2}\rangle \\
&= \frac{1}{4\tau}\sum_{\phi=\pm1}\langle\Psi_{\text{snk},2}|D^{-1} \\
&\otimes D^{-1}X(\phi_\tau)\otimes X(\phi_\tau)D^{-1}\otimes D^{-1}X(\phi_{\tau-1})\otimes X(\phi_{\tau-1})\cdots|\Psi_{\text{src},2}\rangle .
\end{aligned}
\tag{5.50}
$$

I will now write out the components of the matrices explicitly:

$$C_2(\tau) = \frac{1}{4\tau} \sum_{x_1 x_2 x_1' x_2' \cdots y_1 y_2} \sum_{\phi_{x_1} \phi_{x_1'} \cdots = \pm 1} \langle \Psi_{\text{snk},2} | x_1 x_2 \rangle D^{-1}_{x_1 x_1'} D^{-1}_{x_2 x_2'} (\delta_{x_1 x_1'} + \sqrt{g_0} \phi_{x_1} \delta_{x_1 x_1'})$$
$$(\delta_{x_2 x_2'} + \sqrt{g_0} \phi_{x_2} \delta_{x_2 x_2'})$$
$$\times D^{-1}_{x_1' x_1''} D^{-1}_{x_2 x_2''} \cdots \langle y_1 y_2 | \Psi_{\text{src},2} \rangle \,. \tag{5.51}$$

The first (last) piece in angle brackets represents the position space wavefunction created by the sink (source). All ϕ fields in Eq. (5.51) are uncorrelated, so we can perform the sum for each time slice independently. One such sum is given by,

$$\frac{1}{4} \sum_{x_1 x_1' x_2 x_2'} \sum_{\phi_{x_1} \phi_{x_2}} \delta_{x_1 x_1'} \delta_{x_2 x_2'} (1 + \sqrt{g_0} \phi_{x_1} + \sqrt{g_0} \phi_{x_2} + g_0 \phi_{x_1} \phi_{x_2})$$
$$= \sum_{x_1 x_2} (1 + g_0 \delta_{x_1 x_2}) \,, \tag{5.52}$$

where the cross terms vanish upon performing the sum. If we make the following definitions,

$$\langle x_1 x_1' | \mathscr{D}^{-1} | x_2 x_2' \rangle \equiv D^{-1}_{x_1 x_1'} D^{-1}_{x_2 x_2'} \,, \qquad \langle x_1 x_2 | \mathscr{V} | x_1' x_2' \rangle \equiv g_0 \delta_{x_1 x_1'} \delta_{x_2 x_2'} \delta_{x_1 x_2} \,, \tag{5.53}$$

then we can write the two-particle correlation function as,

$$C_2(\tau) = \langle \Psi_{\text{snk},2} | \mathscr{D}^{-1} (1 + \mathscr{V}) \mathscr{D}^{-1} (1 + \mathscr{V}) \cdots \mathscr{D}^{-1} (1 + \mathscr{V}) \mathscr{D}^{-1} | \Psi_{\text{src}} \rangle$$
$$= \langle \Psi_{\text{snk}} | \mathscr{D}^{-1/2} \mathscr{T} \mathscr{D}^{-1/2} | \Psi_{\text{src},2} \rangle \,, \tag{5.54}$$

where I have made the definition

$$\mathscr{T} \equiv \mathscr{D}^{-1/2} (1 + \mathscr{V}) \mathscr{D}^{-1/2} \,. \tag{5.55}$$

Recall from statistical mechanics that correlation functions may be written as τ insertions of the transfer matrix, e^{-H}, acting between two states,

$$C(\tau) = \langle \Psi_{\text{snk},2} | e^{-H\tau} | \Psi_{\text{src},2} \rangle$$
$$= \langle \Psi_{\text{snk},2} | \left[e^{-H} \right]^{\tau} | \Psi_{\text{src},2} \rangle \,. \tag{5.56}$$

Then we may identify $\mathscr{T}$ in Eq. (5.55) as the transfer matrix of the theory, $\mathscr{T} = e^{-H}$. This in turn implies that the logarithm of the eigenvalues of $\mathscr{T}$ give the energies of the two-particle system.

We will now evaluate the transfer matrix in momentum space:

$$
\begin{aligned}
\langle pq|\mathscr{T}|p'q'\rangle &= \sum_{kk'll'} \langle pq|\mathscr{D}^{-1/2}|kl\rangle\langle kl|1+\mathscr{V}|k'l'\rangle\langle k'l'|\mathscr{D}^{-1/2}|p'q'\rangle \\
&= \sum_{kk'll'} \delta_{k'p'}\delta_{l'q'}\delta_{pk}\delta_{ql}\left(\delta_{kk'}\delta_{ll'} + \delta_{k+l,k'+l'}\frac{g_0}{V}\right) \\
&\quad \times \left[\frac{1}{\left(1+\frac{\Delta(p)}{M}\right)\left(1+\frac{\Delta(q)}{M}\right)\left(1+\frac{\Delta(p')}{M}\right)\left(1+\frac{\Delta(q')}{M}\right)}\right]^{1/2} \\
&= \frac{\delta_{pp'}\delta_{qq'} + \frac{g_0}{V}\delta_{p+q,p'+q'}}{\sqrt{\xi(p)\xi(q)\xi(q')\xi(p')}} ,
\end{aligned}
\tag{5.57}
$$

where I have made the definition,

$$
\xi(p) \equiv 1 + \frac{\Delta(q)}{M} . \tag{5.58}
$$

The eigenvalues of the matrix $\mathscr{T}$ may be evaluated numerically to reproduce the entire two-particle spectrum. However, for the moment we only need to set a single coupling, g_0, so one eigenvalue will be sufficient. The largest eigenvalue of the transfer matrix, corresponding to the ground state, may be found using a simple variational analysis.[4] Choosing a simple trial state wavefunction,

$$
\langle pq|\Psi\rangle = \frac{\psi(p)}{\sqrt{V}}\delta_{p,-q} , \tag{5.59}
$$

subject to the normalization constraint,

$$
\frac{1}{V}\sum_p |\psi(p)|^2 = 1 , \tag{5.60}
$$

we now need to maximize the following functional:

$$
\langle\Psi|\mathscr{T}|\Psi\rangle = \left[\frac{1}{V}\sum_p \frac{|\psi(p)|^2}{\xi^2(p)} + \frac{g_0}{V^2}\left|\sum_p \frac{\psi(p)}{\xi(p)}\right|^2 + \lambda\left(1 - \frac{1}{V}\sum_p |\psi(p)|^2\right)\right] , \tag{5.61}
$$

[4] Many thanks to Michael Endres for the following variational argument.

where λ is a Lagrange multiplier enforcing the normalization constraint, and I have used the fact that $\xi(p)$ is symmetric in p to simplify the expression. Taking a functional derivative with respect to $\psi^\dagger(q)$ on both sides gives

$$-\lambda\psi(q) + \frac{\psi(q)}{\xi^2(q)} + \frac{g_0}{V}\sum_p \frac{\psi(p)}{\xi(p)\xi(q)} = 0\,, \tag{5.62}$$

where I have set the expression equal to zero in order to locate the extrema. Rearranging this equation, then taking a sum over q on both sides gives

$$\sum_q \frac{\psi(q)}{\xi(q)} = \sum_q \frac{g_0}{V}\frac{1}{\lambda\xi^2(q)-1}\sum_p \frac{\psi(p)}{\xi(p)}\,, \tag{5.63}$$

finally resulting in

$$1 = \frac{g_0}{V}\sum_q \frac{1}{\lambda\xi^2(q)-1}\,. \tag{5.64}$$

We now have an equation involving two unknowns, λ and g_0. We need a second equation in order to determine these two parameters. We may use the constraint equation to solve for $\psi(p)$, giving

$$\psi(p) = \mathscr{N}\frac{\xi(p)}{\lambda\xi^2(p)-1}\,, \qquad \frac{1}{\mathscr{N}^2} = \frac{1}{V}\sum_p \frac{\xi^2(p)}{[\lambda\xi^2(p)-1]^2}\,. \tag{5.65}$$

Plugging this back in to our transfer matrix we find,

$$\langle\Psi|\mathscr{T}|\Psi\rangle = \lambda\,. \tag{5.66}$$

This tells us that λ is equivalent to the eigenvalue we sought, $E_0 = -\ln\lambda(g_0)$. As a check, we can compare Eqs. (5.47), (5.64) in the unitary limit: $a \to \infty, \lambda \to 1$, giving

$$\frac{1}{g_0} = \frac{M}{2V}\sum_p \frac{1}{\Delta\left(1+\frac{\Delta}{2M}\right)} \tag{5.67}$$

for both equations.

In Sect. 5.2.2.5 we will discuss a simple formalism for determining the exact two particle spectrum in a box for any given scattering phase shift. This will allow us to eliminate certain finite volume systematic errors automatically. The transfer matrix method is also powerful because it gives us access to the entire two particle, finite-volume spectrum. When we discuss improvement in Sect. 5.4.2, we will add more operators and couplings to the interaction in order to match not only the ground state energy we desire, but higher eigenvalues as well. This will allow us to control the interaction between particles with non-zero relative momentum. To gain access to higher eigenvalues, the transfer matrix must be solved numerically, however, this may be accomplished quickly and easily for a finite volume system.

5.3 Calculating Observables

Perhaps the simplest observable to calculate using lattice (or any imaginary time) methods is the ground-state energy. While the two-body system may be solved exactly and used to set the couplings for two-body interactions, correlation functions for N-body systems can then be used to make predictions. However, the transfer matrix for $N \gtrsim 4$ cannot in general be solved exactly, because the dimension of the matrix increases with particle number. For this reason we form instead N-body correlation functions,

$$C_N(\tau) = \frac{1}{Z}\int \mathcal{D}\phi\mathcal{D}\psi^\dagger\mathcal{D}\psi e^{-S[\psi^\dagger,\psi,\phi]}\Psi^{(b)}_{b_1\cdots b_N}(\tau)\Psi^{\dagger(a)}_{a_1\cdots a_N}(0) , \tag{5.68}$$

where

$$\Psi^{(a)\dagger}_{a_1\cdots a_N}(\tau) = \int dx_1\cdots dx_N A^{(a)}(x_1\cdots x_N)\psi_{a_1}(x_1,\tau)\cdots\psi_{a_N}(x_N,\tau) \tag{5.69}$$

is a source for N particles with spin/flavor indices $a_1\cdots a_N$, and a spatial wavefunction $A^{(a)}(x_1\cdots x_N)$. For the moment the only requirement we will make of the wavefunction is that it has non-zero overlap with the ground-state wavefunction (i.e. it must have the correct quantum numbers for the state of interest).

Recall that a correlation function consists of τ insertions of the transfer matrix between source and sink. We can then expand the correlation function in a basis of eigenstates,

$$\begin{aligned} C_N(\tau) &= \frac{1}{Z}\langle\tilde{\Psi}^{(a)}_{a_1\cdots a_N}|e^{-H\tau}|\tilde{\Psi}^{(b)}_{b_1\cdots b_N}\rangle = \frac{1}{Z}\sum_{m,n}\langle\tilde{\Psi}^{(a)}_{a_1\cdots a_N}|m\rangle\langle m|e^{-H\tau}|n\rangle\langle n|\tilde{\Psi}^{(b)}_{b_1\cdots b_N}\rangle \\ &= \sum_m Z^{(a)}_m Z^{*(b)}_m e^{-E_n\tau} , \end{aligned} \tag{5.70}$$

where $Z^{(a)}_m$ is the overlap of wavefunction a with the energy eigenstate m, and E_n is the nth eigenvalue of the Hamiltonian. In the limit of large Euclidean time (zero temperature), the ground state dominates,

$$C_N(\tau) \underset{\tau\to\infty}{\longrightarrow} Z^{(a)}_0 Z^{*(b)}_0 e^{-E_0\tau} , \tag{5.71}$$

with higher excited states exponentially suppressed by $\sim e^{-\Delta_{n0}\tau}$, where $\Delta_{n0} \equiv E_n - E_0$ is the energy splitting between the nth state and the ground state. It should be noted that for a non-relativistic theory the rest masses of the particles do not contribute to these energies, so the ground state energy of a single particle at rest is $E_0 = 0$, in contrast to lattice QCD formulations.

In this way, we can think of the transfer matrix as acting as a filter for the ground state, removing more excited state contamination with each application in time. A common method for determining the ground state energy from a correlation function is to construct the so-called effective mass function,

$$M_{\text{eff}}(\tau) \equiv \ln \frac{C(\tau)}{C(\tau+1)} \underset{\tau\to\infty}{\longrightarrow} E_0 , \tag{5.72}$$

and look for a plateau at long times, whose value corresponds to the ground-state energy.

Once the ground state has been isolated, we can calculate matrix elements with the ground state as follows,

$$\begin{aligned} \langle \Psi^{(a)}_{a_1\cdots a_N}|A(\tau')|\Psi^{(b)}_{b_1\cdots b_N}\rangle &= \sum_{lmnq} \langle \Psi^{(a)}_{a_1\cdots a_N}|l\rangle\langle l|e^{-H(\tau-\tau')}|m\rangle\langle m|A|n\rangle\langle n|e^{-H\tau'}|q\rangle\langle q|\Psi^{(b)}_{b_1\cdots b_N}\rangle \\ &= \sum_{ln} Z_l^{(a)} Z_n^{*(b)} e^{-E_l(\tau-\tau')} e^{-E_n\tau'} \langle m|A|n\rangle . \end{aligned} \tag{5.73}$$

To filter out the ground state, the matrix element insertion A must be placed sufficiently far in time from both source and sink, $\{\Delta_{l0}(\tau-\tau'), \Delta_{n0}\tau'\} \gg 1$,

$$\underset{\tau,\tau'\to\infty}{\longrightarrow} Z_0^{(a)} Z_0^{*(b)} e^{-E_0\tau} \langle 0|A|0\rangle . \tag{5.74}$$

In order to isolate the matrix element and remove unknown Z factors and ground state energies, ratios may be formed with correlation functions at various times, similar to the effective mass function.

Another observable one may calculate using lattice methods is the scattering phase shift between interacting particles. Because all lattice calculations are performed in a finite volume, which cannot accommodate true asymptotic scattering states, direct scattering measurements are not possible. However, a method has been devised by Lüscher which uses finite volume energy shifts to infer the interaction, and therefore, the infinite volume scattering phase shift. The Lüscher method will be discussed further in Sect. 5.4.2.1. Because the inputs into the Lüscher formalism are simply energies, correlation functions may be used in the same way as described above to produce this data.

5.3.1 Signal-to-Noise

Recall that we must use Monte Carlo methods to approximate the partition function using importance sampling,

$$C(\tau) \approx \frac{1}{N_{\text{cfg}}} \sum_{i=1}^{N_{\text{cfg}}} C(\phi_i, \tau) \underset{\tau \to \infty}{\longrightarrow} Z_0 e^{-E_0 \tau} , \tag{5.75}$$

where $C(\phi_i, \tau)$ is the operator for some correlation function of interest evaluated on a single configuration ϕ_i, and the set of all fields, ϕ, are generated according to the appropriate probability distribution. In the long Euclidean time limit we expect that this quantity will give us an accurate value for the ground state energy. As stated previously, if the ensemble is large enough for the central limit theorem to hold, then the error on the mean (noise) will be governed by the sample standard deviation,

$$\sigma_C^2(\tau) = \frac{1}{N_{\text{cfg}}} \left[\sum_{i=1}^{N_{\text{cfg}}} |C(\phi_i, \tau)|^2 - \left| \sum_{i=1}^{N_{\text{cfg}}} C(\phi_i, \tau) \right|^2 \right] . \tag{5.76}$$

As an example of how to estimate the size of the fluctuations relative to the signal, let's consider a single particle correlation function, consisting of a single propagator,

$$\frac{1}{Z_\phi} \int \mathcal{D}\phi P(\phi) \langle \Psi_a | K^{-1}(\phi, \tau) | \Psi_b \rangle \approx \frac{1}{N_{\text{cfg}}} \sum_{i=1}^{N_{\text{cfg}}} K_{ab}^{-1}(\phi_i, \tau) , \tag{5.77}$$

where the indices $\{ab\}$ indicate projection onto the states specified by the source/sink. In the large Euclidean time limit, this object will approach a constant, Z_0, because the ground state energy for a single particle is $E_0 = 0$. For the non-relativistic theory as we have set it up, the matrix K is real so long as $g_0 > 0$ (attractive interaction). The standard deviation is then given by

$$\sigma_{C_1}^2(\tau) = \frac{1}{N_{\text{cfg}}} \left[\sum_{i=1}^{N_{\text{cfg}}} \left(K_{ab}^{-1}(\phi_i, \tau) \right)^2 - \left(\sum_{i=1}^{N_{\text{cfg}}} K_{ab}^{-1}(\phi_i, \tau) \right)^2 \right] . \tag{5.78}$$

The second term on the right hand side of the above equation is simply the square of the single particle correlation function, and will therefore also go to a constant, Z_0^2, for large Euclidean time. To gain an idea of how large the first term of $\sigma_{C_1}^2$ is, let's take a look at a correlation function for one spin up and one spin down particle,

$$C_2(\tau) = \frac{1}{Z} \int \mathcal{D}\phi \mathcal{D}\psi^\dagger \mathcal{D}\psi e^{-S[\psi^\dagger, \psi, \phi]} \psi_\uparrow^{(b)}(\tau) \psi_\downarrow^{(b)}(\tau) \psi_\uparrow^{\dagger(a)}(0) \psi_\downarrow^{\dagger(a)}(0) , \tag{5.79}$$

where I have chosen the same single particle source (sink), $\psi^{(a)}$ ($\psi^{(b)}$), for both particles (this is only allowed for bosons or for fermions with different spin/flavor labels). After integrating out the ψ fields we have

$$C_2(\tau) = \frac{1}{Z_\phi} \int \mathscr{D}\phi P(\phi) K_{ab}^{-1}(\phi, \tau) K_{ab}^{-1}(\phi, \tau) \,, \tag{5.80}$$

which is approximately given by

$$C_2(\tau) \approx \frac{1}{N_{\text{cfg}}} \sum_{i=1}^{N_{\text{cfg}}} \left[K_{ab}^{-1}(\phi_i, \tau)\right]^2 \,. \tag{5.81}$$

This is precisely what we have for the first term on the right hand side of Eq. (5.78). Therefore, this term should be considered a two-particle correlation function, whose long Euclidean time behavior is known. Note that we must interpret this quantity as a two-particle correlation function whose particles are either bosons or fermions with different spin/flavor labels due to the lack of anti-symmetrization.

We may now write the long-time dependence of the variance of the single particle correlator as

$$\sigma_{C_1}^2(\tau) \approx C_2(\tau) - (C_1(\tau))^2 \underset{\tau\to\infty}{\longrightarrow} Z_2 e^{-E_0^{(2)}\tau} - Z_1^2 \,, \tag{5.82}$$

where $E_0^{(2)}$ is the ground state energy of the two-particle system. For a two-body system with an attractive interaction in a finite volume, $E_0^{(2)} < 0$, and we may write

$$\sigma_{C_1}^2(\tau) \underset{\tau\to\infty}{\longrightarrow} Z_2 e^{E_B^{(2)}\tau} - Z_1^2 \,, \tag{5.83}$$

where I've defined $E_B^{(2)} \equiv -E_0^{(2)}$. This tells us that $\sigma_{C_1}^2$, and therefore the noise, grows exponentially with time. We can write the signal-to-noise ratio $\mathscr{R}_{C_1}(\tau)$ as

$$\mathscr{R}_{C_1}(\tau) \equiv \frac{C_1(\tau)}{\frac{1}{\sqrt{N_{\text{cfg}}}}\sigma_{C_1}(\tau)} \underset{\tau\to\infty}{\longrightarrow} \sqrt{N_{\text{cfg}}} \frac{Z_1}{\sqrt{Z_2} e^{E_B^{(2)}\tau/2}} = \sqrt{N_{\text{cfg}}} \frac{Z_1}{\sqrt{Z_2}} e^{-E_B^{(2)}\tau/2} \,, \tag{5.84}$$

where I've dropped the constant term in $\sigma_{C_1}^2$, because it is suppressed in time relative to the exponentially growing term. This expression indicates that the signal-to-noise ratio itself grows exponentially with time, and therefore an exponentially large N_{cfg} will be necessary to extract a signal at large Euclidean time. Unfortunately, large Euclidean time is necessary in order to isolate the ground state.

This exponential signal-to-noise problem is currently the limiting factor in system size for the use of any lattice method for nuclear physics. Here, we will discuss it in some detail because in many cases understanding the physical basis behind the problem can lead to methods for alleviation. One method we can use is to employ knowledge of the wavefunction of the signal and/or the wavefunction of the undesired noise in order to maximize the ratio of Z-factors, $Z_1/\sqrt{Z_2}$. For example, choosing a plane wave source for our single particle correlator gives perfect overlap with the desired signal, but will give poor overlap with the bound state expected in the noise. This leads to what has been referred to as a "golden window" in time where the ground-state dominates before the noise begins to turn on [42]. In general, choosing a perfect source for the signal is not possible, however, a proposal for simultaneously maximizing the overlap with the desired state as well as reducing the overlap with the noise using a variational principle has been proposed in [43, 44]. We will discuss other methods for choosing good interpolating fields in Sect. 5.3.3, in order to allow us to extract a signal at earlier times where the signal-to-noise problem is less severe.

Another situation where understanding of the noise may allow us to reduce the noise is when the auxiliary fields and couplings used to generate the interactions can often be introduced in different ways, for instance, via the density channel vs. the BCS channel as mentioned previously. While different formulations can give the same effective interaction, they may lead to different sizes of the fluctuations. Understanding what types of interactions generate the most noise is therefore crucial. This will become particularly relevant when we discuss adding interactions beyond leading order to our EFT in Sect. 5.5, where different combinations of interactions can be tuned to give the same physical observables.

Let's now discuss what happens to $\sigma^2_{C_1}$ if we have a repulsive interaction ($g_0 < 0$). Because nuclear potentials have repulsive cores, such a scenario occurs for interactions at large energy. Since the auxiliary-field-mediated interaction is given by $\sqrt{g_0}\phi\psi^\dagger\psi$, this implies that the interaction is complex. Our noise is now given by

$$\sigma^2_{C_1}(\tau) = \frac{1}{N_{\text{cfg}}}\sum_{i=1}^{N_{\text{cfg}}} K^{-1}_{ab}(\phi_i,\tau)\left[K^{-1}_{ab}(\phi_i,\tau)\right]^\dagger - |C_1(\tau)|^2 . \tag{5.85}$$

Recall that the single particle propagator can be written

$$K^{-1}(\phi_i,\tau) = D^{-1}X(\phi_{i,\tau})D^{-1}X(\phi_{i,\tau-1})\cdots \qquad X(\phi_{i,\tau}) = 1+\sqrt{g_0}\phi_{i,\tau} . \tag{5.86}$$

The complex conjugate of the propagator then corresponds to taking $\phi \to -\phi$,

$$\left[K^{-1}(\phi_i,\tau)\right]^\dagger = D^{-1}X(-\phi_{i,\tau})D^{-1}X(-\phi_{i,\tau-1}) . \tag{5.87}$$

Again, ϕ fields on different time slices are independent, so we may perform each sum over $\phi = \pm 1$ separately. Each sum that we will encounter in the two-particle correlator consists of the product of $X(\phi_\tau)X(-\phi_\tau)$,

$$\sum_\phi (1+\sqrt{g_0}\phi)(1-\sqrt{g_0}\phi) = 1 - g_0^2 = 1 + |g_0|^2 , \tag{5.88}$$

which is exactly the same as we had for the attractive interaction. This implies that even though the interaction in the theory we're using to calculate the correlation function is repulsive, the noise is controlled by the energy of two particles with an attractive interaction, which we have already investigated. In this particular case for a single particle propagator, the signal-to-noise ratio is the same regardless of the sign of the interaction.[5]

In general, however, signal-to-noise problems for systems with repulsive interactions are exponentially worse than those for attractive interactions. This is because generically the signal-to-noise ratio falls off as,

$$\mathscr{R} \sim e^{-(E_{\mathscr{S}}-E_{\mathscr{N}}/2)\tau}\,, \tag{5.89}$$

where $E_{\mathscr{S}(\mathscr{N})}$ is the ground-state energy associated with the signal (noise). Because the signal corresponds to a repulsive system while the noise corresponds to an attractive system, the energy difference in the exponential will be greater than for a signal corresponding to an attractive system.

5.3.1.1 Sign Problems

A related but generally more insidious problem can occur in formulations having fermion determinants in the probability measure, known as a sign problem. A sign problem occurs when the determinant is complex, for example, in our case of a repulsive interaction. While we were able to eliminate the fermion determinant in one particular formulation, there are situations when having a fermion determinant in the probability measure may be beneficial, for example, when using forms of favorable reweighting, as will be discussed later on, or may be necessary, such as for non-zero chemical potential or finite temperature, when the boundary conditions in time may not be altered. For these reasons, we will now briefly discuss sign problems.

The basic issue behind a sign problem is that a probability measure, by definition, must be real and positive. Therefore, a complex determinant cannot be used for importance sampling. Methods to get around the sign problem often result in exponentially large fluctuations of the observable when calculated on a finite sample, similar to the signal-to-noise problem (the two usually result from the same physical mechanism). One particular method is called reweighting, in which

[5]This argument is somewhat simplified by our particular lattice setup in which we have no fermion determinant as part of the probability measure. For cases where there is a fermion determinant, there will be a mismatch between the interaction that the particles created by the operators see (attractive) and the interaction specified by the determinant used in the probability measure (repulsive). This is known as a partially quenched theory, and is unphysical. However, one may calculate a spectrum using an effective theory in which valence (operator) and sea (determinant) particles are treated differently. Often it is sufficient to ignore the effects from partial quenching because any differences contribute only to loop diagrams and may be suppressed.

a reshuffling occurs between what is considered the "observable" and what is considered the "probability measure". For example, when calculating an observable,

$$\langle \mathscr{O} \rangle = \frac{1}{Z_\phi} \int \mathscr{D}\phi P(\phi)\mathscr{O}(\phi)\,, \tag{5.90}$$

when $P(\phi)$ is complex, we can multiply and divide by the magnitude of $P(\phi)$ in both numerator and denominator,

$$\langle \mathscr{O} \rangle = \frac{\int \mathscr{D}\phi |P(\phi)| \frac{P(\phi)\mathscr{O}(\phi)}{|P(\phi)|}}{\int \mathscr{D}\phi |P(\phi)| \frac{P(\phi)}{|P(\phi)|}}\,, \tag{5.91}$$

as well as multiply and divide by $\tilde{Z}_\phi \equiv \int \mathscr{D}\phi |P(\phi)|$,

$$\langle \mathscr{O} \rangle = \frac{\int \mathscr{D}\phi |P(\phi)| \frac{P(\phi)\mathscr{O}(\phi)}{|P(\phi)|}}{\tilde{Z}_\phi} \Bigg/ \frac{\int \mathscr{D}\phi |P(\phi)| \frac{P(\phi)}{|P(\phi)|}}{\tilde{Z}_\phi} = \langle \mathscr{O}' \rangle_{|P|} / \langle \mathscr{O}'' \rangle_{|P|}\,, \tag{5.92}$$

where

$$\mathscr{O}' \equiv \frac{P(\phi)\mathscr{O}(\phi)}{|P(\phi)|}\,, \qquad \mathscr{O}'' \equiv \frac{P(\phi)}{|P(\phi)|}\,, \tag{5.93}$$

and $\langle \cdots \rangle_{|P|}$ implies that the path integrals in the expectation values use the measure $|P(\phi)|$. The advantage is that now the probability measure used for sampling is real and positive, at the cost of having to calculate two observables, $\mathscr{O}', \mathscr{O}''$. The real disadvantage, however, is that the second observable, $\mathscr{O}''$ corresponds to the complex phase of the original measure, $P(\phi)$, which is highly oscillatory from field configuration to field configuration.

We can measure the size of the fluctuations of the phase of $P(\phi) = [\det K(\phi)]^2$, corresponding to a two-spin (or flavor) theory with a repulsive interaction,

$$\langle \mathscr{O}'' \rangle_{|P|} = \frac{\int \mathscr{D}\phi \det K(\phi) \det K^*(\phi)}{\int \mathscr{D}\phi\, [\det K(\phi)]^2}\,. \tag{5.94}$$

The denominator of the above ratio corresponds to the partition function of the original theory which has two spins of particles interacting via a repulsive interaction. The numerator also corresponds to the partition function of a two-spin theory. However, recall that $K^*(\phi)$ corresponds to a propagator with the opposite sign on the interaction term. Because fermions of the same spin don't interact (Pauli principle), the only interaction in this theory is that between two particles of opposite spin, which we established previously will be an attractive interaction due to the sign flip on $K^*(\phi)$. Thus, the numerator corresponds to the partition function of a two-spin theory with an attractive interaction.

A partition function is simply the logarithm of the free energy, $Z = e^{-\beta F}$. For a system in a finite volume at zero temperature this becomes $Z = e^{-V\mathscr{E}_0}$, where $\mathscr{E}_0$ is the energy density of the ground state of the theory. This implies that

$$\langle \mathscr{O}'' \rangle_{|P|} \underset{\tau\to\infty}{\sim} e^{-V(\mathscr{E}_0^{(\mathrm{rep})} - \mathscr{E}_0^{(\mathrm{att})})}, \tag{5.95}$$

where $\mathscr{E}_0^{(\mathrm{rep})}$ ($\mathscr{E}_0^{(\mathrm{att})}$) is the energy density of the ground state of the repulsive (attractive) theory. Generically, $\mathscr{E}_0^{(\mathrm{att})} \leq \mathscr{E}_0^{(\mathrm{rep})}$, for theories which are identical up to the sign of their interaction. This may be shown using the Cauchy-Schwarz theorem,

$$\langle |\det K(\phi)| \rangle \leq |\langle \det K(\phi) \rangle| \,. \tag{5.96}$$

Therefore, $\langle \mathscr{O}'' \rangle_{|P|}$ will be exponentially small for large Euclidean times so long as $\mathscr{E}_0^{(\mathrm{rep})} \neq \mathscr{E}_0^{(\mathrm{att})}$. The variance, on the other hand, is

$$\langle |\mathscr{O}''|^2 \rangle_{|P|} - |\langle \mathscr{O}'' \rangle_{|P|}|^2 = \langle 1 \rangle - |\langle \mathscr{O}'' \rangle_{|P|}|^2 \underset{\tau\to\infty}{\sim} 1 - e^{-2V(\mathscr{E}_0^{(\mathrm{rep})} - \mathscr{E}_0^{(\mathrm{att})})} \sim 1\,. \tag{5.97}$$

So again, we have an exponentially small signal-to-noise ratio at large Euclidean time for the observable $\mathscr{O}''$. This argument is very similar to our signal-to-noise argument for correlation functions. In general, if a theory has a sign problem there will be a corresponding signal-to-noise problem for correlation functions. The reverse is not always true, however, because reweighting is only necessary when the integration measure is complex, so even if there is a signal-to-noise problem in calculating correlation functions (as there is for an attractive interaction), a sign problem may not arise. Sign problems are in general far more problematic due to the exponential scaling with the volume, and because correlation functions give us the additional freedom of choosing interpolating fields in order to try to minimize the noise. In some cases, however, it may be possible to use knowledge learned from signal-to-noise problems in order to solve or reduce sign problems, and vice-versa [45–47].

5.3.1.2 Noise in Many-Body Systems

Let us now discuss signal-to-noise ratios for N-body correlation functions. First, we'll look at the two-particle case. We have already defined the correlation function for two particles with different spin/flavor labels,

$$C_2(\tau) = \langle \left[K_{ab}^{-1}(\phi_i, \tau)\right]^2 \rangle \,. \tag{5.98}$$

The variance is given by

$$\sigma^2_{C_2}(\tau) = \langle \left[K^{-1}_{ab}(\phi_i, \tau)\right]^4 \rangle - (C_2(\tau))^2 \ . \tag{5.99}$$

It is simple to see that the first term in this expression corresponds to a four-particle correlation function, where each particle has a different flavor/spin index (because there is no anti-symmetrization of the fermion fields). Thus, we can write,

$$\sigma^2_{C_2}(\tau) = C_4(\tau) - (C_2(\tau))^2 \ , \tag{5.100}$$

where $C_4(\tau)$ corresponds to a correlator with four particles having different flavors. This is much like a correlator for an alpha particle in the spin/flavor $SU(4)$ limit, thus, it will be dominated at large times by the binding energy, $E_B^{(4)}$, of a state with a large amount of binding energy per particle. Our signal-to-noise ratio is then,

$$\mathscr{R}_{C_2}(\tau) \underset{\tau\to\infty}{\sim} \frac{e^{E_B^{(2)}\tau}}{e^{E_B^{(4)}\tau/2}} \ , \tag{5.101}$$

where, $E_B^{(4)}/2 > E_B^{(2)}$. Therefore, the signal-to-noise ratio is again falling off exponentially in time; this problem clearly becomes worse as the coupling becomes stronger. Finally, we can consider a many-body correlator composed of a Slater determinant over N single-particle states in a two spin/flavor theory,

$$C_{2N}(\tau) = \langle \left[\det K^{-1}(\phi_i, \tau)\right]^2 \rangle \ . \tag{5.102}$$

The ground state of this correlator will be either a BEC or BCS state, as discussed earlier in Sect. 5.2.1.1. The noise, on the other hand, will be dominated by a system of alpha-like clusters, since the number of flavors in the noise is always double that of the signal, which can bind to form nuclei. The ground-state energy of this bound state will clearly be much lower than that of a dilute BEC/BCS state, and our signal-to-noise ratio will be exponentially small in the large time limit.

In general this pattern continues for fermion correlators with any number of particles, spins, and flavors. This is because doubling the number of flavors reduces the amount of Pauli repulsion in the resulting expression for the variance. Even for bosonic systems signal-to-noise can be a problem, simply as a result of the Cauchy-Schwarz triangle inequality, which tells you that, at best, your signal-to-noise ratio can be 1, corresponding to a non-interacting system. Turning on interactions then generally leads to exponential decay of the signal-to-noise ratio. Signal-to-noise problems also generally scale exponentially with the system size, leading to limitations on system size based on computational resources. Thus, understanding and combatting signal-to-noise problems is paramount to further development in the field.

5.3.2 Statistical Overlap

For the lattice formulations we have thus far explored one generates configurations according to the probability distribution associated with the vacuum. One then introduces sources to create particles, which are considered part of the "observable". However, the configurations which are the most important for creating the vacuum may not necessarily be the most important for the observable one wishes to calculate.

We can look to lattice QCD for a pedagogical example. In QCD, the fermion determinant encodes vacuum bubbles created by quark/anti-quark pairs. According to the tenets of confinement, bubbles with large spacetime area require a large energy to produce, and are therefore highly suppressed in the partition function. When doing importance sampling, small vacuum bubbles will dominate. On the other hand, if we now calculate an observable which introduces particle sources, a configuration involving a large vacuum bubble may become very important to the calculation. This is because the total relevant spacetime area of the given configuration, taking into account the particles created by the sources, can in fact be small (see Fig. 5.7). However, by sampling according to the vacuum probability, this configuration will be missed, skewing the calculation in an unknown manner. The farther the observable takes us from the vacuum, the worse this problem becomes, making this a particularly troublesome issue for many-body calculations.

Such problems are referred to as statistical overlap problems. Another situation where these overlap problems can often occur is when doing reweighting to evade a sign problem, as discussed in Sect. 5.3.1.1. For example, if the distribution being sampled corresponds to a theory with an attractive interaction, but the desired

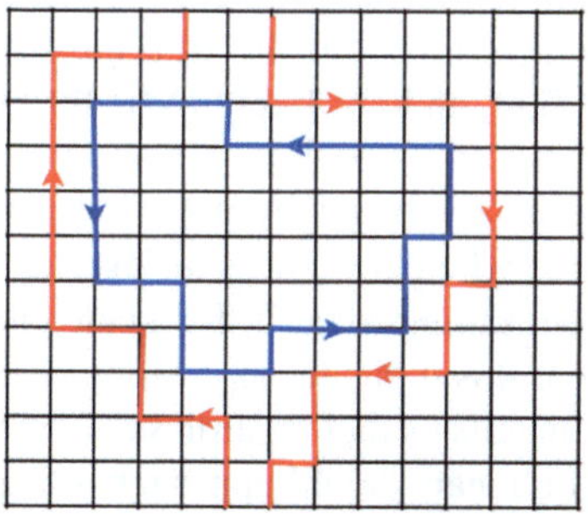

Fig. 5.7 A schematic of an example configuration in LQCD which may lead to a statistical overlap problem. *Red propagators* correspond to valence quarks (quarks created by the sources/sinks in the operator), while *blue* corresponds to sea quarks (vacuum bubbles generated via Monte Carlo). Due to confinement, large bubbles (determined by the area enclosed by the *blue propagator*) are suppressed in the QCD vacuum and thus will likely be thrown out during importance sampling. In the presence of quark sources, however, these configurations are very important in the calculation of the observable (due to the small area enclosed between the *red* and *blue propagators*)

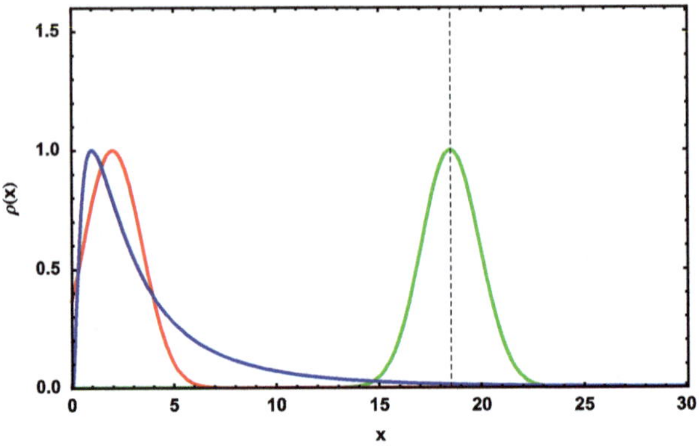

Fig. 5.8 Schematic drawing of a long-tailed probability distribution (*blue*) which leads to an overlap problem. Monte Carlo sampling leads to a sample distribution which is centered around the peak of the underlying distribution (*red*), far from the mean. The ideal probability distribution one would like to sample is narrow and centered around the mean (*green*)

observable has a repulsive interaction, the Monte Carlo sampling will be unlikely to pick up the most relevant configurations, affecting the numerator of Eq. (5.92).

We can understand the problem further by studying probability distributions of observables. While the distribution of the sampled field, ϕ in our case, may be peaked around the mean value of ϕ, the distribution of the observable as calculated over the sample may not be peaked near the true mean of the observable. Such a distribution necessarily has a long tail. Plotting histograms of the values of the observable as calculated over the sample, $\{C(\phi_1), C(\phi_2), \cdots C(\phi_{N_{\text{cfg}}})\}$, can allow us to gain an idea of the shape of the distribution for that observable. An example of a distribution with a statistical overlap problem is plotted in Fig. 5.8. In this case, the peak of the distribution is far from the true mean. Values in the tail of the distribution have small weight, and are likely to be thrown out during importance sampling, skewing the sample mean without a corresponding increase in the error bar. The error bar is instead largely set by the width of the distribution near the peak. One way to determine whether there is an overlap problem is to recalculate the observable on a different sample size; if the mean value fluctuates significantly outside the original error bar this indicates an overlap problem.

The central limit theorem tells us that regardless of the initial distribution we pull from, the distribution of the mean should approach a Gaussian for a large enough sample size, so in principle we should be able to combat an overlap problem by brute force. However, what constitutes a "large enough" sample size is dictated by the shape of the original distribution. The Berry-Esseen theorem [48, 49] can be used to determine that the number of configurations necessary to assume the central limit theorem applies is governed by

$$\sqrt{N_{\text{cfg}}} \sim \frac{\langle \mathscr{X}^3 \rangle}{\langle \mathscr{X}^2 \rangle^{3/2}}, \tag{5.103}$$

where $\langle \mathscr{X}^n \rangle$ is the nth moment of the distribution of an observable, $\mathscr{X}$. Thus, a large skewness, or long tail, increases the number of configurations necessary before the central limit theorem applies, and therefore, to trust an error bar determined by the standard deviation of the distribution of the mean.

One could imagine repeating an argument similar to that made for estimating the variance of our correlation functions in order to estimate the third moment. For example, if our observable is the two-particle correlation function, $C_2(\tau)$, then the third moment will be

$$\langle \mathscr{X}^3 \rangle \sim \langle [K_{ab}(\phi_i, \tau)]^6 \rangle \,, \tag{5.104}$$

corresponding to a correlation function containing six particles of different flavors. Again, increasing the number of flavors generally increases the binding energy per particle of the system, leading to a third moment which is exponentially large compared to the appropriately scaled second moment. This implies that an exponentially large number of configurations will be necessary before the central limit theorem applies to the distribution of the mean of correlation functions calculated using this formulation.

While we mentioned that using reweighting to avoid a sign problem is one situation where overlap problems often occur, it is also possible to use reverse reweighting in order to lessen an overlap problem. Here instead we would like to reweight in order to make the distribution of ϕ have *more* overlap with the configurations that are important for the observable. An example that is commonly used is to include the desired correlation function itself, calculated at some fixed time, to be part of the probability measure. This may be accomplished using ratios of correlators at different times,

$$\frac{C_N(\tau' + \tau)}{C_N(\tau')} = \frac{\int \mathscr{D}\phi \tilde{P}(\phi) \tilde{\mathscr{O}}(\phi, \tau)}{\int \mathscr{D}\phi \tilde{P}(\phi)} \,, \tag{5.105}$$

where

$$\tilde{P}(\phi) \equiv P(\phi) C_N(\tau', \phi) \,, \qquad \tilde{\mathscr{O}}(\phi, \tau) \equiv \frac{C_N(\tau' + \tau, \phi)}{C_N(\tau', \phi)} \,. \tag{5.106}$$

Now the probability distribution incorporates an N-body correlator at one time, τ', and will therefore do a much better job of generating configurations relevant for the N-body correlator at different times. A drawback of this method is that it is much more computationally expensive to require the calculation of propagators for the generation of each configuration. Furthermore, the configurations that are generated will be operator-dependent, so that calculating the correlator C_{N+1} will require the generation of a whole new set of field configurations.

Another method for overcoming a statistical overlap problem is to try to get a more faithful estimate of the mean from the long-tailed distribution itself. To try to better understand the distribution, let's use our signal-to-noise argument to estimate higher moments of the distribution. We can easily estimate the Nth moment of the correlation function for a single particle,

$$\mathcal{M}_N \sim C_N \underset{\tau\to\infty}{\sim} e^{-E_0^{(N)}\tau}, \tag{5.107}$$

where $E_0^{(N)}$ is the ground-state energy of N particles with different flavors. Let's consider the theory to be weakly coupled (small scattering length, $a/L \ll 1$). In this case the two-body interaction dominates and we can use perturbation theory to estimate the energy of two particles in a box: $E_0^{(2)} \approx \frac{4\pi a}{ML^3}$. A weakly coupled system of N particles interacting via the two-body interaction is given by simply counting the number of possible pairs of interacting particles, $E_0^{(N)} \approx N(N-1)\frac{4\pi a}{ML^3}$, leading to the following expression for the moments [50]:

$$\mathcal{M}_N \sim e^{-N(N-1)\frac{4\pi a}{ML^3}}. \tag{5.108}$$

Distributions with the particular N dependence seen in Eq. (5.108) are called log-normal distributions, so named because the distribution of the logarithm of a log-normally distributed quantity is normal. While we derived this expression for theories near weak coupling, there is also evidence that the log-normal distribution occurs for correlators near unitarity as well [51, 52].

The central limit theorem implies that normal distributions occur generically for large sums of random numbers; the same argument leads to the conclusion that log-normal distributions occur for large products of random numbers. Let's think about how correlation functions are calculated on the lattice: particles are created, then propagate through random fields from one time slice to the next until reaching a sink. Each application of the random field is multiplied by the previous one,

$$K^{-1}(\tau) = D^{-1}X(\tau)D^{-1}X(\tau-1)\cdots, \tag{5.109}$$

and then products of these propagators may be used to form correlation functions for multiple particles. Thus, one might expect that in the $\tau \to \infty$ limit (or for large numbers of particles), the distributions of these correlation functions might flow toward the log-normal distribution. More precisely though, each block $X(\tau)$ is actually a matrix of random numbers, and products of random matrices are far less well understand than products of random numbers. Nonetheless, products of random link variables are used to form most observables in nearly all lattice calculations, and approximately log-normal distributions appear to be ubiquitous as well, including in lattice QCD calculations.

If it is $\ln C$ that is nearly Gaussian rather than C, then it may be better to sample $\ln C$ as our observable instead. Without asserting any assumptions about the actual form of the distribution, we can expand around the log-normal distribution using what is known as a cumulant expansion,

$$\ln\langle\mathscr{O}\rangle = \sum_{n=1}^{\infty}\frac{1}{n!}\kappa_n(\ln\mathscr{O})\,, \tag{5.110}$$

where κ_n is the nth cumulant, or connected moment. The cumulants may be calculated using the following recursion relation:

$$\kappa_n(\mathscr{X}) = \langle\mathscr{X}^n\rangle - \sum_{m=1}^{n-1}\binom{n-1}{m-1}\kappa_m(\mathscr{X})\langle\mathscr{X}^{n-m}\rangle\,. \tag{5.111}$$

Note that the expansion in Eq. (5.110) is an exact equality for an observable obeying any distribution. We may now expand the correlation function as

$$\ln\langle C\rangle \underset{\tau\to\infty}{\longrightarrow} -E_0\tau = \langle\ln C\rangle + \frac{1}{2}\left(\langle(\ln C)^2\rangle - \langle\ln C\rangle^2\right) + \frac{1}{6}\kappa_3(\ln C) + \cdots\,. \tag{5.112}$$

Again, this expansion is true for a correlation function obeying any distribution. However, if the distribution of $\ln C$ is exactly log-normal, then $\kappa_{n\geq 3}(\ln C) = 0$. If the distribution is approximately log-normal, then the third and higher cumulants are small corrections, further suppressed in the cumulant expansion by $1/n!$. This suggests that we may cut off the expansion after including a finite number of cumulants without significantly affecting the result (see Fig. 5.9). We may also include the next higher order cumulant in order to estimate any systematic error associated with our cutoff.

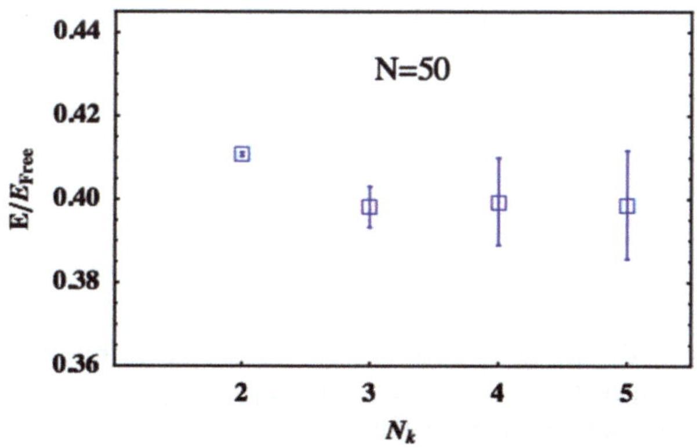

Fig. 5.9 Results for the energy of 50 two-component fermions at unitarity using the cumulant expansion [Eq. (5.110)] cut off at $\mathscr{O}(N_k)$. Figure from [53]

The benefit of using the cumulant expansion to estimate the mean rather than using the standard method is that for a finite sample size, high-order cumulants of $\ln C$ are poorly measured, which is the culprit behind the overlap problem. However, for approximately log-normal distributions these high-order cumulants should be small in the infinite statistics limit. Thus, by not including them in the expansion we do a better job at estimating the true mean on a finite sample size. In other words, by sampling $\ln C$ rather than C, we have shifted the overlap problem into high, irrelevant moments which we may neglect.

The cumulant expansion avoids some of the drawbacks of reweighting, such as greatly increased computational effort in importance sampling. However, the farther the distribution is from log-normal, the higher one must go in the cumulant expansion, which can be particularly difficult to do with noisy data. Thus, for some observables it may be difficult to show convergence of the series on a small sample. Which method is best given the competition between the computational effort used in generating samples via the reweighting method versus the large number of samples which may be required to show convergence of the cumulant expansion is unclear and probably observable dependent.

5.3.3 *Interpolating Fields*

The previous section highlights the importance of gaining access to the ground state as early in time as possible, since the number of configurations required grows exponentially with time. Returning to our expression for the expansion of a correlation function in terms of energy eigenstates,

$$\begin{aligned} C(\tau) &= Z_0 e^{-E_0\tau} + Z_1 e^{-E_1\tau} + \cdots \\ &= Z_0 e^{-E_0\tau}\left[1 + \frac{Z_1}{Z_0} e^{-(E_1-E_0)\tau} + \cdots\right], \end{aligned} \tag{5.113}$$

we see that the condition that must be met in order to successfully suppress the leading contribution from excited state contamination is

$$\tau \gg \frac{\ln\left(\frac{Z_1}{Z_0 E_0}\right)}{E_1 - E_0}, \tag{5.114}$$

where E_0, Z_0 (E_1, Z_1) are the ground (first excited) state energy and wavefunction overlap factor, respectively. Assuming we have properly eliminated excited states corresponding to unwanted quantum numbers through the choice of our source/sink, we have no further control over the energy difference $E_1 - E_0$ in the denominator, because this is set by the theory. Unfortunately, this makes the calculation of many-body observables extremely difficult as this energy splitting can become arbitrarily small due to collective excitations. Therefore, our only recourse is to choose excellent interpolating fields in order to reduce the numerator of Eq. (5.114).

The simplest possible choice for a many-body interpolating field is composed of non-interacting single particle states. A Slater determinant over the included states takes care of fermion antisymmetrization. For example, a correlation function for $N_\uparrow$ ($N_\downarrow$) spin up (spin down) particles can be written,

$$C_{N_\uparrow,N_\downarrow}(\tau) = \langle \det S^\downarrow(\tau) \det S^\uparrow(\tau)\rangle \,, \tag{5.115}$$

where

$$S^\sigma_{ij}(\phi,\tau) \equiv \langle \alpha^\sigma_i | K^{-1}(\phi,\tau) | \alpha^\sigma_j \rangle \,, \tag{5.116}$$

and $\langle \alpha^\sigma_j |$ corresponds to single particle state i with spin σ. As an example, we may use a plane wave basis for the single particle states,

$$|\alpha^\uparrow_j\rangle = |\mathbf{p}_j\rangle \,, \qquad |\alpha^\downarrow_j\rangle = |-\mathbf{p}_j\rangle \,, \tag{5.117}$$

where I've chosen equal and opposite momenta for the different spin labels in order to enforce zero total momentum (this condition may be relaxed to attain boosted systems).

Though the interpolating field chosen in Eq. (5.115) has non-zero overlap with the ground state of interest, if the overlap is small it may take an inordinately long time to remove excited state contributions. Consider a system involving only two-particle correlations, as in our two-spin fermion system, and make the simplification that the ground state consists of non-interacting two-body pairs having wavefunction $\Psi_{\text{2-body}}$, and overlap with a product of two non-interacting single particle states given by

$$\langle \Psi_{\text{2-body}} | \left(|\mathbf{p}\rangle \otimes |-\mathbf{p}\rangle \right) = \epsilon < 1. \tag{5.118}$$

Then the corresponding overlap of the Slater determinant in Eq. (5.115) with the ground state wavefunction scales as

$$\left(\langle \Psi_{\text{2-body}} | \otimes \cdots \otimes \langle \Psi_{\text{2-body}} | \right) \left(|\mathbf{p}_1\rangle \otimes |-\mathbf{p}_1\rangle \otimes \cdots \otimes |\mathbf{p}_N\rangle \otimes |-\mathbf{p}_N\rangle \right) \sim \epsilon^N . \tag{5.119}$$

Thus the overlap of single-particle states with an interacting $2N$-body state is exponentially small with N. This condition worsens for systems with 3- and higher-body correlations.

In order to do a better job we can incorporate two-body correlations into the sinks as follows: first, we construct a two particle propagator,

$$S_{ij}^{\uparrow}\downarrow(\phi,\tau) = \langle\Psi_2|K^{-1}(\phi,\tau)\otimes K^{-1}(\phi,\tau)\left(|\alpha_i^{\uparrow}\rangle\otimes|\alpha_j^{\downarrow}\rangle\right)$$
$$= \sum_{\mathbf{p}}\Psi(\mathbf{p})\langle\mathbf{p}|K^{-1}(\phi,\tau)|\alpha_i^{\uparrow}\rangle\langle-\mathbf{p}|K^{-1}(\phi,\tau)|\alpha_j^{\downarrow}\rangle\,, \qquad (5.120)$$

where $\Psi_2(\mathbf{p})$ is some two-body wavefunction (this process could equally well be performed in position space). As an example, to incorporate BCS pairing, we may use a wavefunction of the form:

$$\Psi_2(\mathbf{p}) \sim \frac{e^{-b|\mathbf{p}|}}{|\mathbf{p}|^2}\,, \qquad (5.121)$$

where b is some parameter which may be tuned to maximize the overlap of the wavefunction. We may also use the wavefunction derived in Eq. (5.65) for a lattice version of such a wavefunction. An example code fragment for implementing such wavefunctions is given here. We show a portion of our c++ code for implementing two types of two-body source vector: Eq. (5.65) (GND) and Eq. (5.121) (PAIR2). Note that these vectors are computed in momentum space. The first operator applied to a source is the kinetic operator, D^{-1}, which is also computed in momentum space.

```
if (two_body_arg.wavefunc_type==WAVEFUNC_TYPE_GND) {

  Dispersion dispersion1(two_body_arg.dispersion_type1,
      two_body_arg.mass1, CUTOFF_TYPE_HARD);
  Dispersion dispersion2(two_body_arg.dispersion_type2,
      two_body_arg.mass2, CUTOFF_TYPE_HARD);

  double xi1;
  double xi2;
  double lambda = two_body_arg.lambda;

  for (int i=0; i<vol; i++) {
    xi1 = 1.0 + dispersion1.Get(i);
    xi2 = 1.0 + dispersion2.Get(i);

    // Here, I use the fermion wave function, as derived in my
        note.
    // Note that psi(p) = xi(p)/(lambda*xi(p)^2-1) is the
        eigenstate of the transfer matrix.
    // The correlation function, however, is given by <final| D
        ^{-1/2} T^N D^{-1/2} |initial>
    // Hence <p|final> = xi(p) psi(p), where xi is the same as
        D^{1/2} in momentum space).
    //
    // wavefunc[i] = xi^2/(lambda*xi^2-1.0);
```

```
        wavefunc[i] = 1.0;              // This way avoids NAN when
            dispersion type is QUADRATIC, or PERFECT
        wavefunc[i] /= lambda - 1.0/(xi1*xi2); // so do it this way
             instead:
      }
    }
  //
    if (two_body_arg.wavefunc_type==WAVEFUNC_TYPE_PAIR2) {

      Dispersion dispersion(DISPERSION_TYPE_QUADRATIC, 1.0,
          CUTOFF_TYPE_HARD);
      double b = two_body_arg.lambda;
      double psq;

      for (int i=0; i<vol; i++) {
        psq = 2.0*dispersion.Get(i);
        if (psq < 1e-15) {
          wavefunc[i] = 0.0; // Omit divergent contribution to wave
               funcion--must be treated separately
        } else {
          wavefunc[i] = exp(- b*sqrt(psq) )/psq;
        }
      }

    }
```

For the full source code see https://github.com/ManyBodyPhysics/LectureNotes Physics/tree/master/Programs/Chapter5-programs.

To ensure Pauli exclusion, it is sufficient to antisymmetrize only the sources, $|\alpha_i\rangle$, leading to the following many-body correlation function,

$$C_{N_\uparrow,N_\downarrow}(\tau) = \langle \det S^{\uparrow\downarrow}(\tau)\rangle \,, \tag{5.122}$$

where the determinant runs over the two sink indices. For correlation functions having an odd number of particles, one may replace a row i of $S^{\uparrow\downarrow}$ with the corresponding row of the single particle object, $S^\uparrow$. The benefit of folding the wavefunction in at the sinks only is an $\mathcal{O}(V^2)$ savings in computational cost: to fold a two-body wavefunction in at both source and sink requires the calculation of propagators from all possible spatial points on the lattice to all possible spatial points in order to perform the resulting double sum.

Higher-body correlations may also be important and can be incorporated using similar methods. However, these will lead to further $\mathcal{O}(V)$ increases in computation time. Finally, the entire system should be projected onto the desired parity, lattice cubic irreducible representation (which we will now briefly discuss), etc. in order to eliminate any contamination from excited states having different quantum numbers.

Table 5.1 Decomposition of the cubic group onto total angular momentum, j

j	Cubic irreps
0	A_1
1	T_1
2	$E + T_2$
3	$A_2 + T_1 + T_2$
4	$A_1 + E + T_1 + T_2$

5.3.3.1 Angular Momentum in a Box

The projection onto the cubic irreps is the lattice equivalent of a partial wave decomposition in infinite volume (and the continuum limit). The cubic group is finite, and therefore has a finite number of irreps, reflecting the reduced rotational symmetry of the box. The eigenstates of the systems calculated on the lattice will have good quantum numbers corresponding to the cubic irreps. When mapping these states onto angular momenta associated with infinite volume, there will necessarily be copies of the same irrep corresponding to the same angular momentum due to the reduced symmetry. This means that the box mixes angular momenta, as displayed in Table 5.1. For example, an energy level calculated in a finite volume that has been projected onto the positive parity A_1 irrep will have overlap with $j = 0, 4, \cdots$. For low energies it may be possible to argue that contributions from high partial waves are kinematically suppressed, since the scattering amplitude scales with p^{2l+1}, but in general the different partial wave contributions must be disentangled using multiple data points from different cubic irreps.

A pedagogical method for projecting two-particle states onto the desired cubic irrep involves first projecting the system onto a particular spin state: for example, a two nucleon system may be projected onto either a spin singlet (symmetric) or spin triplet (anti-symmetric) state. The wavefunctions may then be given an "orbital angular momentum" label by performing a partial projection using spherical harmonics confined to only the allowed rotations in the box. For example, we could fix the position of one of the particles at the origin $(0, 0, 0)$, then displace the second particle to a position (x_0, y_0, z_0). This configuration will be labeled by the wavefunction $\psi_{s,m_s}[(x_0, y_0, z_0)]$, where s, m_s are the total and z-component of the spin. We can then perform the partial projection,

$$\tilde{\psi}_{l,m_l;s,m_s} = \sum_i Y_{l,m_l}\left[R_i(x_0, y_0, z_0)\right] \psi_{s,m_s}\left[R_i(x_0, y_0, z_0)\right] , \tag{5.123}$$

where the R_i are cubic rotation matrices. Essentially, the set $R_i(x, y, z)$ correspond to all possible lattice vectors of the same magnitude. For example, if our original vector was $(1, 0, 0)$, then we would sum over the set of displacements $\{(\pm 1, 0, 0), (0, \pm 1, 0), (0, 0, \pm 1)\}$. I want to emphasize that the l, m_l are only wavefunction labels and do not correspond to good quantum numbers due to the reduced rotational symmetry.

Now that the wavefunctions have spin and orbital momentum labels, these may be combined into total angular momentum labels j, m_j using the usual Clebsch-Gordan coefficients. Finally, these wavefunctions are projected onto cubic irreps using so-called subduction matrices [54]. As an example, a wavefunction labeled with $j = 2$ (having five possible m_j labels) will have overlap with two cubic irreps, T_2, E. The subduction matrices are:

$$T_2 : \overbrace{\begin{pmatrix} 0 & 1 & 0 & 0 & 0 \\ 1/\sqrt{2} & 0 & 0 & 0 & -1/\sqrt{2} \\ 0 & 0 & 0 & 1 & 0 \end{pmatrix}}^{m_j=-2,-1,0,1,2} , \qquad E : \begin{pmatrix} 0 & 0 & 1 & 0 & 0 \\ 1/\sqrt{2} & 0 & 0 & 0 & 1/\sqrt{2} \end{pmatrix} . \tag{5.124}$$

Note that the T_2 irrep has three degenerate states, while the E irrep has two, matching the total of five degenerate states for $j = 2$ in infinite volume.

Using this method for projection onto the cubic irreps has several benefits, including ease of bookkeeping and extension to higher-body systems using pairwise combinations onto a given j, m_j, followed by subduction of the total resulting wavefunction. Furthermore, in cases where more than one partial wave has overlap onto the chosen cubic irrep, wavefunctions with different partial wave labels may have different overlap onto the ground- and excited states of the system. Therefore, they can be used as a handle for determining the best source for the state of interest. We will discuss methods for using multiple sources for disentangling low-lying states and allowing for measurements at earlier times in the next subsection.

5.3.4 Analysis Methods

Having done our best to come up with interpolating wavefunctions, we can attempt to extract the ground state energy (and possibly excited state energies) earlier in time by performing multiple exponential fits to take into account any remaining excited state contamination. Using the known functional form for the correlator,

$$y(\tau) = \sum_{n}^{\Lambda} Z_n e^{-E_n \tau} , \tag{5.125}$$

where Λ is a cutoff in the number of exponentials included in the fit, we may perform a correlated χ^2 minimization,

$$\chi^2_\Lambda = \sum_{\tau,\tau'} [C(\tau) - y(\tau)] \left(\mathscr{C}^{-1}\right)_{\tau\tau'} \left[C(\tau') - y(\tau')\right] , \tag{5.126}$$

where $\mathscr{C}$ is the covariance matrix taking into account the correlation between different time steps. Because the correlation function at a given time is built directly upon the correlation function for the previous time step, there is large correlation between times that must be taken into account.

We can go further by noting that correlation functions formed using different sources, but having the same quantum numbers, will lead to the same spectrum in Eq. (5.125), but with different overlap factors, Z_n. Thus, the χ^2 minimization can be expanded to include different sources s, with only a modest increase in the number of parameters to be fit. Different sources may be produced, for example, by varying some parameter in the wavefunction, such as b in Eq. (5.121), through a different basis of non-interacting single particle states, such as plane waves vs. harmonic oscillator states, or through different constructions of the same cubic irrep, as discussed in the previous subsection. The resulting χ^2 minimization is

$$y_s(\tau) = \sum_n^{\Lambda} Z_n^{(s)} e^{-E_n\tau}\,, \chi^2_{\Lambda} = \sum_{\tau,\tau',s,s'} \left[C_s(\tau) - y_s(\tau)\right] \left(\mathscr{C}^{-1}\right)^{ss'}_{\tau\tau'} \left[C_{s'}(\tau') - y_{s'}(\tau')\right]\,, \tag{5.127}$$

where the covariance matrix now takes into account the correlation between different sources calculated on the same ensembles.

In general, multiple parameter fits require high precision from the data in order to extract several parameters. The use of priors through Bayesian analysis techniques may be beneficial in some circumstances when performing multi-exponential fits to noisy data.

A more elegant approach using a set of correlation functions created using different operators is based on a variational principle [55, 56]. A basic variational argument proceeds as follows [57]: starting with some set of operators $\mathscr{O}_i$ which produce states $|\phi_i\rangle = \mathscr{O}_i|0\rangle$ from the vacuum, we can evolve the state to some time τ_0, $|\tilde{\phi}_i\rangle = e^{-\tau_0 H/2}|\phi_i\rangle$ in order to eliminate the highest excited states, but leaving a finite set of states contributing to the correlation function. We would like to find some wavefunction $|\psi\rangle = \sum_{i=1}^{N} \alpha_i|\tilde{\phi}_i\rangle$ which is a linear combination of our set of operators parameterized by $\{\alpha_i\}$, that maximizes the following quantity for $\tau > \tau_0$:

$$\lambda_0(\tau,\tau_0) = \underset{\{\alpha_i\}}{\text{Max}} \frac{\langle\psi|e^{-(\tau-\tau_0)H}|\psi\rangle}{\langle\psi|\psi\rangle}\,, \tag{5.128}$$

so that

$$\lambda_0(\tau,\tau_0) \approx e^{-E_0(\tau-\tau_0)}\,. \tag{5.129}$$

A powerful method for finding the appropriate linear combination of states satisfying the variational principle uses a generalized eigenvalue problem (GEVP). For this method we form a matrix of correlation functions using all combinations of sources and sinks formed from a set of operators,

$$C_{ij}(\tau) = \langle \mathcal{O}_i(\tau)\mathcal{O}_j^*(0)\rangle = \sum_n e^{-E_n\tau} Z_i^{(n)} Z_j^{(n)} . \tag{5.130}$$

The GEVP may be stated as:

$$C(\tau)v_n(\tau,\tau_0) = \lambda_n(\tau,\tau_0)C(\tau_0)v_n(\tau,\tau_0) , \tag{5.131}$$

where v_n (λ_n) are a set of eigenvectors (eigenvalues) to be determined as follows: assume we choose τ_0 to be far out enough in time such that only N states contribute to the correlation function,

$$C_{ij}(\tau) = \sum_n^N e^{-E_n\tau} Z_i^{(n)} Z_j^{(n)} . \tag{5.132}$$

Let's introduce a set of dual vectors $u_i^{(n)}$ such that

$$\sum_i u_i^{(n)} Z_i^{(m)} = \delta_{mn} . \tag{5.133}$$

Applying u_i to C_{ij} gives

$$\sum_j C_{ij}(\tau)u_j^{(m)} = \sum_j \sum_n e^{-E_n\tau} Z_i^{(n)} Z_j^{(n)} u_j^{(m)} = e^{-E_m\tau} Z_i^{(m)} . \tag{5.134}$$

Going back to our original GEVP, Eq. (5.131),

$$C(\tau)u^{(m)} = \lambda_m(\tau,\tau_0)C(\tau_0)u^{(m)} , \tag{5.135}$$

we can now identify,

$$\lambda_m(\tau,\tau_0) = e^{-E_m(\tau-\tau_0)} . \tag{5.136}$$

Thus, the energies may be found from the eigenvalues of the matrix, $C^{-1}(\tau_0)C(\tau)$. Solving this GEVP gives us access to not only the ground state, but some of the lowest excited states as well.

Any remaining contributions from states corresponding to $E_n, n > N$ can be shown to be exponentially suppressed as $e^{-(E_{N+1}-E_n)\tau_0}$, where E_{N+1} is the first state neglected in the analysis. We should define a new effective mass function to study the time dependence of each of the extracted states,

$$E_n^{(\text{eff})}(\tau,\tau_0) \equiv \ln \frac{\lambda_n(\tau,\tau_0)}{\lambda_n(\tau+1,\tau_0)} , \tag{5.137}$$

and look for a plateau,

$$\lim_{\tau \to \infty} E_n^{(\mathrm{eff})}(\tau, \tau_0) = E_n \,, \tag{5.138}$$

to indicate convergence to the desired state. The reference time τ_0 may be chosen to optimize this convergence, and should generally be close to the beginning of the plateau of the standard effective mass.

The GEVP method works very well in many situations and has been used extensively for LQCD spectroscopy. The main determining factor on the applicability of the method is whether one is able to construct a basis of operators which encapsulates the full low-lying spectrum sufficiently well. One major drawback is that the GEVP assumes a symmetric correlator matrix, meaning that the same set of operators must be used at both source and sink. As discussed in Sect. 5.3.3, this may be difficult to do numerically due to increases in computational time which scale with the volume when projecting onto a given wavefunction (unless the wavefunction is simply a delta function; however, this operator generally has extremely poor overlap with any physical states of interest). This is particularly a problem for noisy systems where large amounts of statistics are necessary.

There are a few alternatives to the GEVP which do not require a symmetric correlator matrix, such as the generalized pencil of functions (GPof) method [58–60], and the matrix Prony method [61, 62]. We will now briefly discuss the latter, following the discussion of Beane et al. [61].

The Prony method uses the idea of a generalized effective mass,

$$M_{\tau_0}^{(\mathrm{eff})}(\tau) = \frac{1}{\tau_0} \ln \frac{C(\tau)}{C(\tau + \tau_0)} \underset{\tau \to \infty}{\longrightarrow} E_0 \,, \tag{5.139}$$

for some, in principle arbitrary, offset τ_0. Because the correlator $C(\tau)$ is a sum of exponentials, it follows certain recursion relations. As an example, for times where only a single exponential contributes we have,

$$C(\tau + \tau_0) + \alpha C(\tau) = 0 \,. \tag{5.140}$$

Plugging in our single exponential for the correlator we can solve for α, then plug it back in to our original expression,

$$\begin{aligned} e^{-E_0 \tau_0} + \alpha &= 0 \\ \longrightarrow C(\tau - \tau_0) - e^{E_0 \tau_0} C(\tau) &= 0 \,. \end{aligned} \tag{5.141}$$

Solving for the ground state energy gives us the same expression as the generalized effective mass at large times,

$$E_0 = \frac{1}{\tau_0} \ln \frac{C(\tau)}{C(\tau + \tau_0)} \,. \tag{5.142}$$

This recursion relation may be generalized for times with contributions from multiple states using the correlation function at different time separations,

$$C(\tau + \tau_0 k) + \alpha_k C(\tau + \tau_0(k-1)) + \cdots + \alpha_1 C(\tau) = 0 . \tag{5.143}$$

We can now generalize this method for a set of correlation functions produced using different operators. Let $C_i(\tau)$ be an N-component vector of correlation functions corresponding to different sources and/or sinks. The correlators then obey the following matrix recursion relation,

$$MC(\tau + \tau_0) - VC(\tau) = 0 , \tag{5.144}$$

for some matrices, M, V, to be determined. Assume the correlator has contributions from Λ states,

$$C(\tau) = \sum_n^{\Lambda} \alpha_n u_n \lambda_n^{-\tau} , \tag{5.145}$$

where $\lambda_n = e^{E_n}$, and u_n is a normalized vector, then we have the following modified GEVP,

$$Mu = \lambda^{\tau_0} Vu . \tag{5.146}$$

A solution for M and V may be found by applying $\sum_{t=\tau}^{\tau+t_W} C(t)^T$ to both sides of Eq. (5.144),

$$M \sum_{t=\tau}^{\tau+t_W} C(t+\tau_0)C(t)^T - V \sum_{t=\tau}^{\tau+t_W} C(t)C(t)^T = 0 , \tag{5.147}$$

leading to the solution,

$$M = \left[\sum_{t=\tau}^{\tau+t_W} C(t+\tau_0)C(t)^T \right]^{-1} , \qquad V = \left[\sum_{t=\tau}^{\tau+t_W} C(t)C(t)^T \right] . \tag{5.148}$$

The parameter t_W is essentially free and may be tuned for optimization, but must obey $t_W \geq \Lambda - 1$ in order to ensure that the matrices are full rank. The λ_n may then be found from the eigenvalues of $V^{-1}M$.

Here we have only used a single recursion relation, which is useful for finding the ground state at earlier times than traditional methods. However, this method is generally less effective for calculating excited states than the symmetric GEVP described previously. It may be possible to construct higher order recursion relations for the matrix Prony method in order to get more reliable access to excited states.

5.4 Systematic Errors and Improvement

5.4.1 *Improving the Kinetic Energy Operator*

The first systematic effect we will examine comes from the discretization of the kinetic operator, first discussed in Sect. 5.2.2. In this section I will show the lattice spacing dependence explicitly so that we may see how discretization errors scale. The kinetic term depends on the definition of the Laplacian operator, which we originally defined to be,

$$\nabla_L^2 f_j = \sum_{k=1,2,3} \frac{1}{b_s^2}\left[f_{j+\hat{k}} + f_{j-\hat{k}} - 2f_j\right] , \tag{5.149}$$

leading to the following kinetic term in momentum space,

$$\Delta(p) = \frac{1}{b_s^2}\sum_i \sin^2\frac{b_s p_i}{2} \approx -\frac{p^2}{2} + \frac{p^4}{24}b_s^2 + \cdots . \tag{5.150}$$

The transfer matrix for the non-interacting system is given by

$$\mathscr{T} = e^{-b_\tau H} = 1 + b_\tau \frac{\Delta(p)}{M} , \tag{5.151}$$

leading to the energy,

$$E = \frac{p^2}{2M} + \mathscr{O}\left(\frac{p^4}{M}b_s^2\right) . \tag{5.152}$$

Therefore, discretization errors in this observable appear at $\mathscr{O}\left(b_s^2\right)$ using this particular discretization. To be more precise, the errors scale with the dimensionless combination $(pb_s)^2$, reflecting the fact that the errors grow as higher momentum scales are probed. As we will discuss in Sect. 5.5, small lattice spacings can lead to computational difficulties beyond the obvious scaling with the number of lattice sites, and taking the continuum limit may prove to be quite difficult. Therefore, it would be beneficial to have an improved operator whose discretization errors come in at a higher order in pb_s. One way to determine such an operator is to examine the relation between the finite difference and the continuum derivative in more detail using a Taylor expansion of the finite difference operator acting on a generic function, $f(x)$,

$$f(x+b_s) - f(x) = b_s f'(x) + \frac{b_s^2}{2}f''(x) + \frac{b_s^3}{6}f'''(x) + \frac{b_s^4}{24}f''''(x) + \cdots . \tag{5.153}$$

Using this expansion, the expression we used previously for the discretized Laplacian can be written,

$$\nabla_L^2 f(x) = \frac{1}{b_s^2}\left(f(x+b_s) + f(x-b_s) - 2f(x)\right) = f''(x) + \frac{b_s^2}{12} f''''(x) + \cdots . \tag{5.154}$$

We see that the leading error comes in at $\mathcal{O}(b_s^2)$, as expected. One method for eliminating the leading error is to add terms involving multiple hops,

$$\tilde{\nabla}_L^2 f(x) = \frac{1}{b_s^2}\left(f(x+b_s) + f(x-b_s) - 2f(x) + c_1 f(x+2b_s) + c_2 f(x-2b_s)\right) , \tag{5.155}$$

where c_1, c_2 must be fixed in such a way as to eliminate the leading error. From symmetry, we must have $c_1 = c_2$. We can then Taylor expand these new terms in our action, and determine the resulting energy as a function of c_1,

$$E(c_1) = \frac{p^2}{2M} + h(c_1)\frac{p^4}{M} b_s^2 + \cdots . \tag{5.156}$$

By solving $h(c_1) = 0$ for c_1, discretization errors will only enter at $\mathcal{O}(b_s^4)$, implying a faster approach to the continuum as b_s is decreased. Perhaps more importantly, in cases where decreasing the lattice spacing is difficult or impossible, the resulting systematic errors at finite lattice spacing will be significantly reduced.

This is our first, very simple, example of improvement. A more general method for improving the action in order to reduce discretization effects utilizes an EFT-like approach [53, 63–66]: we add higher dimension operators consistent with the symmetries of the theory and having unknown coefficients. The coefficients are then fixed by matching onto known physical quantities. The dimension of the operator added determines the order at which discretization errors have been eliminated.

In principle, one would need an infinite number of operators in order to eliminate all discretization errors. We are, of course, limited in the number of displacements we can add, as in Eq. (5.155), by the number of lattice sites. Therefore, the best possible kinetic operator, utilizing all possible spatial hops allowed by the lattice, may still only exactly reproduce the non-interacting spectrum up to the momentum cutoff set by the edge of the first Brillouin zone. Because the kinetic operator Δ is diagonal in momentum space, we may determine this "perfect" operator directly by setting the transfer matrix,

$$\mathcal{T} = 1 + \frac{b_\tau \Delta(p)}{M} = e^{-\frac{b_\tau p^2}{2M}} , \tag{5.157}$$

up to a cutoff, leading to the operator,

$$\Delta_{\text{perf}}(p) = M\left(e^{\frac{b_\tau p^2}{2M}} - 1\right), \qquad p < \frac{\pi}{b_s}. \tag{5.158}$$

While this operator is simple in momentum space, it is highly non-local in position space, as expected, and would be unwieldy to use in a typical lattice calculation. However, another benefit of having a non-relativistic formulation with a separable interaction is that the form of the propagator,

$$\begin{aligned} K^{-1}(\tau) &= D^{-1}X(\tau)D^{-1}X(\tau-1)\cdots D^{-1} \\ &= D^{-1}X(\tau)K^{-1}(\tau-1), \end{aligned} \tag{5.159}$$

suggests that the kinetic (D^{-1}) and interaction (X) operators may each be applied separately in whatever basis is most convenient. So, we may choose to start with a source in momentum space (which is often preferable), then apply an exact kinetic operator, D^{-1}, also in momentum space, perform a FFT to position space, hit the resulting vector with the X operator, which is most easily specified in position space, FFT again back to momentum space to perform a kinetic operation, and so on until finally the sink is applied. Example code for calculating various forms of inverse kinetic operator in momentum space is shown here in terms of a c++ code fragment for computing various lattice Laplacian operators: Eq. (5.150) (STANDARD), Eq. (5.158) (PERFECT), as well as a simple quadratic in momentum (QUADRATIC). Note that these are computed in momentum space, and they may be used to calculate the kinetic operator D^{-1}, then directly applied to the momentum space vectors discussed above.

```
for(int i=0; i<GJP.Vol(); i++) {

  z = i%z_sites;
  Y = i/z_sites;
  y = Y%y_sites;
  x = Y/y_sites;

  if (dispersion_type==DISPERSION_TYPE_STANDARD) {
    //
    // \Delta(p) = 2 \sum_j \sin^2(p_j/2)
    //
    sx = sin( (x+bc[0]) * PI / x_sites );
    sy = sin( (y+bc[1]) * PI / y_sites );
    sz = sin( (z+bc[2]) * PI / z_sites );
    dispersion[i] = 2.0 * ( sx*sx + sy*sy + sz*sz );
    dispersion[i] /= mass;
  }
  if ( (dispersion_type==DISPERSION_TYPE_QUADRATIC)
  ||(dispersion_type==DISPERSION_TYPE_PERFECT) ) {
```

```
// additional code not displayed

  sx = (x + bc[0]) / (double)x_sites;
  sy = (y + bc[1]) / (double)y_sites;
  sz = (z + bc[2]) / (double)z_sites;
  psq = (sx*sx + sy*sy + sz*sz)*4.0*PISQ;

  //---- Cutoff, beyond which we take psq = infinity
  if ( (cutoff_type==CUTOFF_TYPE_HARD)&&(psq > GJP.CutoffSq()
      ) ) {
    dispersion[i] = numeric_limits<double>::infinity();
  } else {
    if (dispersion_type==DISPERSION_TYPE_QUADRATIC) {
        dispersion[i] = psq/(2.0*mass); }
    if (dispersion_type==DISPERSION_TYPE_PERFECT) {
        dispersion[i] = ( exp( psq/(2.0*mass) ) -1.0 ); }
  }
}
```

The benefit to using the FFT repeatedly rather than simply converting the kinetic operator into position space is that modern FFT libraries are highly optimized and cheap to use. For comparison, if we used the "perfect" kinetic operator in position space it would be a dense $V \times V$ matrix. The operation of applying such an object to a V-dimensional vector,

$$D^{-1}(x)|\psi(x)\rangle, \tag{5.160}$$

scales like V^2. On the other hand, using the FFT to convert the V-dimensional vector to momentum space, then applying a diagonal matrix to it,

$$D^{-1}(p)\left(\text{FFT}|\psi(x)\rangle = |\tilde{\psi}(p)\rangle\right), \tag{5.161}$$

scales like $V \log V$. This is a method referred to as "Fourier acceleration" (see e.g. [67–70]).

For formulations lacking separability of the kinetic and interaction operations, this method cannot generally be applied. In such cases, the kinetic operator should be kept relatively sparse in position space. Such a condition disfavors the use of Eq. (5.158) for a more modestly improved operator, composed of only a few spatial displacements, using the method outlined in the beginning of this section.

5.4.2 Improving the Interaction

To discuss systematic errors and improvement of the interaction, we will focus on systems tuned to unitarity. Because unitarity corresponds to a conformal fixed-point, the systems we will study only depend on a single scale, the density, n. The finite lattice spacing necessarily breaks this conformal symmetry, and we can consider dependence on any new scales to stem from systematic errors. Systems having multiple intrinsic scales contain more complicated dependences of systematic errors, and will be discussed later on.

Recall that the scattering phase shift for two particles at unitarity is,

$$p \cot \delta = 0 , \tag{5.162}$$

implying that the inverse scattering length, effective range, and all other shape parameters vanish. In Sect. 5.2.2.5, we discussed how to tune the two-particle coupling in order to reproduce infinite scattering length. The lattice, however, naturally induces an effective range for the interactions, which have been generated via auxiliary fields extending across a lattice link, of size b_s. In order to improve the interaction and eliminate the unwanted effective range contribution stemming from discretization, we may add a higher-order interaction operator,

$$\sum_{\mathbf{x}} \sqrt{g_2} \phi \psi_{\mathbf{x}}^\dagger \nabla_L^2 \psi_{\mathbf{x}} , \tag{5.163}$$

recalculate the scattering amplitude, A, as a function of g_0, g_2, and tune g_2 to eliminate the r_0 term in the effective range expansion. In principle, one may further generalize the interaction operator,

$$\mathscr{L}_{\text{int}} = \sum_n \sqrt{g_{2n}} \phi \psi^\dagger \nabla_L^{2n} \psi , \tag{5.164}$$

where we will now suppress spacetime indices, and use the g_{2n} to tune away successive terms in the effective range expansion. In practice this may be difficult because the interaction is generally no longer separable, so that loops can't be summed analytically. An easier method may be to use the transfer matrix, as we did in Sect. 5.2.2, to determine the two particle energy spectrum in a box, then tune the couplings in order to reproduce the desired energies. The target energies may be determined for systems obeying any known physical scattering phase shift using an approach known as the Lüscher method, which we will now briefly review.

5.4.2.1 Lüscher's Method

Lüscher's method [71, 72] was originally developed as a tool for extracting physical scattering phase shifts from finite volume, Euclidean space observables produced by lattice QCD. The concept of asymptotic "in" and "out" scattering states does not exist in a finite volume, making direct scattering "experiments" impossible on the lattice. Furthermore, the issue of analytic continuation from Euclidean to Minkowski time is a tricky one, particularly when utilizing stochastic techniques. Thus, Lüscher proposed utilizing a different observable, finite volume energy shifts, and inferring the infinite volume scattering phase shift that would lead to the observed finite volume spectrum. In this section, we will largely follow the discussion in [73].

First let's recap how to calculate the infinite volume s-wave scattering phase shift in our effective theory assuming the following generic tree-level interaction: $\mathscr{L}_2 = \sum_n g_{2n} p^{2n}$. The scattering amplitude is given by,

$$A_\infty = \frac{\sum_n g_{2n} p^{2n}}{1 - \sum_n g_{2n} p^{2n} I_0^\infty} = \frac{4\pi}{M} \frac{1}{p \cot \delta - ip}, \tag{5.165}$$

where I will now include the super/subscript "∞" to indicate infinite volume quantities, and I_0^∞ is defined as,

$$I_0^\infty = \int \frac{d^3 q}{(2\pi)^3} \frac{1}{E - q^2/M}. \tag{5.166}$$

Note that I have assumed that the interaction is separable in deriving Eq. (5.165). This would not be possible using a momentum cutoff as a regulator, so we will use dimensional regularization for this integral. By investigating the inverse scattering amplitude,

$$A_\infty^{-1} = \frac{1}{\sum_n g_{2n} p^{2n}} - I_0^\infty = \frac{M}{4\pi}(p \cot \delta - ip), \tag{5.167}$$

we can identify

$$\sum_n g_{2n} p^{2n} = \left[I_0^\infty + \frac{M}{4\pi}(p \cot \delta - ip) \right]^{-1}. \tag{5.168}$$

the quantity on the right can be expanded using the effective range expansion; the couplings are then determined by the scattering parameters, as we have seen previously.

Now that we have a relation between the couplings and the physical scattering parameters, let's now use this same effective theory to determine its finite volume spectrum. In a finite volume, there is no continuum of scattering states, but rather a discrete spectrum corresponding to poles in the finite volume analogue of the scattering amplitude, A_{FV},

$$\text{Re}\left[A_{\text{FV}}^{-1}\right] = 0. \tag{5.169}$$

Because the imposition of a finite volume can affect only the IR behavior of the theory, the interactions, and therefore the couplings, g_{2n}, remain unchanged. Any differences come from loops, where intermediate particles may go on shell and explore the finite boundary. Therefore, our finite volume analogue of the scattering amplitude may be written, where

$$A_{\rm FV}^{-1} = \frac{1}{\sum_n g_{2n} p^{2n}} - I_0^{\rm FV} \,, \tag{5.170}$$

where the loop integral has been replaced by a finite volume sum over the allowed quantized momenta in a box,

$$I_0^{\rm FV} = \frac{1}{L^3} \sum_{\mathbf{n}}^{\Lambda} \frac{1}{E - \left(\frac{2\pi n}{L}\right)^2 / M} \,. \tag{5.171}$$

Again, because the couplings are unchanged by the finite volume we are free to use Eq. (5.168) to replace them with the physical infinite volume phase shift, resulting in,

$$A_{\rm FV}^{-1} = \frac{M}{4\pi}(p \cot\delta - ip) + I_0^{\infty} - I_0^{\rm FV} \,. \tag{5.172}$$

This leads to the eigenvalue equation,

$$\text{Re}\left[A_{\rm FV}^{-1}\right] = \frac{M}{4\pi} p \cot\delta + \text{Re}\left[I_0^{\infty} - I_0^{\rm FV}\right] = 0 \,. \tag{5.173}$$

I have specified taking the real part of the inverse amplitude merely for calculational simplicity; this quantity is, in fact, already purely real because there are no integrals, and therefore, no $i\epsilon$ prescription. Furthermore, the difference between the infinite volume integral and the finite volume sum must be finite because the two encode the same UV behavior. Finally, we have the result,

$$p \cot\delta = \frac{4\pi}{M}\left[-\frac{M}{4\pi^2 L} \sum_{\mathbf{n}}^{\Lambda} \frac{1}{\left(\frac{pL}{2\pi}\right)^2 - n^2} - \frac{M\Lambda}{\pi L}\right] = \frac{1}{\pi L} S(\eta) \,, \tag{5.174}$$

where $\eta \equiv \left(\frac{pL}{2\pi}\right)^2$, and

$$S(\eta) \equiv \sum_{\mathbf{n}}^{\Lambda} \frac{1}{n^2 - \eta} - 4\pi\Lambda \,, \tag{5.175}$$

is related to the Riemann zeta function. The cutoff on the sum, Λ, may be interpreted as an upper limit on the allowed momenta due to the finite lattice spacing, however, in practice it is taken to ∞ so that discretization and finite volume effects may be separately accounted for (note that we haven't used our lattice propagators in this derivation, which would be necessary for a proper treatment of discretization

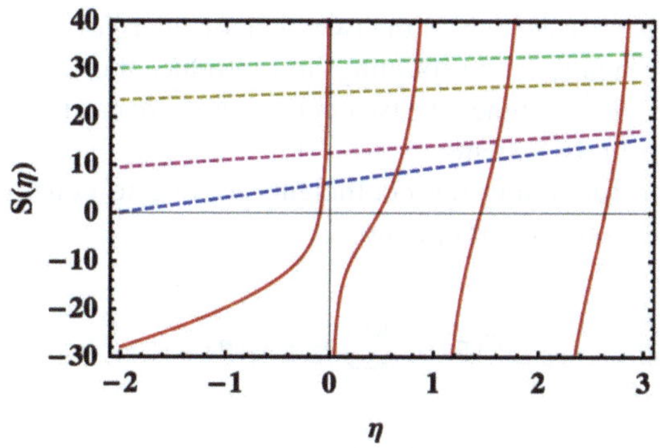

Fig. 5.10 $S(\eta)$ (*solid red*) and $\pi Lp \cot\delta$ (*dashed*) as a function of $\eta \equiv \left(\frac{pL}{2\pi}\right)^2$. The $\pi Lp \cot\delta$ correspond to $r_0/a = -0.1$, for the following volumes: $L/|a| = 2$ (*blue*), $L/|a| = 4$ (*pink*), $L/|a| = 8$ (*yellow*), $L/|a| = 10$ (*green*). The energy eigenstates for the corresponding volumes are given by the intercepts of $S(\eta)$ with the *dashed lines*. Figure from [41]

effects). Values of momenta which solve this eigenvalue equation for a given phase shift and volume correspond to the predicted finite volume spectrum. This is illustrated in Fig. 5.10, where the function $S(\eta)$ has been plotted, along with several representative phase shifts, corresponding to positive and negative scattering lengths. The locations of the intersections give the energy eigenvalues for that volume. The poles of the S function give the locations of the energies of a non-interacting system in a box, while the zeroes give the energies for systems at unitarity.

Many extensions of Lüscher's method exist for more complicated systems, such as multi-channel processes [74–81], higher partial waves [82–84], moving frames [85, 86], moving bound states [87, 88], asymmetric boxes [89, 90], and three-body systems [91–94], as well as perturbative expansions for many-boson systems [95–97]. Formulations for general systems involving two nucleons may be found in [98, 99]. These formulations have been successfully applied in Lattice QCD for the determination of scattering phase shifts of nucleon-nucleon [2, 5–8, 12, 13, 16, 19, 23, 100], meson-meson [101–119], meson-baryon [120–123], and hyperon-nucleon [14, 124, 125] systems.

5.4.2.2 Applying Lüscher's Method to Tune the Two-Body Couplings

The prescription for a lattice QCD calculation of nucleon-nucleon phase shifts is to start with quark interpolating fields to create a two nucleon correlation function, measure a set of finite volume energies, then use the eigenvalue equation, Eq. (5.174), to infer the infinite volume two nucleon phase shift that produces those energies. For our lattice EFT, however, two nucleon phase shifts are used as input into the coefficients in the Lagrangian. Thus, we can use the Lüscher method in reverse to calculate what we expect the two nucleon energies in a box to be given a known phase shift, then tune the couplings to reproduce those same energies in our lattice calculations. Having tuned the two-body sector, we can then make predictions about 3- and higher-body systems.

Our prescription for tuning the coefficients will be to construct the two-body transfer matrix with some set of operators,

$$\mathscr{G}(\mathbf{p}) = \sum_{n}^{\Lambda_n} g_{2n} \mathscr{O}_{2n}(\mathbf{p}) \,, \tag{5.176}$$

which satisfy the low energy expansion $\mathscr{O}_{2n}(\mathbf{p}) = \mathbf{p}^{2n}\left[1 + \mathscr{O}(\mathbf{p}^2)\right]$ at low momenta, and should be chosen to depend only on the relative momentum of the two particle system in order to ensure Galilean invariance. This is important so that once the interaction is tuned boosted pairs of particles will see the same interaction. A convenient choice for the operators is given by,

$$\mathscr{O}_{2n}(\mathbf{p}) = M^n \left(1 - e^{-\hat{\mathbf{p}}^2/M}\right)^n \,, \tag{5.177}$$

where $\hat{\mathbf{p}}$ is taken to be a periodic function of $\mathbf{p}$ and satisfies the relation $\hat{\mathbf{p}}^2 = \mathbf{p}^2\theta(\Lambda - |\mathbf{p}|) + \Lambda^2\theta(|\mathbf{p}| - \Lambda)$ for $\mathbf{p}$ in the first Brillouin zone. Here we show a C++ code fragment for calculating the interaction given in Eq. (5.176), using the operators equation (5.177), given some set of input coefficients interaction_arg.couplings[Λ_n]. Note that this operator is calculated in momentum space. It may be applied directly to the momentum space vector resulting from the first operation of the kinetic operator, D^{-1}. A FFT must then be performed before applying the random auxiliary field, ϕ_x. A final FFT must then be performed to return to momentum space before applying the next operation of D^{-1} in order to propagate the system forward in time.

```
double xi1;
double xi2;

double psq;
double mass = 2.0 * kinetic_arg1.mass * kinetic_arg2.mass; //
    reduced mass
mass /= kinetic_arg1.mass + kinetic_arg2.mass; // reduced mass

double PSQ = GJP.Cutoff();
PSQ *= PSQ;
double XI1 = exp( PSQ/(2.0*kinetic_arg1.mass));
double XI2 = exp( PSQ/(2.0*kinetic_arg2.mass));

for(int i=0; i<vol; i++) {

  xi1 = 1.0 + dispersion1.Get(i);
  if (xi1 > XI1) { xi1 = XI1; }
```

```
    xi2 = 1.0 + dispersion2.Get(i);
    if (xi2 > XI2) { xi2 = XI2; }

    switch (interaction_arg.interaction_type) {
      case INTERACTION_TYPE_NONE:
        ERR.NotImplemented(fname,"Interaction type
            INTERACTION_TYPE_NONE.");
        break;
      case INTERACTION_TYPE_ONEMINUSXIINVSQ:
        psq = 1.0 - 1.0 /(xi1*xi2);
        break;
      case INTERACTION_TYPE_XISQMINUSONE:
        psq = xi1*xi2 - 1.0;
        break;
      default:
        ERR.NotImplemented(fname,"Unrecognized interaction_type."
            );
    }

    psq *= mass;

    //---- Evaluate the interaction as a Taylor series in p^2: O
        (\,p) = sum_n C_n p^(2n)
    interaction[i] = 0.0;
    for (int j=0; j<interaction_arg.num_couplings; j++) {
      interaction[i] += interaction_arg.couplings[j]*pow(psq,j);
    }

    //---- Make sure O(p) is non-negative before taking the
        square root!
    if (interaction[i]<0.0) {
      ERR.General(fname, "Interaction is less the zero; cannot
          take square root.");
    }

    //---- Normalize operator, etc..
    interaction[i] *= 2.0 * TWOPI / mass; // Just a convention
    interaction[i] = sqrt(interaction[i]); // Division by vol
        because FFT is not normalized
    interaction[i] /= vol; // Division by vol because FFT is not
        normalized
    interaction[i] *= dt; // Controls overall sign and step size
        of the interaction
  }
```

The transfer matrix may then be diagonalized numerically to determine the energy eigenvalues. The g_{2n} should then be tuned until the energies match the first Λ_n eigenvalues given by the Lüscher method. This process serves a dual purpose: tuning multiple couplings helps reduce lattice spacing effects like the effective range, as we discussed previously, and also takes into account finite volume effects by correctly translating the exact infinite volume phase shifts into a finite volume.

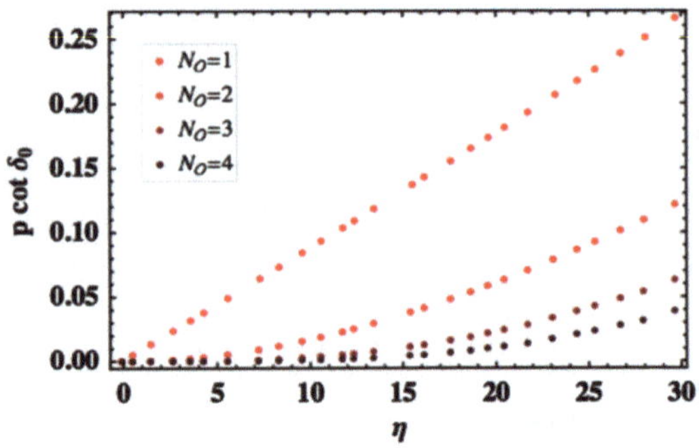

Fig. 5.11 Effective scattering phase shifts $p\cot\delta$ vs. η produced by a set of contact interactions of the form in Eq. (5.176), with $N_{\mathcal{O}}$ coefficients tuned to unitarity. Figure from [126]

The process of tuning for the case of unitarity is illustrated in Fig. 5.11. Here, $N_{\mathcal{O}}$ coefficients have been tuned to correctly reproduce the first $N_{\mathcal{O}}$ Lüscher eigenvalues. The entire two-body spectrum is then calculated using these coefficients, and the resulting energies are plugged back into Eq. (5.174) to determine the effective phase shift seen by pairs of particles with different momenta. To be truly at unitarity, we should have $p\cot\delta = 0$ for all momenta. Clearly, tuning more coefficients brings us closer to unitarity for larger and larger momenta. This is particularly important for calculations involving many-body systems, where the average momentum grows with the density, $\langle p\rangle \sim n^{1/3}$.

A quantitative prediction can be made for the error remaining in higher, untuned two-body energy levels [126]. Assuming $N_{\mathcal{O}}$ terms in the effective range expansion have been tuned to zero,

$$p\cot\delta \sim r_{N_{\mathcal{O}}-1}p^{2N_{\mathcal{O}}} = \left(\frac{2\pi}{L}\right)^{2N_{\mathcal{O}}} r_{N_{\mathcal{O}}-1}\eta^{N_{\mathcal{O}}}\,, \tag{5.178}$$

we can then use Lüscher's relation for the first untuned eigenvalue η_k,

$$\left(\frac{2\pi}{L}\right)^{2N_{\mathcal{O}}} r_{N_{\mathcal{O}}-1}\eta_k^{N_{\mathcal{O}}} = \frac{1}{\pi L}S(\eta_k)\,. \tag{5.179}$$

Let's suppose η_k^* is the eigenvalue one would expect in the true unitary limit. We can then Taylor expand the function $S(\eta_k)$ around η_k^*,

$$S(\eta_k) \approx c_k(\eta_k - \eta_k^*)\,, \tag{5.180}$$

where c_k is the slope near η_k^*. The error is then estimated as,

$$\frac{\eta_k}{\eta_k^*} - 1 \approx \frac{\pi L}{\eta_k^* c_k}\left(\frac{2\pi}{L}\right)^{2N_{\mathcal{O}}} r_{N_{\mathcal{O}}-1}\left(\eta_k^*\right)^{N_{\mathcal{O}}} \sim \mathcal{O}\left(L^{1-2N_{\mathcal{O}}}\right) \sim \mathcal{O}\left((b_s n^{1/3})^{2N_{\mathcal{O}}-1}\right)\,, \tag{5.181}$$

where on the right I have rewritten the scaling with the volume as a scaling with the density to remind you that though the errors scale with the volume, these are not actually finite volume errors we are investigating, but discretization effects scaling with the dimensionless quantity $b_s n^{1/3} \sim b_s/L_{phys} = 1/L$ for systems at unitarity. The Lüscher method takes into account finite volume effects automatically.

5.4.3 Scaling of Discretization Errors for Many-Body Systems

Having tuned our two-body interaction, we can now also predict the scaling of errors that we should expect to find in an N-body calculation. Let us suppose that the first untuned operator contains at most $2N_{\mathcal{O}}$ derivatives,

$$\mathcal{O}_{2N_{\mathcal{O}}} \sim (\psi\psi)^\dagger \, \psi\nabla^{2N_{\mathcal{O}}}\psi \,. \tag{5.182}$$

The leading error results when any pair of particles interacts via this operator, and should scale with the dimension of this operator.

To determine the operator dimension, first let me briefly recap how scaling dimensions are determined in a non-relativistic theory (see [38] for more details). We expect the action, S, to be a dimensionless quantity, so we will consider the action for a non-interacting theory to determine how the fields and derivatives must scale,

$$S = \int d\tau d^3x \psi^\dagger \left(\partial_\tau - \frac{\nabla^2}{2M}\right)\psi \,. \tag{5.183}$$

First, note that the mass, M, carries zero scaling dimension in a non-relativistic theory because it is considered to be much larger than any scale of interest. Then, from the expression in parentheses, we see that time and space must scale differently, $[\partial_\tau] = 2[\nabla]$. Using the convention $[\nabla] = 1$, we can then determine that the dimension of the fermion field must be $[\psi] = 3/2$.

Now let us return to the operator, Eq. (5.182), and determine its scaling dimension relative to the energy,

$$\left[(\psi\psi)^\dagger \, \psi\nabla^{2N_{\mathcal{O}}}\psi\right] - \left[\psi^\dagger\partial_\tau\psi\right] = (6 + 2N_{\mathcal{O}}) - (5) = 1 + 2N_{\mathcal{O}} \,. \tag{5.184}$$

This indicates that the error from such an operator will scale as $\sim \mathcal{O}(b_s p)^{1+2N_{\mathcal{O}}}$, or $\sim \mathcal{O}\left((b_s n^{1/3})^{1+2N_{\mathcal{O}}}\right)$ for unitary fermions. This is similar scaling that we saw for higher two-body states, however, here the dependence on the number of particles is also important.

One may in principle tune as many operators as possible in order to perfect the interaction for higher energies. In practice, however, as more and more operators are tuned, the coefficients in front of higher dimensional operators which are still untuned can become very large. This can cause interactions seen by pairs of particles far in the tail of the momentum distribution to generate large errors. Thus, similar to the case of the kinetic operator, there is a limit to how "perfect" the interaction can be made.

On the other hand, these s-wave two-body interactions are not the only possible errors that are induced by the lattice, so we should not expect to see much improvement by tuning more operators corresponding to errors which are higher order than the leading operator which is not accounted for. For example, an unfortunate consequence of our tuning program is the introduction of interactions in the p-wave channel, as well as in higher partial waves. While a simple interaction which is point-like in space has no p-wave contribution, the introduction of spatial derivatives in our tuning operators gives rise to these new p-wave interactions. The leading p-wave operator has the form,

$$\mathcal{O}_{p\text{-wave}} \sim \psi^\dagger \nabla \psi \cdot \psi^\dagger \nabla \psi \ , \tag{5.185}$$

and induces errors at $\mathcal{O}\left((b_s n^{1/3})^3\right)$. In order to cancel this operator we could in principle add a ϕ field which carries momentum and carry out a similar program for tuning the coefficients as we used for the s-wave interaction. This destroys the separability of our interaction, however, and may be difficult to implement, in addition to introducing a new source of noise.

In general, we can determine all possible sources of discretization error as well as their scaling using a method referred to as the Symanzik effective action [53, 63–66]. The basic procedure begins through considering any possible operators (that have not been explicitly tuned) which are allowed by the symmetry of the theory. Because these operators may only be induced through discretization and must disappear in the continuum limit, they should be multiplied by the lattice spacing raised to the appropriate scaling dimension of the operator. We can then determine at what order in b_s, relative to the energy, we can expect systematic errors to arise.

Let's take a look another interesting operator which arises due to discretization, corresponding to a three-body interaction. While there can be no point-like 3-body interaction in the continuum limit for 2-component fermions due to the Pauli exclusion principle, three particles separated by a lattice spacing may interact via ϕ-field exchange because they don't all lie on the same spacetime point. Thus, we should include in our Symanzik effective action an operator,

$$\mathcal{O}_{3\text{-body}} \sim (\psi\psi\psi)^\dagger \psi\psi\psi \ . \tag{5.186}$$

Naïvely, the dimension of this operator is 9, and therefore should contribute errors of $\mathcal{O}\left((b_s n^{1/3})^4\right)$. So far, all of the operators we've discussed obey this simple scaling, corresponding to naïve dimensional analysis. However, our theory is strongly interacting, which can in general lead to large anomalous dimensions of certain operators.

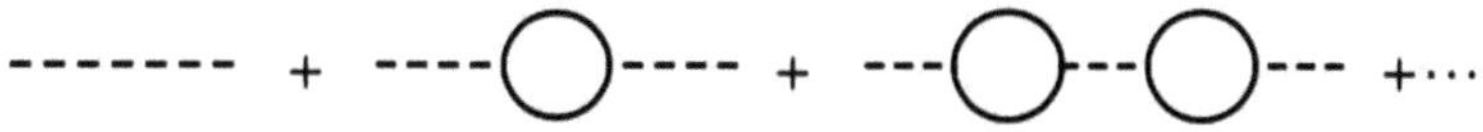

Fig. 5.12 Propagator for the bosonic field ϕ, dressed by fermionic loops

As an example, let's consider the scaling dimension of a very basic operator, the field ϕ. The canonical (non-interacting) dimension for a generic bosonic field in a non-relativistic theory can be deduced by looking at the kinetic term in the action,

$$S_{\text{kin}} = \int d\tau d^3x \nabla^2 \phi^2 \,, \tag{5.187}$$

leading to a scaling dimension, $[\phi] = 3/2$. However, once interactions with the ψ fields are included, the ϕ propagator is renormalized through loop diagrams (see Fig. 5.12). For a non-perturbative interaction, we must sum all possible loop diagrams. However, there is a simpler way to determine the scaling dimension of the strongly interacting ϕ field. The key is to recognize that near unitarity the ϕ field represents a bound state of two ψ fields at threshold. We can therefore write ϕ as a local operator,

$$\phi(x) = \lim_{x \to y} |x - y| \psi^\dagger(x) \psi(y) \,, \tag{5.188}$$

where $|x - y|$ must be included to ensure that matrix elements of the operator are finite (the wavefunction for two particles at unitarity must scale as $|x-y|^{-1}$ at short distances [127]). Using our previous analysis for the scaling dimension of the ψ field, we find,

$$[\phi]_{\text{int}} = 2 \,, \tag{5.189}$$

which implies a very strong wavefunction renormalization.

In general it can be very difficult to calculate anomalous dimensions directly in a non-perturbative fashion. However, for non-relativistic conformal field theories (CFT), there exists an operator-state correspondence (similar to an ADS/CFT correspondence), which relates the scaling dimension of an operator in the CFT (e.g. for unitary fermions) to the energy of the corresponding state in a harmonic potential [127]. For example, we have already determined the dimension of the field ψ to be 3/2, and the energy of a single fermion in a harmonic potential with oscillator frequency ω is $3/2\omega$. The energy of two unitary fermions in a harmonic potential is 2ω, corresponding to the dimension of the ϕ field, $[\phi] = 2$.

Returning now to our 3-body operator, we can use numerical results for the energy of three fermions in a total $l = 0$ state in a harmonic potential [128, 129] to determine that,

$$[\psi\psi\psi] = 4.67 \,. \tag{5.190}$$

The error-inducing operator in the Symanzik effective action both creates and destroys this 3-body state, resulting in

$$\left[(\psi\psi\psi)^\dagger\,\psi\psi\psi\right] = 9.34\,. \tag{5.191}$$

The relative error in the energy will then be $\mathcal{O}\left(L^{-(9.34-5)}\right) = \mathcal{O}\left(L^{-4.34}\right)$.

It turns out that the ground state of three fermions in a harmonic potential is actually not the s-wave state, but a p-wave state with energy $\sim 4.27\omega$. Thus, we should expect an additional systematic error corresponding to a 3-body p-wave operator that contributes at $\mathcal{O}\left(L^{-3.55}\right)$ [130]. Finally, at approximately the same order as the 3-body s-wave there is a 2-body d-wave operator (four derivatives) with zero anomalous dimension, and therefore contributing at $\mathcal{O}\left(L^{-5}\right)$.

While certainly only the leading error, $\left(\mathcal{O}\left(L^{-3}\right)\right)$, will dominate very close to the continuum limit, at a finite lattice spacing we have just demonstrated that there are several sources of error scaling with very similar powers of the lattice spacing. If we wish to eliminate discretization errors through extrapolation to the continuum limit, we must include all possible non-negligible contributions in our extrapolation function. For example, we could employ the following function:

$$E(L) = E_0\left[1 + aL^{-3} + bL^{-3.55} + cL^{-4.34} + dL^{-5} + \cdots\right], \tag{5.192}$$

and fit the coefficients $\{a, b, c, d\}$ using data at several volumes, in order to extract the continuum energy, E_0 [53].

5.4.4 Additional Sources of Systematic Error

It should be pretty clear by now that understanding and controlling systematic errors can be quite complicated, even for conformal systems! For more complex systems with contributions from multiple scales, such as nuclei, things become even messier. As a simple example of a system with more than one scale we can consider trapping our unitary fermions in a harmonic potential, which will allow us to discuss finite volume errors that are not accounted for by the Lüscher method. This is clearly relevant for cold atom experiments, which utilize traps, but may also be useful for calculating the energies needed to use the operator-state correspondence discussed in the previous subsection.

The new characteristic length scale contributed by the introduction of the harmonic trap is given by the size of the trap, L_0. We now have two different dimensionless quantities which determine the scaling of systematic errors due to discretization, b_s/L_0, and finite volume, L_0/L_{phys}, individually. To determine the size of discretization errors we may use the Symanzik effective action method as previously described, with the average momentum scale replaced by $n^{1/3} \to N^{1/3}/L_0$. Finite volume errors may be estimated by examining the long distance behavior of the wavefunction of the system of interest, where distortions due to the finite boundary can occur. For a system in a harmonic trap with local interactions, wavefunctions behave as Gaussians at large distance, so we might consider using a function $E(L_{\text{phys}}) = E_0\left(1 + ae^{-(L_0/L_{\text{phys}})^2}\right)$ to extrapolate to the infinite volume limit.

For the case of nuclei, which are bound states whose wavefunctions fall off exponentially at long distance, we might expect systematic errors to scale as $e^{-R/L_{\text{phys}}}$, where R is the characteristic size of the bound state. In general, one may also need to consider effects from interactions between images produced due to the periodic boundary conditions. For example, if the interaction between images is mediated at long distances by the exchange of a light particle, such as a pion, then we might expect systematic errors to fall off exponentially with $\sim (m_\pi L_{\text{phys}})$. Note that this type of finite volume effect is not accounted for by the Lüscher formalism; this is because in order to derive Eq. (5.174) we had to assume that all interactions were point-like.

Finally, we should briefly discuss systematic errors associated with temporal discretization. These tend to be far less worrisome for zero temperature results for several reasons. The first is due to the relation $b_\tau = \frac{b_s^2}{M}$ for non-relativistic theories, indicating that temporal discretization errors are of lower order than spatial discretization errors. Furthermore, our tuning method for improving the kinetic and interaction operators also translates into an improved temporal derivative operator. The lattice temporal derivative is given by the finite difference,

$$\partial_\tau \psi \sim \psi_{\tau+1} - \psi_\tau \sim (\mathscr{T} - 1)\,\psi_\tau\,, \tag{5.193}$$

where on the right hand side I have used the knowledge that the transfer matrix $\mathscr{T}$ is our time-translation operator. By perfecting the transfer matrix with our tuning method, we are in turn perfecting the single time hop operation, thereby reducing temporal discretization errors.

We also have the freedom to use the anisotropy parameter M to tune the temporal lattice spacing to be intrinsically smaller than the spatial lattice spacing. However, it should be noted that because the temperature is controlled by the physical Euclidean time length, $1/(b_\tau N_\tau)$, increasing the anisotropy parameter M will necessitate an increase in the number of temporal lattice points to reach the zero temperature limit. On the other hand, having a finer temporal lattice spacing may also help to better resolve plateaus occurring within a short "golden window" before the noise begins to set in, due to the increase in the number of points available for fitting. For this reason, anisotropic lattices are sometimes used in lattice QCD for noisy systems. However, points corresponding to a finer temporal lattice spacing are also more correlated, so it is currently unclear whether anisotropic lattices are actually beneficial for resolving noisy signals.

5.5 Beyond Leading Order EFT

The first step away from unitarity and toward real nuclear physics that we can easily take is to introduce a four-component nucleon field, N, containing two flavors of spin up and spin down fermions. The nucleons have two allowed s-wave scattering channels, ${}^1S_0$ and ${}^3S_1$, which should be tuned independently (breaking the approximate $SU(4)$ symmetry between the nucleons) to give the physical nucleon-nucleon scattering lengths. One possible way to achieve this is to introduce two four-fermion interactions corresponding to,

$$\mathscr{L}_{\text{int}} = -\frac{1}{2}g_S\left(N^\dagger N\right)^2 - \frac{1}{2}g_T\left(N^\dagger \boldsymbol{\sigma} N\right)^2 , \tag{5.194}$$

where σ_i is a Pauli matrix acting on the spin indices, and g_S, g_T are couplings for the spin singlet and spin triplet channel, respectively. The lattice version of this interaction requires the introduction of two independent auxiliary fields, ϕ_S, ϕ_T. One possibility is,

$$\mathscr{L}_{\text{int}}^{(L)} = \sqrt{g_S}\phi_S N^\dagger N + \sqrt{g_T}\phi_T\boldsymbol{\sigma}\cdot N^\dagger \boldsymbol{\sigma} N . \tag{5.195}$$

There are, in fact, many ways to implement the same interactions, and the different implementations will affect the signal-to-noise ratios of observables. For example, one could imagine having one of the ϕ fields couple to both channels equally (the $SU(4)$ limit), tuned to give the scattering length of the more attractive channel,${}^3S_1$, then adding a second auxiliary field coupling only to the ${}^1S_0$ channel and tuning this coupling to be repulsive, making this channel more weakly attractive as desired. As we learned in Sect. 5.3.1, repulsive interactions cause severe sign and noise problems, so this would clearly be a poor choice of implementation.

Let's look at the signal-to-noise ratio for a two-particle correlator in the ${}^1S_0$ channel using the interaction shown above, Eq. (5.195), where neither interaction is repulsive, but their relative strengths are different. The signal goes like,

$$\langle K_n^{\uparrow-1}(\tau)K_n^{\downarrow-1}(\tau)\rangle \sim e^{-E_0^{({}^1S_0)}\tau} . \tag{5.196}$$

while the noise is given by,

$$\sigma^2 \sim \langle K_n^{\uparrow-1}(\tau)K_n^{\downarrow-1}(\tau)\rangle K_{n'}^{\uparrow-1}(\tau)K_{n'}^{\downarrow-1}(\tau)\rangle \sim e^{E_B^{(4)}\tau} , \tag{5.197}$$

where n' denotes a particle of different flavor from n, and $E_B^{(4)}$ is the binding energy of a four particle, four flavor state. This causes a signal-to-noise problem which is similar to our original two-body correlator, however, in this case the problem is exacerbated by the fact that particles in Eq. (5.197) having different flavor index interact through the most attractive channel, ${}^3S_1$. This results in a greater disparity between the energies governing the signal and the noise, leading to more severe exponential decay of the signal-to-noise ratio. Unequal interactions can also lead to problems with reweighting methods designed to alleviate an overlap problem if the desired reweighting factor is no longer real or positive.

One method, devised by the Bonn-Raleigh group (for a review, see e.g. [131]), for avoiding the extra noise caused by unequal interactions in the two s-wave channels, is to use an $SU(4)$ symmetric transfer matrix, $\mathcal{T}_{SU(4)}$, to evolve the system for several time steps before applying the full asymmetric transfer matrix. This process may be thought of as utilizing several applications of $\mathcal{T}_{SU(4)}$ in order to produce a better interpolating wavefunction from some initial guess wavefunctions, $\Psi_{i,f}$, which is then used as a source for the correlation function,

$$C(\tau) = \langle \Psi_f | \mathcal{T}_{SU(4)}^{\tau'} \mathcal{T}^{\tau} \mathcal{T}_{SU(4)}^{\tau'} | \Psi_i \rangle = \langle \tilde{\Psi}_f | \mathcal{T}^{\tau} | \tilde{\Psi}_i \rangle , \quad (5.198)$$

where $|\tilde{\Psi}_i\rangle \equiv \mathcal{T}_{SU(4)}^{\tau'} |\Psi_i\rangle$. Using this method reduces the number of times the noisier $\mathcal{T}$ must be used because the system begins in a state that is already closer to the true ground state.

Another method used by the same group to reduce noise is to perform a Fierz transformation on the four-fermion interactions in order to define interactions with more symmetric couplings [132]. Using the identity,

$$\left(N^\dagger N\right)^2 = -\frac{1}{2}\left(N^\dagger \boldsymbol{\sigma} N\right)^2 - \frac{1}{2}\left(N^\dagger \boldsymbol{\tau} N\right)^2 , \quad (5.199)$$

we can rewrite the four-fermion interactions, Eq. (5.194), to give the following,

$$\tilde{\mathcal{L}}_{\text{int}} = -\frac{1}{2} g_0 \left(N^\dagger N\right)^2 - \frac{1}{2} g_I \left(N^\dagger \boldsymbol{\tau} N\right)^2 , \quad (5.200)$$

where τ_i is a Pauli matrix acting on the flavor components of N, and the couplings $g_{0,I}$ are related to the original couplings by,

$$g_0 = g_S - 2g_T , \qquad g_I = -g_T . \quad (5.201)$$

5.5.1 Tuning the Effective Range

The method outlined in Sect. 5.2.2.5 was devised as a way to allow us to tune our couplings to reproduce any physical scattering phase shift using the Lüscher finite volume method. We were able to successfully tune the system to unitarity, where the effective range and all higher shape parameters vanish. For nucleon scattering, the effective ranges in the s-wave channels are given roughly by the Compton wavelength of the pion, so the next logical step in our quest toward nuclear physics should be to try to tune our coefficients to give the physical effective ranges.

Unfortunately, a problem arises for producing a non-zero effective range non-perturbatively using point-like interactions in combination with a lattice regulator.

The choice of regulator is relevant when attempting to perform non-perturbative calculations because EFTs in general are non-renormalizable. However, they should be renormalizable order by order in perturbation theory, because at each order we introduce a new operator having the correct dimensions and symmetries to act as a counterterm, absorbing infinities from loops containing lower order interactions. Lattice methods incorporate the Lagrangian of the theory non-perturbatively, effectively summing the entire subset of diagrams for each interaction. In principle, such a formulation may also require the introduction of an infinite number of counterterms to absorb the divergences from all loop diagrams.

In certain cases, however, this situation can be avoided. An example is our non-perturbative tuning of the scattering length. Recall that all bubble diagrams involving only the coupling g_0 were separable; this allowed us to write the non-perturbative scattering amplitude as a geometric sum, and we were able to absorb all loop divergences into the single coupling, g_0. The condition of separability for loop diagrams containing interactions which carry momenta is dependent on the choice of regulator. Our choice of a lattice regulator, which is similar to a momentum cutoff, leads to a bound, known as the Wigner bound, on the allowed effective ranges one can access non-perturbatively [133–135].

Because the general tuning method introduced in Sect. 5.2.2.5 involves the numerical calculation of the transfer matrix, understanding the Wigner bound in this context is difficult. To better illustrate the issue, let's attempt to tune the effective range instead using the first method for tuning, outlined in Sect. 5.2.1.2. This method involves calculating the scattering amplitude and tuning the couplings to match the desired scattering parameters directly from the effective range expansion.

We will again calculate a sum of bubble diagrams, however, we must now include an interaction of the form $\mathscr{L}_{\text{int}} \sim g_2 \psi^\dagger \nabla^2 \psi$, which we would like to use to tune the effective range. We will largely follow the discussion of Phillips et al. [136]. A generic integral from one of these diagrams will have the form,

$$I_{2n} = \frac{1}{2\pi^2} \int dq \frac{q^{2+2n}}{E - q^2/M}, \tag{5.202}$$

where $n = 0, 1, 2$, depending on which of the two interactions we have at the two vertices. Since we are interested in the renormalizability of the scattering amplitude, we will separate out the divergent pieces of such an integral by expanding around $q \to \infty$,

$$I_{2n} = \frac{1}{2\pi^2} \int dq \left[Mq^{2n} - EM \int dq \frac{q^{2n-2}}{E - q^2/M} \right], \tag{5.203}$$

and investigate the integrals using different regularization schemes. The above relation may be iterated for a given n until the remaining integral is finite. The lowest order integral that we will need is given by,

$$I_0 = -\frac{1}{2\pi^2}\int dq \frac{q^2}{E - q^2/M} \,. \tag{5.204}$$

We evaluated this integral previously using a cutoff, $\pi\Lambda/2$, to find,

$$I_0 = \frac{M}{4\pi}\left[\Lambda + iME\right] \qquad \text{(cutoff)} \,. \tag{5.205}$$

Using dimensional regularization (dim reg), on the other hand, eliminates power-law divergences, so the result becomes,

$$I_0 = \frac{M}{4\pi} iME \qquad \text{(dim reg)} \,. \tag{5.206}$$

The other two integrals we will need have two and four additional powers of the momentum. Using our relation, Eq. (5.203), we can write,

$$I_2 = MEI_0 - \lambda_2 \,, \tag{5.207}$$

where

$$\lambda_2 = \frac{M}{2\pi^2}\int dq q^2 = \begin{cases} -\frac{M\pi}{48}\Lambda^3 & \text{cutoff} \\ 0 & \text{dim reg} \end{cases} \,, \tag{5.208}$$

and

$$I_4 = MEI_2 - \lambda_4 \,, \tag{5.209}$$

where

$$\lambda_4 = \frac{M}{2\pi^2}\int dq q^4 = \begin{cases} -\frac{M\pi^3}{320}\Lambda^5 & \text{cutoff} \\ 0 & \text{dim reg} \end{cases} \,. \tag{5.210}$$

From these results we see that dim reg leads to a separable interaction because each of the integrals can be written in terms of I_0 times some overall factor. On the other hand, the cutoff introduces new terms which cannot be factorized.

In order to evaluate the scattering amplitude more generally for a non-separable interaction we must solve a matrix equation. We will set this up by noting that the interaction can be written,

$$V(p,p') = \sum_{i,j=0}^{1} p'^{2i} v_{ij} p^{2j} \,, \tag{5.211}$$

where

$$v = \begin{pmatrix} g_0 & g_2 \\ g_2 & 0 \end{pmatrix} . \tag{5.212}$$

The amplitude is then,

$$A = -\sum_{i,j=0}^{1} (ME)^{i+j} a_{ij} , \tag{5.213}$$

where

$$a = v + v\mathcal{I} a , \qquad \mathcal{I} = \begin{pmatrix} I_0 & I_2 \\ I_2 & I_4 \end{pmatrix} . \tag{5.214}$$

We can now solve for a,

$$a = [1 - v\mathcal{I}]^{-1} v = \frac{1}{\lambda} \begin{pmatrix} g_0 + g_2^2 I_4 & g_2(1 - g_2 I_2) \\ g_2(1 - g_2 I_2) & g_2^2 I_0 \end{pmatrix} , \tag{5.215}$$

where

$$\lambda \equiv 1 - g_0 I_0 - 2g_2 I_2 + g_2^2 (I_2^2 - I_0 I_4) . \tag{5.216}$$

Finally, we have

$$\begin{aligned} \frac{1}{A} &= -\frac{(g_2\lambda_2 - 1)^2}{g_0 + g_2[ME(2 - g_2\lambda_2) + g_2\lambda_4]} + I_0 \\ &= \frac{M}{4\pi} \left(-1/a + 1/2r_0 ME - i\sqrt{ME} \right) , \end{aligned} \tag{5.217}$$

where I have used the effective range expansion for the inverse scattering amplitude on the right hand side.

This expression may be used to determine the couplings $g_{0,2}$ in terms of the effective range parameters, a, r_0, by expanding the left hand side in powers of ME, and comparing the resulting coefficients to the corresponding parameters in the effective range expansion. The leading order is,

$$\left.\frac{1}{A}\right|_{E=0} = -\frac{(g_2\lambda_2 - 1)^2}{g_0 + g_2^2\lambda_4} + I_0|_{E=0} = -\frac{M}{4\pi a} , \tag{5.218}$$

while the next order gives,

$$\left[\frac{\partial}{\partial(ME)} \frac{1}{A} \right]_{E=0} \frac{g_2 \left(I_0|_{E=0} + \frac{M}{4\pi a} \right)^2)(2 - g_2\lambda_2)}{(g_2\lambda_2 - 1)^2} = \frac{M}{8\pi} r_0 . \tag{5.219}$$

Using these two expressions and the above relations for λ_n and I_0, we can derive the following dependence of the effective range on the couplings for a theory regularized using dim reg,

$$r_0 = \frac{Mg_2}{\pi a^2} \,. \tag{5.220}$$

Because the effective range is proportional to the coupling g_2, it can be tuned arbitrarily. Thus, as expected from the separability of the interaction, there are no issues with renormalizability when using dim reg.

Let us now see what happens for the case of a cutoff. The relation becomes,

$$\begin{aligned} r_0 &= \frac{8\pi}{M}\left(\frac{M}{4\pi a} + I_0|_{E=0}\right)^2 \left[\frac{1}{(g_2\lambda_2 - 1)^2\lambda_2} - \frac{1}{\lambda_2}\right] \\ &= \frac{M}{2\pi}(1/a + \Lambda)^2 \left[-\frac{1}{\left(g_2 \frac{M\pi}{48}\Lambda^3 - 1\right)^2 \frac{M\pi}{48}\Lambda^3} + \frac{48}{M\pi\Lambda^3}\right] . \end{aligned} \tag{5.221}$$

We should now attempt to remove the cutoff by taking, $\Lambda \to \infty$,

$$r_0 \underset{\Lambda\to\infty}{\longrightarrow} -\frac{\frac{M}{2\pi}\Lambda^2}{(g_2 \frac{M\pi}{48}\Lambda^3 - 1)^2 \frac{M\pi}{48}\Lambda^3} \,, \tag{5.222}$$

where I have kept the first term in square brackets in Eq. (5.221) because there g_2 may be renormalized to absorb factors of Λ. Because g_2 must be real to ensure a Hermitian Hamiltonian, this expression shows that if we attempt to remove the cutoff of the theory, we are only allowed to tune $r_0 \leq 0$.

More generally, Wigner showed that for any potential which obeys $V(r, r') \to 0$ for $r, r' > R$ sufficiently quickly for some characteristic radius R, then

$$r_0 \leq 2\left(R - \frac{R^2}{a} + \frac{R^3}{3a^2}\right) . \tag{5.223}$$

For a potential generated using delta function interactions and a momentum cutoff, $R \sim 1/\Lambda$, and we arrive at our expression $r_0 \leq 0$.

In our lattice formulation the interactions are generated by an auxiliary field extending across a single time link, so that $R \sim b_s$. Therefore, if we try to tune r_0 non-perturbatively via the inclusion of such interactions in the Lagrangian, we are limited to $r_0 \lesssim b_s$. This was not a problem when we considered unitarity, since at this point $r_0 = 0$. For nuclear physics, this bound restricts us to tuning the effective range to be smaller than the lattice spacing, implying that there is no continuum limit to the theory. On the other hand, the theory we are attempting to simulate is only an effective theory of nucleons, valid up to a physical cutoff. Thus, so long as we do not attempt to probe physics beyond scales of order $\sim 1/r_0$ there will be no inconsistencies. This is clearly a limitation, however, and also restricts our ability to vary the lattice spacing when studying discretization effects.

One possibility for avoiding this restriction is to include the effective range contribution to observables perturbatively, keeping the renormalizability of the effective theory intact. Perturbative corrections may be added by expanding the transfer matrix,

$$\mathcal{T} \approx e^{-H_0 b_\tau} - b_\tau \delta H e^{-H_0 b_\tau} , \tag{5.224}$$

where $H = H_0 + \delta H$ is the full Hamiltonian and δH is the piece we wish to treat perturbatively. Multiple insertions of δH may be included to reach higher orders in the effective theory.

5.5.2 Including Pions

If we wish to probe energies of order the pion mass we must include pions explicitly into the effective theory. Unfortunately, pions are notoriously difficult to include in a consistent power counting scheme. Here, we will only briefly outline some of the issues related to power counting for pion contributions.

The KSW expansion proposed that pion exchange be treated as a series of perturbative corrections to the leading order pionless EFT [32–34]. In this case, a tree level one pion exchange (1PE) diagram may be given by Fleming et al. [137],

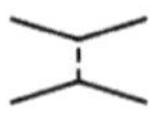

$$\sim \frac{g_A^2}{2f_\pi^2} f\left(\frac{p}{m_\pi}\right) , \tag{5.225}$$

where solid lines represent nucleons, dashed lines represent pions, g_A is the axial coupling, f_π is the pion decay constant, and $f(p/m_\pi)$ is a dimensionless function. By comparison, at one loop there is a box diagram,

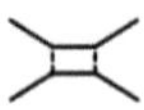

$$\sim \left(\frac{g_A^2}{2f_\pi^2}\right)^2 \frac{Mm_\pi}{4\pi} \tilde{f}\left(\frac{p}{m_\pi}\right) . \tag{5.226}$$

Note that the factor of the nucleon mass, a large energy scale for the effective theory, comes from diagrams in which intermediate nucleons can go on-shell. This implies that an expansion parameter for the set of ladder diagrams is approximately,

$$\frac{g_A^2 M m_\pi}{8\pi f_\pi^2} \sim 0.5 , \tag{5.227}$$

and that the expansion may converge very slowly. In practice, the convergence for this formulation might be acceptable in the $^1S_0$ scattering channel, but is poor in the spin triplet channel. This is likely due to the singular tensor force contribution to the two-nucleon potential in this channel, which we will discuss in a moment [137].

Weinberg's formulation for nuclear EFT involves summing a subset of diagrams non-perturbatively, by first calculating a nucleon-nucleon potential diagrammatically, then using this potential to solve the Schrodinger equation. In doing so, we can take into account higher orders in a perturbative expansion that breaks down or converges slowly. For the pions, tree-level pion exchange diagrams give rise to the following 1PE potential [36]:

$$V_{\text{1PE}}(\mathbf{r}) = \left(\frac{g_A}{2f_\pi}\right)^2 \boldsymbol{\tau}_1 \cdot \boldsymbol{\tau}_2 \left[m_\pi^2 \frac{e^{-m_\pi r}}{12\pi r} \left(S_{12}(\hat{r}) \left(1 + \frac{3}{m_\pi r} + \frac{3}{(m_\pi r)^2}\right) + \boldsymbol{\sigma}_1 \cdot \boldsymbol{\sigma}_2 \right) - \frac{1}{3} \boldsymbol{\sigma}_1 \cdot \boldsymbol{\sigma}_2 \delta^3(r) \right] , \tag{5.228}$$

where $S_{12} = 3\boldsymbol{\sigma}_1 \cdot \hat{r} \boldsymbol{\sigma}_2 \cdot \hat{r} - \boldsymbol{\sigma}_1 \cdot \boldsymbol{\sigma}_2$.

The most divergent part of this potential, scaling like $\sim 1/r^2$, comes from the tensor force in the spin triplet channel. Attractive potentials which scale as r^{-n} for $n \geq 2$ are referred to as singular potentials. Particles sitting in a singular potential eventually fall toward the center with infinite velocity, which is clearly unphysical. Thus, singular potentials can only be defined with an explicit cutoff that cannot be removed. Particles generally sit near this cutoff, rendering the system sensitive to the short-range details of the choice of boundary condition. Therefore, systems involving singular potentials are generally model dependent and we can no longer have a true effective theory because the cutoff cannot be removed.

The reason such a singular potential arises is similar to that which led to the Wigner bound in the previous section. Again, we are attempting to sum a subset of diagrams in an effective theory non-perturbatively, which cannot in general be assumed to be a renormalizable process. In practice, nuclear theorists using so-called chiral potentials are generally able to demonstrate that the cutoff dependence is small so long as the cutoff is only varied within a particular range, typically $\Lambda \sim 300$–$1000\,\text{MeV}$. Therefore, if we wish to include pions non-perturbatively in our lattice theory we should keep this in mind as it implies a restriction on the allowed lattice spacings, just as we found for the non-perturbative inclusion of effective range contributions.

Pion fields may be added directly to our lattice Lagrangian in a straightforward way. The incorporation of dynamical pions, however, will likely complicate importance sampling by introducing noise and/or sign problems, and adds complexity to the Monte Carlo algorithms. Fortunately fully dynamical pions are unnecessary; all we actually seek is the addition of a term in the Lagrangian which generates the tree level diagrams between a single pion and two nucleons. The lattice formulation then non-perturbatively accounts for all possible loop diagrams involving this pion-nucleon interaction. Diagrams involving vacuum pion loops, pion self-energies, etc. are higher order in our chiral expansion and can be included perturbatively if necessary.

One possible implementation utilized by the Bonn-Raleigh group is to use static pion auxiliary fields, $\pi^{(I)}_{\mathbf{x},\tau}$, with isospin I, and the following action [131, 132]:

$$S_{\pi\pi} = \left(\frac{m_\pi^2}{2} + 3\right) \sum_{\mathbf{x},\tau,I} \pi^{(I)}_{\mathbf{x},\tau} \pi^{(I)}_{\mathbf{x},\tau} - \sum_{\mathbf{x},\tau,I,k} \pi^{(I)}_{\mathbf{x},\tau} \pi^{(I)}_{\mathbf{x}+\hat{k},\tau} . \qquad (5.229)$$

Because the pions are derivatively coupled to the nucleons, the interaction term should behave like,

$$S_{\pi NN} \sim \frac{g_A}{2f_\pi} \sum_{I,k} \left[\pi^{(I)}_{\mathbf{x}+\hat{k}} - \pi^{(I)}_{\mathbf{x}-\hat{k}}\right] \psi^\dagger_{\mathbf{x}} \psi_{\mathbf{x}} , \qquad (5.230)$$

(see [131] for more details on the particular interaction chosen). The pions have been chosen to only couple to the nucleons through spatial displacements. This simplifies the analysis by eliminating the renormalization of the nucleon mass through nucleon self-energy diagrams such as:

Then we can simply utilize the physical value, $M \sim 938\,\mathrm{MeV}$, for the nucleon mass. These pions therefore act instantaneously, much the same way as they do in a pion potential picture.

5.5.3 3- and Higher-Body Interactions

Naïve dimensional analysis dictates that the leading three-body interaction should be suppressed relative to the two-body interaction by $\mathcal{O}(L^3)$. We should be more cautious by this point, since we have seen dimensional analysis fail in previous cases for strongly interacting systems. For that reason, we will now inspect the three-body system more carefully.

To begin, we will consider a system of three particles interacting via only the simplest, leading order two-body contact interaction. We will follow the discussion of Braaten et al. [138]. Let us assume that all three particles carry different quantum numbers, as they do for the triton and ^{3}He, and that all pairs of particles interact via the same two-body coupling, g_0. To calculate the three-particle scattering amplitude for a strongly coupled system we must iterate this interaction non-perturbatively, as we did for the two-particle system.

A useful trick for calculating this quantity is the addition of a bosonic dimer field, ϕ, coupling to two fermion particles, ψ. This allows us to rewrite the three-particle scattering amplitude in the form of a two-particle scattering amplitude. The dimer propagator must be fully dressed by fermion loop bubbles and can be written diagrammatically as shown in Fig. 5.13. This bubble sum is essentially the same as

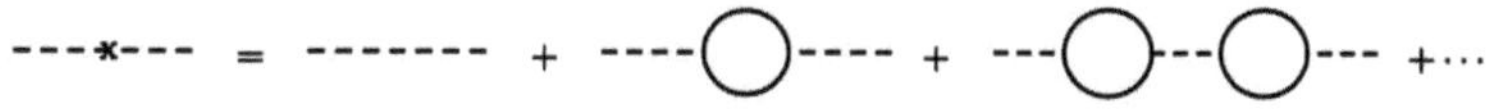

Fig. 5.13 Dressed propagator for the bosonic dimer field, ϕ

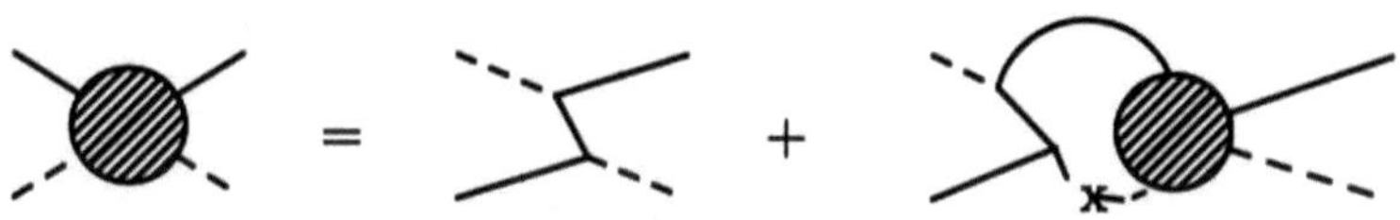

Fig. 5.14 Full three-particle scattering amplitude written in terms of a two-particle amplitude for a fermion scattering with a dimer field. Here we have only included two-body interactions, with no explicit three-body contact interaction

the one we have encountered several times before in these lectures. However, we must now allow external momentum, $(p_0, \mathbf{p})$ to flow through the diagrams, leading to the following dressed propagator for the dimer field,

$$D_0(p_0, \mathbf{p}) = \frac{1}{1 - g_0\, I_0|_{E=p_0-p^2/M}} = \frac{1/a - \Lambda}{1/a + i\sqrt{Mp_0 - p^2 - i\epsilon}}, \qquad (5.231)$$

where I've used the results from Sect. 5.2.1.2 to rewrite the coupling in terms of the scattering length, a, and the cutoff, Λ. We see that the dimer propagator has a pole at $p_0 = \frac{p^2}{M} - \frac{1}{Ma^2}$, corresponding to a (virtual) bound state for (negative) positive scattering length with energy $E_B = \frac{1}{Ma^2}$.

Using this dimer field, we can write the full three-body scattering amplitude, A_3, as an integral equation, shown in Fig. 5.14. To simplify the expression, we can set the ψ fields to be on-shell, so that all off-shell properties are absorbed into the dimer propagator. The amplitude can then be written,

$$\begin{aligned} A_3(p,k;E,p^2/M) = &-\frac{g_0}{E - p^2/M - k^2/M - (p+k)^2/M + i\epsilon} \\ &+\frac{8\pi i}{g_0}\int \frac{d^4q}{(2\pi)^4}\left(\frac{g_0}{E - p^2/M - q_0 - (p+q)^2/M + i\epsilon}\right) \\ &\times \left(\frac{1}{q_0 - q^2/M + i\epsilon}\right)\left(\frac{A_3(q,k;E,q_0)}{1/a + i\sqrt{M(E-q_0) + q^2 - i\epsilon}}\right), \end{aligned} \qquad (5.232)$$

known as the Skorniakov-Ter-Martirosian (STM) integral equation. Integrating over q_0 and projecting the system onto the s-wave channel gives (see [138] for more details),

$$
\begin{aligned}
\tilde{A}_3(p,k;E) = & \frac{1}{apk}\ln\left(\frac{p^2+pk+k^2-ME-i\epsilon}{p^2-pk+k^2-ME-i\epsilon}\right) \\
& + \frac{1}{4\pi^2}\int^{\Lambda} dq\frac{q}{p}\ln\left(\frac{p^2+pq+q^2-ME-i\epsilon}{p^2-pq+q^2-ME-i\epsilon}\right) \\
& \frac{\tilde{A}_3(q,k;E)}{-1/a+\sqrt{3q^2-ME-i\epsilon}} \,.
\end{aligned}
\tag{5.233}
$$

For large scattering length (strong interaction) we have,

$$
\tilde{A}_3(p,k;E) \underset{a\to\infty}{\longrightarrow} \frac{1}{4\pi^2}\int^{\Lambda} dq\frac{q}{p}\ln\left(\frac{p^2+pq+q^2-ME-i\epsilon}{p^2-pq+q^2-ME-i\epsilon}\right)\frac{\tilde{A}_3(q,k;E)}{\sqrt{3q^2-ME-i\epsilon}}
\tag{5.234}
$$

This integral contains divergences, which may be renormalized by adding an explicit three-body coupling, H. To absorb the divergences, the coupling must have the following dependence on the momentum cutoff, Λ [139–141]:

$$
H(\Lambda) = \frac{\cos\left[s_0\ln(\Lambda/\Lambda_*)+\tan^{-1}s_0\right]}{\cos\left[s_0\ln(\Lambda/\Lambda_*)-\tan^{-1}s_0\right]} \,,
\tag{5.235}
$$

where $s_0 \sim 1.006$ is a constant, and Λ_* is some reference scale which may be set by a three-body observable, such as the triton binding energy, or the neutron-deuteron scattering length.

There are two remarkable things to note here: the first is that this result for the scattering amplitude is only a leading order result, yet we had to introduce a three-body coupling in order to renormalize the theory. This illustrates another case where naïve dimensional analysis does not work, because the three-body coupling contributes at the same order as the two-body coupling. The second is the running of the coupling $H(\Lambda)$, plotted on a logarithmic scale in Fig. 5.15. We see that the coupling, and therefore also observables depending on the coupling, displays a log-periodic discrete scaling symmetry, related to the so-called Efimov effect. This property arises for systems obeying a potential at the threshold of singularity, $\sim 1/r^2$, as can be shown to occur for our three-body system using hyperspherical coordinates [142, 143].

Because the three-body interaction has been demonstrated to be relevant at leading order, we should in general include it non-perturbatively to our lattice theory by adding an interaction term to the Lagrangian such as,

$$
C_3\phi_3\psi_\tau^\dagger\psi_{\tau+1} \,,
\tag{5.236}
$$

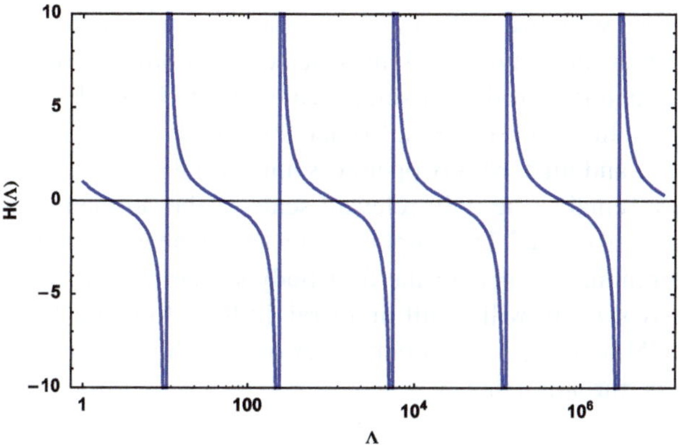

Fig. 5.15 Running of the three-body contact interaction $H(\Lambda)$ at unitarity vs. the momentum cutoff, Λ, showing log-periodicity

where C_3 is tuned to reproduce some three-body observable, and $\phi_3 \in Z_3$ (cube roots of 1). However, ϕ_3 is necessarily a complex field, will induce severe noise and/or sign problems. The interaction may alternatively be introduced via multiple Z_2 interactions, but the noise problem remains.

Figure 5.15 is important for our discussion because it shows how the three-body coupling runs as we change the lattice spacing. The larger the coupling, the worse the noise/sign problem will be. The solution chosen by the Bonn-Raleigh group is to tune the ratio b_τ/b_s until a chosen three-body observable is sufficiently well-described by tuning only the two-body interactions. This implies that the three-body interaction is small at this point, and can then be regarded as a higher-order correction and included perturbatively. A drawback to this approach is that we can no longer use the anisotropy parameter as a knob for probing temporal discretization errors. Because the spatial lattice spacing may also already be restricted by the condition of renormalizability of any pion or effective range contributions to the Lagrangian, we have forfeited most of our ability to demonstrate that discretization errors are under control.

Another possibility for reducing the contribution from the three-body interaction might be to change the short-distance behavior of the two-body sector in another way. For example, tuning different numbers of two-body interaction coefficients (Sect. 5.2.2.5) or changing the discretization of the kinetic operator will shift the reference scale Λ^*, giving us a different value for $H(\Lambda)$ at a fixed lattice spacing.

Finally, given that the three-body sector required a reshuffling of the orders in perturbation theory at strong coupling, should we expect the same for higher N-body interactions? Fortunately it has been fairly well established that four- and higher body operators are not necessary to renormalize the theory at leading order and are therefore irrelevant. This means that we may treat four- and higher-body interactions as perturbative corrections.

This is observed via the so-called Tjon line (see, e.g. [144]). Recall that while the two-body system at unitarity has no intrinsic scale, in order to describe the three-body system we had to introduce a single scale, Λ_*, to be set by some three-body observable. Once this scale is set, all other three-body observables may then be predicted. If four- and higher-body operators appear only at higher orders, then this three-body scale remains the only relevant scale in the problem, and observables must be proportional to Λ_*.[6] This implies that varying the three-body parameter Λ_*, in a plot of the binding energy for the four-body system versus the binding energy of the three-body system, will result in a straight line. Any non-linear dependence on higher-order N-body operators contributes only within the error band predicted at this order in perturbation theory.

5.5.4 Final Considerations

Perhaps the most worrisome issue we have discussed is the inability to take the continuum limit due to interactions that are included non-pertubatively and which generate new non-zero scales beyond the scattering length. The lattice spacing must also be kept reasonably large for another reason mentioned previously, related to numerical stability: if the lattice spacing becomes too small, the system will begin to probe the repulsive core of the two-body potential, leading to sign and/or signal-to-noise problems.

Though we may not have the ability to vary the lattice spacing by significant amounts, we must still prove that our results do not depend strongly on the short-distance details of the action. This can be demonstrated instead by changing the discretization of derivatives in the action, using more or less improvement of the interaction, etc., and showing that the results do not change significantly [131].

Showing convergence of the EFT for the lattice results is also a major concern, particularly since we have no single power-counting scheme that is known to converge in all channels even in the continuum theory. One possible indication of issues with convergence in the current Bonn-Raleigh method is the need for a significant repulsive four-body interaction in order to stabilize four- and higher-body systems, which seem prone to forming four-body clusters on a single lattice site. This is akin to the particles falling to the bottom of a singular potential, and may be related to the particular tuning of the three-body interaction. However, once

[6]This single scale is also critical for the appearance of the log-normal distribution in correlators near unitarity, where the moments are given by

$$\mathcal{M}_N \sim e^{-E_{\text{N-body}}\tau} \sim e^{-f(N)\Lambda_*\tau}\,. \tag{5.237}$$

Numerical evidence was shown in [51] that $f(N)$ has the expected form for the log-normal distribution.

this interaction has been set the convergence of the results appears to be relatively stable.

Possibly the biggest open issues to be resolved are the sign/noise problems and proving convergence to the ground (or desired excited) state. Noise problems have restricted most calculations of nuclear systems to nuclei in (or near) the alpha ladder, where approximate $SU(4)$ symmetry applies. New theories and/or algorithms would be enormously helpful in this arena. The engineering of better sources or methods for extracting the desired states might be particularly beneficial for both the reduction of noise and to eliminate the need for performing long temporal extrapolations.

Despite these limitations there have been enormous successes for lattice EFT for few- and many-body states both for systems at unitarity and nuclei. As an example, at unitarity the energies of up to 50 two-component fermions have been calculated with errors comparable to state-of-the-art Green's Function Monte Carlo calculations [53, 126, 145–148]. The Raleigh-Bonn group has calculated properties of nuclei up to $A = 28$ [149–153]. Particularly exciting is their investigation of the structure of the Hoyle state, a key component of the triple alpha process necessary for Carbon production in stars [154–158].

5.6 Reading Assignments and Exercises

5.1 Much of these lecture notes follow this review: arXiv:1208.6556. There you will also find more information about algorithms. The following is an excellent pedagogical introduction to EFT's by David B. Kaplan: arXiv:nucl-th/0510023.

5.2 Explore the cumulant expansion using a toy model [145]:

$$C(\tau, \phi) = \prod_{i=1}^{\tau}(1 + g\phi_i) , \tag{5.238}$$

for $0 \leq g \leq 1$ and $\phi \in [-1, 1]$. The true mean of the correlator should be $\langle C(\tau, \phi)\rangle = 1$, corresponding to $E_0 = 0$. Compare the cumulant expansion cut off at various orders on a finite sample size to the mean calculated using standard methods as the sample size is varied.

5.3 Reading: D. Lee: arXiv:0804.3501 [131] G.P. Lepage: Analysis of algorithms for lattice field theory [1].

5.4 Add a term

$$c\psi_\tau^\dagger \nabla_L^2 \psi_{\tau-1} \tag{5.239}$$

to the simple interaction, Eq. (5.41), and derive an analytic expression for tuning the couplings, g_0 and c in order to eliminate the effective range contribution. You may use either the scattering amplitude or the transfer matrix method.

5.5 Write numerical code (Mathematica will suffice) to solve the transfer matrix for two particles for a chosen set of coefficients, g_{2n} [Eq. (5.176)], using $L = 32, M = 5$, and tune your coefficients to match the first few expected Lüscher eigenvalues at unitarity. Compare your results with those in Table II of [126].

Acknowledgements The author would like to thank Michael Endres, David B. Kaplan, and Jong-Wan Lee for extensive discussions, and especially M. Endres for the development of and permission to use this code. AN was supported in part by U.S. DOE grant No. DE-SC00046548.

Appendix

Compilation and Running the Code

This code requires the use of the FFTW library, which you may download and install from fftw.org. The script "create_lib.sh" should be run first from the head directory. Once this script is successful, you may go into the production directory, modify the script "create_binary.sh" to reflect your path to the FFTW library, and compile by running this script. The executable created is called "a.out", which should be run without specifying any additional parameters in the command line. Input parameters are specified in the files included in the "arg" folder. The parameters for each file are described in the header "arg.h". The codes can be downloaded from the link https://github.com/ManyBodyPhysics/LectureNotesPhysics/tree/master/Programs/Chapter5-programs/.

Output is created in the folder "results". The file gives a list of the values (real part listed first, imaginary second) of the two-particle correlation function calculated at different values of Euclidean time, on a set of auxiliary field configurations. The organization of the output is as follows:

$$\begin{array}{ccccccc}
\mathrm{Re}[C(\phi_1,\tau_1)] & \mathrm{Im}[C(\phi_1,\tau_1)] & \mathrm{Re}[C(\phi_1,\tau_2)] & \mathrm{Im}[C(\phi_1,\tau_2)] & \cdots & \mathrm{Re}[C(\phi_1,\tau_{N_\tau})] & \mathrm{Im}[C(\phi_1,\tau_{N_\tau})] \\
\mathrm{Re}[C(\phi_2,\tau_1)] & \mathrm{Im}[C(\phi_2,\tau_1)] & \mathrm{Re}[C(\phi_2,\tau_2)] & \mathrm{Im}[C(\phi_2,\tau_2)] & \cdots & \mathrm{Re}[C(\phi_2,\tau_{N_\tau})] & \mathrm{Im}[C(\phi_2,\tau_{N_\tau})] \\
 & & & \vdots & & & \\
\mathrm{Re}[C(\phi_{N_{\mathrm{cfg}}},\tau_1)] & \mathrm{Im}[C(\phi_{N_{\mathrm{cfg}}},\tau_1)] & \mathrm{Re}[C(\phi_{N_{\mathrm{cfg}}},\tau_2)] & \mathrm{Im}[C(\phi_{N_{\mathrm{cfg}}},\tau_2)] & \cdots & \mathrm{Re}[C(\phi_{N_{\mathrm{cfg}}},\tau_{N_\tau})] & \mathrm{Im}[C(\phi_{N_{\mathrm{cfg}}},\tau_{N_\tau})]
\end{array}$$

where N_τ and N_{cfg} are the total number of time steps, specified in "do.arg", and total number of configurations, specified in "evo.arg", respectively. To calculate the correlation function at a given time, τ, average over all values: $C(\tau) = \sum_i (\mathrm{Re}\,[C(\phi_i,\tau)] + i\,\mathrm{Im}\,[C(\phi_i,\tau)])$.

Exercises

5.6 Set the first value in the file "interaction.arg" to a coupling of your choice, and the remaining couplings to 0. Use the long time behavior of the effective mass function, $\ln \frac{C(\tau)}{C(\tau+1)} \xrightarrow[\tau\to\infty]{} E_0$ (see Sect. 5.3), to determine the ground state energy for your choice of coupling, g. Compare this with what you expect from Eq. (5.64), using the relation $\lambda = e^{-E_0}$, as the number of lattice points is increased. You may test the improved interaction, Sect. 5.2.2.5, using coefficients calculated from your code developed in Prob. 4 by setting multiple couplings in the "interaction.arg" file. Be careful to set the dispersion relation in "kinetic.arg" to match the one used in setting up your transfer matrix for the tuning.

5.7 Add a harmonic potential by setting the parameters in potential.arg. The three numerical values correspond to the spring constant, κ, for the x, y, z-directions. Set the interaction coefficients to correspond to unitarity, then find the energies of two unitary fermions in a harmonic trap, exploring and removing finite volume and discretization effects by varying the parameters, $L, L_0 = (\kappa M)^{-1/4}$, and performing extrapolations in these quantities if necessary. Compare your result to the expected value of 2ω, where $\omega = \sqrt{\kappa/M}$, and the mass M is set in the file "kinetic.arg".

5.8 Construct sources for three fermions in an $l = 0$ and $l = 1$ state and find the lowest energies corresponding to each state at unitarity. Which l corresponds to the true ground state of this system?

References

1. G.P. Lepage, *From Actions to Answers, Proceedings of the 1989 Theoretical Advanced Study Institute (TASI)* (1989)
2. E. Berkowitz, T. Kurth, A. Nicholson, B. Joo, E. Rinaldi, M. Strother, P.M. Vranas, A. Walker-Loud, Phys. Lett. B **765**, 285–292 (2017)
3. T. Kurth, E. Berkowitz, E. Rinaldi, P. Vranas, A. Nicholson, M. Strother, A. Walker-Loud, *Proceedings, 33rd International Symposium on Lattice Field Theory (Lattice 2015)* (2015)
4. A. Nicholson, E. Berkowitz, E. Rinaldi, P. Vranas, T. Kurth, B. Joo, M. Strother, A. Walker-Loud, *Proceedings, 33rd International Symposium on Lattice Field Theory (Lattice 2015)* (2015)
5. K. Orginos, A. Parreno, M.J. Savage, S.R. Beane, E. Chang, W. Detmold, Phys. Rev. D **92**, 114512 (2015)
6. W. Detmold, K. Orginos, A. Parreno, M.J. Savage, B.C. Tiburzi, S.R. Beane, E. Chang, Phys. Rev. Lett. **116**, 112301 (2016)
7. E. Chang, W. Detmold, K. Orginos, A. Parreno, M.J. Savage, B.C. Tiburzi, S.R. Beane, Phys. Rev. D **92**, 114502 (2015)
8. S.R. Beane, E. Chang, W. Detmold, K. Orginos, A. Parreño, M.J. Savage, B.C. Tiburzi, Phys. Rev. Lett. **115**, 132001 (2015)
9. S.R. Beane, E. Chang, S.D. Cohen, W. Detmold, H.W. Lin, K. Orginos, A. Parreño, M.J. Savage, Phys. Rev. D **91**, 114503 (2015)
10. S.R. Beane, E. Chang, S. Cohen, W. Detmold, H.W. Lin, K. Orginos, A. Parreno, M.J. Savage, B.C. Tiburzi, Phys. Rev. Lett. **113**, 252001 (2014)

11. S.R. Beane et al., Phys. Rev. C **88**, 024003 (2013)
12. S.R. Beane, E. Chang, S.D. Cohen, W. Detmold, H.W. Lin, T.C. Luu, K. Orginos, A. Parreno, M.J. Savage, A. Walker-Loud, Phys. Rev. D **87**, 034506 (2013)
13. S.R. Beane, E. Chang, W. Detmold, H.W. Lin, T.C. Luu, K. Orginos, A. Parreno, M.J. Savage, A. Torok, A. Walker-Loud, Phys. Rev. D **85**, 054511 (2012)
14. S.R. Beane, W. Detmold, H.W. Lin, T.C. Luu, K. Orginos, M.J. Savage, A. Torok, A. Walker-Loud, Phys. Rev. D **81**, 054505 (2010)
15. T. Yamazaki, *Proceedings, 33rd International Symposium on Lattice Field Theory (Lattice 2015)* (2015)
16. T. Yamazaki, K.I. Ishikawa, Y. Kuramashi, A. Ukawa, Phys. Rev. D **92**, 014501 (2015)
17. T. Yamazaki, K.I. Ishikawa, Y. Kuramashi, A. Ukawa, PoS **LATTICE2013**, 230 (2014)
18. T. Yamazaki, K.I. Ishikawa, Y. Kuramashi, A. Ukawa, PoS **LATTICE2012**, 143 (2012)
19. T. Yamazaki, K.I. Ishikawa, Y. Kuramashi, A. Ukawa, Phys. Rev. D **86**, 074514 (2012)
20. T. OPTdoi et al., *12th International Conference on Hypernuclear and Strange Particle Physics (HYP 2015)*, Sendai, 7–12 Sept 2015
21. T. OPTdoi et al., *Proceedings, 33rd International Symposium on Lattice Field Theory (Lattice 2015)* (2015)
22. N. Ishii, S. Aoki, T. Hatsuda, Phys. Rev. Lett. **99**, 022001 (2007)
23. K. Murano, N. Ishii, S. Aoki, T. OPTdoi, T. Hatsuda, Y. Ikeda, T. Inoue, H. Nemura, K. Sasaki, Phys. Lett. B **735**, 19 (2014)
24. S. Aoki, PoS **Hadron2013**, 020 (2013)
25. K. Murano, Few Body Syst. **54**, 1105 (2013)
26. N. Ishii, S. Aoki, T. OPTdoi, T. Hatsuda, Y. Ikeda, T. Inoue, K. Murano, H. Nemura, K. Sasaki, Phys. Lett. B **712**, 437 (2012)
27. T. Inoue, N. Ishii, S. Aoki, T. OPTdoi, T. Hatsuda, Y. Ikeda, K. Murano, H. Nemura, K. Sasaki, Prog. Theor. Phys. **124**, 591 (2010)
28. S.R. Beane, W. Detmold, T.C. Luu, K. Orginos, A. Parreno, M.J. Savage, A. Torok, A. Walker-Loud, Phys. Rev. D **80**, 074501 (2009)
29. T. OPTdoi, S. Aoki, T. Hatsuda, Y. Ikeda, T. Inoue, N. Ishii, K. Murano, H. Nemura, K. Sasaki, Prog. Theor. Phys. **127**, 723 (2012)
30. S. Weinberg, Phys. Lett. B **251**, 288 (1990)
31. S. Weinberg, Nucl. Phys. B **363**, 3 (1991)
32. D.B. Kaplan, M.J. Savage, M.B. Wise, Nucl. Phys. B **478**, 629 (1996)
33. D.B. Kaplan, M.J. Savage, M.B. Wise, Phys. Lett. B **424**, 390 (1998)
34. D.B. Kaplan, M.J. Savage, M.B. Wise, Nucl. Phys. B **534**, 329 (1998)
35. E. Epelbaum, H.W. Hammer, U.G. Meißner, Rev. Mod. Phys. **81**, 1773 (2009)
36. E. Epelbaum (2010). arXiv:1001.3229
37. R. Machleidt, D.R. Entem, Phys. Rep. **503**, 1 (2011)
38. D.B. Kaplan, Five lectures on effective field theory (2005). arXiv:Nucl-th/0510023
39. S. Giorgini, L. Pitaevskii, S. Stringari, Rev. Mod. Phys. **80**, 1215 (2008)
40. I. Block, J. Dalibard, W. Zwerger, Rev. Mod. Phys. **80**, 885 (2008)
41. J.E. Drut, A.N. Nicholson, J. Phys. G **40**, 043101 (2013)
42. S.R. Beane, W. Detmold, K. Orginos, M.J. Savage, Prog. Part. Nucl. Phys. **66**, 1 (2011)
43. W. Detmold, M.G. Endres, Phys. Rev. D **90**, 034503 (2014)
44. W. Detmold, M.G. Endres, PoS **LATTICE2014**, 170 (2015)
45. D. Grabowska, D.B. Kaplan, A.N. Nicholson, Phys. Rev. D **87**, 014504 (2013)
46. A.N. Nicholson, D. Grabowska, D.B. Kaplan (2012) [J. Phys. Conf. Ser. **432**, 012032 (2013)]
47. M.G. Endres, D.B. Kaplan, J.W. Lee, A.N. Nicholson, PoS **LATTICE2011**, 017 (2011)
48. A.C. Berry, Trans. Am. Math. Soc. **49**, 122 (1941)
49. C.G. Esseen, Arkiv for matematik, astronomi och fysik A **28**, 1 (1942)
50. T. DeGrand, Phys. Rev. D **86**, 014512 (2012)
51. A.N. Nicholson, Phys. Rev. Lett. **109**, 073003 (2012)
52. A.N. Nicholson, EPJ Web Conf. **113**, 03019 (2016)
53. M.G. Endres, D.B. Kaplan, J.W. Lee, A.N. Nicholson, Phys. Rev. A **87**, 023615 (2013)

54. J.J. Dudek, R.G. Edwards, M.J. Peardon, D.G. Richards, C.E. Thomas, Phys. Rev. D **82**, 034508 (2010)
55. C. Michael, I. Teasdale, Nucl. Phys. B **215**, 433 (1983)
56. M. Lüscher, U. Wolff, Nucl. Phys. B **339**, 222 (1990)
57. B. Blossier, M. Della Morte, G. von Hippel, T. Mendes, R. Sommer, J. High Energy Phys. **04**, 094 (2009)
58. C. Aubin, K. Orginos, PoS **LATTICE2011**, 148 (2011)
59. Y. Hua, T. Sarkar, IEEE Trans. Antennas Propag. **37**, 229 (1989)
60. T. Sarkar, O. Pereira, IEEE Antennas Propag. Mag. **37**, 48 (1995)
61. S.R. Beane, W. Detmold, T.C. Luu, K. Orginos, A. Parreno et al., Phys. Rev. D **79**, 114502 (2009)
62. G.T. Fleming, S.D. Cohen, H.W. Lin, V. Pereyra, Phys. Rev. D **80**, 074506 (2009)
63. K. Symanzik, *Recent Developments in Gauge Theories* (Plenum, New York, 1980)
64. K. Symanzik, *Mathematical Problems in Theoretical Physics* (Springer, New York, 1982)
65. K. Symanzik, Nucl. Phys. B **226**, 187 (1983)
66. K. Symanzik, Nucl. Phys. B **226**, 205 (1983)
67. G. Batrouni, A. Hansen, M. Nelkin, Phys. Rev. Lett. **57**, 1336 (1986)
68. C. Davies, G. Batrouni, G. Katz, A. Kronfeld, P. Lepage, P. Rossi, B. Svetitsky, K. Wilson, J. Stat. Phys. **43**, 1073 (1986)
69. C.T.H. Davies, G.G. Batrouni, G.R. Katz, A.S. Kronfeld, G.P. Lepage, K.G. Wilson, P. Rossi, B. Svetitsky, Phys. Rev. D **37**, 1581 (1988)
70. G. Katz, G. Batrouni, C. Davies, A. Kronfeld, P. Lepage, P. Rossi, B. Svetitsky, K. Wilson, Phys. Rev. D **37**, 1589 (1988)
71. M. Lüscher, Commun. Math. Phys. **105**, 153 (1986)
72. M. Lüscher, Nucl. Phys. B **354**, 531 (1991)
73. S.R. Beane, P.F. Bedaque, A. Parreno, M.J. Savage, Phys. Lett. B **585**, 106 (2004)
74. R.A. Briceno, Z. Davoudi, Phys. Rev. D **88**(9), 094507 (2013)
75. M.T. Hansen, S.R. Sharpe, Phys. Rev. D **86**, 016007 (2012)
76. N. Li, S.Y. Li, C. Liu, Phys. Rev. D **90**, 034509 (2014)
77. R.A. Briceno, Phys. Rev. D **89**, 074507 (2014)
78. R.A. Briceño, M.T. Hansen, Phys. Rev. D **94**(1), 013008 (2016)
79. R.A. Briceño, M.T. Hansen, Phys. Rev. D **92**, 074509 (2015)
80. R.A. Briceño, M.T. Hansen, A. Walker-Loud, PoS **LATTICE2014**, 095 (2015)
81. R.A. Briceno, M.T. Hansen, A. Walker-Loud, Phys. Rev. D **91**, 034501 (2015)
82. T. Luu, M.J. Savage, Phys. Rev. D **83**, 114508 (2011)
83. S. Koenig, D. Lee, H.W. Hammer, Phys. Rev. Lett. **107**, 112001 (2011)
84. S. Koenig, D. Lee, H.W. Hammer, Ann. Phys. **327**, 1450 (2012)
85. K. Rummukainen, S.A. Gottlieb, Nucl. Phys. B **450**, 397 (1995)
86. C. Kim, C. Sachrajda, S.R. Sharpe, Nucl. Phys. B **727**, 218 (2005)
87. S. Bour, S. Koenig, D. Lee, H.W. Hammer, U.G. Meissner, Phys. Rev. D **84**, 091503 (2011)
88. Z. Davoudi, M.J. Savage, Phys. Rev. D **84**, 114502 (2011)
89. X. Li, C. Liu, Phys. Lett. B **587**, 100 (2004)
90. X. Feng, X. Li, C. Liu, Phys. Rev. D **70**, 014505 (2004)
91. M.T. Hansen, S.R. Sharpe, Phys. Rev. D **93**, 096006 (2016)
92. M.T. Hansen, S.R. Sharpe, Phys. Rev. D **92**, 114509 (2015)
93. M.T. Hansen, S.R. Sharpe, Phys. Rev. D **90**, 116003 (2014)
94. R.A. Briceno, Z. Davoudi, Phys. Rev. D **87**, 094507 (2013)
95. S.R. Beane, W. Detmold, M.J. Savage, Phys. Rev. D **76**, 074507 (2007)
96. W. Detmold, M.J. Savage, Phys. Rev. **D77**, 057502 (2008)
97. B. Smigielski, J. Wasem, Phys. Rev. D **79**, 054506 (2009)
98. R.A. Briceno, Z. Davoudi, T.C. Luu, Phys. Rev. D **88**, 034502 (2013)
99. R.A. Briceno, Z. Davoudi, T. Luu, M.J. Savage, Phys. Rev. D **88**, 114507 (2013)
100. S.R. Beane, P.F. Bedaque, K. Orginos, M.J. Savage, Phys. Rev. Lett. **97**, 012001 (2006)
101. D.J. Wilson, J.J. Dudek, R.G. Edwards, C.E. Thomas, Phys. Rev. D **91**(5), 054008 (2015)

102. D.J. Wilson, R.A. Briceno, J.J. Dudek, R.G. Edwards, C.E. Thomas, Phys. Rev. D **92**(9), 094502 (2015)
103. J.J. Dudek, R.G. Edwards, C.E. Thomas, Phys. Rev. D **86**, 034031 (2012)
104. J.J. Dudek, R.G. Edwards, C.E. Thomas, Phys. Rev. D **87**, 034505 (2013)
105. J.J. Dudek, R.G. Edwards, C.E. Thomas, D.J. Wilson, Phys. Rev. Lett. **113**, 182001 (2014)
106. S.R. Beane, E. Chang, W. Detmold, H.W. Lin, T.C. Luu, K. Orginos, A. Parreno, M.J. Savage, A. Torok, A. Walker-Loud, Phys. Rev. D **85**, 034505 (2012)
107. S. Aoki et al., Phys. Rev. D **76**, 094506 (2007)
108. S. Aoki et al., Phys. Rev. D **84**, 094505 (2011)
109. C. Pelissier, A. Alexandru, Phys. Rev. D **87**, 014503 (2013)
110. X. Feng, K. Jansen, D.B. Renner, Phys. Rev. D **83**, 094505 (2011)
111. A. Martínez Torres, E. Oset, S. Prelovsek, A. Ramos, J. High Energy Phys. **05**, 153 (2015)
112. D.R. Bolton, R.A. Briceno, D.J. Wilson, Phys. Lett. B **757**, 50–56 (2016)
113. R.A. Briceno, J.J. Dudek, R.G. Edwards, C.J. Shultz, C.E. Thomas, D.J. Wilson, Phys. Rev. Lett. **115**, 242001 (2015)
114. C.B. Lang, L. Leskovec, D. Mohler, S. Prelovsek, Phys. Rev. D **86**, 054508 (2012)
115. S. Prelovsek, L. Leskovec, C.B. Lang, D. Mohler, Phys. Rev. D **88**, 054508 (2013)
116. C.B. Lang, L. Leskovec, D. Mohler, S. Prelovsek, R.M. Woloshyn, Phys. Rev. D **90**, 034510 (2014)
117. C.B. Lang, D. Mohler, S. Prelovsek, R.M. Woloshyn, Phys. Lett. B **750**, 17 (2015)
118. C.B. Lang, D. Mohler, S. Prelovsek, M. Vidmar, Phys. Rev. D **84**, 054503 (2011) [Erratum: Phys. Rev. D **89**(5), 059903 (2014)]
119. R.A. Briceno, J.J. Dudek, R.G. Edwards, C.J. Shultz, C.E. Thomas, D.J. Wilson, Phys. Rev. D **93**(11), 114508 (2016)
120. V. Verduci, C.B. Lang, PoS **LATTICE2014**, 121 (2014)
121. C.B. Lang, V. Verduci, Phys. Rev. D **87**, 054502 (2013)
122. A. Torok, S.R. Beane, W. Detmold, T.C. Luu, K. Orginos et al., Phys. Rev. D **81**, 074506 (2010)
123. W. Detmold, A. Nicholson, Phys. Rev. D **93**(11), 114511 (2016)
124. S.R. Beane, E. Chang, S.D. Cohen, W. Detmold, H.W. Lin, T.C. Luu, K. Orginos, A. Parreno, M.J. Savage, A. Walker-Loud, Phys. Rev. Lett. **109**, 172001 (2012)
125. S.R. Beane, P.F. Bedaque, T.C. Luu, K. Orginos, E. Pallante, A. Parreno, M.J. Savage, Nucl. Phys. A **794**, 62 (2007)
126. M.G. Endres, D.B. Kaplan, J.W. Lee, A.N. Nicholson, Phys. Rev. A **84**, 043644 (2011)
127. Y. Nishida, D.T. Son, Phys. Rev. D **76**, 086004 (2007)
128. D. Blume, J. von Stecher, C.H. Greene, Phys. Rev. Lett. **99**, 233201 (2007)
129. D. Blume, K.M. Daily, C. R. Phys. **12**, 86 (2011)
130. F. Werner, Y. Castin, Phys. Rev. Lett. **97**, 150401 (2006)
131. D. Lee, Prog. Part. Nucl. Phys. **63**, 117 (2009)
132. B. Borasoy, E. Epelbaum, H. Krebs, D. Lee, U.G. Meißner, Eur. Phys. J. A **31**, 105 (2007)
133. E.P. Wigner, Phys. Rev. **98**, 145 (1955)
134. D.R. Phillips, T.D. Cohen, Phys. Lett. B **390**, 7 (1997)
135. T.D. Cohen, Phys. Rev. C **55**, 67 (1997)
136. D.R. Phillips, S.R. Beane, T.D. Cohen, Ann. Phys. **263**, 255 (1998)
137. S. Fleming, T. Mehen, I.W. Stewart, Nucl. Phys. A **677**, 313 (2000)
138. E. Braaten, H.W. Hammer, Phys. Rep. **428**, 259 (2006)
139. P.F. Bedaque, H.W. Hammer, U. van Kolck, Nucl. Phys. A **646**, 444 (1999)
140. P.F. Bedaque, H.W. Hammer, U. van Kolck, Phys. Rev. Lett. **82**, 463 (1999)
141. S.R. Beane, P.F. Bedaque, L. Childress, A. Kryjevski, J. McGuire, U. van Kolck, Phys. Rev. A **64**, 042103 (2001)
142. V.N. Efimov, Phys. Lett. B **33**, 563 (1970)
143. V.N. Efimov, Sov. J. Nucl. Phys. **12**, 589 (1971)
144. H.W. Hammer, L. Platter, Ann. Rev. Nucl. Part. Sci. **60**, 207 (2010)
145. M.G. Endres, D.B. Kaplan, J.W. Lee, A.N. Nicholson, Phys. Rev. Lett. **107**, 201601 (2011)

146. M.G. Endres, D.B. Kaplan, J.W. Lee, A.N. Nicholson, PoS **Lattice2010**, 182 (2010)
147. J.W. Lee, M.G. Endres, D.B. Kaplan, A.N. Nicholson, PoS **Lattice2010**, 197 (2010)
148. A.N. Nicholson, M.G. Endres, D.B. Kaplan, J.W. Lee, PoS **Lattice2010**, 206 (2010)
149. E. Epelbaum, H. Krebs, D. Lee, U.G. Meissner, Eur. Phys. J. A **45**, 335 (2010)
150. E. Epelbaum, H. Krebs, D. Lee, U.G. Meißner, Phys. Rev. Lett. **104**, 142501 (2010)
151. E. Epelbaum, H. Krebs, T.A. Lähde, D. Lee, U.G. Meißner, G. Rupak, Phys. Rev. Lett. **112**(10), 102501 (2014)
152. T.A. Lahde, E. Epelbaum, H. Krebs, D. Lee, U.G. Meißner, G. Rupak (2013). [PoS LATTICE **2013**, 231 (2014)]
153. T.A. Lähde, E. Epelbaum, H. Krebs, D. Lee, U.G. Meißner, G. Rupak, Phys. Lett. B **732**, 110 (2014)
154. E. Epelbaum, H. Krebs, D. Lee, U.G. Meißner, Phys. Rev. Lett. **106**, 192501 (2011)
155. E. Epelbaum, H. Krebs, T. Lähde, D. Lee, U.G. Meißner, Phys. Rev. Lett. **109**, 252501 (2012)
156. E. Epelbaum, H. Krebs, T.A. Lähde, D. Lee, U.G. Meißner, Phys. Rev. Lett. **110**, 112502 (2013)
157. E. Epelbaum, H. Krebs, T.A. Lahde, D. Lee, U.G. Meißner, Eur. Phys. J. A **49**, 82 (2013)
158. T.A. Lahde, E. Epelbaum, H. Krebs, D. Lee, U.G. Meißner, G. Rupak, Pramana **83**, 651 (2014)

Chapter 6
Lattice Methods and the Nuclear Few- and Many-Body Problem

Dean Lee

6.1 Introduction

This chapter builds upon the general overview of lattice methods for effective field theory of the previous chapter. We discuss the theory and algorithms used in lattice simulations of nuclear few and many body systems. We show the exact equivalence of the Grassmann path integral, transfer matrix operator, Grassmann path integral with auxiliary fields, and transfer matrix operator with auxiliary fields. Along with our analysis we include several coding examples and a number of exercises for the calculations of few- and many-body systems at leading order in chiral effective field theory.

Effective field theory (EFT) provides a theoretical framework for organizing low-energy interactions in powers of particle momenta. Chiral effective field theory applies this framework to the low-energy interactions of protons and neutrons while explicitly including the interactions of pions [1–9]. Pions are qualitatively different from other mesons since they become massless in the limit of massless quarks, thereby producing long-range exchange interactions. The low-energy expansion of chiral EFT is organized in powers of Q, where Q denotes the typical momentum of the nucleons as well as explicit factors of the pion mass. The most important interactions are called leading order (LO) or $O(Q^0)$. The next most important contributions are next-to-leading order (NLO) or $O(Q^2)$. The terms after this are next-to-next-to-leading order (NNLO) or $O(Q^3)$, and so on.

Lattice EFT refers generally to lattice simulations based upon the framework of effective field theory. There are a few reviews in the literature which discuss current methods used in lattice effective field theory [10, 11] as well as the discussion in the

D. Lee (✉)
Department of Physics, North Carolina State University, Raleigh, NC 27695, USA
e-mail: dean_lee@ncsu.edu

M. Hjorth-Jensen et al. (eds.), *An Advanced Course in Computational Nuclear Physics*, Lecture Notes in Physics 936, DOI 10.1007/978-3-319-53336-0_6

previous chapter of this volume. Many different phenomena can be studied in lattice EFT using the same lattice action. In principle all systematic errors are introduced up front when defining the low-energy effective theory, as opposed to the particular computational scheme used to calculate observables.

Lattice EFT has been aided by efficient lattice methods developed for lattice QCD and condensed matter applications. The methods include Markov Chain Monte Carlo techniques, auxiliary fields [12, 13], pseudofermion methods [14], and non-local updating schemes such as the hybrid Monte Carlo algorithm[15–17]. Lattice EFT was first used in studies of infinite nuclear matter [18] and infinite neutron matter with and without explicit pions [19–22]. The method has also been used to study light nuclei in pionless EFT [23] and chiral EFT at leading order [24]. There have been further studies of neutron matter [25–27] and light nuclei [28, 29], and there have been several applications to nuclear structure and nuclear clustering [30–35] as well as recent work on nuclear scattering and reactions [36–38].

6.2 Recent Applications

We review here several recent applications of lattice effective field theory to nuclear systems. In [34], the first ab initio evidence is presented for a tetrahedral alpha-cluster structure of the ground state of ^{16}O. The first excited 0^+ state of ^{16}O is found to be a planar or square arrangement of alpha clusters. The evidence for these geometric arrangements come from the strong overlap between nuclear states and initial state configurations with these alpha-cluster geometries.

In Table 6.1 we presented the energies of the low-lying even parity states of oxygen-16. The columns labeled "LO(2N)" and "NNLO(2N)" show the energies at each order using the two-nucleon force only. The column labeled "+3N" also includes the 3NF, which first appears at NNLO. The column "+4N$_{\rm eff}$" includes an "effective" 4N force, and the column "Exp" gives the empirical energies. This "effective" 4N force was introduced in [33] as a proxy measure of unknown systematic errors responsible for overbinding in lattice chiral effective field theory calculations with increasing numbers of nucleons. This tendency towards overbinding has also been noted in other nuclear structure calculations [39, 40].

Table 6.1 Lattice results and experimental energies for the lowest even-parity states of ^{16}O in MeV

J_n^p	LO (2N)	NNLO (2N)	+3N	+4N$_{\rm eff}$	Exp
0_1^+	−147.3(5)	−121.4(5)	−138.8(5)	−131.3(5)	−127.62
0_2^+	−145(2)	−116(2)	−136(2)	−123(2)	−121.57
2_1^+	−145(2)	−116(2)	−136(2)	−123(2)	−120.70

The errors include statistical Monte Carlo errors and uncertainties due to the extrapolation to infinite Euclidean time

In order to understand the source of this overbinding, the problem was revisited again in [35]. In that work numerical evidence from ab initio lattice simulations showed that the problem appears related to the fact that the nuclear forces reside near a quantum phase transition. Using lattice effective field theory, Monte Carlo simulations were performed for systems with up to 20 nucleons. For even and equal numbers of protons and neutrons, a first-order transition was found at zero temperature from a Bose-condensed gas of alpha particles to a nuclear liquid. Whether one has an alpha-particle gas or nuclear liquid is determined by the strength of the alpha-alpha interactions, and the alpha-alpha interactions depend on the strength and locality of the nucleon-nucleon interactions. This insight is useful in improving calculations of nuclear structure and important astrophysical reactions involving alpha capture on nuclei. These findings also provide a tool to probe the structure of alpha cluster states such as the Hoyle state responsible for the production of carbon in red giant stars and point to a connection between nuclear states and the universal physics of bosons at large scattering length.

Processes such as the scattering of alpha particles, the triple-alpha reaction, and alpha capture play an important role in stellar nucleosynthesis. In particular, alpha capture on carbon determines the ratio of carbon to oxygen during helium burning and impacts the following carbon, neon, oxygen, and silicon burning stages. In these reactions the elastic scattering of alpha particles and alpha-like nuclei (nuclei with even and equal numbers of protons and neutrons) are important for understanding background and resonant scattering contributions. In [38] the first ab initio calculations of the scattering of two alpha particles were performed using a technique called the adiabatic projection method. These calculations represent a significant algorithmic improvement since the calculations presented in [38] scale roughly quadratically with the number of nucleons and opens a gateway to scattering and reactions involving heavier nuclei.

6.3 Scattering on the Lattice

At any given order in the chiral EFT expansion, there will be short-range interaction coefficients which depend on the chosen regularization of the large-momentum divergences. On the lattice this regularization is provided by the lattice spacing, unless some additional regularization is applied to the lattice interactions. In order to set the values of the short-range two-nucleon interaction coefficients, we make a comparison of nucleon-nucleon scattering on the lattice with experimental scattering data. The extension to three-nucleon interaction coefficients is also required at NNLO, and that procedure on the lattice has been discussed in [28].

As discussed in the previous chapter, Lüscher [41–43] has shown that the finite-volume energy levels for a two-body system in a periodic cubic box are related to the infinite-volume scattering matrix. While the method is very useful at low momenta, it can become less accurate at higher momenta and higher orbital angular momenta. Also spin-orbit coupling and partial-wave mixing are difficult to measure accurately using Lüscher's method due to scattering artifacts produced by the cubic periodic boundary. An alternative approach has been developed to measure phase shifts for particles on the lattice using a spherical wall boundary [44, 45].

In this approach, a hard spherical wall boundary is imposed on the relative separation between the two particles. This wall is placed at some chosen radius R_{wall}, and it removes copies of the interactions produced by the periodic lattice. Working in the center-of-mass frame, we solve the time-independent Schrödinger equation as a function of the relative separation between the particles and compute spherical standing waves which vanish at $r = R_{\text{wall}}$. At values of r beyond the range of the interaction, the spherical standing waves can be written as a superposition of products of spherical harmonics and spherical Bessel functions,

$$[\cos\delta_\ell \cdot j_\ell(kr) - \sin\delta_\ell \cdot y_\ell(kr)]\, Y_{\ell,\ell_z}(\theta,\phi). \tag{6.1}$$

Here k is the relative momentum between the scattering particles, and δ_ℓ is the phase shift for partial wave ℓ. We can extract k from the energy of the standing wave, and the phase shift δ_ℓ is determined by setting the wave function in Eq. (6.1) to zero at the wall boundary.

When the total intrinsic spin of the two nucleons is nonzero, spin-orbit coupling generates mixing between partial waves. In this case the standing wave at the wall boundary is decomposed into spherical harmonics and coupled-channel equations are solved to extract the phase shifts and mixing angles. The spherical wall method was used to calculate phase shifts and mixing angle for low-energy nucleon-nucleon scattering [25]. Recently the spherical wall approach has been improved in accuracy and computational efficiency [46]. In the improved approach one projects onto spherical harmonics Y_{ℓ,ℓ_z} with angular momentum quantum numbers ℓ, ℓ_z. In this manner one constructs radial position states for a given partial wave,

$$|r\rangle^{\ell,\ell_z} = \sum_{\mathbf{r}'} Y_{\ell,\ell_z}(\hat{\mathbf{r}}')\delta_{r,|\mathbf{r}'|}|\mathbf{r}'\rangle. \tag{6.2}$$

We require that r is less than half the box length $L/2$. Using this technique we are essentially constructing a radial position basis for each partial wave.

It is also useful to introduce auxiliary potentials in the region lying just in front of the spherical wall boundary [46]. The auxiliary potential is a spherical attractive well that is positioned in front of the spherical wall boundary. We can tune to any scattering energy by adjusting the depth of the well. For systems with partial wave mixing due to spin-orbit coupling, we also include a Hermitian but imaginary off-diagonal auxiliary potential for the two coupled channels. This breaks time reversal symmetry, and the resulting standing wave solutions now have both real and imaginary parts that are linearly independent. From the real and imaginary solutions one can determine the scattering phase shifts and mixing angle at any given value of the scattering energy.

This spherical wall approach has been used together with a technique called the adiabatic projection method to study nuclear scattering and reactions on the lattice. The adiabatic projection method [35, 38, 47–49] is a general framework that produces a low-energy effective theory for clusters of particles which becomes exact in the limit of large projection time. For the case of two-cluster scattering, we consider a set of two cluster states $|\mathbf{R}\rangle$ labeled by the spatial separation vector $\mathbf{R}$. The initial wave functions are wave packets which, for large $|\mathbf{R}|$, factorize into a product of two individual clusters,

$$|\mathbf{R}\rangle = \sum_{\mathbf{r}} |\mathbf{r}+\mathbf{R}\rangle_1 \otimes |\mathbf{r}\rangle_2. \tag{6.3}$$

The summation over $\mathbf{r}$ is required to produce states with total momentum equal to zero. We bin the initial cluster states together according to radial distance and angular momentum. In this manner, we form radial position states with projected angular momentum quantum numbers, which we label $|R\rangle^{\ell,\ell_z}$.

The next step is to multiply by powers of the transfer matrix in order to form "dressed" cluster states. This produces states that approximately span the set of low-energy cluster-cluster scattering states in our periodic box. We discuss the transfer matrix formalism in detail later in this chapter. After n_t time steps, we have the dressed cluster states

$$|R\rangle_{n_t}^{\ell,\ell_z} = M^{n_t}|R\rangle^{\ell,\ell_z}. \tag{6.4}$$

These dressed cluster states are then used to compute matrix elements of the transfer matrix M,

$$[M_{n_t}]_{R',R}^{\ell,\ell_z} = {}_{n_t}^{\ell,\ell_z}\langle R'|M|R\rangle_{n_t}^{\ell,\ell_z}. \tag{6.5}$$

Since such states are not orthogonal, we also compute a norm matrix

$$[N_{n_t}]_{R',R}^{\ell,\ell_z} = {}_{n_t}^{\ell,\ell_z}\langle R'|R\rangle_{n_t}^{\ell,\ell_z}. \tag{6.6}$$

The "radial adiabatic transfer matrix" is defined as the matrix product

$$\left[M_{n_t}^a\right]_{R',R}^{\ell,\ell_z} = \left[N_{n_t}^{-\frac{1}{2}} M_{n_t} N_{n_t}^{-\frac{1}{2}}\right]_{R',R}^{\ell,\ell_z}, \tag{6.7}$$

and the scattering phase shifts can then be determined from the standing waves of the radial adiabatic transfer matrix.

6.4 Lattice Formalisms

Throughout our discussion of the lattice formalism we use dimensionless parameters and operators corresponding with physical values times the appropriate power of the spatial lattice spacing a. In our notation the three-component integer vector $\mathbf{n}$

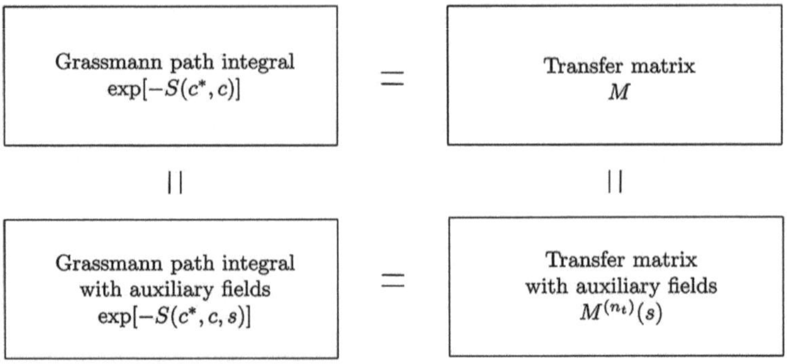

Fig. 6.1 A schematic diagram of the different lattice formulations, namely, the Grassmann path integral, transfer matrix operator, Grassmann path integral with auxiliary fields, and transfer matrix operator with auxiliary fields

labels the lattice sites of a three-dimensional periodic lattice with dimensions L^3. The spatial lattice unit vectors are denoted $\hat{\mathbf{l}} = \hat{\mathbf{1}}, \hat{\mathbf{2}}, \hat{\mathbf{3}}$. We use n_t to label lattice steps in the temporal direction, and L_t denotes the total number of lattice time steps. The temporal lattice spacing is given by a_t, and $\alpha_t = a_t/a$ is the ratio of the temporal to spatial lattice spacing. We also define $h = \alpha_t/(2m)$, where m is the nucleon mass in lattice units. In Fig. 6.1 we show a diagram of the four different but exactly equivalent lattice formulations that we discuss, the Grassmann path integral, transfer matrix operator, Grassmann path integral with auxiliary fields, and transfer matrix operator with auxiliary fields.

6.4.1 Grassmann Path Integral

We define the lattice action starting from the lattice Grassmann path integral action without auxiliary fields. This is the simplest formulation in which to derive the lattice Feynman rules. We let c and c^* be anticommuting Grassmann fields for the nucleons. In our notation c is a column vector composed of the spin-isospin nucleon degrees of freedom c_i, while c^* is a row vector of the components c_i^*. The Grassmann fields are periodic with respect to the spatial extent of the L^3 lattice,

$$c_i(\mathbf{n} + L\hat{1}, n_t) = c_i(\mathbf{n} + L\hat{2}, n_t) = c_i(\mathbf{n} + L\hat{3}, n_t) = c_i(\mathbf{n}, n_t), \tag{6.8}$$

$$c_i^*(\mathbf{n} + L\hat{1}, n_t) = c_i^*(\mathbf{n} + L\hat{2}, n_t) = c_i^*(\mathbf{n} + L\hat{3}, n_t) = c_i^*(\mathbf{n}, n_t), \tag{6.9}$$

and antiperiodic along the temporal direction,

$$c_i(\mathbf{n}, n_t + L_t) = -c_i(\mathbf{n}, n_t), \tag{6.10}$$

$$c_i^*(\mathbf{n}, n_t + L_t) = -c_i^*(\mathbf{n}, n_t). \tag{6.11}$$

We write $DcDc^*$ as shorthand for the integral measure,

$$DcDc^* = \prod_{\mathbf{n},n_t,i} dc_i(\mathbf{n}, n_t) dc_i^*(\mathbf{n}, n_t). \tag{6.12}$$

We use the usual convention for Grassmann integration,

$$\int dc_i(\mathbf{n}, n_t) = \int dc_i^*(\mathbf{n}, n_t) = 0, \tag{6.13}$$

$$\int dc_i(\mathbf{n}, n_t) c_i(\mathbf{n}, n_t) = \int dc_i^*(\mathbf{n}, n_t) c_i^*(\mathbf{n}, n_t) = 1 \text{ (no sum on } i). \tag{6.14}$$

We consider the Grassmann path integral

$$\mathscr{Z} = \int DcDc^* \exp\left[-S\left(c^*, c\right)\right], \tag{6.15}$$

where the lattice action can be broken into a free part and interacting part,

$$S(c^*, c) = S_{\text{free}}(c^*, c) + S_{\text{int}}(c^*, c). \tag{6.16}$$

The free part is the free non-relativistic nucleon action, which is

$$S_{\text{free}}(c^*, c) = \sum_{\mathbf{n},n_t} c^*(\mathbf{n}, n_t) \left[c(\mathbf{n}, n_t + 1) - c(\mathbf{n}, n_t)\right] + \alpha_t \sum_{n_t} K^{(n_t)}(c^*, c), \tag{6.17}$$

where

$$K^{(n_t)}(c^*, c) = \sum_{k=0,1,2,\cdots} (-1)^k \frac{w_k}{2m} \sum_{\mathbf{n},\hat{\mathbf{l}}} c^*(\mathbf{n}, n_t) \left[c(\mathbf{n} + k\hat{\mathbf{l}}, n_t) + c(\mathbf{n} - k\hat{\mathbf{l}}, n_t)\right], \tag{6.18}$$

and the hopping coefficients w_k correspond to a hopping parameter expansion of the squared momentum,

$$P^2(\mathbf{p}) = 2 \sum_{k=0,1,2,\cdots} \sum_{l=1,2,3} (-1)^k w_k \cos\left(kp_l\right). \tag{6.19}$$

Table 6.2 Hopping coefficients w_k for several lattice actions

Coefficient	Standard	$O(a^2)$-improved	$O(a^4)$-improved
w_0	1	5/4	49/36
w_1	1	4/3	3/2
w_2	0	1/12	3/20
w_3	0	0	1/90

The hopping coefficients are chosen to match the continuum relation

$$P^2(\mathbf{p}) = \mathbf{p}^2, \tag{6.20}$$

up to some chosen level of lattice discretization error. The hopping coefficients w_k for a few different lattice actions are shown in Table 6.2.

6.4.2 *Transfer Matrix Operator*

Let $a_i(\mathbf{n})$ and $a_i^\dagger(\mathbf{n})$ denote fermion annihilation and creation operators for the nucleon component i at lattice site $\mathbf{n}$. The shorthand $a(\mathbf{n})$ represents a column vector of nucleon components $a_i(\mathbf{n})$, and $a^\dagger(\mathbf{n})$ represents a row vector of components $a_i^\dagger(\mathbf{n})$. We can write any Grassmann path integral with instantaneous interactions as the trace of a product of operators using the identity [50, 51]

$$\begin{aligned}
&\mathrm{Tr}\left\{: F_{L_t-1}\left[a^\dagger(\mathbf{n}'), a(\mathbf{n})\right] : \times \cdots \times : F_0\left[a^\dagger(\mathbf{n}'), a(\mathbf{n})\right] :\right\} \\
&= \int DcDc^* \exp\left\{\sum_{n_t=0}^{L_t-1} \sum_{\mathbf{n},i} c_i^*(\mathbf{n}, n_t)\left[c_i(\mathbf{n}, n_t) - c_i(\mathbf{n}, n_t+1)\right]\right\} \\
&\times \prod_{n_t=0}^{L_t-1} F_{n_t}\left[c^*(\mathbf{n}', n_t), c(\mathbf{n}, n_t)\right],
\end{aligned} \tag{6.21}$$

where $c_i(\mathbf{n}, L_t) = -c_i(\mathbf{n}, 0)$.

Let us define the free non-relativistic lattice Hamiltonian

$$H_{\text{free}}(a^\dagger, a) = \sum_{k=0,1,2,\cdots} (-1)^k \frac{w_k}{2m} \sum_{\mathbf{n},\hat{\mathbf{l}}} a^\dagger(\mathbf{n})\left[a(\mathbf{n}+k\hat{\mathbf{l}}) + a(\mathbf{n}-k\hat{\mathbf{l}})\right]. \tag{6.22}$$

We write the interaction term as $H_{\text{int}}(a^\dagger, a)$, so that our total Hamiltonian is

$$H(a^\dagger, a) = H_{\text{free}}(a^\dagger, a) + H_{\text{int}}(a^\dagger, a). \tag{6.23}$$

Using the correspondence Eq. (6.21), we can rewrite the path integral $\mathscr{Z}$ defined in Eq. (6.15) as a transfer-matrix partition function,

$$\mathscr{Z} = \mathrm{Tr}\left(M^{L_t}\right), \tag{6.24}$$

where M is the normal-ordered transfer matrix operator

$$M =: \exp\left[-H(a^\dagger, a)\alpha_t\right] : . \tag{6.25}$$

Roughly speaking, the transfer matrix operator is the exponential of the Hamiltonian operator over one Euclidean lattice time step. In order to satisfy the identity Eq. (6.21), the exact definition of the transfer matrix is the normal-ordered exponential as defined in Eq. (6.25).

In this transfer matrix formalism, one can do simulations of nucleons using Monte Carlo, and this would essentially be a lattice version of diffusion or Green's function Monte Carlo [52]. Visually one can view the nucleons as interacting with each other while diffusing in space with each time step, as indicated in Fig. 6.2. At leading order in chiral effective field theory, the interactions include two independent S-wave contact interactions and the exchange of pions. We discuss these interactions in detail in the following.

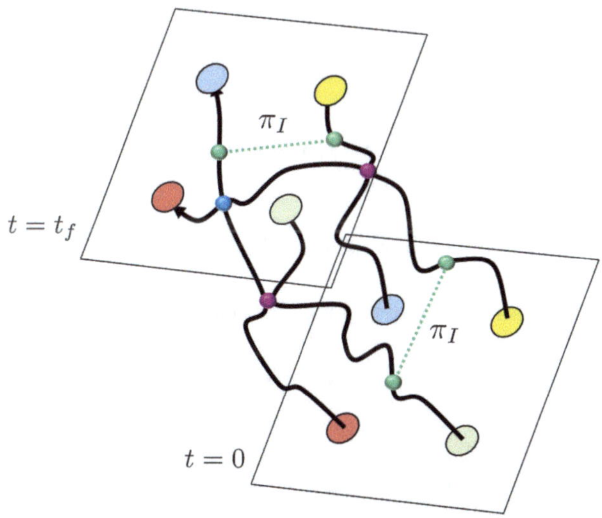

Fig. 6.2 A sketch showing nucleons which evolve with each time step. At leading order in chiral effective field theory, the interactions include two contact interactions and the exchange of pions

6.4.3 Grassmann Path Integral with Auxiliary Field

We assume that there exists an integral relation that allows us to write $\exp\left[-S_{\text{int}}(c^*, c)\right]$ as an integral over auxiliary fields. The purpose of the auxiliary field transformation is to decouple the interactions among the nucleons. Instead the interactions will be between the nucleons and the auxiliary fields.

We illustrate using the interactions that appear at leading order in chiral effective field theory. For pedagogical purposes we discuss the simplest possible implementation of the leading order action on the lattice. We first consider a zero-range contact interaction which is independent of nucleon spin and isospin. The action has the form

$$S_{\text{int}}^{C}(c^*, c) = \alpha_t \frac{C}{2} \sum_{\mathbf{n}, n_t} \left[c^*(\mathbf{n}, n_t) c(\mathbf{n}, n_t)\right]^2 . \tag{6.26}$$

We can write this as

$$\exp\left[-S_{\text{int}}^{C}(c^*, c)\right] = \int Ds \, \exp\left[-S_{ss}(s) - S_s(c^*, c, s)\right] \tag{6.27}$$

for auxiliary field $s(\mathbf{n}, n_t)$, where

$$S_{ss}(s) = \frac{1}{2} \sum_{\mathbf{n}, n_t} s^2(\mathbf{n}, n_t), \tag{6.28}$$

$$S_s(c^*, c, s) = \sqrt{-C\alpha_t} \sum_{\mathbf{n}, n_t} s(\mathbf{n}, n_t) c^*(\mathbf{n}, n_t) c(\mathbf{n}, n_t). \tag{6.29}$$

In our definition of the integration measure Ds, we include a factor of $1/\sqrt{2\pi}$ for each degree of freedom.

Next we consider an isospin-dependent contact interaction

$$S_{\text{int}}^{C'}(c^*, c) = \alpha_t \frac{C'}{2} \sum_{\mathbf{n}, n_t, I} \left[c^*(\mathbf{n}, n_t) \tau_I c(\mathbf{n}, n_t)\right]^2 , \tag{6.30}$$

where τ_I for $I = 1, 2, 3$ are the Pauli matrices in isospin space. Then we can use

$$\exp\left[-S_{\text{int}}^{C'}(c^*, c)\right] = \int \prod_I Ds_I \exp\left[-S_{s_I s_I}(s_I) - S_{s_I}(c^*, c, s_I)\right] \tag{6.31}$$

for auxiliary fields $s_I(\mathbf{n}, n_t)$ where

$$S_{s_I s_I}(s_I) = \frac{1}{2} \sum_{\mathbf{n}, n_t, I} s_I^2(\mathbf{n}, n_t), \tag{6.32}$$

$$S_{s_I}(c^*, c, s_I) = \sqrt{-C'\alpha_t} \sum_{\mathbf{n}, n_t, I} s_I(\mathbf{n}, n_t) c^*(\mathbf{n}, n_t) \tau_I c(\mathbf{n}, n_t). \tag{6.33}$$

Finally we work with the one-pion exchange potential (OPEP). In this case the pion acts much like the auxiliary fields. However there are also spatial correlations in the quadratic part of the pion action and a gradient coupling between the pions and nucleons. The one-pion exchange interaction on the lattice can written as

$$\exp\left[-S_{\text{int}}^{\text{OPEP}}(c^*, c)\right] = \int \prod_I D\pi_I \exp\left[-S_{\pi_I \pi_I}(\pi_I) - S_{\pi_I}(c^*, c, \pi_I)\right]. \tag{6.34}$$

The free pion action is

$$S_{\pi_I \pi_I}(\pi_I) = \frac{1}{2}\alpha_t m_\pi^2 \sum_{\mathbf{n}, n_t, I} \pi_I^2(\mathbf{n}, n_t)$$
$$+ \frac{1}{2}\alpha_t \sum_{k=0,1,2,\cdots} (-1)^k w_k \sum_{\mathbf{n}, n_t, I, \hat{\mathbf{l}}} \pi_I(\mathbf{n}, n_t) \left[\pi_I(\mathbf{n} + k\hat{\mathbf{l}}, n_t) + \pi_I(\mathbf{n} - k\hat{\mathbf{l}}, n_t)\right], \tag{6.35}$$

with the coefficient w_k as defined in Table 6.2 and m_π is the pion mass. At leading order we do not consider any isospin-breaking effects. The pion coupling to the nucleon is

$$S_{\pi_I}(c^*, c, \pi_I) = \frac{g_A \alpha_t}{2 f_\pi} \sum_{\mathbf{n}, n_t, l, I} \Delta_k \pi_I(\mathbf{n}, n_t) c^*(\mathbf{n}, n_t) \sigma_k \tau_I c(\mathbf{n}, n_t), \tag{6.36}$$

where σ_l for $l = 1, 2, 3$ are the Pauli matrices in spin space and

$$\Delta_l \pi_I(\mathbf{n}, n_t) = \frac{1}{2} \sum_{k=1,2,\cdots} (-1)^{k-1} o_k \left[\pi_I(\mathbf{n} + k\hat{\mathbf{l}}, n_t) - \pi_I(\mathbf{n} - k\hat{\mathbf{l}}, n_t)\right], \tag{6.37}$$

with coefficients o_k corresponding to a hopping parameter expansion of the momentum,

$$P(p_l) = \sum_{k=1,2,\cdots} (-1)^{k-1} o_k \sin(k p_l). \tag{6.38}$$

Table 6.3 Hopping coefficients o_k for several lattice actions

Coefficient	Standard	$O(a^2)$-improved	$O(a^4)$-improved
o_1	1	4/3	3/2
o_2	0	1/6	3/10
o_3	0	0	1/30

Here g_A is the axial-vector coupling constant, and f_π is the pion decay constant. The hopping coefficients can be chosen to match the continuum result

$$P(p_l) = p_l. \tag{6.39}$$

The hopping coefficients o_k for a few different lattice actions are shown in Table 6.3.

6.4.4 Transfer Matrix Operator with Auxiliary Field

Using the equivalence in Eq. (6.21), we can write $\mathscr{Z}$ as the trace of a product of transfer matrix operators which depend on the auxiliary field,

$$\mathscr{Z} = \int Ds \prod_I (Ds_I D\pi_I) \exp\left[-S_{ss}(s) - S_{s_I s_I}(s_I) - S_{\pi_I \pi_I}(\pi_I)\right] \mathrm{Tr}\left\{M^{(L_t-1)} \cdots M^{(0)}\right\}. \tag{6.40}$$

The transfer matrix at time step n_t is given by

$$M^{(n_t)} = :\exp\left[-H^{(n_t)}(a^\dagger, a, s, s_I, \pi_I)\alpha_t\right]:, \tag{6.41}$$

where

$$\begin{aligned} H^{(n_t)}(a^\dagger, a, s, s_I, \pi_I)\alpha_t &= H_{\text{free}}(a^\dagger, a)\alpha_t + S_s^{(n_t)}(a^\dagger, a, s) + S_{s_I}^{(n_t)}(a^\dagger, a, s_I) \\ &\quad + S_{\pi_I}^{(n_t)}(a^\dagger, a, \pi_I), \end{aligned} \tag{6.42}$$

and

$$S_s^{(n_t)}(a^\dagger, a, s) = \sqrt{-C\alpha_t} \sum_{\mathbf{n}} s(\mathbf{n}, n_t) a^\dagger(\mathbf{n}) a(\mathbf{n}), \tag{6.43}$$

$$S_{s_I}^{(n_t)}(a^\dagger, a, s_I) = \sqrt{-C'\alpha_t} \sum_{\mathbf{n},I} s_I(\mathbf{n}, n_t) a^\dagger(\mathbf{n}) \tau_I a(\mathbf{n}), \tag{6.44}$$

$$S_{\pi_I}^{(n_t)}(a^\dagger, a, \pi_I) = \frac{g_A \alpha_t}{2f_\pi} \sum_{\mathbf{n},k,I} \Delta_k \pi_I(\mathbf{n}, n_t) a^\dagger(\mathbf{n}) \sigma_k \tau_I a(\mathbf{n}). \tag{6.45}$$

6.5 Projection Monte Carlo

Let us consider a system with A nucleons. We can create a general single-nucleon state using creation operators acting on the vacuum with coefficient function $f(\mathbf{n})$. We write $f(\mathbf{n})$ as a column vector in the space of nucleon spin and isospin components, and the single-nucleon state can be written as

$$|f\rangle = \sum_{\mathbf{n}} a^{\dagger}(\mathbf{n}) f(\mathbf{n}) |0\rangle . \tag{6.46}$$

For our projection Monte Carlo calculation we take our A-body initial state to be a Slater determinant of single nucleon states,

$$|f_1, \cdots, f_A\rangle = \left[\sum_{\mathbf{n}} a^{\dagger}(\mathbf{n}) f_1(\mathbf{n})\right] \cdots \left[\sum_{\mathbf{n}} a^{\dagger}(\mathbf{n}) f_A(\mathbf{n})\right] |0\rangle . \tag{6.47}$$

We use the same construction for the A-body final state.

For the purposes of coding the projection Monte Carlo calculation, it is convenient to view the identical nucleons as having a hidden index $j = 1, \cdots, A$ that makes all of the nucleons distinguishable. If we antisymmetrize all physical states over this extra index then all physical observables are exactly recovered. So our initial state $|f_1, \cdots, f_A\rangle$ becomes

$$\begin{aligned} &\frac{1}{\sqrt{A!}} \sum_{P} \left[\sum_{\mathbf{n}} a^{\dagger}_{[P(1)]}(\mathbf{n}) f_1(\mathbf{n})\right] \cdots \left[\sum_{\mathbf{n}} a^{\dagger}_{[P(A)]}(\mathbf{n}) f_A(\mathbf{n})\right] |0\rangle \\ &= \frac{1}{\sqrt{A!}} \sum_{P'} \operatorname{sgn}(P') \left[\sum_{\mathbf{n}} a^{\dagger}_{[1]}(\mathbf{n}) f_{P'(1)}(\mathbf{n})\right] \cdots \left[\sum_{\mathbf{n}} a^{\dagger}_{[A]}(\mathbf{n}) f_{P'(A)}(\mathbf{n})\right] |0\rangle , \end{aligned} \tag{6.48}$$

where the summations are over all permutations P, and sgn is the sign of the permutation. With these hidden indices our normal-ordered auxiliary-field transfer matrix $M^{(n_t)}$ becomes

$$\left[1 - H^{(n_t)}(a^{\dagger}_{[1]}, a_{[1]}, s, s_I, \pi_I)\alpha_t\right] \cdots \left[1 - H^{(n_t)}(a^{\dagger}_{[A]}, a_{[A]}, s, s_I, \pi_I)\alpha_t\right] \tag{6.49}$$

We see that the higher-order powers of the exponential vanish due to normal ordering.

In the projection Monte Carlo calculation we compute the amplitude

$$Z(n_t) = \langle f_1, \cdots, f_A | M^{(n_t-1)} \cdots M^{(0)} | f_1, \cdots, f_A\rangle \tag{6.50}$$

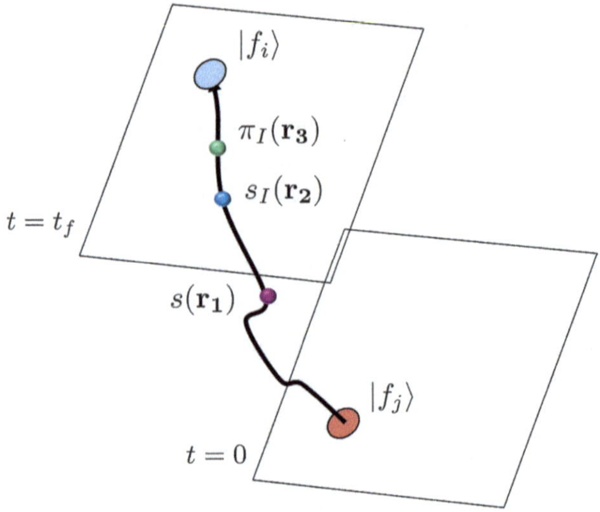

Fig. 6.3 A sketch showing the worldline for a single nucleon with a background of pion fields and auxiliary fields

for $n_t = L_t$ and $n_t = L_t - 1$. In the limit of large L_t the amplitudes will be dominated by the state with the lowest energy E_0 and nonzero overlap with $|f_1, \cdots, f_A\rangle$. In this limit the ratio $Z(n_t)/Z(n_t - 1)$ will converge to $\exp(-E_0\alpha_t)$ from above.

Each nucleon evolves as a particle in a fluctuating background of auxiliary fields and pion fields. The original interactions are reproduced after integrating over the fluctuating auxiliary and pion fields. For a simulation with A nucleons, the amplitude for a given configuration of pion and auxiliary fields is proportional to the determinant of an $A \times A$ matrix $\mathbf{M}$. The entries of $\mathbf{M}_{ij}$ are single nucleon worldline amplitudes for a nucleon starting at state $|f_j\rangle$ at $t = 0$ and ending at state $|f_i\rangle$ at $t = t_f = L_t\alpha_t$. This is shown in Fig. 6.3.

In the following we show sample code 6.1 in the Fortran programming language which calculates the auxiliary-field transfer matrix multiplications on the left starting from the single-nucleon initial states. We show only the terms which arise from the free-nucleon transfer matrix and the auxiliary field s.

Code Listing 6.1 Sample code calculating the auxiliary-field transfer matrix multiplications on the left starting from the single-nucleon initial states

```
DO nt = nt1+1, nt2
  DO np = 0,num-1
    DO nz = 0,L-1; DO ny = 0,L-1; DO nx = 0,L-1; DO ni = 0,1; DO
         ns = 0,1

     zvecs(nx,ny,nz,nt,ns,ni,np) = zvecs(nx,ny,nz,nt-1,ns,ni,np) &
        * (1.D0-6.D0*w0_N*h+CDSQRT(-c0*atovera*(1.D0,0.D0))*s(nx,ny
            ,nz,nt-1))
```

```
      zvecs(nx,ny,nz,nt,ns,ni,np) = zvecs(nx,ny,nz,nt,ns,ni,np) &
        + w1_N*h*zvecs(MOD(nx+1,L),ny,nz,nt-1,ns,ni,np) &
        + w1_N*h*zvecs(MOD(nx-1+L,L),ny,nz,nt-1,ns,ni,np) &
        + w1_N*h*zvecs(nx,MOD(ny+1,L),nz,nt-1,ns,ni,np) &
        + w1_N*h*zvecs(nx,MOD(ny-1+L,L),nz,nt-1,ns,ni,np) &
        + w1_N*h*zvecs(nx,ny,MOD(nz+1,L),nt-1,ns,ni,np) &
        + w1_N*h*zvecs(nx,ny,MOD(nz-1+L,L),nt-1,ns,ni,np)

      IF (improveN >= 1) THEN
        zvecs(nx,ny,nz,nt,ns,ni,np) = zvecs(nx,ny,nz,nt,ns,ni,np) &
          - w2_N*h*zvecs(MOD(nx+2,L),ny,nz,nt-1,ns,ni,np) &
          - w2_N*h*zvecs(MOD(nx-2+L,L),ny,nz,nt-1,ns,ni,np) &
          - w2_N*h*zvecs(nx,MOD(ny+2,L),nz,nt-1,ns,ni,np) &
          - w2_N*h*zvecs(nx,MOD(ny-2+L,L),nz,nt-1,ns,ni,np) &
          - w2_N*h*zvecs(nx,ny,MOD(nz+2,L),nt-1,ns,ni,np) &
          - w2_N*h*zvecs(nx,ny,MOD(nz-2+L,L),nt-1,ns,ni,np)
      END IF

      IF (improveN == 2) THEN
        zvecs(nx,ny,nz,nt,ns,ni,np) = zvecs(nx,ny,nz,nt,ns,ni,np) &
          + w3_N*h*zvecs(MOD(nx+3,L),ny,nz,nt-1,ns,ni,np) &
          + w3_N*h*zvecs(MOD(nx-3+L,L),ny,nz,nt-1,ns,ni,np) &
          + w3_N*h*zvecs(nx,MOD(ny+3,L),nz,nt-1,ns,ni,np) &
          + w3_N*h*zvecs(nx,MOD(ny-3+L,L),nz,nt-1,ns,ni,np) &
          + w3_N*h*zvecs(nx,ny,MOD(nz+3,L),nt-1,ns,ni,np) &
          + w3_N*h*zvecs(nx,ny,MOD(nz-3+L,L),nt-1,ns,ni,np)
      END IF

    END DO; END DO; END DO; END DO; END DO
  END DO
END DO
```

Similarly, we now show sample code 6.2 which calculates the auxiliary-field transfer matrix multiplications on the right starting from the single-nucleon final states. Again we present only the terms arising from the free-nucleon transfer matrix and the auxiliary field s.

Code Listing 6.2 Sample code calculating the auxiliary-field transfer matrix multiplications on the right starting from the single-nucleon final states

```
DO nt = nt2,nt1+1,-1
  DO np = 0,num-1
    DO nz = 0,L-1; DO ny = 0,L-1; DO nx = 0,L-1; DO ni = 0,1; DO
        ns = 0,1

      zdualvecs(nx,ny,nz,nt-1,ns,ni,np) &
        = zdualvecs(nx,ny,nz,nt,ns,ni,np) &
        * (1.D0-6.D0*w0_N*h+CDSQRT(-c0*atovera*(1.D0,0.D0))*s(nx,
            ny,nz,nt-1))

      zdualvecs(nx,ny,nz,nt-1,ns,ni,np) &
```

```
        = zdualvecs(nx,ny,nz,nt-1,ns,ni,np) &
        + w1_N*h*zdualvecs(MOD(nx+1,L),ny,nz,nt,ns,ni,np) &
        + w1_N*h*zdualvecs(MOD(nx-1+L,L),ny,nz,nt,ns,ni,np) &
        + w1_N*h*zdualvecs(nx,MOD(ny+1,L),nz,nt,ns,ni,np) &
        + w1_N*h*zdualvecs(nx,MOD(ny-1+L,L),nz,nt,ns,ni,np) &
        + w1_N*h*zdualvecs(nx,ny,MOD(nz+1,L),nt,ns,ni,np) &
        + w1_N*h*zdualvecs(nx,ny,MOD(nz-1+L,L),nt,ns,ni,np)

      IF (improveN >= 1) THEN
        zdualvecs(nx,ny,nz,nt-1,ns,ni,np) &
          = zdualvecs(nx,ny,nz,nt-1,ns,ni,np) &
          - w2_N*h*zdualvecs(MOD(nx+2,L),ny,nz,nt,ns,ni,np) &
          - w2_N*h*zdualvecs(MOD(nx-2+L,L),ny,nz,nt,ns,ni,np) &
          - w2_N*h*zdualvecs(nx,MOD(ny+2,L),nz,nt,ns,ni,np) &
          - w2_N*h*zdualvecs(nx,MOD(ny-2+L,L),nz,nt,ns,ni,np) &
          - w2_N*h*zdualvecs(nx,ny,MOD(nz+2,L),nt,ns,ni,np) &
          - w2_N*h*zdualvecs(nx,ny,MOD(nz-2+L,L),nt,ns,ni,np)
      END IF

      IF (improveN == 2) THEN
        zdualvecs(nx,ny,nz,nt-1,ns,ni,np) &
        = zdualvecs(nx,ny,nz,nt-1,ns,ni,np) &
        + w3_N*h*zdualvecs(MOD(nx+3,L),ny,nz,nt,ns,ni,np) &
        + w3_N*h*zdualvecs(MOD(nx-3+L,L),ny,nz,nt,ns,ni,np) &
        + w3_N*h*zdualvecs(nx,MOD(ny+3,L),nz,nt,ns,ni,np) &
        + w3_N*h*zdualvecs(nx,MOD(ny-3+L,L),nz,nt,ns,ni,np) &
        + w3_N*h*zdualvecs(nx,ny,MOD(nz+3,L),nt,ns,ni,np) &
        + w3_N*h*zdualvecs(nx,ny,MOD(nz-3+L,L),nt,ns,ni,np)
      END IF

    END DO; END DO
  END DO; END DO; END DO
END DO
```

In the following we show sample code 6.3 where these transfer matrix product multiplications are called as subroutines and used to compute the determinant and inverse of the matrix of single-nucleon amplitudes **M**.

Code Listing 6.3 Sample code where transfer matrix product multiplications are called and used to compute the determinant and inverse of the matrix of single-nucleon amplitudes

```
CALL getzvecs(s,sI,zvecs,zwave,Lt,0,pion,ztau2x2,n_f)
CALL getzdualvecs(s,sI,zdualvecs,zdualwave,Lt,0,pion,ztau2x2,n_f)
CALL getinvcorr(zvecs,zdualvecs,zldeter,zcorrmatrix,zcorrinv,Lt)

aldeterabs = DBLE(zldeter)
zdeterphase = CDEXP((0.D0,1.D0)*DIMAG(zldeter))
act = bose - aldeterabs
```

6.6 Importance Sampling

We do importance sampling according to the positive measure

$$|Z(L_t)| \exp\left[-S_{ss}(s) - S_{s_I s_I}(s_I) - S_{\pi_I \pi_I}(\pi_I)\right], \tag{6.51}$$

and use hybrid Monte Carlo to do global updates of the auxiliary and pion fields. The hybrid Monte Carlo (HMC) algorithm [15–17] is efficient in quickly generating decorrelated configurations for each auxiliary and pion field. Here we describe the updating algorithm for the s field. The updating of the s_I and π_I fields proceed in a very similar fashion. In general terms, the HMC algorithm can be described by means of a probability weight $P(s)$

$$P(s) \propto \exp[-V(s)], \tag{6.52}$$

where $V(s)$ is in general a non-local function of the field $s(\mathbf{n}, n_t)$, and a molecular dynamics (MD) Hamiltonian,

$$H(s,p) \equiv \frac{1}{2} \sum_{\mathbf{n},n_t} [p_s(\mathbf{n}, n_t)]^2 + V(s). \tag{6.53}$$

Classical Hamiltonian dynamics is introduced by defining the momentum $p_s(\mathbf{n}, n_t)$ conjugate to $s(\mathbf{n}, n_t)$.

Given an arbitrary initial configuration $s^0(\mathbf{n}, n_t)$, the conjugate momentum is chosen from a random Gaussian distribution according to

$$P[p_s^0(\mathbf{n}, n_t)] \propto \exp\left\{-\frac{1}{2}\left[p_s^0(\mathbf{n}, n_t)\right]^2\right\}, \tag{6.54}$$

after which the Hamiltonian equations of motion are integrated numerically with a small but nonzero step size $\varepsilon_{\text{step}}$. This method begins with a "half-step" forward in the conjugate momentum,

$$\tilde{p}_s^0(\mathbf{n}, n_t) = p_s^0(\mathbf{n}, n_t) - \frac{\varepsilon_{\text{step}}}{2}\left[\frac{\partial V(s)}{\partial s(\mathbf{n}, n_t)}\right]_{s=s^0}, \tag{6.55}$$

followed by repeated updates of s and $\tilde{p}_s$ according to

$$s^{i+1}(\mathbf{n}, n_t) = s^i(\mathbf{n}, n_t) + \varepsilon_{\text{step}} \tilde{p}_s^i(\mathbf{n}, n_t), \quad \tilde{p}_s^{i+1}(\mathbf{n}, n_t) = \tilde{p}_s^i(\mathbf{n}, n_t) - \varepsilon_{\text{step}}\left[\frac{\partial V(s)}{\partial s(\mathbf{n}, n_t)}\right]_{s=s^{i+1}}, \tag{6.56}$$

for a specified number of steps N_{step}. This is followed by an additional half-step backward in $\tilde{p}_s$ given by

$$p_s^{N_{\text{step}}}(\mathbf{n}, n_t) = \tilde{p}_s^{N_{\text{step}}}(\mathbf{n}, n_t) + \frac{\varepsilon_{\text{step}}}{2}\left[\frac{\partial V(s)}{\partial s(\mathbf{n}, n_t)}\right]_{s=s^0}. \tag{6.57}$$

For algorithmic efficiency the length of such an MD "trajectory" should be taken large enough to ensure decorrelation between successive configurations of the auxiliary field. The evolved configuration is then subjected to a "Metropolis test" against a random number $r \in [0, 1)$. The new configuration is accepted if

$$r < \exp\left[-H(s^{N_{\text{step}}}, p_s^{N_{\text{step}}}) + H(s^0, p_s^0)\right]. \tag{6.58}$$

It should be noted that although H is in principle conserved in the MD evolution, the truncation error of the leapfrog method introduces a systematic error. The Metropolis test eliminates the need for extrapolation in $\varepsilon_{\text{step}}$.

In our case $\exp[-V(s)]$ has the form

$$|Z(L_t)| \exp\left[-S_{ss}(s) - S_{s_I s_I}(s_I) - S_{\pi_I \pi_I}(\pi_I)\right], \tag{6.59}$$

where $Z(L_t)$ is the determinant of an $A \times A$ matrix of single-nucleon amplitudes $\mathbf{M}$. The derivative of V is then computed using

$$\begin{aligned}
\frac{\partial V(s)}{\partial s(\mathbf{n}, n_t)} &= \frac{\partial S_{ss}(s)}{\partial s(\mathbf{n}, n_t)} - \frac{\partial \text{Re}\left[\ln(\det \mathbf{M})\right]}{\partial s(\mathbf{n}, n_t)} \\
&= \frac{\partial S_{ss}(s)}{\partial s(\mathbf{n}, n_t)} - \text{Re}\left[\frac{1}{\det \mathbf{M}} \sum_{k,l} \frac{\partial \det \mathbf{M}}{\partial \mathbf{M}_{kl}} \frac{\partial \mathbf{M}_{kl}}{\partial s(\mathbf{n}, n_t)}\right] \\
&= \frac{\partial S_{ss}(s)}{\partial s(\mathbf{n}, n_t)} - \text{Re}\left[\sum_{k,l} \mathbf{M}_{lk}^{-1} \frac{\partial \mathbf{M}_{kl}}{\partial s(\mathbf{n}, n_t)}\right].
\end{aligned} \tag{6.60}$$

In the following we show sample code 6.4 calculating the quadratic part of the action due to the auxiliary fields and pion fields,

$$\begin{aligned}
&\frac{1}{2}\sum_{\mathbf{n},n_t} [p_s(\mathbf{n}, n_t)]^2 + \frac{1}{2}\sum_{\mathbf{n},n_t,I} [p_{s_I}(\mathbf{n}, n_t)]^2 + \frac{1}{2}\sum_{\mathbf{n},n_t,I} [p_{\pi_I}(\mathbf{n}, n_t)]^2 + S_{ss}(s) + S_{s_I s_I}(s_I) \\
&+ S_{\pi_I \pi_I}(\pi_I).
\end{aligned} \tag{6.61}$$

In the code we have found it convenient to rescale the pion field by a factor of $\sqrt{q_\pi}$ where

$$q_\pi = \alpha_t \left(m_\pi^2 + 6w_0\right). \tag{6.62}$$

Code Listing 6.4 Sample code calculating the quadratic part of the action due to the auxiliary fields and pion fields

```
bose = 0.D0
DO nt = 0,Lt-1
  DO nz = 0,L-1; DO ny = 0,L-1; DO nx = 0,L-1
    bose = bose &
      + s(nx,ny,nz,nt)**2.D0/2.D0 &
      + p_s(nx,ny,nz,nt)**2.D0/2.D0
    DO iso = 1,3
      bose = bose &
        + sI(nx,ny,nz,nt,iso)**2.D0/2.D0 &
        + p_sI(nx,ny,nz,nt,iso)**2.D0/2.D0 &
        + pion(nx,ny,nz,nt,iso)**2.D0/2.D0 &
        + atovera/qpi3*pion(nx,ny,nz,nt,iso)*( &
        - w1_P*pion(MOD(nx+1,L),ny,nz,nt,iso) &
        - w1_P*pion(nx,MOD(ny+1,L),nz,nt,iso) &
        - w1_P*pion(nx,ny,MOD(nz+1,L),nt,iso) &
        + w2_P*pion(MOD(nx+2,L),ny,nz,nt,iso) &
        + w2_P*pion(nx,MOD(ny+2,L),nz,nt,iso) &
        + w2_P*pion(nx,ny,MOD(nz+2,L),nt,iso) &
        - w3_P*pion(MOD(nx+3,L),ny,nz,nt,iso) &
        - w3_P*pion(nx,MOD(ny+3,L),nz,nt,iso) &
        - w3_P*pion(nx,ny,MOD(nz+3,L),nt,iso)) &
        + p_pion(nx,ny,nz,nt,iso)**2.D0/2.D0
    END DO
  END DO; END DO; END DO
END DO
```

In following we show sample code 6.5 which calculates

$$\left[\frac{\partial V(s)}{\partial s(\mathbf{n}, n_t)}\right]_{s=s^0} \tag{6.63}$$

and uses it to compute the half-step forward in the conjugate momentum,

$$\tilde{p}_s^0(\mathbf{n}, n_t) = p_s^0(\mathbf{n}, n_t) - \frac{\varepsilon_{\text{step}}}{2}\left[\frac{\partial V(s)}{\partial s(\mathbf{n}, n_t)}\right]_{s=s^0}. \tag{6.64}$$

Code Listing 6.5 Sample code computing derivative with respect to the auxiliary field and half-step forward in the conjugate momentum

```
DO npart1 = 0,n_f-1; DO npart2 = 0,n_f-1
  zdcorrmatrix(npart2,npart1) = 0.D0
  DO ni = 0,1; DO ns = 0,1
    zdcorrmatrix(npart2,npart1) = &
      zdcorrmatrix(npart2,npart1) + &
      zdualvecs(nx,ny,nz,nt+1,ns,ni,npart2) &
      *zvecs(nx,ny,nz,nt,ns,ni,npart1) &
      *CDSQRT(-c0*atovera*(1.D0,0.D0))/L**3
  END DO; END DO
```

```
END DO; END DO

dVds(nx,ny,nz,nt) = s(nx,ny,nz,nt)

DO npart1 = 0,n_f-1; DO npart2 = 0,n_f-1
  dVds(nx,ny,nz,nt) = dVds(nx,ny,nz,nt) &
    - DBLE(zdcorrmatrix(npart2,npart1) &
    *zcorrinv(npart1,npart2))
END DO; END DO

p_sHMC(nx,ny,nz,nt,0) = &
  p_s(nx,ny,nz,nt) - 0.5D0*eHMC*dVds(nx,ny,nz,nt)
```

In following code 6.6 we show an example code which performs the Metropolis test against a random number $r \in [0,1)$, with the new configuration being accepted if

$$r < \exp\left[-H(s^{N_{\text{step}}}, p_s^{N_{\text{step}}}) + H(s^0, p_s^0)\right]. \tag{6.65}$$

Code Listing 6.6 Sample code which performs the Metropolis acceptance test

```
IF (ntrial .eq. 1 .or. grnd() .lt. DEXP(-actnew+act)) THEN

  accept = accept + 1.

  DO nt = 0,Lt-1
    DO nz = 0,L-1; DO ny = 0,L-1; DO nx = 0,L-1
      s(nx,ny,nz,nt) = snew(nx,ny,nz,nt)
    END DO; END DO; END DO
  END DO

  DO nt = 0,Lt-1
    DO nz = 0,L-1; DO ny = 0,L-1; DO nx = 0,L-1
      DO iso = 1,3
        sI(nx,ny,nz,nt,iso) = sInew(nx,ny,nz,nt,iso)
        pion(nx,ny,nz,nt,iso) = pionnew(nx,ny,nz,nt,iso)
      END DO
    END DO; END DO; END DO
  END DO

  aldeterabs = aldeternewabs
  zdeterphase = zdeternewphase

END IF
```

Although the Monte Carlo importance sampling uses only the absolute value of the amplitude, the complex phase of the amplitude is treated as an observable and is collected with each configuration of the auxiliary and pion fields.

6.7 Exercises

6.1 Write a lattice hybrid Monte Carlo code which performs updates of the lattice action according to only the quadratic part of the action due to the auxiliary fields and pions,

$$\frac{1}{2}\sum_{\mathbf{n},n_t}[p_s(\mathbf{n},n_t)]^2+\frac{1}{2}\sum_{\mathbf{n},n_t,I}[p_{s_I}(\mathbf{n},n_t)]^2+\frac{1}{2}\sum_{\mathbf{n},n_t,I}[p_{\pi_I}(\mathbf{n},n_t)]^2+S_{ss}(s)+S_{s_Is_I}(s_I)$$
$$+S_{\pi_I\pi_I}(\pi_I). \tag{6.66}$$

Verify that the change in the action produced by the hybrid Monte Carlo update is scaling quadratically in the step size, $\varepsilon_{\text{step}}$, in the limit $\varepsilon_{\text{step}} \to 0$ with $N_{\text{step}}\varepsilon_{\text{step}}$ held fixed.

6.2 Write a function or subroutine that generates initial/final single-nucleon states on the lattice corresponding to a Slater-determinant state with one neutron spin-up and one neutron spin-down, both with zero momentum.

6.3 Write a function or subroutine that generates initial/final single-nucleon states on the lattice corresponding to a Slater-determinant state with one proton spin-up and one neutron spin-up, both with zero momentum.

6.4 Write a function or subroutine that generates initial/final single-nucleon states on the lattice corresponding to a Slater-determinant state of four nucleons—proton spin-up, proton spin-down, neutron spin-up, and neutron spin-down—each with zero momentum.

6.5 Write a function or subroutine that extends the sample code 6.1 to repeatedly multiply the auxiliary-field transfer matrix on the left starting from the initial single-nucleon wave functions. Include the contributions from the auxiliary fields s and s_I as well as the pion field π_I.

6.6 Write a function or subroutine that extends the sample code 6.2 to repeatedly multiply the auxiliary-field transfer matrix on the right starting from the final single-nucleon wave functions. Include the contributions from the auxiliary fields s and s_I as well as the pion field π_I.

6.7 Use the Slater-determinant states constructed in Problems 6.2–6.4 as initial and final states. In each case apply the functions or subroutines written in Problems 6.5 and 6.6 with all coupling constants set to zero. Verify that in each case the initial/final state is the ground state of the non-interacting system with energy equal to zero.

6.8 Use the Slater-determinant states constructed in Problems 6.2–6.4 as initial and final states. Using the functions or subroutines written in Problems 6.5 and 6.6, extend the sample code 6.5 to compute the derivatives of $V(s)$ with respect to $s(\mathbf{n},n_t)$, $s_I(\mathbf{n},n_t)$, and $\pi_I(\mathbf{n},n_t)$.

6.9 Take the code you have written for Problem 6.8 and complete the remaining steps needed to do hybrid Monte Carlo updates for s, s_I, and π_I. Verify that the change in the action produced by the hybrid Monte Carlo update is scaling quadratically in $\varepsilon_{\mathrm{step}}$ in the limit $\varepsilon_{\mathrm{step}} \to 0$ with $N_{\mathrm{step}}\varepsilon_{\mathrm{step}}$ held fixed.

6.10 Take the code you have written for Problem 6.9 and complete the remaining steps needed to calculate the energy of the ground state by computing the ratio of the amplitudes $Z(L_t)/Z(L_t - 1)$.

6.8 Codes and Benchmarks

Complete versions of the codes discussed in this chapter and developed in the exercises can be found online via this link. In order to run the codes, one must first copy the corresponding initial/final wavefunctions (waveinit_1S0.f90, waveinit_3S1.f90, or waveinit_He4.f90) into the file waveinit.f90 used by the main program nuclei.f90. The number of nucleons is controlled by the parameter n_f in input.f90 and must correspond to the number of nucleons in waveinit.f90.

As an example we show the beginning of the input file input.f90 for a two nucleon state with spatial lattice spacing $a = 1/(100\,\mathrm{MeV})$, temporal lattice spacing $a_t = 1/(150\,\mathrm{MeV})$, box size $L = 4a$, and Euclidean time extent $L_t = 6a_t$. We use an $O(a^4)$-improved lattice action for the nucleon hopping coefficients, $O(a^0)$-improved lattice action for the pion hopping coefficients, and $O(a^0)$-improved lattice action for the pion-nucleon coupling. The coefficient of the ${}^1S_0$ contact interaction is tuned to the physical ${}^1S_0 n-p$ scattering length and is $-5.615 \times 10^{-5}\,\mathrm{MeV}^{-2}$. The coefficient of the ${}^3S_1$ contact interaction is tuned to the deuteron binding energy at infinite volume and is $-6.543 \times 10^{-5}\,\mathrm{MeV}^{-2}$.

Code Listing 6.7 Parameter declarations at the beginning of the file input.f90

```
parameter(n_f = 2)
parameter(L = 4)
parameter(Lt = 6)
parameter(cutoff = 100.D0, temporalcutoff = 150.D0)
parameter(improveN = 2)
parameter(improveP = 0)
parameter(improveD = 0)
parameter(c1S0_phys = -5.615D-5)
parameter(c3S1_phys = -6.543D-5)
```

Using these values for the parameters of the lattice action, we now present some benchmark values which can be used to test the nuclear lattice simulations in the two-nucleon system. The values presented in these benchmarks are computed using exact calculations of the two-nucleon transfer matrix. They provide a useful independent check that there are no errors in the Monte Carlo simulations. In Table 6.4 we show the energies for the ${}^1S_0$ spin combination of two nucleons. The initial state is one neutron spin-up and one neutron spin-down, both at zero

Table 6.4 Benchmark energies for the $^1S_0$ spin combination of two nucleons

L_t	Energy (MeV)
2	−1.0915
4	−1.3987
6	−1.6209
8	−1.7929
10	−1.9296
12	−2.0398
14	−2.1291
16	−2.2018
18	−2.2610
20	−2.3094

The initial state is one neutron spin-up and one neutron spin-down, both at zero momentum, for $L = 4a$ and various values of L_t

Table 6.5 Benchmark energies for the $^3S_1$ spin combination of two nucleons

L_t	Energy (MeV)
2	−1.4446
4	−2.0400
6	−2.4774
8	−2.8331
10	−3.1341
12	−3.3925
14	−3.6151
16	−3.8069
18	−3.9718
20	−4.1132

The initial state is one proton spin-up and one neutron spin-up, both at zero momentum, for $L = 4a$ and various values of L_t

momentum, for $L = 4a$ and various values of L_t. The energies are extracted by computing the ratio of amplitudes $Z(L_t)/Z(L_t - 1)$ and setting equal to $\exp(-E\alpha_t)$.

We show the energies for the $^3S_1$ spin combination of two nucleons in Table 6.5. The initial state is one proton spin-up and one neutron spin-up, both at zero momentum, for $L = 4a$ and various values of L_t. The energies are extracted by computing the ratio of amplitudes $Z(L_t)/Z(L_t - 1)$ and setting equal to $\exp(-E\alpha_t)$.

Acknowledgements The author is grateful for discussions with Amy Nicholson and Morten Hjorth-Jensen. He is also greatly indebted to his collaborators Jose Alarcón, Dechuan Du, Serdar Elhatisari, Evgeny Epelbaum, Nico Klein, Hermann Krebs, Timo Lähde, Ning Li, Bing-nan Lu, Thomas Luu, Ulf-G. Meißner, Alexander Rokash, and Gautam Rupak. Partial financial support provided by the U.S. Department of Energy (DE-FG02-03ER41260). Computational resources were provided by the Jülich Supercomputing Centre.

References

1. S. Weinberg, Phys. Lett. B **251**, 288 (1990)
2. S. Weinberg, Nucl. Phys. B **363**, 3 (1991)
3. C. Ordonez, U. van Kolck, Phys. Lett. B **291**, 459 (1992)
4. C. Ordonez, L. Ray, U. van Kolck, Phys. Rev. Lett. **72**, 1982 (1994)
5. U. van Kolck, Phys. Rev. C **49**, 2932 (1994)
6. E. Epelbaum, W. Glöckle, U.G. Meißner, Phys. Lett. B **439**, 1 (1998)
7. E. Epelbaum, W. Glöckle, U.G. Meißner, Nucl. Phys. A **637**, 107 (1998)
8. P.F. Bedaque, U. van Kolck, Ann. Rev. Nucl. Part. Sci. **52**, 339 (2002)
9. E. Epelbaum, H.W. Hammer, U.G. Meißner, Rev. Mod. Phys. **81**, 1773 (2009)
10. D. Lee, Prog. Part. Nucl. Phys. **63**, 117 (2009)
11. J.E. Drut, A.N. Nicholson, J. Phys. G Nucl. Part. Phys. **40**, 043101 (2013)
12. J. Hubbard, Phys. Rev. Lett. **3**, 77 (1959)
13. R.L. Stratonovich, Sov. Phys. Dokl. **2**, 416 (1958)
14. D.H. Weingarten, D.N. Petcher, Phys. Lett. B **99**, 333 (1981)
15. R.T. Scalettar, D.J. Scalapino, R.L. Sugar, Phys. Rev. B **34**, 7911 (1986)
16. S. Gottlieb, W. Liu, D. Toussaint, R.L. Renken, R.L. Sugar, Phys. Rev. D **35**, 2531 (1987)
17. S. Duane, A.D. Kennedy, B.J. Pendleton, D. Roweth, Phys. Lett. B **195**, 216 (1987)
18. H.M. Müller, S.E. Koonin, R. Seki, U. van Kolck, Phys. Rev. C **61**, 044320 (2000)
19. D. Lee, B. Borasoy, T. Schäfer, Phys. Rev. C **70**, 014007 (2004)
20. D. Lee, T. Schäfer, Phys. Rev. C **72**, 024006 (2005)
21. D. Lee, T. Schäfer, Phys. Rev. C **73**, 015201 (2006)
22. D. Lee, T. Schäfer, Phys. Rev. C **73**, 015202 (2006)
23. B. Borasoy, H. Krebs, D. Lee, U.G. Meißner, Nucl. Phys. A **768**, 179 (2006)
24. B. Borasoy, E. Epelbaum, H. Krebs, D. Lee, U.G. Meißner, Eur. Phys. J. A **31**, 105 (2007)
25. B. Borasoy, E. Epelbaum, H. Krebs, D. Lee, U.G. Meißner, Eur. Phys. J. A **35**, 343 (2008)
26. B. Borasoy, E. Epelbaum, H. Krebs, D. Lee, U.G. Meißner, Eur. Phys. J. A **35**, 357 (2008)
27. G. Wlazłowski, J.W. Holt, S. Moroz, A. Bulgac, K.J. Roche, Phys. Rev. Lett. **113**(18), 182503 (2014)
28. E. Epelbaum, H. Krebs, D. Lee, U.G. Meißner, Eur. Phys. J. A **41**, 125 (2009)
29. E. Epelbaum, H. Krebs, D. Lee, U.G. Meißner, Phys. Rev. Lett. **104**, 142501 (2010)
30. E. Epelbaum, H. Krebs, D. Lee, U.G. Meißner, Phys. Rev. Lett. **106**, 192501 (2011)
31. E. Epelbaum, H. Krebs, T. Lähde, D. Lee, U.G. Meißner, Phys. Rev. Lett. **109**, 252501 (2012)
32. E. Epelbaum, H. Krebs, T.A. Lähde, D. Lee, U.G. Meißner, Phys. Rev. Lett. **110**, 112502 (2013)
33. T.A. Lähde, E. Epelbaum, H. Krebs, D. Lee, U.G. Meißner, G. Rupak, Phys. Lett. B **732**, 110 (2014)
34. E. Epelbaum, H. Krebs, T.A. Lähde, D. Lee, U.G. Meißner, G. Rupak, Phys. Rev. Lett. **112**(10), 102501 (2014)
35. S. Elhatisari, et al., Nuclear binding near a quantum phase transition. Phys. Rev. Lett. **117**(13), 132501 (2016). doi:10.1103/PhysRevLett.117.132501
36. G. Rupak, D. Lee, Phys. Rev. Lett. **111**(3), 032502 (2013)
37. G. Rupak, P. Ravi, Phys. Lett. B **741**, 301 (2014)
38. S. Elhatisari, D. Lee, G. Rupak, E. Epelbaum, H. Krebs, T.A. Lähde, T. Luu, U.G. Meißner, Nature **528**, 111 (2015). doi:10.1038/nature16067
39. A. Ekström, G.R. Jansen, K.A. Wendt, G. Hagen, T. Papenbrock, B.D. Carlsson, C. Forssén, M. Hjorth-Jensen, P. Navrátil, W. Nazarewicz, Phys. Rev. C **91**, 051301 (2015)
40. G. Hagen, T. Papenbrock, M. Hjorth-Jensen, D.J. Dean, Rep. Prog. Phys. **77**, 096302 (2014)
41. M. Lüscher, Commun. Math. Phys. **104**, 177 (1986)
42. M. Lüscher, Commun. Math. Phys. **105**, 153 (1986)
43. M. Lüscher, Nucl. Phys. B **354**, 531 (1991)
44. B. Borasoy, E. Epelbaum, H. Krebs, D. Lee, U.G. Meißner, Eur. Phys. J. A **34**, 185 (2007)

45. J. Carlson, V. Pandharipande, R. Wiringa, Nucl. Phys. A **424**(1), 47 (1984)
46. B.N. Lu, T.A. Lähde, D. Lee, U.G. Meißner, Precise determination of lattice phase shifts and mixing angles. Phys. Lett. B **760**, 309–313 (2016). doi:10.1016/j.physletb.2016.06.081
47. M. Pine, D. Lee, G. Rupak, Eur. Phys. J. A **49**, 151 (2013)
48. S. Elhatisari, D. Lee, Phys. Rev. C **90**(6), 064001 (2014)
49. A. Rokash, M. Pine, S. Elhatisari, D. Lee, E. Epelbaum, H. Krebs, Scattering cluster wave functions on the lattice using the adiabatic projection method. Phys. Rev C **92**(5), 054612 (2015). doi:10.1103/PhysRevC.92.054612
50. M. Creutz, Phys. Rev. D **38**, 1228 (1988)
51. M. Creutz, Found. Phys. **30**, 487 (2000)
52. J. Carlson, S. Gandolfi, F. Pederiva, S.C. Pieper, R. Schiavilla, K.E. Schmidt, R.B. Wiringa, Rev. Mod. Phys. **87**, 1067 (2015). doi:10.1103/RevModPhys.87.1067

Chapter 7
Ab Initio Methods for Nuclear Structure and Reactions: From Few to Many Nucleons

Giuseppina Orlandini

7.1 Introduction: Theory, Model, Method

The importance of studying nuclei lays in the fact that they are the most common manifestation of the strong interaction at *low*-energy (order of MeV), a regime where the fundamental theory, Quantum Chromo-Dynamics (QCD), is non-perturbative.

Describing nuclei as an assembly of interacting protons and neutrons corresponds to choosing the *effective* degrees of freedom (d.o.f) most relevant at that energy. This idea comes from observing that just protons and/or neutrons emerge, when energies of a few MeV are transferred to a nuclear system. Since such degrees of freedom have a comparatively much larger mass (about one GeV), we are allowed to adopt a non relativistic quantum mechanical framework to describe nuclear properties. Then, any observable we would like to account for, will require solving the Schrödinger equation, governed by a many-nucleon Hamiltonian. In other words we will need to solve the so called *non relativistic quantum many-body problem*.

In this context we will adopt the word *Theory* referred exclusively to non relativistic quantum mechanics (NRQM), in Schrödinger, Heisenberg or Interaction representation. The word **Model** will be used in connection to the choice of the d.o.f. and of their mutual interaction, namely the potential part of the Hamiltonian. It is clear that any nuclear *Model* (in the above acceptation) must have its roots in QCD, and that the Hamiltonian will have to share all its symmetries with QCD. These lectures will not deal with the problem of establishing the best *model* for a

G. Orlandini (✉)
Dipartimento di Fisica, Università di Trento, Via Sommarive, 14, I-38123 Trento, Italy

Trento Institute for Fundamental Physics and Applications - I.N.F.N, Via Sommarive, 14, I-38123 Trento, Italy
e-mail: giuseppina.orlandini@unitn.it

M. Hjorth-Jensen et al. (eds.), *An Advanced Course in Computational Nuclear Physics*, Lecture Notes in Physics 936, DOI 10.1007/978-3-319-53336-0_7

"realistic" potential (this subject is extensively treated elsewhere in this book), here we limit ourselves to consider it as an input for our problem.

Summarizing, we want to calculate observables within our defined *Theory*, with an input nuclear *Model*, which is both rooted in QCD and *realistic* enough to accurately reproduce (at least) the nucleon-nucleon scattering data (χ-square per datum close to 1). However, in addition, we want to be able to control the degree of *accuracy* of the method used to solve the NRQM many-body problem, namely we want to determine the theoretical accuracy on the value of an observable. This is done by benchmarking different results obtained by different methods, using the same input. All this is what characterizes an ab initio approach. Comparing the ab initio results to data we can learn about the degree of reliability of the nuclear *Model*. In this way we will be able to predict new nuclear observables, as well as to give other fields (e.g. astrophysics) the needed nuclear information, complemented by the degree of accuracy of the many-body method used.

7.2 The Non-relativistic Quantum Mechanical Many-Nucleon Problem

The non-relativistic quantum dynamics of a system of A nucleons, supposed to have equal masses m, is governed by the nuclear Hamiltonian H, which consists of kinetic energy T and potential V:

$$H = T + V = \sum_{i=1}^{A} \frac{\mathbf{p}_i^2}{2m} + \sum_{i<j}^{A} V_{ij} + \sum_{i<j<k}^{A} V_{ijk} + \ldots . \tag{7.1}$$

In the equation above $\mathbf{p}_i$ is the momentum of the ith nucleon in a general laboratory system, and V_{ij} and V_{ijk} denote the nucleon-nucleon (NN) potential V_{NN}, the three-nucleon potential V_{NNN}, etc., respectively. Notice that the reason why in general the nuclear Hamiltonian should contain many-body potentials is due to the fact the nucleons are *effective* degrees of freedom. The Chiral Effective Field Theory approach to the nuclear potential [1] shows that nuclear forces obey a hierarchy: forces of more and more many-body nature appear at higher and higher order in a perturbative expansion. In our discussion we will restrict to two- and three-body potentials, which appear to be the most relevant for common observables (in general the inclusion of three-body potentials represents a technical challenge for most ab initio approaches).

Our problem consists in solving the Schrödinger equation,

$$(H - E_n)|\Psi_n\rangle = 0\,, \tag{7.2}$$

where E_n and $|\Psi_n\rangle$ denote the eigenenergies and eigenfunctions of H, respectively. The spectrum of H, represented by the infinite set of eigenenergies is discrete below, and continuous above, the first break-up threshold E_{th}.

In order to solve the Schrödinger equation one has to supply proper boundary conditions. For $E < E_{\text{th}}$ the wave function represents a bound state and thus it is described by a square integrable (localized) function. This characteristic leads to major technical simplifications, compared to the case $E \geq E_{\text{th}}$, where the asymptotic boundary conditions pose serious problems, especially when $A > 2$ (many-body scattering problem).

There are different ways to tackle the NRQM problem for bound states or for the continuum. Very often a reformulation of the problem allows a practical solution, which seems impossible otherwise. It is this possibility that generates the richness of methods in many-body theory.

7.2.1 Translation and Galileian Invariance

A correct approach to the non relativistic many-nucleon problem should fulfill two fundamental symmetries, namely those related to translational and Galileian invariance. One can easily show that the corresponding conserved quantities are center of mass (CM) momentum $\mathbf{p}_{CM} = \sum_i^A \mathbf{p}_i$ and CM position $\mathbf{R}_{CM} = \frac{1}{A}\sum_i^A \mathbf{r}_i$, respectively. Therefore the correct nuclear Hamiltonian must commute with those operators. This can be achieved if one rewrites Eq. (7.1) in terms of the Jacobi vectors, i.e. the A independent (normalized) vectors given by $\mathbf{R}_{CM}$ and

$$\begin{aligned}
\boldsymbol{\eta}_1 &= \sqrt{\frac{A-1}{A}}\Big(\mathbf{r}_1 - \frac{1}{A-1}(\mathbf{r}_2 + \mathbf{r}_3 + \cdots + \mathbf{r}_A)\Big) \\
\boldsymbol{\eta}_2 &= \sqrt{\frac{A-2}{A-1}}\Big(\mathbf{r}_2 - \frac{1}{A-2}(\mathbf{r}_3 + \mathbf{r}_4 + \cdots + \mathbf{r}_A)\Big) \\
&\cdots \\
\boldsymbol{\eta}_N &= \sqrt{\frac{1}{2}}\Big(\mathbf{r}_{A-1} - \mathbf{r}_A\Big), \qquad N = A-1\,;
\end{aligned} \tag{7.3}$$

together with their conjugate momenta $\mathbf{P}_{CM}$ and $\boldsymbol{\pi}_1, \boldsymbol{\pi}_2, \ldots \boldsymbol{\pi}_N$.

In terms of Jacobi vectors the Hamiltonian in (7.1) becomes

$$H = H_{CM} + \mathscr{H} = \frac{\mathbf{P}_{CM}^2}{2Am} + \sum_{i=1}^{A-1} \frac{\boldsymbol{\pi}_i^2}{2m} + V(\boldsymbol{\eta}_1, \boldsymbol{\eta}_2 \ldots \boldsymbol{\eta}_{A-1})\,, \tag{7.4}$$

and the translation and Galileian invariant Hamiltonian of our interest is $\mathscr{H}$, which commutes both with $\mathbf{P}_{CM}$ and $\mathbf{R}_{CM}$. An interesting remark is in order here: any potential, even when it is limited to have a two-(or three-)body character, becomes an unseparated function of the $(A-1)$-coordinate in $\mathscr{H}$. This means that the system, when correctly expressed in terms of relative coordinates only, contains a correlation among the constituents, which goes beyond the dynamical one. One can call it a *CM correlation*. The latter can be easily understood from the fact that the movement of one particle will affect all the others, since the CM momentum and position remain fixed (conserved).

7.3 Classification of Ab Initio Approaches for Ground-State Calculations

As it was stated above the NRQM problem can be formulated in different ways. Therefore one can classify the ab initio methods in terms of just such different formulations, grouping them in different classes:

- The Faddeev-Yakubowski (FY) method,
- Methods based on the variational theorem,
- Methods based on similarity transformations,
- Quantum Monte Carlo methods.

In the following we will concentrate in particular on two of the methods rooted in the variational theorem. However, in the following a brief summary of the main peculiarities characterizing each group will be given. A more extensive description of the methods can be found in the quoted original references. A recent review can be found in [2].

7.3.1 The Faddeev-Yakubowski (FY) Method

The very nice feature of the FY method is that it is formulated in a way that it is applicable to both bound and scattering states. This method starts from the Lipmann-Schwinger (LS) reformulation of the Schrödinger equation and therefore deals with integral equations instead of differential equations. Today it is well known that a direct application of LS-type equations to the scattering problem for a system with more than two particles does not lead to a unique solution. However, for quite some time it was not clear how such a unique solution could be obtained. It was in 1961 that in his seminal work for the three-body system [3] Faddeev showed how the problem can be solved. He derived the right set of coupled integral equations which have taken his name: the Faddeev equations. Yakubowsky [4] generalized the approach, in principle to any number of particles. However, the number of coupled equations to solve becomes prohibitive for more than four particles.

A sort of variation of the FY equations are those introduced by Alt, Grassberger and Sandhas (AGS equations) [5] who looked for possible further reductions of the FY problem. Assuming a separable form for the NN t-matrix leads to one-variable integral equations, which are much simpler to solve than the FY equations.

7.3.2 *Methods Based on the Variational Theorem (Diagonalization Methods)*

The diagonalization methods are based on the Rayleigh-Ritz variational theorem [6, 7]. This theorem, which is very profitably applied every time the solution of some useful equation renders stationary some proper functional, finds a large application in quantum mechanics. In particular, one can show that the solution of the Schrödinger equation (for a state with finite norm) renders stationary the energy functional

$$E[\Psi] = \frac{\langle\Psi|H|\Psi\rangle}{\langle\Psi|\Psi\rangle} . \tag{7.5}$$

An important lemma complements the fundamental variational theorem, stating that the value of the energy functional calculated with any trial function is always greater than the ground-state energy and equal to it, only when the trial function coincides with the exact ground-state wave function. This means that one can find the ground state energy of a system by solving a minimization problem

$$\delta E[\Psi] = 0 . \tag{7.6}$$

Numerous approaches use this variational principle to find the ground-state energy of a many-body system. The approach is efficient if the trial function has a parametrized functional form that is both convenient and suitable to the problem to be solved. The various variational approaches differ by the choice of the trial function. One very well known classical example is the Hartree-Fock method, where the trial function is a Slater determinant. In this case, however, it is not possible to give a theoretical estimate of how far the Hartree-Fock energy is from the correct result. Also the use of a parametrized functional form for the trial function and the minimization with respect to the parameters does not allow a theoretical estimate of the error. More sophisticated approaches exist like the resonating group method [8, 9], where the trial function is chosen according to a cluster picture of the system or the variational Monte Carlo (VMC) technique where the trial function reflects the form of the potential, see the presentation in Chap. 9.

A more systematic approach consists in choosing the trial function Ψ_T as an expansion on a complete (or over complete) set of square integrable functions ϕ_n

that respect the symmetries of the Hamiltonian:

$$|\Psi_T(N)\rangle = \sum_{n=1}^{N} c_n|\phi_n\rangle \,. \tag{7.7}$$

In this case the minimization procedure corresponds to finding the solution of a (generalized) eigenvalue problem

$$(\mathbf{H} - E\,\mathbf{M})\,C = 0\,, \tag{7.8}$$

where $\mathbf{H}$ and $\mathbf{M}$ are $N \times N$ Hermitean matrices of the Hamiltonian ($H_{nm} = \langle\phi_n|H|\phi_m\rangle$) and overlap integrals of the basis functions ($M_{nm} = \langle\phi_n|\phi_m\rangle$), while C represents the N-component vector formed by the linear parameters c_n. With growing N the size of the Hamiltonian matrix, represented on the chosen basis, increases and the true ground-state energy is approached from above. The basis can be complete as the hyperspherical harmonics (HH) or the harmonic oscillator (HO) basis, or over-complete. In principle the true result would be obtained only for an infinite number of basis functions, however, the convergence of the smallest energy obtained after the diagonalization for large enough N gives the ground state energy. An estimate of the error can also be given, related to the convergence pattern. One can consider this as an ab initio result.

Here one should also mention another interesting variational method, namely the stochastic variational method (SVM) [10, 11]. Here again the variational procedure does not proceed systematically by the diagonalization of a larger and larger Hamiltonian matrix, but in a stochastic way (trial and error), obtaining nevertheless rather good results when compared to other approaches [12].

Variational approaches also allow to obtain the wave function corresponding to the minimal energy, which can then be used to calculate other ground-state observables. However, one has to remember that the difference between the exact value of the energy and that obtained with the trial function Ψ_T which minimizes the energy functional, is an infinitesimal of higher order than the difference between the true wave function and Ψ_T. Therefore one should expect a slower convergence and less accuracy for such observables.

7.3.2.1 The Hyperspherical Harmonics (HH) Method

The HH method is a variational method where the trial function is written as an expansion on the *hyper*-spherical harmonics (HH) basis. The HH are the generalization of the spherical harmonics Y_{lm}. In fact as the latter represent a basis for the relative wave function of a two-body system, the HH represent a general basis for the internal wave function of an A-body system. Because of this, they are expressed in terms of the hyperspherical coordinates which are defined by a transformation of the Jacobi vectors.

Let us remember that the set of A Jacobi vectors is composed by $\mathbf{R}_{CM}$ and the $N = A - 1$ relative vectors $\boldsymbol{\eta}_1, \ldots, \boldsymbol{\eta}_N$ in Eq. (7.3), for a total of $3A$ coordinates. The hyperspherical coordinates are defined by further transforming the $3N$ coordinates $\boldsymbol{\eta}_1, \ldots, \boldsymbol{\eta}_N$ as follows: the $2N$ polar angles θ_i and ϕ_i of the $\boldsymbol{\eta}_i \equiv (\eta_i, \theta_i, \phi_i)$ are left unaffected by the transformation. The remaining N hyperspherical coordinates η_i are expressed in terms of one *hyper*-radius ρ_N and $(N-1)$ *hyper*-angles α_n defined by

$$\sin\alpha_n = \frac{|\eta_n|}{\rho_n}\,; \qquad \rho_n^2 = \sum_{i=1}^{n} \eta_i^2\,, \quad n = 2, \ldots, N\,. \tag{7.9}$$

A very interesting feature of the hyperspherical coordinates is that, when expressed in such coordinates, the A-body kinetic energy operator of A nucleons of equal masses is a sum of two terms ($\rho^2 \equiv \rho_N^2$) [13]

$$T = T_\rho + T_K(\rho)\,, \quad \text{with} \qquad T_\rho = -\frac{1}{2m}\Delta_\rho\,, \qquad T_K(\rho) = \frac{1}{2m}\frac{\mathbf{K}_N^2}{\rho^2}\,, \tag{7.10}$$

namely it has a form which is in perfect analogy to the three-dimensional case, with a *hyper*-radial dependent Laplacian T_ρ and a *hyper*-centrifugal barrier $T_K(\rho)$.

The *hyper*-angular momentum operator $\mathbf{K}_N$ depends on all the $(3N-1)$ angles (denoted by $\hat{\Omega}_{[N]}$) and has a rather complicated form. But the main point here is that the HH are the orthonormal eigenfunctions $\mathcal{Y}_{[K_N]}(\hat{\Omega}_{[N]})$ of $\mathbf{K}_N^2$

$$\mathbf{K}_N^2\, \mathcal{Y}_{[K_N]}(\hat{\Omega}_{[N]}) = K_N(K_N + 3N - 2)\mathcal{Y}_{[K_N]}(\hat{\Omega}_{[N]})\,. \tag{7.11}$$

As one sees the eigenvalues are expressed in terms of the quantum number K_N. The subscript $[K_N]$ stands for the total set of $(3N-1)$ quantum numbers corresponding to commuting operators, namely the hyperangular momenta $\mathbf{K}_N^2, \mathbf{K}_{N-1}^2, \ldots \mathbf{K}_2^2$ relative to the subsets of $N, N-1\ldots 2\eta$-coordinates, the angular momenta relative to each of the N Jacobi coordinates $\mathbf{l}_N^2, \mathbf{l}_{N-1}^2, \ldots \mathbf{l}_1^2$, the total angular momenta $\mathbf{L}_N^2, \mathbf{L}_{N-1}^2, \ldots \mathbf{L}_2^2$ of the same subsets of $N, N-1\ldots 2\eta$-coordinates, and the third component of the total angular momentum L^z.

The $\mathcal{Y}_{[K_N]}(\hat{\Omega}_{[N]})$ are good basis functions for the hyperangular part of the A-body internal wave function, however, one also needs good basis functions for the hyperradial part of the wave function. A suitable choice are the orthogonal Laguerre polynomials, because of their exponential weight function, reproducing the correct asymptotic behavior of the wave function.

The basis obtained by the product of Laguerre polinomials and HH is a translation invariant CM "correlated" basis (all particles are connected to each other!) and has good asymptotic conditions, therefore one can expect a faster convergence with respect to using a translation invariant HO basis. However, just because of the mentioned correlation, the basis presents big difficulties when coping with the Pauli principle, a problem also common to the translation invariant HO basis. Based only on intuition one can guess that permutations of particles will lead to different definitions of the Jacobi coordinates and consequently of the hyperspherical coordinates. It would be a miracle if the HH would have definite permutational symmetries. And in fact they do not, but they possess different components of the irreducible representations of the symmetry group S_A. Even when this problem is overcome (see [14–17]), as the number of particles increases the convergence becomes rather slow.

In order to speed up the convergence two ways have been followed: the Correlated Hyperspherical Harmonics expansion (CHH) and the Hyperspherical Harmonics expansion with Effective Interaction (EIHH). The latter will be explained in Sect. 7.4.2. The main idea of the CHH approach consists in acting on the bare HH functions with a Jastrow operator $\hat{J}$ embodying the short range correlation due to the repulsive part of the potential. Such a repulsion leads to high momentum components in the wave function which is responsible for the slow convergence of the bare HH expansion. The correlation operator $\hat{J}$ takes the form

$$\hat{J} = \mathscr{S} \prod_{i<j} \sum_{s,t} f_{st}(r_{ij}) P_{st}(i,j) \tag{7.12}$$

where $P_{st}(i,j)$ are projection operators onto nucleon pairs (ij) with spin s and isospin t and $\mathscr{S}$ is a particle symmetrization operator. This method has been applied only to A=3,4 systems, since the loss of orthonormality of the CHH limits its efficiency. In fact calculating the matrix elements of the potential requires $3N$ dimensional integrals. The reason why this is not the case for the uncorrelated HH basis is explained in the following.

When expressed in HH coordinates the invariant Hamiltonian $\mathscr{H}$ is

$$\mathscr{H} = T_\rho + T_K(\rho) + V(\rho, \hat{\Omega}_{[N]}) . \tag{7.13}$$

However, supposing for simplicity that the potential has a two-body character $\sum_{i<j} V_{ij}$ (but the present argument can be easily extended to three-body potentials) its matrix element on antisymmetric functions will be the sum of $A(A-1)$ identical integrals with, say, $i = A$ and $j = A - 1$. This means that V will be a function only of the Jacobi vector η_N, namely $V(\sqrt{2}\rho \sin \alpha_N, \theta_N, \phi_N)$, the recursive construction of the HH allows then to use the orthonormality condition of the $\mathscr{Y}_{[K_{N-1}]}(\hat{\Omega}_{[N-1]})$ and reduce the calculation of the matrix element of the potential to a (at most) four-dimensional integral, for any number of particles. When the orthonormality condition of the HH is lost, like in the CHH case, this is no longer true and one is left with 3N-dimensional integrals.

7.3.3 *Methods Based on Similarity Transformations*

Another reformulation of the quantum mechanical many-body problem is based on the use of similarity transformations [18–21]. In this case one considers that the following mean value

$$E_0 = \langle \Psi_0 | H | \Psi_0 \rangle \,, \tag{7.14}$$

where $|\Psi_0\rangle$ is the ground state of the Hamiltonian H, is invariant under similarity transformations e^S, i.e.

$$E_0 = \langle \Psi_0 | e^{-S} e^S H e^{-S} e^S | \Psi_0 \rangle \equiv \langle \bar{\Phi} | \bar{H} | \Phi \rangle \tag{7.15}$$

with

$$|\Phi\rangle = e^S |\Psi_0\rangle \,, \qquad |\bar{\Phi}\rangle = e^{-S^\dagger} |\Psi_0\rangle \,, \qquad \bar{H} = e^S H e^{-S} \,. \tag{7.16}$$

At this point one may consider a subspace P of the Hilbert space with eigenprojector $\hat{P}$ given by

$$\hat{P} = \sum_{n=1}^{N} |\phi_n\rangle\langle\phi_n| \,, \tag{7.17}$$

where the $|\phi_n\rangle$ are eigenfunctions of some well known Hamiltonian (e.g. H_{HO}). Indicating by $\hat{Q} = I - \hat{P}$ the corresponding eigenprojector on the residual space, one can write Eq. (7.15) as

$$E = \langle \bar{\Phi} | (\hat{P} + \hat{Q}) \bar{H} (\hat{P} + \hat{Q}) | \Phi \rangle = \langle \bar{\Phi} | \hat{P}\bar{H}\hat{P} + \hat{P}\bar{H}\hat{Q} + \hat{Q}\bar{H}\hat{P} + \hat{Q}\bar{H}\hat{Q} | \Phi \rangle \,. \tag{7.18}$$

If the following decoupling condition is satisfied

$$\hat{Q}\bar{H}\hat{P} = \hat{Q} e^S H e^{-S} \hat{P} = 0 \,, \tag{7.19}$$

one has

$$E = \langle \bar{\Phi} | \hat{P}\bar{H}\hat{P} | \Phi \rangle \,. \tag{7.20}$$

This means that if one solves the decoupling equation (7.19) it is possible, in principle, to determine S and therefore calculate E_0 as the mean value of the *effective* operator $\bar{H}$ on the P-space.

Notice that, while in the *bare* Hamiltonian H the operators may have a two- or three-body nature, the *effective* operator $\bar{H}$ will be in principle an A-body operator. So, in general the operator S, which generates the similarity transformation, may be written as a combination of operators of any $n \leq A$-body nature. It is clear that in actual calculations one has to apply some restrictions on the number of these

n-body operators. In this respect the similarity transformation approach has been used in two ways:

- in the so called Coupled Cluster (CC) method [22, 23] (discussed in the next chapter) to calculate the ground state energies and radii of A-body nuclei ($n \leq 3$, known as CCSD and CCSDT);
- to construct effective two- or three-body potentials, as described in the next Sect. 7.3.3.1, in order to accelerate convergence in the variational diagonalization approaches using the HO and the HH basis (see Sects. 7.4.1 and 7.4.2).

7.3.3.1 The Similarity Transformation Method for Effective Interactions

As was stressed in the previous section, the similarity transformation leads to an effective Hamiltonian which is an A-body operator. To avoid this complication, an approximation is made. It consists in first finding only a two-body effective interaction $\tilde{V}_{ij}^{[2,\text{eff}]}$ which is then used to replace the *bare* interaction term V_{ij}. This approach is often referred to as the Lee-Suzuki (LS) method [21]. The effective interaction $V_{ij}^{[2,\text{eff}]}$ is obtained by applying the decoupling condition of Eq. (7.19) to a two-nucleon Hamiltonian $H^{[2]}$ that arises from $\mathscr{H}$ by restricting the kinetic and potential operators to two nucleons only (e.g. $i = A$ and $j = A - 1$). In this simple case the decoupling condition can be solved. In fact the two-body problem can be fully solved and in this case one has the knowledge of both the P and the Q space.

Once the effective interaction $V_{12}^{[2,\text{eff}]}$ is obtained one replaces it in the $\sum_{i<j} V_{ij}$. The replacement in the potential term of the effective interaction $V_{ij}^{[2,\text{eff}]}$ makes the approach no longer variational. The n-body terms neglected in the full effective Hamiltonian, could either increase or decrease the binding energy. On the other hand, one has an important result: as the P_A-space is increased the result has to converge to the exact solution. This may be illustrated in a pictorial way as in Fig. 7.1. At each P_A, since the similarity transformation transfers information from the Q_2 space to the P_2 space, there is much less information left out. Consequently the convergence on E_0 is much faster. When P_A is sufficiently large, so that it covers almost the whole Hilbert space, the effective interaction practically coincides with the bare one, and one has an accurate result. From the figure one can infer that the exact result could be reached also applying the similarity transformation to the three-body and then four-body Hamiltonian etc., namely systematically applying the similarity transformation to move the information from the larger Q_3 space into the P_3 space, from the Q_4 space into the P_4 space, etc.

However, this is much more problematic and impossible in practice, since one would need to know the entire n-body spectrum to construct the n-body effective interaction. Of course if three-body forces are present in the original Hamiltonian one has to apply the procedure at least up to $n = 3$.

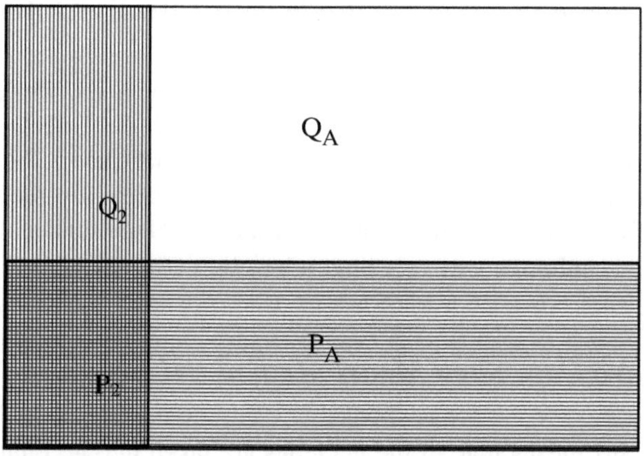

Fig. 7.1 The various P and Q spaces relevant for the construction of the two-body effective interaction (see text)

7.3.4 Monte Carlo Methods

The Monte Carlo (MC) methods are based on a formulation of the quantum mechanical many-body problem which is suited to a stochastic approach. These methods are the Green's Function MC (GFMC) [24], Diffusion MC (DMC) [25] and Auxiliary Field Diffusion MC (AFDMC) [26], the Chiral Effective Field Theory on a Lattice (LCEFT) [27], the Monte Carlo Shell Model Diagonalization (MCSMD) [28–30], and the Variational Monte Carlo (VMC) [31]. The GFMC, DMC and AFDMC methods are based on the path integral formulation of quantum mechanics (there are small differences between the GFMC and DMC methods so that in the literature they are often interchanged). The LCEFT is a DMC approach, except that the dynamical degrees of freedom are nucleon and pion fields rather than particles. Both GFMC and LCEFT methods are based on the Euclidean time (imaginary time) evolution of the system. The MCSMD, although inspired by the imaginary time formulation, is effectively a variational method. Starting from an imaginary time evolved trial function, after some manipulation, MCSMD leads to an expression that suggests a way of constructing a variational shell model basis. In this case the imaginary time is just one of the non-linear parameters. The VMC method is a fully variational method. Here the MC technique is used to evaluate the many-dimensional energy functional integrals. The variational wave function obtained with this method usually serves as starting trial function for the GFMC imaginary time evolution.

An extensive treatment of Monte Carlo approaches can be found in Chap. 9 of this text.

7.4 Two Diagonalization Methods with Effective Interactions

In this section we will describe two diagonalization methods, which have much in common: the No Core Shell Model (NCSM) and the Effective Interaction Hyperspherical Harmonic (EIHH) methods. They both make use of the LS similarity transformation method (see Sect. 7.3.3.1), in order to speed up the convergence. They differ only by the choice of the basis on which the Hamiltonian is diagonalized and, for $A > 4$, also for the way they treat translation invariance.

7.4.1 The No Core Shell Model Method (NCSM)

The name NCSM means that all the nucleons are taken into account explicitly as degrees of freedom, namely it is not assumed that there is an inert core, like in the traditional shell model. The NCSM couples the advantage of the shell model (i.e. working in a HO basis) with the accuracy of an ab initio approach.

In the literature two versions of the NCSM exist, which differ in the treatment of translation invariance. In one version, the Hamiltonian of Eq. (7.1) is modified [32] by adding a harmonic oscillator CM Hamiltonian $H_{\rm CM}$ to the intrinsic Hamiltonian $\mathscr{H}$

$$H_\Omega^{[A]} = \mathscr{H} + H_{\rm CM}^{\rm HO} = \mathscr{H} + \frac{\mathbf{p}_{\rm CM}^2}{2Am} + \frac{Am}{2}\Omega^2\mathbf{R}_{\rm CM}^2 \tag{7.21}$$

$$= \sum_{i=1}^{A}\left[\frac{\mathbf{p}_i^2}{2m} + \frac{1}{2}m\Omega^2\mathbf{r}_i^2\right] + \sum_{i<j=1}^{A}\left[V_{ij} - \frac{m\Omega^2}{2A}(\mathbf{r}_i - \mathbf{r}_j)^2\right] \tag{7.22}$$

$$\equiv \sum_{i=1}^{A} h_i^{\rm HO} + \sum_{i<j=1}^{A} \tilde{V}_{ij}\,, \tag{7.23}$$

where $\tilde{V}_{jk}$ is a modified potential which depends on both the HO frequency Ω and the nuclear system via the mass number A:

$$\tilde{V}_{ij} = \left[V_{ij} - \frac{m\Omega^2}{2A}(\mathbf{r}_i - \mathbf{r}_j)^2\right]. \tag{7.24}$$

Of course the added center of mass HO term has no influence on the internal motion. Therefore the ground-state energy E of $\mathscr{H}$ is obtained by subtraction of the CM ground-state energy $3\hbar\Omega/2$ from the ground-state energy $E_\Omega^{[A]}$ of $H_\Omega^{[A]}$.

In order to obtain an accurate result the calculation of $E_\Omega^{[A]}$ should be performed with the $\tilde{V}$ of Eq. (7.24) in a finite model space P_A spanned by all the A-body HO Slater determinants formed by filling the single-particle HO eigenstates with $N \leq N_{\max}$ (N is the total number of single-particle HO quanta) and increasing P_A, namely $N_{\max}$, up to convergence. However, since the convergence is very slow one can speed it up using a $\tilde{V}^{[2,\text{eff}]}$. This is obtained as described in Sect. 7.3.3.1, namely applying the SL procedure to a two-body internal Hamiltonian obtained from (7.22) by restricting the sums to two nucleons only (e.g. nucleons A and $A-1$), keeping however the original mass number A in the interaction term $\tilde{V}$:

$$\mathscr{H}_{\Omega}^{[2]} = \left[\frac{\pi^2}{2m} + \frac{1}{2}m\Omega^2\eta^2\right] + \tilde{V}_{A(A-1)} \equiv \mathscr{H}_{\text{HO}}^{[2]} + \tilde{V}_{A(A-1)}\,. \tag{7.25}$$

The effective Hamiltonian $\mathscr{H}_{\text{eff}}^{[2]}$ is determined in the P_2-space ($\hat{P}_2 + \hat{Q}_2 = I_2$), a subspace of P, via the two-body transformation operator $S^{[2]} = \hat{Q}_2 S^{[2]} \hat{P}_2$. Then by subtracting $\mathscr{H}_{\text{HO}}^{[2]}$ from $\mathscr{H}_{\text{eff}}^{[2]}$ the two-body effective interaction is obtained i.e.

$$\tilde{V}_{12}^{[2,\text{eff}]} = \mathscr{H}_{\text{eff}}^{[2]} - \mathscr{H}_{\text{HO}}^{[2]}\,. \tag{7.26}$$

The obtained $\tilde{V}_{ij}^{[2,\text{eff}]}$ is then used in Eq. (7.23). As is clear from [20], this procedure is equivalent to (1) limiting the similarity operator S of Sect. 7.3.3 to a two-body operator $S^{[2]}$ and (2) truncating the effective Hamiltonian at the two-body operator level. When the diagonalization of the Hamiltonian is performed with the two-body effective interaction the NCSM is no longer variational. In fact, as was already stated above the real effective interaction obtained by a full similarity transformation at a fixed P-space is an A-body interaction. The neglected $n > 2$-body terms could either increase or decrease the binding energy. However, as the P-space increases the result converges to the exact solution, as was already discussed in Sect. 7.3.3.1.

Another version of the NCSM exists [33], where the problem is formulated directly in terms of Jacobi coordinates and the A-body basis is the translationally invariant HO basis. However, it is restricted to $A = 3, 4$, because of the same complications generated by the Pauli principle, as those was mentioned in Sect. 7.4.2.

7.4.2 *The Hyperspherical Harmonics Method with Effective Interaction (EIHH)*

The idea is very similar to that of the NCSM approach, but with the following two differences: one is that the P-space is defined by a maximal value $K_{\max}$ of the grand angular quantum number K_N ($N = A-1$), and the second is that the EIHH two-body Hamiltonian undergoing the similarity transformation, $H_A^{[2]}$, is a *quasi* two-body Hamiltonian, because it contains information about the dynamics of the entire A-body system via an hyperradial dependence.

To be more explicit, the HH *quasi* two-body Hamiltonian is chosen as

$$H_A^{[2]}(\rho) = T_K(\rho) + V_{A,A-1}(\sqrt{2}\rho\sin\alpha_N\,\theta_N, \phi_N)\,, \tag{7.27}$$

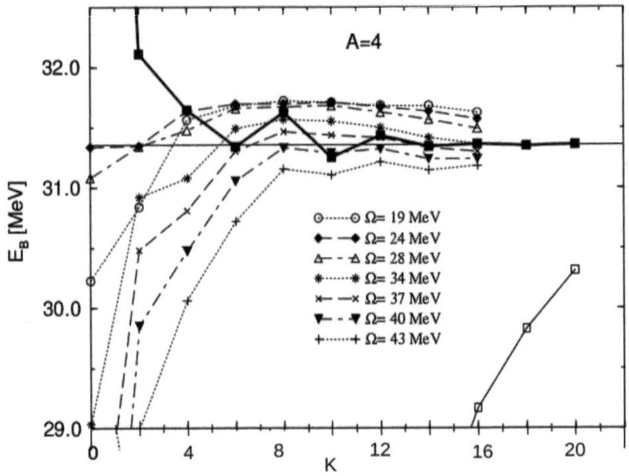

Fig. 7.2 Comparison between EIHH (*full squares*), pure HH (*open squares*) and NCSM results obtained with different values of the HO parameter

where $T_K(\rho)$ is the collective hyperspherical kinetic energy of the entire A-body system [see Eqs. (7.10)] and we make explicit only the spatial dependence of V on two particles (chosen to be A and $A-1$, consistent with Jacobi coordinates constructed in normalized reversed order, like those in Eq. (7.3).

The transformation is applied separately for each value of the hyperradius ρ, therefore the effective interaction becomes a function of ρ. In addition to its ρ-dependence, the HH effective interaction also depends on some quantum numbers of the residual system. The additional many-body information contained in the HH effective interaction (obtained by subtracting the hypercentrifugal term) is responsible for leading to a fast convergence of the HH expansion as can be seen in an example reported in Fig. 7.2 from [34]. In this figure the binding energy of ^{4}He was obtained using a central NN potential. One can notice the enormous advantage in the convergence pattern of the EIHH results in comparison to the HH ones. It is also interesting to notice the comparison with the NCSM results, which are spread around the EIHH one, depending on the HO parameter, as well as the non-variational behavior of the convergence pattern.

7.5 Excited States

By diagonalizing the Hamiltonian matrix on a given finite basis of square integrable functions one gets a spectrum of N eigenstates Ψ_i with eigenvalues E_i. If the trial function has the same quantum numbers of the ground state of the nucleus, the lowest E_i corresponds to the ground-state energy. The other solutions correspond to discrete excited states of the system, with the same quantum numbers as the ground state. Getting convergence for the energy of such excited states requires in general larger matrices. To find discrete excited states with quantum numbers different from those of the ground state one has to implement the proper quantum numbers into the basis set.

Few-nucleon systems have very few, if any, discrete excited states, in fact most of them lie in the continuum. Continuum eigenvalues are larger than the threshold energy E_{th}, corresponding to their ground state energy plus the nucleon separation energy (the proton separation energy is always smaller than that of a neutron if the Coulomb force is considered). The energy states larger than E_{th} obtained by diagonalization do not have the proper continuum boundary conditions. Proceeding in such a way one obtains a fake discretization of the continuum.

In order to avoid finding the continuum solutions of the Schrödinger equation, corresponding to solve the many-body scattering problem, one can try again to reformulate the quantum mechanical problem in an accessible way. This will be done in the next sections, referring in particular to observables that require the knowledge of such continuum states.

7.5.1 *Response Functions to Perturbative Probes*

Here we focus on a particular family of reactions involving states in the continuum: we deal with nuclear reactions on light systems induced by perturbations. Typical examples are electroweak (e.w.) reactions like electron or neutrino scattering on nuclei or nuclear photoabsorption.

The strength of the interaction Hamiltonian H_{int} between an electromagnetic or weak probe and the nucleus is very weak when compared to the strong interaction among the nucleons. Therefore the cross section can be calculated in first order perturbation theory, using the Fermi Golden rule. This means that the cross section will contain the transition rate proportional to the square modulus of the matrix element $|\langle f|H_{int}|i\rangle|^2$ with $|i\rangle$ and $|f\rangle$ the initial and final states of the nucleus, as well as an energy conserving δ-function

$$\sigma \propto |\langle f|H_{int}|i\rangle|^2 \delta(\omega - E_f + E_i) , \tag{7.28}$$

where ω is the energy transferred by the probe to the nucleus ($\hbar = c = 1$) and E_f, E_i are the energies of the nuclear final and initial states, generally the ground state $|0\rangle$ and one of its $|n\rangle$ eigenstates (we suppose that in ω the energy that has served to recoil the nucleus has been subtracted). In general the interaction H_{int} can be described as the product of the "current" densities inside the nucleus and the field generated by the probe, e.g. the charge density $\rho(\mathbf{r})$ and the electromagnetic scalar potential $\varphi(\mathbf{r})$ generated by the electrons in an electron scattering experiment

$$H_{int} = \int d\mathbf{r}\, \rho(\mathbf{r})\, \varphi(\mathbf{r}) \tag{7.29}$$

In this example the nuclear charge density is due to the presence of the protons

$$\rho(r) = \sum_{j=1}^{Z} e\,\delta(\mathbf{r}-\mathbf{r}_j) = e\sum_{j=1}^{A}\delta(\mathbf{r}r-\mathbf{r}_j)\frac{1+\tau_j^3}{2}\,. \tag{7.30}$$

The initial and final states will be the product of the CM wave functions and the internal ones, which are antisymmetric and translation invariant

$$|i\rangle = |\Phi_i(\mathbf{R}_{CM})\rangle|\psi_i(\boldsymbol{\eta}_1\ldots\boldsymbol{\eta}_{A-1})\rangle \tag{7.31}$$

$$|f\rangle = |\Phi_f(\mathbf{R}_{CM})\rangle|\psi_f(\boldsymbol{\eta}_1\ldots\boldsymbol{\eta}_{A-1})\rangle \tag{7.32}$$

Therefore

$$\begin{aligned}\langle f|H_{int}|i\rangle &= e\int d\mathbf{r}\,\varphi(\mathbf{r})\int d\mathbf{R}_{CM}e^{-i\mathbf{P}_f\cdot\mathbf{R}_{CM}}\,e^{i\mathbf{P}_i\cdot\mathbf{R}_{CM}}\\ &\times\int d\boldsymbol{\eta}_1\ldots d\boldsymbol{\eta}_{A-1}\psi_f^*(\boldsymbol{\eta}_1\ldots\boldsymbol{\eta}_{A-1};\sigma_1^z\ldots\sigma_A^z;\tau_1^z\ldots\tau_A^z)\\ &\times\sum_{j=1}^{A}\delta(\mathbf{r}-\mathbf{r}_j)\frac{1+\tau_j^3}{2}\psi_i(\boldsymbol{\eta}_1\ldots\boldsymbol{\eta}_{A-1};\sigma_1^z\ldots\sigma_A^z;\tau_1^z\ldots\tau_A^z)\,.\end{aligned} \tag{7.33}$$

Because the ψ are antisymmetric the matrix element above is a sum of A equal integrals, with the charge density operator limited to only one element of the sum (e.g. j=1).

At this point it is necessary to express $\delta(\mathbf{r}-\mathbf{r}_1)$ in terms of the integration variables. Using the definition of the Jacobi coordinate $\boldsymbol{\eta}_1$ as in (7.3)

$$\boldsymbol{\eta}_1 = \sqrt{\frac{A}{A-1}}(\mathbf{r}_1-\mathbf{R}_{CM}) \tag{7.34}$$

the matrix element in (7.33) becomes

$$\begin{aligned}\langle f|H_{int}|i\rangle &= Ae\int d\mathbf{r}\,\varphi(\mathbf{r})\int d\mathbf{R}_{CM}e^{i\mathbf{r}\cdot\mathbf{R}_{CM}}\\ &\times\int d\boldsymbol{\eta}_1\ldots d\boldsymbol{\eta}_{A-1}\psi_f^*(\boldsymbol{\eta}_1\ldots\boldsymbol{\eta}_{A-1};\sigma_1^z\ldots\sigma_A^z;\tau_1^z\ldots\tau_A^z)\\ &\times\sum_{j=1}^{A}\delta\left(\mathbf{r}-\mathbf{R}_{CM}-\sqrt{\frac{A-1}{A}}\boldsymbol{\eta}_1\right)\frac{1+\tau_1^3}{2}\psi_i(\boldsymbol{\eta}_1\ldots\boldsymbol{\eta}_{A-1};\sigma_1^z\ldots\sigma_A^z;\tau_1^z\ldots\tau_A^z)\,.\end{aligned} \tag{7.35}$$

After performing the integral in $d\mathbf{R}_{CM}$ one has

$$\langle f|H_{int}|i\rangle = Ae \int d\mathbf{r}\, \varphi(\mathbf{r})\, e^{i\mathbf{q}\cdot\mathbf{r}} \int d\boldsymbol{\eta}_1 \ldots d\boldsymbol{\eta}_{A-1}$$
$$\times \psi_f^*(\boldsymbol{\eta}_1 \ldots \boldsymbol{\eta}_{A-1}; \sigma_1^z \ldots \sigma_A^z; \tau_1^z \ldots \tau_A^z)\, e^{i\mathbf{q}\cdot\sqrt{\frac{A-1}{A}}\boldsymbol{\eta}_1}\, \frac{1+\tau_1^3}{2}$$
$$\psi_i(\boldsymbol{\eta}_1 \ldots \boldsymbol{\eta}_{A-1}; \sigma_1^z \ldots \sigma_A^z; \tau_1^z \ldots \tau_A^z)\,. \tag{7.36}$$

One can notice that in the matrix element one has a factorization in two terms: the Fourier transform of the field

$$\varphi(\mathbf{q}) = \int d\mathbf{r}\, \varphi(\mathbf{r})\, e^{i\mathbf{q}\cdot\mathbf{r}} \tag{7.37}$$

and the Fourier transform of the so called *proton transition density*, defined as

$$\rho_{i,f}^p(\boldsymbol{\eta}_1) = \int d\boldsymbol{\eta}_2 \ldots d\boldsymbol{\eta}_{A-1} \psi_f^*(\boldsymbol{\eta}_1 \ldots \boldsymbol{\eta}_{A-1}; \sigma_1^z \ldots \sigma_A^z; \tau_1^z \ldots \tau_A^z)$$
$$\sum_j \frac{1+\tau_j^3}{2}\, \psi_i(\boldsymbol{\eta}_1 \ldots \boldsymbol{\eta}_{A-1}; \sigma_1^z \ldots \sigma_A^z; \tau_1^z \ldots \tau_A^z)\,. \tag{7.38}$$

Since $\sqrt{\frac{A-1}{A}}\boldsymbol{\eta}_1 = \mathbf{r}_1 - \mathbf{R}_{CM} \equiv \mathbf{r}'$ one can write the final expression as

$$\langle f|H_{int}|i\rangle = e\varphi(\mathbf{q})\sqrt{\frac{A}{A-1}} \int d\mathbf{r}'\, e^{i\mathbf{q}\cdot\mathbf{r}'}\, \rho_{i,f}^p(\mathbf{r}')\,. \tag{7.39}$$

Some remarks are in order here:

- if $\psi_i = \psi_f = \psi_0$, namely the nucleus recoils, but does not excite, the cross section is called *elastic* and will be proportional to the square modulus of what is called the *charge form factor*. This is the Fourier transform of the average charge distribution with respect to the center of mass $\rho_{0,0}^p(r')$;
- if $\psi_i \neq \psi_f$ the cross section is called *inelastic* and one has the Fourier transform of the transition density $\rho_{i,f}^p(\mathbf{r}')$ (called the *transition form factor*);
- the matrix element entering in the cross section is involving an integral of the wave functions. This means that in principle one does not need to know the whole detailed wave function, but only an integral of it (an infinite number of different wave functions will have the same integral!).

The previous derivation has been done for the case of electron scattering on a nucleus, considering only the interaction between the charge density and the electromagnetic scalar field. Similar results are obtained in case of other kinds of interactions, e.g. between the nuclear current density and the vector field or axial currents and axial field for neutrino scattering. What changes is the form of the transition form factors which will not result from the charge density operator, but from other kinds of operators.

From what has been illustrated above we can conclude what is the main ingredient of the cross section for perturbative *inclusive* experiments. Such experiments are those where the only experimentally controlled quantities are the energy ω and the momentum $\mathbf{q}$ transferred to the nucleus, while nothing is known on what has happened to it (in how many fragments it has possibly broken). The cross section will be proportional to the so called *structure* or *response* function $S(\mathbf{q}, \omega)$

$$S(\mathbf{q}, \omega) = \sum_{n=0}^{\infty} |\langle n|G|0\rangle|^2 \delta(\omega - E_n + E_0) . \tag{7.40}$$

The operator G denotes the general operator that interacts with the field created by the probe.

Notice that in (7.40), because of the presence of the energy conserving δ function, we have been allowed to introduce the sum on all the eigenstates of the nuclear Hamiltonian. Of course, depending whether the energy transferred to the system corresponds to the discrete or the continuum part of the spectrum, that sum may be intended as an integral. Moreover it has to be noticed that for each energy E_n in the continuum one has the degeneration given by the different possible “asymptotic channels”, namely the possibility that such a many-body system breaks in many different sets of fragments at the same energy.

In the following we will concentrate on the inelastic part of the response function, namely $S(\mathbf{q}, \omega)$ where the term $n = 0$ in the sum is excluded

$$R(\mathbf{q}, \omega) = \sum_{n\neq 0}^{\infty} |\langle n|G|0\rangle|^2 \delta(\omega - E_n + E_0) . \tag{7.41}$$

The reason is that the elastic part requires only the knowledge of the ground state of the system and its calculation can be done with one of the methods described above. Here, instead we want to face the inelastic scattering problem, and in particular the situation when the energy is large enough to break the system and continuum states $|n\rangle$ are in the game. To this aim we define a fluctuation operator $\Theta = G - \langle 0|G|0\rangle$. Since $\langle 0|\Theta|0\rangle = 0$ one can write

$$R(\mathbf{q}, \omega) = \sum_{n=0}^{\infty} |\langle n|\Theta|0\rangle|^2 \delta(\omega - E_n + E_0) , \tag{7.42}$$

where the complete sum (or the integral) on all states has been recovered. Having a complete sum on the Hamiltonian eigenstates is crucial as it will be clear below.

7.6 Integral Transform Approaches

Integral transform approaches allow to calculate inclusive response functions, avoiding to calculate the wave functions in the continuum. One starts from the consideration, already mentioned above, that the amount of information contained in the wave function is redundant with respect to the transition matrix elements needed, since the latter involve their integrals. Therefore, one tries to avoid the difficult task of solving the Schrödinger equation for positive energies and one concentrates instead on $R(\mathbf{q}, \omega)$, directly.

An integral transform of $R(\omega)$ (here and in the following we drop the dependence on $\mathbf{q}$) is defined as

$$\Phi(\sigma) = \int K(\sigma, \omega)\, R(\omega)\, d\omega\,, \tag{7.43}$$

with a smooth kernel K. Performing the integral in $d\omega$ and applying the closure property of Hamiltonian eigenstates $\sum_{n=0}^{\infty} |n\rangle\langle n| = I$ (see the remark at the end of Sect. 7.5.1) one has

$$\Phi(\sigma) = \langle 0|\Theta^{\dagger}\hat{K}\left(\sigma, (\hat{H} - E_0)\right)\, \Theta|0\rangle\,. \tag{7.44}$$

From Eq. (7.44) one can see that the calculation of $\Phi(\sigma)$ seems to require in principle the knowledge of the ground state only. However, the possibility to actually calculate $\Phi(\sigma)$ depends on how complicate is the operator $\hat{K}(\sigma, \hat{H})$. Moreover, in order to access the quantity of interest, namely $R(\omega)$ an inversion of the transform is necessary.

7.6.1 *Sum Rules*

Before proceeding towards discussing useful kernels for our aim, let us recall that the so called "method of moments" has something to do with this approach. The moments of $R(\omega)$, seen as a probability distribution, would be the $\Phi(\sigma)$ obtained by the kernel $K(\sigma, \omega) = \omega^{\sigma}$ with σ integer (they are also known as *sum rules*).

$$\Phi(\sigma) \equiv m_{\sigma} = \int d\omega\, \omega^{\sigma} R(\omega) = \langle 0|\Theta^{\dagger}(H - E_0)^{\sigma}\, \Theta|0\rangle\,. \tag{7.45}$$

For integer positive σ the moments $m(\sigma)$ can also be written as mean values on the ground state of multiple commutators or anticommutators of the Hamiltonian with the Θ operator (see e.g [35]). As already stressed they can be evaluated using the knowledge of the ground state only. The knowledge of only few moments can give information about some gross features of $R(\omega)$, like its normalization (m_0) the mean

ω value (m_1/m_0) etc. However, the real problem of such an approach is that the very detailed knowledge of $R(\omega)$ would require the calculation of a large number of m_σ, arising then the problem of their existence, since the high energy behavior of $R(\omega)$ is in general unknown. Cumulant expansion approaches [36] have been suggested in order to get a lot of information on $R(\omega)$ by the knowledge of only the first few m_σ, however with limited success.

7.6.2 *Integral Transform with the Laplace Kernel*

There is a kernel that is used extensively in many different contexts. This is the Laplace kernel $e^{-\sigma\,\omega}$. Before coming to that let's first consider the Fourier transform of $R(\omega)$

$$F(t) = \int d\omega\, e^{i\omega t}\, R(\omega)\,, \tag{7.46}$$

one can again make use of closure and write it as

$$F(t) = \langle 0|\, \Theta^\dagger\, e^{i(H-E_0)t}\, \Theta\, |0\rangle\,, \tag{7.47}$$

or, in Heisemberg representation

$$F(t) = \langle 0|\, \Theta^\dagger(t)\, \Theta(0)\, |0\rangle\,. \tag{7.48}$$

This is a complex quantity of the real time variable t. However, if one extends its domain in the complex plane, one realizes that $F^*(t) = F(-t^*)$ and for imaginary time, i.e. $t = i\tau$, the function $F(\tau)$ is real. Since $R(\omega) = 0$ for $\omega \leq 0$, $F(\tau)$ corresponds to its Laplace transform. It turns out that GFMC or DMC methods are suitable to calculate $F(\tau)$. However, one is left with the thorny problem of the inversion of the Laplace transform.

It is well known that the inversion of a Laplace transform is a typical *ill posed problem*. In order to explain in simpler terms what an *ill posed problem* means in practice, consider two examples for $R(\omega)$ as plotted in the left panel of Fig. 7.3. Suppose that the dashed curve is the real one and the full line a wrong one. If you look at their relative Laplace transforms $\Lambda(\sigma)$ in the right panel, you will notice that they can fall both within a possible numerical error of the size shown in the figure, so that no inversion algorithm can discriminate with absolute certitude between the right and wrong $R(\omega)$. This is due to the fact that the exponential kernel tends to wash out rapidly any information at high ω. Moreover errors in the transform can generate oscillations (see discussion in Sect. 7.6.6).

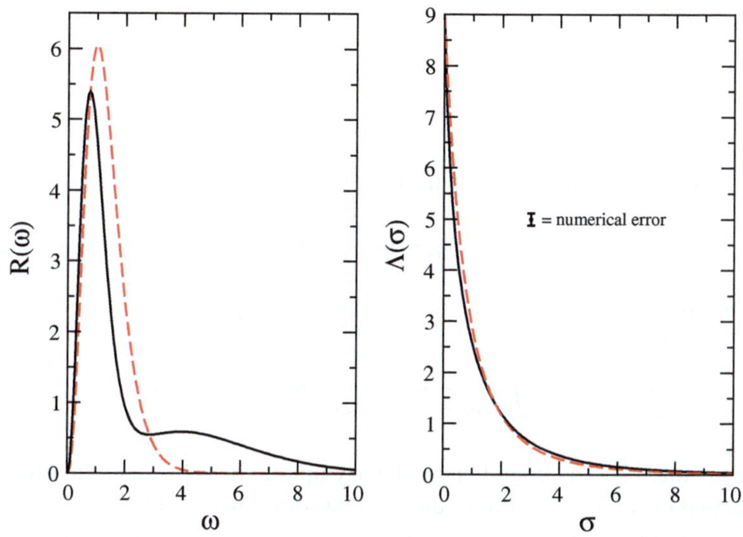

Fig. 7.3 *Left panel*: two different $R(\omega)$. *Right panel*: their relative Laplace transforms $\Lambda(\sigma)$. See discussion in the text

7.6.3 Integral Transform with the Lorentzian Kernel

One may ask what would be the "perfect" kernel, in the sense of a kernel that returns a transform which reproduces the features of $R(\omega)$. It would trivially be $\delta(\sigma - \omega)$. This is of course of no use. However, probably a representation of the δ-function could be an "almost perfect" kernel. This is the idea that was pursued in [37], when the Lorentz Integral Transform (LIT) was proposed. It is clear that a good kernel has not only to be able to reproduce the features of the original $R(\omega)$ in the transformed function, in order to make the inversion procedure easier and reliable, but it has also to allow the calculation of $\Phi(\sigma)$ in practice. In the following it will be illustrated that if the kernel is a Lorentzian function (one of the representations of the δ-function) this is just the case.

The (normalized) Lorentzian kernel can be defined for a complex $\sigma = \sigma_R + i\sigma_I$ as

$$K_L(\sigma, \omega) = \frac{\sigma_I}{\pi} \frac{1}{(\omega - \sigma_R)^2 + \sigma_I^2} = \frac{\sigma_I}{\pi} \frac{1}{\omega - \sigma_R - i\sigma_I} \frac{1}{\omega - \sigma_R + i\sigma_I}. \tag{7.49}$$

With such a kernel Eq. (7.43) becomes

$$L(\sigma_R, \sigma_I) = \frac{\sigma_I}{\pi} \langle 0 | \Theta^\dagger \frac{1}{\hat{H} - E_0 - \sigma_R - i\sigma_I} \frac{1}{\hat{H} - E_0 - \sigma_R + i\sigma_I} \Theta | 0 \rangle . \tag{7.50}$$

If one defines

$$|\tilde{\Psi}\rangle \equiv \sqrt{\frac{\sigma_I}{\pi}}\,\frac{1}{\hat{H}-E_0-\sigma_R+i\sigma_I}\,\Theta|\,0\,\rangle\,, \tag{7.51}$$

the LIT of $R(\omega)$ corresponds to the norm of this $|\tilde{\Psi}\rangle$, namely

$$L(\sigma_R,\sigma_I) = \langle\tilde{\Psi}|\tilde{\Psi}\rangle\,. \tag{7.52}$$

The most important feature of $|\tilde{\Psi}\rangle$ is that its norm is finite since

$$L(\sigma_R,\sigma_I) = \frac{\sigma_I}{\pi}\int\frac{1}{(\omega-\sigma_R)^2+\sigma_I^2}\,R(\omega)\,d\omega\ <\ \infty \tag{7.53}$$

This means that $|\tilde{\Psi}\rangle$ has bound-state like asymptotic behavior and therefore it can be calculated using one or more of the methods described in Sect. 7.3.

In the literature there are examples of LIT calculations with the Faddeev method for inclusive electron scattering[38] and photoabsorption on ^{3}He and ^{3}H [39], the CHH method, again for electron scattering [40] and photoabsorption [41] of three-nucleon systems, the NCSM method for photoabsorption of ^{4}He [42], the EIHH method for photoabsorption of ^{4}He [43], ^{6}He and ^{6}Li [44], ^{7}Li [45], for neutrino scattering on ^{4}He [46] and more recently with CC on photoabsorption of ^{16}O and ^{22}O [47].

Now we will illustrate how the Lanczos method can be particularly useful to calculate $L(\sigma_R,\sigma_I)$ in practice. One easily realizes that, because of Eqs. (7.50)–(7.53), one has

$$\langle\tilde{\Psi}|\tilde{\Psi}\rangle = \frac{1}{\pi}\,Im\left[\,\langle\,0\,|\,\Theta^\dagger\,\frac{1}{\hat{H}-E_0-\sigma_R-i\sigma_I}\,\Theta|\,0\,\rangle\,\right]. \tag{7.54}$$

Due to the finiteness of the norm of $|\tilde{\Psi}\rangle$ one can represent the Hamiltonian on a basis of localized functions. After a Lanczos diagonalization, $L(\sigma_R,\sigma_I)$ will appear as

$$\begin{aligned} L(\sigma_R,\sigma_I) &= \frac{1}{\pi}\,Im\left[\sum_\mu\langle\,0\,|\,\Theta^\dagger\,\frac{1}{\epsilon_\mu-E_0-\sigma_R-i\sigma_I}\,\Theta|\,0\,\rangle\right] \\ &= \frac{\sigma_I}{\pi}\sum_\mu\frac{|\langle\,\mu\,|\Theta|\,0\,\rangle|^2}{(\epsilon_\mu-E_0-\sigma_R)^2+\sigma_I^2}\,, \end{aligned} \tag{7.55}$$

namely a superposition of Lorentzians functions of width σ_I, centered on $\epsilon_\mu-E_0$ (ϵ_μ is the Hamiltonian matrix eigenvalue) and weighted by the strength of the transition between the ground state and the Hamiltonian matrix eigenvectors $|\mu\rangle$, induced by

Θ. In principle the "exact result" is reached when the Hamiltonian matrix is large enough to let $L(\sigma_R, \sigma_I)$ become a converged curve.

A few remarks are in order here:

- the range of σ_R, where it is convenient to calculate $L(\sigma_R, \sigma_I)$, is connected to the ω-range of interest for $R(\omega)$. In fact one should remember that the kernel is a representation of the δ-function and $L(\sigma_R, \sigma_I)$ resembles $R(\omega)$ more and more as σ_I becomes small;
- the choice of σ_I is determined by the kind of resolution that one wants to have on $R(\omega)$. If σ_I is of the same order as the experimental resolution one does not even need to invert the transform and one can compare $L(\sigma_R, \sigma_I)$ to data, directly;
- even in case that an inversion is necessary, σ_I is related to the the kind of resolution one wishes to have for $R(\omega)$. A crucial quantity to look at is the average distance between contiguous ϵ_μ. In fact for σ_I larger than this average distance the convergence of $L(\sigma_R, \sigma_I)$ is rather easy to reach, since in this case the different Lorentzians that compose $L(\sigma_R, \sigma_I)$ [see Eq. (7.55)] overlap smoothly. The inversion of a well converged result is safe. However, if the inverted result shows oscillations with wavelength smaller that σ_I, one has to check the result, calculating the LIT up to convergence with an as small σ_I (see discussion in Sect. 7.6.6).
- when σ_I is smaller than the average distance between contiguous ϵ_ν, $L(\sigma_R, \sigma_I)$ will appear as a set of separated peaks, which in most cases move around as the matrix increases. In fact enlarging the matrix does not necessarily mean that the density of ϵ_μ increases in the σ_R region of interest. Only if this is the case one could reach a convergence in $L(\sigma_R, \sigma_I)$. The choice of an appropriate basis in representing the Hamiltonian plays an essential role for this problem, as it was shown in [48].

An extensive review of the LIT method, together with more examples of applications can be found in [49].

7.6.4 Integral Transform with the Sumudu Kernel

One of the advantages of the LIT method is that the kernel is a representation of the δ-function. This makes the transform resemble $R(\omega)$, reducing in this way the difficulties in the inversion considerably. On the other hand, the necessity to diagonalize, up to convergence of the transform, an Hamiltonian matrix built on a many-body basis, seems to limit its applicability to rather small A. One could ask whether it is possible to find a kernel that is both a representation of the δ-function and that allows to calculate the transform with one of the Monte Carlo methods, which are rather powerful also for a rather large number of particles. Such a kernel has been proposed in [50], see also Chap. 9 here. It is a combination of Sumudu kernels, (or more simply of exponentials) and reads

$$K_P(\sigma,\omega) = N\left[\frac{e^{-\mu\frac{\omega}{\sigma}}}{\sigma} - \frac{e^{-\nu\frac{\omega}{\sigma}}}{\sigma}\right]^P, \tag{7.56}$$

where

$$\mu = \frac{ln[b] - ln[a]}{b-a}a; \quad \nu = \frac{ln[b] - ln[a]}{b-a}b, \tag{7.57}$$

and the parameters P, a, b are integer numbers with $b > a$. The normalization constant N is a function of P, a, b such that $\int d\sigma K_P(\sigma,\omega) = 1$.

Independent on the choice of a and b the kernel $K_P(\sigma,\omega)$ converges to $\delta(\omega-\sigma)$ in the $P \to \infty$ limit. For a finite P, at each value of ω the kernel has a finite width which depends on P and represents the *resolution* with which one can study $R(\omega)$. The maximum of $\sigma K_P(\sigma,\omega)$ is at $\omega = \sigma$, therefore one can choose both the energy range of interest (the σ values) and the resolution (the larger is P, the higher is the resolution). This makes the transform with such a kernel extremely flexible, similarly to the case of the LIT method.

Using the binomial expansion the kernel becomes a linear combination of the so-called Sumudu transform kernels [50]

$$K_P(\sigma,\omega) = \frac{N}{\sigma}\sum_{k=0}^{P}\binom{P}{k}(-1)^k e^{-\ln(b/a)[\frac{a}{b-a}P+k]\frac{\omega}{\sigma}}. \tag{7.58}$$

This expression makes it clear how to calculate the transform by quantum Monte Carlo. In fact by operating the usual substitution $\omega \to \hat{H}$ [see Eqs. (7.43)–(7.44)], one has a simple linear combination of imaginary-time propagators

$$K_P(\sigma,\hat{H}) = \frac{N}{\sigma}\sum_{k=0}^{P}\binom{P}{k}(-1)^k e^{-\tau_{Pk}\hat{H}}, \tag{7.59}$$

taken at different imaginary-time points

$$\tau_{Pk} = \ln(b/a)[\frac{a}{b-a}P + k]/\sigma. \tag{7.60}$$

Until now this transform has only been applied to the case of bosons (liquid Helium) [50]. It would be desirable to investigate its application for nuclear systems.

7.6.5 Integral Transform with the Stieltjes Kernel

The integral transform with the Stieltjes kernel, $K(\sigma, \omega) = \frac{1}{\omega+\sigma}$ and $\sigma > 0$, is given by

$$\mathscr{S}(\sigma) = \int d\omega \frac{1}{(\omega+\sigma)} R(\omega) = \langle 0 | \Theta^\dagger \frac{1}{\hat{H} - E_0 + \sigma} \Theta | 0 \rangle . \quad (7.61)$$

It was introduced long ago [51] to study reactions in the continuum. It has had, however, a limited application, since it shares the same problems as the Laplace transform. In fact the kernel has a similar decreasing behavior in σ, even if much smoother than the exponential, washing out information at higher energies, useful for the success of the inversion algorithms. In [52] the test on the electron scattering longitudinal response function has shown that its inversion can lead to rather large uncertainties, even in presence of rather small errors in the calculation of the transform.

There is, however, an interesting application of such a transform to calculate a dynamical quantity, which in general requires the knowledge of the continuum part of the spectrum. This quantity is the *generalized* polarizability (e.g. dipole polarizability or magnetic susceptibility) of a nucleus in response to a constant perturbative field.

Applying time dependent perturbation theory one can show that these polarizabilities α_Θ are related to the inverse moment m_{-1} of $R(\omega)$ as

$$\alpha_\Theta = 2m_{-1} = 2\langle 0 | \Theta^\dagger \frac{1}{\hat{H} - E_0} \Theta | 0 \rangle , \quad (7.62)$$

where Θ represents the operator relevant for the kind of probe used. Comparing Eqs. (7.61) and (7.62) it is clear that the polarizability corresponds to the limit for vanishing σ of $\mathscr{S}(\sigma)$. It turns out that this limit represents a viable way to calculate the polarizability. In fact, one notices that the polarizability can be written as

$$\alpha_\Theta = 2\langle 0 | \Theta^\dagger | \tilde{\Phi} \rangle \quad (7.63)$$

where $| \tilde{\Phi} \rangle$ is the solution of the Schrödinger-like equation with a source

$$(\hat{H} - E_0 + \sigma)|\tilde{\Phi}\rangle = \Theta |0\rangle \quad (7.64)$$

One notices that $\tilde{\Phi}$ is a function of finite norm, since

$$\langle \tilde{\Phi} | \tilde{\Phi} \rangle = \int d\omega \frac{1}{(\omega+\sigma)^2} R(\omega) < \infty. \quad (7.65)$$

Therefore representing the Hamiltonian and $\tilde{\Phi}$ on bound state-like functions, Eq. (7.64) becomes a linear matrix equation to be solved up to convergence in the size of the matrix. Repeating the calculation for smaller and smaller σ one can plot $\mathscr{S}(\sigma)$ as a function of σ and read α_{Θ} as the extrapolated value at $\sigma = 0$. A first application of this procedure for the calculation of the dipole polarizability of light system can be found in [53].

7.6.6 Methods of Inversion

A crucial part of the integral transform method is the inversion of the transforms. The inversion has to be made with care, since errors in the transform can generate spurious oscillations in $R(\omega)$. To illustrate this let us consider a well defined $R(\omega)$ to which we add a term $\Delta_\nu R(\omega)$ oscillating at a high frequency ν. Such a term leads to an additional $\Delta_\nu \Phi(\sigma)$ in the transform of Eq. (7.43). One should realize that, for any amplitude of the oscillation, $\Delta_\nu \Phi$ decreases with increasing ν. This means that for some value of ν the term $\Delta_\nu \Phi$ may be smaller than the size of the errors in the calculation. Therefore in this case $\Delta_\nu R$ cannot be discriminated. In principle, by reducing the error in the calculation one can push the frequency of the undiscriminated $\Delta_\nu R$ to higher and higher values, so to render their spuriosity manifest.

At this point something should be said regarding the algorithms which are commonly used to invert the transforms. The Laplace transforms are obtained by Monte Carlo methods and are affected by statistical errors. In this case the inversion algorithms are necessarily based on Bayes theorem. Therefore, one gets the probability of a solution based on known input conditions. Because of the highly ill-posed nature of the Laplace transform inversion it may be necessary to use many input conditions to obtain a highly probable solution. The best known algorithm is the Maximum Entropy method[54] with its many sophisticated versions.

The calculation of the LIT is much less affected by numerical errors of statistical nature (they are generally tiny) and much more by the systematic errors related to estimations of the convergence quality. Therefore in this case it is more convenient to use another algorithm called the *regularization method* [55]. This has led to very safe inversion results. Alternative inversion methods of the same nature are discussed in [56].

A LIT inversion method that has been largely used with success consists in making the following ansatz for the response function

$$R(\omega') = \sum_{n=1}^{N_{max}} c_n \chi_n(\omega', \alpha_i) , \tag{7.66}$$

with $\omega' = \omega - E_{th}$, where E_{th} is the break–up threshold energy into the continuum (calculable with bound state methods). The χ_n are given functions with nonlinear parameters α_i. A basis set that has rather frequently been used to invert the LIT inversions is

$$\chi_n(\omega', \alpha_i) = \omega'^{\alpha_1} \exp(-\frac{\alpha_2 \omega'}{n}) . \tag{7.67}$$

In addition also possible information on narrow levels could be incorporated easily into the set χ_n. Substituting such an expansion into the right hand side of (7.53) one obtains (here too the σ_I dependence is omitted)

$$L(\sigma_R) = \sum_{n=1}^{N_{max}} c_n \tilde{\chi}_n(\sigma_R, \alpha_i) , \tag{7.68}$$

where

$$\tilde{\chi}_n(\sigma_R, \alpha_i) = \int_0^\infty \frac{\chi_n(\omega', \alpha_i)}{(\omega' - \sigma_R)^2 + \sigma_I^2} . \tag{7.69}$$

For given α_i the linear parameters c_n are determined from a least–square best fit of $L(\sigma_R)$ of Eq. (7.68) to the calculated $L(\sigma_R)$ for a number of σ_R points much larger than N_{max}.

For every value of N_{max} the overall best fit is selected and then the procedure is repeated for $N'_{max} = N_{max} + 1$ till a stability of the inverted response is obtained and taken as inversion result. A further increase of N_{max} will eventually reach a point, where the inversion becomes unstable leading typically to random oscillations. The reason is that $L(\sigma_R)$ of Eq. (7.55) is not determined precisely enough so that a randomly oscillating $R(\omega)$ leads to a better fit than the true response. If the accuracy in the determination of $L(\sigma_R)$ is increased, one may include more basis functions in the expansion (7.68).

The LIT method has to be understood as an approach with a *controlled resolution*. If one expects that $R(\omega)$ has structures of width Γ, then the LIT resolution parameter σ_I should be similar in size. Then it is sufficient to determine the corresponding LIT with a moderately high precision, and the inversion should lead to reliable results for $R(\omega)$, if in fact no structures with a width smaller than Γ are present. If, however, there is a reason to believe that $R(\omega)$ exhibits such smaller structures one should reduce σ_I accordingly and perform again a calculation of the Lorentz integral transform with the same relative precision as before. Such a calculation is of course more expensive than the previous one with larger σ_I, but in principle one can reduce the LIT resolution parameter σ_I more and more.

The advantage of the LIT approach as compared with a conventional approach is evident. In the LIT case one makes the calculation with the proper resolution, while in a conventional calculation an infinite resolution (corresponding to $\sigma_I = 0$) is requested, which often makes such a calculation not feasible.

There are several tests of the reliability of the inversion. First of all, performing the calculation at different σ_I one has to obtain the same stable result for $R(\omega)$ from the inversion. If σ_I is too small the solution reaches its asymptotic behavior to zero very slowly, therefore for $\sigma_I < \sigma_I^{min}$ one may have convergence problems, turning into large errors for the LIT. As already said above, this will show up in unphysical oscillations in $R(\omega)$. This means that the stable result obtained with $\sigma_I \geq \sigma_I^{min}$ is the correct one, at that resolution.

Another test is the control of the moments, in fact the moments of $R(\omega)$ can be calculating both integrating $R(\omega)$ or by Eq. (7.45), which needs only the knowledge of the ground state.

7.7 Conclusion

From what has been written in these short lecture notes it should be clear that they are only a partial presentation of the amount of work that the theoretical nuclear physicists have done in the last two decades in the attempt to account for nuclear observables from first principles, for a number of particles that exceeds the classical few-body definition, traditionally limited to A=2,3.

In describing the methods, here only the main points have been exposed, leaving out the complicated formalism of the more detailed presentation needed by possible practitioners. The reference to the numerous original works should compensate this lack.

Very active theoretical research is still going on in this field, especially in the attempt to find unifying approaches for structure and reactions and to reach regions of the nuclear data chart still unexplored. Fortunately this research is accompanied by an as active experimental activity, which produces observables that at the same time need a theoretical explanation and constitute the reference for testing the models and the methods. All that shows the relevance of the ab initio approaches in nuclear physics as the necessary bridge between QCD, the fundamental theory of strong interaction and nuclear phenomena, many of which are at the basis of the evolution of the Universe.

References

1. E. Epelbaum, U.-G. Meißner, Ann. Rev. Nucl. Part. Sci. **62**, 159 (2012)
2. W. Leidemann, G. Orlandini, Prog. Part. Nucl. Phys. **68**, 158 (2013)
3. L.D. Faddeev, Sov. Phys. JETP **12**, 1014 (1961)
4. O. Yakubowsky, Yad. Fiz. **5**, 1312 (1967) [Sov. J. Nucl. Phys. **5**, 937]
5. E.O. Alt, P. Grassberger, W. Sandhas, Nucl. Phys. B **2**, 167 (1967); P. Grassberger, W. Sandhas, Nucl. Phys. B **2**, 181 (1967)
6. J.M. Rayleigh, Philos. Trans. **161**, 77 (1870)
7. W. Ritz, J. Reine Angew. Math. **135**, 1 (1909)
8. K. Wildermuth, Y.C. Tang, *A Unified Theory of the Nucleus* (Vieweg, Braunschweig, 1977); Y.C. Tang, M. Lemere, D.R. Thompson, Phys. Rep. **47**, 169 (1978)

9. H.M. Hofmann, G.M. Hale, Nucl. Phys. A **613**, 69 (1997); Phys. Rev. C **77**, 044002 (2008)
10. V.I. Kukulin, V.M. Krasnopolsk, J. Phys. **3**, 795 (1977)
11. Y. Suzuki, K. Varga, *Stochastic Variational Approach to Quantum Mechanical Few-Body Problems* (Springer, Berlin, 1998)
12. H. Kamada et al., Phys. Rev. C **64**, 044001 (2001)
13. J.O. Hirschfelder, J. Dahler, Proc. Natl. Acad. Sci. U. S. A. **42**, 363 (1956)
14. A. Novoselsky, J. Katriel, Phys. Rev. A **49**, 833 (1994)
15. N. Barnea, A. Novoselsky, Ann. Phys. (N.Y.) **256**, 192 (1997); Phys. Rev. A **57**, 48 (1998)
16. M. Gattobigio, A. Kievsky, M. Viviani, Phys. Rev. C **83**, 024001 (2011)
17. S. Deflorian, N. Barnea, W. Leidemann, G. Orlandini, Few-Body Syst. **54**, 1879 (2013)
18. S. Okubo, Prog. Theor. Phys. **12**, 603 (1954)
19. F. Coester, H. Kümmel, Nucl. Phys. **17**, 477 (1960)
20. J. da Providencia, C.M. Shakin, Ann. Phys. **30**, 95 (1964)
21. K. Suzuki, S.Y. Lee, Prog. Theor. Phys. **64**, 2091 (1980)
22. H. Kümmel, K.H. Lührmann, J.G. Zabolitzky, Phys. Rep. **36**, 1 (1978)
23. D.J. Dean, M. Hjorth-Jensen, Phys. Rev. C **69**, 054320 (2004)
24. M.H. Kalos, Phys. Rev. **128**, 1791 (1962)
25. J. Carlson, Phys. Rev. C **36**, 2026 (1987)
26. K.E. Schmidt, S. Fantoni, Phys. Lett. B **446**, 99 (1999)
27. D. Lee, Prog. Part. Nucl. Phys. **64**, 117 (2009)
28. S.E. Koonin, D.J. Dean, K. Langanke, Phys. Rep. **278**, 1 (1997)
29. M. Honma, T. Mizusaki, T. Otsuka, Phys. Rev. Lett. **75**, 1284 (1995)
30. T. Otsuka et al., Prog. Part. Nucl. Phys. **47**, 319 (2001)
31. S.C. Pieper, R.B. Wiringa, Ann. Rev. Nucl. Part. Sci. **51**, 53 (2001)
32. A. de Shalit, H. Feshbach, *Theoretical Nuclear Physics: Nuclear Structure* (Wiley, New York, 1974)
33. P. Navratil, B.R. Barrett, Phys. Rev. C **57**, 562 (1998)
34. N. Barnea, W. Leidemann, G. Orlandini, Phys. Rev. C **61**, 054001 (2000)
35. G. Orlandini, M. Traini, Rep. Prog. Phys. **54**, 257 (1991)
36. R. Rosenfelder, Ann. Phys. **128**, 188 (1980)
37. V.D. Efros, W. Leidemann, G. Orlandini, Phys. Lett. B **338**, 130 (1994)
38. S. Martinelli et al., Phys. Rev. C **52**, 1778 (1995)
39. J. Golak, R. Skibinski, W. Glöckle, H. Kamada, A. Nogga, H. Witala, V.D. Efros, W. Leidemann, G. Orlandini, E.L. Tomusiak, Nucl. Phys. **A707**, 365 (2002)
40. V.D. Efros, W. Leidemann, G. Orlandini, E.L. Tomusiak, Phys. Rev. C **69**, 044001 (2004)
41. N. Barnea, W. Leidemann, G. Orlandini, V.D. Efros, E.L. Tomusiak, Few-Body Syst. C **39**, 1 (2006)
42. I. Stetcu, B.R. Barrett, P. Navrátil, J.P. Vary, Phys. Rev. C **71**, 044325 (2005)
43. D. Gazit, S. Bacca, N. Barnea, W. Leidemann, G. Orlandini, Phys. Rev. Lett. **96**, 112301 (2006)
44. S. Bacca, N. Barnea, W. Leidemann, G. Orlandini, Phys. Rev. C **69**, 057001 (2004)
45. S. Bacca, H. Arenhoevel, N. Barnea, W. Leidemann, G. Orlandini, Phys. Lett. B **603**, 159 (2004)
46. D. Gazit, N. Barnea, Nucl. Phys. A **790**, 356 (2007)
47. S. Bacca, N. Barnea, G. Hagen, M. Miorelli, G. Orlandini, T. Papenbrock, Phys. Rev. C **90**, 064619 (2014)
48. W. Leidemann, Phys. Rev. C **91**, 054001 (2015)
49. V.D. Efros, W. Leidemann, G. Orlandini, N. Barnea, J. Phys. G Nucl. Part. Phys. **34**, R459 (2007)
50. A. Roggero, F. Pederiva, G. Orlandini, Phys. Rev. B **88**, 094302 (2013)
51. V.D. Efros, Sov. J. Nucl. Phys. **41**, 949 (1985)
52. V.D. Efros, W. Leidemann, G. Orlandini, Few-Body Syst. **14**, 151 (1993)

53. M. Miorelli, S. Bacca, N. Barnea, G. Hagen, G.R. Jansen, G. Orlandini, T. Papenbrock, Phys. Rev. C **94**, 034317. http://journals.aps.org/prc/abstract/10.1103/PhysRevC.94.034317
54. E.T. Jaynes, Information theory and statistical mechanics, in *Statistical Physics*, ed. by K. Ford (Benjamin, New York, 1963), p. 181
55. A.N. Tikhonov, V.Y. Arsenin, *Solutions of Ill–Posed Problems* (Winston, Washington, 1977)
56. D. Andreasi, W. Leidemann, C. Reiss, M. Schwamb, Eur. Phys. J. A **24**, 361 (2005)

Chapter 8
Computational Nuclear Physics and Post Hartree-Fock Methods

Justin G. Lietz, Samuel Novario, Gustav R. Jansen, Gaute Hagen, and Morten Hjorth-Jensen

8.1 Introduction

Studies of dense baryonic matter are of central importance to our basic understanding of the stability of nuclear matter, spanning from matter at high densities and temperatures to matter as found within dense astronomical objects like neutron stars. An object like a neutron star offers an intriguing interplay between nuclear processes and astrophysical observables, spanning many orders of magnitude in density and several possible compositions of matter, from the crust of the star to a possible quark matter phase in its interior, see for example [1–8] for discussions. A central issue in studies of infinite nuclear matter is the determination of the equation of state (EoS), which can in turn be used to determine properties like the mass range, the

J.G. Lietz (✉) • S. Novario
Department of Physics and Astronomy and National Superconducting Cyclotron Laboratory, Michigan State University, East Lansing, MI, USA
e-mail: lietz@nscl.msu.edu; novarios@nscl.msu.edu

G.R. Jansen
National Center for Computational Sciences and Physics Division, Oak Ridge National Laboratory, Oak Ridge, TN, USA
e-mail: jansengr@ornl.gov

G. Hagen
Physics Division, Oak Ridge National Laboratory, Oak Ridge, TN, USA

Department of Physics and Astronomy, University of Tennessee, Knoxville, TN, USA
e-mail: hageng@ornl.gov

M. Hjorth-Jensen
National Superconducting Cyclotron Laboratory and Department of Physics and Astronomy, Michigan State University, East Lansing, Michigan, USA

Department of Physics, University of Oslo, Oslo, Norway
e-mail: hjensen@msu.edu

M. Hjorth-Jensen et al. (eds.), *An Advanced Course in Computational Nuclear Physics*, Lecture Notes in Physics 936, DOI 10.1007/978-3-319-53336-0_8

mass-radius relationship, the thickness of the crust and the rate by which a neutron star cools down over time. The EoS is also an important ingredient in studies of the energy release in supernova explosions.

The determination and our understanding of the EoS for dense nuclear matter is intimately linked with our capability to solve the nuclear many-body problem. In particular, to be able to provide precise constraints on the role of correlations beyond the mean field is crucial for improved and controlled calculations of the EoS. In recent years, there has been a considerable algorithmic development of first principle (or ab initio) methods for solving the nuclear many-body problem. Linked with a similar progress in the derivation of nuclear forces based on effective field theory (EFT), see Chaps. 4, 5 and 6 of the present text and [9–16], we are now in a situation where reliable results can be provided at different levels of approximation. The nuclear Hamiltonians which are now used routinely in nuclear structure and nuclear matter calculations, include both nucleon-nucleon (*NN*) and three-nucleon forces (3NFs) derived from EFT, see for example [5, 5, 17–26]. Parallel to the development of nuclear forces from EFT, which employ symmetries of quantum chromodynamics, there are recent and promising approaches to derive the EoS using forces constrained from lattice quantum chromodynamics calculations [27], see Chaps. 2 and 3 of the present text.

Theoretical studies of nuclear matter and the pertinent EoS span back to the very early days of nuclear many-body physics. These early developments are nicely summarized in for example the review of Day [28] from 1967. These early state-of-the-art calculations were performed using what is known as Brueckner-Bethe-Goldstone theory [29, 30], see for example [8, 20, 21, 31–33] for recent reviews and developments. In these calculations, mainly particle-particle correlations were summed to infinite order. Other correlations were often included in a perturbative way. A systematic inclusion of other correlations in a non-perturbative way are nowadays accounted for in many-body methods like coupled cluster theory [17, 19, 23, 34–38] (this chapter), various Monte Carlo methods [39–46] (Chap. 9), Green's function approaches [26, 32, 47, 48] (Chap. 11) and similarity renormalization group methods [24, 49] (Chap. 10), just to mention a few of the actual many-body methods which are used for nuclear matter studies. Many of these methods are discussed in detail in this and the following chapters.

The aim of this part of the present lecture notes (comprising this chapter and the three subsequent ones) is to provide the necessary ingredients for performing studies of neutron star matter (or matter in β-equilibrium) and symmetric nuclear matter. We will mainly focus on pure neutron matter, but the framework and formalism can easily be extended to other dense and homogeneous fermionic systems such as the electron gas in two and three dimensions. The introductory material we present here forms also the basis for the next three chapters, starting with the definition of the single-particle basis and our Hamiltonians as well as Hartree-Fock theory. For infinite matter, due to the translational invariance of the Hamiltonian, the single-particle basis, in terms of plane waves, is unchanged under Hartree-Fock calculations.

Neutron star matter at densities of $0.1\,\mathrm{fm}^{-3}$ and greater, is often assumed to consist mainly of neutrons, protons, electrons and muons in beta equilibrium. However, other baryons like various hyperons may exist, as well as possible mesonic condensates and transitions to quark degrees of freedom at higher densities [3, 4, 8, 50]. In this chapter we limit ourselves to matter composed of neutrons only. Furthermore, we will also consider matter at temperatures much lower than the typical Fermi energies.

In the next section we present some of the basic quantities that enter the different many-body methods discussed in this and the three subsequent chapters. All these methods start with some single-particle basis states, normally obtained via the solution of mean-field approaches like Hartree-Fock theory. Contributions from correlations beyond such a mean-field basis to selected observables, are then obtained via a plethora of many-body methods. These methods represent different mathematical algorithms used to solve either Schrödinger's or Dirac's equations for many interacting fermions. After the definitions of our basis states, we derive the Hartree-Fock equations in the subsequent section and move on with many-body perturbation theory, full configuration interaction theory and coupled cluster theory. Monte Carlo methods, Green's function theory approaches and Similarity Renormalization group approaches are discussed in the subsequent three chapters.

The strengths and weaknesses of these methods are discussed throughout these chapters, with applications to either a simple pairing model and/or pure neutron matter. Our focus will be on pure neutron matter, starting with a simple model for the interaction between nucleons. This allows us to focus on pedagogical and algorithmic aspects of the various many-body methods, avoiding thereby the full complexity of nuclear forces. If properly written however, the codes can easily be extended to include models of the nuclear interactions based on effective field theory (see Chaps. 4, 5 and 6 of the present text) and other baryon species than just neutrons. In our conclusions we point back to models for nuclear forces and their links to the underlying theory of the strong interaction discussed in the first chapters of this book, bridging thereby the gap between the theory of nuclear Hamiltonians and many-body methods.

8.2 Single-Particle Basis, Hamiltonians and Models for the Nuclear Force

8.2.1 Introduction to Nuclear Matter and Hamiltonians

Although our focus here and in the coming chapters is on neutron matter only, our formalism lends itself easily to studies of nuclear matter with a given proton fraction and electrons. In this section we outline some of the background details, with a focus on the calculational basis and the representation of a nuclear Hamiltonian.

Neutron star matter is not composed of only neutrons. Rather, matter is composed of various baryons and leptons in chemical and charge equilibrium. The equilibrium conditions are governed by the weak processes (normally referred to as the processes for β-equilibrium)

$$b_1 \to b_2 + l + \bar{\nu}_l \qquad b_2 + l \to b_1 + \nu_l,$$

where b_1 and b_2 refer to different types of baryons, for example a neutron and a proton. The symbol l is either an electron or a muon and $\bar{\nu}_l$ and ν_l their respective anti-neutrinos and neutrinos. Leptons like muons appear at a density close to nuclear matter saturation density ρ_0, the latter being

$$\rho_0 \approx 0.16 \pm 0.02 \text{ fm}^{-3},$$

with a corresponding energy per baryon $\mathscr{E}_0$ for symmetric nuclear matter at saturation density of

$$\mathscr{E}_0 = B/A = -15.6 \pm 0.2 \text{ MeV}.$$

The energy per baryon is the central quantity in the present studies. From the energy per baryon, we can define the pressure P which counteracts the gravitational forces and hinders the gravitational collapse of a neutron star. The pressure is defined through the relation

$$P = \rho^2 \frac{\partial \mathscr{E}}{\partial \rho} = \rho \frac{\partial \varepsilon}{\partial \rho} - \varepsilon,$$

where ε is the energy density and ρ the density. Similarly, the chemical potential for particle species i is given by

$$\mu_i = \left(\frac{\partial \varepsilon}{\partial \rho_i} \right).$$

In calculations of properties of neutron star matter in β-equilibrium, we need to calculate the energy per baryon $\mathscr{E}$ for several proton fractions x_p. The proton fraction corresponds to the ratio of protons as compared to the total nucleon number (Z/A). It is defined as

$$x_p = \frac{\rho_p}{\rho},$$

where $\rho = \rho_p + \rho_n$ is the total baryonic density if neutrons and protons are the only baryons present (the subscripts used here, n, p, e, μ, refer to neutrons, protons, electrons and muons, respectively). If this is the case, the total Fermi momentum k_F and the Fermi momenta k_{Fp}, k_{Fn} for protons and neutrons, respectively, are related to the total nucleon density n by

$$\rho = \frac{2}{3\pi^2}k_F^3 = x_p\rho + (1-x_p)\rho = \frac{1}{3\pi^2}k_{Fp}^3 + \frac{1}{3\pi^2}k_{Fn}^3.$$

The energy per baryon will thus be labelled as $\mathscr{E}(\rho, x_p)$. The quantity $\mathscr{E}(\rho, 0)$ refers then to the energy per baryon for pure neutron matter while $\mathscr{E}(\rho, \frac{1}{2})$ is the corresponding value for symmetric nuclear matter.

An important ingredient in the discussion of the EoS and the criteria for matter in β-equilibrium is the so-called symmetry energy $\mathscr{S}(\rho)$, defined as the difference in energy for symmetric nuclear matter and pure neutron matter

$$\mathscr{S}(\rho) = \mathscr{E}(\rho, x_p = 0) - \mathscr{E}(\rho, x_p = 1/2).$$

If we expand the energy per baryon in the case of nucleonic degrees of freedom only in the proton concentration x_p about the value of the energy for SNM ($x_p = \frac{1}{2}$), we obtain,

$$\mathscr{E}(\rho, x_p) = \mathscr{E}(n, x_p = \frac{1}{2}) + \frac{1}{2}\frac{d^2\mathscr{E}}{dx_p^2}(\rho)\left(x_p - 1/2\right)^2 + \ldots,$$

where the term $d^2\mathscr{E}/dx_p^2$ is to be associated with the symmetry energy $\mathscr{S}(\rho)$ in the empirical mass formula. If we assume that higher order derivatives in the above expansion are small, then through the conditions for β-equilibrium (see for example [8]) we can define the proton fraction by the symmetry energy as

$$\hbar c\left(3\pi^2\rho x_p\right)^{1/3} = 4\mathscr{S}(\rho)\left(1 - 2x_p\right),$$

where the electron chemical potential is given by $\mu_e = \hbar c k_F$, i.e. ultrarelativistic electrons are assumed. Thus, the symmetry energy is of paramount importance for studies of neutron star matter in β-equilibrium. One can extract information about the value of the symmetry energy at saturation density ρ_0 from systematic studies of the masses of atomic nuclei. However, these results are limited to densities around ρ_0 and for proton fractions close to $\frac{1}{2}$. Typical values for $\mathscr{S}(\rho)$ at ρ_0 are in the range 27–38 MeV [51]. For densities greater than ρ_0 it is more difficult to get a reliable information on the symmetry energy, and thereby the related proton fraction.

With this background, we are now ready to define our basic inputs and approximations to the various many-body theories discussed in this chapter and the three subsequent ones. We will assume that the interacting part of the Hamiltonian can be approximated by a two-body interaction. This means that our Hamiltonian is written as the sum of a one-body part and a two-body part

$$\hat{H} = \hat{H}_0 + \hat{H}_I = \sum_{i=1}^{A} \hat{h}_0(x_i) + \sum_{i<j}^{A} \hat{v}(r_{ij}),$$

with

$$\hat{H}_0 = \sum_{i=1}^{A} \hat{h}_0(x_i),$$

being the so-called unperturbed part of the Hamiltonian defined by the one-body operator

$$\hat{h}_0(x_i) = \hat{t}(x_i) + \hat{u}_{\rm ext}(x_i),$$

where $\hat{t}$ represents the kinetic energy and x_i represents both spatial and spin degrees of freedom. For many-body calculations of finite nuclei, the external potential $u_{\rm ext}(x_i)$ is normally approximated by a harmonic oscillator or Woods-Saxon potential. For atoms, the external potential is defined by the Coulomb interaction an electron feels from the atomic nucleus. However, other potentials are fully possible, such as one derived from the self-consistent solution of the Hartree-Fock equations to be discussed below. Since we will work with infinite matter and plane wave basis states, the one-body operator is simply given by the kinetic energy operator. Finally, the term $\hat{H}_I$ represents the residual two-body interaction

$$\hat{H}_I = \sum_{i<j}^{A} \hat{v}(r_{ij}).$$

Our Hamiltonian is invariant under the permutation (interchange) of two particles. Since we deal with fermions however, the total wave function is anti-symmetric and we assume that we can approximate the exact eigenfunction for say the ground state with a Slater determinant

$$\Phi_0(x_1, x_2, \ldots, x_A, \alpha, \beta, \ldots, \sigma) = \frac{1}{\sqrt{A!}} \left| \begin{array}{cccc} \psi_\alpha(x_1) & \psi_\alpha(x_2) & \ldots\ldots & \psi_\alpha(x_A) \\ \psi_\beta(x_1) & \psi_\beta(x_2) & \ldots\ldots & \psi_\beta(x_A) \\ \ldots & \ldots & \ldots\ldots & \ldots \\ \ldots & \ldots & \ldots\ldots & \ldots \\ \psi_\sigma(x_1) & \psi_\sigma(x_2) & \ldots\ldots & \psi_\sigma(x_A) \end{array} \right|,$$

where x_i stand for the coordinates and spin values of particle i and $\alpha, \beta, \ldots, \gamma$ are quantum numbers needed to describe remaining quantum numbers.

The single-particle function $\psi_\alpha(x_i)$ are eigenfunctions of the one-body Hamiltonian h_0, that is

$$\hat{h}_0(x_i)\psi_\alpha(x_i) = \left(\hat{t}(x_i) + \hat{u}_{\rm ext}(x_i)\right) \psi_\alpha(x_i) = \varepsilon_\alpha \psi_\alpha(x_i).$$

The energies ε_α are the so-called non-interacting single-particle energies, or unperturbed energies. The total energy is in this case the sum over all single-particle energies, if no two-body or more complicated many-body interactions are present.

The properties of the determinant lead to a straightforward implementation of the Pauli principle since no two particles can be at the same place (two columns being the same in the above determinant) and no two particles can be in the same state (two rows being the same). As a practical matter, however, Slater determinants beyond $N = 4$ quickly become unwieldy. Thus we turn to the occupation representation or second quantization to simplify calculations. For a good introduction to second quantization see for example [35, 52–54].

We start with a set of orthonormal single-particle states $\{\psi_\alpha(x)\}$. To each single-particle state $\psi_\alpha(x)$ we associate a creation operator $a^\dagger_\alpha$ and an annihilation operator a_α. When acting on the vacuum state $|0\rangle$, the creation operator $a^\dagger_\alpha$ causes a particle to occupy the single-particle state $\psi_\alpha(x)$

$$\psi_\alpha(x) \to a^\dagger_\alpha|0\rangle.$$

But with multiple creation operators we can occupy multiple states

$$\psi_\alpha(x)\psi_\beta(x')\psi_\delta(x'') \to a^\dagger_\alpha a^\dagger_\beta a^\dagger_\delta|0\rangle.$$

Now we impose anti-symmetry by having the fermion operators satisfy the anti-commutation relations

$$a^\dagger_\alpha a^\dagger_\beta + a^\dagger_\beta a^\dagger_\alpha = \left\{a^\dagger_\alpha, a^\dagger_\beta\right\} = 0,$$

with the consequence that

$$a^\dagger_\alpha a^\dagger_\beta = -a^\dagger_\beta a^\dagger_\alpha.$$

Because of this property, we obtain $a^\dagger_\alpha a^\dagger_\alpha = 0$, enforcing the Pauli exclusion principle. Thus we can represent a Slater determinant using creation operators as

$$a^\dagger_\alpha a^\dagger_\beta a^\dagger_\delta \ldots |0\rangle,$$

where each index $\alpha, \beta, \delta, \ldots$ has to be unique.

We will now find it convenient to define a Fermi (F) level and introduce a new reference vacuum. The Fermi level is normally defined in terms of all occupied single-particle states below a certain single-particle energy. With the definition of a Fermi level we define our ansatz for the ground state, represented by a Slater determinant Φ_0. We will throughout the rest of this text use creation and annihilation operators to represent quantum mechanical operators and states. It means that our compact representation for a given Slater determinant in Fock space is

$$\Phi_0 = |i_1 \dots i_A\rangle = a_{i_1}^\dagger \dots a_{i_A}^\dagger |0\rangle$$

where $|0\rangle$ is the true vacuum and we have defined the creation and annihilation operators as

$$a_p^\dagger|0\rangle = |p\rangle, \quad a_p|q\rangle = \delta_{pq}|0\rangle$$

with the anti-commutation relations

$$\delta_{pq} = \left\{a_p, a_q^\dagger\right\},$$

and

$$\left\{a_p^\dagger, a_q\right\} = \left\{a_p, a_q\right\} = \left\{a_p^\dagger, a_q^\dagger\right\} = 0.$$

We can rewrite the ansatz for the ground state as

$$|\Phi_0\rangle = \prod_{i \le F} a_i^\dagger |0\rangle,$$

where we have introduced the shorthand labels for states below the Fermi level F as $i, j, \dots \le F$. For single-particle states above the Fermi level we reserve the labels $a, b, \dots > F$, while the labels $p, q, \dots$ represent any possible single-particle state.

Since our focus is on infinite systems, the one-body part of the Hamiltonian is given by the kinetic energy operator only. In second quantization it is defined as

$$\hat{H}_0 = \hat{T} = \sum_{pq} \langle p|\hat{t}|q\rangle a_p^\dagger a_q,$$

where the matrix elements $\langle p|\hat{t}|q\rangle$ represent the expectation value of the kinetic energy operator (see the discussion below as well). The two-body interaction reads

$$\hat{H}_I = \hat{V} = \frac{1}{4} \sum_{pqrs} \langle pq|\hat{v}|rs\rangle_{AS} a_p^\dagger a_q^\dagger a_s a_r,$$

where we have defined the anti-symmetrized matrix elements

$$\langle pq|\hat{v}|rs\rangle_{AS} = \langle pq|\hat{v}|rs\rangle - \langle pq|\hat{v}|sr\rangle.$$

We can also define a three-body operator

$$\hat{V}_3 = \frac{1}{36} \sum_{pqrstu} \langle pqr|\hat{v}_3|stu\rangle_{AS} a_p^\dagger a_q^\dagger a_r^\dagger a_u a_t a_s,$$

with the anti-symmetrized matrix element

$$\langle pqr|\hat{v}_3|stu\rangle_{AS} = \langle pqr|\hat{v}_3|stu\rangle + \langle pqr|\hat{v}_3|tus\rangle + \langle pqr|\hat{v}_3|ust\rangle - \langle pqr|\hat{v}_3|sut\rangle - \langle pqr|\hat{v}_3|tsu\rangle - \langle pqr|\hat{v}_3|uts\rangle.$$

In this and the forthcoming chapters we will limit ourselves to two-body interactions at most. Throughout this chapter and the subsequent three we will drop the subscript AS and use only anti-symmetrized matrix elements.

Using the ansatz for the ground state $|\Phi_0\rangle$ as new reference vacuum state, we need to redefine the anticommutation relations to

$$\{a_p^\dagger, a_q\} = \delta_{pq},\ p, q \leq F,$$

and

$$\{a_p, a_q^\dagger\} = \delta_{pq},\ p, q > F.$$

It is easy to see that

$$a_i|\Phi_0\rangle = |\Phi_i\rangle \neq 0, \qquad a_a^\dagger|\Phi_0\rangle = |\Phi^a\rangle \neq 0,$$

and

$$a_i^\dagger|\Phi_0\rangle = 0 \qquad a_a|\Phi_0\rangle = 0.$$

With the new reference vacuum state the Hamiltonian can be rewritten as, see Problem 8.1,

$$\hat{H} = E_{\mathrm{Ref}} + \hat{H}_N,$$

with the reference energy defined as the expectation value of the Hamiltonian using the reference state Φ_0

$$E_{\mathrm{Ref}} = \langle\Phi_0|\hat{H}|\Phi_0\rangle = \sum_{i\leq F}\langle i|\hat{h}_0|i\rangle + \frac{1}{2}\sum_{ij\leq F}\langle ij|\hat{v}|ij\rangle,$$

and the new normal-ordered Hamiltonian (all creation operators to the left of the annihilation operators) is defined as

$$\hat{H}_N = \sum_{pq} \langle p|\hat{h}_0|q\rangle \left\{a_p^\dagger a_q\right\} + \frac{1}{4}\sum_{pqrs}\langle pq|\hat{v}|rs\rangle \left\{a_p^\dagger a_q^\dagger a_s a_r\right\} + \sum_{pq,i\leq F} \langle pi|\hat{v}|qi\rangle \left\{a_p^\dagger a_q\right\}, \tag{8.1}$$

where the curly brackets represent normal-ordering with respect to the new reference vacuum state. The normal-ordered Hamiltonian can be rewritten in terms of a new one-body operator and a two-body operator as

$$\hat{H}_N = \hat{F}_N + \hat{V}_N,$$

with

$$\hat{F}_N = \sum_{pq} \langle p|\hat{f}|q\rangle \left\{a_p^\dagger a_q\right\}, \tag{8.2}$$

where

$$\langle p|\hat{f}|q\rangle = \langle p|\hat{h}_0|q\rangle + \sum_{i\leq F}\langle pi|\hat{v}|qi\rangle.$$

The last term on the right hand side represents a medium modification to the single-particle Hamiltonian due to the two-body interaction. Finally, the two-body interaction is given by

$$\hat{V}_N = \frac{1}{4}\sum_{pqrs}\langle pq|\hat{v}|rs\rangle \left\{a_p^\dagger a_q^\dagger a_s a_r\right\}.$$

8.2.2 *Single-Particle Basis for Infinite Matter*

Infinite nuclear or neutron matter is a homogeneous system and the one-particle wave functions are given by plane wave functions normalized to a volume Ω for a box with length L (the limit $L \to \infty$ is to be taken after we have computed various expectation values)

$$\psi_{\mathbf{k}\sigma}(\mathbf{r}) = \frac{1}{\sqrt{\Omega}} \exp{(i\mathbf{kr})}\xi_\sigma$$

where **k** is the wave number and ξ_σ is the spin function for either spin up or down nucleons

$$\xi_{\sigma=+1/2} = \begin{pmatrix} 1 \\ 0 \end{pmatrix} \quad \xi_{\sigma=-1/2} = \begin{pmatrix} 0 \\ 1 \end{pmatrix}.$$

As an interesting aside, the recent works of Binder et al. [55] and McElvain and Haxton [56] offer new perspectives on the construction of effective Hamiltonians and choices of basis functions.

We focus first on the kinetic energy operator. We assume that we have periodic boundary conditions which limit the allowed wave numbers to

$$k_i = \frac{2\pi n_i}{L} \quad i = x, y, z \quad n_i = 0, \pm 1, \pm 2, \ldots$$

The operator for the kinetic energy can be written as

$$\hat{T} = \sum_{\mathbf{p}\sigma_p} \frac{\hbar^2 k_p^2}{2m} a_{\mathbf{p}\sigma_p}^\dagger a_{\mathbf{p}\sigma_p}.$$

When using periodic boundary conditions, the discrete-momentum single-particle basis functions (excluding spin and/or isospin degrees of freedom) result in the following single-particle energy

$$\varepsilon_{n_x,n_y,n_z} = \frac{\hbar^2}{2m}\left(\frac{2\pi}{L}\right)^2 \left(n_x^2 + n_y^2 + n_z^2\right) = \frac{\hbar^2}{2m}\left(k_{n_x}^2 + k_{n_y}^2 + k_{n_z}^2\right),$$

for a three-dimensional system with

$$k_{n_i} = \frac{2\pi n_i}{L}, \quad n_i = 0, \pm 1, \pm 2, \ldots,$$

We will select the single-particle basis such that both the occupied and unoccupied single-particle states have a closed-shell structure. This means that all single-particle states corresponding to energies below a chosen cutoff are included in the basis. We study only the unpolarized spin phase, in which all orbitals are occupied with one spin-up and one spin-down fermion (neutrons and protons in our case). With the kinetic energy rewritten in terms of the discretized momenta we can set up a list (assuming identical particles one and including spin up and spin down solutions) of single-particle energies with momentum quantum numbers such that $n_x^2+n_y^2+n_z^2 \leq 3$, as shown in for example Table 8.1.

Continuing in this way we get for $n_x^2 + n_y^2 + n_z^2 = 4$ a total of 12 additional states, resulting in 66 as a new magic number. For the lowest six energy values the degeneracy in energy gives us 2, 14, 38, 54, 66 and 114 as magic numbers. These numbers will then define our Fermi level when we compute the energy in

Table 8.1 Total number of particle filling $N_{\uparrow\downarrow}$ for various $n_x^2 + n_y^2 + n_z^2$ values for one spin-1/2 fermion species

$n_x^2 + n_y^2 + n_z^2$	n_x	n_y	n_z	$N_{\uparrow\downarrow}$
0	0	0	0	2
1	−1	0	0	
1	1	0	0	
1	0	−1	0	
1	0	1	0	
1	0	0	−1	
1	0	0	1	14
2	−1	−1	0	
2	−1	1	0	
2	1	−1	0	
2	1	1	0	
2	−1	0	−1	
2	−1	0	1	
2	1	0	−1	
2	1	0	1	
2	0	−1	−1	
2	0	−1	1	
2	0	1	−1	
2	0	1	1	38
3	−1	−1	−1	
3	−1	−1	1	
3	−1	1	−1	
3	−1	1	1	
3	1	−1	−1	
3	1	−1	1	
3	1	1	−1	
3	1	1	1	54

Borrowing from nuclear shell-model terminology, filled shells corresponds to all single-particle states for one $n_x^2+n_y^2+n_z^2$ value being occupied. For matter with both protons and neutrons, the filling degree increased with a factor of 2

a Cartesian basis. When performing calculations based on many-body perturbation theory, coupled cluster theory or other many-body methods, we need then to add states above the Fermi level in order to sum over single-particle states which are not occupied.

If we wish to study infinite nuclear matter with both protons and neutrons, the above magic numbers become $4, 28, 76, 108, 132, 228, \ldots$.

Every number of particles for filled shells defines also the number of particles to be used in a given calculation. The number of particles can in turn be used to define the density ρ (or the Fermi momentum) of the system via

$$\rho = g\frac{k_F^3}{6\pi^2},$$

where k_F is the Fermi momentum and the degeneracy g, which is two for one type of spin-$1/2$ particles and four for symmetric nuclear matter. With the density defined and having fixed the number of particles A and the Fermi momentum k_F, we can define the length L of the box used with periodic boundary contributions via the relation

$$V = L^3 = \frac{A}{\rho}.$$

With L we can to define the spacing between various k-values given by

$$\Delta k = \frac{2\pi}{L}.$$

Here, A is the number of nucleons. If we deal with the electron gas only, this needs to be replaced by the number of electrons N. Exercise 8.4 deals with the set up of a program that establishes the single-particle basis for nuclear matter calculations with input a given number of nucleons and a user specified density or Fermi momentum.

8.2.3 Two-Body Interaction

As mentioned above, we will employ a plane wave basis for our calculations of infinite matter properties. With a cartesian basis we can calculate directly the various matrix elements. However, a cartesian basis represents an approximation to the thermodynamical limit. In order to compare the stability of our basis with results from the thermodynamical limit, it is convenient to rewrite the nucleon-nucleon interaction in terms of a partial wave expansion. This will allow us to compute the Hartree-Fock energy of the ground state in the thermodynamical limit (with the caveat that we need to limit the number of partial waves). In order to find the expressions for the Hartree-Fock energy in a partial wave basis, we will find it convenient to rewrite our two-body force in terms of the relative and center-of-mass motion momenta.

The direct matrix element, with single-particle three-dimensional momenta $\mathbf{k}_p$, spin σ_p and isospin τ_p, is defined as

$$\langle \mathbf{k}_p\sigma_p\tau_p\mathbf{k}_q\sigma_q\tau_q|\hat{v}|\mathbf{k}_r\sigma_r\tau_r\mathbf{k}_s\sigma_s\tau_s\rangle,$$

or in a more compact form as $\langle \mathbf{pq}|\hat{v}|\mathbf{rs}\rangle$ where the boldfaced letters $\mathbf{p}$ etc represent the relevant quantum numbers, here momentum, spin and isospin. Introducing the relative momentum

$$\mathbf{k} = \frac{1}{2}\left(\mathbf{k}_p - \mathbf{k}_q\right),$$

and the center-of-mass momentum

$$\mathbf{K} = \mathbf{k}_p + \mathbf{k}_q,$$

we have

$$\langle \mathbf{k}_p\sigma_p\tau_p\mathbf{k}_q\sigma_q\tau_q|\hat{v}|\mathbf{k}_r\sigma_r\tau_r\mathbf{k}_s\sigma_s\tau_s\rangle = \langle \mathbf{k}\mathbf{K}\sigma_p\tau_p\sigma_q\tau_q|\hat{v}|\mathbf{k}'\mathbf{K}'\sigma_r\tau_r\sigma_s\tau_s\rangle.$$

The nucleon-nucleon interaction conserves the total momentum and charge, implying that the above uncoupled matrix element reads

$$\begin{aligned}\langle \mathbf{k}\mathbf{K}\sigma_p\tau_p\sigma_q\tau_q|\hat{v}|\mathbf{k}'\mathbf{K}'\sigma_r\tau_r\sigma_s\tau_s\rangle &= \delta_{T_z,T_z'}\delta(\mathbf{K}-\mathbf{K}')\langle \mathbf{k}T_zS_z = (\sigma_a+\sigma_b)|\hat{v}|\mathbf{k}'T_zS_z' \\ &= (\sigma_c+\sigma_d)\rangle,\end{aligned}$$

where we have defined the isospin projections $T_z = \tau_p + \tau_q$ and $T_z' = \tau_r + \tau_s$. Defining $\hat{v} = \hat{v}(\mathbf{k},\mathbf{k}')$, we can rewrite the previous equation in a more compact form as

$$\begin{aligned}\delta_{T_z,T_z'}\delta(\mathbf{K}-\mathbf{K}')\langle \mathbf{k}T_zS_z = (\sigma_p+\sigma_q)|\hat{v}|\mathbf{k}'T_zS_z' = (\sigma_r+\sigma_s)\rangle \\ = \delta_{T_z,T_z'}\delta(\mathbf{K}-\mathbf{K}')\langle T_zS_z|\hat{v}(\mathbf{k},\mathbf{k}')|T_zS_z'\rangle.\end{aligned}$$

These matrix elements can in turn be rewritten in terms of the total two-body quantum numbers for the spin S of two spin-1/2 fermions as

$$\langle \mathbf{k}T_zS_z|\hat{v}(\mathbf{k},\mathbf{k}')|\mathbf{k}'T_zS_z'\rangle = \sum_{SS'}\langle\frac{1}{2}\sigma_p\frac{1}{2}\sigma_q|SS_z\rangle\langle\frac{1}{2}\sigma_r\frac{1}{2}\sigma_s|S'S_z'\rangle\langle \mathbf{k}T_zSS_z|\hat{v}(\mathbf{k},\mathbf{k}')|\mathbf{k}T_zS'S_z'\rangle$$

The coefficients $\langle\frac{1}{2}\sigma_p\frac{1}{2}\sigma_q|SS_z\rangle$ are so-called Clebsch-Gordan recoupling coefficients. We will assume that our interactions conserve charge. We will refer to $T_z = 0$ as the *pn* (proton-neutron) channel, $T_z = -1$ as the *pp* (proton-proton) channel and $T_z = 1$ as the *nn* (neutron-neutron) channel.

The nucleon-nucleon force is often derived and analyzed theoretically in terms of a partial wave expansion. A state with linear momentum **k** can be written in terms of spherical harmonics Y_{lm} as

$$|\mathbf{k}\rangle = \sum_{l=0}^{\infty}\sum_{m=-l}^{l} \imath^l Y_{lm}\langle\hat{k}|klm_l\rangle.$$

In terms of the relative and center-of-mass momenta $\mathbf{k}$ and $\mathbf{K}$, the potential in momentum space is related to the nonlocal operator $V(\mathbf{r}, \mathbf{r}')$ by

$$\langle \mathbf{k}'\mathbf{K}'|\hat{v}|\mathbf{k}'\mathbf{K}\rangle = \int d\mathbf{r}d\mathbf{r}'e^{-\imath\mathbf{k}'\mathbf{r}'}V(\mathbf{r}',\mathbf{r})e^{\imath\mathbf{k}\mathbf{r}}\delta(\mathbf{K},\mathbf{K}').$$

We will assume that the interaction is spherically symmetric and use the partial wave expansion of the plane waves in terms of spherical harmonics. This means that we can separate the radial part of the wave function from its angular dependence. The wave function of the relative motion is described in terms of plane waves as

$$e^{\imath\mathbf{k}\mathbf{r}} = \langle \mathbf{r}|\mathbf{k}\rangle = 4\pi\sum_{lm}\imath^l j_l(kr)Y^*_{lm}(\hat{\mathbf{k}})Y_{lm}(\hat{\mathbf{r}}),$$

where j_l is a spherical Bessel function and Y_{lm} the spherical harmonic. This partial wave basis is useful for defining the operator for the nucleon-nucleon interaction, which is symmetric with respect to rotations, parity and isospin transformations. These symmetries imply that the interaction is diagonal with respect to the quantum numbers of total angular momentum J, spin S and isospin T. Using the above plane wave expansion, and coupling to final J, S and T we get

$$\langle \mathbf{k}'|V|\mathbf{k}\rangle = (4\pi)^2\sum_{JM}\sum_{lm}\sum_{l'm'}\imath^{l+l'}Y^*_{lm}(\hat{\mathbf{k}})Y_{l'm'}(\hat{\mathbf{k}}')\mathscr{C}^{l'SJ}_{m'M_SM}\mathscr{C}^{lSJ}_{mM_SM}\langle k'l'STJM|V|klSTJM\rangle,$$

where we have defined

$$\langle k'l'STJM|V|klSTJM\rangle = \int j_{l'}(k'r')\langle l'STJM|V(r',r)|lSTJM\rangle j_l(kr)r'^2dr'r^2dr.$$

We have omitted the momentum of the center-of-mass motion $\mathbf{K}$ and the corresponding orbital momentum L, since the interaction is diagonal in these variables.

The interaction we will use for these calculations is a semirealistic nucleon-nucleon potential known as the Minnesota potential [57] which has the form, $V_\alpha(r) = V_\alpha\exp(-\alpha r^2)$. The spin and isospin dependence of the Minnesota potential is given by,

$$V(r) = \frac{1}{2}\left(V_R + \frac{1}{2}\left(1+P^\sigma_{12}\right)V_T + \frac{1}{2}\left(1-P^\sigma_{12}\right)V_S\right)\left(1-P^\sigma_{12}P^\tau_{12}\right),$$

where $P^\sigma_{12} = \frac{1}{2}(1+\sigma_1\cdot\sigma_2)$ and $P^\tau_{12} = \frac{1}{2}(1+\tau_1\cdot\tau_2)$ are the spin and isospin exchange operators, respectively. A Fourier transform to momentum space of the radial part $V_\alpha(r)$ is rather simple, see Problem 8.5, since the radial depends only on the magnitude of the relative distance and thereby the relative momentum $\mathbf{q} = \frac{1}{2}\left(\mathbf{k}_p - \mathbf{k}_q - \mathbf{k}_r + \mathbf{k}_s\right)$. Omitting spin and isospin dependencies, the momentum

Table 8.2 Parameters used to define the Minnesota interaction model [57]

α	V_α in MeV	κ_α in fm^{-2}
R	200	1.487
T	178	0.639
S	91.85	0.465

space version of the interaction reads

$$\langle \mathbf{k}_p\mathbf{k}_q|V_\alpha|\mathbf{k}_r\mathbf{k}_s\rangle = \frac{V_\alpha}{L^3}\left(\frac{\pi}{\alpha}\right)^{3/2} \exp\left(\frac{-q^2}{4\alpha}\right)\delta_{\mathbf{k}_p+\mathbf{k}_q,\mathbf{k}_r+\mathbf{k}_s}$$

The various parameters defining the interaction model used in this work are listed in Table 8.2.

8.2.4 *Models from Effective Field Theory for the Two- and Three-Nucleon Interactions*

During the past two decades it has been demonstrated that chiral effective field theory represents a powerful tool to deal with hadronic interactions at low energy in a systematic and model-independent way (see [14–16, 18, 58–62]). Effective field theories (EFTs) are defined in terms of effective Lagrangians which are given by an infinite series of terms with increasing number of derivatives and/or nucleon fields, with the dependence of each term on the pion field prescribed by the rules of broken chiral symmetry. Applying this Lagrangian to a particular process, an unlimited number of Feynman graphs can be drawn. Therefore, a scheme is needed that makes the theory manageable and calculable. This scheme which tells us how to distinguish between large (important) and small (unimportant) contributions is chiral perturbation theory (ChPT). Chiral perturbation theory allows for an expansion in terms of $(Q/\Lambda_\chi)^\nu$, where Q is generic for an external momentum (nucleon three-momentum or pion four-momentum) or a pion mass, and $\Lambda_\chi \sim$ 1 GeV is the chiral symmetry breaking scale. Determining the power ν has become known as power counting.

Nuclear potentials are defined as sets of irreducible graphs up to a given order. The power ν of a few-nucleon diagram involving A nucleons is given in terms of naive dimensional analysis by:

$$\nu = -2 + 2A - 2C + 2L + \sum_i \Delta_i\,, \tag{8.3}$$

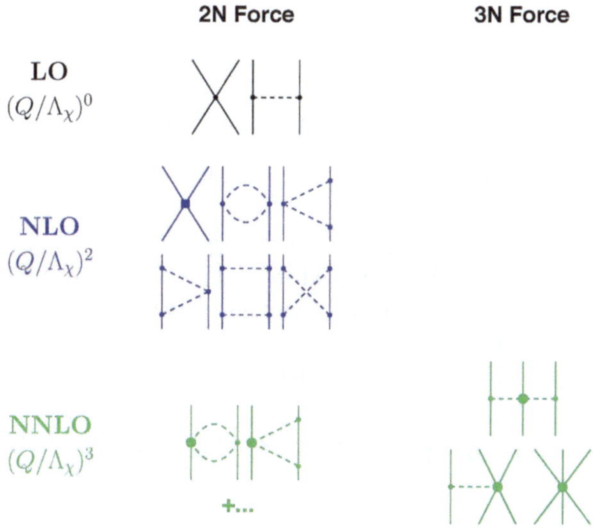

Fig. 8.1 Nuclear forces in ChPT up to NNLO. *Solid lines* represent nucleons and *dashed lines* pions. *Small dots*, *large solid dots*, and *solid squares* denote vertices of index $\Delta_i = 0$, 1, and 2, respectively

with

$$\Delta_i \equiv d_i + \frac{n_i}{2} - 2\,,$$

where A labels the number of nucleons, C denotes the number of separately connected pieces and L the number of loops in the diagram; d_i is the number of derivatives or pion-mass insertions and n_i the number of nucleon fields (nucleon legs) involved in vertex i; the sum runs over all vertices contained in the diagram under consideration. Note that $\Delta_i \geq 0$ for all interactions allowed by chiral symmetry. In this work we will focus on the simple Minnesota model discussed above. It is however possible, see also the exercises, to include two- and three-nucleon forces at order NNLO, as indicated in Fig. 8.1.

Below we revisit briefly the formalism and results presented in [62]. For further details on chiral effective field theory and nuclear interactions, see for example [15, 16, 18, 61, 62] For an irreducible *NN* diagram ("two-nucleon potential", $A = 2$, $C = 1$), Eq. (8.3) collapses to

$$\nu = 2L + \sum_i \Delta_i\,.$$

Thus, in terms of naive dimensional analysis or "Weinberg counting" [58], the various orders of the irreducible graphs which define the chiral *NN* potential are given by (see Fig. 8.1)

$$\begin{aligned} V_{\rm LO} &= V_{\rm ct}^{(0)} + V_{1\pi}^{(0)} \\ V_{\rm NLO} &= V_{\rm LO} + V_{\rm ct}^{(2)} + V_{1\pi}^{(2)} + V_{2\pi}^{(2)} \\ V_{\rm NNLO} &= V_{\rm NLO} + V_{1\pi}^{(3)} + V_{2\pi}^{(3)} \end{aligned}$$

where the superscript denotes the order ν of the low-momentum expansion, LO stands for leading order, NLO for next-to-leading order and NNLO stands for next-to-next-to leading order. Contact potentials carry the subscript "ct" and pion-exchange potentials can be identified by an obvious subscript.

The charge-independent one-pion-exchange (1PE) potential reads

$$V_{1\pi}(\mathbf{k}', \mathbf{k}) = -\frac{g_A^2}{4f_\pi^2}\, \boldsymbol{\tau}_1 \cdot \boldsymbol{\tau}_2\, \frac{\boldsymbol{\sigma}_1 \cdot \mathbf{q}\, \boldsymbol{\sigma}_2 \cdot \mathbf{q}}{q^2 + m_\pi^2}\,, \tag{8.4}$$

where $\mathbf{k}'$ and $\mathbf{k}$ represent the final and initial nucleon momenta in the center-of-mass system (CMS) and $\mathbf{q} \equiv \mathbf{k}' - \mathbf{k}$ is the momentum transfer; $\boldsymbol{\sigma}_{1,2}$ and $\boldsymbol{\tau}_{1,2}$ are the spin and isospin operators of nucleon 1 and 2; g_A, f_π, and m_π denote axial-vector coupling constant, the pion decay constant, and the pion mass, respectively. Since higher order corrections contribute only to mass and coupling constant renormalizations and since, on shell, there are no relativistic corrections, the on-shell 1PE has the form of Eq. (8.4) to all orders.

It is well known that for high-precision *NN* potentials, charge dependence is important. To take into account the charge dependence of the 1PE contribution we define a pion-mass dependent 1PE by

$$V_{1\pi}(m_\pi) \equiv -\frac{g_A^2}{4f_\pi^2}\, \frac{\boldsymbol{\sigma}_1 \cdot \mathbf{q}\, \boldsymbol{\sigma}_2 \cdot \mathbf{q}}{q^2 + m_\pi^2}\,.$$

The 1PE for proton-proton (*pp*) and neutron-neutron (*nn*) terms are then given by

$$V_{1\pi}^{(pp)}(\mathbf{k}', \mathbf{k}) = V_{1\pi}^{(nn)}(\mathbf{k}', \mathbf{k}) = V_{1\pi}(m_{\pi^0})\,,$$

while for the neutron-proton (*np*) part we have

$$V_{1\pi}^{(np)}(\mathbf{k}', \mathbf{k}) = -V_{1\pi}(m_{\pi^0}) + (-1)^{T+1}\, 2\, V_{1\pi}(m_{\pi^\pm})\,,$$

where T denotes the isospin of the two-nucleon system. The pion masses are defined as $m_{\pi^0} = 134.9766\,\text{MeV}$ and $m_{\pi^\pm} = 139.5702\,\text{MeV}$. For the leading-order, next-to-leading order and NNLO, we refer the reader to [15, 62]. The final interaction at order NNLO is multiplied with the following factors [15],

$$\widehat{V}(\mathbf{k}',\mathbf{k}) \equiv \frac{1}{(2\pi)^3}\sqrt{\frac{M_N}{E_{p'}}}\,V(\mathbf{p}',\mathbf{p})\,\sqrt{\frac{M_N}{E_p}}$$

with $E_p = \sqrt{M_N^2 + p^2}$ and where the factor $1/(2\pi)^3$ is just added for convenience. The potential $\widehat{V}$ satisfies the nonrelativistic Lippmann-Schwinger (LS) equation, see [15] for discussions,

$$\widehat{T}(\mathbf{k}',\mathbf{k}) = \widehat{V}(\mathbf{k}',\mathbf{k}) + \int d^3p''\,\widehat{V}(\mathbf{k}',\mathbf{k}'')\,\frac{M_N}{k^2 - k''^2 + i\epsilon}\,\widehat{T}(\mathbf{k}'',\mathbf{k}).$$

In pp scattering, we use $M_N = M_p = 938.2720\,\text{MeV}$, and in nn scattering, $M_N = M_n = 939.5653\,\text{MeV}$. Moreover, the on-shell momentum is simply

$$p^2 = \frac{1}{2}M_N T_{\text{lab}},$$

where T_{lab} denotes the kinetic energy of the incident nucleon in the laboratory system ("Lab. Energy"). For np scattering, we have the relations

$$M_N = \frac{2M_pM_n}{M_p + M_n} = 938.9182\,\text{MeV, and}$$

$$p^2 = \frac{M_p^2 T_{\text{lab}}(T_{\text{lab}} + 2M_n)}{(M_p + M_n)^2 + 2T_{\text{lab}}M_p},$$

which are based upon relativistic kinematics.

Iteration of $\hat{V}$ in the Lippman-Schwinger equation discussed above, requires cutting $\hat{V}$ off for high momenta to avoid infinities. This is consistent with the fact that ChPT is a low-momentum expansion which is valid only for momenta $Q \ll \Lambda_\chi \approx 1\,\text{GeV}$. Therefore, the potential $\hat{V}$ is multiplied with the regulator function $f(k',k)$,

$$\hat{V}(\mathbf{k}',\mathbf{k}) \longmapsto \hat{V}(\mathbf{k}',\mathbf{k})f(k',k)$$

with

$$f(p',p) = \exp[-(p'/\Lambda)^{2n} - (p/\Lambda)^{2n}],$$

as a possible example.

Up to NNLO in chiral perturbation theory there are, in addition to the two-body interaction diagrams discussed above, also a few three-body interaction diagrams, see Fig. 8.1. In chiral perturbation theory, the orders are generated systematically, and at a given chiral order the number of Feynman diagrams is finite. Consistency requires that a calculation includes all diagrams which are present at the chosen order. There are in total five contact terms that determine the strength of the NNLO three-nucleon force (3NF); c_1, c_3, and c_4 are associated with the three-body two-pion-exchange (2PE) diagram, c_D and c_E determine the strength of the one-pion-exchange plus contact (1PE) diagram and the pure contact (CNT) diagram, respectively. References [16, 63] give an extensive discussions of these terms.

8.3 Hartree-Fock Theory

Hartree-Fock (HF) theory is an algorithm for finding an approximative expression for the ground state of a given Hamiltonian. The basic ingredients contain a single-particle basis $\{\psi_\alpha\}$ defined by the solution of the following eigenvalue problem

$$\hat{h}^{\mathrm{HF}}\psi_\alpha = \varepsilon_\alpha \psi_\alpha,$$

with the Hartree-Fock Hamiltonian defined as

$$\hat{h}^{\mathrm{HF}} = \hat{t} + \hat{u}_{\mathrm{ext}} + \hat{u}^{\mathrm{HF}}.$$

The term $\hat{u}^{\mathrm{HF}}$ is a single-particle potential to be determined by the HF algorithm. The HF algorithm means to select $\hat{u}^{\mathrm{HF}}$ in order to have

$$\langle \hat{H} \rangle = E^{\mathrm{HF}} = \langle \Phi_0^{HF} | \hat{H} | \Phi_0^{HF} \rangle,$$

as a local minimum with a Slater determinant Φ_0^{HF} being the ansatz for the ground state. The variational principle ensures that $E^{\mathrm{HF}} \geq E_0$, with E_0 representing the exact ground state energy.

We will show that the Hartree-Fock Hamiltonian $\hat{h}^{\mathrm{HF}}$ equals our definition of the operator $\hat{f}$ discussed in connection with the new definition of the normal-ordered Hamiltonian, that is we have, for a specific matrix element

$$\langle p|\hat{h}^{\mathrm{HF}}|q\rangle = \langle p|\hat{f}|q\rangle = \langle p|\hat{t} + \hat{u}_{\mathrm{ext}}|q\rangle + \sum_{i\leq F} \langle pi|\hat{V}|qi\rangle,$$

meaning that

$$\langle p|\hat{u}^{\mathrm{HF}}|q\rangle = \sum_{i\leq F} \langle pi|\hat{V}|qi\rangle.$$

The so-called Hartree-Fock potential $\hat{u}^{\mathrm{HF}}$ adds an explicit medium dependence due to the summation over all single-particle states below the Fermi level F. It brings also in an explicit dependence on the two-body interaction (in nuclear physics we can also have complicated three- or higher-body forces). The two-body interaction, with its contribution from the other bystanding fermions, creates an effective mean field in which a given fermion moves, in addition to the external potential $\hat{u}_{\text{ext}}$ which confines the motion of the fermion. For systems like nuclei or infinite nuclear matter, there is no external confining potential. Nuclei and nuclear matter are examples of self-bound systems, where the binding arises due to the intrinsic nature of the strong force. For nuclear systems thus, there would be no external one-body potential in the Hartree-Fock Hamiltonian.

Another possibility is to expand the single-particle functions in a known basis and vary the coefficients, that is, the new single-particle wave function is written as a linear expansion in terms of a fixed chosen orthogonal basis (for example the well-known harmonic oscillator functions or the hydrogen-like functions etc). We define our new Hartree-Fock single-particle basis by performing a unitary transformation on our previous basis (labelled with Greek indices) as

$$\psi_p^{HF} = \sum_{\lambda} C_{p\lambda}\phi_{\lambda}. \tag{8.5}$$

In this case we vary the coefficients $C_{p\lambda}$. If the basis has infinitely many solutions, we need to truncate the above sum. We assume that the basis ϕ_{λ} is orthogonal. A unitary transformation keeps the orthogonality, as discussed in Problem 8.6 below.

It is normal to choose a single-particle basis defined as the eigenfunctions of parts of the full Hamiltonian. The typical situation consists of the solutions of the one-body part of the Hamiltonian, that is we have

$$\hat{h}_0\phi_{\lambda} = \epsilon_{\lambda}\phi_{\lambda}.$$

For infinite nuclear matter $\hat{h}_0$ is given by the kinetic energy operator and the states are given by plane wave functions. Due to the translational invariance of the two-body interaction, the Hartree-Fock single-particle eigenstates are also given by the same functions. For infinite matter thus, it is only the single-particle energies that change when we solve the Hartree-Fock equations.

The single-particle wave functions $\phi_{\lambda}(\mathbf{r})$, defined by the quantum numbers λ and $\mathbf{r}$ are defined as the overlap

$$\phi_{\lambda}(\mathbf{r}) = \langle \mathbf{r}|\lambda\rangle.$$

In our discussions we will use our definitions of single-particle states above and below the Fermi (F).

We use Greek letters to refer to our original single-particle basis. The expectation value for the energy with the ansatz Φ_0 for the ground state reads (see Problem 8.7, with application to infinite nuclear matter)

$$E[\Phi_0] = \sum_{\mu\le F}\langle\mu|h|\mu\rangle + \frac{1}{2}\sum_{\mu,\nu\le F}\langle\mu\nu|\hat{v}|\mu\nu\rangle.$$

Now we are interested in defining a new basis defined in terms of a chosen basis as defined in Eq. (8.5). We define the energy functional as

$$E[\Phi^{HF}] = \sum_{i\leq F} \langle i|h|i\rangle + \frac{1}{2}\sum_{ij\leq F} \langle ij|\hat{v}|ij\rangle, \tag{8.6}$$

where Φ^{HF} is the new Slater determinant defined by the new basis of Eq. (8.5).

Using Eq. (8.5) we can rewrite Eq. (8.6) as

$$E[\Psi] = \sum_{i\leq F}\sum_{\alpha\beta} C^*_{i\alpha} C_{i\beta} \langle \alpha|h|\beta\rangle + \frac{1}{2}\sum_{ij\leq F}\sum_{\alpha\beta\gamma\delta} C^*_{i\alpha} C^*_{j\beta} C_{i\gamma} C_{j\delta} \langle \alpha\beta|\hat{v}|\gamma\delta\rangle. \tag{8.7}$$

In order to find the variational minimum of the above functional, we introduce a set of Lagrange multipliers, noting that since $\langle i|j\rangle = \delta_{i,j}$ and $\langle\alpha|\beta\rangle = \delta_{\alpha,\beta}$, the coefficients $C_{i\gamma}$ obey the relation

$$\langle i|j\rangle = \delta_{i,j} = \sum_{\alpha\beta} C^*_{i\alpha} C_{i\beta} \langle\alpha|\beta\rangle = \sum_{\alpha} C^*_{i\alpha} C_{i\alpha},$$

which allows us to define a functional to be minimized that reads

$$F[\Phi^{HF}] = E[\Phi^{HF}] - \sum_{i\leq F} \epsilon_i \sum_{\alpha} C^*_{i\alpha} C_{i\alpha}. \tag{8.8}$$

Minimizing with respect to $C^*_{i\alpha}$ (the equations for $C^*_{i\alpha}$ and $C_{i\alpha}$ can be written as two independent equations) we obtain

$$\frac{d}{dC^*_{i\alpha}}\left[E[\Phi^{HF}] - \sum_{j} \epsilon_j \sum_{\alpha} C^*_{j\alpha} C_{j\alpha}\right] = 0,$$

which yields for every single-particle state i and index α (recalling that the coefficients $C_{i\alpha}$ are matrix elements of a unitary matrix, or orthogonal for a real symmetric matrix) the following Hartree-Fock equations

$$\sum_{\beta} C_{i\beta} \langle\alpha|h|\beta\rangle + \sum_{j\leq F}\sum_{\beta\gamma\delta} C^*_{j\beta} C_{j\delta} C_{i\gamma} \langle\alpha\beta|\hat{v}|\gamma\delta\rangle = \epsilon_i^{HF} C_{i\alpha}.$$

We can rewrite this equation as (changing dummy variables)

$$\sum_{\beta}\left\{\langle\alpha|h|\beta\rangle + \sum_{j\leq F}\sum_{\gamma\delta} C^*_{j\gamma} C_{j\delta} \langle\alpha\gamma|\hat{v}|\beta\delta\rangle\right\} C_{i\beta} = \epsilon_i^{HF} C_{i\alpha}.$$

Note that the sums over Greek indices run over the number of basis set functions (in principle an infinite number).

Defining

$$h_{\alpha\beta}^{HF} = \langle\alpha|h|\beta\rangle + \sum_{j\leq F}\sum_{\gamma\delta} C_{j\gamma}^{*}C_{j\delta}\langle\alpha\gamma|\hat{v}|\beta\delta\rangle,$$

we can rewrite the new equations as

$$\sum_{\beta} h_{\alpha\beta}^{HF} C_{i\beta} = \epsilon_i^{HF} C_{i\alpha}. \quad (8.9)$$

The latter is nothing but a standard eigenvalue problem. Our Hartree-Fock matrix is thus

$$\hat{h}_{\alpha\beta}^{HF} = \langle\alpha|\hat{h}_0|\beta\rangle + \sum_{j\leq F}\sum_{\gamma\delta} C_{j\gamma}^{*}C_{j\delta}\langle\alpha\gamma|\hat{v}|\beta\delta\rangle.$$

The Hartree-Fock equations are solved in an iterative way starting with a guess for the coefficients $C_{j\gamma} = \delta_{j,\gamma}$ and solving the equations by diagonalization till the new single-particle energies ϵ_i^{HF} do not change anymore by a user defined small quantity.

Normally we assume that the single-particle basis $|\beta\rangle$ forms an eigenbasis for the operator $\hat{h}_0$, meaning that the Hartree-Fock matrix becomes

$$\hat{h}_{\alpha\beta}^{HF} = \epsilon_\alpha\delta_{\alpha,\beta} + \sum_{j\leq F}\sum_{\gamma\delta} C_{j\gamma}^{*}C_{j\delta}\langle\alpha\gamma|\hat{v}|\beta\delta\rangle.$$

8.3.1 Hartree-Fock Algorithm with Simple Python Code

The equations are often rewritten in terms of a so-called density matrix, which is defined as

$$\rho_{\gamma\delta} = \sum_{i\leq F}\langle\gamma|i\rangle\langle i|\delta\rangle = \sum_{i=1}^{N} C_{i\gamma}C_{i\delta}^{*}. \quad (8.10)$$

It means that we can rewrite the Hartree-Fock Hamiltonian as

$$\hat{h}_{\alpha\beta}^{HF} = \epsilon_\alpha\delta_{\alpha,\beta} + \sum_{\gamma\delta}\rho_{\gamma\delta}\langle\alpha\gamma|V|\beta\delta\rangle.$$

It is convenient to use the density matrix since we can precalculate in every iteration the product of the eigenvector components C.

The Hartree-Fock equations are, in their simplest form, solved in an iterative way, starting with a guess for the coefficients $C_{i\alpha}$. We label the coefficients as $C_{i\alpha}^{(n)}$, where the superscript n stands for iteration n. To set up the algorithm we can proceed as follows.

1. We start with a guess $C_{i\alpha}^{(0)} = \delta_{i,\alpha}$. Alternatively, we could have used random starting values as long as the vectors are normalized. Another possibility is to give states below the Fermi level a larger weight. We construct then the density matrix and the Hartree-Fock Hamiltonian.
2. The Hartree-Fock matrix simplifies then to

$$\hat{h}_{\alpha\beta}^{HF}(0) = \epsilon_\alpha \delta_{\alpha,\beta} + \sum_{\gamma\delta} \rho_{\gamma\delta}^{(0)} \langle \alpha\gamma|V|\beta\delta\rangle.$$

   Solving the Hartree-Fock eigenvalue problem yields then new eigenvectors $C_{i\alpha}^{(1)}$ and eigenvalues $\epsilon_i^{\mathrm{HF}}(1)$.
3. With the new eigenvectors we can set up a new Hartree-Fock potential

$$\sum_{\gamma\delta} \rho_{\gamma\delta}^{(1)} \langle \alpha\gamma|V|\beta\delta\rangle.$$

   The diagonalization with the new Hartree-Fock potential yields new eigenvectors and eigenvalues.
4. This process is continued till a user defined test is satisfied. As an example, we can require that

$$\frac{\sum_p |\epsilon_i^{(n)} - \epsilon_i^{(n-1)}|}{m} \leq \lambda,$$

   where λ is a small number defined by the user ($\lambda \sim 10^{-8}$ or smaller) and p runs over all calculated single-particle energies and m is the number of single-particle states.

The following simple Python program implements the above algorithm using the density matrix formalism outlined above. We have omitted the functions that set up the single-particle basis and the anti-symmetrized two-body interaction matrix elements. These have to be provided, see https://github.com/ManyBodyPhysics/LectureNotesPhysics/tree/master/Programs/Chapter8-programs/python for full code and matrix elements.

```
# We skip here functions that set up the one- and two-body parts
     of the Hamiltonian
# These functions need to be defined by the user. The two-body
    interaction below is
# calculated by calling the function TwoBodyInteraction(alpha,
    gamma,beta,delta)
# Similarly, the one-body part is computed by the function
    singleparticleH(alpha)
# We have omitted specific quantum number tests as well (isospin
     conservation,
# momentum conservation etc)
import numpy as np
from decimal import Decimal

if __name__ == '__main__':

  """ Star HF-iterations, preparing variables and density matrix
       """

      """ Coefficients for setting up density matrix, assuming
          only one along the diagonals """
  C = np.eye(spOrbitals) # HF coefficients
      DensityMatrix = np.zeros([spOrbitals,spOrbitals])
      for gamma in range(spOrbitals):
         for delta in range(spOrbitals):
            sum = 0.0
            for i in range(Nparticles):
               sum += C[gamma][i]*C[delta][i]
            DensityMatrix[gamma][delta] = Decimal(sum)
      maxHFiter = 100
      epsilon = 1.0e-5
      difference = 1.0
  hf_count = 0
  oldenergies = np.zeros(spOrbitals)
  newenergies = np.zeros(spOrbitals)
  while hf_count < maxHFiter and difference > epsilon:
         HFmatrix = np.zeros([spOrbitals,spOrbitals])
    for alpha in range(spOrbitals):
      for beta in range(spOrbitals):
                      """ Setting up the Fock matrix using the
                          density matrix and antisymmetrized two-
                          body interaction """
                sumFockTerm = 0.0
                      for gamma in range(spOrbitals):
                         for delta in range(spOrbitals):
                            sumFockTerm += DensityMatrix[gamma][
                                delta]*
                                       TwoBodyInteraction(alpha,
                                          gamma,beta,delta)
                      HFmatrix[alpha][beta] = Decimal(sumFockTerm
                          )
                      """ Adding the one-body term """
```

```
                    if beta == alpha: HFmatrix[alpha][alpha] +=
                        singleparticleH(alpha)
    spenergies, C = np.linalg.eigh(HFmatrix)
            """ Setting up new density matrix """
            DensityMatrix = np.zeros([spOrbitals,spOrbitals])
            for gamma in range(spOrbitals):
                for delta in range(spOrbitals):
                    sum = 0.0
                    for i in range(Nparticles):
                        sum += C[gamma][i]*C[delta][i]
                    DensityMatrix[gamma][delta] = Decimal(sum)
    newenergies = spenergies
            """ Brute force computation of difference between
                previous and new sp HF energies """
            sum =0.0
            for i in range(spOrbitals):
                sum += (abs(newenergies[i]-oldenergies[i]))/
                    spOrbitals
            difference = sum
            oldenergies = newenergies
            print "Single-particle energies, ordering may have
                changed "
            for i in range(spOrbitals):
                print('{0:4d}  {1:.4f}'.format(i, Decimal(
                    oldenergies[i])))
    hf_count += 1
```

We end this section by rewriting the ground state energy by adding and subtracting $\hat{u}^{HF}$. Using anti-symmetrized two-body matrix elements we have

$$E_0^{HF} = \langle \Phi_0 | \hat{H} | \Phi_0 \rangle = \sum_{i \le F}^{A} \langle i | \hat{h}_0 + \hat{u}^{HF} | i \rangle + \frac{1}{2} \sum_{i \le F}^{A} \sum_{j \le F}^{A} \langle ij | \hat{v} | ij \rangle - \sum_{i \le F}^{A} \langle i | \hat{u}^{HF} | i \rangle,$$

which results in

$$E_0^{HF} = \sum_{i \le F}^{A} \varepsilon_i^{HF} + \frac{1}{2} \sum_{i \le F}^{A} \sum_{j \le F}^{A} \langle ij | \hat{v} | ij \rangle - \sum_{i \le F}^{A} \langle i | \hat{u}^{HF} | i \rangle.$$

Our single-particle states $ijk \ldots$ are now single-particle states obtained from the solution of the Hartree-Fock equations.

Using our definition of the Hartree-Fock single-particle energies we obtain then the following expression for the total ground-state energy

$$E_0^{HF} = \sum_{i \le F}^{A} \varepsilon_i - \frac{1}{2} \sum_{i \le F}^{A} \sum_{j \le F}^{A} \langle ij | \hat{v} | ij \rangle.$$

This equation demonstrates that the total energy is not given as the sum of the individual single-particle energies.

8.4 Full Configuration Interaction Theory

Full configuration theory (FCI), which represents a discretized variant of the continuous eigenvalue problem, allows for, in principle, an exact (to numerical precision) solution of Schrödinger's equation for many interacting fermions or bosons with a given basis set. This basis set defines an effective Hilbert space. For fermionic problems, the standard approach is to define an upper limit for the set of single-particle states. As an example, if we use the harmonic oscillator one-body Hamiltonian to generate an orthogonal single-particle basis, truncating the basis at some oscillator excitation energy provides thereby an upper limit. Similarly, truncating the maximum values of $n_{x,y,z}$ for plane wave states with periodic boundary conditions, yields a similar upper limit. Table 8.1 lists several possible truncations to the basis set in terms of the single-particle energies as functions of $n_{x,y,z}$. This single-particle basis is then used to define all possible Slater determinants which can be constructed with a given number of fermions A. The total number of Slater determinants determines thereafter the dimensionality of the Hamiltonian matrix and thereby an effective Hilbert space. If we are able to set up the Hamiltonian matrix and solve the pertinent eigenvalue problem within this basis set, FCI provides numerically exact solutions to all states of interest for a given many-body problem. The dimensionality of the problem explodes however quickly. To see this it suffices to consider the total number of Slater determinants which can be built with say N neutrons distributed among n single-particle states. The total number is

$$\binom{n}{N} = \frac{n!}{(n-N)!N!}.$$

As an example, for a model space which comprises the first four major harmonic oscillator shells only, that is the $0s$, $0p$, $1s0d$ and $1p0f$ shells we have 40 single particle states for neutrons and protons. For the eight neutrons of oxygen-16 we would then have

$$\binom{40}{8} = \frac{40!}{(32)!8!} \sim 8 \times 10^7,$$

possible Slater determinants. Multiplying this with the number of proton Slater determinants we end up with approximately $d \sim 10^{15}$ possible Slater determinants and a Hamiltonian matrix of dimension $10^{15} \times 10^{15}$, an intractable problem if we wish to diagonalize the Hamiltonian matrix. The dimensionality can be reduced if we look at specific symmetries, however these symmetries will never reduce the problem to dimensionalities which can be handled by standard eigenvalue solvers.

These are normally lumped into two main categories, direct solvers for matrices of dimensionalities which are smaller than $d \sim 10^5$, and iterative eigenvalue solvers (when only selected states are being sought after) for dimensionalities up to $10^{10} \times 10^{10}$.

Due to its discreteness thus, the effective Hilbert space will always represent an approximation to the full continuous problem. However, with a given Hamiltonian matrix and effective Hilbert space, FCI provides us with true benchmarks that can convey important information on correlations beyond Hartree-Fock theory and various approximative many-body methods like many-body perturbation theory, coupled cluster theory, Green's function theory and the Similarity Renormalization Group approach. These methods are all discussed in this text. Assuming that we can diagonalize the Hamiltonian matrix, and thereby obtain the exact solutions, this section serves the aim to link the exact solution obtained from FCI with various approximative methods, hoping thereby that eventual differences can shed light on which correlations play a major role and should be included in the above approximative methods. The simple pairing model discussed in Problem 8.10 is an example of a system that allows us to compare exact solutions with those defined by many-body perturbation theory to a given order in the interaction, coupled cluster theory, Green's function theory and the Similarity Renormalization Group (SRG).

In order to familiarize the reader with these approximative many-body methods, we start with the general definition of the full configuration interaction problem.

We have defined the ansatz for the ground state as

$$|\Phi_0\rangle = \left(\prod_{i\le F} \hat{a}_i^\dagger\right)|0\rangle,$$

where the variable i defines different single-particle states up to the Fermi level. We have assumed that we have A nucleons and that the chosen single-particle states are eigenstates of the one-body Hamiltonian $\hat{h}_0$ (defining thereby an orthogonal basis set). A given one-particle-one-hole ($1p1h$) state can be written as

$$|\Phi_i^a\rangle = \hat{a}_a^\dagger \hat{a}_i|\Phi_0\rangle,$$

while a $2p2h$ state can be written as

$$|\Phi_{ij}^{ab}\rangle = \hat{a}_a^\dagger \hat{a}_b^\dagger \hat{a}_j \hat{a}_i|\Phi_0\rangle,$$

and a general $ApAh$ state as

$$|\Phi_{ijk\ldots}^{abc\ldots}\rangle = \hat{a}_a^\dagger \hat{a}_b^\dagger \hat{a}_c^\dagger \ldots \hat{a}_k \hat{a}_j \hat{a}_i|\Phi_0\rangle.$$

As before, we use letters $ijkl\ldots$ for states below the Fermi level and $abcd\ldots$ for states above the Fermi level. A general single-particle state is given by letters $pqrs\ldots$.

We can then expand our exact state function for the ground state as

$$|\Psi_0\rangle = C_0|\Phi_0\rangle + \sum_{ai} C_i^a|\Phi_i^a\rangle + \sum_{abij} C_{ij}^{ab}|\Phi_{ij}^{ab}\rangle + \cdots = (C_0 + \hat{C})|\Phi_0\rangle,$$

where we have introduced the so-called correlation operator

$$\hat{C} = \sum_{ai} C_i^a \hat{a}_a^\dagger \hat{a}_i + \sum_{abij} C_{ij}^{ab} \hat{a}_a^\dagger \hat{a}_b^\dagger \hat{a}_j \hat{a}_i + \ldots$$

Since the normalization of Ψ_0 is at our disposal and since C_0 is by assumption not zero, we may arbitrarily set $C_0 = 1$ with corresponding proportional changes in all other coefficients. Using this so-called intermediate normalization we have

$$\langle \Psi_0 | \Phi_0 \rangle = \langle \Phi_0 | \Phi_0 \rangle = 1,$$

resulting in

$$|\Psi_0\rangle = (1 + \hat{C})|\Phi_0\rangle.$$

We rewrite

$$|\Psi_0\rangle = C_0|\Phi_0\rangle + \sum_{ai} C_i^a|\Phi_i^a\rangle + \sum_{abij} C_{ij}^{ab}|\Phi_{ij}^{ab}\rangle + \ldots,$$

in a more compact form as

$$|\Psi_0\rangle = \sum_{PH} C_H^P \Phi_H^P = \left(\sum_{PH} C_H^P \hat{A}_H^P\right)|\Phi_0\rangle,$$

where H stands for $0, 1, \ldots, n$ hole states and P for $0, 1, \ldots, n$ particle states. The operator $\hat{A}_H^P$ represents a given set of particle-hole excitations. For a two-particle-to-hole excitation this operator is given by $\hat{A}_{2h}^{2p} = \hat{a}_a^\dagger \hat{a}_b^\dagger \hat{a}_j \hat{a}_i$. Our requirement of unit normalization gives

$$\langle \Psi_0 | \Psi_0 \rangle = \sum_{PH} |C_H^P|^2 = 1,$$

and the energy can be written as

$$E = \langle \Psi_0 | \hat{H} | \Psi_0 \rangle = \sum_{PP'HH'} C_H^{*P} \langle \Phi_H^P | \hat{H} | \Phi_{H'}^{P'} \rangle C_{H'}^{P'}.$$

The last equation is normally solved by diagonalization, with the Hamiltonian matrix defined by the basis of all possible Slater determinants. A diagonalization is equivalent to finding the variational minimum of

$$\langle\Psi_0|\hat{H}|\Psi_0\rangle - \lambda\langle\Psi_0|\Psi_0\rangle,$$

where λ is a variational multiplier to be identified with the energy of the system.

The minimization process results in

$$0 = \delta\left[\langle\Psi_0|\hat{H}|\Psi_0\rangle - \lambda\langle\Psi_0|\Psi_0\rangle\right] \tag{8.11}$$

$$= \sum_{P'H'}\left\{\delta[C_H^{*P}]\langle\Phi_H^P|\hat{H}|\Phi_{H'}^{P'}\rangle C_{H'}^{P'} + C_H^{*P}\langle\Phi_H^P|\hat{H}|\Phi_{H'}^{P'}\rangle\delta[C_{H'}^{P'}] - \lambda(\delta[C_H^{*P}]C_{H'}^{P'} + C_H^{*P}\delta[C_{H'}^{P'}]\right\}. \tag{8.12}$$

Since the coefficients $\delta[C_H^{*P}]$ and $\delta[C_{H'}^{P'}]$ are complex conjugates it is necessary and sufficient to require the quantities that multiply with $\delta[C_H^{*P}]$ to vanish.

This leads to

$$\sum_{P'H'}\langle\Phi_H^P|\hat{H}|\Phi_{H'}^{P'}\rangle C_{H'}^{P'} - \lambda C_H^P = 0,$$

for all sets of P and H.

If we then multiply by the corresponding C_H^{*P} and sum over PH we obtain

$$\sum_{PP'HH'} C_H^{*P}\langle\Phi_H^P|\hat{H}|\Phi_{H'}^{P'}\rangle C_{H'}^{P'} - \lambda\sum_{PH}|C_H^P|^2 = 0,$$

leading to the identification $\lambda = E$. This means that we have for all PH sets

$$\sum_{P'H'}\langle\Phi_H^P|\hat{H} - E|\Phi_{H'}^{P'}\rangle = 0. \tag{8.13}$$

An alternative way to derive the last equation is to start from

$$(\hat{H} - E)|\Psi_0\rangle = (\hat{H} - E)\sum_{P'H'} C_{H'}^{P'}|\Phi_{H'}^{P'}\rangle = 0,$$

and if this equation is successively projected against all Φ_H^P in the expansion of Ψ, we end up with Eq. (8.13).

If we are able to solve this equation by numerical diagonalization in a large Hilbert space (it will be truncated in terms of the number of single-particle states included in the definition of Slater determinants), it can then serve as a benchmark for other many-body methods which approximate the correlation operator $\hat{C}$. Our pairing model discussed in Problem 8.10 is an example of a system which can be diagonalized exactly, providing thereby benchmarks for different approximative methods.

To better understand the meaning of possible configurations and the derivation of a Hamiltonian matrix, we consider here a simple example of six fermions. We assume we can make an ansatz for the ground state with all six fermions below the Fermi level. We label this state as a zero-particle-zero-hole state $0p-0h$. With six nucleons we can make at most $6p-6h$ excitations. If we have an infinity of single particle states above the Fermi level, we will obviously have an infinity of say $2p-2h$ excitations. Each specific way to distribute the particles represents a configuration. We will always have to truncate the basis of single-particle states. This gives us a finite number of possible Slater determinants. Our Hamiltonian matrix would then look like (where each block which is marked with an x can contain a large quantity of non-zero matrix elements) as shown here if the Hamiltonian contains at most a two-

	$0p-0h$	$1p-1h$	$2p-2h$	$3p-3h$	$4p-4h$	$5p-5h$	$6p-6h$
$0p-0h$	x	x	x	0	0	0	0
$1p-1h$	x	x	x	x	0	0	0
$2p-2h$	x	x	x	x	x	0	0
$3p-3h$	0	x	x	x	x	x	0
$4p-4h$	0	0	x	x	x	x	x
$5p-5h$	0	0	0	x	x	x	x
$6p-6h$	0	0	0	0	x	x	x

body interaction, as demonstrated in Problem 8.8. If we use a so-called canonical Hartree-Fock basis [35], this corresponds to a particular unitary transformation where matrix elements of the type $\langle 0p-0h|\hat{H}|1p-1h\rangle = \langle\Phi_0|\hat{H}|\Phi_i^a\rangle = 0$. With a canonical Hartree-Fock basis our Hamiltonian matrix reads

	$0p-0h$	$1p-1h$	$2p-2h$	$3p-3h$	$4p-4h$	$5p-5h$	$6p-6h$
$0p-0h$	$\tilde{x}$	0	$\tilde{x}$	0	0	0	0
$1p-1h$	0	$\tilde{x}$	$\tilde{x}$	$\tilde{x}$	0	0	0
$2p-2h$	$\tilde{x}$	$\tilde{x}$	$\tilde{x}$	$\tilde{x}$	$\tilde{x}$	0	0
$3p-3h$	0	$\tilde{x}$	$\tilde{x}$	$\tilde{x}$	$\tilde{x}$	$\tilde{x}$	0
$4p-4h$	0	0	$\tilde{x}$	$\tilde{x}$	$\tilde{x}$	$\tilde{x}$	$\tilde{x}$
$5p-5h$	0	0	0	$\tilde{x}$	$\tilde{x}$	$\tilde{x}$	$\tilde{x}$
$6p-6h$	0	0	0	0	$\tilde{x}$	$\tilde{x}$	$\tilde{x}$

If we do not make any truncations in the possible sets of Slater determinants (many-body states) we can make by distributing A nucleons among n single-particle states, we call such a calculation for a full configuration interaction (FCI) approach. If we make truncations, we have several different possibilities to reduce the dimensionality of the problem. A well-known example is the standard nuclear shell-model. For the nuclear shell model we define an effective Hilbert space with respect to a given core. The calculations are normally then performed for all many-body states that can be constructed from the effective Hilbert spaces. This approach requires a properly defined effective Hamiltonian. Another possibility to constrain the dimensionality of the problem is to truncate in the number of excitations. As an example, we can limit the possible Slater determinants to only $1p-1h$ and $2p-2h$ excitations. This is called a configuration interaction calculation at the level of singles and doubles excitations. If we truncate at the level of three-particle-three-hole excitations we end up with singles, doubles and triples excitations. Such truncations reduce considerably the size of the Hamiltonian matrices to be diagonalized, but can lead to so-called unlinked contributions, and thereby wrong results, for a given expectation value [34]. A third possibility is to constrain the number of excitations by an energy cutoff. This cutoff defines a maximum excitation energy. The maximum excitation energy is normally given by the sum of single-particle energies defined by the unperturbed one-body part of the Hamiltonian. A commonly used basis in nuclear physics is the harmonic oscillator. The cutoff in energy is then defined by the maximum number of harmonic oscillator excitations. If we do not define a core, this defines normally what is called the no-core shell-model approach, see for example [25, 64].

8.4.1 A Non-practical Way of Solving the Eigenvalue Problem

For reasons to come (links with coupled cluster theory and many-body perturbation theory), we will rewrite Eq. (8.13) as a set of coupled non-linear equations in terms of the unknown coefficients C_H^P. To obtain the eigenstates and eigenvalues in terms of non-linear equations is less efficient than using standard eigenvalue solvers [65]. However, this digression serves the scope of linking full configuration interaction theory with approximative solutions to the many-body problem.

To see this, we look at the contributions arising from

$$\langle\Phi_H^P| = \langle\Phi_0|$$

in Eq. (8.13), that is we multiply with $\langle\Phi_0|$ from the left in

$$(\hat{H}-E)\sum_{P'H'} C_{H'}^{P'}|\Phi_{H'}^{P'}\rangle = 0.$$

If we assume that we have a two-body operator at most, the Slater-Condon rule for a two-body interaction, see Problem 8.8, results in an expression for the correlation energy in terms of C_i^a and C_{ij}^{ab} only, namely

$$\langle\Phi_0|\hat{H}-E|\Phi_0\rangle+\sum_{ai}\langle\Phi_0|\hat{H}-E|\Phi_i^a\rangle C_i^a+\sum_{abij}\langle\Phi_0|\hat{H}-E|\Phi_{ij}^{ab}\rangle C_{ij}^{ab}=0,$$

or

$$E-E_{\rm Ref}=\Delta E=\sum_{ai}\langle\Phi_0|\hat{H}|\Phi_i^a\rangle C_i^a+\sum_{abij}\langle\Phi_0|\hat{H}|\Phi_{ij}^{ab}\rangle C_{ij}^{ab},$$

where the energy $E_{\rm Ref}$ is the reference energy and ΔE defines the so-called correlation energy. The single-particle basis functions could result from a Hartree-Fock calculation or they could be the eigenstates of the one-body operator that defined the non-interacting part of the Hamiltonian.

In our Hartree-Fock discussions, we have already computed the matrix $\langle\Phi_0|\hat{H}|\Phi_i^a\rangle$ and $\langle\Phi_0|\hat{H}|\Phi_{ij}^{ab}\rangle$. If we are using a Hartree-Fock basis we have $\langle\Phi_0|\hat{H}|\Phi_i^a\rangle=0$ and we are left with a *correlation energy* given by

$$E-E_{\rm Ref}=\Delta E^{HF}=\sum_{abij}\langle\Phi_0|\hat{H}|\Phi_{ij}^{ab}\rangle C_{ij}^{ab}.$$

Inserting the various matrix elements we can rewrite the previous equation as

$$\Delta E=\sum_{ai}\langle i|\hat{f}|a\rangle C_i^a+\sum_{abij}\langle ij|\hat{v}|ab\rangle C_{ij}^{ab}. \tag{8.14}$$

This equation determines the correlation energy but not the coefficients C. We need more equations. Our next step is to set up

$$\langle\Phi_i^a|\hat{H}-E|\Phi_0\rangle+\sum_{bj}\langle\Phi_i^a|\hat{H}-E|\Phi_j^b\rangle C_j^b+\sum_{bcjk}\langle\Phi_i^a|\hat{H}-E|\Phi_{jk}^{bc}\rangle C_{jk}^{bc}$$
$$+\sum_{bcdjkl}\langle\Phi_i^a|\hat{H}-E|\Phi_{jkl}^{bcd}\rangle C_{jkl}^{bcd}=0,$$

as this equation will allow us to find an expression for the coefficients C_i^a through

$$\langle i|\hat{f}|a\rangle+\langle\Phi_i^a|\hat{H}|\Phi_i^a\rangle C_i^a+\sum_{bj\neq ai}\langle\Phi_i^a|\hat{H}|\Phi_j^b\rangle C_j^b+\sum_{bcjk}\langle\Phi_i^a|\hat{H}|\Phi_{jk}^{bc}\rangle C_{jk}^{bc}$$
$$+\sum_{bcdjkl}\langle\Phi_i^a|\hat{H}|\Phi_{jkl}^{bcd}\rangle C_{jkl}^{bcd}=EC_i^a. \tag{8.15}$$

We see that on the right-hand side we have the energy E. This leads to a non-linear equation in the unknown coefficients since the coefficients appear also in the definition of the correlation energy of Eq. (8.14). These equations are normally solved iteratively, that is we start with a guess for the coefficients C_i^a. A common choice is to use perturbation theory as a starting point for the unknown coefficients. For the one-particle-one-hole coefficients, the wave operator (see Sect. 8.5) to first order in the interaction is given by

$$C_i^a = \frac{\langle i|\hat{f}|a\rangle}{\epsilon_i - \epsilon_a}.$$

The observant reader will however see that we need an equation for C_{jk}^{bc} and C_{jkl}^{bcd} and more complicated particle-hole excitations as well. To find the equations for these coefficients we need then to continue our multiplications from the left with the various Φ_H^P terms.

For C_{jk}^{bc} we have

$$0 = \langle \Phi_{ij}^{ab}|\hat{H} - E|\Phi_0\rangle + \sum_{kc}\langle \Phi_{ij}^{ab}|\hat{H} - E|\Phi_k^c\rangle C_k^c \tag{8.16}$$

$$+ \sum_{cdkl}\langle \Phi_{ij}^{ab}|\hat{H} - E|\Phi_{kl}^{cd}\rangle C_{kl}^{cd} + \sum_{cdeklm}\langle \Phi_{ij}^{ab}|\hat{H} - E|\Phi_{klm}^{cde}\rangle C_{klm}^{cde}$$

$$+ \sum_{cdefklmn}\langle \Phi_{ij}^{ab}|\hat{H} - E|\Phi_{klmn}^{cdef}\rangle C_{klmn}^{cdef}. \tag{8.17}$$

We can isolate the coefficients C_{kl}^{cd} in a similar way as we did for the coefficients C_i^a. A standard choice for the first iteration is to use again perturbation theory to first order in the interaction and set

$$C_{ij}^{ab} = \frac{\langle ij|\hat{v}|ab\rangle}{\epsilon_i + \epsilon_j - \epsilon_a - \epsilon_b}.$$

At the end we can rewrite our solution of the Schrödinger equation in terms of a series coupled equations for the coefficients C_H^P. This is a very cumbersome way of solving a many-body problem. However, by using this iterative scheme we can illustrate how we can compute the various terms in the wave operator or correlation operator $\hat{C}$. We will later identify the calculation of the various terms C_H^P as parts of different many-body approximations to full configuration interaction theory.

8.4.2 *Short Summary*

If we can directly diagonalize large matrices, full configuration interaction theory is the method of choice since we obtain all eigenvectors and eigenvalues. The eigenvectors are obtained directly from the coefficients C_H^P which result from the diagonalization. We can then compute expectation values of other operators, as well as transition probabilities. Moreover, correlations are easy to understand in terms of contributions to a given operator beyond the Hartree-Fock contribution. For larger dimensionalities d, with $d > 10^5$, iterative methods [65] like Lanczos' [66] or Davidson's [67, 68] algorithms are frequently used. These methods yield, with a finite number of iteration, only a subset of all eigenvalues of interest. Lanczos' algorithm converges to the extreme values, yielding the lowest-lying and highest-lying eigenstates, see for example [65] for a proof.

With the eigenvectors we can compute the correlation energy, which is defined as (with a two-body Hamiltonian)

$$\Delta E = \sum_{ai} \langle i|\hat{f}|a\rangle C_i^a + \sum_{abij} \langle ij|\hat{v}|ab\rangle C_{ij}^{ab}.$$

The energy of the ground state is then

$$E = E_{\text{Ref}} + \Delta E.$$

However, as we have seen, even for a small case like the four first major shells and oxygen-16 with 16 active nucleons, the dimensionality becomes quickly intractable. If we wish to include single-particle states that reflect weakly bound systems, we need a much larger single-particle basis. We need thus approximative methods that sum specific correlations to infinite order. All these methods start normally with a Hartree-Fock basis as the calculational basis. In the next section we discuss one of these possible approximative methods, namely many-body perturbation theory.

8.5 Many-Body Perturbation Theory

We assume here that we are only interested in the non-degenerate ground state of a given system and expand the exact wave function in terms of a series of Slater determinants

$$|\Psi_0\rangle = |\Phi_0\rangle + \sum_{m=1}^{\infty} C_m |\Phi_m\rangle,$$

where we have assumed that the true ground state is dominated by the solution of the unperturbed problem, that is

$$\hat{H}_0|\Phi_0\rangle = W_0|\Phi_0\rangle.$$

The state $|\Psi_0\rangle$ is not normalized and we employ again intermediate normalization via $\langle\Phi_0|\Psi_0\rangle = 1$.

The Schrödinger equation is given by

$$\hat{H}|\Psi_0\rangle = E|\Psi_0\rangle,$$

and multiplying the latter from the left with $\langle\Phi_0|$ gives

$$\langle\Phi_0|\hat{H}|\Psi_0\rangle = E\langle\Phi_0|\Psi_0\rangle = E,$$

and subtracting from this equation

$$\langle\Psi_0|\hat{H}_0|\Phi_0\rangle = W_0\langle\Psi_0|\Phi_0\rangle = W_0,$$

and using the fact that the operators $\hat{H}$ and $\hat{H}_0$ are hermitian results in

$$\Delta E = E - W_0 = \langle\Phi_0|\hat{H}_I|\Psi_0\rangle, \tag{8.18}$$

which is an exact result. This resembles our previous definition of the correlation energy except that the reference energy is now defined by the unperturbed energy W_0. The reader should contrast this equation to our previous definition of the correlation energy

$$\Delta E = \sum_{ai}\langle i|\hat{f}|a\rangle C_i^a + \sum_{abij}\langle ij|\hat{v}|ab\rangle C_{ij}^{ab},$$

and the total energy

$$E = E_{\mathrm{Ref}} + \Delta E,$$

where the reference energy is given by

$$E_{\mathrm{Ref}} = \langle\Phi_0|\hat{H}|\Phi_0\rangle.$$

Equation (8.18) forms the starting point for all perturbative derivations. However, as it stands it represents nothing but a mere formal rewriting of Schrödinger's equation and is not of much practical use. The exact wave function $|\Psi_0\rangle$ is unknown. In order to obtain a perturbative expansion, we need to expand the exact wave function in terms of the interaction $\hat{H}_I$.

Here we have assumed that our model space defined by the operator $\hat{P}$ is one-dimensional, meaning that

$$\hat{P} = |\Phi_0\rangle\langle\Phi_0|,$$

and

$$\hat{Q} = \sum_{m=1}^{\infty} |\Phi_m\rangle\langle\Phi_m|.$$

We can thus rewrite the exact wave function as

$$|\Psi_0\rangle = (\hat{P} + \hat{Q})|\Psi_0\rangle = |\Phi_0\rangle + \hat{Q}|\Psi_0\rangle.$$

Going back to the Schrödinger equation, we can rewrite it as, adding and a subtracting a term $\omega|\Psi_0\rangle$ as

$$\left(\omega - \hat{H}_0\right)|\Psi_0\rangle = \left(\omega - E + \hat{H}_I\right)|\Psi_0\rangle,$$

where ω is an energy variable to be specified later.

We assume also that the resolvent of $\left(\omega - \hat{H}_0\right)$ exits, that is it has an inverse which defines the unperturbed Green's function as

$$\left(\omega - \hat{H}_0\right)^{-1} = \frac{1}{\left(\omega - \hat{H}_0\right)}.$$

We can rewrite Schrödinger's equation as

$$|\Psi_0\rangle = \frac{1}{\omega - \hat{H}_0}\left(\omega - E + \hat{H}_I\right)|\Psi_0\rangle,$$

and multiplying from the left with $\hat{Q}$ results in

$$\hat{Q}|\Psi_0\rangle = \frac{\hat{Q}}{\omega - \hat{H}_0}\left(\omega - E + \hat{H}_I\right)|\Psi_0\rangle,$$

which is possible since we have defined the operator $\hat{Q}$ in terms of the eigenfunctions of $\hat{H}_0$.

Since these operators commute we have

$$\hat{Q}\frac{1}{\left(\omega - \hat{H}_0\right)}\hat{Q} = \hat{Q}\frac{1}{\left(\omega - \hat{H}_0\right)} = \frac{\hat{Q}}{\left(\omega - \hat{H}_0\right)}.$$

With these definitions we can in turn define the wave function as

$$|\Psi_0\rangle = |\Phi_0\rangle + \frac{\hat{Q}}{\omega - \hat{H}_0}\left(\omega - E + \hat{H}_I\right)|\Psi_0\rangle.$$

This equation is again nothing but a formal rewrite of Schrödinger's equation and does not represent a practical calculational scheme. It is a non-linear equation in two unknown quantities, the energy E and the exact wave function $|\Psi_0\rangle$. We can however start with a guess for $|\Psi_0\rangle$ on the right hand side of the last equation.

The most common choice is to start with the function which is expected to exhibit the largest overlap with the wave function we are searching after, namely $|\Phi_0\rangle$. This can again be inserted in the solution for $|\Psi_0\rangle$ in an iterative fashion and if we continue along these lines we end up with

$$|\Psi_0\rangle = \sum_{i=0}^{\infty} \left\{ \frac{\hat{Q}}{\omega - \hat{H}_0} \left(\omega - E + \hat{H}_I \right) \right\}^i |\Phi_0\rangle,$$

for the wave function and

$$\Delta E = \sum_{i=0}^{\infty} \langle \Phi_0 | \hat{H}_I \left\{ \frac{\hat{Q}}{\omega - \hat{H}_0} \left(\omega - E + \hat{H}_I \right) \right\}^i |\Phi_0\rangle,$$

which is now a perturbative expansion of the exact energy in terms of the interaction $\hat{H}_I$ and the unperturbed wave function $|\Psi_0\rangle$.

In our equations for $|\Psi_0\rangle$ and ΔE in terms of the unperturbed solutions $|\Phi_i\rangle$ we have still an undetermined parameter ω and a dependency on the exact energy E. Not much has been gained thus from a practical computational point of view.

In Brilluoin-Wigner perturbation theory [35] it is customary to set $\omega = E$. This results in the following perturbative expansion for the energy ΔE

$$\Delta E = \sum_{i=0}^{\infty} \langle \Phi_0 | \hat{H}_I \left\{ \frac{\hat{Q}}{\omega - \hat{H}_0} \left(\omega - E + \hat{H}_I \right) \right\}^i |\Phi_0\rangle \tag{8.19}$$

$$= \langle \Phi_0 | \left(\hat{H}_I + \hat{H}_I \frac{\hat{Q}}{E - \hat{H}_0} \hat{H}_I + \hat{H}_I \frac{\hat{Q}}{E - \hat{H}_0} \hat{H}_I \frac{\hat{Q}}{E - \hat{H}_0} \hat{H}_I + \dots \right) |\Phi_0\rangle. \tag{8.20}$$

This expression depends however on the exact energy E and is again not very convenient from a practical point of view. It can obviously be solved iteratively, by starting with a guess for E and then solve till some kind of self-consistency criterion has been reached.

Defining $e = E - \hat{H}_0$ and recalling that $\hat{H}_0$ commutes with $\hat{Q}$ by construction and that $\hat{Q}$ is an idempotent operator $\hat{Q}^2 = \hat{Q}$, we can rewrite the denominator in the above expansion for ΔE as

$$\hat{Q} \frac{1}{\hat{e} - \hat{Q}\hat{H}_I\hat{Q}} = \hat{Q} \left[\frac{1}{\hat{e}} + \frac{1}{\hat{e}} \hat{Q}\hat{H}_I\hat{Q} \frac{1}{\hat{e}} + \frac{1}{\hat{e}} \hat{Q}\hat{H}_I\hat{Q} \frac{1}{\hat{e}} \hat{Q}\hat{H}_I\hat{Q} \frac{1}{\hat{e}} + \dots \right] \hat{Q}.$$

Inserted in the expression for ΔE we obtain

$$\Delta E = \langle \Phi_0 | \hat{H}_I + \hat{H}_I \hat{Q} \frac{1}{E - \hat{H}_0 - \hat{Q}\hat{H}_I\hat{Q}} \hat{Q}\hat{H}_I | \Phi_0 \rangle .$$

In Rayleigh-Schrödinger (RS) perturbation theory [35] we set $\omega = W_0$ and obtain the following expression for the energy difference

$$\Delta E = \sum_{i=0}^{\infty} \langle \Phi_0 | \hat{H}_I \left\{ \frac{\hat{Q}}{W_0 - \hat{H}_0} \left(\hat{H}_I - \Delta E \right) \right\}^i | \Phi_0 \rangle \tag{8.21}$$

$$\langle \Phi_0 | \left(\hat{H}_I + \hat{H}_I \frac{\hat{Q}}{W_0 - \hat{H}_0} (\hat{H}_I - \Delta E) + \hat{H}_I \frac{\hat{Q}}{W_0 - \hat{H}_0} (\hat{H}_I - \Delta E) \frac{\hat{Q}}{W_0 - \hat{H}_0} (\hat{H}_I - \Delta E) + \dots \right) | \Phi_0 \rangle . \tag{8.22}$$

The operator $\hat{Q}$ commutes with $\hat{H}_0$ and since ΔE is a constant we obtain that

$$\hat{Q} \Delta E | \Phi_0 \rangle = \hat{Q} \Delta E | \hat{Q} \Phi_0 \rangle = 0.$$

Inserting this result in the expression for the energy gives us

$$\Delta E = \langle \Phi_0 | \left(\hat{H}_I + \hat{H}_I \frac{\hat{Q}}{W_0 - \hat{H}_0} \hat{H}_I + \hat{H}_I \frac{\hat{Q}}{W_0 - \hat{H}_0} (\hat{H}_I - \Delta E) \frac{\hat{Q}}{W_0 - \hat{H}_0} \hat{H}_I + \dots \right) | \Phi_0 \rangle .$$

We can now perturbatively expand this expression in terms of the interaction $\hat{H}_I$, which is assumed to be small. We obtain then

$$\Delta E = \sum_{i=1}^{\infty} \Delta E^{(i)},$$

with the following expression for $\Delta E^{(i)}$

$$\Delta E^{(1)} = \langle \Phi_0 | \hat{H}_I | \Phi_0 \rangle ,$$

which is just the contribution to first order in perturbation theory,

$$\Delta E^{(2)} = \langle \Phi_0 | \hat{H}_I \frac{\hat{Q}}{W_0 - \hat{H}_0} \hat{H}_I | \Phi_0 \rangle ,$$

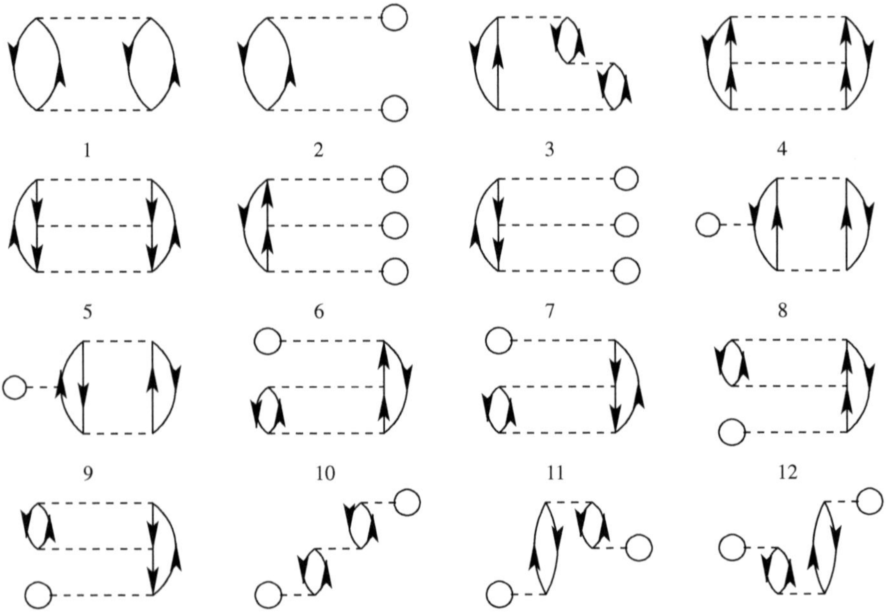

Fig. 8.2 Linked anti-symmetrized Goldstone diagrams which enter the definition of the ground-state correlation energy ΔE_0 to third order in the interaction. We have not included the first-order contribution

which is the contribution to second order and

$$\Delta E^{(3)} = \langle \Phi_0 | \hat{H}_I \frac{\hat{Q}}{W_0 - \hat{H}_0} \hat{H}_I \frac{\hat{Q}}{W_0 - \hat{H}_0} \hat{H}_I \Phi_0 \rangle$$

$$- \langle \Phi_0 | \hat{H}_I \frac{\hat{Q}}{W_0 - \hat{H}_0} \langle \Phi_0 | \hat{H}_I | \Phi_0 \rangle \frac{\hat{Q}}{W_0 - \hat{H}_0} \hat{H}_I | \Phi_0 \rangle,$$

being the third-order contribution. There exists a formal theory for the calculation of ΔE_0, see for example [35]. According to the well-known Goldstone linked-diagram theory, the energy shift ΔE_0 is given exactly by the diagrammatic expansion shown in Fig. 8.2, where ground state diagrams to third order are listed. This theory is a linked-cluster perturbation expansion for the ground state energy of a many-body system, and applies equally well to both nuclear matter and closed-shell nuclei such as the doubly magic nucleus ^{40}Ca. We assume the reader is familiar with the standard rules for deriving and setting up the analytical expressions for various Feymann-Goldstone diagrams [35]. In an infinite system like nuclear matter or the homogenous electron gas, all diagrams with so-called Hartree-Fock insertions like diagrams (2), (6), (7), (10–16) are zero due to lack of momentum conservation. They would also be zero in case a canonical [35] Hartree-Fock basis is employed.

Using the above standard diagram rules, the various diagrams contained in Fig. 8.2 can be readily calculated (in an uncoupled scheme). Diagram (1) results in

$$(1) = \frac{1}{2^2}\sum_{ij\leq F}\sum_{ab>F}\frac{\langle ij|\hat{v}|ab\rangle\langle ab|\hat{v}|ij\rangle}{\varepsilon_i+\varepsilon_j-\varepsilon_a-\varepsilon_b}, \tag{8.23}$$

while diagram (2) is zero due to lack of momentum conservation. We have two factors of $1/2$ since there are two equivalent pairs of fermions (two particle states and two hole states) starting at the same vertex and ending at the same vertex. The expression for diagram (3) is

$$(3) = \sum_{ijk\leq k_F}\sum_{abc>F}\frac{\langle ij|\hat{v}|ab\rangle\langle bk|\hat{v}|ic\rangle\langle ac|\hat{v}|ik\rangle}{(\varepsilon_i+\varepsilon_j-\varepsilon_a-\varepsilon_b)(\varepsilon_i+\varepsilon_k-\varepsilon_a-\varepsilon_c)}. \tag{8.24}$$

Diagrams (4) and (5) read

$$(4) = \frac{1}{2^3}\sum_{ij\leq F}\sum_{abcd>F}\frac{\langle ij|\hat{v}|cd\rangle\langle cd|\hat{v}|ab\rangle\langle ab|\hat{v}|ij\rangle}{(\varepsilon_i+\varepsilon_j-\varepsilon_c-\varepsilon_d)(\varepsilon_i+\varepsilon_j-\varepsilon_a-\varepsilon_b)}, \tag{8.25}$$

$$(5) = \frac{1}{2^3}\sum_{ijkl\leq F}\sum_{ab>F}\frac{\langle ab|\hat{v}|kl\rangle\langle kl|\hat{v}|ij\rangle\langle ij|\hat{v}|ab\rangle}{(\varepsilon_i+\varepsilon_j-\varepsilon_a-\varepsilon_b)(\varepsilon_k+\varepsilon_l-\varepsilon_a-\varepsilon_b)}, \tag{8.26}$$

where the factor $(1/2)^3$ arises due to three equivalent pairs of lines starting and ending at the same vertex. The last two contributions have an even number of hole lines and closed loops, resulting thus in a positive sign. In Problem 8.9, you are asked to calculate the expressions for diagrams like (8) and (9) in the above figure.

In the expressions for the various diagrams the quantity ε denotes the single-particle energies defined by H_0. The steps leading to the above expressions for the various diagrams are rather straightforward. Though, if we wish to compute the matrix elements for the interaction $\hat{v}$, a serious problem arises. Typically, the matrix elements will contain a term $V(|\mathbf{r}|)$ which represents the interaction potential V between two nucleons, where $\mathbf{r}$ is the internucleon distance. All modern models for V have a strong short-range repulsive core. Hence, matrix elements involving $V(|\mathbf{r}|)$, will result in large (or infinitely large for a potential with a hard core) and repulsive contributions to the ground-state energy. A perturbative expansion in terms of such interaction matrix elements may thus lead to a slowly converging expansion. A standard recipe to circumvent such problems has been to sum up a selected class of correlations. We discuss such possibilities in Sect. 8.6.

8.5.1 Interpreting the Correlation Energy and the Wave Operator

In Sect. 8.4 we showed that we could rewrite the exact state function for the ground state as a linear expansion in terms of all possible Slater determinants. We expanded our exact state function for the ground state as

$$|\Psi_0\rangle = C_0|\Phi_0\rangle + \sum_{ai} C_i^a|\Phi_i^a\rangle + \sum_{abij} C_{ij}^{ab}|\Phi_{ij}^{ab}\rangle + \cdots = (C_0 + \hat{C})|\Phi_0\rangle,$$

where we introduced the so-called correlation operator

$$\hat{C} = \sum_{ai} C_i^a \hat{a}_a^\dagger \hat{a}_i + \sum_{abij} C_{ij}^{ab} \hat{a}_a^\dagger \hat{a}_b^\dagger \hat{a}_j \hat{a}_i + \ldots$$

In a shell-model calculation, the unknown coefficients in $\hat{C}$ are the eigenvectors that result from the diagonalization of the Hamiltonian matrix.

How can we use perturbation theory to determine the same coefficients? Let us study the contributions to second order in the interaction, namely

$$\Delta E^{(2)} = \langle \Phi_0 | \hat{H}_I \frac{\hat{Q}}{W_0 - \hat{H}_0} \hat{H}_I | \Phi_0 \rangle.$$

This contribution will also be discussed in connection with the development of a many-body program for nuclear matter, as well as the simple pairing model of Problem 8.10. The intermediate states given by $\hat{Q}$ can at most be of a $2p-2h$ nature if we have a two-body Hamiltonian. This means that to second order in perturbation theory we can at most have $1p - 1h$ and $2p - 2h$ excitations as intermediate states. When we diagonalize, these contributions are included to infinite order. This means that in order to include such correlations to higher order in the interaction, we need to go to higher-orders in perturbation theory.

If we limit the attention to a Hartree-Fock basis, we have that $\langle \Phi_0 | \hat{H}_I | 2p - 2h \rangle$ is the only contribution since matrix elements involving $\langle \Phi_0 | \hat{H}_I | 1p - 1h \rangle$ are zero and the contribution to the energy from second order in Rayleigh-Schrödinger perturbation theory reduces to

$$\Delta E^{(2)} = \frac{1}{4} \sum_{abij} \langle ij | \hat{v} | ab \rangle \frac{\langle ab | \hat{v} | ij \rangle}{\epsilon_i + \epsilon_j - \epsilon_a - \epsilon_b}.$$

Here we have used the results from Problem 8.8. If we compare this to the correlation energy obtained from full configuration interaction theory with a Hartree-Fock basis, we found that

$$E - E_{\text{Ref}} = \Delta E = \sum_{abij} \langle ij | \hat{v} | ab \rangle C_{ij}^{ab},$$

where the energy E_{Ref} is the reference energy and ΔE defines the so-called correlation energy.

We see that if we set

$$C_{ij}^{ab} = \frac{1}{4}\frac{\langle ab|\hat{v}|ij\rangle}{\epsilon_i + \epsilon_j - \epsilon_a - \epsilon_b},$$

we have a perfect agreement between configuration interaction theory and many-body perturbation theory. However, configuration interaction theory includes $2p-2h$ (and more complicated ones as well) correlations to infinite order. In order to make a meaningful comparison we would at least need to sum such correlations to infinite order in perturbation theory. The last equation serves however as a very useful comparison between configuration interaction theory and many-body perturbation theory. Furthermore, for our nuclear matter studies, one-particle-one-hole intermediate excitations are zero due to the requirement of momentum conservation in infinite systems. These two-particle-two-hole correlations can also be summed to infinite order and a particular class of such excitations are given by two-particle excitations only. These represent in case of nuclear interactions, which are strongly repulsive at short interparticle distances, a physically intuitive way to understand the renormalization of nuclear forces. Such correlations are easily computed by simple matrix inversion techniques and have been widely employed in nuclear many-body theory. Summing up two-particle excitations to infinite order leads to an effective two-body interaction which renormalizes the short-range part of the nuclear interactions.

In summary, many-body perturbation theory introduces order-by-order specific correlations and we can make comparisons with exact calculations like those provided by configuration interaction theory. The advantage of for example Rayleigh-Schrödinger perturbation theory is that at every order in the interaction, we know how to calculate all contributions. The two-body matrix elements can for example be tabulated or computed on the fly. However, many-body perturbation theory suffers from not being variational and there is no guarantee that higher-order terms will improve the order-by-order convergence. It is also extremely tedious to compute terms beyond third order, in particular if one is interested in effective valence space interactions. There are however classes of correlations which can be summed up to infinite order in the interaction. The hope is that such correlations can mitigate specific convergence issues, although there is no a priori guarantee thereof. Examples are the so-called TDA and RPA classes of diagrams [52–54], as well as the resummation of two-particle-two-hole correlations discussed in Chap. 11. If we limit ourselves to the resummation of two-particle correlations only, these lead us to the so-called G-matrix resummation of diagrams, see for example [28]. There are however computationally inexpensive methods which sum larger classes of correlations to infinite order in the interaction. This leads us to Sect. 8.6 and the final many-body method of this chapter, coupled cluster theory.

8.6 Coupled Cluster Theory

Coester and Kümmel [69–71] developed the ideas that led to coupled cluster theory in the late 1950s. The correlated wave function of a many-body system $\mid \Psi\rangle$ can be formulated as an exponential of correlation operators T acting on a reference state $\mid \Phi\rangle$,

$$\mid \Psi\rangle = \exp\left(\hat{T}\right) \mid \Phi\rangle .$$

We will discuss how to define the operators later in this work. This simple ansatz carries enormous power. It leads to a non-perturbative many-body theory that includes summation of ladder diagrams [30], ring diagrams [72], and an infinite-order generalization of many-body perturbation theory [73]. Developments and applications of coupled cluster theory took different routes in chemistry and nuclear physics. In quantum chemistry, coupled cluster developments and applications have proven to be extremely useful, see for example the review by Barrett and Musial as well as the recent textbook by Shavitt and Bartlett [35]. Many previous applications to nuclear physics struggled with the repulsive character of the nuclear forces and limited basis sets used in the computations [71]. Most of these problems have been overcome during the last decade and coupled cluster theory is one of the computational methods of preference for doing nuclear physics, with applications ranging from light nuclei to medium-heavy nuclei, see for example the recent reviews [17, 19, 23, 38].

8.6.1 A Quick Tour of Coupled Cluster Theory

The ansatz for the ground state is given by

$$|\Psi_0\rangle = |\Psi_{CC}\rangle = e^{\hat{T}}|\Phi_0\rangle = \left(\sum_{n=1}^{A}\frac{1}{n!}\hat{T}^n\right)|\Phi_0\rangle ,$$

where A represents the maximum number of particle-hole excitations and $\hat{T}$ is the cluster operator defined as

$$\hat{T} = \hat{T}_1 + \hat{T}_2 + \ldots + \hat{T}_A$$

$$\hat{T}_n = \left(\frac{1}{n!}\right)^2 \sum_{\substack{i_1,i_2,\ldots i_n\\ a_1,a_2,\ldots a_n}} t_{i_1 i_2\ldots i_n}^{a_1 a_2\ldots a_n} a_{a_1}^\dagger a_{a_2}^\dagger \ldots a_{a_n}^\dagger a_{i_n} \ldots a_{i_2} a_{i_1} .$$

The energy is given by

$$E_{\mathrm{CC}} = \langle \Phi_0 | \overline{H} | \Phi_0 \rangle,$$

where $\overline{H}$ is a similarity transformed Hamiltonian

$$\overline{H} = e^{-\hat{T}} \hat{H}_N e^{\hat{T}}$$
$$\hat{H}_N = \hat{H} - \langle \Phi_0 | \hat{H} | \Phi_0 \rangle.$$

The coupled cluster energy is a function of the unknown cluster amplitudes $t_{i_1 i_2 \ldots i_n}^{a_1 a_2 \ldots a_n}$, given by the solutions to the amplitude equations

$$0 = \langle \Phi_{i_1 \ldots i_n}^{a_1 \ldots a_n} | \overline{H} | \Phi_0 \rangle. \tag{8.27}$$

In order to set up the above equations, the similarity transformed Hamiltonian $\overline{H}$ is expanded using the Baker-Campbell-Hausdorff expression,

$$\overline{H} = \hat{H}_N + \left[\hat{H}_N, \hat{T} \right] + \frac{1}{2} \left[\left[\hat{H}_N, \hat{T} \right], \hat{T} \right] + \ldots + \frac{1}{n!} \left[\ldots \left[\hat{H}_N, \hat{T} \right], \ldots \hat{T} \right] + \ldots \tag{8.28}$$

and simplified using the connected cluster theorem [35]

$$\overline{H} = \hat{H}_N + \left(\hat{H}_N \hat{T} \right)_c + \frac{1}{2} \left(\hat{H}_N \hat{T}^2 \right)_c + \cdots + \frac{1}{n!} \left(\hat{H}_N \hat{T}^n \right)_c + \ldots$$

We will discuss parts of the derivation below. For the full derivation of these expressions, see for example [35].

A much used approximation is to truncate the cluster operator $\hat{T}$ at the $n = 2$ level. This defines the so-called singles and doubles approximation to the coupled cluster state function, normally shortened to CCSD. The coupled cluster wavefunction is now given by

$$|\Psi_{CC}\rangle = e^{\hat{T}_1 + \hat{T}_2} |\Phi_0\rangle$$

where

$$\hat{T}_1 = \sum_{ia} t_i^a a_a^\dagger a_i$$
$$\hat{T}_2 = \frac{1}{4} \sum_{ijab} t_{ij}^{ab} a_a^\dagger a_b^\dagger a_j a_i.$$

The amplitudes t play a role similar to the coefficients C in the shell-model calculations. They are obtained by solving a set of non-linear equations similar to those discussed above in connection with the configuration interaction theory discussion, see Eqs. (8.15) and (8.16).

In our configuration interaction theory discussion the correlation energy is defined as (with a two-body Hamiltonian)

$$\Delta E = \sum_{ai} \langle i|\hat{f}|a\rangle C_i^a + \sum_{abij} \langle ij|\hat{v}|ab\rangle C_{ij}^{ab}.$$

We can obtain a similar expression for the correlation energy using coupled cluster theory. Using Eq. (8.28) we can write the expression for the coupled cluster ground state energy as an infinite sum over nested commutators

$$\begin{aligned} E_{CC} = \langle \Phi_0| \Big(&\hat{H}_N + \left[\hat{H}_N, \hat{T}\right] + \frac{1}{2}\left[\left[\hat{H}_N, \hat{T}\right], \hat{T}\right] \\ &+ \frac{1}{3!}\left[\left[\left[\hat{H}_N, \hat{T}\right], \hat{T}\right], \hat{T}\right] \\ &+ \frac{1}{4!}\left[\left[\left[\left[\hat{H}_N, \hat{T}\right], \hat{T}\right], \hat{T}\right], \hat{T}\right] + \ldots \Big) |\Phi_0\rangle . \end{aligned}$$

One can show that this infinite series truncates naturally at a given order of nested commutators [35]. Let us demonstrate briefly how we can construct the expressions for the correlation energy by approximating $\hat{T}$ at the CCSD level, that is $\hat{T} \approx \hat{T}_1 + \hat{T}_2$. The first term is zero by construction

$$\langle \Phi_0| \hat{H}_N |\Phi_0\rangle = 0.$$

The second term can be split into the following contributions

$$\langle \Phi_0| \left[\hat{H}_N, \hat{T}\right] |\Phi_0\rangle = \langle \Phi_0| \left(\left[\hat{F}_N, \hat{T}_1\right] + \left[\hat{F}_N, \hat{T}_2\right] + \left[\hat{V}_N, \hat{T}_1\right] + \left[\hat{V}_N, \hat{T}_2\right] \right) |\Phi_0\rangle .$$

Let us start with $\left[\hat{F}_N, \hat{T}_1\right]$, where the one-body operator $\hat{F}_N$ is defined in Eq. (8.2). In the equations below we employ the shorthand $f_q^p = \langle p|\hat{f}|q\rangle$. We write out the commutator as

$$\begin{aligned} \left[\hat{F}_N, \hat{T}_1\right] &= \sum_{pqia} \left(f_q^p \left\{a_p^\dagger a_q\right\} t_i^a \left\{a_a^\dagger a_i\right\} - t_i^a \left\{a_a^\dagger a_i\right\} f_q^p \left\{a_p^\dagger a_q\right\} \right) \\ &= \sum_{pqia} f_q^p t_i^a \left(\left\{a_p^\dagger a_q\right\} \left\{a_a^\dagger a_i\right\} - \left\{a_a^\dagger a_i\right\} \left\{a_p^\dagger a_q\right\} \right) . \end{aligned}$$

We have kept here the curly brackets that indicate that the chains of operators are normal ordered with respect to the new reference state. If we consider the second set of operators and rewrite them with curly brackets (bringing back the normal ordering) we have

$$\left\{a_a^\dagger a_i\right\}\left\{a_p^\dagger a_q\right\} = \left\{a_a^\dagger a_i a_p^\dagger a_q\right\} = \left\{a_p^\dagger a_q a_a^\dagger a_i\right\}$$

$$\begin{aligned}
\left\{a_p^\dagger a_q\right\}\left\{a_a^\dagger a_i\right\} &= \left\{a_p^\dagger a_q a_a^\dagger a_i\right\} \\
&+ \left\{\overline{a_p^\dagger a_q a_a^\dagger a_i}\right\} + \left\{a_p^\dagger \overline{a_q a_a^\dagger} a_i\right\} \\
&+ \left\{\overline{a_p^\dagger \overline{a_q a_a^\dagger} a_i}\right\} \\
&= \left\{a_p^\dagger a_q a_a^\dagger a_i\right\} + \delta_{qa}\left\{a_p^\dagger a_i\right\} + \delta_{pi}\left\{a_q a_a^\dagger\right\} + \delta_{qa}\delta_{pi}.
\end{aligned}$$

We can then rewrite the two sets of operators as

$$\left\{a_p^\dagger a_q\right\}\left\{a_a^\dagger a_i\right\} - \left\{a_a^\dagger a_i\right\}\left\{a_p^\dagger a_q\right\} = \delta_{qa}\left\{a_p^\dagger a_i\right\} + \delta_{pi}\left\{a_q a_a^\dagger\right\} + \delta_{qa}\delta_{pi}.$$

Inserted into the original expression, we arrive at the explicit value of the commutator

$$\left[\hat{F}_N, \hat{T}_1\right] = \sum_{pai} f_a^p t_i^a \left\{a_p^\dagger a_i\right\} + \sum_{qai} f_q^i t_i^a \left\{a_q a_a^\dagger\right\} + \sum_{ai} f_a^i t_i^a.$$

We are now ready to compute the expectation value with respect to our reference state. Since the two first terms require the ground state linking to a one-particle-one-hole state, the first two terms are zero and we are left with

$$\langle\Phi_0|\left[\hat{F}_N, \hat{T}_1\right]|\Phi_0\rangle = \sum_{ai} f_a^i t_i^a. \tag{8.29}$$

The two first terms will however contribute to the calculation of the Hamiltonian matrix element which connects the ground state and a one-particle-one-hole excitation.

Let us next look at the term $\left[\hat{F}_N, \hat{T}_2\right]$. We have

$$\begin{aligned}
\left[\hat{F}_N, \hat{T}_2\right] &= \left[\sum_{pq} f_q^p \left\{a_p^\dagger a_q\right\}, \frac{1}{4}\sum_{ijab} t_{ij}^{ab}\left\{a_a^\dagger a_b^\dagger a_j a_i\right\}\right] \\
&= \frac{1}{4}\sum_{\substack{pq\\ ijab}} f_q^p t_{ij}^{ab}\left(\left\{a_p^\dagger a_q\right\}\left\{a_a^\dagger a_b^\dagger a_j a_i\right\} - \left\{a_a^\dagger a_b^\dagger a_j a_i\right\}\left\{a_p^\dagger a_q\right\}\right).
\end{aligned}$$

The last set of operators can be rewritten as

$$\left\{a_a^\dagger a_b^\dagger a_j a_i\right\}\left\{a_p^\dagger a_q\right\} = \left\{a_a^\dagger a_b^\dagger a_j a_i a_p^\dagger a_q\right\}$$

$$= \left\{a_p^\dagger a_q a_a^\dagger a_b^\dagger a_j a_i\right\}$$

$$\begin{aligned}
\left\{a_p^\dagger a_q\right\} \left\{a_a^\dagger a_b^\dagger a_j a_i\right\} &= \left\{a_p^\dagger a_q a_a^\dagger a_b^\dagger a_j a_i\right\} + \left\{\overline{a_p^\dagger a_q a_a^\dagger a_b^\dagger a_j} a_i\right\} + \left\{\overline{a_p^\dagger a_q a_a^\dagger a_b^\dagger a_j a_i}\right\} \\
&+ \left\{a_p^\dagger \overline{a_q a_a^\dagger} a_b^\dagger a_j a_i\right\} + \left\{a_p^\dagger \overline{a_q a_a^\dagger a_b^\dagger} a_j a_i\right\} + \left\{\overline{a_p^\dagger \overline{a_q a_a^\dagger} a_b^\dagger a_j} a_i\right\} \\
&+ \left\{\overline{a_p^\dagger \overline{a_q a_a^\dagger} a_b^\dagger a_j a_i}\right\} + \left\{\overline{a_p^\dagger \overline{a_q a_a^\dagger a_b^\dagger} a_j} a_i\right\} + \left\{\overline{a_p^\dagger \overline{a_q a_a^\dagger a_b^\dagger} a_j a_i}\right\} \\
&= \left\{a_p^\dagger a_q a_a^\dagger a_b^\dagger a_j a_i\right\} - \delta_{pj} \left\{a_q a_a^\dagger a_b^\dagger a_i\right\} + \delta_{pi} \left\{a_q a_a^\dagger a_b^\dagger a_j\right\} \\
&+ \delta_{qa} \left\{a_p^\dagger a_b^\dagger a_j a_i\right\} - \delta_{qb} \left\{a_p^\dagger a_a^\dagger a_j a_i\right\} - \delta_{pj}\delta_{qa} \left\{a_b^\dagger a_i\right\} \\
&+ \delta_{pi}\delta_{qa} \left\{a_b^\dagger a_j\right\} + \delta_{pj}\delta_{qb} \left\{a_a^\dagger a_i\right\} - \delta_{pi}\delta_{qb} \left\{a_a^\dagger a_j\right\}.
\end{aligned}$$

We can then rewrite the two sets of operators as

$$\begin{aligned}
&\left(\left\{a_p^\dagger a_q\right\} \left\{a_a^\dagger a_b^\dagger a_j a_i\right\} - \left\{a_a^\dagger a_b^\dagger a_j a_i\right\} \left\{a_p^\dagger a_q\right\} \right) \\
&= -\delta_{pj} \left\{a_q a_a^\dagger a_b^\dagger a_i\right\} + \delta_{pi} \left\{a_q a_a^\dagger a_b^\dagger a_j\right\} + \delta_{qa} \left\{a_p^\dagger a_b^\dagger a_j a_i\right\} \\
&- \delta_{qb} \left\{a_p^\dagger a_a^\dagger a_j a_i\right\} - \delta_{pj}\delta_{qa} \left\{a_b^\dagger a_i\right\} + \delta_{pi}\delta_{qa} \left\{a_b^\dagger a_j\right\} + \delta_{pj}\delta_{qb} \left\{a_a^\dagger a_i\right\} \\
&- \delta_{pi}\delta_{qb} \left\{a_a^\dagger a_j\right\},
\end{aligned}$$

which, when inserted into the original expression gives us

$$\begin{aligned}
\left[\hat{F}_N, \hat{T}_2\right] &= \frac{1}{4} \sum_{\substack{pq \\ abij}} f_q^p t_{ij}^{ab} \Big(- \delta_{pj} \left\{a_q a_a^\dagger a_b^\dagger a_i\right\} + \delta_{pi} \left\{a_q a_a^\dagger a_b^\dagger a_j\right\} \\
&+ \delta_{qa} \left\{a_p^\dagger a_b^\dagger a_j a_i\right\} - \delta_{qb} \left\{a_p^\dagger a_a^\dagger a_j a_i\right\} - \delta_{pj}\delta_{qa} \left\{a_b^\dagger a_i\right\} \\
&+ \delta_{pi}\delta_{qa} \left\{a_b^\dagger a_j\right\} + \delta_{pj}\delta_{qb} \left\{a_a^\dagger a_i\right\} - \delta_{pi}\delta_{qb} \left\{a_a^\dagger a_j\right\} \Big).
\end{aligned}$$

After renaming indices and changing the order of operators, we arrive at the explicit expression

$$\left[\hat{F}_N, \hat{T}_2\right] = \frac{1}{2}\sum_{qijab} f^i_q t^{ab}_{ij} \left\{a_q a^\dagger_a a^\dagger_b a_j\right\} + \frac{1}{2}\sum_{pijab} f^p_a t^{ab}_{ij} \left\{a^\dagger_p a^\dagger_b a_j a_i\right\} + \sum_{ijab} f^i_a t^{ab}_{ij} \left\{a^\dagger_b a_j\right\}.$$

In this case we have two sets of two-particle-two-hole operators and one-particle-one-hole operators and all these terms result in zero expectation values. However, these terms are important for the amplitude equations. In a similar way we can compute the terms involving the interaction $\hat{V}_N$. We obtain then

$$\begin{aligned}
\langle \Phi_0 | \left[\hat{V}_N, \hat{T}_1\right] | \Phi_0 \rangle &= \langle \Phi_0 | \left[\frac{1}{4}\sum_{pqrs} \langle pq | \hat{v} | rs \rangle \left\{a^\dagger_p a^\dagger_q a_s a_r\right\}, \sum_{ia} t^a_i \left\{a^\dagger_a a_i\right\}\right] | \Phi_0 \rangle \\
&= \frac{1}{4}\sum_{\substack{pqr\\sia}} \langle pq | \, | rs \rangle \, t^a_i \, \langle \Phi_0 | \left[\left\{a^\dagger_p a^\dagger_q a_s a_r\right\}, \left\{a^\dagger_a a_i\right\}\right] | \Phi_0 \rangle \\
&= 0,
\end{aligned}$$

and

$$\begin{aligned}
&\langle \Phi_0 | \left[\hat{V}_N, \hat{T}_2\right] | \Phi_0 \rangle \\
&= \langle \Phi_0 | \left[\frac{1}{4}\sum_{pqrs} \langle pq | \hat{v} | rs \rangle \left\{a^\dagger_p a^\dagger_q a_s a_r\right\}, \frac{1}{4}\sum_{ijab} t^{ab}_{ij} \left\{a^\dagger_a a^\dagger_b a_j a_i\right\}\right] | \Phi_0 \rangle \\
&= \frac{1}{16}\sum_{\substack{pqr\\sijab}} \langle pq | \hat{v} | rs \rangle \, t^{ab}_{ij} \, \langle \Phi_0 | \left[\left\{a^\dagger_p a^\dagger_q a_s a_r\right\}, \left\{a^\dagger_a a^\dagger_b a_j a_i\right\}\right] | \Phi_0 \rangle \\
&= \frac{1}{16}\sum_{\substack{pqr\\sijab}} \langle pq | \hat{v} | rs \rangle \, t^{ab}_{ij} \, \langle \Phi_0 | \Big(\left\{a^\dagger_p a^\dagger_q a_s a_r a^\dagger_a a^\dagger_b a_j a_i\right\} + \left\{a^\dagger_p a^\dagger_q a_s a_r a^\dagger_a a^\dagger_b a_j a_i\right\} \\
&\quad \left\{a^\dagger_p a^\dagger_q a_s a_r a^\dagger_a a^\dagger_b a_j a_i\right\} + \left\{a^\dagger_p a^\dagger_q a_s a_r a^\dagger_a a^\dagger_b a_j a_i\right\} \Big) | \Phi_0 \rangle \\
&= \frac{1}{4}\sum_{ijab} \langle ij | \hat{v} | ab \rangle \, t^{ab}_{ij}.
\end{aligned}$$

The final contribution to the correlation energy comes from the non-linear terms with the amplitudes squared. The contribution from $\hat{T}^2$ is given by

$$
\begin{aligned}
&\langle \Phi_0| \frac{1}{2}\left(\hat{V}_N \hat{T}_1^2\right)|\Phi_0\rangle \\
&= \frac{1}{8}\sum_{pqrs}\sum_{ijab}\langle pq|\hat{v}|rs\rangle t_i^a t_j^b \langle\Phi_0|\left(\{a_p^\dagger a_q^\dagger a_s a_r\}\{a_a^\dagger a_i\}\left\{a_b^\dagger a_j\right\}\right)_c|\Phi_0\rangle \\
&= \frac{1}{8}\sum_{pqrs}\sum_{ijab}\langle pq|\hat{v}|rs\rangle t_i^a t_j^b \langle\Phi_0| \\
&\left(\left\{a_p^\dagger a_q^\dagger a_s a_r a_a^\dagger a_i a_b^\dagger a_j\right\} + \left\{a_p^\dagger a_q^\dagger a_s a_r a_a^\dagger a_i a_b^\dagger a_j\right\} + \left\{a_p^\dagger a_q^\dagger a_s a_r a_a^\dagger a_i a_b^\dagger a_j\right\}\right. \\
&\left. + \left\{a_p^\dagger a_q^\dagger a_s a_r a_a^\dagger a_i a_b^\dagger a_j\right\}\right)|\Phi_0\rangle \\
&= \frac{1}{2}\sum_{ijab}\langle ij|\hat{v}|ab\rangle t_i^a t_j^b .
\end{aligned}
$$

Collecting all terms we have the final expression for the correlation energy with a two-body interaction given by

$$
\Delta E = \sum_{ai}\langle i|\hat{f}|a\rangle t_i^a + \frac{1}{2}\sum_{ijab}\langle ij|\hat{v}|ab\rangle t_i^a t_j^b + \frac{1}{4}\sum_{ijab}\langle ij|\hat{v}|ab\rangle t_{ij}^{ab} . \tag{8.30}
$$

We leave it as a challenge to the reader to derive the corresponding equations for the Hamiltonian matrix elements of Eq. (8.27).

There are several interesting features with the coupled cluster equations. With a truncation like CCSD or even with the inclusion of triples (CCSDT), we can include to infinite order correlations based on one-particle-one-hole and two-particle-two-hole contributions. We can include a large basis of single-particle states, normally not possible in standard FCI calculations. Typical FCI calculations for light nuclei $A \leq 16$ can be performed in at most some few harmonic oscillator shells. For heavier nuclei, at most two major shells can be included due to too large dimensionalities. However, coupled cluster theory is non-variational and if we want to find properties of excited states, additional calculations via for example equation of motion methods are needed [17, 35]. If correlations are strong, a single-reference ansatz may not be the best starting point and a multi-reference approximation is needed [74]. Furthermore, we cannot quantify properly the error we make when truncations are made in the cluster operator.

8.6.2 The CCD Approximation

We will now approximate the cluster operator $\hat{T}$ to include only $2p-2h$ correlations. This leads to the so-called CCD approximation, that is

$$\hat{T} \approx \hat{T}_2 = \frac{1}{4}\sum_{abij} t_{ij}^{ab} a_a^\dagger a_b^\dagger a_j a_i,$$

meaning that we have

$$|\Psi_0\rangle \approx |\Psi_{CCD}\rangle = \exp\left(\hat{T}_2\right)|\Phi_0\rangle.$$

Inserting these equations in the expression for the computation of the energy we have, with a Hamiltonian defined with respect to a general reference vacuum

$$\hat{H} = \hat{H}_N + E_{\mathrm{ref}},$$

with

$$\hat{H}_N = \sum_{pq}\langle p|\hat{f}|q\rangle a_p^\dagger a_q + \frac{1}{4}\sum_{pqrs}\langle pq|\hat{v}|rs\rangle a_p^\dagger a_q^\dagger a_s a_r,$$

we obtain that the energy can be written as

$$\langle\Phi_0|\exp\left(-\hat{T}_2\right)\hat{H}_N\exp\left(\hat{T}_2\right)|\Phi_0\rangle = \langle\Phi_0|\hat{H}_N(1+\hat{T}_2)|\Phi_0\rangle = E_{CCD}.$$

This quantity becomes

$$E_{CCD} = E_{\mathrm{ref}} + \frac{1}{4}\sum_{abij}\langle ij|\hat{v}|ab\rangle t_{ij}^{ab},$$

where the latter is the correlation energy from this level of approximation of coupled cluster theory. Similarly, the expression for the amplitudes reads (see Problem 8.13)

$$\langle\Phi_{ij}^{ab}|\exp\left(-\hat{T}_2\right)\hat{H}_N\exp\left(\hat{T}_2\right)|\Phi_0\rangle = 0.$$

These equations can be reduced to (after several applications of Wick's theorem), for all $i > j$ and all $a > b$,

$$
\begin{aligned}
0 = \langle ab|\hat{v}|ij\rangle + \left(\epsilon_a + \epsilon_b - \epsilon_i - \epsilon_j\right) t_{ij}^{ab} + \frac{1}{2}\sum_{cd}\langle ab|\hat{v}|cd\rangle t_{ij}^{cd} + \frac{1}{2}\sum_{kl}\langle kl|\hat{v}|ij\rangle t_{kl}^{ab} \\
+ \hat{P}(ij|ab)\sum_{kc}\langle kb|\hat{v}|cj\rangle t_{ik}^{ac} + \frac{1}{4}\sum_{klcd}\langle kl|\hat{v}|cd\rangle t_{ij}^{cd} t_{kl}^{ab} + \hat{P}(ij)\sum_{klcd}\langle kl|\hat{v}|cd\rangle t_{ik}^{ac} t_{jl}^{bd} \\
- \frac{1}{2}\hat{P}(ij)\sum_{klcd}\langle kl|\hat{v}|cd\rangle t_{ik}^{dc} t_{lj}^{ab} - \frac{1}{2}\hat{P}(ab)\sum_{klcd}\langle kl|\hat{v}|cd\rangle t_{lk}^{ac} t_{ij}^{db},
\end{aligned}
\tag{8.31}
$$

where we have defined

$$\hat{P}\,(ab) = 1 - \hat{P}_{ab},$$

where $\hat{P}_{ab}$ interchanges two particles occupying the quantum numbers a and b. The operator $\hat{P}(ij|ab)$ is defined as

$$\hat{P}(ij|ab) = (1 - \hat{P}_{ij})(1 - \hat{P}_{ab}).$$

The single-particle energies ϵ_p are normally taken to be Hartree-Fock single-particle energies. Recall also that the unknown amplitudes t_{ij}^{ab} represent anti-symmetrized matrix elements, meaning that they obey the same symmetry relations as the two-body interaction, that is

$$t_{ij}^{ab} = -t_{ji}^{ab} = -t_{ij}^{ba} = t_{ji}^{ba}.$$

The two-body matrix elements are also anti-symmetrized, meaning that

$$\langle ab|\hat{v}|ij\rangle = -\langle ab|\hat{v}|ji\rangle = -\langle ba|\hat{v}|ij\rangle = \langle ba|\hat{v}|ji\rangle.$$

The non-linear equations for the unknown amplitudes t_{ij}^{ab} are solved iteratively. We discuss the implementation of these equations below.

8.6.3 Approximations to the Full CCD Equations

It is useful to make approximations to the equations for the amplitudes. These serve as important benchmarks when we are to develop a many-body code. The standard method for solving these equations is to set up an iterative scheme where method's like Newton's method or similar root searching methods are used to find the amplitudes, see for example [75].

Iterative solvers need a guess for the amplitudes. A good starting point is to use the correlated wave operator from perturbation theory to first order in the interaction. This means that we define the zeroth approximation to the amplitudes as

$$(t_{ij}^{ab})^{(0)} = \frac{\langle ab|\hat{v}|ij\rangle}{\left(\epsilon_i + \epsilon_j - \epsilon_a - \epsilon_b\right)},$$

leading to our first approximation for the correlation energy at the CCD level to be equal to second-order perturbation theory without $1p - 1h$ excitations, namely

$$\Delta E_{\mathrm{CCD}}^{(0)} = \frac{1}{4}\sum_{abij}\langle ij|\hat{v}|ab\rangle (t_{ij}^{ab})^{(0)} = \frac{1}{4}\sum_{abij}\frac{\langle ij|\hat{v}|ab\rangle\langle ab|\hat{v}|ij\rangle}{\left(\epsilon_i + \epsilon_j - \epsilon_a - \epsilon_b\right)}.$$

With this starting point, we are now ready to solve Eq. (8.31) iteratively. Before we attack the full equations, it is however instructive to study a truncated version of the equations. We will first study the following approximation where we take away all terms except the linear terms that involve the single-particle energies and the two-particle intermediate excitations, that is

$$0 = \langle ab|\hat{v}|ij\rangle + \left(\epsilon_a + \epsilon_b - \epsilon_i - \epsilon_j\right) t_{ij}^{ab} + \frac{1}{2}\sum_{cd}\langle ab|\hat{v}|cd\rangle t_{ij}^{cd}. \tag{8.32}$$

In the above and following equations we have dropped the subscript which indicates the number of iterations. Setting the single-particle energies for the hole states equal to an energy variable $\omega = \epsilon_i + \epsilon_j$, Eq. (8.32) reduces to the well-known equations for the so-called G-matrix, widely used in infinite matter and finite nuclei studies, see for example [8, 28]. The equation can then be reordered and solved by matrix inversion. To see this let us define the following quantity

$$\tau_{ij}^{ab} = (\omega - \epsilon_a - \epsilon_b)\, t_{ij}^{ab},$$

and inserting

$$1 = \frac{(\omega - \epsilon_c - \epsilon_d)}{(\omega - \epsilon_c - \epsilon_d)},$$

in the intermediate sums over cd in Eq. (8.32), we can rewrite the latter equation as

$$\tau_{ij}^{ab}(\omega) = \langle ab|\hat{v}|ij\rangle + \frac{1}{2}\sum_{cd}\langle ab|\hat{v}|cd\rangle \frac{1}{\omega - \epsilon_c - \epsilon_d}\tau_{ij}^{cd}(\omega),$$

where we have inserted an explicit energy dependence via the parameter ω. This equation, transforming a two-particle configuration into a single index, can be rewritten as a matrix inversion problem. Alternatively, the same equation can be solved by iteration. Solving the equations for a fixed energy ω allows us to compare directly with results from Green's function theory when only two-particle intermediate states are included.

To solve Eq. (8.32), we start with a guess for the unknown amplitudes, normally using the wave operator defined by first order in perturbation theory, leading to a zeroth-order approximation for the correlation energy given by second-order perturbation theory. A simple approach to the solution of Eq. (8.32), is thus to

1. Start with a guess for the amplitudes and compute the zeroth approximation to the correlation energy.
2. Use the ansatz for the amplitudes to solve Eq. (8.32) via for example your root-finding method of choice (Newton's method or modifications thereof can be used) and continue these iterations till the correlation energy does not change more than a prefixed quantity λ; $\Delta E_{\mathrm{CCD}}^{(i)} - \Delta E_{\mathrm{CCD}}^{(i-1)} \leq \lambda$.
3. It is common during the iterations to scale the amplitudes with a parameter α, with $\alpha \in (0, 1]$ as $t^{(i)} = \alpha t^{(i)} + (1-\alpha)t^{(i-1)}$.

The next approximation is to include the two-hole term in Eq. (8.31), a term which allows us to make a link with Green's function theory with two-particle and two-hole correlations discussed in Chap. 11. This means that we solve

$$0 = \langle ab|\hat{v}|ij\rangle + \left(\epsilon_a + \epsilon_b - \epsilon_i - \epsilon_j\right) t_{ij}^{ab} + \frac{1}{2}\sum_{cd}\langle ab|\hat{v}|cd\rangle t_{ij}^{cd} + \frac{1}{2}\sum_{kl}\langle kl|\hat{v}|ij\rangle t_{kl}^{ab}. \tag{8.33}$$

This equation is solved the same way as we would do for Eq. (8.32). The final step is then to include all terms in Eq. (8.31).

8.7 Developing a Numerical Project

A successful numerical project relies on us having expertise in several scientific and engineering disciplines. We need a thorough understanding of the relevant scientific domain to ask the right questions and interpret the results, but the tools we use require a proficiency in mathematics to develop models and work out analytical results, in numerics to choose the correct algorithms, in computer science to understand what can go wrong with our algorithms when the problem is discretized and solved on a digital computer, and in software engineering to develop and maintain a computer program that solves our problem.

Independent of your scientific background, you are probably also educated in mathematics and numerics. Unfortunately, the computer science and software engineering aspects of computing are often neglected and thought of as skills you pick up along the way. This is a problem for many reasons. First, running a numerical project is very similar to running a physical experiment. The codes we develop are the blueprints the compiler uses to build the experiment from the components of the computer. It is unthinkable to publish results from a physical experiment without a thorough understanding of the experimental equipment. Second, the blueprints are not only used to tell the compiler what to build, but also by humans to understand what is being built, how to fix it if something goes wrong, and how to improve it. If the blueprints are not properly written and readable, human understanding is lost. Last, components of an experiment are always tested individually to establish tolerances and that they work according to specification. In software engineering, this corresponds to writing testable code where you can be confident of the quality of each piece. These are skills many writers of scientific software never learn and as a consequence many numerical experiments are not properly understood and are never independently verified.

In this section we will focus on some key tools and strategies that we feel are important for developing and running a numerical experiment. Our main concerns are that our results can be validated, independently verified, and run efficiently. In addition, we will discuss tools that make the whole process somewhat easier. We will cover testing, tracking changes with version control software, public code repositories, and touch upon simple profiling tools to guide the optimization process. Finally we will present a numerical project where we have developed a code to calculate properties of nuclear matter using coupled-cluster theory. Here, we will make extensive use of the simple pairing model of Problem 8.10. This model allows for benchmarks against exact results. In addition, it provides analytical answers to several approximations, from perturbation theory to specific terms in the solution of the coupled cluster equations, the in-medium similarity renormalization group approach of Chap. 10 and the Green's function approach of Chap. 11.

8.7.1 *Validation and Verification*

The single most important thing in a numerical experiment is to get the correct answer. A close second is to be confident that the answer is correct and why. A lucky coincidence must be distinguishable from a consistently correct result. The only way to do this is to validate the code by writing and running tests—lots of tests. Ideally, every aspect of a code should be tested and it should be possible to run the tests automatically. As most of you have probably experienced, it's very easy to introduce errors into a project. And very often, symptoms of the errors are not visible where the errors have been introduced. By having a large set of automated tests and running them often, symptoms of errors can be discovered quickly and the errors tracked down while recent changes are fresh in memory.

As scientists we are trained to validate our methods and findings. By applying the same rigorous process to our software, we can achieve the same level of confidence in our code as we have for the rest of our work. We advocate testing at three distinct levels. Let's start with discussing validation tests, as this is the type of test you are probably most familiar with. In a validation test, your application is run as in production mode and the test fails if it cannot reproduce a known result. The known result could be a published benchmark, a simplified model where analytical results are available, an approximate result from a different method, or even an earlier result from the same code. We will discuss this type of testing further in Sect. 8.7.4.

Analogous to testing individual components in a physical experiment, is a type of test called a unit test. This is a very fine-grained test that will typically only test a class or a procedure, or even a small part of your code at a time. This is where you test that a data structure has been correctly filled, that an algorithm works appropriately, that a file has been read correctly, and basically every other component test you can think of. It does take a little more work to setup as testing needs to be done outside of your normal program flow. Typically this involves writing different executables that create the necessary dependencies before testing a component. The advantages of writing unit tests are many. First, because you know that the individual pieces of your code work independently, you will achieve a higher degree of confidence in your results. Second, you will develop a programming style that favors highly decoupled units because such units are easier to test. This allows talking about the code at a higher level of abstraction, which helps understanding. Last, your tests become the documentation of how your code is supposed to be used. This might not seem important while your are actively working on a project, but it will be invaluable down the line when you want to add new features. Also, when you share your code as part of the scientific process, these tests will be the way your peers will start to understand your work. This means that your final production code will also include various tests.

While validation tests test your code at the coarsest level, and unit tests test your code at the finest level, integration tests test how your components work together. If, for example, your program solves differential equations as parts of a larger problem, the components that make up your differential-equation solver can be tested alone. If your solver can solve a set of representative problems that either have analytical solutions or can be worked out using some other tool, you can be more confident that it will work on your specific problem. Moreover, writing integration tests pushes you to develop more general components. Instead of writing a routine that only solves the differential equations you need, you write a solver that can solve many different types of differential equations. This allows your components to be reused in other projects and by other people.

To many, this rigorous approach to testing software might seem like a waste of time. Our view is that testing software is crucial to the scientific process and we should strive to apply the same level of rigour to our software as we do to every other aspect of our work. On a more pragmatic level, you can either spend your time writing tests and make sure your components work, or you can spend your time debugging when something goes wrong and worry that your results are not valid. We definitely prefer, from own and other people's experience, the first approach.

8.7.2 Tracking Changes

If you're not using a tool to track the changes you make to your code, now is the time to start. There are several tools available, but the authors are using git (https://git-scm.com/), an open-source version-control system that can run on Linux, OSX, and Windows. By tracking changes, it is easier to correct a mistake when it inevitably creeps into the code. It is possible to go back to a previously validated version and by using branches, you can work on different versions of the code simultaneously. For example, you can create a production branch where everything is validated and ready to run, and you can create a development branch to implement new features. There are also code repositories where you can store a copy of your code for free, without worrying about things getting lost. The source codes discussed in this book are hosted on for example GitHub (https://github.com/), which uses git to track all changes to the code. By using a service like this, it is easier to synchronize code between multiple machines. Multiple developers can work on the same code at the same time and share changes without worrying about losing contributions. It can also become the official public repository of your software to enable your peers to verify your work. The software discussed in this chapter is available from our GitHub repository https://github.com/ManyBodyPhysics/LectureNotesPhysics/tree/master/Programs/Chapter8-programs/.

8.7.3 Profile-Guided Optimization

The aim of this subsection is to discuss in more detail how we can make the computations discussed in connection with Eqs. (8.52) and (8.53) more efficient using physical constraints, algorithm improvements, and parallel processing. For pedagogical reasons, we will use the MBPT parts of the program due to their simplicity while still containing the important elements of a larger, more complicated CCD calculation. The codes can be found at the github link https://github.com/ManyBodyPhysics/LectureNotesPhysics/tree/master/Programs/Chapter8-programs/cpp/MBPT/src. We will demonstrate the use of a simple profiler to help guide our development efforts. Our starting points are naive implementations of many-body perturbation theory to second (MBPT2) and third order (MBPT3) in the interaction. For reference, we calculate properties of nuclear matter and construct our Hamiltonian in a free-wave basis using the Minnesota [57] potential discussed in Sect. 8.2.3. As the model is not as important as the performance in this section, we postpone a discussion of the model to Sect. 8.7.4.

Code Listing 8.1 Trivial implementation of a MBPT2 diagram

```
  double energy = 0.0;
  for(int i = 0; i < modelspace.indhol; ++i){
    for(int j = 0; j < modelspace.indhol; ++j){
      if(i == j){ continue; }
      for(int a = modelspace.indhol; a < modelspace.indtot; ++a){
        for(int b = modelspace.indhol; b < modelspace.indtot; ++b)
            {
          if(a == b){ continue; }
          energy0 = potential->get_element(modelspace.qnums, i, j,
                a, b);
          energy0 *= energy0;
          energy0 /= (modelspace.qnums[i].energy + modelspace.
               qnums[j].energy -
                       modelspace.qnums[a].energy - modelspace.
                            qnums[b].energy);
          energy += energy0;
        }
      }
    }
  }
  energy *= 0.25;
  return energy;
}
```

Listing 8.1 shows a possible early implementation to solve Eq. (8.52) from MBPT2. This function has a loop over all single-particle indices and calls the V_Minnesota function via the get_element function to calculate the two-body interaction for each set of indices. The energy denominators are calculated from the single-particle energies stored in the modelspace structure and partial results are accumulated into the energy variable. This function represents a straightforward implementation of MBPT2. We normally recommend, when developing a code, to write the first implementation in a way which is as close as possible to the mathematical expressions, in this particular case Eq. (8.52).

Table 8.3 shows the total execution time for this application for different model spaces (defined by the number of single-particle states) and number of particles on a local workstation. Your runtimes will be different. Our goals are converged calculations of pure neutron matter as well as nuclear matter, where the number of states and the number of protons and neutrons goes to infinity. It suffices to say that we cannot reach our goals with this code.

We want to decrease the run time of this application, but it can be difficult to decide where we should spend our time improving this code. Our first approach is to observe what goes on inside the program. For that we will use one the simplest possible profiling tools called gprof (https://sourceware.org/binutils/docs/gprof/). Alternatively, software like Valgrind is also highly recommended http://valgrind.org. If you are using integrated development environments (IDEs) like Qt https://www.qt.io/, performance and debugging tools are integrated with the IDE.

Table 8.3 Total runtime for the MBPT2 implementation in Listing 8.1 for different model spaces and particle numbers

Number of states	Number of protons	Number of neutrons	Runtime (s) Listing 8.1
342	0	2	< 0.01
	0	14	0.15
	0	38	1.00
	0	54	1.81
684	2	2	0.88
	14	14	2.43
	38	38	15.7
	54	54	28.2
1598	0	2	0.04
	0	14	3.32
	0	38	25.0
	0	54	58.9
3196	2	2	0.88
	14	14	54.3
	38	38	399
	54	54	797

Table 8.4 Flat profile for the MBPT2 implementation in Listing 8.1 using 1598 states calculating pure neutron matter with 54 neutrons

```
Flat profile:

Each sample counts as 0.01 seconds.
  %    cumulative   self              self     total
 time    seconds   seconds    calls  ms/call  ms/call  name
 58.01     26.26    26.26 2523524846     0.00     0.00  V_Minnesota(...)
 39.64     44.21    17.95                              mbpt2V00::getEnergy()
  2.35     45.27     1.07                              spinExchangeTerm(...)
```

To use gprof the code must first be compiled and linked with the -pg flag. This flag enables the collection of runtime information so that a call graph and a profile can be constructed when your program is run.

Table 8.4 shows the top few lines of the flat profile generated for MBPT2 version in Listing 8.1. The leftmost column shows the percentage of run time spent in the different functions and it shows that about 58% of the time is spent calculating the potential while about 40% is spent in the loops in the actual MBPT2 function. The remaining part is spent in the spinExchangeTerm function which is called from the potential function. Even though the application spends most of its time generating the potential, we don't want to spend too much time on improving this code. We use the Minnesota potential for testing and benchmark purposes only. For more realistic calculations, one should employ the chiral interaction models discussed earlier. It is, however, possible to reduce the number of times this function is called. The 4th column in Table 8.4 shows that for this particular instance, the potential function was called 2.5 billion times. However, due to known symmetries of the nuclear interaction we know that most of these calls result in matrix elements that are zero. If we can exploit this structure to reduce the number of calls to the potential function we will greatly reduce the total run time of this program. The details of how this is done is presented in Sect. 8.7.4.

Code Listing 8.2 Block-sparse implementation of a MBPT2 diagram

```
double mbpt2V02::getEnergy() {
  double energy = 0.0;
  double energy0;
  int nhh, npp, i, j, a, b;
  for(int chan = 0; chan < channels.size; ++chan){
    nhh = channels.nhh[chan];
    npp = channels.npp[chan];
    if(nhh*npp == 0){ continue; }

    for(int hh = 0; hh < nhh; ++hh){
      i = channels.hhvec[chan][2*hh];
      j = channels.hhvec[chan][2*hh + 1];
      for(int pp = 0; pp < npp; ++pp){
        a = channels.ppvec[chan][2*pp];
        b = channels.ppvec[chan][2*pp + 1];
        energy0 = V_Minnesota(modelspace, i, j, a, b, L);
        energy0 *= energy0;
        energy0 /= (modelspace.qnums[i].energy + modelspace.qnums[j].energy -
                    modelspace.qnums[a].energy - modelspace.qnums[b].energy);
        energy += energy0;
      }
    }
  }
  energy *= 0.25;
  return energy;

}
```

Listing 8.2 shows a version of this code where the potential function is not called when we know that the matrix element is zero. This code loops over channels, which are the dense blocks of the full interaction. We have pre-computed the two-body configurations allowed in each channel and store them in the channels structure. The potential is computed in the same way as before, but for fewer combinations of indices.

The profile in Table 8.5 shows that the potential function is now only called 2.5 million times, a reduction of three orders of magnitude. Table 8.6 summarizes the execution times of these two versions of MBPT2.

Code Listing 8.3 Block-sparse implementation of a MBPT3 diagram

```
double mbpt3V02::getEnergy() {
  double energy = 0.0;
```

Table 8.5 Flat profile for the MBPT2 implementation in Listing 8.2 using 1598 states calculating pure neutron matter with 54 neutrons

```
Flat profile:

Each sample counts as 0.01 seconds.
  %   cumulative   self              self     total
 time   seconds   seconds    calls  ms/call  ms/call  name
 66.69      0.08     0.08  2520526     0.00     0.00  V_Minnesota(...)
 16.67      0.10     0.02  4770508     0.00     0.00  Chan_2bInd(...)
  8.34      0.11     0.01        1    10.00    10.00  Build_Model_Space(...)
  8.34      0.12     0.01        1    10.00    30.01  Setup_Channels_MBPT(...)
```

Table 8.6 Total runtime for different MBPT2 implementations for different model spaces

Number of states	Number of protons	Number of neutrons	Runtime (s)	
			Listing 8.1	Listing 8.2
342	0	2	< 0.01	< 0.01
	0	14	0.15	< 0.01
	0	38	1.00	0.03
	0	54	1.81	0.05
684	2	2	0.88	< 0.01
	14	14	2.43	0.04
	38	38	15.7	0.19
	54	54	28.2	0.31
1598	0	2	0.04	< 0.01
	0	14	3.32	0.03
	0	38	25.0	0.23
	0	54	58.9	0.44
3196	2	2	0.88	< 0.01
	14	14	54.3	0.21
	38	38	399	1.40
	54	54	797	2.67

```
double energy0, energy1;
int nhh, npp, i, j, a, b, c, d;
for(int chan = 0; chan < channels.size; ++chan){
  nhh = channels.nhh[chan];
  npp = channels.npp[chan];
  if(nhh*npp == 0){ continue; }

  for(int hh = 0; hh < nhh; ++hh){
    i = channels.hhvec[chan][2*hh];
    j = channels.hhvec[chan][2*hh + 1];
    for(int ab = 0; ab < npp; ++ab){
      a = channels.ppvec[chan][2*ab];
      b = channels.ppvec[chan][2*ab + 1];
      energy0 = V_Minnesota(modelspace, i, j, a, b, L);
```

```
      energy0 /= (modelspace.qnums[i].energy + modelspace.qnums[
          j].energy -
                  modelspace.qnums[a].energy - modelspace.qnums[b
                      ].energy);
      for(int cd = 0; cd < npp; ++cd){
        c = channels.ppvec[chan][2*cd];
        d = channels.ppvec[chan][2*cd + 1];
        energy1 = V_Minnesota(modelspace, a, b, c, d, L);
        energy1 *= V_Minnesota(modelspace, c, d, i, j, L);
        energy1 /= (modelspace.qnums[i].energy + modelspace.
            qnums[j].energy -
                    modelspace.qnums[c].energy - modelspace.qnums[
                        d].energy);
        energy += energy0*energy1;
      }
    }
  }
}
energy *= 0.125;
return energy;
}
```

Table 8.7 Flat profile for the MBPT3 implementation in Listing 8.3 using 3196 states calculating nuclear matter with 14 protons and 14 neutrons

```
Flat profile:

Each sample counts as 0.01 seconds.
  %   cumulative   self              self     total
 time   seconds   seconds    calls  ms/call  ms/call  name
 90.76    177.21   177.21 6068445596     0.00     0.00  V_Minnesota(...)
  8.76    194.31    17.11                             mbpt3V02::getEnergy()
```

Listing 8.3 shows an implementation of MBPT3 that uses a block-sparse representation of the interaction. Compared to MBPT2 it loops over two additional particle indices which increases the computational complexity by several orders of magnitude. However, we are now calculating the interaction many more times than what is necessary. The profile in Table 8.7 shows that we calculated over six billion matrix elements. By moving the construction of the interaction out of the main loops and storing the elements, we can eliminate these redundant calls to the potential function at the expense of using memory to store the elements.

Code Listing 8.4 Block-sparse implementation of a MBPT3 diagram with interaction stored in memory

```
double mbpt3V05::getEnergy() {
  double energy = 0.0;
  double *V1, *S1, *V2, *S2;
  char N = 'N';
  char T = 'T';
  double fac0 = 0.0;
  double fac1 = 1.0;
  int nhh, npp, i, j, a, b, c, d, idx;
  double energy0;
  for(int chan = 0; chan < channels.size; ++chan){
    nhh = channels.nhh[chan];
    npp = channels.npp[chan];
    if(nhh*npp == 0){ continue; }

    V1 = new double[nhh * npp];
    S1 = new double[nhh * npp];
    V2 = new double[npp * npp];
    S2 = new double[nhh * nhh];
    for(int ab = 0; ab < npp; ++ab){
      a = channels.ppvec[chan][2*ab];
      b = channels.ppvec[chan][2*ab + 1];
      for(int hh = 0; hh < nhh; ++hh){
        i = channels.hhvec[chan][2*hh];
        j = channels.hhvec[chan][2*hh + 1];
        idx = hh * npp + ab;
        energy0 = V_Minnesota(modelspace, i, j, a, b, L);
        energy0 /= (modelspace.qnums[i].energy + modelspace.qnums[j].
             energy -
                   modelspace.qnums[a].energy - modelspace.qnums[b].
                         energy);
        V1[idx] = energy0;
      }
      for(int cd = 0; cd < npp; ++cd){
        c = channels.ppvec[chan][2*cd];
        d = channels.ppvec[chan][2*cd + 1];
        idx = ab * npp + cd;
        V2[idx] = V_Minnesota(modelspace, a, b, c, d, L);
      }
    }

    RM_dgemm(V1, V2, S1, &nhh, &npp, &npp, &fac1, &fac0, &N, &N);
    RMT_dgemm(S1, V1, S2, &nhh, &nhh, &npp, &fac1, &fac0, &N, &T);
    delete V1; delete V2; delete S1;
    for(int hh = 0; hh < nhh; ++hh){
      energy += S2[nhh*hh + hh];
    }
```

```
    delete S2;
  }
  energy *= 0.125;

  return energy;
}
```

Listing 8.4 shows the new version of the code that stores matrix elements of the interaction. The explicit summation to calculate the energy can now be done by using matrix products by calling the BLAS (Basic Linear Algebra Subprograms) [76] dgemm wrappers `RM_dgemm` and `RMT_dgemm`. Note that the code has grown more complicated for every new optimization we have introduced. This increases the possibility of introducing errors significantly. It is a good thing we have tests to make sure that the results haven't changed between versions. Table 8.8 shows the profile for this version and we have reduced considerably the number of calls to the function which sets up the interaction.

Table 8.9 summarizes the execution times so far.

Table 8.8 shows that the potential function is still the most expensive function in our program, but we would like to get a more detailed profile of this function. The code to calculate the potential is filled with calls to the exponential function which is part of the standard library. Since we have linked to the standard library dynamically, gprof is not able to show time spent in these functions. We can get a little bit more detail by linking statically. This is done by introducing the static flag to the compiler. Table 8.10 shows the new profile.

The profile is now a lot more busy and it shows a longer runtime than the previous profile. This is because gprof doesn't sample time spent in dynamically linked libraries. The total runtime in this profile corresponds better with Table 8.9, but it is also more difficult to read. What is clear is that the call to the function labelled `__ieee754_exp_avx` takes up almost 70% of the total run time. This function represents the calls to the exponential function in the potential code. If we can reduce the number of evaluations of the exponential function, we can further reduce the run time of this application. We leave that as an exercise to the reader.

The next level of optimization that we will discuss here is the introduction of parallelism. Most modern computers have more than one cpu core available for

Table 8.8 Flat profile for the MBPT3 implementation in Listing 8.4 using 3196 states calculating nuclear matter with 14 protons and 14 neutrons

```
Flat profile:

Each sample counts as 0.01 seconds.
  %   cumulative   self              self     total
 time   seconds   seconds    calls  ms/call  ms/call  name
 91.16     14.59    14.59 484191644     0.00     0.00  V_Minnesota(...)
  7.68     15.82     1.23                             mbpt3V05::getEnergy()
```

Table 8.9 Total runtime for the MBPT3 implementation in Listing 8.3 for different model spaces

Number of states	Number of protons	Number of neutrons	Runtime (s)	
			Listing 8.3	Listing 8.4
342	0	2	0.08	0.02
	0	14	2.38	0.31
	0	38	9.20	0.49
684	2	2	1.05	0.18
	14	14	27.8	2.00
	38	38	107	3.18
1598	0	2	1.81	0.46
	0	14	80.0	10.5
	0	38	456	29.8
3196	2	2	23.1	4.00
	14	14	884	69.8
	38	38	$> 10^3$	190

Table 8.10 Flat profile for the MBPT3 implementation in Listing 8.4 compiled with the static flag enabled using 3196 states calculating nuclear matter with 14 protons and 14 neutrons

```
Flat profile:

Each sample counts as 0.01 seconds.
  %   cumulative   self              self     total
 time   seconds   seconds    calls  ms/call  ms/call  name
 68.06     44.68    44.68                             __ieee754_exp_avx
 21.26     58.64    13.96 484191644     0.00     0.00  V_Minnesota(...)
  5.37     62.17     3.53                             exp
  2.18     63.60     1.43                             mbpt3V05::getEnergy()
  0.94     64.22     0.62                             dgemm_otcopy
  0.55     64.58     0.36                             dgemm_kernel
  0.53     64.93     0.35                             __mpexp_fma4
  0.25     65.09     0.17                             __floor_sse41
```

computation, but the codes we have presented so far will only run on one of these cores. The simplest way to make this code run in parallel is to introduce OpenMP (http://www.openmp.org/) directives. This will split the work between multiple execution streams that all share the same view of memory. Listing 8.5 shows a new version of the MBPT3 function where we have introduced OpenMP directives in lines 19 and 21. The first line marks the start of a parallel region and defines which variables the cores can share and which must be duplicated. The second line defines a parallel loop, where each core is responsible for only a section of the loop. As long as we have enough work in the outermost loop, this strategy will work quite well as shown in Fig. 8.3. Here we show the total run time of this code using different number of cores compared to the best run times we could have gotten with this approach if our parallel regions scaled perfectly with the number of cores. In reality this never happens. In this particular case, we could have made

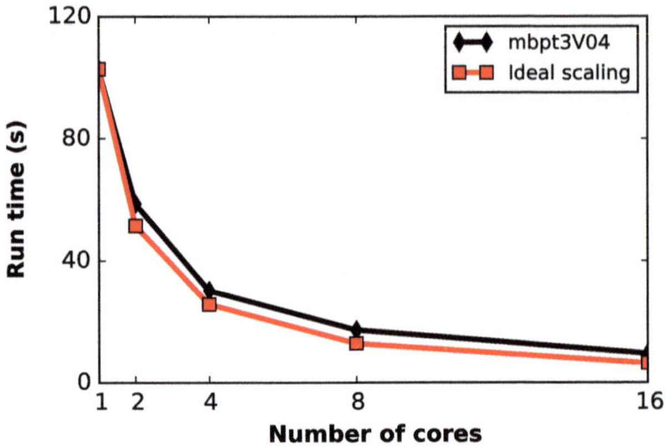

Fig. 8.3 Total runtime for the mbpt3 code in Listing 8.5 using different number of OpenMP threads on a 16 code node of a Cray XK7

the potential function more cache friendly. With this version the different cores are fighting each other for access to memory and cache. This reduces performance somewhat.

Code Listing 8.5 Block-sparse implementation of a MBPT3 diagram with interaction stored in memory and openmp directives

```
double mbpt3V04::getEnergy() {
  double energy = 0.0;
  double *V1, *S1, *V2, *S2;
  char N = 'N';
  char T = 'T';
  double fac0 = 0.0;
  double fac1 = 1.0;
  int nhh, npp, i, j, a, b, c, d, idx;
  double energy0;
  for(int chan = 0; chan < channels.size; ++chan){
    nhh = channels.nhh[chan];
    npp = channels.npp[chan];
    if(nhh*npp == 0){ continue; }

    V1 = new double[nhh * npp];
    S1 = new double[nhh * npp];
    V2 = new double[npp * npp];
    S2 = new double[nhh * nhh];
    #pragma omp parallel shared(V1, V2) private(i, j, a, b, c, d,
          idx, energy0)
    {
      #pragma omp for schedule(static)
      for(int ab = 0; ab < npp; ++ab){
        a = channels.ppvec[chan][2*ab];
        b = channels.ppvec[chan][2*ab + 1];
        for(int hh = 0; hh < nhh; ++hh){
          i = channels.hhvec[chan][2*hh];
          j = channels.hhvec[chan][2*hh + 1];
          idx = hh * npp + ab;
```

```
          energy0 = V_Minnesota(modelspace, i, j, a, b, L);
          energy0 /= (modelspace.qnums[i].energy + modelspace.
              qnums[j].energy -
                      modelspace.qnums[a].energy - modelspace.qnums[
                          b].energy);
          V1[idx] = energy0;
        }
        for(int cd = 0; cd < npp; ++cd){
          c = channels.ppvec[chan][2*cd];
          d = channels.ppvec[chan][2*cd + 1];
          idx = ab * npp + cd;
          V2[idx] = V_Minnesota(modelspace, a, b, c, d, L);
        }
      }
    }

    RM_dgemm(V1, V2, S1, &nhh, &npp, &npp, &fac1, &fac0, &N, &N);
    RMT_dgemm(S1, V1, S2, &nhh, &nhh, &npp, &fac1, &fac0, &N, &T)
        ;
    delete V1; delete V2; delete S1;
    for(int hh = 0; hh < nhh; ++hh){
      energy += S2[nhh*hh + hh];
    }
    delete S2;
  }
  energy *= 0.125;

  return energy;
}
```

In this function we have used the matrix-matrix multiplication function *dgemm* of BLAS [76]. Finally, the above codes can easily be extended upon by including MPI [77, 78] and/or a mix of OpenMP and MPI commands for distributed memory architectures. We leave this as a challenge to the reader. The coding practices and examples developed in this section, are reused in our development of the coupled cluster code discussed in the next section. There we discuss however in more detail how to develop an efficient coupled cluster code for infinite matter, with a focus on validation and verification and simplifications of the equations.

8.7.4 Developing a CCD Code for Infinite Matter

This section focuses on writing a working CCD code from scratch. Based on the previous discussion, what follows serves also the scope of outlining how to start a larger numerical project. We will in particular pay attention to possible benchmarks that can be used to validate our codes.

We will assume that you have opted for a specific mathematical method for solving Schrödinger's equation. Here the mathematics is given by the CCD equations. Our basic steps can then be split up as follows

- Write a first version of the CCD code which is as close as possible to the mathematics of your equations. In this stage we will not focus on high-performance computing aspects and code efficiency. This will mimick our discussion of many-body perturbation theory, in particular the calculation of the correlation energy to second order discussed above.
- Try to find possible benchmarks you can test your code against. In our case, the pairing model serves as an excellent testcase.
- With a functioning code that reproduces possible analytical and/or numerical results, we can start to analyze our code. In particular, if there are mathematical operations which can be simplified and/or can be represented in simpler ways etc. The modified code can hopefully reduce memory needs and time spent on computations. The usage of specific symmetries of the interaction will turn out particularly useful.

In this specific section, we will try to follow the above three steps, with less attention on speed and numerical efficiency. Our aim is to have a code which passes central tests and can be properly validated and verified. If you are familiar with high-performance computing topics, you are obviously not limited to follow the basic steps outlined here. However, when developing a numerical project we have often found it easier and less error-prone to start with the basic mathematical expressions. With a first functioning code, we will delve into high-performance computing topics. A good read on developing numerical projects and clear code is Martin's recent text [79]. We recommend it highly and have borrowed many ideas and coding philosophies therefrom.

We start with implementing the CCD equations as they stand here

$$
\begin{aligned}
\left(\epsilon_i+\epsilon_j-\epsilon_a-\epsilon_b\right) t_{ij}^{ab} &= \langle ab|\hat{v}|ij\rangle + \frac{1}{2}\sum_{cd}\langle ab|\hat{v}|cd\rangle t_{ij}^{cd} + \frac{1}{2}\sum_{kl}\langle kl|\hat{v}|ij\rangle t_{kl}^{ab} \\
&+ \hat{P}(ij|ab)\sum_{kc}\langle kb|\hat{v}|cj\rangle t_{ik}^{ac} + \frac{1}{4}\sum_{klcd}\langle kl|\hat{v}|cd\rangle t_{ij}^{cd} t_{kl}^{ab} \\
&+ \hat{P}(ij)\sum_{klcd}\langle kl|\hat{v}|cd\rangle t_{ik}^{ac} t_{jl}^{bd} - \frac{1}{2}\hat{P}(ij)\sum_{klcd}\langle kl|\hat{v}|cd\rangle t_{ik}^{dc} t_{lj}^{ab} \\
&- \frac{1}{2}\hat{P}(ab)\sum_{klcd}\langle kl|\hat{v}|cd\rangle t_{lk}^{ac} t_{ij}^{db},
\end{aligned}
\tag{8.34}
$$

for all $i<j$ and all $a<b$, using the standard notation that $a,b,\ldots$ are particle states and $i,j,\ldots$ are hole states. The CCD correlation energy is given by

$$
\Delta E_{CCD} = \frac{1}{4}\sum_{ijab}\langle ij|\hat{v}|ab\rangle\, t_{ij}^{ab}. \tag{8.35}
$$

One way to solve these equations, is to write Eq. (8.34) as a series of iterative nonlinear algebraic equations

$$
\begin{aligned}
t_{ij}^{ab(n+1)} = \frac{1}{\epsilon_{ij}^{ab}}\Big(&\langle ab|\hat{v}|ij\rangle + \frac{1}{2}\sum_{cd}\langle ab|\hat{v}|cd\rangle t_{ij}^{cd(n)} + \frac{1}{2}\sum_{kl}\langle kl|\hat{v}|ij\rangle t_{kl}^{ab(n)} \\
&+ \hat{P}(ij|ab)\sum_{kc}\langle kb|\hat{v}|cj\rangle t_{ik}^{ac(n)} \\
&+ \frac{1}{4}\sum_{klcd}\langle kl|\hat{v}|cd\rangle t_{ij}^{cd(n)} t_{kl}^{ab(n)} + \hat{P}(ij)\sum_{klcd}\langle kl|\hat{v}|cd\rangle t_{ik}^{ac(n)} t_{jl}^{bd(n)} \\
&- \frac{1}{2}\hat{P}(ij)\sum_{klcd}\langle kl|\hat{v}|cd\rangle t_{ik}^{dc(n)} t_{lj}^{ab(n)} - \frac{1}{2}\hat{P}(ab)\sum_{klcd}\langle kl|\hat{v}|cd\rangle t_{lk}^{ac(n)} t_{ij}^{db(n)}\Big),
\end{aligned}
\tag{8.36}
$$

for all $i < j$ and all $a < b$, where $\epsilon_{ij}^{ab} = \left(\epsilon_i + \epsilon_j - \epsilon_a - \epsilon_b\right)$, and $t_{ij}^{ab(n)}$ is the t amplitude for the nth iteration of the series. This way, given some starting guess $t_{ij}^{ab(0)}$, we can generate subsequent t amplitudes that converge to some value. Discussions of the mathematical details regarding convergence will be presented below; for now we will mainly focus on implementing these equations into a computer program and assume convergence. In pseudocode, the function that updates the t amplitudes looks like

```
CCD_Update()
for i ∈ {0, N_Fermi − 1} do
    for j ∈ {0, N_Fermi − 1} do
        for a ∈ {N_Fermi, N_sp − 1} do
            for b ∈ {N_Fermi, N_sp − 1} do
                sum ← TBME[index(a, b, i, j)]
                for c ∈ {N_Fermi, N_sp − 1} do
                    for d ∈ {N_Fermi, N_sp − 1} do
                        sum ← sum + 0.5 × TBME[index(a, b, c, d)] ×
t_amplitudes_old[index(c, d, i, j)]
                    end for
                end for
                ...
                sum ← sum + (all other terms)
                ...
                energy_denom = SP_energy[i]+SP_energy[j]-
SP_energy[a]-SP_energy[b]
            t_amplitudes[index(a, b, i, j)] = sum/energy_denom
```

(continued)

```
            end for
        end for
    end for
end for
```

Here we have defined N_{Fermi} to be the fermi level while N_{sp} is the total number of single particle (s.p.) states, indexed from 0 to $N_{sp} - 1$. At the most basic level, the CCD equations are just the addition of many products containing t_{ij}^{ab} amplitudes and two-body matrix elements (TBMEs) $\langle ij|\hat{v}|ab\rangle$. Care should thus be placed into how we store these objects. These are objects with four indices and a sensible first implementation of the CCD equations would be to create two four-dimensional arrays to store the objects. However, it is often more convenient to work with simple one-dimensional arrays instead. The function *index*() maps the four indices onto one index so that a one-dimensional array can be used. An example of a brute force implementation of such a function is

```
function INDEX(p, q, r, s)
    return p × N_sp^3 + q × N_sp^2 + r × N_sp + s
end function
```

Because elements with repeated indices vanish, $t_{ii}^{ab} = t_{ij}^{aa} = 0$ and $\langle pp|\hat{v}|rs\rangle = \langle pq|\hat{v}|rr\rangle = 0$, data structures using this index function will contain many elements that are automatically zero. This means that we need to discuss more efficient storage strategies later. Notice also that we are looping over all indices i, j, a, b, rather than the restricted indices. This means that we are doing redundant work. The reason for presenting the equations this way is merely pedagogical. When developing a program, we would recommend to write a code which is as close as possible to the mathematical expressions. The first version of our code will then often be slow, as discussed in Sect. 8.7.3. Below we will however unrestrict these indices in order to achieve a better speed up of our code.

The goal of our code is to calculate the correlation energy, ΔE_{CCD}, meaning that after each iteration of our equations, we use our newest t amplitudes to update the correlation energy

$$\Delta E_{CCD}^{(n)} = \frac{1}{4} \sum_{ijab} \langle ij|\hat{v}|ab\rangle \, t_{ij}^{ab(n)}. \tag{8.37}$$

We check that our result is converged by testing whether the most recent iteration has changed the correlation energy by less than some tolerance threshold η,

$$\eta > |\Delta E_{CCD}^{(n+1)} - \Delta E_{CCD}^{(n)}|. \tag{8.38}$$

The basic structure of the iterative process looks like

```
while (abs(energy_Diff) > tolerance) do
    CCD_Update()
    correlation_Energy ← CCD_Corr_Energy()
    energy_Diff ← correlation_Energy - correlation_Energy_old
    correlation_Energy_old ← correlation_Energy
    t_amplitudes_old ← t_amplitudes
end while
```

Prior to this algorithm, the t amplitudes should be initialized, $t_{ij}^{ab(0)}$. A particularly convenient choice, as discussed above, is to use many-body perturbation theory for the wave operator with

$$t_{ij}^{ab(0)} = \frac{\langle ab|\hat{v}|ij\rangle}{\epsilon_{ij}^{ab}}, \tag{8.39}$$

which results in the correlation energy

$$\Delta E_{CCD}^{(1)} = \frac{1}{4}\sum_{ijab}\frac{\langle ij|\hat{v}|ab\rangle\langle ab|\hat{v}|ij\rangle}{\epsilon_{ij}^{ab}}. \tag{8.40}$$

This is the familiar result from many-body perturbation theory to second order (MBPT2). It is a very useful result, as one iteration of the CCD equations can be ran, and checked against MBPT2 to give some confidence that everything is working correctly. Additionally, running a program using a minimal test case is another useful way to make sure that a program is working correctly. For this purpose, we turn our attention to the simple pairing model Hamiltonian of Problem 8.10,

$$\hat{H}_0 = \delta\sum_{p\sigma}(p-1)a_{p\sigma}^{\dagger}a_{p\sigma} \tag{8.41}$$

$$\hat{V} = -\frac{1}{2}g\sum_{pq}a_{p+}^{\dagger}a_{p-}^{\dagger}a_{q-}a_{q+} \tag{8.42}$$

which represents a basic pairing model with p levels, each having a spin degeneracy of 2. The form of the coupled cluster equations uses single-particle states that are

Table 8.11 Single-particle states and their quantum numbers and their energies from Eq. (8.43)

State label	p	$2s_z$	E	Type
0	1	1	$-g/2$	Hole
1	1	-1	$-g/2$	Hole
2	2	1	$1-g/2$	Hole
3	2	-1	$1-g/2$	Hole
4	3	1	2	Particle
5	3	-1	2	Particle
6	4	1	3	Particle
7	4	-1	3	Particle

The degeneracy for every quantum number p is equal to 2 due to the two possible spin values

eigenstates of the Hartree-Fock operator, $(\hat{u}+\hat{u}_{\mathrm{HF}})\,|p\rangle = \epsilon_p|p\rangle$. In the pairing model, this condition is already fulfilled. All we have to do is define the lowest N_{Fermi} states as holes and redefine the single-particle energies,

$$\epsilon_q = h_{qq} + \sum_i \langle qi|\hat{v}|qi\rangle\,. \tag{8.43}$$

To be more specific, let us look at the pairing model with four particles and eight single-particle states. These states (with $\delta = 1.0$) could be labeled as shown in Table 8.11. The Hamiltonian matrix for this four-particle problem with no broken pairs is defined by six possible Slater determinants, one representing the ground state and zero-particle-zero-hole excitations $0p-0h$, four representing various $2p-2h$ excitations and finally one representing a $4p-4h$ excitation. Problem 8.10 gives us for this specific problem

$$H = \begin{bmatrix} 2\delta-g & -g/2 & -g/2 & -g/2 & -g/2 & 0 \\ -g/2 & 4\delta-g & -g/2 & -g/2 & -0 & -g/2 \\ -g/2 & -g/2 & 6\delta-g & 0 & -g/2 & -g/2 \\ -g/2 & -g/2 & 0 & 6\delta-g & -g/2 & -g/2 \\ -g/2 & 0 & -g/2 & -g/2 & 8\delta-g & -g/2 \\ 0 & -g/2 & -g/2 & -g/2 & -g/2 & 10\delta-g \end{bmatrix}$$

The python program (included for pedagogical purposes only) at https://github.com/ManyBodyPhysics/LectureNotesPhysics/tree/master/Programs/Chapter8-programs/python/mbpt.py diagonalizes the above Hamiltonian matrix for a given span of interaction strength values, performing a full configuration interaction calculation. It plots the correlation energy, that is the difference between the ground state energy and the reference energy. Furthermore, for the pairing model we have added results from perturbation theory to second order (MBPT2) and third order in the interaction MBPT3. Second order perturbation theory includes diagram (2) of Fig. 8.2 while MBPT3 includes diagrams (3), (4), (5), (8) and (9) as well. Note that diagram (3) is zero for the pairing model and that diagrams (8) and (9) do not contribute either if we work with a canonical Hartree-Fock basis. In the case of the simple pairing model it is easy to calculate ΔE_{MBPT2} analytically. This is a very useful check of our codes since this analytical expression can also be used to check our first CCD iteration. We restate this expression here but restrict the sums over single-particle states

$$\Delta E_{MBPT2} = \frac{1}{4}\sum_{abij} \frac{\langle ij|\hat{v}|ab\rangle \langle ab|\hat{v}|ij\rangle}{\epsilon_{ij}^{ab}} = \sum_{a<b,i<j} \frac{\langle ij|\hat{v}|ab\rangle \langle ab|\hat{v}|ij\rangle}{\epsilon_{ij}^{ab}}$$

For our pairing example we obtain the following result

$$\Delta E_{MBPT2} = \frac{\langle 01|\hat{v}|45\rangle^2}{\epsilon_{01}^{45}} + \frac{\langle 01|\hat{v}|67\rangle^2}{\epsilon_{01}^{67}} + \frac{\langle 23|\hat{v}|45\rangle^2}{\epsilon_{23}^{45}} + \frac{\langle 23|\hat{v}|67\rangle^2}{\epsilon_{23}^{67}},$$

which translates into

$$\Delta E_{MBPT2} = -\frac{g^2}{4}\left(\frac{1}{4+g} + \frac{1}{6+g} + \frac{1}{2+g} + \frac{1}{4+g}\right).$$

This expression can be used to check the results for any value of g and provides thereby an important test of our codes. Figure 8.4 shows the resulting correlation energies for the exact case, MBPT2 and MBPT3. We note from the above program

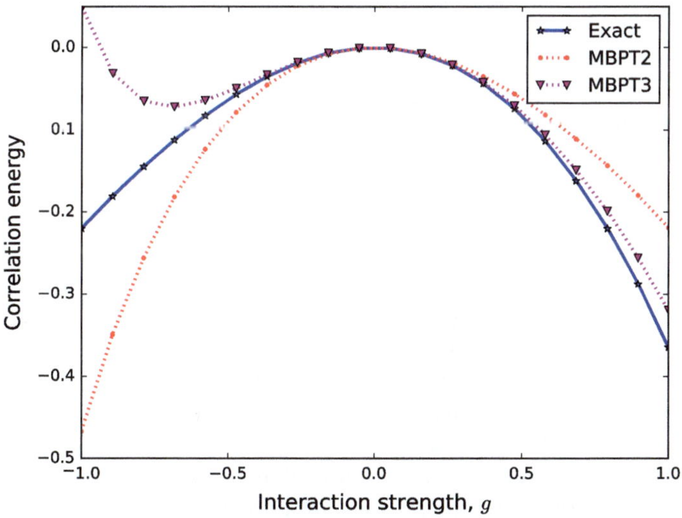

Fig. 8.4 Correlation energy for the pairing model with exact diagonalization, MBPT2 and perturbation theory to third order MBPT3 for a range of interaction values. A canonical Hartree-Fock basis has been employed in all MBPT calculations

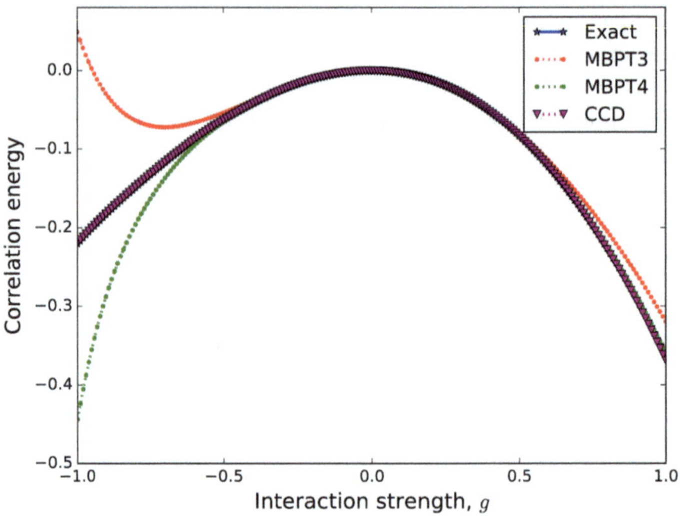

Fig. 8.5 Correlation energy for the pairing model with exact diagonalization, CCD and perturbation theory to third (MBPT3) and fourth order (MBPT4) for a range of interaction values

that we have coded the expressions for the various diagrams following strictly the mathematical expressions of for example Eqs. (8.24)–(8.26). This means that for every diagram we loop explicitly over every single-particle state. The python program linked to above is included mainly for pedagogical reasons. As we have already seen, this approach is extremely inefficient from a computational point of view. In our discussions of MBPT, as well as for CCD code, we rewrite the computations of most diagrams in terms of efficient matrix-matrix multiplications or matrix-vector multiplications. Figure 8.4 shows us that the approximation to both second and third order are very good when the interaction strength is small and contained in the interval $g \in [-0.5, 0.5]$, but as the interaction gets stronger in absolute value the agreement with the exact reference energy for MBPT2 and MBPT3 worsens. We also note that the third-order result is actually worse than the second order result for larger values of the interaction strength, indicating that there is no guarantee that higher orders in many-body perturbation theory may reduce the relative error in a systematic way. Adding fourth order contributions as shown in Fig. 8.5 for negative interaction strengths gives a better result than second and third order. The fourth order contributions include also four-particle-four-hole correlations. However, the disagreement for stronger interaction values hints at the possibility that many-body perturbation theory may not converge order by order. Perturbative studies of nuclear systems may thus be questionable, unless selected contributions that soften the interactions are properly softened. We note also the non-variational character of many-body perturbation theory, with results at different levels of many-body perturbation theory either overshooting or undershooting the true ground state correlation energy. The coupled cluster results are included in

Table 8.12 Coupled cluster and MBPT2 results for the simple pairing model with eight single-particle levels and four spin 1/2 fermions for different values of the interaction strength g

g	E_{ref}	ΔE_{MBPT2}	ΔE_{CCD}
−1.0	3	−0.46667	−0.21895
−0.5	2.5	−0.08874	−0.06306
0.0	2	0	0
0.5	1.5	−0.06239	−0.08336
1.0	1	−0.21905	−0.36956

Fig. 8.5 where we display the difference between the exact correlation energy and the correlation energy obtained with many-body perturbation theory to third order. Coupled cluster theory with doubles only shows a very good agreement with the exact results. For larger values of g one will however observe larger discrepancies. In Table 8.12 we list for the sake of completeness also our coupled cluster results at the CCD level for the same system. The $g = -1.0$ case diverges without implementing iterative mixing. Sometimes iterative solvers run into oscillating solutions, and mixing can help the iterations break this cycle.

$$t^{(i)} = \alpha t^{(i)}_{no\_mixing} + (1 - \alpha)t^{(i-1)} \tag{8.44}$$

Once a working pairing model has been implemented, improvements can start to be made, all the while using the pairing model to make sure that the code is still working and giving correct answers. Realistic systems will be much larger than this small pairing example.

One limitation that will be ran into while trying to do realistic CCD calculations is that of memory. The four-indexed two-body matrix elements (TBMEs) and t-amplitudes have to store a lot of elements, and the size of these arrays can quickly exceed the available memory on a machine. If a calculation wants to use 500 single-particle basis states, then a structure like $\langle pq|v|rs\rangle$ will need a length of 500 for each of its four indices, which means it will have $500^4 = 625 \times 10^8$ elements. To get a handle on how much memory is used, consider the elements as double-precision floating point type. One double takes up 8 bytes of memory. This specific array would take up $8 \times 625 \times 10^8$ bytes = 5000×10^8 bytes = 500 GB of memory. Most personal computers in 2016 have 4–8 GB of RAM, meaning that this calculation would be way out of reach. There are supercomputers that can handle applications using 500 GB of memory, but we can quickly reduce the total memory required by applying some physical arguments. In addition to vanishing elements with repeated indices, mentioned above, elements that do not obey certain symmetries are also zero. Almost all realistic two-body forces preserve some quantities due to symmetries in the interaction. For example, an interaction with rotational symmetry will conserve angular momentum. This means that a two-body ket state $|rs\rangle$, which has some set of quantum numbers, will retain quantum numbers corresponding to the interaction symmetries after being acted on by $\hat{v}$. This state is then projected onto $|pq\rangle$ with its own set of quantum numbers. Thus $\langle pq|v|rs\rangle$ is only non-zero if

$|pq\rangle$ and $|rs\rangle$ share the same quantum numbers that are preserved by $\hat{v}$. In addition, because the cluster operators represent excitations due to the interaction, t_{ij}^{ab} is only non-zero if $|ij\rangle$ has the same relevant quantum numbers as $|ab\rangle$.

To take advantage of this, these two-body ket states can be organized into symmetry "channels" of shared quantum numbers. In the case of the pairing model, the interaction preserves the total spin projection of a two-body state, $S_z = s_{z1} + s_{z2}$. The single particle states can have spin of $+1/2$ or $-1/2$, so there can be three two-body channels with $S_z = -1, 0, +1$. These channels can then be indexed with a unique label in a similar way to the single particle index scheme. In more complicated systems, there will be many more channels involving multiple symmetries, so it is useful to create a data structure that stores the relevant two-body quantum numbers to keep track of the labeling scheme.

It is more efficient to use two-dimensional array data structures, where the first index refers to the channel number and the second refers to the element within that channel. So to access matrix elements or t amplitudes, you can loop over the channels first, then the indices within that channel. To get an idea of the savings using this block diagonal structure, let's look at a case with a plane wave basis, with three momentum and one spin quantum numbers, with an interaction that conserves linear momentum in all three dimensions, as well as the total spin projection. Using 502 basis states, the TBME's require about 0.23 Gb of memory in block diagonal form, which is an enormous saving from the 500 Gb mentioned earlier in the naïve storage scheme.

Since the calculation of all zeros can now be avoided, improvements in speed and memory will now follow. To get a handle on how these CCD calculations are implemented we need only to look at the most expensive sum in Eq. (8.34). This corresponds to the sum over $klcd$. Since this sum is repeated for all $i < j$ and $a < b$, it means that these equations will scale as $\mathscr{O}(n_p^4 n_h^4)$. However, they can be rewritten using intermediates as

$$
\begin{aligned}
0 = \langle ab|\hat{v}|ij\rangle &+ \hat{P}(ab)\sum_c \langle b|\chi|c\rangle \langle ac|t|ij\rangle - \hat{P}(ij)\sum_k \langle k|\chi|j\rangle \langle ab|t|ik\rangle \\
&+ \frac{1}{2}\sum_{cd} \langle ab|\chi|cd\rangle \langle cd|t|ij\rangle + \frac{1}{2}\sum_{kl} \langle ab|t|kl\rangle \langle kl|\chi|ij\rangle \\
&+ \hat{P}(ij)\hat{P}(ab)\sum_{kc} \langle ac|t|ik\rangle \langle kb|\chi|cj\rangle ,
\end{aligned} \tag{8.45}
$$

for all i, j, a, b, the reason why these indices are now unrestricted will be explained later. The intermediates χ are defined as

$$
\langle b|\chi|c\rangle = \langle b|f|c\rangle - \frac{1}{2}\sum_{kld} \langle bd|t|kl\rangle \langle kl|v|cd\rangle , \tag{8.46}
$$

$$
\langle k|\chi|j\rangle = \langle k|f|j\rangle + \frac{1}{2}\sum_{cdl} \langle kl|v|cd\rangle \langle cd|t|jl\rangle , \tag{8.47}
$$

$$\langle kl|\chi|ij\rangle = \langle kl|v|ij\rangle + \frac{1}{2}\sum_{cd}\langle kl|v|cd\rangle\,\langle cd|t|ij\rangle\,, \tag{8.48}$$

$$\langle kb|\chi|cj\rangle = \langle kb|v|cj\rangle + \frac{1}{2}\sum_{dl}\langle kl|v|cd\rangle\,\langle db|t|lj\rangle\,, \tag{8.49}$$

$$\langle ab|\chi|cd\rangle = \langle ab|v|cd\rangle\,. \tag{8.50}$$

With the introduction of the above intermediates, the CCD equations scale now as $\mathcal{O}(n_h^2 n_p^4)$, which is an important improvement. This is of course at the cost of computing the intermediates at the beginning of each iteration, where the most expensive one, $\langle kb|\chi|cj\rangle$ scales as $\mathcal{O}(n_h^3 n_p^3)$. To further speed up these computations, we see that these sums can be written in terms of matrix-matrix multiplications. It is not obvious how to write all of these sums in such a way, but it is useful to first recall that the expression for the multiplication of two matrices $\hat{C} = \hat{A} \times \hat{B}$ can be written as

$$C_{ij} = \sum_k A_{ik} \times B_{kj}. \tag{8.51}$$

We see then that Eq. (8.48) can be written as

$$\langle K|\chi|I\rangle = \langle K|v|I\rangle + \frac{1}{2}\sum_C \langle K|v|C\rangle\,\langle C|t|I\rangle$$

by mapping the two index pairs $kl \rightarrow K, ij \rightarrow I, cd \rightarrow C$. The sum looks now like a matrix-matrix multiplication. This is useful because there are packages like BLAS (Basic Linear Algebra Subprograms) [76] which have extremely fast implementations of matrix-matrix multiplication, as discussed in connection with the listing 8.5. The simplest example to consider is the expression for the correlation energy from MBPT2. We rewrite

$$\Delta E_{MBPT2} = \frac{1}{4}\sum_{abij}\frac{\langle ij|\hat{v}|ab\rangle\,\langle ab|\hat{v}|ij\rangle}{\epsilon_{ij}^{ab}}, \tag{8.52}$$

by defining the matrices $\hat{A}$ and $\hat{B}$ with new indices $I = (ij)$ and $A = (ab)$. The individual matrix elements of these matrices are

$$A_{IA} = \langle I|\hat{v}|A\rangle,$$

and

$$B_{AI} = \frac{\langle A|\hat{v}|I\rangle}{\epsilon_I^A}.$$

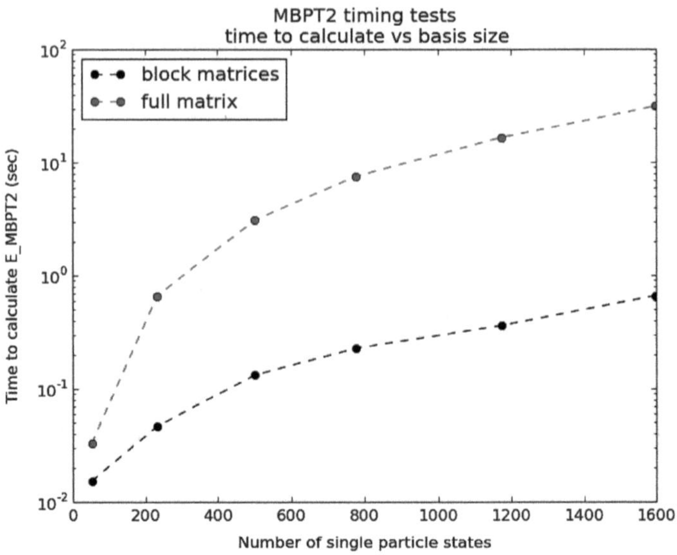

Fig. 8.6 MBPT2 contribution to the correlation for pure neutron matter with $N = 14$ neutrons and periodic boundary conditions. Up to approximately 1600 single-particle states have been included in the sums over intermediate states in Eqs. (8.52) and (8.53)

We can then rewrite the correlation energy from MBPT2 as

$$\Delta E_{MBPT2} = \frac{1}{4}\hat{A} \times \hat{B}. \tag{8.53}$$

Figure 8.6 shows the difference between the brute force summation over single-particle states of Eq. (8.52) and the smarter set up in terms of indices including two-body configurations only, that is Eq. (8.53). In these calculations we have only considered pure neutron matter with $N = 14$ neutrons and a density $n = 0.08\,\mathrm{fm}^{-3}$ and plane wave single-particle states with periodic boundary conditions, allowing for up to 1600 single-particle basis states. The Minnesota interaction model [57] has been used in these calculations. With 40 single-particle shells, see Table 8.1 for example, we have in total 2713 single-particle states. Using the smarter algorithm the final calculation time is 2.4 s (this is the average time from ten numerical experiments). The total time using the brute force summation over single-particle indices is 100.6 s (again the average of ten numerical experiments), resulting in a considerable speed up. It is useful to dissect the final time in terms of different operations. For the smarter algorithm most of the time is spent setting up the matrix elements for the two-body channels and to load the matrix elements. The final matrix-matrix multiplication takes only 1% of the total time. For the brute force algorithm, the multiplication and summation over the various single-particle states takes almost half of the total time. Here we have deliberately only focused on the difference between the two ways of computing Eqs. (8.52) and (8.53). We have, on purpose, not performed a proper timing analysis since this was done in the previous subsection. In this section we have chosen to focus on the development of a program which produces the correct results. As mentioned above, the true elephant in the room, in terms of computational time, is our computation of the matrix elements of the nucleon-nucleon potential. We have deliberately omitted the time spent on setting up the interactions here. For the rest of this section we will focus on various physics applications of our newly developed CCD code.

With the definition of the intermediates and appropriate matrix-matrix multiplications and a working CCD program, we can move on to more realistic cases. One such case is infinite nuclear matter using a plane-wave basis. These states are solutions to the free-particle Hamiltonian,

$$\frac{-\hbar^2}{2m}\nabla^2\phi(\mathbf{x}) = \epsilon\,\phi(\mathbf{x})\,. \tag{8.54}$$

For a finite basis, as discussed earlier, we approximate the problem by constructing a box with sides of length L, which quantizes the momentum, and impose periodic boundary conditions in each direction by requiring $\phi(x_i) = \phi(x_i + L)$.

The first step in calculating infinite matter is to construct a model space by finding every single-particle state relevant to a given problem. In our case, this amounts to looping over the quantum numbers for spin, isospin, and the three momentum directions. To control the model space size, the momentum can be truncated to give a cubic space, where $n_i \leq n_{\max}$, or a spherical space, where $n_x^2 + n_y^2 + n_z^2 \leq N_{\max}$. The number of single-particle states in a cubic space increases rapidly with $n_{\max}$ compared to the spherical case with $N_{\max}$. For example, in pure neutron matter a cubic space with $n_{\max} = 3$ has 668 states while the spherical space with $N_{\max} = 17$ has 610 states. Therefore, the spherical case will be used for the rest of the calculations here. The loop increases in energy by counting the number of shells, so states can be 'filled' by labeling the first P proton and N neutron states as holes. The following loop is for pure neutron matter and requires the number of neutrons, N and density, $\rho = N/L^3$, as input. Symmetric nuclear matter requires an extra loop over isospin.

$n = 0$
for shell $\in \{0, \ldots, N_{\max}\}$ **do**
  for $-\sqrt{N_{\max}} \leq n_x \leq \sqrt{N_{\max}}$ **do**
    for $-\sqrt{N_{\max}} \leq n_y \leq \sqrt{N_{\max}}$ **do**
      for $-\sqrt{N_{\max}} \leq n_z \leq \sqrt{N_{\max}}$ **do**
        for $s_z \in \{-\frac{1}{2}, \frac{1}{2}\}$ **do**
          if $n_x^2 + n_y^2 + n_z^2 =$ shell **then**
            Energy $= \frac{4\pi^2\hbar^2}{2m} \times$ shell
            if $n < N$ **then**

(continued)

```
                    type = "hole"
                else
                    type = "particle"
                end if
                STATES ← (n, n_x, n_y, n_z, s_z, Energy, type)
                n ← n + 1
            end if
        end for
    end for
  end for
 end for
end for
```

The next step is to build every two-body state in the model space and separate them by their particle/hole character and combined quantum numbers. While each single-particle state is unique, two-body states can share quantum numbers with other members of a particular two-body channel. These channels allow us to remove matrix elements and cluster amplitudes that violate the symmetries of the interaction. This reduces greatly the size and speed of the calculation. Our structures will depend on direct two-body channels, T, where the quantum numbers are added, and cross two-body channels, X, where the quantum numbers are subtracted. Before filling the channels, it is helpful to order them with an index function which returns a unique index for a given set of two-body quantum numbers. Without an index function, one has to loop over all the channels for each two-body state, adding a substantial amount of time to this algorithm. An example of an index function for the direct channels in symmetric nuclear matter is, for $N_x = n_{x,1} + n_{x,2}$, $N_y = n_{y,1} + n_{y,2}$, $N_z = n_{z,1} + n_{z,2}$, $S_z = s_{z,1} + s_{z,2}$, $T_z = t_{z,1} + t_{z,2}$, $m = 2\lfloor\sqrt{N_{\max}}\rfloor$, and $M = 2m + 1$,

$$\text{Ind}\left(N_x, N_y, N_z, S_z, T_z\right) = 2\left(N_x + m\right) M^3 + 2\left(N_y + m\right) M^2 + 2\left(N_z + m\right) M + 2\left(S_z + 1\right) + \left(T_z + 1\right). \tag{8.55}$$

This function, which can also be used for the cross-channel index function, is well suited for a cubic model space but can be applied in either case. An additional restriction for two-body states is that they must be composed of two different states to satisfy the Pauli-exclusion principle.

```
for sp1 ∈ STATES do
    for sp2 ∈ STATES do
        if sp1 ≠ sp2 then
            N_i ← n_{i,1} + n_{i,2}
            S_z ← s_{z,1} + s_{z,2}
            T_z ← t_{z,1} + t_{z,2}
            i_dir ← Ind (N_x, N_y, N_z, S_z, T_z)
            T ← (sp1, sp2, i_dir)
            N'_i ← n_{i,1} − n_{i,2}
            S'_z ← s_{z,1} − s_{z,2}
            T'_z ← t_{z,1} − t_{z,2}
            i_cross ← Ind (N'_x, N'_y, N'_z, S'_z, T'_z)
            X ← (sp1, sp2, i_cross)
        end if
    end for
end for
```

From the cross channels, one can construct the cross channel complements, X', where $X\,(pq) \equiv X'\,(qp)$. Also from the direct channels, one can construct one-body, and corresponding three-body, channels for each single-particle state, K by finding every combination of two two-body states within a direct channel that contains that single particle state, $T\,(pq) = T\,(rs) \Rightarrow K_p \leftarrow (qrs)$.

```
for Chan ∈ T do
    for tb1 ∈ Chan do
        for tb2 ∈ Chan do
            K ← tb1_1
            K_{tb1_1} ← tb1_2, tb2_1, tb2_2
        end for
    end for
end for
```

These different structures can be further categorized by a two-body state's particle-hole character, $\langle pp|t|hh\rangle$, $\langle hh|v|hh\rangle$, $\langle pp|v|pp\rangle$, $\langle hh|v|pp\rangle$, and $\langle hp|v|hp\rangle$, which greatly simplifies the matrix-matrix multiplications of the CCD iterations by indexing the summed variables in a systematic way. Summations are constructed by placing two structures next to each other in such a way that the inner summed indices are of the same channel. The resulting structure is indexed by the outer channels as shown here for several of the intermediates defined above

$$\langle b|\chi|c\rangle = \langle b|f|c\rangle - \frac{1}{2}\sum_{kld} \langle bd|t|kl\rangle \langle kl|v|cd\rangle \rightarrow f_c^b\,(K\,(b)\,,K\,(c))$$

$$- \frac{1}{2} t_{kl}^{bd}\,(K\,(b)\,,K_b\,(kld))\, v_{cd}^{kl}\,(K_c\,(kld)\,,K\,(c))\,,$$

$$\langle kl|\chi|ij\rangle = \langle kl|v|ij\rangle + \frac{1}{2}\sum_{cd} \langle kl|v|cd\rangle \langle cd|t|ij\rangle \rightarrow v_{ij}^{kl}\,(T\,(kl)\,,T\,(ij))$$

$$+ \frac{1}{2} v_{cd}^{kl}\,(T\,(kl)\,,T\,(cd))\, t_{ij}^{cd}\,(T\,(cd)\,,T\,(ij))\,,$$

$$\langle kb|\chi|cj\rangle = \langle kb|v|cj\rangle + \frac{1}{2}\sum_{dl} \langle kl|v|cd\rangle \langle db|t|lj\rangle \rightarrow v_{cj}^{kb}\,(X\,(kc)\,,X\,(jb))$$

$$+ \frac{1}{2} v_{cd}^{kl}\,(X\,(kc)\,,X\,(dl))\, t_{lj}^{db}\,(X\,(dl)\,,X\,(jb))\,,$$

$$\langle ab|\chi|cd\rangle = \langle ab|v|cd\rangle \rightarrow v_{cd}^{ab}\,(T\,(ab)\,,T\,(cd))\,,$$

$$\sum_c \langle b|\chi|c\rangle \langle ac|t|ij\rangle \rightarrow \chi_c^b\,(K\,(b)\,,K\,(c)) \cdot t_{ij}^{ac}\,(K\,(c)\,,K_c\,(ija))\,,$$

$$\sum_k \langle k|\chi|j\rangle \langle ab|t|ik\rangle \rightarrow \chi_j^k\,(K\,(j)\,,K\,(k)) \cdot t_{ik}^{ab}\,(K\,(c)\,,K_c\,(ija))\,,$$

$$\sum_{cd} \langle ab|\chi|cd\rangle \langle cd|t|ij\rangle \rightarrow \chi_{cd}^{ab}\,(T\,(ab)\,,T\,(cd)) \cdot t_{ij}^{cd}\,(T\,(cd)\,,T\,(ij))\,,$$

$$\sum_{kl} \langle ab|t|kl\rangle \langle kl|\chi|ij\rangle \rightarrow t_{kl}^{ab}\,(T\,(ab)\,,T\,(kl)) \cdot \chi_{ij}^{kl}\,(T\,(kl)\,,T\,(ij))\,,$$

and finally

$$\sum_{kc} \langle ac|t|ik\rangle \langle kb|\chi|cj\rangle = \sum_{kc} \langle ai^{-1}|t|kc^{-1}\rangle \langle kc^{-1}|\chi|jb^{-1}\rangle$$

$$\rightarrow t_{ik}^{ac}\,(X\,(ia)\,,X\,(kc)) \cdot \chi_{cj}^{kb}\,(X\,(kc)\,,X\,(jb))\,.$$

Figure 8.7 displays the convergence of the energy per particle for pure neutron matter as function of number particles for the CCD approximation with the Minnesota interaction model [57] for different with $N_{max} = 20$. Similarly, Fig. 8.8 shows the convergence in terms of different model space sizes $N_{max} = 20$ with a fixed number of neutrons $N = 114$. We see from Fig. 8.8 that at the CCD level and neutron matter only there is a good convergence with $N_{max} = 25$. These results depends however on the type of interaction and many-body approximation.

In these calculations we approximated our problem with periodic boundary conditions, $\phi(x_i) = \phi(x_i + L)$, but we could have chosen anti-periodic boundary

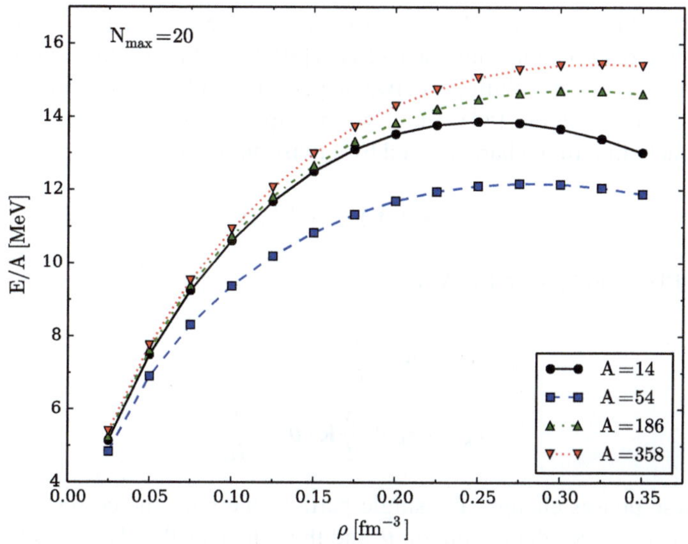

Fig. 8.7 Energy per particle of pure neutron matter computed in the CCD approximation with the Minnesota interaction model [57] for different numbers of particles with $N_{max} = 20$

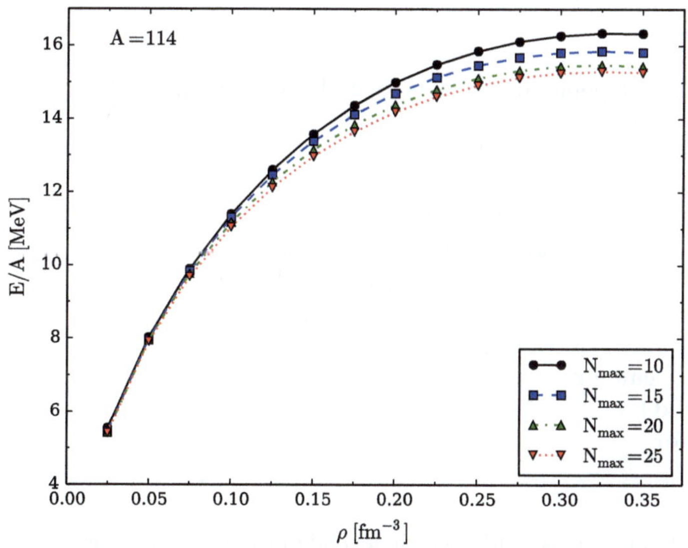

Fig. 8.8 Energy per particle of pure neutron matter computed in the CCD approximation with the Minnesota interaction model [57] for different model space sizes with A = 114

conditions, $\phi(x_i) = -\phi(x_i + L)$. The difference between these two shows how the correlation energy contains finite-size effects [80–83]. One solution to this problem is by integrating over solutions between periodic and anti-periodic conditions, known as twist-averaging [84]. First, we multiply the single-particle states by a phase for each direction, characterized by a twist-angle, θ_i.

$$\phi_{\mathbf{k}}(\mathbf{x} + \mathbf{L}) \rightarrow e^{i\boldsymbol{\theta}} \phi_{\mathbf{k}}(\mathbf{x}) \tag{8.56}$$

$\theta_i = 0$ for PBC and $\theta_i = \pi$ for APBC

$$\mathbf{k} \rightarrow \mathbf{k} + \frac{\boldsymbol{\theta}}{L} \tag{8.57}$$

$$\epsilon_{\mathbf{k}} \rightarrow \epsilon_{\mathbf{k}} + \frac{\pi}{L}\mathbf{k} \cdot \boldsymbol{\theta} + \frac{\pi^2}{L^2} \tag{8.58}$$

Adding these phases changes the single-particle energies, the correction of which disappear as $L \rightarrow \infty$, depending on $\boldsymbol{\theta}$ and thus changes the shell structure so that hole states can jump up to particle states and *vice versa*. It is thence necessary to fill hole states separately for each $\boldsymbol{\theta}$. Integration over a quantity is approximated by a weighted sum, such as Gauss-Legendre quadrature, over the quantity for each set of twist angles. The algorithm becomes then

Build mesh points and weights for each direction i: $\{\theta_i, w_i\}$
$E_{\text{twist}} = 0$
for $(\theta_x, w_x) \in \{\theta_x, w_x\}$ **do**
    for $(\theta_y, w_y) \in \{\theta_y, w_y\}$ **do**
        for $(\theta_z, w_z) \in \{\theta_z, w_z\}$ **do**
            Build Basis States with $k_i \rightarrow k_i + \frac{\theta_i}{L}$
            Order States by Energy and Fill Holes
            Get Result E (T,HF,CCD)
            $E_{\text{twist}} = E_{\text{twist}} + \frac{1}{\pi^3} w_x w_y w_z E$
        end for
    end for
end for

This technique gives results which depend much less on the particle number, but requires a full calculation for each set of twist angles, which can grow very quickly. For example, using 10 twist angles in each direction requires 1000 calculations. To see the effects of twist averaging, it is easy to calculate the kinetic energy per particle and the Hartree-Fock energy per particle, which avoids the full CCD calculation. These calculations can be compared to the exact values for infinite matter, which are calculated by integrating the relevant values up to the fermi surface. The kinetic energy is given by

$$\mathrm{T}_{\mathrm{inf}} = \frac{3\hbar^2 k_f^2}{10m},$$

while the potential energy to first order (corresponding to the Hartree-Fock contribution) reads

$$\mathrm{HF}_{\mathrm{inf}} = \frac{1}{(2\pi)^6} \frac{L^3}{2\rho} \int_0^{k_f} d\mathbf{k}_1 \int_0^{k_f} d\mathbf{k}_2 \, \langle \mathbf{k}_1 \mathbf{k}_2 | \hat{v} | \mathbf{k}_1 \mathbf{k}_2 \rangle \, .$$

Figure 8.9 shows the CCD kinetic energy of pure neutron matter computed with the Minnesota interaction model [57] as a function of the number of particles for both periodic boundary conditions (PBC) and twist-averaged boundary conditions (TABC5). We see clearly that the twist-averaged boundary conditions soften the dependence on finite size effects. Similarly, Fig. 8.10 displays the corresponding Hartree-Fock energy (the reference energy) obtained with Minnesota interaction using both periodic and twist-average boundary conditions.

The results show again a weaker dependence on finite size effects.

With all these ingredients, we can now compute the final CCD energy and thereby the equation of state for infinite neutron matter. Figure 8.11 displays the total CCD energy (including the reference energy) as well as the reference energy obtained with the Minnesota interaction model. The computations have been performed with

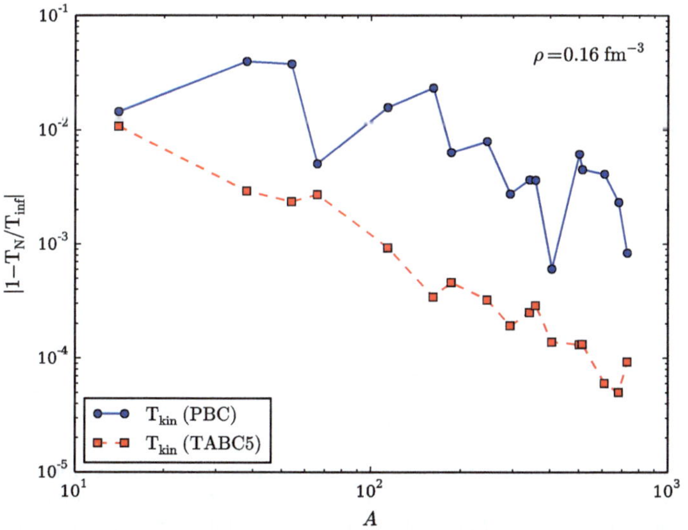

Fig. 8.9 Finite-size effects in the kinetic energy of pure neutron matter computed with the Minnesota interaction model [57] as a function of the number of particles for both periodic boundary conditions (PBC) and twist-averaged boundary conditions (TABC5)

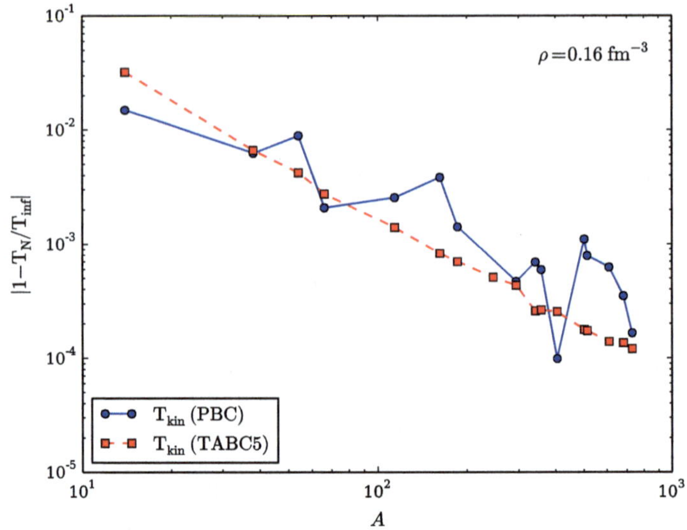

Fig. 8.10 Finite-size effects in the Hartree-Fock energy of pure neutron matter computed with the Minnesota interaction model [57] as a function of the number of particles for both periodic boundary (PBC) conditions and twist-averaged boundary conditions (TABC5)

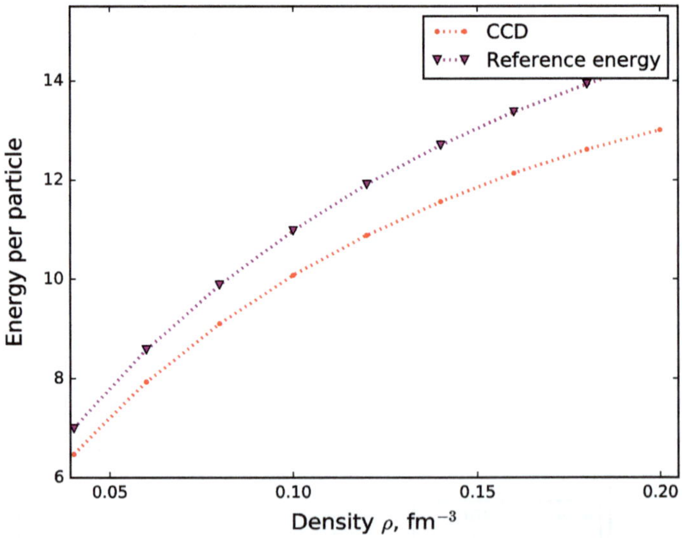

Fig. 8.11 Energy per particle for pure neutron matter as function of density from coupled cluster calculations with doubles correlations only. The reference energy is included for comparison. The results have been obtained with Minnesota interaction model using periodic boundary conditions and $N = 66$ neutrons and single-particle states up to $N_{max} = 36$, resulting in a total of 2377 single-particle states

Table 8.13 CCD and MBPT2 results for infinite neutron matter with $N = 66$ neutrons and a maximum number of single-particle states constrained by $N_{max} = 36$

Density ρ fm^{-3}	E_{MBPT2}	E_{CCD}
0.04	6.472	6.468
0.06	7.919	7.932
0.08	9.075	9.136
1.0	9.577	10.074
1.2	10.430	10.885
1.4	11.212	11.565
1.6	11.853	12.136
1.8	12.377	12.612
2.0	12.799	13.004

$N = 66$ neutrons and a maximum number of single-particle states constrained by $N_{max} = 36$. This corresponds to 2377 single-particle states.

We see from this figure that the correlations brought by coupled cluster theory are at the order of 10% roughly of the reference energy. It means that for this system (neutrons only) with the Minnesota potential, higher-order correlations can most likely be treated in a perturbative way. Many-body perturbation theory to second order gives results which are very close to our CCD results, as seen in Table 8.13. For low densities we observe a good agreement while higher densities bring in particle-particle correlations that become more important as the density increases. Coupled cluster theory sums to infinite order for example particle-particle correlations and with increasing densities this is reflected in differences between the two many-body approximations. The above results agree well with the recent coupled cluster calculations of [36, 37], obtained with interaction models from effective field theory. With the inclusion of proton correlations as well as other potential models we may expect larger differences between different methods and interactions. In Chaps. 9, 10 and 11 we compare the above results with those obtained with Monte Carlo methods, the in-medium renormalization group approach and Green's function theory.

The discussions hitherto have focused on the development of an efficient and flexible many-body code. The codes have been structured to allow users and developers to study and add different physical systems, spanning from the simple pairing model to quantum dots and infinite nuclear matter. Structuring the codes in for example an abstract system class and a solver class allows an eventual user to study different physical systems and add new many-body solvers without having to write a totally new program. As demonstrated in Chap. 10, with few additions one can add the widely popular similarity renormalization group method.

Till now we have limited our discussion to the construction of a many-body code following the underlying mathematical expressions, including elements like how to structure a code in terms of functions, how to modularize the code, how to develop classes and how to verify and validate our computations (our checks provided by the simple pairing model and many-body perturbation theory to second order) against known results. With the rewriting of our equations in terms of efficient matrix and vector operations we have also shown how to make our code more efficient. Matrix and vector operations can easily be parallelized, as demonstrated in our discussion of many-body perturbation theory to second and third order. Such algorithmic improvements are necessary in order to be able to study complicated physical systems. Our codes can also be easily parallelized in order to run on shared memory architectures using either OpenMP [85] and/or MPI/OpenMPI [77, 78].

We conclude this section by summarizing and emphasizing some topics we feel can help in structuring a large computational project. Amongst these, the validation and verification of the correctness of the employed algorithms and programs are central issues which can save you a lot of time when developing a numerical project. In the discussions above we used repeatedly the simple pairing model of Problem 8.10. This model allows for benchmarks against exact results. In addition, it provides analytical answers to several approximations, from perturbation theory to specific terms in the solution of the coupled cluster equations, the in-medium similarity renormalization group approach of Chap. 10 and the Green's function approach of Chap. 11.

It is also important to develop an understanding of how our algorithms can go wrong and how they can be implemented in order to run as efficiently as possible. When working on a numerical project it is important to keep in mind that computing covers numerical as well as symbolic computing and paper and pencil solutions. Furthermore, version control is something we strongly recommend. It does not only save you time in case you struggle with odd errors in a new version of your code. It allows you also to make science reproducible. Making your codes available to a larger audience and providing proper benchmarks allows fellow scientists to test what you have developed, and perhaps come with considerable improvements and/or find flaws or errors you were not aware of. Sharing your codes using for example modern version control software makes thus your science reproducible and adds in a natural way a sound ethical scientific element to what you develop. In this chapter we have thus provided several code examples, hoping they can serve as good examples.

8.8 Conclusions

In this chapter we have presented many of the basic ingredients that enter theoretical studies of infinite nuclear matter, with possible extensions to the homogeneous electron gas in two and three dimensions or other quantum mechanical systems. We have focused on the construction of a single-particle and many-body basis appropriate for such systems, as well as introducing post Hartree-Fock many-body methods like full configuration interaction theory, many-body perturbation theory and coupled cluster theory. The results here, albeit being obtained with a simpler model for the nuclear forces, can easily be extended to more complicated and realistic models for nuclear interactions and to include other many-body methods. We have however, for pedagogical reasons, tried to keep the theoretical inputs to the various many-body methods as simple as possible. The reader should however, with the inputs from Chaps. 2–6, be able to have a better understanding of nuclear forces and how these can be derived from the underlying theory for the strong force and effective field theory. The last exercise in this chapter replaces the simple Minnesota potential with realistic interactions from effective field theory.

The subsequent Chaps. 9, 10 and 11 show how many of the theoretical concepts and code elements discussed in this chapter can be used to add other many-body methods, without having to develop a new numerical project. With a proper modularization and flexible classes, we can add new physical systems as well as new many-body methods. The codes which have been developed in this chapter can be reused in the development and analysis of the in-medium similarity renormalization group approach of Chap. 10 or the Green's function approach in Chap. 11. Similarly, the theoretical concepts we have developed in this chapter, such as the definition of a single-particle basis using plane wave functions and correlations from many-body perturbation theory or coupled cluster theory, can be used in Chaps. 9, 10 and 11 as well. Chapter 9 for example, uses results from coupled cluster theory in order to provide better ways to constrain the Jastrow factor, which accounts for correlations beyond a mean-field picture, in Monte Carlo calculations.

8.9 Exercises

8.1 Show that the one-body part of the Hamiltonian

$$\hat{H}_0 = \sum_{pq} \langle p| \hat{h}_0 |q\rangle a_p^\dagger a_q$$

can be written, using standard annihilation and creation operators, in normal-ordered form as

$$\hat{H}_0 = \sum_{pq} \langle p| \hat{h}_0 |q\rangle \{a_p^\dagger a_q\} + \sum_i \langle i| \hat{h}_0 |i\rangle$$

Explain the meaning of the various symbols. Which reference vacuum has been used?

8.2 Show that the two-body part of the Hamiltonian

$$\hat{H}_I = \frac{1}{4} \sum_{pqrs} \langle pq| \hat{v} |rs\rangle a_p^\dagger a_q^\dagger a_s a_r$$

can be written, using standard annihilation and creation operators, in normal-ordered form as

$$\hat{H}_I = \frac{1}{4}\sum_{pqrs} \langle pq|\, \hat{v}\, |rs\rangle\, a_p^\dagger a_q^\dagger a_s a_r$$
$$= \frac{1}{4}\sum_{pqrs} \langle pq|\, \hat{v}\, |rs\rangle \left\{a_p^\dagger a_q^\dagger a_s a_r\right\} + \sum_{pqi} \langle pi|\, \hat{v}\, |qi\rangle \left\{a_p^\dagger a_q\right\} + \frac{1}{2}\sum_{ij} \langle ij|\, \hat{v}\, |ij\rangle$$

Explain again the meaning of the various symbols.

8.3 Derive the normal-ordered form of the threebody part of the Hamiltonian.

$$\hat{H}_3 = \frac{1}{36}\sum_{\substack{pqr\\stu}} \langle pqr|\, \hat{v}_3\, |stu\rangle\, a_p^\dagger a_q^\dagger a_r^\dagger a_u a_t a_s,$$

and specify the contributions to the two-body, one-body and the constant part.

8.4 Develop a program which sets up a single-particle basis for nuclear matter calculations with input a given number of nucleons and a user specified density or Fermi momentum. Follow the setup discussed in Table 8.1. You need to define the number of particles A and the density of the system using

$$\rho = g\frac{k_F^3}{6\pi^2}.$$

Here you can either define the density itself or the Fermi momentum k_F. With the density/Fermi momentum and a fixed number of nucleons A, we can define the length L of the box used with our periodic boundary contributions via the relation

$$V = L^3 = \frac{A}{\rho}.$$

We can then can use L to define the spacing between various k-values, that is

$$\Delta k = \frac{2\pi}{L}.$$

8.5 The interaction we will use for these calculations is a semirealistic nucleon-nucleon potential known as the Minnesota potential [57] which has the form, $V_\alpha(r) = V_\alpha \exp-(\alpha r^2)$. The spin and isospin dependence of the Minnesota potential is given by,

$$V(r) = \frac{1}{2}\left(V_R + \frac{1}{2}\left(1 + P_{12}^\sigma\right) V_T + \frac{1}{2}\left(1 - P_{12}^\sigma\right) V_S\right)\left(1 - P_{12}^\sigma P_{12}^\tau\right),$$

where $P^{\sigma}_{12} = \frac{1}{2}(1+\sigma_1\cdot\sigma_2)$ and $P^{\tau}_{12} = \frac{1}{2}(1+\tau_1\cdot\tau_2)$ are the spin and isospin exchange operators, respectively. Show that a Fourier transform to momentum space results in

$$\langle \mathbf{k}_p\mathbf{k}_q|V_\alpha|\mathbf{k}_r\mathbf{k}_s\rangle = \frac{V_\alpha}{L^3}\left(\frac{\pi}{\alpha}\right)^{3/2}\exp\frac{-q^2}{4\alpha}\delta_{\mathbf{k}_p+\mathbf{k}_q,\mathbf{k}_r+\mathbf{k}_s}.$$

Write thereafter a function which sets up the full anti-symmetrized matrix elements for the Minnesota potential using the parameters listed in Table 8.2.

8.6 Consider a Slater determinant built up of orthogonal single-particle orbitals ψ_λ, with $\lambda = 1,2,\dots,A$.

The unitary transformation

$$\psi_a = \sum_\lambda C_{a\lambda}\phi_\lambda,$$

brings us into the new basis. The new basis has quantum numbers $a = 1,2,\dots,A$. Show that the new basis is orthogonal.

(a) Show that the new Slater determinant constructed from the new single-particle wave functions can be written as the determinant based on the previous basis and the determinant of the matrix C.
(b) Show that the old and the new Slater determinants are equal up to a complex constant with absolute value unity. Hint: C is a unitary matrix.

8.7 Use the ansatz for the ground state in second quantization

$$|\Phi_0\rangle = \left(\prod_{i<F}\hat{a}_i^\dagger\right)|0\rangle,$$

where the index i defines different single-particle states up to the Fermi level, to calculate using Wick's theorem (see the appendix) the expectation value

$$E[\Phi_0] = E_{\mathrm{Ref}} = \sum_{i\le F}^{A}\langle i|\hat{h}_0|i\rangle + \frac{1}{2}\sum_{ij\le F}^{A}\langle ij|\hat{v}|ij\rangle.$$

Insert thereafter the plane wave function basis for the various single-particle states and show that the above energy can be written as

$$E[\Phi_0] = \sum_{i\le F}^{A}\langle k_i|\hat{t}|k_i\rangle + \frac{1}{2}\sum_{ij\le F}^{A}\langle k_ik_j|\hat{v}|k_ik_j\rangle,$$

where we use the shorthand $|k_i\rangle = |\mathbf{k}_i,\sigma_i,\tau_{z_i}\rangle$ for the single-particle states in three dimensions.

Replace then the discrete sums with integrals, that is

$$\sum_i \to \sum_{\sigma_i}\sum_{\tau_{z_i}} \frac{L^3}{(2\pi)^3}\int_0^{\mathbf{k}_F} d\mathbf{k},$$

and show that the energy per particle A can be written as (for symmetric nuclear matter)

$$\frac{E_{\mathrm{Ref}}}{A} = \frac{3\hbar^2 k_F^2}{10M_N} + \frac{1}{2n}\frac{L^3}{(2\pi)^6}\sum_{\sigma_i\sigma_j}\sum_{\tau_{z_i}\tau_{z_j}}\int_0^{\mathbf{k}_F} d\mathbf{k}_i \int_0^{\mathbf{k}_F} d\mathbf{k}_j \langle k_i k_j|\hat{v}|k_i k_j\rangle,$$

with the density $n = V/A = L^3/A$.

Find the following expression for pure neutron matter. Use the Minnesota interaction and try to simplify the above six-dimensional integral for pure neutron matter (Hint: the interaction depends only the momentum transfer squared and fix one of the momentum integrations along the z-axis. Integrate out the dependence on the various angles).

Finally, write a program which computes the above energy for pure neutron matter using the Minnesota potential.

8.8 We will assume that we can build various Slater determinants using an orthogonal single-particle basis ψ_λ, with $\lambda = 1, 2, \ldots, A$.

The aim of this exercise is to set up specific matrix elements that will turn useful when we start our discussions of different many-body methods. In particular you will notice, depending on the character of the operator, that many matrix elements will actually be zero.

Consider three A-particle Slater determinants $|\Phi_0$, $|\Phi_i^a\rangle$ and $|\Phi_{ij}^{ab}\rangle$, where the notation means that Slater determinant $|\Phi_i^a\rangle$ differs from $|\Phi_0\rangle$ by one single-particle state, that is a single-particle state ψ_i is replaced by a single-particle state ψ_a. It will later be interpreted as a so-called one-particle-one-hole excitation. Similarly, the Slater determinant $|\Phi_{ij}^{ab}\rangle$ differs by two single-particle states from $|\Phi_0\rangle$ and is normally thought of as a two-particle-two-hole excitation.

Define a general one-body operator $\hat{F} = \sum_i^A \hat{f}(x_i)$ and a general two-body operator $\hat{G} = \sum_{i>j}^A \hat{g}(x_i, x_j)$ with g being invariant under the interchange of the coordinates of particles i and j.

(a)

$$\langle\Phi_0|\hat{F}|\Phi_0\rangle,$$

and

$$\langle\Phi_0|\hat{G}|\Phi_0\rangle.$$

(b) Find thereafter

$$\langle \Phi_0 | \hat{F} | \Phi_i^a \rangle,$$

and

$$\langle \Phi_0 | \hat{G} | \Phi_i^a \rangle,$$

(c) Finally, find

$$\langle \Phi_0 | \hat{F} | \Phi_{ij}^{ab} \rangle,$$

and

$$\langle \Phi_0 | \hat{G} | \Phi_{ij}^{ab} \rangle.$$

(d) What happens with the two-body operator if we have a transition probability of the type

$$\langle \Phi_0 | \hat{G} | \Phi_{ijk}^{abc} \rangle,$$

where the Slater determinant to the right of the operator differs by more than two single-particle states?

(e) With an orthogonal basis of Slater determinants Φ_λ, we can now construct an exact many-body state as a linear expansion of Slater determinants, that is, a given exact state

$$\Psi_i = \sum_{\lambda=0}^{\infty} C_{i\lambda} \Phi_\lambda.$$

In all practical calculations the infinity is replaced by a given truncation in the sum.

If you are to compute the expectation value of (at most) a two-body Hamiltonian for the above exact state

$$\langle \Psi_i | \hat{H} | \Psi_i \rangle,$$

based on the calculations above, which are the only elements which will contribute? [there is no need to perform any calculation here, use your results from exercises (a), (b), and (c)].

These results simplify to a large extent shell-model calculations.

8.9 Write down the analytical expressions for diagrams (8) and (9) in Fig. 8.2 and discuss whether these diagrams should be accounted for or not in the calculation of the energy per particle of infinite matter. If a Hartree-Fock basis is used, should these diagrams be included? Show also that diagrams (2), (6)–(7) and (10)–(16) are zero in infinite matter due to the lack of momentum conservation.

8.10 We present a simplified Hamiltonian consisting of an unperturbed Hamiltonian and a so-called pairing interaction term. It is a model which to a large extent mimicks some central features of atomic nuclei, certain atoms and systems which exhibit superfluidity or superconductivity. To study this system, we will use a mix of many-body perturbation theory (MBPT), Hartree-Fock (HF) theory and full configuration interaction (FCI) theory. The latter will also provide us with the exact answer. When setting up the Hamiltonian matrix you will need to solve an eigenvalue problem.

We define first the Hamiltonian, with a definition of the model space and the single-particle basis. Thereafter, we present the various exercises (some of them are solved).

The Hamiltonian acting in the complete Hilbert space (usually infinite dimensional) consists of an unperturbed one-body part, $\hat{H}_0$, and a perturbation $\hat{V}$.

We limit ourselves to at most two-body interactions and our Hamiltonian is represented by the following operators

$$\hat{H} = \sum_{\alpha\beta} \langle\alpha|h_0|\beta\rangle a_\alpha^\dagger a_\beta + \frac{1}{4}\sum_{\alpha\beta\gamma\delta} \langle\alpha\beta|V|\gamma\delta\rangle a_\alpha^\dagger a_\beta^\dagger a_\delta a_\gamma,$$

where $a_\alpha^\dagger$ and a_α etc. are standard fermion creation and annihilation operators, respectively, and $\alpha\beta\gamma\delta$ represent all possible single-particle quantum numbers. The full single-particle space is defined by the completeness relation

$$\hat{\mathbf{1}} = \sum_{\alpha=1}^{\infty} |\alpha\rangle\langle\alpha|.$$

In our calculations we will let the single-particle states $|\alpha\rangle$ be eigenfunctions of the one-particle operator $\hat{h}_0$. Note that the two-body part of the Hamiltonian contains anti-symmetrized matrix elements.

The above Hamiltonian acts in turn on various many-body Slater determinants constructed from the single-basis defined by the one-body operator $\hat{h}_0$. As an example, the two-particle model space $\mathcal{P}$ is defined by an operator

$$\hat{P} = \sum_{\alpha\beta=1}^{m} |\alpha\beta\rangle\langle\alpha\beta|,$$

where we assume that $m = \dim(\mathscr{P})$ and the full space is defined by

$$\hat{P} + \hat{Q} = \hat{\mathbf{1}},$$

with the projection operator

$$\hat{Q} = \sum_{\alpha\beta=m+1}^{\infty} |\alpha\beta\rangle\langle\alpha\beta|,$$

being the complement of $\hat{P}$.

Our specific model consists of N doubly-degenerate and equally spaced single-particle levels labelled by $p = 1, 2, \ldots$ and spin $\sigma = \pm 1$. We write the Hamiltonian as

$$\hat{H} = \hat{H}_0 + \hat{V},$$

where

$$\hat{H}_0 = \delta \sum_{p\sigma} (p-1) a_{p\sigma}^{\dagger} a_{p\sigma}$$

and

$$\hat{V} = -\frac{1}{2} g \sum_{pq} a_{p+}^{\dagger} a_{p-}^{\dagger} a_{q-} a_{q+}.$$

Here, H_0 is the unperturbed Hamiltonian with a spacing between successive single-particle states given by δ, which we will set to a constant value $\delta = 1$ without loss of generality. The two-body operator $\hat{V}$ has one term only. It represents the pairing contribution and carries a constant strength g.

The indices $\sigma = \pm$ represent the two possible spin values. The interaction can only couple pairs and excites therefore only two particles at the time.

(a) Show that the unperturbed Hamiltonian $\hat{H}_0$ and $\hat{V}$ commute with both the spin projection $\hat{S}_z$ and the total spin $\hat{S}^2$, given by

$$\hat{S}_z := \frac{1}{2} \sum_{p\sigma} \sigma a_{p\sigma}^{\dagger} a_{p\sigma}$$

and

$$\hat{S}^2 := \hat{S}_z^2 + \frac{1}{2}(\hat{S}_+ \hat{S}_- + \hat{S}_- \hat{S}_+),$$

where

$$\hat{S}_{\pm} := \sum_{p} a^{\dagger}_{p\pm} a_{p\mp}.$$

This is an important feature of our system that allows us to block-diagonalize the full Hamiltonian. We will focus on total spin $S = 0$. In this case, it is convenient to define the so-called pair creation and pair annihilation operators

$$\hat{P}^{+}_{p} = a^{\dagger}_{p+} a^{\dagger}_{p-},$$

and

$$\hat{P}^{-}_{p} = a_{p-} a_{p+},$$

respectively.

(b) Show that you can rewrite the Hamiltonian (with $\delta = 1$) as

$$\hat{H} = \sum_{p\sigma} (p-1) a^{\dagger}_{p\sigma} a_{p\sigma} - \frac{1}{2} g \sum_{pq} \hat{P}^{+}_{p} \hat{P}^{-}_{q}.$$

(c) Show also that Hamiltonian commutes with the product of the pair creation and annihilation operators. This model corresponds to a system with no broken pairs. This means that the Hamiltonian can only link two-particle states in so-called spin-reversed states.

(d) Construct thereafter the Hamiltonian matrix for a system with no broken pairs and total spin $S = 0$ for the case of the four lowest single-particle levels. Our system consists of four particles only that can occupy four doubly degenerate single-particle states. Our single-particle space consists of only the four lowest levels $p = 1, 2, 3, 4$. You need to set up all possible Slater determinants. Find all eigenvalues by diagonalizing the Hamiltonian matrix. Vary your results for values of $g \in [-1, 1]$. We refer to this as the exact calculation. Comment the behavior of the ground state as function of g.

8.11 (a) We will now set up the Hartree-Fock equations by varying the coefficients of the single-particle functions. The single-particle basis functions are defined as

$$\psi_p = \sum_{\lambda} C_{p\lambda} \psi_{\lambda}.$$

where in our case $p = 1, 2, 3, 4$ and $\lambda = 1, 2, 3, 4$, that is the first four lowest single-particle orbits. Set up the Hartree-Fock equations for this system by varying the coefficients $C_{p\lambda}$ and solve them for values of $g \in [-1, 1]$. Comment your results and compare with the exact solution. Discuss also which diagrams in Fig. 8.2 that can be affected by a Hartree-Fock basis. Compute the total binding energy using a Hartree-Fock basis and comment your results.

(b) We will now study the system using non-degenerate Rayleigh-Schrödinger perturbation theory to third order in the interaction. If we exclude the first order contribution, all possible diagrams (so-called anti-symmetric Goldstone diagrams) are shown in Fig. 8.2.

Based on the form of the interaction, which diagrams contribute to the binding energy of the ground state? Write down the expressions for the diagrams that contribute and find the contribution to the ground state energy as function $g \in [-1, 1]$. Comment your results. Compare these results with those you obtained from the exact diagonalization with and without the $4p - 4h$ state. Discuss your results for a canonical Hartree-Fock basis and a non-canonical Hartree-Fock basis.

Diagram 1 in Fig. 8.2 represents a second-order contribution to the energy and a so-called $2p - 2h$ contribution to the intermediate states. Write down the expression for the wave operator in this case and compare the possible contributions with the configuration interaction calculations without the $4p - 4h$ Slater determinant. Comment your results for various values of $g \in [-1, 1]$.

We limit now the discussion to the canonical Hartree-Fock case only. To third order in perturbation theory we can produce diagrams with $1p - 1h$, $2p - 2h$ and $3p - 3h$ intermediate excitations as shown in

Define first linked and unlinked diagrams and explain briefly Goldstone's linked diagram theorem. Based on the linked diagram theorem and the form of the pairing Hamiltonian, which diagrams will contribute to third order?

Calculate the energy to third order with a canonical Hartree-Fock basis for $g \in [-1, 1]$ and compare with the full diagonalization case in exercise (b). Discuss the results.

8.12 This project serves as a continuation of the pairing model with the aim being to solve the same problem but now by developing a program that implements the coupled cluster method with double excitations only. In doing so you will find it convenient to write classes which define the single-particle basis and the Hamiltonian. Your functions that solve the coupled cluster equations will then just need to set up variables which point to interaction elements and single-particle states with their pertinent quantum numbers.

(a) Explain why no single excitations are involved in this model.
(b) Set up the coupled cluster equations for doubles excitations and convince yourself about their meaning and correctness.
(c) Write a class which holds single-particle data like specific quantum numbers, single-particle Hamiltonian etc. Write also a class which sets up and stores two-body matrix elements defined by the single-particle states. Write thereafter functions/classes which implement the coupled cluster method with doubles only.

(d) Compare your results with those from the exact diagonalization with and without the $4p4h$ excitation. Compare also your results to perturbation theory at different orders, in particular to second order. Discuss your results.

8.13 Derive the amplitude equations of Eq. (8.31) starting with

$$0 = \langle \Phi_{i_1 \ldots i_n}^{a_1 \ldots a_n} | \overline{H} | \Phi_0 \rangle .$$

8.14 Replace the Minnesota interaction model with realistic models for nuclear forces based on effective field theory. In particular replace the Minnesota interaction with the low-order (LO) contribution which includes a contact term and a one-pion exchange term only. The expressions are discussed in Sect. 8.2.4 and Eq. (8.4). Reference [62] contains a detailed compilation of all terms up to order NNLO, with tabulated values for all constants. When adding realistic interaction models we recommend that you use the many-body perturbation theory codes to second order in the interaction, see the code link at https://github.com/ManyBodyPhysics/LectureNotesPhysics/tree/master/Programs/Chapter8-programs/cpp/MBPT2/src/.

8.10 Solutions to Selected Exercises

8.1 To solve this problem, we start by introducing the shorthand label for single-particle states below the Fermi level F as $i, j, \ldots \leq F$. For single-particle states above the Fermi level we reserve the labels $a, b, \ldots > F$, while the labels $p, q, \ldots$ represent any possible single particle state. Using the ansatz for the ground state $|\Phi_0$ as new reference vacuum state, the anticommutation relations are

$$\left\{a_p^\dagger, a_q\right\} = \delta_{pq}, p, q \leq F,$$

and

$$\left\{a_p, a_q^\dagger\right\} = \delta_{pq},\ p, q > F.$$

It is easy to see then that

$$a_i|\Phi_0\rangle = |\Phi_i\rangle \neq 0, \quad a_a^\dagger|\Phi_0\rangle = |\Phi^a\rangle \neq 0,$$

and

$$a_i^\dagger|\Phi_0\rangle = 0 \quad a_a|\Phi_0\rangle = 0.$$

We can then rewrite the one-body Hamiltonian as

$$\begin{aligned}
\hat{H}_0 &= \sum_{pq} \langle p|\, \hat{h}_0 \,|q\rangle \, a_p^\dagger a_q \\
&= \sum_{pq} \langle p|\, \hat{h}_0 \,|q\rangle \left\{a_p^\dagger a_q\right\} + \delta_{pq\in i} \sum_{pq} \langle p|\, \hat{h}_0 \,|q\rangle \\
&= \sum_{pq} \langle p|\, \hat{h}_0 \,|q\rangle \left\{a_p^\dagger a_q\right\} + \sum_{i} \langle i|\, \hat{h}_0 \,|i\rangle ,
\end{aligned}$$

where the curly brackets represent normal-ordering with respect to the new vacuum state. Withe respect to the new vacuum reference state, the

8.2 Using our anti-commutation rules, Wick's theorem discussed in the appendix and the definition of the creation and annihilation operators from the previous problem, we can rewrite the set of creation and annihilation operators of

$$\hat{H}_I = \frac{1}{4} \sum_{pqrs} \langle pq|\, \hat{v} \,|rs\rangle \, a_p^\dagger a_q^\dagger a_s a_r$$

as

$$\begin{aligned}
a_p^\dagger a_q^\dagger a_s a_r &= \left\{a_p^\dagger a_q^\dagger a_s a_r\right\} \\
&+ \left\{a_p^\dagger \overline{a_q^\dagger a_s} a_r\right\} + \left\{a_p^\dagger \overline{a_q^\dagger a_s a_r}\right\} + \left\{\overline{a_p^\dagger a_q^\dagger a_s} a_r\right\} \\
&+ \left\{\overline{a_p^\dagger a_q^\dagger a_s a_r}\right\} + \left\{\overline{a_p^\dagger \overline{a_q^\dagger a_s} a_r}\right\} + \left\{\overline{a_p^\dagger \overline{a_q^\dagger a_s a_r}}\right\} \\
&= \left\{a_p^\dagger a_q^\dagger a_s a_r\right\} + \delta_{qs\in i} \left\{a_p^\dagger a_r\right\} - \delta_{qr\in i} \left\{a_p^\dagger a_s\right\} - \delta_{ps\in i} \left\{a_q^\dagger a_r\right\} \\
&+ \delta_{pr\in i} \left\{a_q^\dagger a_s\right\} + \delta_{pr\in i}\delta_{qs\in i} - \delta_{ps\in i}\delta_{qr\in i}.
\end{aligned}$$

Inserting the redefinition of the creation and annihilation operators with respect to the new vacuum state, we have

$$\begin{aligned}
\hat{H}_I &= \frac{1}{4} \sum_{pqrs} \langle pq|\, \hat{v} \,|rs\rangle \, a_p^\dagger a_q^\dagger a_s a_r \\
&= \frac{1}{4} \sum_{pqrs} \langle pq|\, \hat{v} \,|rs\rangle \left\{a_p^\dagger a_q^\dagger a_s a_r\right\} + \frac{1}{4} \sum_{pqrs} \Big(\delta_{qs\in i} \langle pq|\, \hat{v} \,|rs\rangle \left\{a_p^\dagger a_r\right\} \\
&- \delta_{qr\in i} \langle pq|\, \hat{v} \,|rs\rangle \left\{a_p^\dagger a_s\right\} - \delta_{ps\in i} \langle pq|\, \hat{v} \,|rs\rangle \left\{a_q^\dagger a_r\right\} \\
&+ \delta_{pr\in i} \langle pq|\, \hat{v} \,|rs\rangle \left\{a_q^\dagger a_s\right\} + \delta_{pr\in i}\delta_{qs\in i} - \delta_{ps\in i}\delta_{qr\in i} \Big)
\end{aligned}$$

$$
\begin{aligned}
&= \frac{1}{4}\sum_{pqrs} \langle pq|\, \hat{v}\, |rs\rangle \left\{a_p^\dagger a_q^\dagger a_s a_r\right\} \\
&+ \frac{1}{4}\sum_{pqi} \Big(\langle pi|\, \hat{v}\, |qi\rangle - \langle pi|\, \hat{v}\, |iq\rangle - \langle ip|\, \hat{v}\, |qi\rangle + \langle ip|\, \hat{v}\, |iq\rangle \Big) \left\{a_p^\dagger a_q\right\} \\
&+ \frac{1}{4}\sum_{ij} \Big(\langle ij|\, \hat{v}\, |ij\rangle - \langle ij|\, \hat{v}\, |ji\rangle \Big) \\
&= \frac{1}{4}\sum_{pqrs} \langle pq|\, \hat{v}\, |rs\rangle \left\{a_p^\dagger a_q^\dagger a_s a_r\right\} + \sum_{pqi} \langle pi|\, \hat{v}\, |qi\rangle \left\{a_p^\dagger a_q\right\} + \frac{1}{2}\sum_{ij} \langle ij|\, \hat{v}\, |ij\rangle \,.
\end{aligned}
$$

Summing up, we obtain a two-body part defined as

$$
\hat{V}_N = \frac{1}{4}\sum_{pqrs} \langle pq|\, \hat{v}\, |rs\rangle \left\{a_p^\dagger a_q^\dagger a_s a_r\right\},
$$

a one-body part given by

$$
\hat{F}_N = \sum_{pqi} \langle pi|\, \hat{v}\, |qi\rangle \left\{a_p^\dagger a_q\right\},
$$

and finally the so-called reference energy

$$
E_{\mathrm{ref}} = \frac{1}{2}\sum_{ij} \langle ij|\, \hat{v}\, |ij\rangle \,.
$$

which is the energy expectation value for the reference state. Thus, our normal-ordered Hamiltonian with at most a two-body nucleon-nucleon interaction is defined as

$$
\hat{H}_N = \frac{1}{4}\sum_{pqrs} \langle pq|\, \hat{v}\, |rs\rangle \left\{a_p^\dagger a_q^\dagger a_s a_r\right\} + \sum_{pq} f_q^p \left\{a_p^\dagger a_q\right\} = \hat{V}_N + \hat{F}_N,
$$

with

$$
\hat{F}_N = \sum_{pq} f_q^p \left\{a_p^\dagger a_q\right\},
$$

and

$$
\hat{V}_N = \frac{1}{4}\sum_{pqrs} \langle pq|\, \hat{v}\, |rs\rangle \left\{a_p^\dagger a_q^\dagger a_s a_r\right\},
$$

where

$$f_q^p = \langle p|\,\hat{h}_0\,|q\rangle + \sum_i \langle pi|\,\hat{v}\,|qi\rangle$$

8.4 The following python code sets up the quantum numbers for both infinite nuclear matter and neutron matter employing a cutoff in the value of n. The full code can be found at https://github.com/ManyBodyPhysics/LectureNotesPhysics/tree/master/Programs/Chapter8-programs/python/spstatescc.py.

```
from numpy import *

nmax =2
nshell = 3*nmax*nmax
count = 1
tzmin = 1

print "Symmetric nuclear matter:"
print "a, nx, ny, nz, sz, tz, nx^2 + ny^2 + nz^2"
for n in range(nshell):
   for nx in range(-nmax,nmax+1):
       for ny in range(-nmax,nmax+1):
         for nz in range(-nmax, nmax+1):
            for sz in range(-1,1+1):
               tz = 1
               for tz in range(-tzmin,tzmin+1):
                  e = nx*nx + ny*ny + nz*nz
                  if e == n:
                     if sz != 0:
                        if tz != 0:
                           print count, " ",nx," ",ny, " ",nz,"
                               ",sz," ",tz," ",e
                           count += 1


nmax =1
nshell = 3*nmax*nmax
count = 1
tzmin = 1
print "------------------------------------"
print "Neutron matter:"
print "a, nx, ny, nz, sz, nx^2 + ny^2 + nz^2"
for n in range(nshell):
   for nx in range(-nmax,nmax+1):
       for ny in range(-nmax,nmax+1):
         for nz in range(-nmax, nmax+1):
            for sz in range(-1,1+1):
               e = nx*nx + ny*ny + nz*nz
               if e == n:
                  if sz != 0:
                     print count, " ",nx," ",ny, " ",sz," ",tz,"
                         ",e
                     count += 1
```

Acknowledgements We are much indebted to Carlo Barbieri, Scott Bogner, Alex Brown, David Dean, Heiko Hergert, Dean Lee, Titus Morris, Thomas Papenbrock, Nathan Parzuchowski, Piotr Piecuch and Fei Yuan for many discussions on many-body theory. Computational resources were provided by Michigan State University and the Norwegian Notur project (Supercomputing grant NN2977K). This work was supported by NSF Grant No. PHY-1404159 (Michigan State University), by the Oce of Nuclear Physics, U.S. Department of Energy, under grants DE-FG02-96ER40963, DE-SC0008499 (NUCLEI SciDAC collaboration) and the Field Work Proposal ERKBP57 at Oak Ridge National Laboratory (ORNL).

This research used also resources of the Oak Ridge Leadership Computing Facility located at ORNL, which is supported by the Office of Science of the Department of Energy under Contract No. DE-AC05-00OR22725.

Appendix, Wick's Theorem

Wick's theorem is based on two fundamental concepts, namely *normal ordering* and *contraction*. The normal-ordered form of $\hat{A}\hat{B}..\hat{X}\hat{Y}$, where the individual terms are either a creation or annihilation operator, is defined as

$$\left\{\hat{A}\hat{B}..\hat{X}\hat{Y}\right\} \equiv (-1)^p \,[\text{creation operators}] \cdot [\text{annihilation operators}]\,. \tag{8.59}$$

The p subscript denotes the number of permutations that is needed to transform the original string into the normal-ordered form. A contraction between to arbitrary operators $\hat{X}$ and $\hat{Y}$ is defined as

$$\overbrace{\hat{X}\hat{Y}} \equiv \langle 0|\hat{X}\hat{Y}|0\rangle. \tag{8.60}$$

It is also possible to contract operators inside a normal ordered products. We define the original relative position between two operators in a normal ordered product as p, the so-called permutation number. This is the number of permutations needed to bring one of the two operators next to the other one. A contraction between two operators with $p \neq 0$ inside a normal ordered is defined as

$$\left\{\overbrace{\hat{A}\hat{B}..\hat{X}}\hat{Y}\right\} = (-1)^p \left\{\overbrace{\hat{A}\hat{B}}..\hat{X}\hat{Y}\right\}. \tag{8.61}$$

In the general case with m contractions, the procedure is similar, and the prefactor changes to

$$(-1)^{p_1+p_2+..+p_m}. \tag{8.62}$$

Wick's theorem states that every string of creation and annihilation operators can be written as a sum of normal ordered products with all possible ways of contractions,

$$\hat{A}\hat{B}\hat{C}\hat{D}..\hat{R}\hat{X}\hat{Y}\hat{Z} = \left\{\hat{A}\hat{B}\hat{C}\hat{D}..\hat{R}\hat{X}\hat{Y}\hat{Z}\right\} \tag{8.63}$$

$$+\sum_{[1]}\left\{\overline{\hat{A}\hat{B}}\hat{C}\hat{D}..\hat{R}\hat{X}\hat{Y}\hat{Z}\right\} \tag{8.64}$$

$$+\sum_{[2]}\left\{\overline{\hat{A}\hat{B}\hat{C}\hat{D}}..\hat{R}\hat{X}\hat{Y}\hat{Z}\right\} \tag{8.65}$$

$$+\dots \tag{8.66}$$

$$+\sum_{[\frac{N}{2}]}\left\{\overline{\hat{A}\hat{B}\hat{C}\hat{D}}..\overline{\hat{R}\hat{X}\hat{Y}\hat{Z}}\right\}. \tag{8.67}$$

The $\sum_{[m]}$ means the sum over all terms with m contractions, while $\left[\frac{N}{2}\right]$ means the largest integer that not do not exceeds $\frac{N}{2}$ where N is the number of creation and annihilation operators. When N is even,

$$\left[\frac{N}{2}\right] = \frac{N}{2}, \tag{8.68}$$

and the last sum in Eq. (8.63) is over fully contracted terms. When N is odd,

$$\left[\frac{N}{2}\right] \neq \frac{N}{2}, \tag{8.69}$$

and none of the terms in Eq. (8.63) are fully contracted.

An important extension of Wick's theorem allow us to define contractions between normal-ordered strings of operators. This is the so-called generalized Wick's theorem,

$$\left\{\hat{A}\hat{B}\hat{C}\hat{D}..\right\}\left\{\hat{R}\hat{X}\hat{Y}\hat{Z}..\right\} = \left\{\hat{A}\hat{B}\hat{C}\hat{D}..\hat{R}\hat{X}\hat{Y}\hat{Z}\right\} \tag{8.70}$$

$$+\sum_{[1]}\left\{\overline{\hat{A}\hat{B}\hat{C}\hat{D}..\hat{R}}\hat{X}\hat{Y}\hat{Z}\right\} \tag{8.71}$$

$$+\sum_{[2]}\left\{\hat{A}\hat{B}\hat{C}\hat{D}..\hat{R}\hat{X}\hat{Y}\hat{Z}\right\} \tag{8.72}$$

$$+\ldots \tag{8.73}$$

Turning back to the many-body problem, the vacuum expectation value of products of creation and annihilation operators can be written, according to Wick's theorem in Eq. (8.63), as a sum over normal ordered products with all possible numbers and combinations of contractions,

$$\langle 0|\hat{A}\hat{B}\hat{C}\hat{D}..\hat{R}\hat{X}\hat{Y}\hat{Z}|0\rangle = \langle 0|\left\{\hat{A}\hat{B}\hat{C}\hat{D}..\hat{R}\hat{X}\hat{Y}\hat{Z}\right\}|0\rangle \tag{8.74}$$

$$+\sum_{[1]}\langle 0|\left\{\hat{A}\hat{B}\hat{C}\hat{D}..\hat{R}\hat{X}\hat{Y}\hat{Z}\right\}|0\rangle \tag{8.75}$$

$$+\sum_{[2]}\langle 0|\left\{\hat{A}\hat{B}\hat{C}\hat{D}..\hat{R}\hat{X}\hat{Y}\hat{Z}\right\}|0\rangle \tag{8.76}$$

$$+\ldots \tag{8.77}$$

$$+\sum_{[\frac{N}{2}]}\langle 0|\left\{\hat{A}\hat{B}\hat{C}\hat{D}..\hat{R}\hat{X}\hat{Y}\hat{Z}\right\}|0\rangle. \tag{8.78}$$

All vacuum expectation values of normal ordered products without fully contracted terms are zero. Hence, the only contributions to the expectation value are those terms that *is* fully contracted,

$$\langle 0|\hat{A}\hat{B}\hat{C}\hat{D}..\hat{R}\hat{X}\hat{Y}\hat{Z}|0\rangle = \sum_{[all]}\langle 0|\left\{\hat{A}\hat{B}\hat{C}\hat{D}..\hat{R}\hat{X}\hat{Y}\hat{Z}\right\}|0\rangle \tag{8.79}$$

$$= \sum_{[all]}\hat{A}\hat{B}\hat{C}\hat{D}..\hat{R}\hat{X}\hat{Y}\hat{Z}. \tag{8.80}$$

To obtain fully contracted terms, Eq. (8.68) must hold. When the number of creation and annihilation operators is odd, the vacuum expectation value can be set to zero at once. When the number is even, the expectation value is simply the sum of terms with all possible combinations of fully contracted terms. Observing that the only contractions that give nonzero contributions are

$$a_\alpha a_\beta^\dagger = \delta_{\alpha\beta}, \tag{8.81}$$

the terms that contribute are reduced even more.

Wick's theorem provides us with an algebraic method for easy determination of the terms that contribute to the matrix element.

References

1. S.L. Shapiro, S.A. Teukolsky, *Black Holes, White Dwarfs, and Neutron Stars: The Physics of Compact Objects* (Wiley, New York, 1983)
2. J.M. Lattimer, M. Prakash, Astrophys. J. **550**, 426 (2001)
3. J.M. Lattimer, M. Prakash, Phys. Rep. **442**, 109 (2007)
4. A.W. Steiner, J.M. Lattimer, E.F. Brown, Astrophys J. **722**, 33 (2010)
5. A.W. Steiner, S. Gandolfi, Phys. Rev. Lett. **108**, 081102 (2012)
6. J.M. Lattimer, Ann. Rev. Nucl. Part. Sci. **62**, 485 (2012)
7. F. Weber, *Pulsars as Astrophysical Laboratories for Nuclear and Particle Physics* (Institute of Physics Publishing, London, 1999)
8. H. Heiselberg, M. Hjorth-Jensen, Phys. Rep. **328**, 237 (2000)
9. S. Weinberg, Phys. Lett. B **251**, 288 (1990)
10. S. Weinberg, Nucl. Phys. B **363**, 3 (1991)
11. C. Ordonez, U. van Kolck, Phys. Lett. B **291**, 459 (1992)
12. C. Ordonez, L. Ray, U. van Kolck, Phys. Rev. Lett. **72**, 1982 (1994)
13. U. van Kolck, Phys. Rev. C **49**, 2932 (1994)
14. U. van Kolck, Prog. Part. Nucl. Phys. **43**, 337 (1999)
15. R. Machleidt, D. Entem, Phys. Rep. **503**, 1 (2011)
16. E. Epelbaum, H.W. Hammer, U.G. Meißner, Rev. Mod. Phys. **81**, 1773 (2009)
17. G. Hagen, T. Papenbrock, M. Hjorth-Jensen, D.J. Dean, Rep. Prog. Phys. **77**, 096302 (2014)
18. A. Ekström, G.R. Jansen, K.A. Wendt, G. Hagen, T. Papenbrock, B.D. Carlsson, C. Forssén, M. Hjorth-Jensen, P. Navrátil, W. Nazarewicz, Phys. Rev. C **91**, 051301 (2015)
19. G. Hagen, M. Hjorth-Jensen, G.R. Jansen, T. Papenbrock, Phys. Scr. **91**, 063006 (2016)
20. L. Coraggio, J.W. Holt, N. Itaco, R. Machleidt, L.E. Marcucci, F. Sammarruca, Phys. Rev. C **89**, 044321 (2014)
21. F. Sammarruca, L. Coraggio, J.W. Holt, N. Itaco, R. Machleidt, L.E. Marcucci, Phys. Rev. C **91**, 054311 (2015)
22. R. Roth, S. Binder, K. Vobig, A. Calci, J. Langhammer, P. Navrátil, Phys. Rev. Lett. **109**, 052501 (2012)
23. S. Binder, P. Piecuch, A. Calci, J. Langhammer, P. Navrátil, R. Roth, Phys. Rev. C **88**, 054319 (2013)
24. H. Hergert, S. Bogner, T. Morris, A. Schwenk, K. Tsukiyama, Phys. Rep. **621**, 165 (2016)
25. P. Navratil, S. Quaglioni, G. Hupin, C. Romero-Redondo, A. Calci, Phys. Scr. **91**, 053002 (2016)
26. A. Carbone, A. Polls, A. Rios, Phys. Rev. C **88**, 044302 (2013)
27. T. Inoue, S. Aoki, T. Doi, T. Hatsuda, Y. Ikeda, N. Ishii, K. Murano, H. Nemura, K. Sasaki, Phys. Rev. Lett. **111**, 112503 (2013)
28. B.D. Day, Rev. Mod. Phys. **39**, 719 (1967)
29. K.A. Brueckner, C.A. Levinson, H.M. Mahmoud, Phys. Rev. **95**, 217 (1954)
30. K.A. Brueckner, Phys. Rev. **100**, 36 (1955)
31. M. Baldo, G.F. Burgio, Rep. Prog. Phys. **75**, 026301 (2012)
32. M. Baldo, A. Polls, A. Rios, H.J. Schulze, I. Vidaña, Phys. Rev. C **86**, 064001 (2012)
33. L. Coraggio, J.W. Holt, N. Itaco, R. Machleidt, F. Sammarruca, Phys. Rev. C **87**, 014322 (2013)
34. R.J. Bartlett, M. Musiał, Rev. Mod. Phys. **79**, 291 (2007)
35. I. Shavitt, R.J. Bartlett, *Many-Body Methods in Chemistry and Physics* (Cambridge University Press, Cambridge, 2009)

36. G. Baardsen, A. Ekström, G. Hagen, M. Hjorth-Jensen, Phys. Rev. C **88**, 054312 (2013)
37. G. Hagen, T. Papenbrock, A. Ekström, K.A. Wendt, G. Baardsen, S. Gandolfi, M. Hjorth-Jensen, C.J. Horowitz, Phys. Rev. C **89**, 014319 (2014)
38. G. Hagen, G.R. Jansen, T. Papenbrock, Phys. Rev. Lett. **117**, 172501 (2016)
39. J. Carlson, J. Morales, V.R. Pandharipande, D.G. Ravenhall, Phys. Rev. C **68**, 025802 (2003)
40. S. Gandolfi, A.Y. Illarionov, K.E. Schmidt, F. Pederiva, S. Fantoni, Phys. Rev. C **79**, 054005 (2009)
41. S. Gandolfi, A.Y. Illarionov, F. Pederiva, K.E. Schmidt, S. Fantoni, Phys. Rev. C **80**, 045802 (2009)
42. A. Gezerlis, J. Carlson, Phys. Rev. C **81**, 025803 (2010)
43. S. Gandolfi, J. Carlson, S. Reddy, Phys. Rev. C **85**, 032801 (2012)
44. A. Lovato, O. Benhar, S. Fantoni, K.E. Schmidt, Phys. Rev. C **85**, 024003 (2012)
45. A. Gezerlis, I. Tews, E. Epelbaum, S. Gandolfi, K. Hebeler, A. Nogga, A. Schwenk, Phys. Rev. Lett. **111**, 032501 (2013)
46. J. Carlson, S. Gandolfi, F. Pederiva, S.C. Pieper, R. Schiavilla, K.E. Schmidt, R.B. Wiringa, Rev. Mod. Phys. **87**, 1067 (2015)
47. V. Somà, A. Cipollone, C. Barbieri, P. Navrátil, T. Duguet, Phys. Rev. C **89**, 061301 (2014)
48. W. Dickhoff, C. Barbieri, Prog. Part. Nucl. Phys. **52**, 377 (2004)
49. T.D. Morris, N.M. Parzuchowski, S.K. Bogner, Phys. Rev. C **92**, 034331 (2015)
50. M. Prakash, I. Bombaci, M. Prakash, P.J. Ellis, J.M. Lattimer, R. Knorren, Phys. Rep. **280**, 1 (1997)
51. C.J. Horowitz, E.F. Brown, Y. Kim, W.G. Lynch, R. Michaels, A. Ono, J. Piekarewicz, M.B. Tsang, H.H. Wolter, J. Phys. G **41**, 093001 (2014)
52. J.P. Blaizot, G. Ripka, *Quantum Theory of Finite Systems* (MIT press, Cambridge, 1986)
53. W. Dickhoff, D. Van Neck, *Many-Body Theory Exposed!: Propagator Description of Quantum Mechanics in Many-Body Systems*, 2nd edn. (World Scientific, New Jersey, 2008)
54. R.D. Mattuck, *A guide to Feynman Diagrams in the Many-Body Problem*, 2nd edn. (Dover, New York, 1992)
55. S. Binder, A. Ekström, G. Hagen, T. Papenbrock, K.A. Wendt, Phys. Rev. C **93**, 044332 (2016)
56. K.S. McElvain, W.C. Haxton, Nuclear Physics without High-Momentum Potentials: Direct Construction of the Effective Interaction from Scattering Observables (2016). ArXiv:1607.06863
57. D. Thompson, M. Lemere, Y. Tang, Nucl. Phys. A **286**, 53 (1977)
58. S. Weinberg, Phys. Lett. B **363**, 288 (1990)
59. C. Ordóñez, L. Ray, U. van Kolck, Phys. Rev. C **53**, 2086 (1996)
60. U. van Kolck, Phys. Rev. C **49**, 2932 (1994)
61. A. Ekström, G. Baardsen, C. Forssén, G. Hagen, M. Hjorth-Jensen, G.R. Jansen, R. Machleidt, W. Nazarewicz, T. Papenbrock, J. Sarich, S.M. Wild, Phys. Rev. Lett. **110**, 192502 (2013)
62. B.D. Carlsson, A. Ekström, C. Forssén, D.F. Strömberg, G.R. Jansen, O. Lilja, M. Lindby, B.A. Mattsson, K.A. Wendt, Phys. Rev. X **6**, 011019 (2016)
63. E. Epelbaum, A. Nogga, W. Glöckle, H. Kamada, U.G. Meißner, H. Witała, Phys. Rev. C **66**, 064001 (2002)
64. P. Navrátil, S. Quaglioni, I. Stetcu, B.R. Barrett, J. Phys. G **36**(8), 083101 (2009)
65. G. Golub, C. Van Loan, *Matrix Computations* (John Hopkins University Press, Baltimore, 1996)
66. C. Lanczos, J. Res. Natl. Bur. Stand. **45**, 255 (1950)
67. E.R. Davidson, Comput. Phys. Commun. **53**, 49 (1989)
68. E.R. Davidson, Comput. Phys. **5**, 519 (1993)
69. F. Coester, Nucl. Phys. **7**, 421 (1958)
70. F. Coester, H. Kümmel, Nucl. Phys. **17**, 477 (1960)
71. H. Kümmel, K.H. Lührmann, J.G. Zabolitzky, Phys. Rep. **36**, 1 (1978)
72. B.H. Brandow, Rev. Mod. Phys. **39**, 771 (1967)
73. R.J. Bartlett, Ann. Rev. Phys. Chem. **32**, 359 (1981)

74. G.R. Jansen, M.D. Schuster, A. Signoracci, G. Hagen, P. Navrátil, Phys. Rev. C **94**, 011301 (2016)
75. A. Baran, A. Bulgac, M.M. Forbes, G. Hagen, W. Nazarewicz, N. Schunck, M.V. Stoitsov, Phys. Rev. C **78**, 014318 (2008)
76. L.S. Blackford, J. Demmel, I. Du, G. Henry, M. Heroux, L. Kaufman, A. Lumsdaine, A. Petitet, R.C. Whaley, ACM Trans. Math. Softw. **28**, 135 (2002)
77. W. Gropp, E. Lusk, A. Skjellum, *Using MPI* (MIT Press, Cambridge, 1999)
78. E. Gabriel, G.E. Fagg, G. Bosilca, T. Angskun, J.J. Dongarra, J.M. Squyres, V. Sahay, P. Kambadur, B. Barrett, A. Lumsdaine, R.H. Castain, D.J. Daniel, R.L. Graham, T.S. Woodall, in *Proceedings, 11th European PVM/MPI Users' Group Meeting*, Budapest (2004), p. 97
79. R.C. Martin, *Clean Code: A Handbook of Agile Software Craftsmanship* (Prentice Hall, Upper Saddle River, 2015)
80. C. Gros, Zeit. Phys. B **86**, 359 (1992)
81. C. Gros, Phys. Rev. B **53**, 6865 (1996)
82. J.J. Shepherd, A. Grüneis, G.H. Booth, G. Kresse, A. Alavi, Phys. Rev. B **86**, 035111 (2012)
83. J.J. Shepherd, J. Chem. Phys. **145**, 031104 (2016)
84. C. Lin, F.H. Zong, D.M. Ceperley, Phys. Rev. E **64**, 016702 (2001)
85. B. Chapman, G. Jost, R. van der Pas, *Using OpenMP. Portable Shared Memory Parallel Programming* (MIT Press, Cambridge, 2008)

Chapter 9
Variational and Diffusion Monte Carlo Approaches to the Nuclear Few- and Many-Body Problem

Francesco Pederiva, Alessandro Roggero, and Kevin E. Schmidt

9.1 Monte Carlo Methods in Quantum Many-Body Physics

9.1.1 Expectations in Quantum Mechanics

In the previous chapters the authors pointed out in several different ways that the non-relativistic quantum many-body problem is equivalent to the solution of a very complicated differential equation, the many-body Schrödinger equation.

As it was illustrated, in the few-body case ($A < 6$) it possible to find compute exact solutions. At the very least, one can expand the eigenfunctions on a basis set including $\mathscr{M}$ elements, diagonalize the Hamiltonian matrix, and try to reach convergence as a function of $\mathscr{M}$. Unfortunately, this procedure becomes more and more expensive when the number of bodies A increases. There are many ingenuous ways to improve the speed of convergence and the quality of the results. The price to pay often is the introduction of more or less controlled approximations.

F. Pederiva (✉)
Physics Department, University of Trento, and INFN-TIFPA, via Sommarive 14, 38123 Trento, Italy
e-mail: francesco.pederiva@unitn.it

A. Roggero
Institute for Nuclear Theory, University of Washington, Seattle, WA 98195-1550, USA
e-mail: roggero@uw.edu

K.E. Schmidt
Department of Physics and Astronomy, Arizona State University, Tempe, AZ 85283-1506, USA
e-mail: kevin.schmidt@asu.edu

M. Hjorth-Jensen et al. (eds.), *An Advanced Course in Computational Nuclear Physics*, Lecture Notes in Physics 936, DOI 10.1007/978-3-319-53336-0_9

All these approaches have one common feature: they end up with some closed expression for the eigenfunctions. However, we should remember that the wavefunction per se is not an observable. In order to make predictions to be compared with experiments, we only need a way to compute *expectations* of operators $\hat{O}$ describing the observables we are interested in.

Given a many-body Hamiltonian $\hat{H}$, we might want, for instance, to look for the ground state eigenfunction and eigenvalue. This means that we want to solve the following equation:

$$\hat{H}|\Psi_0\rangle = E_0|\Psi_0\rangle. \tag{9.1}$$

At this point we to provide a representation of the Hilbert space in term of some basis set. This set will be denoted as $\{|X\rangle\}$. Its elements could be eigenstates of the position or of the momentum operators, or eigenstates of a simpler Hamiltonian of which we know the exact spectrum. In order to make the notation less cumbersome, we will assume that the quantum numbers X characterizing the basis states are in the continuum. In the case of a discrete spectrum, integrals in the following have to be replaced by sums over all their possible values, without any loss of generality. As an example, X could include the positions or the momenta of A nucleons, and their spin and isospin values.

All the physical information we need about the time-independent problem is then included in integrals of the form:

$$\langle O\rangle \equiv \langle\Psi_0|\hat{O}\Psi_0\rangle = \frac{\int dXdX'\langle\Psi_0|X\rangle\langle X|\hat{O}|X'\rangle\langle X'|\Psi_0\rangle}{\int dX|\langle X|\Psi_0\rangle|^2}. \tag{9.2}$$

These integrals are apparently as hard to solve as the Schrödinger equation itself, even if we had access to the explicit form of the wavefunction. Is there any real gain in reformulating the problem this way?

We can first notice that expectations can in general be written in a slightly different form, independent of the nature of the operator $\hat{O}$:

$$\langle O\rangle = \frac{\int dX|\langle X|\Psi_0\rangle|^2\frac{\langle X|\hat{O}\Psi_0\rangle}{\langle X|\Psi_0\rangle}}{\int dX|\langle X|\Psi_0\rangle|^2}. \tag{9.3}$$

For the moment we will just assume that the quotient appearing at numerator of the expectation is always well defined, and we will later discuss this aspect in more detail. The standard quantum mechanical interpretation of the wavefunction tells us that the quantity:

$$P[X] = \frac{|\langle X|\Psi_0\rangle|^2}{\int dX|\langle X|\Psi_0\rangle|^2}, \tag{9.4}$$

is the probability density of finding the system in the state $|X\rangle$ labeled by the set of quantum numbers X. Thereby, the expectation integral has the general form:

$$\langle O\rangle = \int dXP[X]\frac{\langle X|\hat{O}\Psi_0\rangle}{\langle X|\Psi_0\rangle}, \tag{9.5}$$

i.e. the average of what we will call the *local* operator $O_{loc} \equiv \frac{\langle X|\hat{O}\Psi_0\rangle}{\langle X|\Psi_0\rangle}$ weighted with the probability of finding the system in a given state $|X\rangle$. Integrals like that in Eq. (9.5) have a direct physical interpretation. In a measurement process what we would observe is essentially the result of a *sampling process* of $P[X]$. The expectation of our operator is approximated by:

$$\langle O\rangle \simeq \frac{1}{M}\sum_{k=1}^{M} O(X_k), \tag{9.6}$$

where M is the number of measurements performed, and $O(X_k)$ is a shorthand notation to indicate the value assumed by the observable $\hat{O}$ in the state labeled by the quantum numbers X_k. The laws of statistics also give us a way of estimating a *statistical* error on $\langle O\rangle$, and we know that the error decreases by increasing the number of measurements.

There is here an important point to notice: in a physical measurement process we have *no direct knowledge of the wavefunction*, we just *sample* its squared modulus!

This argument suggests that if we had a numerical way of sampling the squared modulus of a wavefunction, we could in principle compute expectations and make comparisons with experiments without needing an explicit expression of the wavefunction itself. Quantum Monte Carlo methods aim exactly at solving the many-body Schrödinger equation by sampling its solutions, eventually without any need of an explicit analytical form.

The remainder of this chapter will be organized as follows. First we will discuss how to perform calculations based on an accurate, explicit ansatz for the wavefunction of an A-body system interacting via a purely central potential, exploiting the variational principle of quantum mechanics (Variational Monte Carlo methods). Then we will discuss how to sample the exact ground state of the system by projecting it out of an initial ansatz (Projection Monte Carlo methods). Finally, we will see how these methods need to be extended when we are interested in studying Hamiltonians that have an explicit dependence on the spin and isospin states of the particles, as it happens for the modern interactions employed in nuclear physics.

9.2 Variational Wavefunctions and VMC for Central Potentials

9.2.1 Coordinate Space Formulation

As previously discussed, we are in principle free to choose any representation of the Hilbert space of the system we like, in order to compute expectations. The most convenient choice, for a system of particles interacting via a purely central potential, with no explicit dependence on the spin or isospin state, is to use the eigenstates of the position operator. If $R = \mathbf{r}_1, \ldots \mathbf{r}_A$ are the coordinates of the A (identical)[1] particles of mass m constituting the system, we have that:

$$|X\rangle \equiv |R\rangle \tag{9.7}$$

with the normalization:

$$\langle R'|R\rangle = \delta(R - R'). \tag{9.8}$$

Notice that we are here considering a $3A$-dimensional Cartesian space, without decomposing it in the product of A 3-dimensional spaces. In this representation the wavefunction is simply given by:

$$\langle R|\Psi_0\rangle \equiv \Psi_0(R) = \Psi_0(\mathbf{r}_1, \ldots \mathbf{r}_A). \tag{9.9}$$

The Hamiltonian instead reads:

$$\hat{H} = \sum_{i=1}^{A} \frac{p_i^2}{2m} + V(\mathbf{r}_1, \ldots \mathbf{r}_A), \tag{9.10}$$

or

$$\hat{H} = \int dR |R\rangle \left[-\frac{\hbar^2}{2m} \sum_{i=1}^{A} \nabla_i^2 + V(\mathbf{r}_1, \ldots \mathbf{r}_A) \right] \langle R|, \tag{9.11}$$

where V is the interparticle potential. Substituting this form into Eq. (9.1), operating from the left with $\langle R|$ gives the Schrödinger differential equation

$$\left[-\frac{\hbar^2}{2m} \sum_{i=1}^{A} \nabla_i^2 + V(\mathbf{r}_1, \ldots \mathbf{r}_A) \right] \Psi_0(R) = E_0 \Psi_0(R). \tag{9.12}$$

[1] We will always refer to systems of identical particle throughout the text. The generalization to mixtures is normally straightforward, and it will not be discussed here.

We will often use the same symbol for the Hilbert space operator and its differential form and write this simply as $\hat{H}\Psi_0(R) = E_0\Psi_0(R)$; whether the operator or differential form is used can be discerned readily from context. In this representation the states of the Hilbert space are sampled by sampling the particle positions from the squared modulus of the wavefunction $|\Psi_0(R)|^2$.

9.2.2 Variational Principle and Variational Wavefunctions

As already seen in the previous chapters, one of the possible ways to approximate a solution of the many-body Schrödinger equation is to exploit the variational principle. Given a *trial state* $|\Psi_T\rangle$, the following inequality holds:

$$E_T = \frac{\langle \Psi_T|\hat{H}\Psi_T\rangle}{\langle \Psi_T|\Psi_T\rangle} \geq E_0, \tag{9.13}$$

where E_0 is the ground state eigenvalue of the Hamiltonian $\hat{H}$. The equality holds if and only if $|\Psi_T\rangle = |\Psi_0\rangle$. The variational principle holds for the ground state, but also for excited states, provided that $|\Psi_T\rangle$ is orthogonal to all the eigenstates having eigenvalue lower than that of the state one wants to approximate.

In coordinate space the formulation of the variational principle can be directly transformed in a form equivalent to that of Eq. (9.5):

$$E_T = \frac{\int dR|\Psi_T(R)|^2 \frac{\hat{H}\Psi_T(R)}{\Psi_T(R)}}{\int dR|\Psi_T(R)|^2} \geq E_0, \tag{9.14}$$

where $\frac{\hat{H}\Psi_T(R)}{\Psi_T(R)}$ is called the *local energy*. Contrary to what happens in functional minimization approaches (such as the Hartree-Fock method), the variational principle is used to determine the best trial wavefunction within a class defined by some proper ansatz. The wavefunction will depend on a set of *variational parameters* $\{\alpha\}$. The solution of the variational problem will therefore be given by the solution of the Euler problem:

$$\frac{\delta E_T(\{\alpha\})}{\delta\{\alpha\}} = 0. \tag{9.15}$$

This means that in order to find the variational solution to the Schrödinger problem we need to evaluate many times the integral of Eq. (9.14) using different values of the variational parameters, and find the minimum trial eigenvalue.

9.2.3 Monte Carlo Evaluation of Integrals

The integral in Eq. (9.14) is in general defined in a $3A$-dimensional space. Since particles interact, we expect that the solution cannot be expressed as a product of single particle functions, and therefore the integral cannot be factorized in a product of simpler integrals. In this sense, the problem is strictly analogous to that of a classical gas at finite temperature $\beta = 1/K_BT$. In that case, given a classical Hamiltonian $H(p,q) = \sum_{i=1}^{A} \frac{p_i^2}{2m} + V(q_1 \dots q_A)$, the average energy of the system is given by:

$$E = \frac{3A}{2} K_B T + \frac{1}{Z} \int dq_1 \cdots dq_A V(q_1 \cdots q_A) e^{-\beta V(q_1 \cdots q_A)}, \tag{9.16}$$

where

$$Z \equiv \int dq_1 \cdots dq_A e^{-\beta V(q_1 \cdots q_A)} \tag{9.17}$$

is the *configurational partition function* of the system. Also in this case the integral to be evaluated is of the same form as Eq. (9.14). We can distinguish in the integrand the product of a *probability density*:

$$P(q_1 \dots q_A) = \frac{e^{-\beta V(q_1 \cdots q_A)}}{Z}, \tag{9.18}$$

and a function to be integrated which is the potential energy V. For classical systems we have a quite intuitive way of proceeding, which is at the basis of statistical mechanics. If we are able to compute (or measure) the potential for some given set of particle coordinates, and we average over many different configurations (sets of particle positions), we will obtain the estimate of the potential energy we need.

This fact can be easily formalized by making use of the Central Limit Theorem. Given a probability density $P[X]$ defined in a suitable event space X, let us consider an arbitrary function $F(X)$. One can define a stochastic variable:

$$S_N(F) = \frac{1}{N} \sum_{i=1}^{N} F(X_i), \tag{9.19}$$

where the events X_i are assumed to be *statistically independent*, and are distributed according to $P[X]$. The stochastic variable $S_N(F)$ will in turn have its own probability density $P[S_N]$, which in general depends on the index N. The Central Limit Theorem states that for large N the probability density $P[S_N]$ will be a Gaussian, namely:

$$\lim_{N\to\infty} P[S_N] = \frac{1}{\sqrt{2\pi\sigma_N^2(F)}} \exp\left\{-\frac{(S_N - \langle F\rangle)^2}{2\sigma_N^2(F)}\right\}, \tag{9.20}$$

where we define the expectation of F as:

$$\begin{aligned} \langle F\rangle &= \int P[X]F(X)dX, \\ \langle F^2\rangle &= \int P[X]F^2(X)dX, \end{aligned} \tag{9.21}$$

and

$$\sigma_N^2(F) = \frac{1}{N}\left[\langle F^2\rangle - \langle F\rangle^2\right] \tag{9.22}$$

is the variance of the Gaussian. The reported average is estimated as $S_N(F)$, while $\langle F^2\rangle - \langle F\rangle^2$ is estimated by $\frac{N}{N-1}\left[S_N(F^2) - S_N^2(F)\right]$. This well known result is at the basis of all measurement theory. Averages over a set of measurements of a system provide the correct expectation of the measured quantity with an error that can be in turn estimated, and that decreases with the square root of the number of measurements N.

This result is very important from the point of view of numerical evaluation of integrals. If we had a way to numerically sample an arbitrary probability density $P[X]$, we could easily estimate integrals like that in Eq. (9.14). The statistical error associated with the estimate would decrease as the square root of the sampled points *regardless of the dimensionality of the system.*

For a classical system, configurations might be generated by solving Newton's equations, possibly adding a thermostat in order to be consistent with the canonical averaging. However, this is not certainly possible for a quantum system. The solution is to use an artificial dynamics, provided that it generates (at least in some limit) configurations that are distributed according to the probability density we want to use. Once again, in order to simplify the following description we will work in the space of the coordinates of the A particles, but the argument can be generalized to arbitrary spaces.

A very detailed description of what follows in this section can be found in the book of Kalos and Whitlock [1] and references therein.

We start defining a *transition matrix* $T_k(R_{k+1} \leftarrow R_k)$ expressing the probability that in the k-th step of the dynamics the system moves from the configuration R to a configuration R'. If at the first step the system is in a configuration R_0, sampled from an arbitrary distribution $P_0[R_0]$, the probability density of finding the system in a configuration R_1 at the next step will be given by:

$$P_1[R_1] = \int dR_0 P_0[R_0] T_0(R_1 \leftarrow R_0). \qquad (9.23)$$

We the introduce an integral operator $\hat{T}_0$ such that:

$$P_1[R_1] = \hat{T}_0 P[R_0]. \qquad (9.24)$$

With this notation, the probability density of the configuration at an arbitrary step k will become:

$$P_k[R_k] = \hat{T}_{k-1} P[R_{k-1}] = \hat{T}_{k-1} \cdots \hat{T}_1 \hat{T}_0 P_0[R_0]. \qquad (9.25)$$

The sequence of stochastic variables R_k generated at each step of this procedure is called a *Markov Chain*. Let us assume that $\hat{T}_k$ does not depend on the index k. What we will generate is then a *stationary* Markov Chain, for which the probability density generated at each step will only depend on the transition matrix and the probability density of the first element. In fact:

$$P_k[R_k] = \hat{T} P[R_{k-1}] = \hat{T} \cdots \hat{T}\hat{T} P_0[R_0] = \hat{T}^k P_0[R_0]. \qquad (9.26)$$

Under these assumptions one might wonder if the sequence is convergent (in functional sense), i.e. if a limiting probability density $P_\infty[R]$ exists. It is interesting to notice that if such function exists, it has to be an eigenvector of the integral operator $\hat{T}$. In fact, since we assume $\hat{T}$ to be independent of k we have:

$$\lim_{k\to\infty} \hat{T} P_k[R_k] = \lim_{k\to\infty} P_{k+1}[R_{k+1}]$$

$$\hat{T} P_\infty[R] = P_\infty[R].$$

It is also easy to realize that the eigenvalue is indeed 1. In fact, let us consider the general relation:

$$\hat{T} P_\infty[R] = \gamma P_\infty[R]. \qquad (9.27)$$

The recursive application of $\hat{T}$ would give:

$$\hat{T}^k P_\infty[R] = \gamma^k P_\infty[R]. \qquad (9.28)$$

If $\gamma \neq 1$ we would lose the normalization property of $P\infty[R]$.

These properties of stationary Markov chains can be exploited to sample a generic probability density $P[R]$. In fact, if we can determine the transition operator that has as eigenvector a *given* $P_\infty[R]$, a repeated application of such operator to an *arbitrary* initial distribution of points will eventually generate a chain in which each element is distributed according to $P_\infty[R]$. There is a simple recipe to construct such transition operator. We will assume that we have at hand a transition operator $\hat{\bar{T}}$ that we can sample (it could be as simple as a uniform probability within a given volume). We will split the searched transition operator in the product of $\hat{\bar{T}}$ and an unknown factor $\hat{A}$ that we will call "acceptance probability", defined in such a way that:

$$\hat{\bar{T}}\hat{A} = \hat{T}. \tag{9.29}$$

In order for the system to preserve its equilibrium state once the probability distribution is reached, we expect that the dynamics described by the random walk will not change the density of sampled points anywhere in the events space. Transitions carrying away from a state R to anywhere must be balanced by transitions leading from anywhere to the same state R:

$$\int dR' P(R) T(R' \leftarrow R) = \int dR' P(R') T(R \leftarrow R') \,. \tag{9.30}$$

One way to enforce this condition is to impose the more stringent *detailed balance* condition, which requires the *integrands* in Eq. (9.30) be equal:

$$P(R) T(R' \leftarrow R) = P(R') T(R \leftarrow R') \,. \tag{9.31}$$

The detailed balance condition can be in turn recast into a requirement on the acceptance probability. In fact:

$$\frac{A(R' \leftarrow R)}{A(R \leftarrow R')} = \frac{P(R')}{P(R)} \frac{\bar{T}(R \leftarrow R')}{\bar{T}(R' \leftarrow R)}. \tag{9.32}$$

The quantities on the r.h.s. of Eq. (9.32) are all known. The configuration R' has to be sampled originating in R from the given transition probability $\bar{T}(R \leftarrow R')$. The probability density $P(R)$ is the one we actually want to asymptotically sample. If we interpret the A values to be probabilities to actually keep the transition, then maximizing the possible A values leads to the slightly modified version of Eq. (9.32):

$$A(R' \leftarrow R) = \min\left(\frac{P(R')}{P(R)} \frac{\bar{T}(R \leftarrow R')}{\bar{T}(R' \leftarrow R)}, 1\right). \tag{9.33}$$

This expression is often called the *acceptance ratio*. In practice, it represents the probability according to which we have to *accept* the new configuration as the new member of the Markov chain, rather than keeping the original point as the next point in the chain.[2] Further analysis shows that existence and uniqueness of the correct eigenvalue 1 solution and therefore convergence to the correct distribution will be guaranteed if (1) every allowed state can be reached from any other by a finite sequence of transitions and (2) there are no cycle of states. The latter is guaranteed if there are any transitions that leave the system in the same state, that is any rejections.

There is a case in which Eq. (9.32) further simplifies. If the transition matrix is taken to be symmetric in the arguments R and R', the ratio becomes unity, and one is left with:

$$A(R' \leftarrow R) = \min\left(\frac{P(R')}{P(R)}, 1\right). \tag{9.34}$$

At this point we have all the ingredients to describe an algorithm that performs a Monte Carlo evaluation of an integral such that of Eq. (9.14). In the following we will describe the simplest version, i.e. the so called "Metropolis-Hastings algorithm" [2, 3].

1. Start from an arbitrary configuration of the A particles. If the potential has a strongly repulsive core one has to pay attention to avoid overlapping pairs.
2. Sweep over the coordinates and generate new positions according to some transition probability. A simple choice is a uniform displacement within a cube of side Δ, i.e.:

$$\bar{T}(R' \leftarrow R) = \begin{cases} \frac{1}{\Delta} \text{ if } & |R_i'^{\alpha} - R_i^{\alpha}| < \frac{\Delta}{2} \\ 0 \text{ otherwise} \end{cases} \tag{9.35}$$

with $\alpha = x, y, z$, and $i = 1 \ldots A$. This choice has the advantage of being symmetric. If we imagine to store our configuration in an array $R[0\ldots2][0\ldots A-1]$ the implementation of this step would read:

```
MC_Move()
for i ∈ {0, A − 1} do
    for j ∈ {0, 2} do
        R_new[i][j] ← R[i][j] + (rand() − 0.5) ∗ Δ
    end for
end for
```

[2] The standard jargon refers to this as a "rejection" event. However one has not to be confused: this is the result of a *reversed* move, and generates a new element in the chain coincident with the starting point.

We will assume that the function rand() generates a random number uniformly distributed in [0, 1).

3. At this point we need to evaluate the acceptance ratio. This is easily done with our choice of the transition matrix, since we only need to evaluate the probability densities in R and R':

$$A(R' \leftarrow R) = \min\left(\frac{|\Psi_T(R')|^2}{|\Psi_T(R)|^2}, 1\right). \tag{9.36}$$

4. Next we need to decide whether we keep the proposed configuration as the next element in the chain or if we want to resort to the original one. If we define acc $= A(R' \leftarrow R)$, then:

```
Accept_reject()
ξ = rand()
if acc > ξ then
    R[i][j] ← R_new[i][j]
end if
```

5. According to the Central Limit Theorem, we now need to cumulate the values of the rest of the integrand. In the case of our variational calculation we need to sum up the local energies. Notice that this step has to be taken whatever the result of the procedure described at the previous point. If we want to estimate the statistical error, we also need to cumulate the *square* of the local energy.

```
Acuest()
eloc ← ĤΨ_T(R) / Ψ_T(R)
ecum ← ecum + eloc
ecum2 ← ecum2 + eloc * eloc
```

6. Steps 2–5 need to be repeated N_{steps} times, where N_{steps} must be sufficiently large to provide a small enough statistical error. The final estimate of the energy is given by $\langle E\rangle \pm \Delta E$, where:

$$\langle E\rangle = \frac{1}{N_{steps}} \cdot \text{ecum}$$

$$\Delta E = \sqrt{\frac{1}{N_{steps}-1}\left(\frac{1}{N_{steps}} \cdot \text{ecum2} - \langle E\rangle^2\right)} \tag{9.37}$$

Notice that this algorithm could in principle be used to evaluate arbitrary integrals. In fact, it is always possible to multiply and divide the integrand by a probability density $P(X)$ that can be used to sample the values of X:

$$I = \int F(X)dX = \int P[X]\frac{F(X)}{P[X]} \tag{9.38}$$

9.2.3.1 Autocorrelations

The main hypothesis underlying the Central Limit Theorem is that data used to construct the averages are sampled independently. While in a measurement process this is a quite reasonable assumption, in the case of the computation of an integral by means of any method based on the Markov chain theory (including the Metropolis-Hastings method) this requirement is not satisfied by construction. In fact, data are sampled based on a transition matrix, and the resulting random walk has a certain degree of memory of the past events. What are the consequences of such memory? Let us consider a sequence of points $X_1, X_2, \cdots, X_N$ sampled via the Metropolis algorithm from some probability density $P[X]$. If we assume these data not to be independent, we have to consider the joint probability for the specific realization of the chain in order to estimate the integral of a given function F:

$$I = \frac{1}{N}\sum_{i=1}^{N}\int dX_1, dX_2, \cdots dX_N P[X_1, X_2, \cdots X_N]F(X_i). \tag{9.39}$$

If the samples are independent then $P[X_1, X_2, \cdots X_N] = P[X_1]P[X_2]\cdots P[X_N]$, and we are in the case previously discussed. However, since we can arbitrarily exchange the indexes of the integration variables, we can easily see that the value of I is unchanged despite the presence of correlations. By construction, in a Markov process two consecutive samples will always be correlated to each other. This seems to be inconsistent to the use we want to make of these samples, i.e. to apply the Central Limit Theorem to integration. However, we can hope that after a certain number of steps memory is lost, and data will become effectively independent. Is it possible to estimate this typical *autocorrelation length*? Based on the previous argument one can define a measure of the autocorrelation by looking at the variance of the expectation of F with respect to P:

$$(\Delta I)^2 = \left\langle \frac{1}{N^2}\sum_{i=1}^{N} F(X_i)\sum_{i=1}^{N} F(X_j)\right\rangle - \langle F\rangle^2. \tag{9.40}$$

The corresponding standard deviation is the estimate of the statistical error on the integral of F. The first term can be recast in the following way:

$$\left\langle \frac{1}{N^2}\sum_{i=1}^{N} F(X_i)\sum_{i=1}^{N} F(X_j)\right\rangle$$
$$= \frac{1}{N^2}\sum_{i,j=1}^{N}\int P[X_1, X_2, \cdots X_N]F(X_i)F(X_j)dX_1\cdots dX_N$$
$$= \frac{1}{N^2}\sum_{i,j=1}^{N}\langle F(X_i)F(X_j)\rangle. \tag{9.41}$$

Since the Markov chain is stationary, this quantity is expected to depend only on the difference of the indexes $\tau = i - j$. We will then define an *autocorrelation coefficient*:

$$c(F)_\tau = \frac{\langle F(X_i)F(X_{i+\tau})\rangle - \langle F\rangle^2}{\langle F^2\rangle - \langle F\rangle^2}. \tag{9.42}$$

The coefficient is normalized to the variance $\sigma^2(F)$, in such a way that $C(F)_0 = 1$. Correlation coefficients are related to the average of the product of the F in the following way:

$$\langle F(X_i)F(X_{i+\tau})\rangle = c(F)_\tau\sigma^2(F) + \langle F\rangle^2. \tag{9.43}$$

We can use the previous expression to estimate the error on I:

$$(\Delta I)^2 = \frac{1}{N^2}\sum_{i,j=1}^{N}\langle F(X_i)F(X_j)\rangle - \langle F\rangle^2$$
$$= \frac{1}{N}\sigma^2(F)\sum_{\tau=1}^{N} c(F)_\tau + \langle F\rangle^2 - \langle F\rangle^2 = \frac{\sigma^2(F)}{N}\sum_{\tau=1}^{N} c(F)_\tau. \tag{9.44}$$

As it can be seen the error not only depends on the variance of F, but also on the sum over all the autocorrelation coefficients of F. This is the main consequence of having autocorrelated samples: the statistical error is underestimated by the variance of F, and needs to be corrected by a factor that depends on the autocorrelation length.

Usually the coefficients $c(F)_\tau$ have an exponential decay. If we approximate them as $c(F)_\tau \sim \exp(-\tau/\bar{\tau})$, the sum of the coefficients can be approximated as:

$$\sum_{\tau=1}^{N} c(F)_\tau \sim \int_0^\infty d\tau e^{-\frac{\tau}{\bar{\tau}}} = \bar{\tau}. \tag{9.45}$$

This means that it is sufficient to fit the exponential decay of the autocorrelation coefficients in order to find an estimate of the characteristic autocorrelation length that corrects the estimate of the error on the integral. In particular the correct expression for the error is:

$$\Delta I \simeq \sqrt{\frac{1}{N-1}\sigma^2(F)\bar{\tau}}, \tag{9.46}$$

which has a simple interpretation: We are not generating N independent samples of the variable X during our Markov process, but rather $N/\bar{\tau}$ of them, and this number must be used as the correct count of events for the error estimation.

It is important to be extremely careful about the estimation of autocorrelations. in many cases an underestimation of the statistical errors leads to a wrong interpretation of the results and to wrong physical conclusions.

Autocorrelations also play a crucial role in choosing the step width Δ in the Metropolis-Hastings algorithm. A common criterion is to choose it in such a way that the fraction of accepted moves is about 50%. However, the ideal value is clearly the one minimizing the autocorrelations among samples, and quite often this value corresponds to acceptances of the order 30 or 40%.

Once the value of $\bar{\tau}$ has been estimated, it is possible to organize the calculation in such a way that the statistical error computed by the code is more realistic by using a *reblocking* technique. In practice the values of the quantity to be averaged are summed up in blocks of N_b elements each:

$$F_l^b = \sum_{i=1}^{N_b} F(X_i). \tag{9.47}$$

Then, the F_l^b are used as the data on which performing the computation of the variance and of the standard deviation. If $N_b \gtrsim \bar{\tau}$, the standard deviation will be corrected by the effects of the autocorrelation of the original data. Typically calculations store block values so that the values can be "reblocked" for example by combining pairs of blocks. The estimated error should be unchanged if the blocks are uncorrelated. In addition, the ratio of the block variance to the variance of the original function can be used to estimate the number of independent samples, and therefore the autocorrelation time.

9.2.4 Construction of the Wavefunction and Computational Procedures

When performing a variational calculation, the first step consists of deciding which model wavefunction we intend to use.

First of all we have to take care of the symmetry of the particles. Nucleons are Fermions, and therefore it is necessary to build an antisymmetric wavefunction. If the Hamiltonian does not contain terms acting on the spin or isospin state of a nucleon or of a pair of nucleons, each particle will preserve its own initial state. In this case it is easy to write an antisymmetric wavefunction simply using a product of Slater determinants, one for each species.

To build the determinants one needs some single particle orbitals. There are several possible choices. For nuclei linear combinations of Gaussians or the eigenstates of the harmonic oscillator are definitely an option. Another choice might be that of using orbitals coming from a Hartree-Fock calculation. In this case the orbitals contain some information about the fact that nucleons interact, but there usually is no consistency between the Hamiltonian used to compute the orbitals and the Hamiltonian we are interested in.

The basic starting point is then a wavefunction of the form:

$$\varphi(R) = \det[\phi_j(\mathbf{r}_{p_i\uparrow})]\det[\phi_j(\mathbf{r}_{p_i\downarrow})]\det[\phi_j(\mathbf{r}_{n_i\uparrow})]\det[\phi_j(\mathbf{r}_{n_i\downarrow})], \tag{9.48}$$

If we just limited ourselves to this kind of wavefunction we would miss most of the interesting physics that happens when particles are close together. A seen in the previous chapters, a very important role is played by the *short range correlations*, that should introduce the many-body effects due to repulsion/attraction of particles at short distance. Contrarily to what one does in other methods, such as coupled clusters, in Quantum Monte Carlo calculations it is easier to work with wavefunctions containing *explicit* two-, three- or many-body correlations.

Here we will use the so-called *Jastrow* factor, i.e. a product of two-body functions that helps to reproduce the correlations from the pair-wise potential. The simplest version of a trial wavefunction therefore reads:

$$\Psi_T(R) = \varphi(R)\prod_{i<j}^{A} f(r_{ij}), \tag{9.49}$$

where $R = (\mathbf{r_1}, \cdots, \mathbf{r_A})$, and f is the so called *Jastrow function* (JF). How do we determine the JF? We have some information that we can exploit. In particular we might seek for analytic forms of f that satisfy what is commonly called the *cusp condition*,(see e.g. [4]) i.e. we must have:

$$\frac{\hat{H}f(r_{ij})}{f(r_{ij})} < \infty \tag{9.50}$$

everywhere in space. It is easy to realize that satisfying the cusp condition helps to prevent the local energy from fluctuating too much even in presence of a divergence of the potential, thereby reducing the variance and the statistical error. Usually in nuclear physics problems it is customary to take a further step. Recognizing that at small separations, the many-body Schrödinger equation is dominated by the short-range pair potential, the two-body problem is solved to determine the f. In particular one can solve the following Schrödinger equation in relative coordinates:

$$-\frac{\hbar^2}{2m}\nabla^2 + qV(r)f(r) = \epsilon f(r), \tag{9.51}$$

and impose the boundary condition that the function becomes a constant at a distance h from the origin, where other parts of the Hamiltonian become important. The quantities q and h are two variational parameters. One could in principle consider a third variational parameter in the Jastrow factor by using a modified Jastrow function $\tilde{f}$ such that:

$$\tilde{f}(r) = e^{-b\log f(r_{ij})} \tag{9.52}$$

The function f is usually determined by numerically solving Eq. (9.51) with the Numerov or Runge-Kutta methods. One has to be careful that the resulting table has to be interpolated to compute the function at an arbitrary distance. Therefore it is important to choose an appropriate number of points (usually of the order of a few thousands). Single particle orbitals can also be either tabulated or computed analytically. Tabulation guarantees in general a faster computation at the price of a loss in numerical accuracy.

In the code it is necessary to compute derivatives of the wavefunction in order to estimate the local energy. This can be done either numerically or analytically. A very good test for checking that there are no major mistakes either in the Monte Carlo evaluation of integrals or in the computation of the local energy is to use the so-called *Jackson-Feenberg* identity for the kinetic energy.

The expectation of the kinetic energy is an integral of the form:

$$\langle T\rangle = \frac{-\frac{\hbar^2}{2m}\int_\Omega dR\Psi^*(R)\nabla^2\Psi(R)}{\int_\Omega dR|\Psi(R)|^2}, \tag{9.53}$$

where Ω is the integration volume. Integrating the numerator by parts one gets:

$$\langle T\rangle = \frac{\frac{\hbar^2}{2m}\int_\Omega dR\nabla\Psi^*(R)\cdot\nabla\Psi(R)}{\int_\Omega dR|\Psi(R)|^2} - \frac{\frac{\hbar^2}{2m}\int_{S(\Omega)} dS\Psi^*(R)\nabla\Psi(R)}{\int_\Omega dR|\Psi(R)|^2}. \tag{9.54}$$

The surface term is zero if the wavefunction is well behaved. We are therefore left with the integral:

$$\langle T\rangle = \frac{\frac{\hbar^2}{2m}\int_\Omega dR\nabla\Psi^*(R)\nabla\Psi(R)}{\int_\Omega dR|\Psi(R)|^2} = \frac{\frac{\hbar^2}{2m}\int_\Omega dR|\Psi(R)|^2\frac{\nabla\Psi(R)}{\Psi(R)}\cdot\frac{\nabla\Psi^*(R)}{\Psi^*(R)}}{\int_\Omega dR|\Psi(R)|^2}. \tag{9.55}$$

We can sum Eqs. (9.53) and (9.55), and divide by 2 in order to obtain a new kinetic energy estimator:

$$\langle T\rangle_{JF} = \frac{\frac{\hbar^2}{4m}\int_\Omega dR|\Psi(R)|^2\left[\frac{\nabla\Psi^*(R)}{\Psi^*(R)}\cdot\frac{\nabla\Psi(R)}{\Psi(R)} - \frac{\nabla^2\Psi(R)}{\Psi(R)}\right]}{\int_\Omega dR|\Psi(R)|^2}. \tag{9.56}$$

This is the *Jackson-Feenberg* kinetic energy estimator. It is easy to see that configuration by configuration the value of the integrand of T and T_{JF} are different. However, they have to be the same on average (i.e. always within the current statistical error). The equivalence of the two estimators checks the integration procedure, the correctness of the implementation of the boundary conditions, and the computation of derivatives. If any of these quantities are wrong, the two estimates of the kinetic energy will differ. This is an extremely useful consistency check, and should always be used in a variational calculation.

At this point it is necessary to perform several calculations varying the parameters in the wavefunction, and looking for a minimum of the energy. In the next subsection we will describe algorithms that allow for performing this search in an automatic way. However, when the number of parameters is small, it is also possible in principle to perform a scan on a grid.

In Fig. 9.1, as an example, we report the behavior of the variational energy computed in a ^{4}He nucleus, modeled with a two body Minnesota potential, and a wavefunction containing only a central Jastrow product. The spatial part of the orbital is an s-wave Gaussian with half width equal to 1.1 fm. The energies have been computed for a fixed value of the healing distance $h = 3.1$ fm as a function of the quencher parameter q, keeping fixed the amplitude parameter $b = 1$. Each run consists of an average over 6.4×10^5 samples, preceded by 6.4×10^4 equilibration steps.

As it can be seen, there is a clear minimum of the energy. The minimum can be determined with sufficient accuracy by fitting the resulting curve. A fit with a quadratic function predicts a minimum at $q \sim 1.4$. The corresponding eigenvalue is $E_T = -15.31(4)$ MeV.[3] The procedure should be repeated for different values of all other variational parameters until an absolute minimum is found.

[3]The number in parenthesis indicates the statistical error on the last figure.

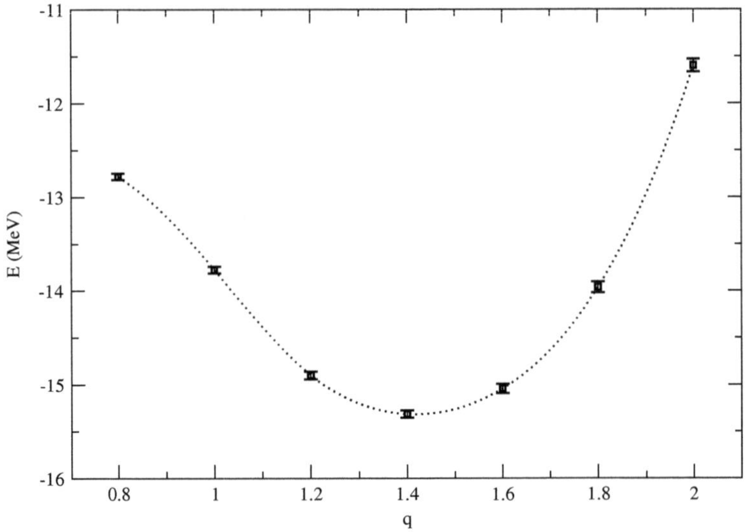

Fig. 9.1 An example of variational minimization of the energy. The estimate of the binding energy of a ^{4}He nucleus described by the Minnesota potential is here plotted as a function of the quencher parameter q, for a fixed value of the healing distance h (see text). The *dotted line* serves as a guide for the eye

The variational wavefunctions can be made arbitrarily richer in structure in order to improve the results, including what our physical intuition suggests as important terms to describe correlations. We will later discuss how to construct trial wavefunctions for realistic nuclear Hamiltonians. A full variational calculation for the ^{4}He nucleus with the Minnesota potential, including Jastrow factors with a spin/isospin dependence would give a binding energy $E_T = -25.52(4)$ MeV. As an example, the optimized Jastrow function for the central channel of the Minnesota potential is shown in Fig. 9.2.

9.2.5 *Wave Function Optimization*

9.2.5.1 Reweighting Methods

The brute-force optimization of the trial wave function becomes quite cumbersome with more than a few parameters. In general the problem is equivalent to searching an absolute minimum in a multi-dimensional space, and does not admit a simple solution. If one is interested in a quick search for local minima, it is possible to compute the gradient of the energy in the parameter space, and use for instance some variant of the steepest descent method. Computation of gradients is based on the so-called "reweighting method". If we have a trial function depending on a set of parameters $\{\alpha\}$, and another depending on a set $\{\alpha + \delta\alpha\}$, it s not necessary to

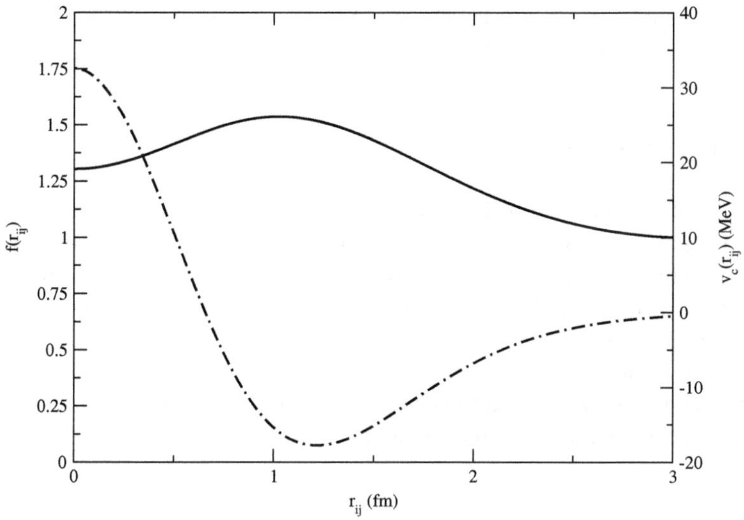

Fig. 9.2 The central channel of the Minnesota potential (*dashed-dotted line*), and the corresponding numerical Jastrow function (*solid line*) evaluated for $h = 3.1$, $q = 1.4$, and $b = 1$

perform two independent calculations to compute the difference (which would also be affected by rather large statistical errors). In fact, the following identity holds):

$$\frac{\int dR|\Psi_T(R,\{\alpha+\delta\alpha\})|^2 O(R)}{\int dR|\Psi_T(R,\{\alpha+\delta\alpha\})|^2} = \frac{\int dR|\Psi_T(R,\{\alpha\})|^2 \frac{|\Psi_T(R,\{\alpha+\delta\alpha\}|^2}{|\Psi_T(R,\{\alpha\})|^2} O(R)}{\int dR|\Psi_T(R,\{\alpha\})|^2 \frac{|\Psi_T(R,\{\alpha+\delta\alpha\})|^2}{|\Psi_T(R,\{\alpha+\delta\alpha\})|^2}} \tag{9.57}$$

It is therefore possible to use the configurations sampled from a trial wavefunction with a given parametrization $\{\alpha\}$ to compute expectations over a wavefunction with a different parametrization $\{\alpha+\delta\alpha\}$ by simply reweighting the values of the operator with the ration between the square moduli of the two wavefunctions:

$$\begin{aligned} \langle O_{\{\alpha+\delta\alpha\}}\rangle &\equiv \frac{\langle\Psi_T(R_k,\{\alpha+\delta\alpha\})|O(R_k)|\Psi_T(R_k,\{\alpha+\delta\alpha\})\rangle}{\langle\Psi_T(R_k,\{\alpha+\delta\alpha\})|\Psi_T(R_k,\{\alpha+\delta\alpha\})\rangle} \\ &= \frac{\sum_k \frac{|\Psi_T(R_k,\{\alpha+\delta\alpha\}|^2}{|\Psi_T(R_k,\{\alpha\}|^2} O(R_k)}{\sum_k \frac{|\Psi_T(R_k,\{\alpha+\delta\alpha\}|^2}{|\Psi_T(R_k,\{\alpha\}|^2}} \end{aligned} \tag{9.58}$$

where the R_k are sampled from $|\Psi_T(R,\{\alpha\})|^2$. Besides the obvious advantage of avoiding multiple calculations to compute the derivatives, the use of this reweighting technique allows direct computation of expectations of the gradients in the parameter space with very high accuracy. The access to gradients opens the way to the use of automated minimization algorithms such as the already mentioned steepest descent method, the Levemberg-Marquardt algorithm [5] or the Linear Method [6] briefly sketched below.

9.2.5.2 Power Method

There is another class of algorithms that have been recently introduced, and based on the power method. We will here discuss in particular the algorithm due to Sandro Sorella [7]. This algorithm was originally discussed in terms of the Lanczos method, but for a single multiplication by his propagator it becomes equivalent to the simpler power method that we discuss here.

For Λ larger than the largest eigenvalue of the eigenvectors contained in $|\psi_n\rangle$, operating with $\Lambda - H$ will multiply the ground state by a larger number than any other state. Therefore iterating the equation

$$|\psi_{n+1}\rangle = (\Lambda - H)|\psi_n\rangle \tag{9.59}$$

will converge to the ground state. One way to implement this is to use a set of test functions (which, in principle, should be complete), $|\phi_m\rangle$. This gives the set of equations

$$\langle\phi_m|\psi_{n+1}\rangle = \langle\phi_m|(\Lambda - H)|\psi_n\rangle \,. \tag{9.60}$$

In his original paper Sorella assumes $|\psi_n\rangle = |\Psi_T\rangle$, and next approximates $|\psi_{n+1}\rangle$ as a linear combination of the original state and the derivatives with respect to the parameters

$$|\psi_{n+1}\rangle \simeq \Delta\alpha_0|\Psi_T\rangle + \sum_{n=1} \Delta\alpha_n \partial_{\alpha_n}|\psi_T\rangle \equiv \sum_{n=0} O^n|\psi_T\rangle\Delta\alpha_n \tag{9.61}$$

and he uses the same functions for $|\phi_m\rangle$, so that

$$|\phi_m\rangle = O^m|\Psi_T\rangle \,. \tag{9.62}$$

When evaluated in the position representation, the O^m for $m > 0$ correspond to multiplying by the derivative of the logarithm of the trial function. Substituting these expressions, and dividing by $\langle\Psi_T|\Psi_T\rangle$, Eq. (9.60) becomes

$$\frac{\langle\Psi_T|O^m(\Lambda - H)|\Psi_T\rangle}{\langle\Psi_T|\Psi_T\rangle} = \sum_{n=0} \frac{\langle\Psi_T|O^m O^n|\Psi_T\rangle}{\langle\Psi_T|\Psi_T\rangle}\Delta\alpha_n \,. \tag{9.63}$$

The expectation values can be calculated and the linear equations solved to get $\Delta\alpha_n$.

Alternatively, the $m = 0$ and $n = 0$ terms can be separated. Writing the trial function expectation of an operator O as $\langle O \rangle$, Eq. (9.63) becomes

$$\langle \Lambda - H \rangle = \Delta\alpha_0 + \sum_{n=1} \langle O^n \rangle \Delta\alpha_n \;\; m = 0 \tag{9.64}$$

$$\langle O^m(\Lambda - H) \rangle = \langle O^m \rangle \Delta\alpha_0 + \sum_{n=1} \langle O^m O^n \rangle \Delta\alpha_n \;\; m > 0\,. \tag{9.65}$$

Substituting Eq. (9.64) into Eq. (9.65) gives

$$\langle O^m(\Lambda - H) \rangle - \langle \Lambda - H \rangle \langle O^m \rangle = \sum_{n=1} \left[\langle O^m O^n \rangle - \langle O^m \rangle \langle O^n \rangle \right] \Delta\alpha_n\,. \tag{9.66}$$

Solving gives $\Delta\alpha_{n>0}$ and Eq. (9.64) then gives the value for $\Delta\alpha_0$.

In either case, the result gives an approximation to the next trial function as a linear combination of the original function and its parameter derivatives. The new parameters are chosen to give this same linear combination as the first two terms in the Taylor series. Since dividing the approximate expression for $|\psi_{n+1}\rangle$ by $\Delta\alpha_0$ gives an expression that is the first two terms in the Taylor series, the new parameters are

$$\alpha_{n>0}^{(\mathrm{new})} = \alpha_n^{(\mathrm{old})} + \frac{\Delta\alpha_{n>0}}{\Delta\alpha_0} \tag{9.67}$$

More recently Toulouse and Umrigar [6] proposed a much more efficient method where the Hamiltonian is diagonalized in the reduced space spanned by the $|\phi_m\rangle$. The parameter variation is then given by the solution of the generalized eigenvalue equation

$$\sum_{n=0} \frac{\langle \Psi_T | O^m H O^n | \Psi_T \rangle}{\langle \Psi_T | \Psi_T \rangle} \Delta\alpha_n = \Delta E \sum_{n=0} \frac{\langle \Psi_T | O^m O^n | \Psi_T \rangle}{\langle \Psi_T | \Psi_T \rangle} \Delta\alpha_n\,. \tag{9.68}$$

with the lowest eigenvalue $\Delta E^m in$:

$$\alpha_{n>0}^{(\mathrm{new})} = \alpha_n^{(\mathrm{old})} + \frac{\Delta\alpha_{n>0}^{min}}{\Delta\alpha_0^{min}}. \tag{9.69}$$

The gradient of the local energy is required for the expectation values appearing in Eq. (9.68), and can be efficiently estimated using the reweighting technique presented in the previous section.

When the parameters are far away from the minimum this approach can be less stable than the previous one giving rise to large parameter variations that invalidate the linear approximation Eq. (9.61). A quick strategy is then to use the solution of Eq. (9.66) early on in the optimization process and then switch to Eq. (9.68) when the resulting norm of the variation is below some threshold.

9.3 Projection Monte Carlo Methods in Coordinate Space

9.3.1 General Formulation

Variational calculations provide only an upper bound for the ground-state eigenvalue of a given Hamiltonian. However, it is possible to use Monte Carlo algorithms to actually solve the Schrödinger equation for an arbitrary number of interacting particles. This class of algorithms is based on the idea of imaginary time propagation.

Let us consider a Hamiltonian $\hat{H}$. The imaginary time evolution of an arbitrary state is defined starting by the standard time-dependent Schrödinger equation:

$$-i\hbar\frac{\partial}{\partial t}|\Psi(t)\rangle = \hat{H}|\Psi(t)\rangle. \tag{9.70}$$

It is possible to Wick rotate, and introduce an *imaginary time* $\tau = \frac{it}{\hbar}$. The time-dependent Schrödinger equation is transformed into an imaginary-time-dependent equation:

$$-\frac{\partial}{\partial \tau}|\Psi(\tau)\rangle = \hat{H}|\Psi(\tau)\rangle, \tag{9.71}$$

where τ is defined as an *inverse energy* that parametrizes the propagation of the quantum state. The formal solution can be written using the imaginary time propagator

$$|\Psi(\tau)\rangle = e^{-\tau\hat{H}}|\Psi(0)\rangle \tag{9.72}$$

It is possible to expand the initial state $|\Psi(0)\rangle$ in eigenstates $|\phi_n\rangle$ of the Hamiltonian itself, such that $\hat{H}|\phi_n\rangle = E_n|\phi_n\rangle$. The imaginary time propagation of $|\Psi(0)\rangle = \sum_n c_n|\phi_n\rangle$ becomes:

$$|\Psi(\tau)\rangle = e^{-\tau\hat{H}}\sum_n c_n|\phi_n\rangle = \sum_n c_n e^{-\tau E_n}|\phi_n\rangle \tag{9.73}$$

Let us now consider the limit of the propagation for $\tau \to \infty$. The coefficients of the expansion $c_n e^{-\tau E_n}$ will either decrease (if $E_n > 0$) or increase (if $E_n < 0$) with the imaginary time, but in the limit the coefficient corresponding to the ground state of $\hat{H}$, i.e. $c_0 e^{-\tau E_0}$ will be dominant. This means that the imaginary time propagator has the interesting property of filtering out of an arbitrary state in the Hilbert space the ground state of a given Hamiltonian, provided that the state is not orthogonal to the ground state to begin with. We want to stress a very important point. The ground state we are referring to is the *mathematical* ground state of the Hamiltonian $\hat{H}$. The *physical* ground state needs to take into account the symmetry of the particles, either bosons or fermions. It is very easy to convince oneself that such mathematical ground state is always a nodeless function (i.e. it is zero nowhere but possibly on the boundaries of the domain of existence of the wavefunction expressed in some representation). This is because the propagator is a positive definite function, at least for a Hamiltonian of the standard form $\hat{H} = \hat{T} + \hat{V}$, where $\hat{T}$ is the kinetic energy of a system of free particles and $\hat{V}$ is a local potential. In this case the eigenvector corresponding to the largest eigenvalue of the propagator is positive definite within the domain that defines the system. The largest eigenvalue of the propagator corresponds to the lowest eigenvalue of $\hat{H}$.

Notice that the imaginary time propagator is hermitian not unitary, and the normalization of the projected ground state is not guaranteed in general. By means of a small change in the propagator definition it is possible to guarantee the normalization of the projected ground state. In fact, let us define the propagator as:

$$|\Psi(\tau)\rangle = e^{-\tau(H-E_0)}|\Psi(0)\rangle. \tag{9.74}$$

It is easy to realize that in this case the amplitude of the component of the initial state along the ground state is preserved (while all other amplitudes decrease exponentially), and therefore the projected state is normalizable.

We will later discuss in detail the implications of these properties as concerns the application of imaginary-time propagation to many-fermion systems.

9.3.2 *Imaginary Time Propagator in Coordinate Representation*

We will focus on a practical implementation of imaginary time propagation, and we will limit ourselves to a system of bosons (or Boltzmannions) which do admit a ground-state wavefunction that is positive definite. We will also consider Hamiltonians of the form mentioned in the previous subsection, in which the interaction is local. In this case the propagator is easily represented in coordinates. Formally we would have:

$$\langle R|\Psi(\tau)\rangle = \int dR' \langle R|e^{-\tau(H-E_0)}|R'\rangle\langle R'|\Psi(0)\rangle, \tag{9.75}$$

where we have inserted a complete set of position eigenstates. The propagator

$$\langle R|e^{-\tau(\hat{H}-E_0)}|R'\rangle \tag{9.76}$$

seems to be still quite difficult to evaluate. However, let us break up the imaginary time interval τ in two equal intervals $\tau/2$. We can write

$$\langle R|e^{-\tau(\hat{H}-E_0)}|R'\rangle = \langle R|e^{-\frac{\tau}{2}(\hat{H}-E_0)}e^{-\frac{\tau}{2}(\hat{H}-E_0)}|R'\rangle, \quad (9.77)$$

since $\hat{H}$ obviously commutes with itself. Inserting a complete set we obtain:

$$\langle R|e^{-\tau(\hat{H}-E_0)}|R'\rangle = \int dR'' \langle R|e^{-\frac{\tau}{2}(\hat{H}-E_0)}|R''\rangle \langle R''|e^{-\frac{\tau}{2}(\hat{H}-E_0)}|R'\rangle, \quad (9.78)$$

This process can be iterated for an arbitrary large number of times M:

$$\langle R|e^{-\tau(\hat{H}-E_0)}|R'\rangle = \int \cdots \int dR'' \cdots dR^M \langle R|e^{-\frac{\tau}{2}(\hat{H}-E_0)}|R''\rangle \cdots \langle R^M|e^{-\frac{\tau}{2}(\hat{H}-E_0)}|R'\rangle. \quad (9.79)$$

Each of the factors in the integrand corresponds to a propagation for a *short* imaginary time $\Delta\tau = \tau/M$. In this case we can split the propagator using the Trotter-Suzuki formula:

$$e^{-\frac{\Delta\tau}{2}(\hat{H}-E_0)} \sim e^{-\frac{\Delta\tau}{2}(\hat{V}-E_0)}e^{-\Delta\tau\hat{T}}e^{-\frac{\Delta\tau}{2}(\hat{V}-E_0)} + o(\Delta\tau^3) \quad (9.80)$$

The representation in coordinates of each factor is known. The factors containing the potential, under the hypotheses made, are diagonal in the coordinates themselves, and simply become:

$$e^{-\frac{\Delta\tau}{2}(\hat{V}-E_0)}|R\rangle = |R\rangle e^{-\frac{\Delta\tau}{2}(V(R)-E_0)}, \quad (9.81)$$

while the kinetic term is the propagator of a set of A free particles obeying the equation:

$$-\frac{\partial}{\partial\tau}\Psi(R,t) = -\frac{\hbar^2}{2m}\nabla^2\Psi(R,t) \quad (9.82)$$

This is a classical *free diffusion* equation. If we interpret $\Psi(R,t)$ as a the density of the A particles, its evolution in time will be given by the well known diffusion law:

$$\Psi(R,t) = \frac{1}{(2\pi\frac{\hbar^2}{m}\Delta\tau)^{\frac{3A}{2}}}\int dR' e^{-\frac{(R-R')^2}{2\frac{\hbar^2}{m}\Delta\tau}}\Psi(R',0). \quad (9.83)$$

The short-time approximation for the propagator, correct at order $\Delta\tau$, will then become:

$$\langle R|e^{-\frac{\Delta\tau}{2}(\hat{H}-E_0)}|R'\rangle \sim \frac{1}{(2\pi\frac{\hbar^2}{m}\Delta\tau)^{\frac{3A}{2}}} e^{-\frac{\Delta\tau}{2}(V(R)-E_0)} e^{-\frac{(R-R')^2}{2\frac{\hbar^2}{m}\Delta\tau}} e^{-\frac{\Delta\tau}{2}(V(R)-E_0)}. \quad (9.84)$$

At this point it is possible to proceed in different ways. By substituting Eq. (9.84) in Eq. (9.79), one obtains an integral in which the integrand is a function of M replicas of the coordinates of the particles in the system. The ground-state expectation value of an operator that is a function of the coordinates can then be computed using on the left and on the right the imaginary time propagation started from an arbitrary state $\Psi(R,0)$. The resulting expression is:

$$\langle\phi_0|O(R)|\phi_0\rangle = \lim_{\tau\to\infty}\langle\Psi(R,\tau)|O(R)|\Psi(R,\tau)\rangle \sim \left(\frac{1}{(2\pi\frac{\hbar^2}{m}\Delta\tau)^{\frac{3A}{2}}}\right)^M \times$$

$$\times\int\int\cdots\int dR\,dR'\cdots dR^M\,\Psi(R,0)e^{-\frac{\Delta\tau}{2}(V(R)-E_0)}e^{-\frac{(R-R')^2}{2\frac{\hbar^2}{m}\Delta\tau}}e^{-\Delta\tau(V(R')-E_0)}\ldots O(R^{M/2})$$

$$\cdots e^{-\Delta\tau(V(R^{M-1})-E_0)}e^{-\frac{(R^{M-1}-R^M)^2}{2\frac{\hbar^2}{m}\Delta\tau}}e^{-\frac{\Delta\tau}{2}(V(R^M)-E_0)}\Psi(R^M,0) \tag{9.85}$$

This expression is reminiscent of a path-integral formulation of the problem. The integral can in principle be computed by means of a Metropolis-like algorithm, and gives the ground-state expectation of an arbitrary observable, provided that the number of slices M used is large enough to guarantee a correct filtering of the ground state. This method is known as Path Integral Ground State Monte Carlo (PIGS-MC) [8].

However, there is a simpler way to implement the imaginary time propagation. Let us expand the initial state from which we want to project the ground state in eigenstates of the position:

$$|\Psi\rangle \simeq \sum_i \Psi(R_i)|R_i\rangle \tag{9.86}$$

We will call each of these points in coordinate space a *walker*, and we will refer to the whole ensemble of points as to the *population* of walkers. If we apply the short-time propagator to each walker, it is easy to understand its effect. We will call the application of the short-time propagator to the walker population an *imaginary time step* (or simply a *time step*). Each time step originates a new *generation* of walkers.

The Gaussian factor in the propagator tells us the probability that a walker positioned in R' is displaced to a new position R. Since the probability density is a Gaussian of variance $\sigma^2 = \frac{\hbar^2}{m}\Delta\tau$, the RMS displacement will be proportional to $\sqrt{\Delta\tau}$ times a constant, which plays the role of a *diffusion constant* D, equal to $\frac{\hbar^2}{m}$. For each coordinate of each particle we need to extract a random number η distributed as:

$$P[\eta] = \frac{1}{\sqrt{2\pi D\Delta\tau}}e^{-\frac{\eta^2}{2D\Delta\tau}} \tag{9.87}$$

and add it to the original coordinate.

```
DMC_Move()
for i ∈ {0, A − 1} do
    for j ∈ {0, 2} do
        R_new[i][j] ← R[i][j] + DΔτ[rgaus()]
    end for
end for
```

The function rgaus() generating normally-distributed random numbers is now universally available as a library routine, but it can easily be implemented starting from a uniform distribution by using the Box-Muller formula. The part of the propagator depending on the potential has a slightly different interpretation. In the classical analogy we could say that the factor $W = e^{-\frac{\Delta\tau}{2}(V(R^M)-E_0)}$ represents the probability of a process to occur by which new points might be created in the time interval $\Delta\tau$ (if $W > 1$) or destroyed (if $W < 1$), or in other words, a process related to the presence of a source or a sink of walkers. W is interpreted as the average number of walkers that this process would generate over time at the position R. As we will later see, this creation/absorption (or *branching*) process is related to the fact that the normalization of the propagated state is not preserved. Since we cannot work with a non-integer number of walkers, we can use the following strategy

1. use the quantity W as a weight for the contribution to the estimates from the walker at a given position. Since in the short-time propagator we have two such factors, one from the initial position and on from the final position of the walker, we can use the product of the two as the total weight:

$$W = \exp\left\{-\Delta\tau\left[\frac{V(R)+V(R')}{2} - E_0\right]\right\} \tag{9.88}$$

   Estimates will be integrals of the form $\langle O\rangle = \int \phi_0(R)O(R)dR$, and they can be computed as:

$$\langle O\rangle = \frac{\sum_l^{N_{wk}} W_{kl}O(R_{kl})}{\sum_l^{N_{wk}} W_{kl}}, \tag{9.89}$$

   where N_{wk} is the number of walkers in a given generation. We will discuss later the specific form of the function O for interesting cases.
2. In order to generate a number of points that is correct on average, we can sample N_{mult}, the number of points to be generated for the next generation, in the following way:

$$N_{mult} = \text{int}(W + \xi), \tag{9.90}$$

where int() is the function truncating the argument to an integer, and ξ is a random number in $[0, 1)$. N_{mult} could be ≥ 1, in which case the next generation will contain *Nmult* copies of the walker, or 0, in which case the walker is suppressed.

The projection of the ground state will be achieved when propagating for a sufficiently long imaginary time. This means that we need to evolve the population of walkers for a large number of time steps, and eventually we will sample a density of points with a distribution *proportional* to the ground-state wavefunction. In the initial stage of the run, the energy and other estimators will have a value that is still strongly biased by the initial state. This means that the initial part of the propagation should be excluded from the averages. There is no automatic recipe to choose how much of the walk should be discarded. Usually it is convenient to monitor some observable (typically energy) and try to see where its value stops having a systematic trend as a function of the imaginary time.

How is the constant E_0 fixed? In principle it should be equal to the ground-state energy. This would mean that we need to know the solution of the problem... before solving it! In practice it is not strictly necessary to use the exact value of E_0, but it is sufficient to use a realistic variational estimate. The value of E_0 can also be used to reduce the fluctuations in the population of walkers due to the branching process, at the cost of introducing additional bias. For example, it is possible to modify the weight of a given configuration in the following way:

$$\tilde{W} = \frac{N_t}{N_g} \exp\left\{ -\Delta\tau \left[\frac{V(R) + V(R')}{2} - E_0 \right] \right\} \tag{9.91}$$

where N_t is a "target" number of walkers in the population and N_g is the number of walkers in the current generation. This modified weight reacts to the variations of the population, increasing or decreasing the weight depending on whether N_g is smaller or larger than N_t, respectively. This modification obviously introduces a bias in the results, since it modifies the propagator. However, this bias will be linearly decreasing with the time-step $\Delta\tau$. The weight can also be rewritten as:

$$\tilde{W} = \exp\left\{ -\Delta\tau \left[\frac{V(R) + V(R')}{2} - \tilde{E} \right] \right\}, \tag{9.92}$$

where

$$\tilde{E} = E_0 + \frac{1}{\Delta\tau} \log\left(\frac{N_t}{N_g} \right) \tag{9.93}$$

Therefore, at each generation the constant can be modified to keep the size of the population under control.

The weight can also be used to estimate the energy. In fact if we take the logarithm of both members of Eq. (9.92) we obtain:

$$\log \tilde{W} = -\Delta\tau \left[\frac{V(R) + V(R')}{2} - \tilde{E} \right] \tag{9.94}$$

from which we obtain:

$$E_0 = \frac{1}{\Delta\tau} \log \left(\frac{N_g \tilde{W}}{N_t} \right) + \frac{V(R) + V(R')}{2}. \tag{9.95}$$

This is the so-called *growth energy* estimator, and it can be used in principle to evaluate the ground-state eigenvalue.

A simpler way of evaluating the energy is to use a test function $\Psi_T(R)$. In this case the idea is to evaluate the following matrix element:

$$\langle E \rangle = \frac{\langle \phi_0 | \hat{H} | \Psi_T \rangle}{\langle \phi_0 | \Psi_T \rangle} = \frac{\int dR \phi_0(R) \hat{H} \Psi_T(R)}{\int dR \phi_0(R) \Psi_T(R)} \tag{9.96}$$

Both numerator and denominator integrals are suitable for Monte Carlo evaluation. The probability density that we sample is $\phi_0(R)$, and the functions to be cumulated following the recipe in Eq. (9.89) are $\hat{H}\Psi_T(R)$ and $\Psi_T(R)$. The latter is necessary whenever $\Psi_T(R)$ is not normalized. We will then have

$$\langle E \rangle = \frac{\langle \hat{H} \Psi_T \rangle}{\langle \Psi_T \rangle}. \tag{9.97}$$

However, due to the hermiticity of the hamiltonian, one has:

$$\langle E \rangle = \frac{\langle \phi_0 | \hat{H} | \Psi_T \rangle}{\langle \phi_0 | \Psi_T \rangle} = \frac{\langle \Psi_T | \hat{H} | \phi_0 \rangle}{\langle \phi_0 | \Psi_T \rangle} = E_0, \tag{9.98}$$

independent of the choice of $\Psi_T(R)$. This is the most practical way to evaluate the energy eigenvalue and its standard deviation. Other observables can be evaluated in a similar way. However the results will always depend on the choice of the test function. We will discuss this aspect later.

A last important remark remains to be made. In devising the algorithm we are making some approximations. First of all the imaginary time propagator is not exact, but is correct only at order $\Delta\tau^2$. This means that for any finite imaginary time step value, the answer will be biased of an amount proportional to $\Delta\tau^2$. The same holds for the population size whenever one wants to apply population control as described above. For any finite target population N_t there will be a bias on the answer of order $1/N_t$. These biases can be corrected by performing several simulations with different values of $\Delta\tau$ and N_t, and then extrapolating to $\Delta\tau \to 0$ and $1/N_t \to 0$. As we will show in the last part of this chapter, methods exist to completely eliminate the time step bias. However, it is possible to reduce the bias with some minor modifications in the propagator and by introducing an acceptance/rejection mechanism (cite CYRUS TIME STEP).

9.3.3 Application to the Harmonic Oscillator

A very simple illustration of the sense of the algorithm can be made by implementing to the one-dimensional harmonic oscillator. We consider the Hamiltonian:

$$\hat{H} = -\frac{1}{2}\frac{\partial^2}{\partial x^2} + \frac{1}{2}x^2 \tag{9.99}$$

The ground-state eigenvalue is $E_0 = \frac{1}{2}$ and the ground-state eigenfunction is the Gaussian $\Psi_0(x) = \frac{1}{\pi^{1/4}}e^{-\frac{x^2}{2}}$. As we have illustrated in the previous section, the propagation can start from any distribution of points with a density not orthogonal to the ground state. A very simple choice in this case is a constant. In Fig. 9.3 we can see how the histogram of the walkers evolves as a function of the imaginary time applying the algorithm described in the previous section, including population control. The initial uniform distribution of walkers in the interval $[-6, 6]$ is transformed into the correct Gaussian density. The mechanism that leads to this

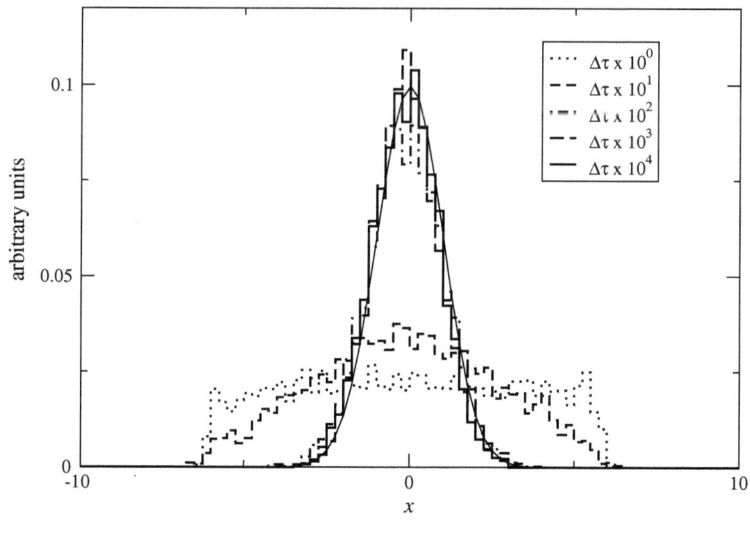

Fig. 9.3 Histogram of the walker population after N DMC imaginary time steps for the harmonic oscillator Hamiltonian described in text. Here we used $\Delta\tau = 10^{-3}$, with a target population of 4000 walkers

result is easy to understand. Any walker finding itself after diffusion in a region where the potential is larger than the eigenvalue will tend to be suppressed, while walkers near the origin will tend to multiply themselves. This will result in a histogram peaked at the origin ad decaying fast to zero when moving away from it.

In order to estimate the energy we need a test function. An approximation to the ground state might be given by the function:

$$\Psi_T(x) = \frac{1}{1+x^2}. \tag{9.100}$$

We can therefore estimate the energy by means of the following quotient [see Eq. (9.97)]:

$$\langle E \rangle = \frac{\sum_i w(x_i)\frac{1-3x_i^2}{(1+x_i^2)^3} + \frac{1}{2}\frac{x_i^2}{1+x_i^2}}{\sum_i w(x_i)\frac{1}{1+x_i^2}}, \tag{9.101}$$

where the sums runs first over all the generations (i.e. the imaginary time steps performed) and then over all the walkers belonging to a given generation.

In Fig. 9.4 we show the logarithm of the energy estimator averaged over each single generation as a function of the imaginary time. As we would expect from the general behavior of the coefficients of the excited states as function of the imaginary

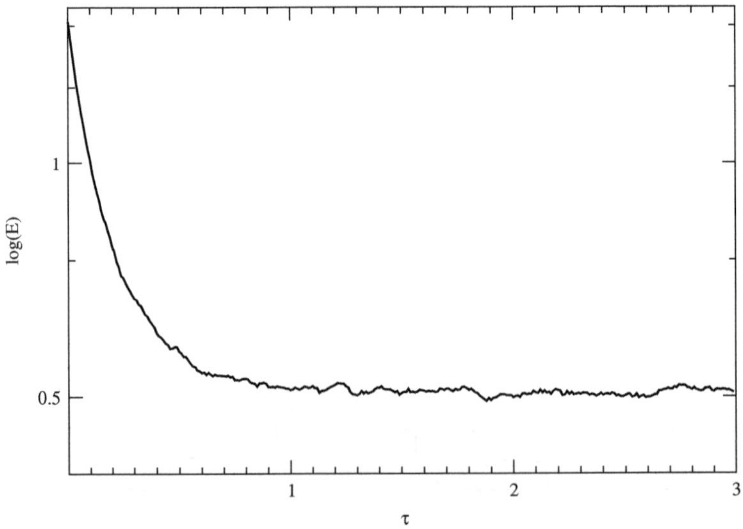

Fig. 9.4 Logarithm of the estimated energy averaged over a single generation as a function of the imaginary time in a run with a population target of 4000 walkers, and with an imaginary time step $\Delta\tau = 10^{-4}$

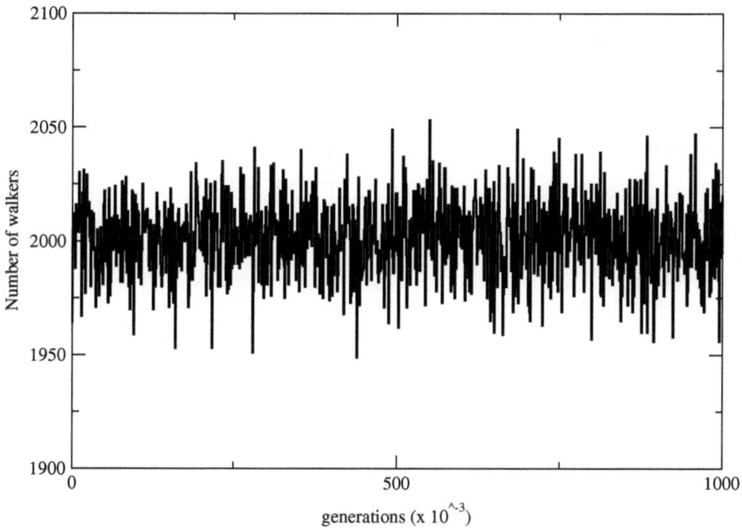

Fig. 9.5 Typical fluctuations of the walker population in a DMC run for the one dimensional harmonic oscillator. The target population in this case is 2000. The imaginary time step is set to $\Delta\tau = 0.075$

time, we see a clear exponential decay of the energy towards the exact eigenvalue. The figure clearly shows how the transient is not made up of a single exponential. The initial state needs includes a large number of excited states, that all need to be projected out before reaching the ground state. In Fig. 9.5 the typical behavior of the fluctuation in the walker number is reported. In the specific case the time step was set to $\Delta\tau = 0.3$. Nevertheless, the walker number never departs from the target by more than 3%. This is the effect of the population control procedure described in the previous subsection. Unfortunately population control alone is not sufficient to guarantee a stable calculation. In presence of particle-particle interactions that diverge at the origin fluctuations in the number of walkers become extremely wide. This is the reason why it is necessary to introduce the so-called importance sampling, that we will discuss in a later section.

Finally, in Fig. 9.6 we show one of the points discussed in the previous section, that is the bias of the result due to the finite imaginary time step. The difference between the energy estimate and the exact eigenvalue is plotted as a function of $\Delta\tau$ for a target population of 2000 walkers and a total of 10^5 generations for each value of $\Delta\tau$. The observed bias is quite small, but well outside of the statistical error. The dependence on $\Delta\tau$ is quadratic, as expected from the analysis of the propagator. Interpolating the data with a function of the form $E = E_0 + \alpha(\Delta\tau)^2$ we predict E_0 to be $(-3 \pm 1) \times 10^{-5}$. As it can be seen there is still a small residual bias due to the finiteness of the population. Further extrapolation would be needed to recover the exact answer.

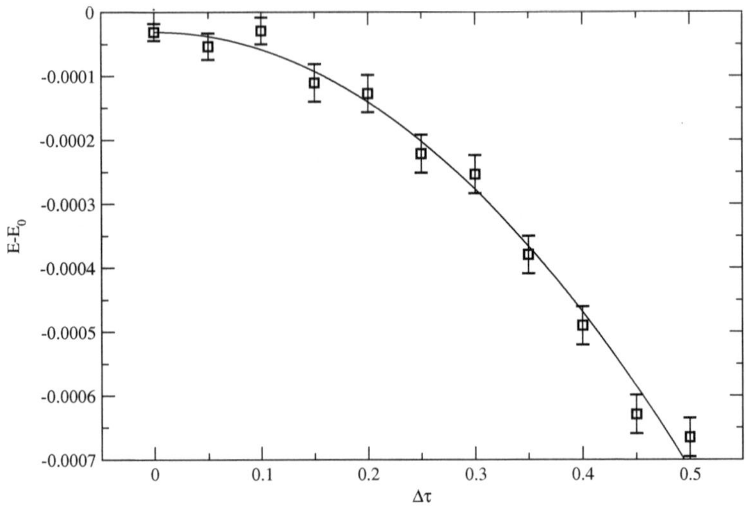

Fig. 9.6 Illustration of the imaginary time-step extrapolation procedure. The energy is computed for different values of $\Delta\tau$, and the results are fitted with a linear function. The intercept will give the correct prediction for the eigenvalue. Notice that the results should still be extrapolated for an infinite population. Here we use a target number of walkers equal to 2000, and the runs consist of 10^5 generations each. Errorbars refer to one standard deviation. The plotted value for $\Delta\tau = 0$ and the corresponding errorbar are obtained from the linear fit of the data

9.3.4 Importance Sampling

The simple diffusion algorithm we have illustrated above suffers of a substantial deficiency when particles interact with a potential having a repulsive or attractive core. Since the free particle diffusion propagator does not have any information about the potential, particles have no restrictions to come close to each other. This means that the weights will suffer of large fluctuations whenever a pair of particles find themselves at short distance. The consequent fluctuations in the population make the computation unmanageable.

The use of an importance function to guide the diffusion process [9] was the key to make Diffusion Monte Carlo algorithms usable. The idea is to give up on the request of sampling the ground-state wavefunction, and rather try to sample a distribution that, asymptotically in imaginary time, is the product of the ground-state wavefunction and of a known function that is the best possible approximation to the ground state obtained, for instance, by means of a variational calculation. We will call this function Ψ_G. Starting from Eq. (9.75) we can multiply both sides by Ψ_G and obtain:

$$\Psi_G(R)\Psi(R,\Delta\tau) = \int dR'G'(R',R,\Delta\tau)\Psi_G(R)\Psi(R',0), \tag{9.102}$$

where we have defined:

$$G'(R,R',\Delta\tau) = \frac{1}{(2\pi\frac{\hbar^2}{m}\Delta\tau)^{\frac{3A}{2}}} e^{-\Delta\tau(V(R')-E_0)} e^{-\frac{(R'-R)^2}{2\frac{\hbar^2}{m}\Delta\tau}}. \tag{9.103}$$

Since all the expressions we have written are correct at order $\Delta\tau$, for our purposes we can assume the equivalence of G' and G. We can multiply and divide the integrand in Eq. (9.102) by $\Psi_G(R')$ to obtain:

$$\Psi_G(R)\Psi(R,\Delta\tau) = \int dR' G'(R',R,\Delta\tau)\frac{\Psi_G(R)}{\Psi_G(R')}\Psi_G(R')\Psi(R',0),. \tag{9.104}$$

In Eq. (9.104) we can identify a new walker density to be propagated, namely:

$$f(R,\tau) = \Psi_G(R)\Psi(R,\tau), \tag{9.105}$$

and the corresponding propagator:

$$\tilde{G}(R,R',\Delta\tau) = G'(R,R',\tau)\frac{\Psi_G(R)}{\Psi_G(R')}. \tag{9.106}$$

The quotient of the wavefunctions can be included in the weight, and provides a correction that prevents the walkers to excessively multiply or die near the divergent points of the potential This point is better illustrated considering the short time limit it is possible to expand the ratio of the guiding functions. At first order in $\Delta\tau$ the result is:

$$\tilde{G}(R,R',\Delta\tau) \simeq G_0(R,R',\tau)\left[1 + \frac{\nabla\Psi_G(R')}{\Psi_G(R')}(R-R') + \cdots\right] \tag{9.107}$$

At the same order we can regard the terms in bracket as the expansion of an exponential and write:

$$\tilde{G}(R,R',\Delta\tau) \simeq G_0(R,R',\tau)e^{\frac{\nabla\Psi_G(R')}{\Psi_G(R')}(R-R')} \tag{9.108}$$

This can be combined with the Gaussian factor in G_0, and by completing the square (which introduces a term at order $\Delta\tau 2$), the propagator is modified as follows:

$$\tilde{G}(R,R',\Delta\tau) \simeq \frac{1}{(2\pi\frac{\hbar^2}{m}\Delta\tau)^{\frac{3A}{2}}} e^{-\frac{\Delta\tau}{2}(V(R')-E_0)} e^{-\frac{(R-R'-\frac{\hbar^2}{m}\Delta\tau\frac{\nabla\Psi_G(R')}{\Psi_G(R')})^2}{2\frac{\hbar^2}{m}\Delta\tau}} e^{-\frac{\Delta\tau}{2}(V(R)-E_0)}. \tag{9.109}$$

The same expansion can be performed to compute the *change in normalization* of the propagated density after a time step. The change in normalization is given by:

$$\mathscr{N} = \int dR\tilde{G}(R, R', \tau), \tag{9.110}$$

i.e. the total weight of the final points R that can be reached starting from R'. Once more we can expand the ratio of the guiding functions in the propagator, but this time up to second order:

$$\tilde{G}(R, R', \Delta\tau) \simeq G_0(R, R', \tau)\left[1 + \frac{\nabla\Psi_G(R')}{\Psi_G(R')}(R - R') \right.$$
$$\left. + \frac{1}{2}\frac{\partial_{i\alpha}\partial_{j\beta}\Psi_G(R')(R - R')_{i\alpha}(R - R')_{j\beta}}{\Psi_G(R')} + \cdots\right] \tag{9.111}$$

Inserting the previous equation in Eq. (9.110) we can see that after integrating over R the terms containing odd powers of $(R - R')$ disappear by parity. We are therefore left with:

$$\mathscr{N} = e^{-\Delta\tau[V(R') - E_0]}\left[1 + \frac{1}{2}\frac{\nabla^2\Psi_G(R')}{\Psi_G(R')}\frac{\hbar^2}{m}\Delta\tau + \cdots\right] \tag{9.112}$$

We can now use the same trick used above to write the expression in square parenthesis as an exponential. The result is:

$$\mathscr{N} = \exp\left[-\Delta\tau\left(V(R') - \frac{\hbar^2}{2m}\frac{\nabla^2\Psi_G(R')}{\Psi_G(R')} - E_0 z\right)\right] \tag{9.113}$$

In the previous expression it is possible to immediately recognize the local energy. In fact, when using importance sampling, the normalization assumes the expression:

$$\mathscr{N} = \exp\left[-\Delta\tau\left(\frac{\hat{H}\Psi_G(R')}{\Psi_G(R')} - E_0\right)\right] \tag{9.114}$$

This is the new form of the weight factor that one needs to compute in order to determine the multiplicity of the walker at a given position. It is immediately clear that the fact that in the exponential we have the difference between the local energy, instead of the potential energy, and the reference eigenvalue E_0 essentially resolves the issue related to the fluctuations of the population related to a divergent behavior of the interaction. In fact, if we knew the exact solution the exponent would be identically zero, and the population would be absolutely stable. However, by means of an accurate variational calculation it is possible to obtain a very good approximation of the ground-state wavefunction, thereby reducing the fluctuations in the population to a minimum.

The algorithm including importance sampling is modified in the following way.

1. For each walker, and for each coordinate perform a "drift" move along the gradient of the guiding function. This displacement is deterministic.

```
DMC_Drift()
for i ∈ {0, A − 1} do
    for j ∈ {0, 2} do
        R_drift[i][j] ← R[i][j] + (∇Ψ_G(R)/Ψ_G(R))|_[i][j] DΔτ
    end for
end for
```

2. Cycle again over coordinates and diffuse the position from R_{drift} as in the non-importance sampled case.
3. Compute the new multiplicity of the walker and the weight to assign to estimators using

$$W = \exp\left[-\Delta\tau\left(\frac{\hat{H}\Psi_G(R')}{\Psi_G(R')} - E_0\right)\right] \tag{9.115}$$

In this way the walkers will asymptotically sample the distribution:

$$f(R) = \Psi_G(R)\phi_0(R). \tag{9.116}$$

This means that it is possible to evaluate integrals of the form:

$$\langle O\rangle = \frac{\int dRf(R)O(R)}{\int dRf(R)}. \tag{9.117}$$

As in the previous case the evaluation of the exact energy eigenvalue can be easily obtained by using the local energy. In fact, the matrix element of the Hamiltonian between the guiding function[4] and the ground-state wavefunction is:

$$\langle E\rangle = \frac{\langle\phi_0|\hat{H}|\Psi_G\rangle}{\langle\phi_0|\Psi_G\rangle} = \frac{\int dRf(R)\frac{\hat{H}\Psi_G(R)}{\Psi_G(R)}}{\int dRf(R)} \tag{9.118}$$

Once more, because of the hermiticity of the Hamiltonian we will have that $\langle E\rangle = E_0$. All other estimators will be matrix elements of the operator between Ψ_G and ϕ_0.

[4]It is always possible to project the energy from a function Ψ_T other than Ψ_G, by introducing a further weighing factor $\frac{\Psi_T}{\Psi_G}$. However this is very rarely used in standard applications.

9.3.5 The Fermion Sign Problem

As we have mentioned, imaginary time propagation projects out of an arbitrary initial state the absolute (mathematical) ground state of a given Hamiltonian $\hat{H}$, which is always a nodeless function. One might correctly object that if the initial state is chosen in such a way not to have any overlap with this ground state, the projection will correctly give back some excited state of $\hat{H}$. More rigorously, if our initial state has components only within a certain subspace of the total Hilbert space, which could be selected, for instance, by the wavefunction symmetry, then imaginary time propagation will end up projecting out the eigenstate with lowest eigenvalue within that given subspace.

This seems to be particularly useful when thinking of applying DMC-like algorithm to the study of many-fermion systems, as the nuclear systems we are interested in. The antisymmetry property of the fermionic ground state suggests that it should be sufficient to start from an arbitrary antisymmetric state $|\Psi_A\rangle$ (provided it is not orthogonal to the fermion ground state) to obtain the sought solution. In fact, one might speculate that antisymmetry itself would guarantee that there is no overlap with the symmetric ground state since the beginning:

$$\begin{aligned}\lim_{\tau\to\infty} e^{-\tau(\hat{H}-E_0^A)}|\Psi_A(0)\rangle &= \sum_n e^{-\tau(E_n-E_0^A)}\langle\phi_n|\Psi_A\rangle|\phi_n\rangle = \\ &= \langle\phi_0^A|\Psi_A\rangle|\phi_0^A\rangle \\ &\quad + \lim_{\tau\to\infty}\langle\phi_0|\Psi_A\rangle|\phi_0\rangle e^{-\tau(E_0-E_0^A)}\end{aligned} \tag{9.119}$$

However, this abstract formulation forgets that eventually we need to *sample a probability density* in order to operate with a Monte Carlo integration, and any excited state will have a wavefunction changing sign somewhere, thereby breaking this requirement. If we had an exact knowledge of the *nodal surface* of the ground state (i.e. of the set of points such that $\phi_0^A(0) = 0$), we could use an antisymmetric function $\Psi_G^A(R)$ having the same nodal surface, and obtain by importance function the required positive definite density to sample:

$$\langle E\rangle = \frac{\langle\phi_0^A|\hat{H}|\Psi_G^A\rangle}{\langle\phi_0^A|\Psi_G^A\rangle} = \frac{\int dR\phi_0^A(R)\Psi_G^A(R)\frac{\hat{H}\Psi_G^A(R)}{\Psi_G^A(R)}}{\int dR\phi_0^A(R)\Psi_G^A(R)}. \tag{9.120}$$

If $\Psi_G^a(R)$ does not have the same nodal surface as $\phi_0^A(R)$, we are once again in trouble.

We might have then the idea of separately sampling the positive and the negative part of the wave function. It is always possible to split an antisymmetric function as:

$$\psi^A(R) = \Psi^+(R) - \Psi^-(R), \tag{9.121}$$

where both Ψ^+ and Ψ^- are positive definite functions. It is easy to see that each one, by linearity, is a solution of the Schrödinger equation with the same eigenvalue as the fermionic ground state. We can call $|R^+\rangle$ the walkers sampling the positive part and $|R^-\rangle$ the walkers sampling the negative part of Ψ^A. The energy expectation could be computed as:

$$E_0^A = \frac{\int dR^+ f^+(R^+) \frac{\hat{H}\Psi_G^A(R^+)}{\Psi_G^A(R^+)} - \int dR^+ f^-(R^-) \frac{\hat{H}\Psi_G^A(R^-)}{\Psi_G^A(R^-)}}{\int dR^+ f^+(R^+)\Psi_G^A(R^+) - \int dR^+ f^-(R^-)\Psi_G^A(R^-)}, \tag{9.122}$$

where $f^{\pm}$, as above, has the meaning of the importance sampled density of walkers. However, once more we have to notice that since both f^+ and f^- will obey the same imaginary time Schrödinger equation, the two densities will both converge to the ground-state density for $\hat{H}$. This means that both the numerator and the denominator of Eq. (9.122) will tend to 0 in the limit $\tau \to \infty$, and the ratio becomes undetermined. The major effect that one can observe during the calculation is that the variance of the energy will become exponentially large, and the integral will be dominated by statistical noise. This is the so called *fermion sign problem*. For some authors there is a prove that the computation of estimates such as Eq. (9.122) is an NP complex problem [10], and a solution will always require computer time that is exponentially increasing with the dimension of the system. However there are hints that by using methods that break this *plus/minus symmetry*, based on correlated dynamics and cancellation methods it is possible to reduce the cost to a polynomial dependence [11, 12].

9.3.5.1 Fixed-Node Approximation

A possible way of circumventing the sign problem in the case in which the antisymmetric ground-state wavefunction has to be real is to use some artificial boundary conditions [13].

We can define a nodal pocket $\Omega(R)$ as the set of points that can be reached from R without crossing the nodal surface at any point. For a standard Hamiltonian we can expect that for any pair of points R', R not on the nodal surface of the wavefunction, there exist a permutation P of the coordinates such that $PR' \in \Omega(R)$. This in turn means that all the space (but for the nodal surface, which has zero measure) can be covered by summing over all the permutations of the points lying in a single nodal pocket $\Omega(R)$. This the so-called *tiling theorem*. The tiling theorem implies that the fermion ground-state eigenvalue of the Schrödinger equation solved inside any $\Omega(R)$ is the same as the eigenvalue of the problem solved on the whole space.

The prove of the tiling theorem is quite simple. If the tiling property does not hold for the antisymmetric ground state $\phi_0^A(R)$, then $\sum_P \Omega(PR)$ will not completely cover the space, leaving out some regions. This means that somewhere there are two regions $\Omega(R_a)$ and $\Omega(R_b)$ that share part of the nodal surface and are not equivalent. It is then possible to construct a function with a lower eigenvalue in the region $\Omega(R_a) \bigcup \Omega(R_b)$ by simply removing the common node and solving for the ground state of $\hat{H}$ within that region. Let us call ϕ^0_{ab} this function. Constructing an antisymmetric function $\Psi_A(R) = \sum_P (-1)^P \phi^0_{ab}$ we will have an antisymmetric function with an eigenvalue lower than that of $\phi^A_0(R)$, thereby violating the assumption that ϕ^0_A is the antisymmetric ground state of $\hat{H}$.

By the same kind of construction it is also possible to prove that the solution of the Schrödinger equation within a given nodal pocket $\Omega(R)$ is always an upper bound of the true antisymmetric eigenvalue, and that the exact result is recovered if and only if the nodal surface of the wavefunction generated by replicating the pocket coincides with that of the exact eigenfunction.

The previous considerations suggest that solving for the ground state of a given Hamiltonian within a nodal pocket $\Omega(R)$ will provide an upper bound of the energy of a many-fermion system, which can in principle can by improved by improving the nodal structure of the test function used to determine the boundary conditions. This is called the *fixed-node approximation.* In order to have zero density at the nodal surface we have to assume that at the border of the nodal pocket an infinite absorbing potential exists, such that walkers never cross that surface. From the point of view of the algorithm this introduces a very tiny modification in the code. We have to remember that we can solve for the ground state in *any* pocket. This means that we do not need to care either of the initial position of the walkers or of the associated sign of the wavefunction. We said that the fixed-node approximation corresponds to modify the Hamiltonian as follows

$$\hat{\hat{H}} = \hat{H} + V_\Omega(R), \tag{9.123}$$

where

$$V_\Omega(R) = \begin{cases} \infty & \text{if } R \in S(\Omega) \\ 0 & \text{otherwise} \end{cases} \tag{9.124}$$

This means that every time the walker crosses the border of the nodal pocket $S(\Omega)$ its weight becomes zero, and the walker is simply canceled from the population. Fixed node calculations are presently very widely employed especially in quantum chemistry and solid state physics applications (for a review of applications to many electron systems see [14]). When the wavefunction needs to be complex it is no longer possible to define a nodal surface, and a different kind of approach has to be used. This will be discussed in the next section concerning the applications to the nuclear physics case.

9.4 Quantum Monte Carlo for Nuclear Hamiltonians in Coordinate Space

9.4.1 General Auxiliary Field Formalism

We begin by looking at the auxiliary field formalism without importance sampling. All such diffusion Monte Carlo methods can be formulated as

$$|\Psi(t+\Delta t)\rangle = \int dXP(X)T(X)|\Psi(t)\rangle \tag{9.125}$$

where X is a set of variables which will become our auxiliary fields, $P(X)$ is a probability density,

$$\begin{aligned} P(X) &\geq 0 \\ \int dXP(X) &= 1\,, \end{aligned} \tag{9.126}$$

and $T(X)$ is an operator that operates in the Hilbert space of $|\Psi(t)\rangle$. We are free to choose the variables X, the probability density $P(X)$, and the operator $T(X)$ subject only to the constraint that the integral gives the desired propagator

$$e^{-(H-E_T)\Delta t} = \int dXP(X)T(X)\,, \tag{9.127}$$

at least in the limit that $\Delta t \to 0$.

In diffusion Monte Carlo methods, we represent the state $|\Psi(t)\rangle$ as a linear combination of basis states which obviously must span the Hilbert space. These can be a complete set. An example is the position eigenstates used for diffusion Monte Carlo for central potentials. They can also form an overcomplete set such as or the position and spin/isospin bases used in the nuclear GFMC method and the position and overcomplete outer product of single particle spinor basis used in AFDMC, or the overcomplete single particle bases used in auxiliary field methods such as those developed by Zhang and coworkers. For either case, we can denote these basis states as possible "walkers." We will denote one of these walker states as $|RS\rangle$ since we will be applying the method to systems where the basis is given by the positions of the particles, R, and a spinor for each spin-isospin of the particles, S.

The state, $|\Psi(t)\rangle$, at time t is represented in diffusion Monte Carlo methods as a linear combination of walker states

$$|\Psi(t)\rangle = \sum_{i=1}^{N_W} w_i|R_iS_i\rangle \tag{9.128}$$

where w_i is a coefficient, often called the weight, and N_W is the number of walkers.

The key ingredient to implementing a diffusion Monte Carlo method is to choose the walker basis and the operator $T(X)$ such that when $T(X)$ operates on a walker basis state, it gives one and only one new walker basis state. That is we want

$$T(X)|RS\rangle = W(X,R,S)|R'S'\rangle \tag{9.129}$$

where $|R'S'\rangle$ is normalized in the same way as $|RS\rangle$, and $W(X,R,S)$ is the change in the normalization from the propagation.

Once we have arranged for Eq. (9.129) to be true, we can implement the diffusion Monte Carlo by starting with $|\Psi(0)\rangle$ written, as in Eq. (9.128), as any, not unreasonable, linear combination of walkers. For each walker, we sample X values from $P(X)$, and use Eq. (9.129) to propagate to a new walker $|R_i'S_i'\rangle$, with a new weight w_i' given by the proportionality constant of Eq. (9.129) multiplied by the original weight w_i. We branch on the magnitude of the weight, so usually, after branching, $w_i' = 1$, where we are ignoring the fermion sign or phase problem for now and assuming that all of the weights are greater than or equal to zero. We will deal with the fermion case below.

9.4.2 *Operator Expectations and Importance Sampling*

9.4.2.1 Mixed Averages

Diffusion Monte Carlo methods efficiently calculate ground-state mixed averages

$$\bar{O}_{\text{mixed}} = \frac{\langle \Psi_T|O|\Psi(t)\rangle}{\langle \Psi_T|\Psi(t)\rangle} \tag{9.130}$$

where $|\Psi_T\rangle$ is trial state. If O is the Hamiltonian, operating on $|\Psi(t)\rangle$ shows that the result is the ground-state energy for large t. For other operators, for which the ground state is not an eigenstate, either approximate extrapolation methods or forward walking or its equivalent must be used to extract the correct ground-state expectation value.

Given a set of walkers as in Eq. (9.128), the mixed estimate can be calculated by

$$\bar{O}_{\text{mixed}} \simeq \frac{\sum_{i=1}^{N_w} w_i \langle \Psi_T|O|R_iS_i\rangle}{\sum_{i=1}^{N_w} w_i \langle \Psi_T|R_iS_i\rangle} \tag{9.131}$$

where the right hand side differs from the correct result because of statistical errors from the sampling which decreases as $N_W^{-1/2}$, and possible population size bias which decreases as N_W^{-1}. Statistical errors can be minimized by reducing the variance through importance sampling. Population bias also can be controlled with importance sampling, and, since it decays faster with population size, can be readily detected and removed by either taking larger numbers of walkers or extrapolation.

Efficient Monte Carlo methods need to have low variance so that the statistical error bars can be made small. For our walker propagation, this means that we should sample new walkers not only corresponding to the weight they will receive from our algorithm, but with this weight multiplied by their expected survival probability. The imaginary time Schrödinger equation is self adjoint, so the optimum importance function is the desired function. Typically, a trial function that can be efficiently evaluated is determined variationally and used as an approximation to the optimum trial function. Usually this trial wave function is used as the importance function. Sometimes a different importance function is used, so we will write this more general case.

9.4.2.2 Importance Sampling

To add importance sampling, we arrange to sample our walkers from a new state which we call $|\Psi_I\Psi(t)\rangle$ such that

$$\langle RS|\Psi_I\Psi(t)\rangle = \langle\Psi_I|RS\rangle\langle RS|\Psi(t)\rangle \tag{9.132}$$

so that

$$|\Psi_I\Psi(t)\rangle = \sum_{i=1}^{N_w} w_i|R_iS_i\rangle \tag{9.133}$$

An alternative way of looking at this is that the sampling probability for the walkers at R_iS_i has been modified so that

$$|\Psi'(t)\rangle = \sum_{i=1}^{N_w} w_i\langle\Psi_I|R_iS_i\rangle^{-1}|R_iS_i\rangle\,. \tag{9.134}$$

Calculating a mixed average now becomes

$$\bar{O}_{\text{mixed}} = \frac{\sum_{i=1}^{N_w} w_i \frac{\langle\Psi_T|R_iS_i\rangle}{\langle\Psi_I|R_iS_i\rangle}\frac{\langle\Psi_T|O|R_iS_i\rangle}{\langle\Psi_T|R_iS_i\rangle}}{\sum_{i=1}^{N_w} w_i \frac{\langle\Psi_T|R_iS_i\rangle}{\langle\Psi_I|R_iS_i\rangle}}\,. \tag{9.135}$$

For the usual case where $|\Psi_I\rangle = |\Psi_T\rangle$, and $w_i = 1$, we have

$$\bar{O}_{\text{mixed}} = \frac{1}{N_w}\sum_{i=1}^{N_w}\frac{\langle\Psi_T|O|R_iS_i\rangle}{\langle\Psi_T|R_iS_i\rangle}\,. \tag{9.136}$$

We substitute Eqs. (9.132) and (9.133) into Eq. (9.125)

$$
\begin{aligned}
|\Psi_I \Psi(t+\Delta t)\rangle &= \sum_{i=1}^{N_w} w_i \int dX P(X) \frac{\langle \Psi_I | R_i' S_i' \rangle}{\langle \Psi_I | R_i S_i \rangle} T(X) |R_i S_i\rangle \\
&= \sum_{i=1}^{N_w} w_i \int dX P(X) \frac{\langle \Psi_I | T(X) | R_i S_i \rangle}{\langle \Psi_I | R_i S_i \rangle} \frac{T(X)}{W(X, R_i, S_i)} |R_i S_i\rangle \quad (9.137)
\end{aligned}
$$

where $|R_i' S_i'\rangle$ is defined as in Eq. (9.129). Notice that the operator $T(X)/W(X, R_i, S_i)$ operating on $|R_i S_i\rangle$ gives a normalized walker. The additional weight of this walker is given by $P(X)\frac{\langle \Psi_I | T(X) | R_i S_i \rangle}{\langle \Psi_I | R_i S_i \rangle}$. We want to minimize fluctuations in this weight factor, and to do this we normalize it and sample from the normalized distribution. The normalization will be the weight.

We write

$$
\begin{aligned}
\mathcal{N} &= \int dX P(X) \frac{\langle \Psi_I | T(X) | R_i S_i \rangle}{\langle \Psi_I | R_i S_i \rangle} \\
&= \frac{\langle \Psi_I | e^{-(H-E_T)\Delta t} | R_i S_i \rangle}{\langle \Psi_I | R_i S_i \rangle} \\
&= e^{-(E_L(R_i, S_i) - E_T)\Delta t} + O(\Delta t^2) \quad (9.138)
\end{aligned}
$$

where the local energy $E_L(R_i, S_i)$ is defined by

$$
E_L(R_i, S_i) = \frac{\langle \Psi_I | H | R_i S_i \rangle}{\langle \Psi_I | R_i S_i \rangle} \quad (9.139)
$$

and we now sample X variables from the normalized distribution

$$
\tilde{P}(X) = \mathcal{N}^{-1} P(X) \frac{\langle \Psi_I | T(X) | R_i S_i \rangle}{\langle \Psi_I | R_i S_i \rangle} . \quad (9.140)
$$

The importance sampled diffusion Monte Carlo in the auxiliary field formalism becomes

$$
|\Psi_I \Psi(t+\Delta t)\rangle = \sum_{i=1}^{N_w} w_i \int dX \tilde{P}(X) e^{-(E_L(R_i, S_i) - E_T)\Delta t} \frac{T(X)}{W(X, R_i, S_i)} |R_i S_i\rangle . \quad (9.141)
$$

We propagate a walker by sampling an X value from $\tilde{P}(X)$, we include the local energy expression in the weight, and construct the new normalized walker position and spin state as $W^{-1}(X, R_i, S_i) T(X) |R_i S_i\rangle$. In each of the equations above, the ratio of the wave function terms gives the walker weight. In Eq. (9.141) these terms have been combined to give a weight that depends on the local energy expectation value.

All of the expectation values and weights contain ratios of trial wave functions so that any normalization factor multiplying the $|RS\rangle$ cancels and any convenient normalization can be used. We can therefore drop the W factors and normalize our walker kets at the end of a step. Typically just the walker positions are stored and the walker spinors are normalized to have magnitude 1.

9.4.2.3 Importance Sampling with a Hubbard-Stratonovich Transformation

We often have Hamiltonians where the Hubbard-Stratonovich transformation

$$e^{\frac{O^2}{2}} = \frac{1}{\sqrt{2\pi}} \int_{-\infty}^{\infty} dx e^{-\frac{x^2}{2}} e^{xO} \tag{9.142}$$

can be used to write a propagator in the form of Eq. (9.129). Examples are writing the kinetic energy as an integral over translations, or writing terms like $\boldsymbol{\sigma}_i \cdot \boldsymbol{\sigma}_j = (\boldsymbol{\sigma}_i + \boldsymbol{\sigma}_j)^2 - 6$ as an integral over spin rotations.

Since we primarily use the Hubbard-Stratonovich transformation to define our auxiliary fields, it is useful to work out how importance sampling can be included within the short-time approximation for this particular case. We begin with a Hamiltonian that is quadratic in a set of N_O operators (which for our nuclear problems will be momentum and spin-isospin operators) O_n,

$$H = \frac{1}{2} \sum_{n=1}^{N_O} \lambda_n O_n^2 \tag{9.143}$$

so that the imaginary time propagator is

$$e^{-H\Delta t} = \int dx \frac{1}{(2\pi)^{N_O/2}} e^{-\frac{1}{2}\sum_{n=1}^{N_O} x_n^2} e^{-i\sum_{n=1}^{N_O} x_n \sqrt{\lambda_n \Delta t} O_n} + O(\Delta t^2) \tag{9.144}$$

where the Δt^2 terms comes from the possible non-commutativity of the O_n.

As before, we choose our walker basis and the operators O_n such that operating on a walker, $|RS\rangle$, with a term sampled from the integrand, gives a result proportional to another walker

$$e^{-i\sum_{n=1}^{N_O} x_n \sqrt{\lambda \Delta t} O_n} |RS\rangle = W(\{x_n\}, R, S)|R'S'\rangle \tag{9.145}$$

where $\{x_n\}$ represents the set of sampled x_n values.

We now sample $\tilde{P}(X)$ which is

$$\begin{aligned}\tilde{P}(X) &= \mathcal{N}^{-1} e^{-\frac{1}{2}\sum_{n=1}^{N_O} x_n^2} \frac{\langle \Psi_T | e^{-i\sum_{n=1}^{N_O} x_n \sqrt{\lambda_n \Delta t} O_n} | RS \rangle}{\langle \Psi_T | RS \rangle} \\ &= \mathcal{N}^{-1} e^{-\frac{1}{2}\sum_{n=1}^{N_O} x_n^2} \left(1 - i \sum_{n=1}^{N_O} x_n \sqrt{\lambda_n \Delta t} \frac{\langle \Psi_T | O_n | RS \rangle}{\langle \Psi_T | RS \rangle} \right. \\ & \left. -\frac{1}{2} \sum_{n=1,m=1}^{N_O} x_n x_m \sqrt{\lambda_m \lambda_n} \Delta t \frac{\langle \Psi_T | O_n O_m | RS \rangle}{\langle \Psi_T | RS \rangle} + \ldots \right)\end{aligned} \tag{9.146}$$

Notice that if we were to expand $T(X)/W(X, R_i, S_i)$ it would have the form $1 + O(x_n \Delta t^{1/2}) + O(x_n x_m \Delta t) + \ldots$. Therefore if we drop terms of order Δt^2, the $O(\Delta t)$ term of $P(X)$ contributes only when it multiplies the 1 term from $T(X)/W(X, R_i, S_i)$. We can therefore integrate it over X without changing the result to this order in Δt. This term cancels the normalization, so that

$$\begin{aligned}\tilde{P}(X) &= e^{-\frac{1}{2}\sum_{n=1}^{N_O} x_n^2} \left(1 - i \sum_{n=1}^{N_O} x_n \sqrt{\lambda_n \Delta t} \frac{\langle \Psi_T | O_n | RS \rangle}{\langle \Psi_T | RS \rangle} + O(\Delta t^{3/2}) \right) \\ &= e^{-\frac{1}{2}\sum_{n=1}^{N_O} x_n^2} e^{-i\sum_{n=1}^{N_O} x_n \sqrt{\lambda_n \Delta t} \frac{\langle \Psi_T | O_n | RS \rangle}{\langle \Psi_T | RS \rangle} + \sum_{n=1}^{N_O} \lambda_n [\frac{\langle \Psi_T | O_n | RS \rangle}{\langle \Psi_T | RS \rangle}]^2} + O(\Delta t^{3/2}) \\ &= \exp\left\{ -\frac{1}{2} \sum_{n=1}^{N_O} \left[x_n + i\sqrt{\lambda_n \Delta t} \frac{\langle \Psi_T | O_n | RS \rangle}{\langle \Psi_T | RS \rangle} \right]^2 \right\}\end{aligned} \tag{9.147}$$

where in the last line, we have written the linear term in x in the exponent, and included a canceling term so that only the linear term survives integration to order Δt.

We sample our expression by sampling x_n from the shifted gaussian (Again, we assume here that $i\sqrt{\lambda_n \Delta t}\langle O_n \rangle$ is real. We will discuss what to do for the complex case below.)

$$x_n = \chi_n - i\sqrt{\lambda_n \Delta t}\langle O_n \rangle \tag{9.148}$$

where χ_n is sampled from a gaussian with unit variance. The new unnormalized ket is

$$|R'S'\rangle = e^{-i\sum_{n=1}^{N_O} x_n \sqrt{\lambda \Delta t} O_n} |RS\rangle \langle \Psi_T | RS \rangle \tag{9.149}$$

and its weight is given by the local energy expression

$$W(R', S') = e^{-[\langle H \rangle - E_T]\Delta t} \tag{9.150}$$

9.4.3 Application to Standard Diffusion Monte Carlo

9.4.3.1 Diffusion Monte Carlo Without Importance Sampling

It is helpful to apply the formalism above to derive the well known central potential diffusion Monte Carlo algorithm [9]. The Hamiltonian is

$$H = \sum_{j=1}^{A} \sum_{\alpha=1}^{3} \frac{p_{j\alpha}^2}{2m} + V(R) \tag{9.151}$$

where $p_{j\alpha}$ and R operate on Hilbert space, and $p_{j\alpha}$ is the α component of the momentum operator for the jth particle. Making the short-time approximation, the propagator can be written as

$$e^{-(H-E_T)\Delta t} = e^{\sum_{j=1}^{A} \sum_{\alpha=1}^{3} \frac{p_{j\alpha}^2}{2m} \Delta t} e^{-[V(R)-E_T]\Delta t} + O(\Delta t^2) \, . \tag{9.152}$$

Since the Hamiltonian does not operate on the spin, we can drop the spin variable from the our walker expressions and take just a position basis $|R\rangle$. Operating with the potential term

$$e^{-[V(R)-E_T]\Delta t} |R_j\rangle = e^{-[V(R_j)-E_T]\Delta t} |R_j\rangle \tag{9.153}$$

clearly satisfies Eq. (9.129). The kinetic energy part of the propagator does not satisfy Eq. (9.129). However, by using the Hubbard-Stratonovich transformation, we can write the kinetic energy in terms of the translation operators $e^{-\frac{i}{\hbar} p_{j\beta} a}$. We introduce the auxiliary field or Hubbard-Stratonovich variables, $x_{j\alpha}$, and write

$$e^{-\sum_{j=1}^{A} \sum_{\alpha=1}^{3} \frac{p_{j\alpha}^2}{2m} \Delta t} = \prod_{j\alpha} \frac{1}{(2\pi)^{3/2}} \int dx_{j\alpha} e^{-\frac{x_{j\alpha}^2}{2}} e^{-\frac{i}{\hbar} \mathbf{p}_{j\alpha} x_{j\alpha} \sqrt{\frac{\hbar^2 \Delta t}{m}}} \tag{9.154}$$

With this definition, X is the set $\{x_{j\alpha}\}$, for the A particles,

$$P(X) = \prod_{j\alpha} \frac{1}{\sqrt{2\pi}} e^{-\frac{x_{j\alpha}^2}{2}} \, , \tag{9.155}$$

and

$$T(X)|R\rangle = e^{-[V(R)-E_T]\Delta t}|R + \Delta R\rangle \tag{9.156}$$

where $R' = R + \Delta R$ is given by translating each particle's position in R

$$r'_{j\alpha} = r_{j\alpha} + x_{j\alpha}\frac{\hbar^2 \Delta t}{m} . \tag{9.157}$$

This is identical to the standard diffusion Monte Carlo algorithm without importance sampling. We move each particle with a gaussian distribution of variance $\frac{\hbar^2\Delta t}{m}$, and include a weight of $e^{-[V(R)-E_T]\Delta t}$. We would then include branching on the weight to complete the algorithm.

While the Hubbard-Stratonovich transformation is the most common, there are many other possibilities. For example, the propagator for the relativistic kinetic energy $\sqrt{p^2c^2 + m^2c^4} - mc^2$ can be sampled by using

$$e^{-\left[\sqrt{p^2c^2+m^2c^4}-mc^2\right]\Delta t} = \int d^3x f(x) e^{-\frac{i}{\hbar}\mathbf{p}\cdot\mathbf{x}} \tag{9.158}$$

with

$$\begin{aligned} f(x) &= \int \frac{d^3p}{(2\pi)^3} e^{\frac{i}{\hbar}\mathbf{p}\cdot\mathbf{x}} e^{-\left[\sqrt{p^2c^2+m^2c^4}-mc^2\right]\Delta t} \\ &= e^{mc^2\Delta t} K_2\left(\frac{mc}{\hbar}\sqrt{x^2 + c^2\Delta t^2}\right) \end{aligned} \tag{9.159}$$

where K_2 is the modified Bessel function of order 2 [15].

9.4.3.2 Importance Sampled Diffusion Monte Carlo in the Auxiliary Field Formulism

We break up the Hamiltonian as a kinetic and potential part. The potential part gives the usual $e^{-V(R)\Delta t}$ weight, and we need to work only with the importance sampled kinetic energy part. The kinetic energy operator is already written as a sum of squares,

$$KE = \sum_{j\alpha} \frac{p_{j\alpha}^2}{2m} \tag{9.160}$$

where j is the particle label and α is the x, y, or z coordinate. We can identify $\lambda_{j\alpha} = m^{-1}$, and $O_{j\alpha} = p_{j\alpha}$. Substituting this into our previous formalism, we have

$$\begin{aligned} i\sqrt{\lambda_{j\alpha}\Delta t}\langle O_{j\alpha}\rangle &= i\sqrt{\frac{\Delta t}{m}}\frac{\langle \Psi_T|p_{j\alpha}|RS\rangle}{\langle \Psi_T|RS\rangle} \\ &= -\sqrt{\frac{\hbar^2\Delta t}{m}}\frac{\partial_{j\alpha}\langle \Psi_T|RS\rangle}{\langle \Psi_T|RS\rangle}\,. \end{aligned} \tag{9.161}$$

The sampled value of $x_{j\alpha}$ will be

$$x_{j\alpha} = \chi_{j\alpha} + \sqrt{\frac{\hbar^2\Delta t}{m}}\frac{\partial_{j\alpha}\langle \Psi_T|RS\rangle}{\langle \Psi_T|RS\rangle} \tag{9.162}$$

where the $\chi_{j\alpha}$ are sampled from a gaussian with unit variance. The new walker will be

$$|R'S'\rangle = e^{-\frac{i}{\hbar}\sum_{j\alpha} x_{j\alpha}\sqrt{\frac{\hbar^2\Delta t}{m}}p_{j\alpha}}|RS\rangle\,. \tag{9.163}$$

Since $e^{-\frac{i}{\hbar}p_{j\alpha}a}$ is the translation operator that translates the ket's $j\alpha$ position coordinate by a. We have

$$\begin{aligned} S' &= S \\ R'_{j\alpha} &= R_{j\alpha} + x_{j\alpha}\sqrt{\frac{\hbar^2\Delta t}{m}} \\ &= R_{j\alpha} + \chi_{j\alpha}\sqrt{\frac{\hbar^2\Delta t}{m}} + \frac{\hbar^2\Delta t}{m}\frac{\partial_{j\alpha}\langle \Psi_T|RS\rangle}{\langle \Psi_T|RS\rangle} \end{aligned} \tag{9.164}$$

which is the standard diffusion Monte Carlo propagation. The weight factor is the local energy.

9.4.4 *Fixed-Phase Importance-Sampled Diffusion Monte Carlo*

The fixed-phase approximation [16] was developed to extend the fixed-node approximation to electrons in a magnetic field where the ground-state wave function is complex. The approximation enforces the trial function's phase as the phase for the calculated ground state. Diffusion Monte Carlo is used to sample the magnitude of the ground state.

If the walker phase has been chosen so that $\langle \Psi_T|R\rangle$ is real, the fixed-phase approximation requires that after propagation $\langle \Psi_T|R'\rangle$ would also be real since an imaginary part would correspond to the calculated ground-state having a different phase than the trial function. Therefore in the implementation of the fixed-phase approximation we discard the imaginary part of the weight of a propagated walker. For an arbitrary initial phase, we discard the imaginary part of the ratio $\frac{\langle\Psi_T|R'\rangle}{\langle\Psi_T|R\rangle}$ which means that the we replace the importance sampled factor in Eq. (9.146) with its real part

$$\frac{\langle\Psi_T|e^{-i\sum_{n=1}^{N_O} x_n\sqrt{\lambda_n\Delta t}O_n}|RS\rangle}{\langle\Psi_T|RS\rangle} \to \mathrm{Re}\left[\frac{\langle\Psi_T|e^{-i\sum_{n=1}^{N_O} x_n\sqrt{\lambda_n\Delta t}O_n}|RS\rangle}{\langle\Psi_T|RS\rangle}\right]. \tag{9.165}$$

The fixed-phase algorithm for propagating a walker is then

1. Propagate to the new position (the spin does not change with a central potential)

$$\begin{aligned} S' &= S \\ R'_{j\alpha} &= R_{j\alpha} + \chi_{j\alpha}\sqrt{\frac{\hbar^2\Delta t}{m}} + \frac{\hbar^2\Delta t}{m}\mathrm{Re}\left[\frac{\partial_{j\alpha}\langle\Psi_T|RS\rangle}{\langle\Psi_T|RS\rangle}\right] \end{aligned} \tag{9.166}$$

2. Include a weight factor for the walker of

$$W = e^{-(\mathrm{Re}\langle H\rangle - E_T)\Delta t} \tag{9.167}$$

This is identical to the fixed-phase algorithm of Ortiz et al.

We will see that similar approximations can be used for our spin-isospin dependent problems.

9.4.5 Application to Quadratic Forms

Quadratic forms in operators that change from one walker to another can be diagonalized to produce the sum of squares needed for Eq. (9.143). That is for

$$H = \frac{1}{2}\sum_{ij} O_i A_{ij} O_j \tag{9.168}$$

with A_{nm} real and symmetric, we can calculate the normalized real eigenvectors and eigenvalues of the matrix A,

$$\sum_j A_{ij}\psi_j^{(n)} = \lambda_n|\psi_i\rangle$$
$$\sum_j \psi_j^{(n)}\psi_j^{(m)} = \delta_{nm}\,. \tag{9.169}$$

The matrix is then

$$A_{ij} = \sum_n \psi_i^{(n)}\lambda_n\psi_j^{(n)} \tag{9.170}$$

and substituting back we have

$$H = \frac{1}{2}\sum_n \lambda_n \mathcal{O}_n^2$$
$$\mathcal{O}_n = \sum_j \psi_j^{(n)} O_j\,, \tag{9.171}$$

which is now in the form of Eq. (9.143).

9.4.6 Auxiliary Field Breakups

There are many possible ways to break up the nuclear Hamiltonian using the auxiliary field formalism. As a concrete example let's look at the spinor propagator when we have a spin-exchange potential between A neutrons

$$V = \sum_{i<j} v^{\sigma}(r_{ij})\boldsymbol{\sigma}_i\cdot\boldsymbol{\sigma}_j \tag{9.172}$$

Taking the operators to be the x, y, and z components of the Pauli operators for each particle, we have a quadratic form in these $3A$ operators. Since walker gives the positions of the neutrons, we know the value of $v^{\sigma}(r_{ij})$ for all pairs. We can then write

$$V = \frac{1}{2}\sum_{ij}^{A} B_{ij}\sigma_{ix}\sigma_{jx} + \frac{1}{2}\sum_{ij}^{A} B_{ij}\sigma_{iy}\sigma_{jy} + \frac{1}{2}\sum_{ij}^{A} B_{ij}\sigma_{iz}\sigma_{jz} \tag{9.173}$$

where $B_{ii} = 0$, and $B_{ij} = v^\sigma(r_{ij})$ for $i \neq j$. Finding the eigenvectors $\psi_i^{(n)}$ and eigenvalues λ_n of the B matrix, we can write

$$
\begin{aligned}
V &= \frac{1}{2}\sum_n \lambda_n(\mathcal{O}_{nx})^2 + \frac{1}{2}\sum_n \lambda_n(\mathcal{O}_{ny})^2 + \frac{1}{2}\sum_n \lambda_n(\mathcal{O}_{nz})^2 \\
\mathcal{O}_{nx} &= \sum_{i=1}^{A} \psi_i^{(n)}\sigma_{ix} \\
\mathcal{O}_{ny} &= \sum_{i=1}^{A} \psi_i^{(n)}\sigma_{iy} \\
\mathcal{O}_{nz} &= \sum_{i=1}^{A} \psi_i^{(n)}\sigma_{iz}\,.
\end{aligned}
\tag{9.174}
$$

Using the Hubbard-Stratonovich transformation would give us $3A$ auxiliary fields.

We can modify this transformation. For example, the diagonal elements of the B matrix are zero. Adding a nonzero diagonal term B_{jj}, would give us additional terms proportional to $\sigma_{jx}^2 = \sigma_{jy}^2 = \sigma_{jz}^2 = 1$, that is, these would be additional purely central terms. Subtracting a corresponding central contribution would then give an identical interaction, but different eigenvectors and therefore different spin rotation operators.

Another alternative would be to look at each term in the sum separately as a quadratic form of two operators. The resulting 2×2 matrices have two eigenvalues and eigenvectors so that

$$
v^\sigma(r_{ij})\sigma_{ix}\sigma_{jx} = \frac{1}{4}v^\sigma(r_{ij})[\sigma_{ix} + \sigma_{jx}]^2 - \frac{1}{4}v^\sigma(r_{ij})[\sigma_{ix} - \sigma_{jx}]^2 \tag{9.175}
$$

and each of the $3A(A-1)/2$ terms would require 2 auxiliary fields or $3A(A-1)$ total. We can reduce the number of auxiliary fields by including diagonal terms to our 2×2 matrix equal to the off diagonal terms. These make the eigenvector $(1, -1)/\sqrt{2}$ have a zero eigenvalue, which then does not contribute

$$
v^\sigma(r_{ij})\sigma_{ix}\sigma_{jx} = \frac{1}{2}v^\sigma(r_{ij})[\sigma_{ix} + \sigma_{jx}]^2 - v^\sigma(r_{ij}) \tag{9.176}
$$

where the second term on the right hand side is a central potential counter term that would be added to the physical central potential, and $3A(A-1)/2$ auxiliary fields would be required. This form could also be derived by expanding the square $[\sigma_{ix} + \sigma_{jx}]^2$.

Each of these breakups gives the same net propagator after integration of the auxiliary fields. If a good importance function is used, and the sampling can be carried out, we would expect the local energy for a complete step to have low variance, and therefore the propagation to have low variance. The trade off then would be the complexity of constructing the operator combination versus the number of auxiliary fields needed in the propagation. In our work to date, we have used the full diagonalization to minimize the number of auxiliary field integrations. The cost of the diagonalization is order A^3 which is the same order as the cost for calculating a Slater determinant for the trial functions we need. However, it is easy to imagine having more complicated Hamiltonians where the cost of full diagonalization would be prohibitive (for example adding Δ degrees of freedom to the nuclei) and a simpler breakup using more auxiliary fields would be more efficient.

The best break up will be the one which optimizes the accuracy and variance of the results for a given amount of computational resources.

9.4.7 AFDMC with the v_6' Potential for Nuclear Matter

The Argonne v_6' potential includes central, spin and isospin exchange, and tensor interactions. Writing out the components, the Hamiltonian is

$$
\begin{aligned}
H = & \sum_{i\alpha} \frac{p_{i\alpha}^2}{2m} + \sum_{i<j} v^c(r_{ij}) + \sum_{i<j,\alpha\beta} \left\{ v^\sigma(r_{ij})\delta_{\alpha\beta} + v^t(r_{ij}) \left[3\hat{\alpha}\cdot\hat{r}_{ij}\hat{\beta}\cdot\hat{r}_{ij} - \delta_{\alpha\beta} \right] \right\} \sigma_{i\alpha}\sigma_{j\beta} \\
& + \sum_{i<j,\alpha\beta\gamma} \left\{ v^{\sigma\tau}(r_{ij})\delta_{\alpha\beta} + v^{t\tau}(r_{ij}) \left[3\hat{\alpha}\cdot\hat{r}_{ij}\hat{\beta}\cdot\hat{r}_{ij} - \delta_{\alpha\beta} \right] \right\} [\sigma_{i\alpha}\tau_{i\gamma}][\sigma_{j\beta}\tau_{j\gamma}] \\
& + \sum_{i<j,\gamma} v^\tau(r_{ij})\tau_{i\gamma}\tau_{j\gamma}
\end{aligned}
\tag{9.177}
$$

where α and β refer to the x, y, and z components and $\hat{\alpha}\hat{\beta}$ are the corresponding unit vectors. We work in a position basis. The potential is quadratic in the 15A spin-isospin operators $\sigma_{i\alpha}$, $\tau_{i\gamma}$, $\sigma_{i\alpha}\tau_{i\gamma}$. Since each spin-isospin operator can rotate the corresponding spin-isospinor the natural basis is the overcomplete basis of the outer product of these spin-isospinors—one for each particle. A walker consists of an overall weight factor, and x, y, and z coordinates and four complex numbers for the components of $|p\uparrow\rangle$, $|p\downarrow\rangle$, $|n\uparrow\rangle$, $|n\downarrow\rangle$ for each of the A particles.

9.4.7.1 The v'_6 Hamiltonian as a Sum of Operator Squares

We now follow Sect. 9.4.5 and define matrices

$$\begin{aligned} C^{\sigma}_{i\alpha,j\beta} &= v^{\sigma}(r_{ij})\delta_{\alpha\beta} + v^{t}(r_{ij})\left[3\hat{\alpha}\cdot\hat{r}_{ij}\hat{\beta}\cdot\hat{r}_{ij} - \delta_{\alpha\beta}\right] \\ C^{\sigma\tau}_{i\alpha,j\beta} &= v^{\sigma\tau}(r_{ij})\delta_{\alpha\beta} + v^{t\tau}(r_{ij})\left[3\hat{\alpha}\cdot\hat{r}_{ij}\hat{\beta}\cdot\hat{r}_{ij} - \delta_{\alpha\beta}\right] \\ C^{\tau}_{i,j} &= v^{\tau}(r_{ij}) \end{aligned} \tag{9.178}$$

which have zero matrix elements when $i = j$. Their eigenvalues and normalized eigenvectors are defined as

$$\begin{aligned} \sum_{j\beta} C^{\sigma}_{i\alpha,j\beta}\psi^{\sigma\ (n)}_{j\beta} &= \lambda^{\sigma}_{n}\psi^{\sigma\ (n)}_{i\alpha} \\ \sum_{j\beta} C^{\sigma\tau}_{i\alpha,j\beta}\psi^{\sigma\ (n)}_{j\beta} &= \lambda^{\sigma\tau}_{n}\psi^{\sigma\tau\ (n)}_{i\alpha} \\ \sum_{j} C^{\tau}_{i,j}\psi^{\tau\ (n)}_{j} &= \lambda^{\tau}_{n}\psi^{\tau\ (n)}_{i} \end{aligned} \tag{9.179}$$

with operator combinations

$$\begin{aligned} O^{\sigma}_{n} &= \sum_{i\alpha}\psi^{\sigma\ (n)}_{i\alpha}\sigma_{i\alpha} \\ O^{\sigma\tau}_{n\beta} &= \sum_{i\alpha}\psi^{\sigma\tau\ (n)}_{i\alpha}\sigma_{i\alpha}\tau_{i\beta} \\ O^{\tau}_{n\alpha} &= \sum_{i}\psi^{\tau\ (n)}_{i}\tau_{i\alpha} \end{aligned} \tag{9.180}$$

The Hamiltonian becomes

$$\begin{aligned} H = &\sum_{i=1}^{A}\sum_{\alpha=1}^{3}\frac{p^2_{i\alpha}}{2m} + \sum_{i<j} v^{c}(r_{ij}) + \frac{1}{2}\sum_{n=1}^{3A}\lambda^{\sigma}_{n}(O^{\sigma}_{n})^2 \\ &+ \frac{1}{2}\sum_{n=1}^{A}\sum_{\alpha=1}^{3}\lambda^{\tau}_{n}(O^{\tau}_{n\alpha})^2 + \frac{1}{2}\sum_{n=1}^{3A}\sum_{\alpha=1}^{3}\lambda^{\sigma\tau}_{n}(O^{\sigma\tau}_{n\alpha})^2 \end{aligned} \tag{9.181}$$

This is, of course, identical to the original Hamiltonian given in Eq. (9.177), but now it is in a form that makes the propagator easy to sample using auxiliary fields.

9.4.7.2 Complex Auxiliary Fields

In realistic nuclear physics problems, the fermion sign problem necessarily becomes a phase problem since conservation of angular momentum requires that flipping a spin changes the orbital angular momentum, which induces an angular phase to the wave function. Various fixed-phase approximations can be used. The Hubbard-Stratonovich transformation integrates the auxiliary field over all real values with a gaussian weight. With importance sampling, the gaussian for x_n is shifted by $i\sqrt{\lambda_n \Delta t}\langle O_n \rangle$ as shown in the last line of Eq. (9.147). As Zhang and Krakauer [17] showed for electronic structure problems, it is equally valid to integrate the auxiliary field over any shifted contour, and by shifting the contour so that x_n becomes complex and takes on the values $x_n = z + i\sqrt{\lambda_n \Delta t}\langle O_n \rangle$, $-\infty < z < \infty$. Integrating over these values does not change the result. However, now this factor is real. We implement the fixed phase approximation by taking the real part of $\langle H \rangle$.

Note that this method cannot be used for the momentum operator. This is because the operator $e^{-\frac{i}{\hbar} p_{j\alpha} a}$ is not bounded if a has an imaginary part. We therefore implement the kinetic energy terms exactly as in the central potential fixed-phase approximation.

There are of course other possible approximations that can be used. The auxiliary fields can be kept real. We find that the approximation is more accurate with Zhang-Krakauer prescription for auxiliary fields for the spin operators.

9.4.7.3 The v_6' Algorithm

We can now give the complete algorithm used for the v_6' potential.

1. We begin with a set of walkers $|R_i S_i\rangle$ which we sample from our trial function magnitude squared, $|\langle RS|\Psi_T\rangle|^2$, with Metropolis Monte Carlo. The walkers consist of the $3A$ coordinates of the A particles, and A 4-component normalized spinors.
2. For each walker in turn we calculate the C^σ, C^τ and $C^{\sigma\tau}$ matrices, their eigenvalues, and their eigenvectors.
3. From the trial function and spinor values we evaluate $\langle \sigma_{j\alpha} \rangle$, $\langle \sigma_{j\alpha}\tau_{j\beta} \rangle$, $\langle \tau_{j\alpha} \rangle$, $\langle p_{j\alpha} \rangle$, and $\langle H \rangle$.
4. We sample the complex values for the spin-isospin auxiliary fields

$$x_n = \chi_n + i\sqrt{\lambda_n \Delta t}\langle O_n \rangle \tag{9.182}$$

and transform our walker spinors using

$$|RS'\rangle = e^{-i\sum_{n=1}^{N_O} x_n \sqrt{\lambda \Delta t} O_n}|RS\rangle \tag{9.183}$$

and normalize the spinors

5. We sample the new positions from

$$r'_{j\alpha} = r_{j\alpha} + \chi_{j\alpha}\sqrt{\frac{\hbar^2 \Delta t}{m}} + \frac{\hbar^2 \Delta t}{m} \mathrm{Re} \frac{\partial_{j\alpha} \langle \Psi_T | RS \rangle}{\langle \Psi_T | RS \rangle} \tag{9.184}$$

6. The weight of the new walker is given by $W = e^{-[\mathrm{Re}\langle H \rangle - E_T]\Delta t}$
7. We branch on the walker weight, taking the number of new walkers to be the integer part of W plus a uniform random value on $(0, 1)$. If the weight W is negative, we discard the walker.

9.4.8 Isospin-Independent Spin-Orbit Interaction

Without isospin exchange, the spin orbit term for particles j and k is

$$\frac{1}{4\hbar} v_{LS}(r_{jk})[(\boldsymbol{r}_j - \boldsymbol{r}_k) \times (\boldsymbol{p}_j - \boldsymbol{p}_k)] \cdot (\boldsymbol{\sigma}_j + \boldsymbol{\sigma}_k) \tag{9.185}$$

We can write the kinetic energy plus spin-orbit interaction Hamiltonian as

$$\sum_{j\alpha} \frac{p_{j\alpha}^2}{2m} + \frac{1}{4\hbar} \sum_{j<k} v_{LS}(r_{jk})[(\boldsymbol{r}_j - \boldsymbol{r}_k) \times (\boldsymbol{p}_j - \boldsymbol{p}_k)] \cdot (\boldsymbol{\sigma}_j + \boldsymbol{\sigma}_k)$$

$$= \sum_{j\alpha} \frac{(p_{j\alpha} + \frac{m}{4\hbar} \sum_{k \neq j} v_{LS}(r_{jk})[(\boldsymbol{\sigma}_j + \boldsymbol{\sigma}_k) \times (\boldsymbol{r}_j - \boldsymbol{r}_k)]_\alpha)^2}{2m} + V_{\mathrm{Counter}}$$

$$V_{\mathrm{Counter}} = -\frac{1}{2m} \sum_{j\alpha} \left[\frac{m}{4\hbar} \sum_{k \neq j} v_{LS}(r_{jk})[(\boldsymbol{\sigma}_j + \boldsymbol{\sigma}_k) \times (\boldsymbol{r}_j - \boldsymbol{r}_k)]_\alpha \right]^2 \tag{9.186}$$

where the counter terms subtract off the unwanted interaction from completing the square. The counter terms do not depend on $\boldsymbol{p}_j$, so they can be included with the rest of the local potential, and will contribute to the drift and the local energy for that part. However, we will see that the local energy part is canceled below (that is the final weight will be just the correct total local energy which does not include the counter terms).

Using the Hubbard-Stratonovich break up with importance sampling, we have $\lambda_{j\alpha} = m^{-1}$, and

$$i\sqrt{\lambda_{j\alpha} \Delta t}\langle O_{j\alpha} \rangle = -\sqrt{\frac{\hbar^2 \Delta t}{m}} \frac{\partial_{j\alpha} \langle \Psi_T | RS \rangle}{\langle \Psi_T | RS \rangle} + i\sqrt{\frac{m \Delta t}{16\hbar^2}} \sum_{k \neq j} [(\langle \boldsymbol{\sigma}_j \rangle + \langle \boldsymbol{\sigma}_k \rangle) \times \boldsymbol{r}_{jk}]_\alpha v_{LS}(r_{jk}) \,. \tag{9.187}$$

The sampled value of $x_{j\alpha}$ will be

$$x_{j\alpha} = \chi_{j\alpha} + \sqrt{\frac{\hbar^2 \Delta t}{m}} \frac{\partial_{j\alpha} \langle \Psi_T | RS \rangle}{\langle \Psi_T | RS \rangle} - i \sqrt{\frac{m \Delta t}{16\hbar^2}} \sum_{k \neq j} [(\langle \boldsymbol{\sigma}_j \rangle + \langle \boldsymbol{\sigma}_k \rangle) \times \boldsymbol{r}_{jk}]_\alpha v_{LS}(r_{jk}) \,. \tag{9.188}$$

where our fixed-phase like approximation will modify this to keep the translation real, so that

$$x_{j\alpha} = \chi_{j\alpha} + \mathrm{Re}\left\{ \sqrt{\frac{\hbar^2 \Delta t}{m}} \frac{\partial_{j\alpha} \langle \Psi_T | RS \rangle}{\langle \Psi_T | RS \rangle} - i \sqrt{\frac{m \Delta t}{16\hbar^2}} \sum_{k \neq j} [(\langle \boldsymbol{\sigma}_j \rangle + \langle \boldsymbol{\sigma}_k \rangle) \times \boldsymbol{r}_{jk}]_\alpha v_{LS}(r_{jk}) \right\} . \tag{9.189}$$

The walker propagator is

$$|R'S'\rangle = e^{-\frac{i}{\hbar} \sum_{j\alpha} x_{j\alpha} \sqrt{\frac{\hbar^2 \Delta t}{m}} p_{j\alpha}} e^{i \sum_{j\alpha} x_{j\alpha} \sqrt{\frac{m \Delta t}{16\hbar^2}} \sum_{k \neq j} [(\boldsymbol{\sigma}_j + \boldsymbol{\sigma}_k) \times \boldsymbol{r}_{jk}]_\alpha v_{LS}(r_{jk})} |RS\rangle \tag{9.190}$$

The local energy term for the spin orbit will contain the kinetic energy, the spin orbit, and the negative of the counter terms. Therefore, the counter term contribution cancels in the weight, and the final weight is the local energy.

9.5 GFMC with Full Spin-Isospin Summation

As mentioned above, current high quality trial wave functions for the coordinate space nuclear Hamiltonians require the same computational complexity to calculate either one or all of the spin-isospin amplitudes at a specified position for the particles. Very roughly for A nucleons, each of which can be a proton or neutron with spin up or down, the number of spin-isospin amplitudes is 4^A. Symmetries can lower this factor but not change its overall exponential character (see Table 9.1).

Typically these calculations are done in either a good charge or good isospin basis. In a good charge basis, with A nucleons, with Z protons, the number of combinations of protons and neutrons is $\frac{A!}{Z!(A-Z)!}$, while the tensor force can flip any of the spins so there are 2^A spin states. The total number of allowed spin-isospin states is the product of these factors. Sometimes the initial calculations are done with a Hamiltonian that conserves isospin and the charge symmetry breaking components are added perturbatively. In this case the number of states can be further reduced. Since $T_z = \frac{2Z-A}{2}$, the number of isospin states T states for a given $T_z \leq T$ is given by the difference in the number of charge states with $T_z = T$ and $T_z = T+1$, which is $\frac{A!}{(\frac{A}{2}-T)!(\frac{A}{2}+T)!} \frac{2T+1}{\frac{A}{2}+T+1}$.

Table 9.1 The number of spin-isospin amplitudes for the ground states of some representative nuclei

Nucleus	Spin	Charge states	Total	Isospin/T reversal
^{4}He	16	6	96	16
^{8}Be	256	70	17,920	1792
^{12}C	4096	924	3,784,704	270,336
^{16}O	65,536	12,870	8.4×10^8	4.7×10^7

Time-reversal invariant states have a further factor of 2 reduction, since in that case, the time reversal operator

$$\mathscr{T} = \left[\prod_{i=1}^{A} \sigma_{ix}\sigma_{iz}\right] K \tag{9.191}$$

relates the amplitudes of the states given by flipping all the spins. Here K is the complex conjugating operator.

To see how this works, we can look at a straightforward generalization of a Jastrow-Slater trial state,

$$|\Psi_T\rangle = \left[\mathscr{S}\prod_{i<j}\sum_{p} f_{ij}^{(p)} O_{ij}^{(p)}\right] |\Phi\rangle \tag{9.192}$$

where $|\Phi\rangle$ is a model state, typically one or a small linear combination of antisymmetric products of single particle orbitals. The p sum is over the same sort of operators as those in the potential (usually operators with gradients are either omitted or kept only at lowest order), with the Jastrow correlations $f_{ij}^{(p)}$ depending only on the spatial operator $|\boldsymbol{r}_i - \boldsymbol{r}_j|$, while the $O_{ij}^{(p)}$ contain spin-isospin operators and the unit vector operators $\frac{\boldsymbol{r}_i - \boldsymbol{r}_j}{|\boldsymbol{r}_i - \boldsymbol{r}_j|}$. The $\mathscr{S}$ is a symmetrizing operator applied to the Jastrow product, since the operators in general do not commute, so that the trial function is properly antisymmetric under interchange.

To form a trial wave function we take the inner product with $\langle RS|$ to obtain $\Psi_T(R, S) = \langle RS|\Psi_T\rangle$. The spatial operators operating to the left on their eigenstate $\langle RS|$ are replaced by their eigenvalues. This leaves just the spin-isospin matrix elements. The model state is evaluated for all possible spin-isospin states as enumerated above, $\langle RS'|\Phi\rangle$. In our spin-isospin basis, each of the operators $\langle S''|O_{ij}^{(p)}|S'\rangle$ is a sparse matrix which can either be tabulated or easily calculated as needed. For example, in the charge basis, acting on a single basis state, the interaction can change the spins of a pair to any of the four values. If the particles of the pair are a neutron and a proton, they can be interchanged. This shows that there are at most either 4 or 8 nonzero entries per row or column of the matrix representation. The construction of the Jastrow product is obtained by these repeated sparse-matrix multiplications.

The symmetrizing operator has the factorial of the number of pairs terms. It would be prohibitive to calculate explicitly. However, the commutator terms are small, so the sum over orders of the operators is done by Monte Carlo sampling.

Since much of the computational time is spent in evaluating the trial wave functions, wave functions that include more complicated correlations as well as alpha particle clustering are often included. The simplest wave function above is adequate for the alpha particle.

A GFMC calculation uses walkers given by positions for all the particles, and amplitudes for each of the possible spin-isospin states in the basis.

In the simplest GFMC implementation, the so-called primitive approximation can be used. Here the propagator is

$$\left[\prod_{i<j} e^{-\frac{1}{2}\tau \sum_p v_{ij}^{(p)}}\right] e^{-\tau \sum_i \frac{p_i^2}{2m}} \left[\prod_{i<j} e^{-\frac{1}{2}\tau \sum_p v_{ij}^{(p)}}\right] \tag{9.193}$$

where the opposite order of the pairs is taken in the two products to minimize the time-step errors. The exponentials of the pair operators can be written as a linear combination of pair operators, and these are then operated on the walker states giving new amplitudes. The kinetic energy term is implemented by sampling a gaussian to give new positions.

9.6 General Projection Algorithms in Fock Space and Non-local Interactions

In recent years, a number of projection algorithms working in a discrete Fock space (configuration space) rather than in coordinate space have been proposed [18–23]. While more similar to more standard many-body techniques like Coupled Cluster (CC) and Many Body Perturbation Theory already covered in previous chapters the adoption of statistical techniques in a configuration space has some advantage. First of all Monte Carlo techniques can be implemented with a much milder scaling with the system size enabling the possibility with a much larger number of basis states that build up the total Hilbert space. Contrary to e.g. CC theory we can ensure that the final QMC estimate for the ground-state energy would be an upper bound of the true eigenvalue, thus providing useful benchmark results. Also, working on a finite many-body space allows practical calculations with non-local interactions, like those developed within the Chiral Effective Field Theory approach to nuclear forces, in a far more controllable way than not with the continuous coordinate-space formulation exposed so far (as was done in [24]). Finally, another great advantage of performing the Monte Carlo on a discrete Hilbert space is the possibility to devise an efficient strategy to reduce the impact of the sign-problem by using cancellation techniques [18–20] in an analogous fashion to what was sketched at the end of Sect. 9.3.5. Unfortunately we won't have space here to cover these aspects.

9.6.1 Fock Space Formulation of Diffusion Monte Carlo

To set the stage let us take a finite set $\mathscr{S}$ of single-particle (sp) states of size $\mathscr{N}_s$ and consider a general second-quantized fermionic Hamiltonian including two and possibly many-body interactions

$$H = \sum_{\alpha\in\mathscr{S}} \epsilon_\alpha a^\dagger_\alpha a_\alpha + \sum_{\alpha\beta\gamma\delta\in\mathscr{S}} V_{\alpha\beta\gamma\delta} a^\dagger_\alpha a^\dagger_\beta a_\delta a_\gamma + \dots . \tag{9.194}$$

In this expression Greek letter indices indicates sp states (ie. α is a collective label for all sp quantum numbers), the operator $a^\dagger_\alpha$ (a_α) creates (destroys) a particle in the sp state α and the $V_{\alpha\beta\gamma\delta}$ are general (anti-symmetrized) two-body interaction matrix elements:

$$V_{\alpha\beta\gamma\delta} = \langle\alpha\beta|\hat{V}|\gamma\delta\rangle - \langle\alpha\beta|\hat{V}|\gamma\delta\rangle. \tag{9.195}$$

For an N-fermion system the resulting Fock space would be spanned by the full set of N-particle Slater determinants that can be generated using the sp orbitals $\alpha \in \mathscr{S}$. We will denote these Slater-determinants in the occupation number basis by $|\mathbf{n}\rangle$, where $\mathbf{n} \equiv \{n_\alpha\}$ and $n_\alpha = 0, 1$ are occupation number of the single-particle orbital α satisfying $\sum_\alpha n_\alpha = N$. For example in a system composed by two identical fermions and with $\mathscr{N}_s = 4$ available sp states we will write

$$|0110\rangle \equiv a^\dagger_3 a^\dagger_2 |0\rangle \tag{9.196}$$

where $|0\rangle$ is our vacuum state (that can be conveniently set to the Hartree-Fock ground state Φ_{HF}), while $a^\dagger_2$ and $a^\dagger_3$ creates a particle in sp state 2 and 3 respectively.

We can now use these states as a complete basis in our many-body Hilbert space and express a generic state in it as

$$|\Psi\rangle = \sum_{\mathbf{n}} \langle\mathbf{n}|\Psi\rangle|\mathbf{n}\rangle \equiv \sum_{\mathbf{n}} \Psi(\mathbf{n})|\mathbf{n}\rangle \tag{9.197}$$

where the sum is over all possible basis vectors that one can obtain from the $\mathscr{N}_S$ single-particle orbitals.

It is important to notice at this point that no assumption is made on the locality of the interaction, which translates into restrictions on the structure of the tensor $V_{\alpha\beta\gamma\delta}$. This shows already that possible non-local interactions can be cleanly incorporated in the formalism.

As was already introduced in Sect. 9.3.1, the core idea behind a Diffusion Monte Carlo algorithm is to extract ground-state informations on the system by evolving in imaginary-time an initial guess for the lowest eigenstate of the hamiltonian H:

$$\Psi_{\tau+\Delta\tau}(\mathbf{m}) = \sum_{\mathbf{n}} \langle\mathbf{m}|P|\mathbf{n}\rangle \Psi_\tau(\mathbf{n}). \tag{9.198}$$

with a suitable projection operator P [cf. Eq. (9.125) and discussion above]. In order to illustrate how the evolution in (9.198) can be implemented in a stochastic way, it will be useful first to express the matrix elements of P as follows

$$\langle \mathbf{m}|P|\mathbf{n}\rangle = p(\mathbf{m},\mathbf{n})g(\mathbf{n}) \tag{9.199}$$

with

$$g(\mathbf{n}) = \sum_{\mathbf{m}} \langle \mathbf{m}|P|\mathbf{n}\rangle \tag{9.200}$$

and

$$p(\mathbf{m},\mathbf{n}) = \frac{\langle \mathbf{m}|P|\mathbf{n}\rangle}{\sum_{\mathbf{m}} \langle \mathbf{m}|P|\mathbf{n}\rangle}. \tag{9.201}$$

At this point, provided the matrix elements $\langle \mathbf{m}|P|\mathbf{n}\rangle \geq 0$ we can interpret $p(\mathbf{m},\mathbf{n})$ for fixed $\mathbf{n}$ as (normalized) probability distribution for the states $\mathbf{m}$ and $g(\mathbf{n})$ as a weight factor. This is analogous to what was done in Sect. 9.3.2 for the conventional coordinate-space formulations where now p takes the place of the gaussian Eq. (9.87) while g replaces the weight Eq. (9.88).

Imagine now that at a given imaginary-time τ the wave-function Ψ_τ is non-negative in configuration space

$$\Psi_\tau(\mathbf{n}) \geq 0 \forall \mathbf{n}, \tag{9.202}$$

then we can represent it as an ensemble of configurations. Due to the non-negativity of the matrix elements of P, we also have that the evolution described in (9.198) preserves the signs

$$\Psi_{\tau+\Delta\tau}(\mathbf{m}) \geq 0 \ \forall \mathbf{m}. \tag{9.203}$$

This suggests the following procedure for the stochastic imaginary-time evolution:

1. walker starts at configuration $\mathbf{n}$ with weight $w(\mathbf{n})$
2. a new configuration $\mathbf{m}$ is chosen from the probability distribution $p(\mathbf{m},\mathbf{n})$
3. the walker's weight gets rescaled as $w(\mathbf{n}) \to w(\mathbf{m}) = w(\mathbf{n})g(\mathbf{n})$
4. repeat from 1.

In order to improve efficiency one can include a *branching* step where the new configuration in $\mathbf{m}$ is replicated according to its weight as explained in Sect. 9.3.2.

Expectation values of observables can then be estimated as usual [cf. Eq. (9.130)] with the mixed estimator

$$\langle O\rangle_{mixed} = \frac{\langle\Psi_T|O|\Psi(\tau)\rangle}{\langle\Psi_T|\Psi(\tau)\rangle} = \frac{\sum_l^{N_w} w(\mathbf{m}_l)\langle\Psi_T|O|\mathbf{m}_l\rangle}{\sum_l^{N_w} w(\mathbf{m}_l)\Psi_T(\mathbf{m}_l)} \tag{9.204}$$

where Ψ_T is a trial state and the sums run over the walker population of size N_w.

In practice we have to choose some form for the evolution operator that appears in (9.198), a common choice in discrete spaces is on operator very similar to the one already encountered in the discussion of Sect. 9.2.5.2:

$$\begin{aligned}\langle\mathbf{m}|P|\mathbf{n}\rangle &= \langle\mathbf{m}|1 - \Delta\tau\,(H - E_T)|\mathbf{n}\rangle \\ &= \delta_{\mathbf{m},\mathbf{n}} - \Delta\tau\langle\mathbf{m}|H - E_T|\mathbf{n}\rangle\end{aligned} \tag{9.205}$$

where E_T is an energy shift used in the simulation to preserve the norm of the solution (the constant E_0 introduced in Sect. 9.3.2). Convergence to the ground-state by repeated application of the projector P to the initial state $|\Psi_0\rangle$

$$|\Psi_{gs}\rangle = \lim_{M\to\infty} P^M|\Psi_0\rangle \tag{9.206}$$

is guaranteed provided that the eigenvalues of P lie between -1 and 1 in order to ensure the diagonal part remains positive definite. This requirement translates into a condition on the imaginary-time step $\Delta\tau$ which has to satisfy the bound

$$\Delta\tau < 2/(E_{max} - E_{min}) \tag{9.207}$$

where E_{max} and E_{min} are respectively the maximum and minimum eigenvalue of H in our finite basis. This upper bound becomes tighter and tighter as we increase the number of particle N and/or the number of sp-states $\mathcal{N}_s$. As a consequence the number M of iterations needed for convergence to the ground state increases dramatically. A way to deal with this problem is to employ a different algorithm proposed in [25] (see also [26, 27]) that allows us to sample directly from the exponential propagator

$$\langle\mathbf{m}|P|\mathbf{n}\rangle = \langle\mathbf{m}|e^{-\Delta\tau(H-E_T)}|\mathbf{n}\rangle \tag{9.208}$$

in analogy to Eq. (9.74), but now without any limitation on the choice of the imaginary time step $\Delta\tau$ that can be chosen arbitrarily large. We leave the discussion of its details in Sect. 9.6.4.

In our discussion so far we have assumed that the matrix elements on the projector that defines $p(\mathbf{m},\mathbf{n})$ in Eq. (9.201) are actually positive definite. Under general circumstances however this is not the case. This clearly prevents the interpretation of $p(\mathbf{m},\mathbf{n})$ as a probability distribution invalidating the naive approach employed above. In order to circumvent the problem we can use the same idea behind the fixed node (phase) approximation introduced in Sect. 9.3.5.1.

Before continuing it is worth to mention that in principle one can still produce a stochastic evolution by absorbing the signs into the weight factor $g(\mathbf{n})$ while sampling off-diagonal moves using $|\langle\mathbf{m}|P|\mathbf{n}\rangle|$. However as briefly explained in Sect. 9.3.5.1 this is accompanied by an exponential decay of the signal to noise ratio as a function of the total projection time $\tau = M\Delta\tau$. Recently it was shown that by employing an annihilation step in the evolution this problem can be substantially alleviated [18, 20, 21]. At the end however these algorithms have still an exponential scaling with system-size, though with a reduced exponent.

9.6.2 Importance Sampling and Fixed-Phase Approximation

As we just mentioned, we can deal with the sign-problem in a way which is similar to standard coordinate-space QMC: we will use an initial ansatz Φ_T for the ground-state wave-function and use that to constrain the random walk in a region of the many-body Hilbert space where

$$\langle\mathbf{m}|P|\mathbf{n}\rangle \geq 0 \tag{9.209}$$

is satisfied. In order for this scheme to be practical one needs a systematic way for reducing the bias coming from this approximation, e.g. we want the bias to go to zero as the ansatz Φ_T goes towards the ground-state Ψ_{gs}. That's exactly what is done in coordinate-space fixed-node(fixed-phase) QMC simulations presented in the previous sections.

In this derivation we will follow the work in [26, 27] and generalize it to the case of complex-hermitian hamiltonians usually found in nuclear theory. Similarly to what was done in Sect. 9.4.4 the imaginary part of the solution is constrained to be the same of that of the trial wave-function

$$\Re[\Psi^*(\mathbf{n})\Phi_T(\mathbf{n})] = 0 \tag{9.210}$$

for every distribution $\Psi(\mathbf{n})$ sampled in the random walk. In this expression $\Re$ stands for the real part and * is complex-conjugation.

We start by defining for any configurations $\mathbf{n}$ and $\mathbf{m}$ for which $|\Phi_T(\mathbf{n})| \neq 0$ the following quantity:

$$\begin{aligned} \mathfrak{s}_{\mathbf{mn}} &= \text{sign}\, \Re\left[\Phi_T^*(\mathbf{m}) H_{\mathbf{mn}} \Phi_T^*(\mathbf{n})^{-1}\right] \\ &= \text{sign}\, \frac{\Re\left[\Phi_T^*(\mathbf{m}) H_{\mathbf{mn}} \Phi_T(\mathbf{n})\right]}{|\Phi_T(\mathbf{n})|^2} = \mathfrak{s}_{\mathbf{nm}}. \end{aligned} \tag{9.211}$$

Now define a one-parameter family of Hamiltonians $\mathscr{H}_\gamma$ defined over configurations $\mathbf{n}$ (again such that $|\Phi_T(\mathbf{n})| \neq 0$) with off-diagonal matrix elements given by

$$\langle \mathbf{m}|\mathscr{H}_\gamma|\mathbf{n}\rangle = \begin{cases} -\gamma\langle \mathbf{m}|H|\mathbf{n}\rangle & \mathfrak{s}(\mathbf{m},\mathbf{n}) > 0 \\ \langle \mathbf{m}|H|\mathbf{n}\rangle & \text{otherwise} \end{cases}, \tag{9.212}$$

while the diagonal terms are

$$\begin{aligned} \langle \mathbf{n}|\mathscr{H}_\gamma|\mathbf{n}\rangle &= \langle \mathbf{n}|H|\mathbf{n}\rangle + (1+\gamma) \sum_{\substack{\mathbf{m}\neq\mathbf{n} \\ \mathfrak{s}(\mathbf{m},\mathbf{n})>0}} \langle \mathbf{m}|H|\mathbf{n}\rangle \\ &= \langle \mathbf{n}|H|\mathbf{n}\rangle + \sum_{\mathbf{m}} h_{\mathbf{mn}}. \end{aligned} \tag{9.213}$$

In the limit where $\gamma \to -1$ we clearly recover the original Hamiltonian:

$$\mathscr{H}_{\gamma=-1} \equiv H. \tag{9.214}$$

We proceed to define a corresponding family of propagators $\mathscr{P}_\gamma$ for configurations $\mathbf{n}$ with $|\Phi_T(\mathbf{n})| \neq 0$ by

$$\langle \mathbf{m}|\mathscr{P}_\gamma|\mathbf{n}\rangle = \delta_{\mathbf{m},\mathbf{n}} - \Delta\tau \frac{\Re\left[\Phi_T^*(\mathbf{m})\langle \mathbf{m}|\mathscr{H}_\gamma - E_T|\mathbf{n}\rangle\Phi_T(\mathbf{n})\right]}{|\Phi_T(\mathbf{n})|^2}. \tag{9.215}$$

It is clear now that for any $\gamma \geq 0$ we have

$$\langle \mathbf{m}|\mathscr{P}_\gamma|\mathbf{n}\rangle \geq 0 \tag{9.216}$$

and so the propagator $\mathscr{P}$ is, by construction, free from the sign-problem. Performing the corresponding random-walk allows us to filter the state

$$\Phi_T(\mathbf{n})\phi_\gamma^0(\mathbf{n}), \tag{9.217}$$

where now $\phi_\gamma^0(\mathbf{n})$ is the ground-state of the hamiltonian $\mathscr{H}_\gamma$. The ground-state energy E_γ obtained following this procedure can be proved (the proof is left to the Appendix) to be a strict upper bound for the true ground-state energy E_0 of the true hamiltonian H. Moreover, this upper bound is tighter than the variational upper-bound provided by

$$E_T = \frac{\langle \Phi_T|H|\Phi_T\rangle}{\langle \Phi_T|\Phi_T\rangle} \geq E_0. \tag{9.218}$$

As you can show in Problem 9.5 any linear extrapolation of E_γ from any two values $\gamma \geq 0$ to $\gamma = -1$ (which would correspond to the original hamiltonian) also provides an upper-bound on E_{gs} that is tighter than the individual E_γ's. A good compromise between the tightness of the upper-bound and the statistical noise in the extrapolation is to choose two values of γ: 0 and 1, thus giving the following energy estimator:

$$E_{extr} = 2E_{\gamma=0} - E_{\gamma=1} \tag{9.219}$$

To ensure the success of the proposed method a good choice for the importance function $|\Phi_T\rangle$ is critical.

9.6.3 Trial Wave-Functions from Coupled Cluster Ansatz

As have been pointed out before, a crucial role is played by the importance function Φ_T used to impose the constraint. This is especially true if we want to estimate expectation values of operators other than the energy (cf. discussion in Sect. 9.4.2.1).

Fundamental prerequisites for a viable importance function are

1. enough flexibility to be able to account for the relevant correlations in the system
2. availability of an efficient way to evaluate its overlap with states explored during the random walk

Within a Fock space formulation, an excellent choice for Φ_T that satisfy the first requirement is given by the wave function generated in a Coupled Cluster calculation. Starting from a reference state, which usually is the Hartree-Fock solution of the problem, CC theory allows to include dynamical correlations into a new state as

$$|\Psi_{CC}\rangle = e^{\hat{T}}|\Phi_{HF}\rangle. \tag{9.220}$$

In the above equation, correlations are introduced trough the excitation operator $\hat{T}$ which in CC theory is hierarchically divided as

$$\hat{T} = \hat{T}_1 + \hat{T}_2 + \hat{T}_3 + \cdots \tag{9.221}$$

counting the number of creation/annihilation operators that compose them. The first two terms are:

$$\hat{T}_1 = \sum_{\alpha,\beta\in\mathscr{S}} t_\alpha^\beta a_\beta^\dagger a_\alpha \qquad \hat{T}_2 = \frac{1}{4}\sum_{\alpha,\beta,\gamma,\delta\in\mathscr{S}} t_{\alpha\beta}^{\gamma\delta} a_\gamma^\dagger a_\delta^\dagger a_\alpha a_\beta \quad \cdots \tag{9.222}$$

The final state $|\Psi_{CC}\rangle$ will then be uniquely identified by the coefficients t_α^β and $t_{\alpha\beta}^{\gamma\delta}$ corresponding to single and double particle-hole excitations respectively. The exponentiated form of the CC wave-functions enables to effectively include some correlations up to the maximum N-particle N-hole in a relatively compact way.

But is the wave-function in Eq. (9.220) also quick to evaluate? In order to simplify the discussion we will focus here on the case of a homogeneous system that can be described dropping the one-particle-one-hole excitation operator $\hat{T}_1$ in the expansion (which do not contribute due to translational invariance).[5] In this situation the lowest order of CC theory is the Coupled Cluster Doubles (CCD) approximation.

To set the notation, we will express a generic Slater-Determinant state describing an M-particle-M-hole state as

$$|\mathbf{m}\rangle = a^\dagger_{p_1}\ldots, a^\dagger_{p_M} a_{h_1}\ldots, a_{h_M}|\Phi_{HF}\rangle \equiv |\Phi^{p_1,\ldots,p_M}_{h_1,\ldots,h_M}\rangle. \tag{9.223}$$

The required amplitude can then be expressed as a superposition of $M-2$ particle/hole states that can be generated from **m**. Eventually (the proof is tedious but straightforward) one obtains:

$$\langle\mathbf{m}|\Psi_{CC}\rangle = \sum_{\gamma=2}^{M}\sum_{\mu<\nu}^{M}(-1)^{\gamma+\mu+\nu} t^{p_\mu p_\nu}_{h_1 h_\gamma}\Psi^{M-2}_{CC}\begin{pmatrix} p_1,p_2,\ldots,p_{\mu-1},p_{\mu+1},\ldots,p_{\nu-1},p_{\nu+1},\ldots,p_M \\ h_2\ldots,h_{\gamma-1},h_{\gamma+1},\ldots,h_M \end{pmatrix} \tag{9.224}$$

assuming $p_1 < p_2 < \cdots < p_M$ and $h_1 < h_2 < \cdots < h_M$. The normalization is fixed in such a way that $\langle\Phi_{HF}|\Psi_{CC}\rangle = 1$.

One way to implement Eq. (9.224) is for instance trough a recursive function that takes as input some K-particle-K-hole state and returns 1.0 for $K = 0$, the correct amplitude t^{ab}_{ij} for $K = 2$ and for $K > 2$ calls itself again removing two particle and two hole states. Clearly this approach becomes slow when states with large values of K are sampled often during the random walk. Just to give an idea, for calculations of pure neutron matter with soft Chiral EFT interactions we have $K \leq 6$ at densities $\rho \approx 0.08\,\text{fm}^{-3}$ (cf. discussion in [28]) and the calculation can be made very efficient.

Within CC theory the coefficients $t^{\gamma\delta}_{\alpha\beta}$ appearing in the equations above are to be obtained as the self-consistent solutions of the following non-linear equation:

$$\langle\Phi^{\gamma\delta}_{\alpha\beta}|\hat{H}\left(1+\hat{T}_2+\frac{1}{2}\hat{T}_2^2\right)|\Phi_{HF}\rangle = \left(\frac{1}{4}\sum_{\alpha,\beta,\gamma,\delta\in\mathscr{S}}\langle\alpha\beta||\gamma\delta\rangle t^{\gamma\delta}_{\alpha\beta}\right) t^{\gamma\delta}_{\alpha\beta} \tag{9.225}$$

where $\langle\alpha\beta||\gamma\delta\rangle$ are the anti-symmetrized two-body matrix elements of the interaction defined in Eq. (9.195).

Solving Eq. (9.225) is in general a very expensive computational problem and within the fixed-node approach all that matters are the signs in Eq. (9.211). It could then be possible to find cheaper approximate ways to determine the doubles coeffi-

[5]Extension to singlets (p-h states) and triplets (3p-3h states) is simple.

cients $t_{\alpha\beta}^{\gamma\delta}$ while still preserving a good quality in the fixed-node approximation. A quite precise and very cheap approximation that have been used successfully is to obtain the coefficients within second order Moeller–Plesset perturbation theory:

$$t_{\alpha\beta}^{\gamma\delta} = \frac{\langle\alpha\beta||\gamma\delta\rangle}{\eta_\alpha + \eta_\beta - \eta_\gamma - \eta_\delta} \quad \text{with} \quad \eta_i = \epsilon_i + \sum_{k\in\mathcal{S}} \langle ik||ik\rangle \tag{9.226}$$

and ϵ_i are the single particle energies appearing in the one body part of the Hamiltonian Eq. (9.194). This is equivalent to truncating the self-consistent solution of (9.225) after the first iteration.

9.6.4 Propagator Sampling with No Time-Step Error

As we pointed out before, in simulations employing the linear propagator (9.205) raising the dimension of the basis set has a detrimental effect on the efficiency of the algorithm since in order to satisfy the bound Eq. (9.207) we are forced to employ an exceedingly small time step. Moreover, in practice values of τ much smaller than the maximum value are usually employed due to the difficulty in obtaining reliable estimates of E_{max} in realistic situations.

To further complicate the scenario, when lattice fixed-node(fixed-phase) methods are employed this maximum value is reduced even further because the diagonal matrix elements of P gets pushed towards the negative region by the addition of the sign-violating contributions $\sum_{\mathbf{m}} h_{\mathbf{mn}}$ in Eq. (9.213). If this method is used to control the sign-problem additional care has to be devoted in the choice of the time-step, greatly deteriorating the efficiency of the overall scheme.

In a discrete space however we can cope with the problem by using an algorithm firstly introduced by Trivedi and Ceperley [25], which shares similarities with the Domains Green's Function Monte Carlo by Kalos et al. [29]. The idea is to use directly (meaning sample from) the exponential propagator

$$P^{exp}(\tau, \mathbf{m}, \mathbf{n}) = \langle\mathbf{m}|e^{-\tau(H-E_T)}|\mathbf{n}\rangle, \tag{9.227}$$

that clearly has no problem with negative diagonal elements. These schemes usually come with the name of *continuous-time* evolution.

For simplicity let us forget the sign-problem for the time being and imagine we are working with the positive-definite importance-sampled greens function (9.215) with $\gamma = 0$ and the corresponding Hamiltonian $\widetilde{H}$ which then satisfies

$$\widetilde{H}_{\mathbf{m},\mathbf{n}} \leq 0 \qquad \forall\ \mathbf{m} \neq \mathbf{n}. \tag{9.228}$$

Furthermore, we will neglect the energy shift E_T since its addition is straightforward.

Recall that the propagator can be written as a product of a stochastic matrix $\widetilde{p}_{\mathbf{m},\mathbf{n}}$ and a weight factor $\widetilde{g}_{\mathbf{n}}$ (cf. Sect. 9.6.1):

$$\widetilde{P}_{\mathbf{m},\mathbf{n}}(\Delta\tau) = \delta_{\mathbf{m},\mathbf{n}} - \Delta\tau\widetilde{H}_{\mathbf{m},\mathbf{n}} = \widetilde{p}_{\mathbf{m},\mathbf{n}}\widetilde{g}_{\mathbf{n}} \tag{9.229}$$

where the two factors are given by:

$$\begin{aligned} \widetilde{p}_{\mathbf{m},\mathbf{n}} &= \frac{\widetilde{P}_{\mathbf{m},\mathbf{n}}(\Delta\tau)}{\widetilde{g}_{\mathbf{n}}}, \\ \widetilde{g}_{\mathbf{n}} &= \sum_{\mathbf{m}} \widetilde{P}_{\mathbf{m},\mathbf{n}}(\Delta\tau) = 1 - \Delta\tau E_L(\mathbf{n}) \end{aligned} \tag{9.230}$$

and in the last equation we have used the expression for the local energy

$$E_L(\mathbf{n}) = \frac{\langle\Phi_T|H|\mathbf{n}\rangle}{\langle\Phi_T|\mathbf{n}\rangle} = \sum_{\mathbf{m}} \frac{\Phi_T(\mathbf{m})\langle\mathbf{m}|H|\mathbf{n}\rangle}{\Phi_T|\mathbf{n}\rangle} \equiv \sum_{\mathbf{m}} \widetilde{H}_{\mathbf{m},\mathbf{n}}. \tag{9.231}$$

The continuous-time limit is recovered by applying M times $\widetilde{P}(\Delta\tau)$ and letting $\Delta\tau \to 0$ while preserving constant the product $\tau = M\Delta\tau$:

$$\lim_{M\to\infty} \widetilde{P}_{\mathbf{m},\mathbf{n}}(\tau)^M = \lim_{M\to\infty} \left(1 - \frac{\tau}{M}\widetilde{H}_{\mathbf{m},\mathbf{n}}\right)^M = \lim_{\Delta\tau\to 0} \left(1 - \Delta\tau\widetilde{H}_{\mathbf{m},\mathbf{n}}\right)^{\frac{\tau}{\Delta\tau}} = \langle\mathbf{m}|e^{-\tau\widetilde{H}}|\mathbf{n}\rangle. \tag{9.232}$$

Now note that if we let $\Delta\tau \to 0$ the probability to make a diagonal move in a single step among the M will accordingly go to ≈ 1, in fact:

$$P_{diag} = \frac{\widetilde{P}_{\mathbf{n},\mathbf{n}}(\Delta\tau)}{\widetilde{g}_{\mathbf{n}}} = \frac{1 - \Delta\tau\widetilde{H}_{\mathbf{n},\mathbf{n}}}{1 - \Delta\tau E_L(\mathbf{n})} \xrightarrow{\Delta\tau\to 0} 1 \tag{9.233}$$

since the local-energy E_L does not depend on the time step but just on the current configuration $\mathbf{n}$. Accordingly, the probability of making K consecutive diagonal moves will be:

$$\begin{aligned} P^K_{diag} &= \left(\frac{\widetilde{P}_{\mathbf{n},\mathbf{n}}(\Delta\tau)}{\widetilde{g}_{\mathbf{n}}}\right)^K = \left(\frac{1 - \Delta\tau\widetilde{H}_{\mathbf{n},\mathbf{n}}}{1 - \Delta\tau E_L(\mathbf{n})}\right)^K \\ &\xrightarrow{K\to\infty} \exp\left(\tau(E_L(\mathbf{n}) - \widetilde{H}_{\mathbf{n},\mathbf{n}})\right) = \exp\left(\tau\widetilde{H}^{off}_{\mathbf{n}}\right) = f_{\mathbf{n}}(\tau) \end{aligned} \tag{9.234}$$

where we have implicitly defined the off-diagonal sum

$$\widetilde{H}^{off}_{\mathbf{n}} = \sum_{\mathbf{m}\neq\mathbf{n}} \widetilde{H}_{\mathbf{m},\mathbf{n}} < 0 \tag{9.235}$$

and the inequality holds thanks to Eq. (9.228).

The elapsed time between consecutive off-diagonal moves is therefore distributed as an exponential distribution $f_{\mathbf{n}}(\tau)$ with average time given by

$$\int_0^\infty \tau f_{\mathbf{n}}(\tau) = -\frac{1}{\widetilde{H}^{off}_{\mathbf{n}}} = \left|\frac{1}{\widetilde{H}^{off}_{\mathbf{n}}}\right|. \tag{9.236}$$

We can then sample the time when the off-diagonal move happens by using a transformation technique: suppose we have a way to sample values ξ from a uniform distribution $g(\xi) = \text{const}$, due to conservation of probability the samples τ drawn from the wanted $f_{\mathbf{n}}(\tau)$ will satisfy:

$$|f(\tau)d\tau| = |g(\xi)d\xi| \quad \longrightarrow \left|\frac{d\xi(\tau)}{d\tau}\right| = f_{\mathbf{n}}(\tau) \tag{9.237}$$

where τ are the samples drawn from the wanted PDF $f_{\mathbf{n}}$. By solving now equation (9.237) for $\xi(\tau)$ and performing the inversion to $\tau = \tau(\xi)$ we obtain the following relation

$$\tau_\xi = \frac{log(\xi)}{\widetilde{H}^{off}_{\mathbf{n}}}. \tag{9.238}$$

that allows to sample exactly from $f_{\mathbf{n}}$ using only samples from a uniform distribution $\xi \in (0, 1)$.

Walkers undergoing such random walk accumulate weight during the K diagonal-moves as well as from performing the off-diagonal step. The weight coming from the diagonal moves is given by

$$w_{\mathbf{n}} = \widetilde{g}^K_{\mathbf{n}} = (1 - \Delta\tau E_L(\mathbf{n}))^K \xrightarrow{\Delta\tau\to 0} e^{-\tau E_L(\mathbf{n})}. \tag{9.239}$$

For the off-diagonal moves instead we have at least two options for sampling the new state $|\mathbf{m}\rangle$:

- heat-bath sampling:

$$P_1(\mathbf{m}, \mathbf{n}) = \widetilde{H}_{\mathbf{m},\mathbf{n}}/\widetilde{H}^{off}_{\mathbf{n}} \tag{9.240}$$

  1. new configuration $|\mathbf{m}\rangle$ is chosen using the normalized probability P_1
  2. the off-diagonal weight would be $w_{\mathbf{m},\mathbf{n}} = 1$

- uniform sampling:

$$P_2(\mathbf{m}, \mathbf{n}) = 1/N_{conn} \tag{9.241}$$

1. new configuration $|\mathbf{m}\rangle$ is chosen among the N_{conn} states connected to $|\mathbf{n}\rangle$
2. reweight the new walker using $w_{\mathbf{m},\mathbf{n}} = P_1(\mathbf{m},\mathbf{n})/P_2(\mathbf{m},\mathbf{n})$

The first option is clearly more expensive per iteration than the second since an explicit calculation of the off-diagonal sum $\widetilde{H}^{off}_{\mathbf{n}}$ is needed in order to normalize P_1. In the uniform sampling case however the weights $w_{\mathbf{m},\mathbf{n}}$ can have large fluctuations forcing the use of smaller time-steps to keep them under control. In our case since we already need to compute the off-diagonal sum in order to generate the fixed-phase hamiltonian Eqs. (9.212) and (9.213) the heat-bath sampling comes with no additional cost. It is worth noting that other choice can be made that are more efficient when fixed-node(phase) is not employed at all [30] or when the transformation that produces $\mathscr{H}_\gamma$ is carried out only approximately [31].

Finally, in order for the measurements along the path to be unbiased we want to define equidistant "time-slices" along the random walk. In order to this we simply choose a target time-step τ_t at the beginning then for each move we first sample a value of τ_ξ from Eq. (9.238), if $\tau_\xi > \tau_t$ we set $\tau = \tau_t$ and use correspondingly a diagonal move if instead $\tau_\xi < \tau_t$ we have to sample an off-diagonal move. The process is preformed until the sum of all the sampled τ_ξ reaches the target time τ_t. The final algorithm for a single walker at $|\mathbf{n}\rangle$ is then as follows:

```
EXP_Move()
τ = τ_t
loop
    E_L(n) = Σ_m H̃_{m,n}                          ▷ Eq. (9.231)
    H̃^{off}_n = E_L(n) − H̃_{n,n}                  ▷ Eq. (9.235)
    ξ = rand()
    τ_ξ = log(ξ)/H̃^{off}_n                        ▷ Eq. (9.238)
    if τ_ξ ≥ τ then
        w(n) → w(n) exp(−τ E_L(n))
        exit
    end if
    w(n) → w(n) exp(−τ_ξ E_L(n))
    τ → τ − τ_ξ
    m ← HeatBath[P_1, n]                          ▷ Eq. (9.240)
    n → m
end loop
```

where the function *HeatBath*$[P,\mathbf{n}]$ generates a new configuration according to the probability P (e.g. Eq. (9.240)) starting from the current state $\mathbf{n}$. In Problem 9.6 you will try to devise an implementation of this function.

As a final remark, it is evident that the most expensive part of the algorithm is the computation of the local energy $E_L(\mathbf{n}$ since it will require a sum over all states connected to $\mathbf{n}$ from the Hamiltonian and for each one $\mathbf{m}$ of these we have to compute both the matrix element of the Hamiltonian and the overlap with the trial function $\Psi_T(\mathbf{m})$. The use of symmetries to reduce the size of the sum is thus of fundamental importance to reach medium-sized systems. We can show this for the simple case of a homogeneous system with only two-body interactions so that the connected states will be all the possible 2-particle-2-hole excitations that can be obtained from the initial state $|\mathbf{n}\rangle$. Neglecting the construction of the transformed matrix $\widetilde{H}$, we can then implement the calculation of the local energy as

```
EL_calc1()
E_L = 0
for i ∈ occ(n) do
    for j ∈ occ(n) do
        for a ∈ 𝒮 \ occ(n) do
            for b ∈ 𝒮 \ occ(n) do
                |m⟩ = a†_a a†_b a_i a_j |n⟩
                E_L = E_L + H̃_m,n
            end for
        end for
    end for
end for
E_L = E_L/4
```

where $occ(\mathbf{n})$ is the set of single-particle states that are occupied in the initial state $\mathbf{n}$. The above algorithm requires $O(N_{occ}^2 \mathscr{N}_s^2)$ evaluations of the Hamiltonian. Many of these are however equivalent to other ones or just zero. For instance all the terms with $i = j$ or $a = b$ give zero due to the Pauli principle. If we fix an ordering of the single particle orbitals in the many-body states and use anti-symmetrized matrix elements the configurations obtained interchanging e.g. $i \leftrightarrow j$ are equivalent. Finally if both momentum and spin are conserved, given the triple (i, j, a) there exist only one single particle state b allowed. An implementation like

```
EL_calc2()
E_L = 0
for i ∈ occ(n) do
    for j < i ∈ occ(n) do
        for a ∈ 𝒮 \ occ(n) do
            b ← FourthState[i, j, a]
            if b ∈ 𝒮 \ occ(n) and b < a then
                |m⟩ = a†_a a†_b a_i a_j |n⟩
                E_L = E_L + H̃_m,n
```

(continued)

```
            end if
        end for
    end for
end for
```

will take now only $O(N_{occ}^2 \mathcal{N}_s)$ evaluations of the Hamiltonian at most, and with a reduced prefactor with respect to the previous version. The function *FourthState* returns the only single particle state allowed by symmetry.

9.6.5 Results

The combination of imaginary time projection, use of importance function derived from CC calculations and no time-step error propagator make up the algorithm that goes under the name of Configuration Interaction Monte Carlo (CIMC). Actual calculations with CIMC require a substantial amount of CPU time. Here we report some results obtained by making use of a simplified Hamiltonian in which the nucleon-nucleon interaction is described by the Minnesota interaction. The system under investigation is homogeneous pure neutron matter (PNM). In QMC calculations PNM is typically modeled as a periodic system containing A neutrons. The cell size is adjusted in such a way that the average density of the system is ρ.

In Fig. 9.7 we show how the computed energy depends on the number of plane wave shells included in the model space. As it can be seen, it is necessary to pay attention to the convergence of the results, which can strongly depend on the specific details of the system. In this case, for instance, one can easily see how convergence is faster when the density is increased.

In Fig. 9.8 the energy computed by CIMC shown for the same neutron matter model as a function of the density (the so called "Equation of State" of neutron matter) is compared with the coupled theory results with doubles (CCD) only discussed in the previous chapter. In this calculation single-particle states up to $N_{\max} = 36$ have been used. The CIMCC and CCD results are converged to the fifth digit as function of $N_{\max}$. The agreement between the two methods is at the level of the third digit after the decimal point for neutron matter with the Minnesota interaction. This is a striking agreement between such different many-body methods, in particular for larger densities where correlations and contributions from states above and below the Fermi level play a larger role, as seen from the difference between the reference energy and the CIMC and CCD energies. Most likely, there will be larger differences between different many-body methods when proton correlations are brought in, as well as when more realistic interaction models will be used. Such results will be presented elsewhere. In the next two chapters we will add results using two additional many-body methods, the in-medium SRG approach described in Chap. 10 and the Green's function approach of Chap. 11.

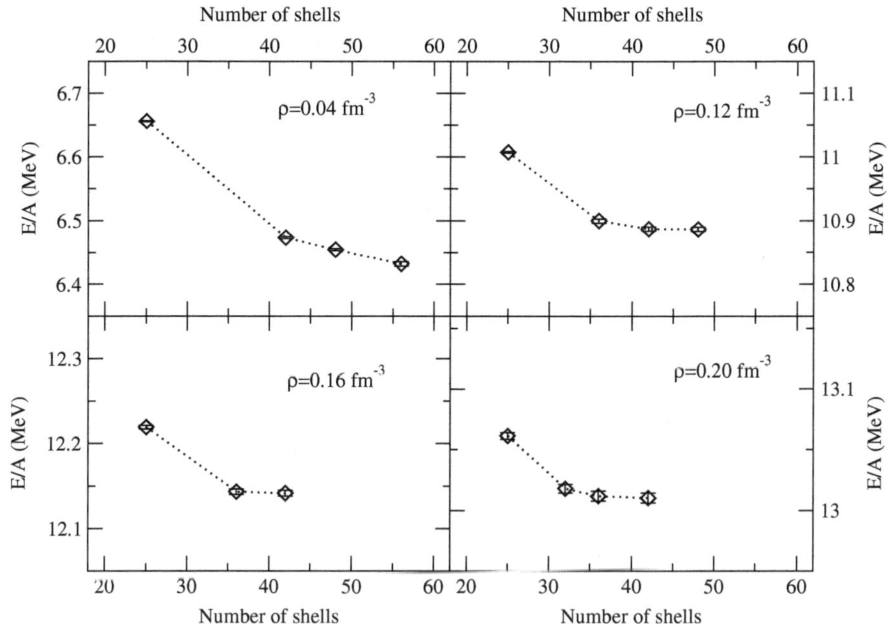

Fig. 9.7 Convergence of the CIMC energies as a function of the number of shells used for a periodic cell of 66 neutrons at different densities

9.7 Conclusions and Perspectives

Quantum Monte Carlo methods are still one of the most powerful tools to attack general many body problems, and in particular the many-nucleon problem. Despite the fact that the Fermion sign problem prevents us so far from having strictly exact results for the solution of the Schrödinger equation, the accuracy that can be reached is very high, and in any cases it constitutes the current benchmark.

Another important general feature of QMC calculations is that they provide a very flexible framework in which it is possible to explore from low temperature condensed helium, to trapped fermions, from atoms and molecules ad solid state devices to nuclei and eventually lattice QCD. It is not rare that technical improvements spread across different disciplines, and the development of the method itself is a common ground that is often the subject of interdisciplinary workshops and conferences.

In the field of nuclear physics it is possible that Fock-space based methods will eventually become the standard. Their main feature is the possibility of dealing with non-local interactions, which makes it possible to extend the use of QMC to the original formulations of ξ-EFT potentials, and a whole class of soft-core interactions that so far have never been used in this context. On the other hand, the availability of more and more accurate versions of the AFDMC codes will open the access of

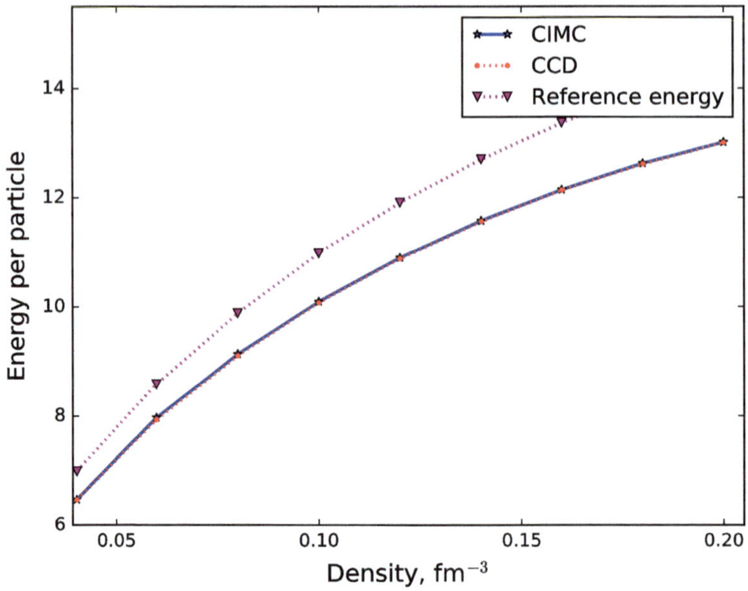

Fig. 9.8 Equation of state of neutron matter modeled as a periodic cell containing A = 66 neutrons using the CIMC method and coupled cluster theory with doubles correlations. Single-particle states up to $N_{\max} = 36$ have been included

accurate studies of the equation of state f neutron and nuclear matter, and of general baryonic matter of extreme importance for astrophysical applications, concerning in particular the physics of neutron stars. The possibility of extending accurate calculations to large A systems is also crucial for understanding the phenomenology of exotic beams.

In this chapter we did not deal with the problem of evaluating excited states and dynamical quantities within a QMC framework. Several methods are nowadays available, mostly based on the evaluation of the Laplace transform of a given response function by means of the calculation of imaginary time correlation functions. Many technical advances have been recently made in this field (see e.g. Refs. [32–35]), and the subject is still under very active investigation.

Finally, the hardest wall to climb remains the solution of the Fermion sign problem. Although there are claims that the problem is NP complete (which is true in general), thereby preventing any solution within standard classical computation, there are hints that many Hamiltonians of interest might admit a viable solution with polynomial scaling in A. This problem would definitely deserve more efforts than those that are presently devoted to its solution.

9.8 Problems

9.1 Evaluate the following integral by means of the Metropolis algorithm

$$I = \int_0^1 e^x - 1\, dx$$

sampling points from

1. P(x) = 1 for $x \in [0, 1]$
2. P(x) = x for $x \in [0, 1]$

- Compare the average and the statistical error for the cases 1) and 2). Which is the best estimate?
- Try to figure out a way to sample a probability density proportional to x^n, and reevaluate the integral I. How is the convergence and the statistical error behaving by increasing n? Try to give an explanation of the result.

9.2 Try to sketch the general proof that for a generic integral I defined as in problem 9.1 the best statistical error is obtained when sampling from a probability density proportional to F(x).

9.3 Consider the one-dimensional Hamiltonian:

$$\hat{H} = \frac{1}{2}\frac{d^2}{dx^2} + \frac{1}{2}x^2$$

and consider the parametrized family of trial solutions $\psi(x, \alpha, \beta) = e^{-\alpha^2 x^2} - \beta$. Compute by means of the Metropolis algorithm the energy and the standard deviation of the energy as a function of α keeping $\beta = 0.01$. Is the minimum found at the same value than for $\beta = 0$? Why?

9.4 Prove that the propagator defined in the integrand of Eq. (9.83) is the Green's function of the differential equation (9.82).

9.5 Show that the given two fixed-phase energies E_a and E_b obtained using the hamiltonians $\mathscr{H}_\gamma$ defined in Eqs. (9.212) and (9.213) with $\gamma = a$ and $\gamma = b$ $(a, b \geq 0)$ the linear extrapolation to $\gamma = -1$ (remember that $\mathscr{H}_{-1} = H$) is still an upper bound. (Hint: show that E_γ is a convex function of the parameter γ).

9.6 Implement the function *HeatBath*[P, **n**] that appears in the algorithm EXP_Move() in Sect. 9.6.4.

9.7 In the case of nuclear Hamiltonians spin is not a conserved quantity. Referring to the calculation of the local energy in Sect. 9.6.4, how would you modify the subroutine EL_calc2 to take into account the non-conservation of spin?

Appendix

In this appendix we give the proof of the upper-bound property for the auxiliary hamiltonians $\mathscr{H}_\gamma$, defined in Sect. 9.6.2, for the general complex-hermitian case (see [27] for the original proof in the real symmetric case). We will concentrate in the simpler case $\gamma = 0$ in Eqs. (9.212) and (9.213), extension to the generic $\gamma \geq 0$ is then straightforward. In what follows we will use the shorthand $\mathscr{H}_{\gamma=0} \equiv \widetilde{H}$. Let $\Psi(\mathbf{n})$ be any arbitrary wave function, our goal is to show that

$$\Re[\langle\Psi|\widetilde{H}|\Psi\rangle] \geq \Re\left[\langle\Psi|H|\Psi\rangle\right] . \tag{9.242}$$

Let us proceed by considering the following difference:

$$\begin{aligned}
\Re[\langle\Psi|\widetilde{H}|\Psi\rangle] - \Re\left[\langle\Psi|H|\Psi\rangle\right] &= \sum_{\mathbf{mn}} \Re\left[\Psi^*(\mathbf{m})(\widetilde{H}_{\mathbf{mn}} - H_{\mathbf{mn}})\Psi(\mathbf{n})\right] \\
&= \sum_{\mathbf{mn}} h_{\mathbf{mn}}|\Psi(\mathbf{n})|^2 + \sum_{\mathbf{m}\neq\mathbf{n}} \Re\left[\Psi^*(\mathbf{m})(\widetilde{H}_{\mathbf{mn}} - H_{\mathbf{mn}})\Psi(\mathbf{n})\right] \\
&= \sum_{\mathbf{n}} \sum_{\mathfrak{s}_{\mathbf{mn}}\neq -} |\Psi(\mathbf{n})|^2 \frac{\Re\left[\Phi_T^*(\mathbf{m})H_{\mathbf{mn}}\Phi_T(\mathbf{n})\right]}{|\Phi_T(\mathbf{n})|^2} - \Re\left[\Psi^*(\mathbf{m})H_{\mathbf{mn}}\Psi(\mathbf{n})\right]
\end{aligned} \tag{9.243}$$

where the second sum is over all $\mathbf{mn}$ pairs such that $\mathfrak{s}_{\mathbf{mn}}$ of (9.211) is positive-definite. The last term can now be rewritten as:

$$\begin{aligned}
\Re\left[\Psi^*(\mathbf{m})H_{\mathbf{mn}}\Psi(\mathbf{n})\right] &= \Re\left[\Psi^*(\mathbf{m})\Phi_T(\mathbf{m})\Phi_T(\mathbf{m})^{-1}H_{\mathbf{mn}}\Phi_T^*(\mathbf{n})^{-1}\Phi_T^*(\mathbf{n})\Psi(\mathbf{n})\right] \\
&= (\Psi^*(\mathbf{m})\Phi(\mathbf{m}))\Re\left[\Phi_T(\mathbf{m})^{-1}H_{\mathbf{mn}}\Phi_T^*(\mathbf{n})^{-1}\right](\Phi_T^*(\mathbf{n})\Psi(\mathbf{n})) \\
&= (\Psi^*(\mathbf{m})\Phi(\mathbf{m}))\Re\left[\frac{\Phi_T^*(\mathbf{m})}{|\Phi_T(\mathbf{m})|^2}H_{\mathbf{mn}}\frac{\Phi_T(\mathbf{n})}{|\Phi_T(\mathbf{n})|^2}\right](\Phi_T^*(\mathbf{n})\Psi(\mathbf{n}))
\end{aligned} \tag{9.244}$$

where in the second step we used the fact that by employing a real propagator we are imposing a fixed-phase constraint, ie $\Im(\Phi_T^*(\mathbf{n})\Psi(\mathbf{n})) = 0$ for every $\mathbf{n}$ explored in the random walk. The equation for the difference becomes:

$$\begin{aligned}
\Re[\langle\Psi|\widetilde{H}|\Psi\rangle] - \Re\left[\langle\Psi|H|\Psi\rangle\right] &= \sum_{\mathbf{mn}} \Re\left[\Psi^*(\mathbf{m})(\widetilde{H}_{\mathbf{mn}} - H_{\mathbf{mn}})\Psi(\mathbf{n})\right] \\
&= \sum_{\mathbf{n}} \sum_{\mathfrak{s}_{\mathbf{mn}}\neq -} \frac{\Re\left[\Phi_T^*(\mathbf{m})H_{\mathbf{mn}}\Phi_T(\mathbf{n})\right]}{|\Phi_T(\mathbf{n})|^2}\left(|\Psi(\mathbf{n})|^2 - \frac{(\Psi^*(\mathbf{m})\Phi_T(\mathbf{m}))(\Phi_T^*(\mathbf{n})\Psi(\mathbf{n}))}{|\Phi(\mathbf{m})|^2}\right) .
\end{aligned} \tag{9.245}$$

Using again the fixed-phase constraint (ie. $(\Phi_T^*(\mathbf{n})\Psi(\mathbf{n})) \equiv (\Phi_T(\mathbf{n})\Psi^*(\mathbf{n}))$) we can rewrite the numerator of the second term as:

$$(\Psi^*(\mathbf{m})\Phi_T(\mathbf{m}))(\Phi_T^*(\mathbf{n})\Psi(\mathbf{n})) = -\frac{1}{2}\big(|\Psi^*(\mathbf{m})\Phi_T(\mathbf{n}) - \Phi_T^*(\mathbf{m})\Psi(\mathbf{n})|^2 \\ - |\Phi_T(\mathbf{n})|^2|\Psi(\mathbf{m})|^2 - |\Phi_T(\mathbf{m})|^2|\Psi(\mathbf{n})|^2\big) \tag{9.246}$$

and then we have:

$$\Re\left[\langle\Psi|\widetilde{H}|\Psi\rangle\right] - \Re\left[\langle\Psi|H|\Psi\rangle\right] = \sum_{\mathbf{mn}} \Re\left[\Psi^*(\mathbf{m})(\widetilde{H}_{\mathbf{mn}} - H_{\mathbf{mn}})\Psi(\mathbf{n})\right]$$
$$= \sum_{\mathbf{n}} \sum_{s_{\mathbf{mn}} \neq -} \frac{\Re\left[\Phi_T^*(\mathbf{m})H_{\mathbf{mn}}\Phi_T(\mathbf{n})\right]}{|\Phi_T(\mathbf{n})|^2}\Big(|\Psi(\mathbf{n})|^2 + \frac{|\Psi^*(\mathbf{m})\Phi_T(\mathbf{n}) - \Phi_T^*(\mathbf{m})\Psi(\mathbf{n})|^2}{2|\Phi_T(\mathbf{m})|^2}$$
$$- \frac{|\Phi_T(\mathbf{n})|^2|\Psi(\mathbf{m})|^2}{2|\Phi_T(\mathbf{m})|^2} - \frac{|\Psi(\mathbf{n})|^2}{2}\Big)$$
$$= \text{(positive terms)}$$
$$+ \sum_{\mathbf{n}} \sum_{s_{\mathbf{mn}} \neq -} \frac{\Re\left[\Phi_T^*(\mathbf{m})H_{\mathbf{mn}}\Phi_T(\mathbf{n})\right]}{2|\Phi_T(\mathbf{n})|^2}\left(|\Psi(\mathbf{n})|^2 - \frac{|\Phi_T(\mathbf{n})|^2|\Psi(\mathbf{m})|^2}{|\Phi_T(\mathbf{m})|^2}\right) \tag{9.247}$$

Now we note that

$$\Re\left[\Phi_T^*(\mathbf{m})H_{\mathbf{mn}}\Phi_T(\mathbf{n})\right] = \Re\left[\Phi_T^*(\mathbf{n})H_{\mathbf{nm}}\Phi_T(\mathbf{m})\right]$$

for a complex-hermitian hamiltonian, we can then express the sums by allowing only unique **mn** combinations:

$$\Re\left[\langle\Psi|\widetilde{H}|\Psi\rangle\right] - \Re\left[\langle\Psi|H|\Psi\rangle\right] = \sum_{\mathbf{mn}} \Re\left[\Psi^*(\mathbf{m})(\widetilde{H}_{\mathbf{mn}} - H_{\mathbf{mn}})\Psi(\mathbf{n})\right]$$
$$= \text{(positive terms)}$$
$$+ \sum_{\mathbf{n}} \sum_{s_{\mathbf{mn}} \neq -}^{\prime} \Re\left[\Phi_T^*(\mathbf{m})H_{\mathbf{mn}}\Phi_T(\mathbf{n})\right]\left(\frac{|\Psi(\mathbf{n})|^2}{2|\Phi_T(\mathbf{n})|^2} + \frac{|\Psi(\mathbf{m})|^2}{2|\Phi_T(\mathbf{m})|^2}\right.$$
$$\left. - \frac{|\Phi_T(\mathbf{n})|^2|\Psi(\mathbf{m})|^2}{2|\Phi_T(\mathbf{m})|^2|\Phi_T(\mathbf{n})|^2} - \frac{|\Phi_T(\mathbf{m})|^2|\Psi(\mathbf{n})|^2}{2|\Phi_T(\mathbf{n})|^2|\Phi_T(\mathbf{m})|^2}\right)$$
$$= \text{(positive terms)} \tag{9.248}$$

which by definition is positive. The extension to the case with $\gamma > 0$ is straightforward since we are basically adding a positive constant to the difference.

References

1. M.H. Kalos, P.A. Whitlock, *Monte Carlo Methods* (Wiley-VCH, Weinheim, 2008)
2. N. Metropolis, A.W. Rosenbluth, M.N. Rosenbluth, A.H. Teller, E. Teller, J. Chem. Phys. **21**, 1087 (1953)
3. W.K. Hastings, Biometrika **57**, 97 (1970)
4. D. Ceperley, G.V. Chester, M.H. Kalos, Phys. Rev. B **16**, 3081 (1977)
5. M.P. Nightingale, C.J. Umrigar, in *Recent Advances in Quantum Monte Carlo Methods*, ed. by W.A. Lester Jr. (World Scientific, Singapore, 1996)
6. J. Toulouse, C.J. Umrigar, J. Chem. Phys. **126**, 084102 (2007)
7. S. Sorella, Phys. Rev. B **64**, 024512 (2001)
8. A. Sarsa, K.E. Schmidt, W.R. Magro, J. Chem. Phys. **113**, 1366 (2000)
9. J.B. Anderson, J. Chem. Phys. **65**, 4121 (1976)
10. M. Troyer, U.-J. Wiese, Phys. Rev. Lett. **94**, 170201 (2005)
11. M.H. Kalos, F. Pederiva, Phys. Rev. Lett. **85**, 3547 (2000)
12. R. Assaraf, M. Caffarel, A. Khelif, J. Phys. A: Math. Theor. **40**, 1181 (2007)
13. D.M. Ceperley, B.J. Alder, Phys. Rev. Lett. **45**, 566 (1980)
14. W.M.C. Foulkes, L. Mitas, R.J. Needs, G. Rajagopal, Rev. Mod. Phys. **73**, 33 (2001)
15. J. Carlson, V.R. Pandharipande, R. Schiavilla, Phys. Rev. C **47**, 484 (1993)
16. G. Ortiz, D.M. Ceperley, R.M. Martin, Phys. Rev. Lett. **71**, 2777 (1993)
17. S. Zhang, H. Krakauer, Phys. Rev. Lett. **90**, 136401 (2003)
18. G.H. Booth, A.J.W. Thom, A. Alavi, J. Chem. Phys. **131**, 054106 (2009)
19. D. Cleland, G.H. Booth, A. Alavi, J. Chem. Phys. **132**, 041103 (2010)
20. F.R. Petruzielo, A.A. Holmes, H.J. Changlani, M.P. Nightingale, C.J. Umrigar, Phys. Rev. Lett. **109**, 230201 (2012)
21. G.H. Booth, A. Grüneis, G. Kresse, A. Alavi, Nature **493**, 365 (2013)
22. A. Mukherjee, Y. Alhassid, Phys. Rev. A **88**, 053622 (2013)
23. A. Roggero, A. Mukherjee, F. Pederiva, Phys. Rev. B **88**, 115138 (2013)
24. A. Roggero, A. Mukherjee, F. Pederiva, Phys. Rev. Lett. **112**, 221103 (2014)
25. N. Trivedi, D.M. Ceperley, Phys. Rev. B **41**, 4552 (1990)
26. S. Sorella, L. Capriotti, Phys. Rev. B **61**, 2599 (2000)
27. D.F.B. ten Haaf, H.J.M. van Bemmel, J.M.J. van Leeuwen, W. van Saarloos, D.M. Ceperley, Phys. Rev. B **51**, 13039 (1995)
28. E. Rrapaj, A. Roggero, J.W. Holt, Phys. Rev. C **93**, 065801 (2016)
29. M.H. Kalos, D. Levesque, L. Verlet, Phys. Rev. A **9**, 2178 (1974)
30. A. Holmes, H.J. Changlani, C.J. Umrigar, J. Chem. Theory Comput. **12**, 1561 (2016)
31. M. Kolodrubetz, B.K. Clark, Phys. Rev. B **86**, 075109 (2012)
32. E. Vitali, M. Rossi, L. Reatto, D.E. Galli, Phys. Rev. B **82**, 174510 (2010)
33. A. Roggero, F. Pederiva, G. Orlandini, Phys. Rev. B **88**, 094302 (2013)
34. A. Lovato, S. Gandolfi, R. Butler, J. Carlson, E. Lusk, S. C. Pieper, R. Schiavilla, Phys. Rev. Lett. **111**, 092501 (2013)
35. A. Lovato, S. Gandolfi, J. Carlson, S.C. Pieper, R. Schiavilla, Phys. Rev. C **91**,062501(R) (2015)

Chapter 10
In-Medium Similarity Renormalization Group Approach to the Nuclear Many-Body Problem

Heiko Hergert, Scott K. Bogner, Justin G. Lietz, Titus D. Morris, Samuel J. Novario, Nathan M. Parzuchowski, and Fei Yuan

10.1 Introduction

Effective Field Theory (EFT) and Renormalization Group (RG) methods have become important tools of modern (nuclear) many-body theory—one need only look at the table of contents of this book to see the veracity of this claim.

Effective Field Theories allow us to systematically take into account the separation of scales when we construct theories to describe natural phenomena. One of the earliest examples that every physics student encounters is the effective force law of gravity near the surface of the Earth: For a mass m at a height h above ground, Newton's force law becomes

$$F(R+h) = G\frac{mM}{(R+h)^2} = m\frac{GM}{R^2}\frac{1}{1+\left(\frac{h}{R}\right)^2} = m\underbrace{\frac{GM}{R^2}}_{\equiv g} + \mathcal{O}\left(\frac{h^2}{R^2}\right), \tag{10.1}$$

where M and R are the mass and radius of the Earth, respectively. Additional examples are the multipole expansion of electric fields [1], which shows that only the moments of an electric charge distribution with characteristic length scale R are resolved by probes with long wave lengths $\lambda \gg R$, or Fermi's theory of beta

H. Hergert (✉) • S.K. Bogner • J.G. Lietz • S.J. Novario • N.M. Parzuchowski • F. Yuan
Department of Physics and Astronomy and National Superconducting Cyclotron Laboratory, Michigan State University, East Lansing, MI, USA
e-mail: hergert@nscl.msu.edu; bogner@nscl.msu.edu; lietz@nscl.msu.edu; novarios@nscl.msu.edu; parzuchowski@frib.msu.edu; yuan@nscl.msu.edu

T.D. Morris
Department of Physics and Astronomy, University of Tennessee, Knoxville, TN, USA

Physics Division, Oak Ridge National Laboratory, Oak Ridge, TN, USA
e-mail: titusmorris@gmail.com

M. Hjorth-Jensen et al. (eds.), *An Advanced Course in Computational Nuclear Physics*, Lecture Notes in Physics 936, DOI 10.1007/978-3-319-53336-0_10

decay [2], which can nowadays be derived from the Standard Model by expanding the propagator of the $W^{\pm}$ bosons that mediate weak processes for small momenta $q \ll M_W = 80\,\text{GeV}$ (in units where $\hbar = c = 1$).

The strong-interaction sector of the Standard Model is provided by Quantum Chromodynamics (QCD), but the description of nuclear observables on the level of quarks and gluons is not feasible, except in the lightest few-nucleon systems (see, e.g., [3] and the chapters on Lattice QCD in this book). The main issue is that QCD is an asymptotically free theory [4, 5], i.e., it is weak and amenable to perturbative methods from Quantum Field Theory at large momentum transfer, but highly non-perturbative in the low-momentum regime which is relevant for nuclear structure physics. A consequence of the latter property is that quarks are confined in baryons and mesons at low momentum or energy scales, which makes those confined particles suitable degrees of freedom for an EFT approach. Chiral EFT, in particular, is constructed in terms of nucleon and pion fields, with some attention now also being given to the lowest excitation mode of the nucleon, namely the Δ resonance. It provides a constructive framework and organizational hierarchy for *NN*, 3*N*, and higher many-nucleon forces, as well as consistent electroweak operators (see, e.g., [6–16]). Since Chiral EFT is a low-momentum expansion, high-momentum (short-range) physics is not explicitly resolved by the theory, but parametrized by the so-called low-energy constants (LECs). In principle, the LECs can be determined by matching calculations of the same observables in chiral EFT and (Lattice) QCD in the overlap region of the two theories. Since such a calculation is currently not feasible, they are fit to experimental data for low-energy QCD observables, typically in the πN, *NN*, and 3*N* sectors [6, 7, 17, 18].

RG methods are natural companions for EFTs, because they smoothly connect theories with different resolution scales and degrees of freedom [19, 20]. Since they were introduced in low-energy nuclear physics around the start of the millennium [21–24], they have provided a systematic framework for formalizing many ideas on the renormalization of nuclear interactions and many-body effects that had been discussed in the nuclear structure community since the 1950s. For instance, soft and hard-core *NN* interactions can reproduce scattering data equally well, but have significantly different saturation properties, which caused the community to all but abandon the former in the 1970s (see, e.g., [25]). What was missing at that time was the recognition of the intricate link between the off-shell *NN* interaction and 3*N* forces that was formally demonstrated for the first time by Polyzou and Glöckle in 1990 [26]. From the modern RG perspective, soft- and hard-core interactions emerge as representations of low-energy QCD at different resolution scales, and the dialing of the resolution scale necessarily leads to induced 3*N* forces, in such a way that observables (including saturation properties) remain invariant under the RG flow (see Sect. 10.2.4.5 and [23, 24]). In conjunction, chiral EFT and nuclear RG applications demonstrate that one cannot treat the *NN*, 3*N*, ... sectors in isolation from each other.

Brueckner introduced the idea of renormalizing the *NN* interaction in the nuclear medium by summing correlations due to the scattering of nucleon pairs to high-energy states into the so-called *G*-matrix, which became the basis of an attempted perturbation expansion of nuclear observables [27–32]. Eventually, the nuclear structure community uncovered severe issues with this approach, like a lack of order-by-order convergence [33–37], and a strong model space dependence in summations over virtual excitations [38]. One of the present authors led a study that revisited this issue, and demonstrated that the G matrix retains significant coupling between low- and high-momentum nodes of the underlying interaction [23], so the convergence issues are not surprising from a modern perspective. In the Similarity Renormalization Group [39, 40] and other modern RG approaches, low- and high-momentum physics are decoupled *properly*, which results in low-momentum interactions that can be treated successfully in finite-order many-body perturbation theory (MBPT) [23, 41–43]. These interactions are not just suited as input for MBPT, but for all methods that work with momentum- or energy-truncated configuration spaces. The decoupling of low- and high-momentum modes greatly improves the convergence behavior of such methods, which can then be applied to heavier and heavier nuclei [44–50].

The idea of decoupling energy scales can also be used to directly tackle the nuclear many-body problem. We implement it in the so-called In-Medium SRG [51–53], which is discussed at length in this chapter. In a nutshell, we will use SRG-like flow equations to decouple physics at different excitation energy scales of the nucleus, and render the Hamiltonian matrix in configuration space block or band diagonal in the process. We will see that this can be achieved while working on the operator level, freeing us from the need to construct the Hamiltonian matrix in a factorially growing basis of configurations. We will show that the IMSRG evolution can also be viewed as a re-organization of the many-body perturbation series, in which correlations that are described explicitly by the configuration space are absorbed into an *RG-improved* Hamiltonian. With an appropriately chosen decoupling strategy, it is even possible to extract specific eigenvalues and eigenstates of the Hamiltonian, which is why the IMSRG qualifies as an ab initio (first principles) method for solving quantum many-body problems.

The idea of using flow equations to solve quantum many-body problems dates back (at least) to Wegner's initial work on the SRG [40] (also see [54] and references therein). In the solid-state physics literature, the approach is also known as continuous unitary transformation theory [55–59]. When we discuss our decoupling strategies for the nuclear many-body problem, it will become evident that the IMSRG is related to Coupled Cluster theory (CC), see also Chap. 8, and a variety of other many-body methods that are used heavily in quantum chemistry (see, e.g., [49, 60–66]). What sets the IMSRG apart from these methods is that the Hamiltonian instead of the wave function is at the center of attention, in the spirit of RG methodology. This seems to be a trivial distinction, but there are practical advantages of this viewpoint, like the ability to simultaneously decouple ground and a number of excited states from the rest of the spectrum (see Sects. 10.3.3 and 10.4.3).

10.1.1 Organization of This Chapter

We conclude our introduction by looking ahead at the remainder of this chapter. In Sect. 10.2, we will introduce the basic concepts of the SRG, and apply it to a pedagogical toy model (Sect. 10.2.2), the pairing Hamiltonian that is also discussed in Chaps. 8, 9, and 11 (Sect. 10.2.3), and last but not least, we will discuss the SRG evolution of modern nuclear interactions (Sect. 10.2.4).

The issue of SRG-induced operators (Sect. 10.2.4.5) will serve as our launching point for the discussion of the IMSRG in Sect. 10.3. First, we will introduce normal-ordering techniques as a means to control the size of induced interaction terms (Sect. 10.3.1). This is followed by the derivation of the IMSRG flow equations, determination of decoupling conditions, and the construction of generators in Sects. 10.3.2–10.3.4. We discuss the essential steps of an IMSRG implementation through the example of a symmetry-unconstrained Python code (Sect. 10.3.5), and use this code to revisit the pairing Hamiltonian in Sect. 10.3.6. In Sect. 10.3.7, we compute the neutron matter equation-of-state in the IMSRG(2) truncation scheme, and compare our result to that of corresponding Coupled Cluster, Quantum Monte Carlo, and Self-Consistent Green's Function results with the same interaction.

Section 10.4 introduces the three major directions of current research: First, we present the Magnus formulation of the IMSRG (Sect. 10.4.1), which is the key to the efficient computation of observables and the construction of approximate version of the IMSRG(3). Second, we give an overview of the multireference IMSRG (MR-IMSRG), which generalizes our framework to arbitrary reference states, and gives us new freedom to manipulate the correlation content of our many-body calculations (Sect. 10.4.2). Third, we will discuss applications of IMSRG-evolved, RG-improved Hamiltonians as input to many-body calculations, in particular for the nuclear (interacting) Shell model and Equation-of-Motion (EOM) methods (Sect. 10.4.3). An outlook on how these three research thrusts interweave concludes the section (Sect. 10.4.4).

In Sect. 10.5, we make some final remarks and close the main body of the chapter in Sect. 10.5. Section 10.6 contains exercises that further flesh out subjects discussed in the preceding sections, as well as outlines for computational projects. Formulas for products and commutators of normal-ordered operators are collected in Appendix.

10.2 The Similarity Renormalization Group

10.2.1 Concept

The basic idea of the Similarity Renormalization Group (SRG) method is quite general: We want to "simplify" our system's Hamiltonian $\hat{H}(s)$ by means of a continuous unitary transformation that is parametrized by a one-dimensional parameter s,

$$\hat{H}(s) = \hat{U}(s)\hat{H}(0)\hat{U}^\dagger(s)\,. \tag{10.2}$$

By convention, $\hat{H}(s = 0)$ is the starting Hamiltonian. To specify what we mean by simplifying $\hat{H}$, it is useful to briefly think of it as a matrix rather than an operator. As in any quantum-mechanical problem, we are primarily interested in finding the eigenstates of $\hat{H}$ by diagonalizing its matrix representation. This task is made easier if we can construct a unitary transformation that renders the Hamiltonian more and more diagonal as s increases. Mathematically, we want to split the Hamiltonian into suitably defined diagonal and off-diagonal parts,

$$\hat{H}(s) = \hat{H}_d(s) + \hat{H}_{od}(s)\,, \tag{10.3}$$

and find $\hat{U}(s)$ so that

$$\hat{H}(s) \underset{s\to\infty}{\longrightarrow} \hat{H}_d(s)\,, \quad \hat{H}_{od}(s) \underset{s\to\infty}{\longrightarrow} 0\,. \tag{10.4}$$

To implement the continuous unitary transformation, we take the derivative of Eq. (10.2) with respect to s to obtain

$$\begin{aligned}\frac{d\hat{H}(s)}{ds} &= \frac{d\hat{U}(s)}{ds}\hat{H}(0)\hat{U}^\dagger(s) + \hat{U}(s)\hat{H}(0)\frac{d\hat{U}^\dagger(s)}{ds}\\ &= \frac{d\hat{U}(s)}{ds}\hat{U}^\dagger(s)\hat{H}(s) + \hat{H}(s)\hat{U}(s)\frac{d\hat{U}^\dagger(s)}{ds}\,.\end{aligned} \tag{10.5}$$

Since $\hat{U}(s)$ is unitary, we also have

$$\frac{d}{ds}\left(\hat{U}(s)\hat{U}^\dagger(s)\right) = \frac{d}{ds}\left(\hat{1}\right) = 0 \quad \Longrightarrow \quad \frac{d\hat{U}(s)}{ds}\hat{U}^\dagger(s) = -\hat{U}(s)\frac{d\hat{U}^\dagger(s)}{ds}\,. \tag{10.6}$$

Defining the anti-Hermitian operator

$$\hat{\eta}(s) \equiv \frac{d\hat{U}(s)}{ds}\hat{U}^\dagger(s) = -\hat{\eta}^\dagger(s)\,, \tag{10.7}$$

we can write the differential equation for the s-dependent Hamiltonian as

$$\frac{d}{ds}\hat{H}(s) = [\hat{\eta}(s), \hat{H}(s)]\,. \tag{10.8}$$

This is the SRG *flow equation* for the Hamiltonian, which describes the evolution of $\hat{H}(s)$ under the action of a dynamical generator $\hat{\eta}(s)$. Since we are considering a *unitary* transformation, the spectrum of the Hamiltonian is preserved.[1] Thus, the SRG is related to so-called isospectral flows, a class of transformations that has been studied extensively in the mathematics literature (see for example [67–71]).

The flow equation (10.8) is the most practical way of implementing an SRG evolution: We can obtain $\hat{H}(s)$ by integrating Eq. (10.8) numerically, without explicitly constructing the unitary transformation itself. Formally, we can also obtain $\hat{U}(s)$ by rearranging Eq. (10.7) into

$$\frac{d}{ds}\hat{U}(s) = \hat{\eta}(s)\hat{U}(s)\,. \tag{10.9}$$

The solution to this differential equation is given by the S-ordered exponential

$$U(s) = \mathscr{S}\exp\int_0^s ds'\,\eta(s')\,, \tag{10.10}$$

because the generator changes dynamically during the flow. This expression is defined equivalently either as a product of infinitesimal unitary transformations,

$$U(s) = \lim_{N\to\infty}\prod_{i=0}^{N} e^{\eta(s_i)\delta s_i}\,,\quad s_{i+1} = s_i + \delta s_i\,,\quad \sum_i \delta s_i = s\,, \tag{10.11}$$

or through a series expansion:

$$U(s) = \sum_n \frac{1}{n!}\int_0^s ds_1 \int_0^s ds_2 \ldots \int_0^s ds_n \mathscr{S}\{\eta(s_1)\ldots\eta(s_n)\}\,. \tag{10.12}$$

Here, the S-ordering operator $\mathscr{S}$ ensures that the flow parameters appearing in the integrands are always in descending order, $s_1 > s_2 > \ldots$. Note that neither Eq. (10.11) nor Eq. (10.12) can be written as a single proper exponential, so we do not obtain a simple Baker-Campbell-Hausdorff expansion of the transformed Hamiltonian. Instead, we would have to use these complicated expressions to construct $\hat{H}(s) = \hat{U}(s)\hat{H}(0)\hat{U}^\dagger(s)$, and to make matters even worse, Eqs. (10.11) and (10.12) depend on the generator at *all intermediate points* of the flow trajectory. The associated storage needs would make numerical applications impractical or entirely unfeasible.

Let us focus on the flow equation (10.8), then, and specify a generator that will transform the Hamiltonian to the desired structure [Eq. (10.4)]. Inspired by the work of Brockett [69] on the so-called double-bracket flow, Wegner [40] proposed the

[1]There are mathematical subtleties due to $\hat{H}(s)$ being an operator that is only bounded from below, and having a spectrum that is part discrete, part continuous (see, e.g., [67, 68]). In practice, we are forced to work with approximate, finite-dimensional matrix representations of $\hat{H}(s)$ in any case.

generator

$$\hat{\eta}(s) \equiv [H_d(s), H_{od}(s)]\,. \tag{10.13}$$

A fixed point of the SRG flow is reached when $\hat{\eta}(s)$ vanishes. At finite s, this can occur if $\hat{H}_d(s)$ and $\hat{H}_{od}(s)$ happen to commute, e.g., due to a degeneracy in the spectrum of $\hat{H}(s)$. A second fixed point at $s \to \infty$ exists if $\hat{H}_{od}(s)$ vanishes as required.

Going back over the discussion, you may notice that we never specified in detail *how* we split the Hamiltonian into diagonal and off-diagonal parts. By "diagonal" we really mean the desired structure of the Hamiltonian, and "off-diagonal" labels the contributions we have to suppress in the limit $s \to \infty$ to obtain that structure. The basic concepts described here are completely general, and we will discuss two examples in which we apply them to the diagonalization of matrices in the following. The renormalization of Hamiltonians (or other operators) is a more specific application of continuous unitary transformations. We make contact with renormalization ideas by imposing a block or band-diagonal structure on the representation of operators in bases that are organized by momentum or energy. This implies a decoupling of low and high momenta or energies in the renormalization group sense. We will conclude this section with a brief discussion of how this SRG decoupling of scales is used to render nuclear Hamiltonians more suitable for ab initio many-body calculations [22, 23, 53, 72, 73].

10.2.2 A Two-Dimensional Toy Problem

In order to get a better understanding of the SRG method, we first consider a simple 2×2 matrix problem that can be solved analytically, and compare the flow generated by Eq. (10.8) with standard diagonalization algorithms like Jacobi's rotation method (see, e.g., [74]).

Let us consider a symmetric matrix H,

$$H \equiv \begin{pmatrix} H_{11} & H_{12} \\ H_{12} & H_{22} \end{pmatrix}. \tag{10.14}$$

and an orthogonal (i.e., unitary and real) matrix U,

$$U = \begin{pmatrix} \cos\gamma & \sin\gamma \\ -\sin\gamma & \cos\gamma \end{pmatrix}, \tag{10.15}$$

that parameterizes a rotation of the basis in which H and U are represented. We want to find an angle γ so that $H' = UHU^T$ is diagonal, and to achieve this, we need to solve

$$(H_{22} - H_{11}) \cos\gamma \, \sin\gamma + H_{12}(\cos^2\gamma - \sin^2\gamma) = 0\,. \tag{10.16}$$

Using the addition theorems $\cos^2\gamma - \sin^2\gamma = \cos(2\gamma)$ and $\cos\gamma\,\sin\gamma = \sin(2\gamma)/2$, we can rewrite this equation as

$$\tan(2\gamma) = \frac{2H_{12}}{H_{11} - H_{22}} \tag{10.17}$$

and obtain

$$\gamma = \frac{1}{2}\tan^{-1}\left(\frac{2H_{12}}{H_{11} - H_{22}}\right) + \frac{k\pi}{2}, \quad k \in \mathbb{Z}, \tag{10.18}$$

where $k\pi/2$ is added due to the periodicity of the tan function. Note that $k = 0$ gives a diagonal matrix of the form

$$H'_{k=0} = \begin{pmatrix} E_1 & 0 \\ 0 & E_2 \end{pmatrix}, \tag{10.19}$$

while $k = 1$ interchanges the diagonal elements:

$$H'_{k=1} = \begin{pmatrix} E_2 & 0 \\ 0 & E_1 \end{pmatrix}. \tag{10.20}$$

Let us now solve the same problem with an SRG flow. We parameterize the Hamiltonian as

$$H(s) = H_d(s) + H_{od}(s) \equiv \begin{pmatrix} E_1(s) & 0 \\ 0 & E_2(s) \end{pmatrix} + \begin{pmatrix} 0 & V(s) \\ V(s) & 0 \end{pmatrix}, \tag{10.21}$$

working in the eigenbasis of $H_d(s)$. We can express $H(s)$ in terms of the 2×2 identity matrix and the Pauli matrices:

$$H(s) = E_+(s)\hat{I} + E_-(s)\sigma_3 + V(s)\sigma_1\,, \tag{10.22}$$

where we have introduced

$$E_\pm(s) \equiv \frac{1}{2}\left(E_1(s) \pm E_2(s)\right)\,. \tag{10.23}$$

The Wegner generator can be determined readily using the algebra of the Pauli matrices, $[\sigma_j, \sigma_k] = 2i\varepsilon_{jkl}\sigma_l$:

$$\eta(s) = [H_d(s), H_{od}(s)] = 2iE_-(s)V(s)\sigma_2\,. \tag{10.24}$$

By evaluating both $\dot{H}(s)$ as well as $[\hat{\eta}(s), \hat{H}(s)]$ and comparing the coefficients of the 2×2 matrices, we obtain the following system of flow equations:

$$\dot{E}_+ = 0 \tag{10.25}$$

$$\dot{E}_- = 4V^2E_- \tag{10.26}$$

$$\dot{V} = -4E_-^2V\,, \tag{10.27}$$

where we have suppressed the flow parameter dependence for brevity. The first flow equation reflects the conservation of the Hamiltonian's trace under unitary transformations,

$$\mathrm{tr}\,(H(s)) = E_1(s) + E_2(s) = E_1(0) + E_2(0) = \mathrm{tr}\,(H(0)) = \mathrm{const.} \tag{10.28}$$

The third flow equation can be rearranged into

$$\frac{1}{V}dV = -4E_-^2\,ds \tag{10.29}$$

and integrated, which yields

$$V(s) = V(0)\exp\left(-4\int_0^s ds'\,E_-^2(s')\right)\,. \tag{10.30}$$

Since $E_-(s)$ is real, the integral is positive for all values of s, and this means that the off-diagonal matrix element will be suppressed exponentially as we evolve $s \to \infty$ (barring singular behavior in $E_\pm(s)$).

To proceed, we introduce new variables Ω and θ:

$$\Omega \equiv \sqrt{E_-^2 + V^2}\,, \quad \tan\frac{\theta}{2} \equiv \frac{V}{E_-}\,. \tag{10.31}$$

Using Eqs. (10.26) and (10.27), we can show that Ω is a flow invariant:

$$\dot{\Omega} = \frac{1}{2\Omega}(2E_-\dot{E}_- + 2V\dot{V}) = 0\,. \tag{10.32}$$

Rewriting $E_-(s)$ and $V(s)$ in terms of Ω and $\theta(s)$, we then have

$$E_-(s) = \Omega\cos\frac{\theta(s)}{2}\,, \quad V(s) = \Omega\sin\frac{\theta(s)}{2}\,. \tag{10.33}$$

Using these expressions, we find that Eqs. (10.26) and (10.27) reduce to a single differential equation for $\theta(s)$:

$$\dot{\theta} = -8\Omega^2 \sin\frac{\theta(s)}{2}\cos\frac{\theta(s)}{2} = -4\Omega^2 \sin\theta\,. \tag{10.34}$$

Bringing $\sin\theta$ to the left-hand side and using

$$\frac{d}{dx}\ln\tan\frac{x}{2} = \frac{1}{\sin x}\,, \tag{10.35}$$

we can integrate the ordinary differential equation (ODE) and obtain

$$\tan\frac{\theta(s)}{2} = \tan\frac{\theta(0)}{2}e^{-4\Omega^2 s}\,. \tag{10.36}$$

At $s = 0$ we have

$$\theta(0) = 2\tan^{-1}\frac{V(0)}{E_-(0)} + 2k\pi = 2\tan^{-1}\frac{2V(0)}{E_1(0)-E_2(0)} + 2k\pi\,, \quad k\in\mathbb{Z}\,, \tag{10.37}$$

which is just four times the angle of the Jacobi rotation that diagonalizes our initial matrix, Eq. (10.18). Likewise $\theta(s)$ is (up to the prefactor) the angle of the Jacobi rotation that will diagonalize the evolved $H(s)$ for $s > 0$. In the limit $s \to \infty$, $\theta(s) \to 0$ because the SRG flow has driven the off-diagonal matrix element to zero and the Hamiltonian is already diagonal.

When we introduced the parameterization (10.33), we chose Ω to be positive, which means that $\theta(s)$ must encode all information on the signs of $E_-(s)$ and $V(s)$. In Fig. 10.1, we show these quantities as a function of $\theta(s)$ over the interval $[0, 4\pi]$. We see that the four possible sign combinations correspond to distinct regions in the interval. We can map any set of initial values—or any point of the flow, really—to a distinct value $\theta(s)$ in this figure, even in limiting cases: For instance, if the diagonal matrix elements are degenerate, $E_1 = E_2$, we have $E_- = 0$, the angle θ will approach $\pm\pi$ and $V(s)/\Omega \to \pm 1$. From this point the SRG flow will drive $E_-(s)$ and $V(s)$ to the nearest fixed point at a multiple of 2π according to the trajectory for $\theta(s)$, Eq. (10.36). The fixed points and flow directions are indicated in the figure.

10.2.3 The Pairing Model

10.2.3.1 Preliminaries

As our second example for diagonalizing matrices by means of SRG flows, we will consider the pairing model that was discussed in the context of Hartree-Fock and beyond-HF methods in Chap. 8. In second quantization, the pairing Hamiltonian is

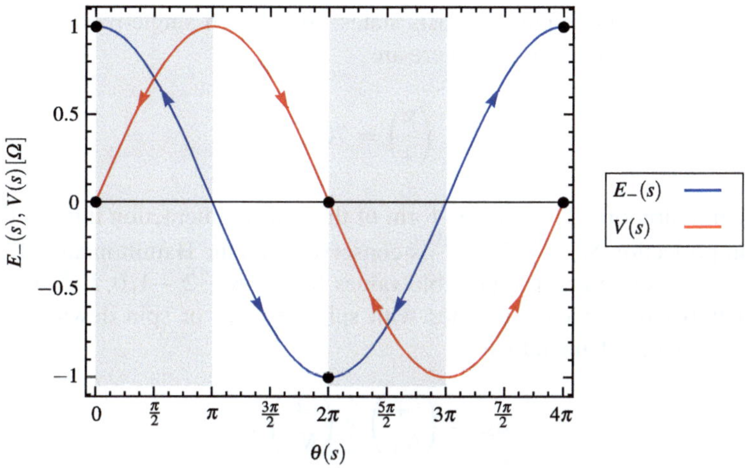

Fig. 10.1 $E_-(s)$ and $V(s)$ as a function of the flowing angle $\theta(s)$, in units of the flow invariant Ω. *Black dots and arrows* the fixed points at $s \to \infty$ and the directions of the SRG flow, respectively, in the domains corresponding to specific sign combinations for $E_-(s)$ and $V(s)$ (see text)

Table 10.1 Single-particle states and their quantum numbers and their energies from Eq. (10.38)

State	p	$2s_z$	ε
0	1	1	0
1	1	−1	0
2	2	1	δ
3	2	−1	δ
4	3	1	2δ
5	3	−1	2δ
6	4	1	3δ
7	4	−1	3δ

The degeneracy for every quantum number p is equal to 2 due to the two possible spin values

given by

$$\hat{H} = \delta \sum_{p\sigma} (p-1) a^\dagger_{p\sigma} a_{p\sigma} - \frac{1}{2} g \sum_{pq} a^\dagger_{p+} a^\dagger_{p-} a_{q-} a_{q+} , \qquad (10.38)$$

where δ controls the spacing of single-particle levels that are indexed by a principal quantum number $p = 1, \ldots, 4$ and their spin projection σ (see Table 10.1), and g the strength of the pairing interaction.

We will consider the case of four particles in eight single-particle states. Following the Full Configuration Interaction (FCI) approach discussed in Chap. 8, we can construct a many-body basis of Slater determinants by placing our particles into the available single-particle basis states. Since each single-particle state can only be occupied by one particle, there are

$$\binom{8}{4} = 70 \tag{10.39}$$

unique configurations. The specific form of the pairing interaction implies that the total spin projection $S_z = \sum_{n=1}^{4} s_z^{(n)}$ is conserved, and the Hamiltonian will have a block diagonal structure. The possible values for S_z are $-2, -1, 0, 1, 2$, depending on the number of particles in states with spin up (N_+) or spin down (N_-). The dimension can be calculated via

$$d_{S_z} = \binom{4}{N_+} \times \binom{4}{N_-}, \tag{10.40}$$

which yields

$$d_{\pm 2} = 1\,, \quad d_{\pm 1} = 16\,, \quad d_0 = 36\,. \tag{10.41}$$

Since the pairing interaction only couples pairs of single-particle states with the same principal quantum number but opposite spin projection, it does not break pairs of particles that occupy such states—in other words, the number of particle pairs is another conserved quantity, which allows us to decompose the S_z blocks into even smaller sub blocks. As in Chap. 8, we consider the $S_z = 0$ sub block with two particle pairs. In this block, the Hamiltonian is represented by the six-dimensional matrix (suppressing block indices)

$$H = \begin{pmatrix} 2\delta - g & -g/2 & -g/2 & -g/2 & -g/2 & 0 \\ -g/2 & 4\delta - g & -g/2 & -g/2 & -0 & -g/2 \\ -g/2 & -g/2 & 6\delta - g & 0 & -g/2 & -g/2 \\ -g/2 & -g/2 & 0 & 6\delta - g & -g/2 & -g/2 \\ -g/2 & 0 & -g/2 & -g/2 & 8\delta - g & -g/2 \\ 0 & -g/2 & -g/2 & -g/2 & -g/2 & 10\delta - g \end{pmatrix}. \tag{10.42}$$

10.2.3.2 SRG Flow for the Pairing Hamiltonian

As in earlier sections, we split the Hamiltonian matrix (10.42) into diagonal and off-diagonal parts:

$$H_d(s) = \mathrm{diag}(E_0(s)\,, \ldots, E_5(s))\,, \quad H_{od}(s) = H(s) - H_d(s)\,, \tag{10.43}$$

with initial values defined by Eq. (10.42). Since $H_d(s)$ is diagonal throughout the flow per construction, the Slater determinants that span our specific subspace are the eigenstates of this matrix. In our basis representation, Eq. (10.8) can be written as

$$\begin{aligned}
\frac{d}{ds}\langle i|\hat{H}|j\rangle &= \sum_k \Big(\langle i|\hat{\eta}|k\rangle\langle k|\hat{H}|j\rangle - \langle i|\hat{H}|k\rangle\langle k|\hat{\eta}|j\rangle\Big) \\
&= -\left(E_i - E_j\right)\langle i|\hat{\eta}|j\rangle + \sum_k \Big(\langle i|\hat{\eta}|k\rangle\langle k|\hat{H}_{od}|j\rangle - \langle i|\hat{H}_{od}|k\rangle\langle k|\hat{\eta}|j\rangle\Big),
\end{aligned} \tag{10.44}$$

where $\langle i|\hat{H}_{od}|i\rangle = 0$ and block indices as well as the s-dependence have been suppressed for brevity. The Wegner generator, Eq. (10.13), is given by

$$\langle i|\hat{\eta}|j\rangle = \langle i|[\hat{H}_d, \hat{H}_{od}]|j\rangle = (E_i - E_j)\langle i|\hat{H}_{od}|j\rangle\,, \tag{10.45}$$

and inserting this into the flow equation, we obtain

$$\frac{d}{ds}\langle i|\hat{H}|j\rangle = -\left(E_i - E_j\right)^2\langle i|\hat{H}_{od}|j\rangle + \sum_k \left(E_i + E_j - 2E_k\right)\langle i|\hat{H}_{od}|k\rangle\langle k|\hat{H}_{od}|j\rangle\,. \tag{10.46}$$

Let us assume that the transformation generated by $\hat{\eta}$ truly suppresses $\hat{H}_{od}$, and consider the asymptotic behavior for large flow parameters $s > s_0 \gg 0$. If $||\hat{H}_{od}(s_0)|| \ll 1$ in some suitable norm, the second term in the flow equation can be neglected compared to the first one. For the diagonal and off-diagonal matrix elements, this implies

$$\frac{dE_i}{ds} = \frac{d}{ds}\langle i|\hat{H}_d|i\rangle = 2\sum_k (E_i - E_k)\langle i|\hat{H}_{od}|k\rangle\langle k|\hat{H}_{od}|i\rangle \approx 0 \tag{10.47}$$

and

$$\frac{d}{ds}\langle i|\hat{H}|j\rangle \approx -\left(E_i - E_j\right)^2\langle i|\hat{H}_{od}|j\rangle\,, \tag{10.48}$$

respectively. Thus, the diagonal matrix elements will be (approximately) constant in the asymptotic region,

$$E_i(s) \approx E_i(s_0)\,, \quad s > s_0\,, \tag{10.49}$$

which in turn allows us to integrate the flow equation for the off-diagonal matrix elements. We obtain

$$\langle i|\hat{H}_{od}(s)|j\rangle \approx \langle i|\hat{H}_{od}(s_0)|j\rangle\, e^{-(E_i-E_j)^2(s-s_0)}\,, \quad s > s_0\,, \tag{10.50}$$

i.e., the off-diagonal matrix elements are suppressed exponentially, as for the 2×2 matrix toy problem discussed in the previous section. If the pairing strength g is sufficiently small so that $\hat{H}^2_{od}(0) \sim \mathcal{O}(g^2)$ can be neglected, we expect to see the exponential suppression of the off-diagonal matrix elements from the very onset of the flow.

Our solution for the off-diagonal matrix elements, Eq. (10.50), shows that the characteristic decay scale for each matrix element is determined by the square of the energy difference between the states it couples, $(\Delta E_{ij})^2 = (E_i - E_j)^2$. Thus, states with larger energy differences are decoupled before states with small energy differences, which means that the Wegner generator generates a proper renormalization group transformation. Since we want to diagonalize (10.42) in the present example, we are only interested in the limit $s \to \infty$, and it does not really matter whether the transformation is an RG or not. Indeed, there are alternative choices for the generator which are more efficient in achieving the desired diagonalization (see Sect. 10.3.4 and [53, 73]). The RG property will matter in our discussion of SRG-evolved nuclear interactions in the next section.

10.2.3.3 Implementation

We are now ready to discuss the implementation of the SRG flow for the pairing Hamiltonian. The main numerical task is the integration of the flow equations, a system of first order ODEs. Readers who are interested in learning the nuts-and-bolts details of implementing an ODE solver are referred to the excellent discussion in [75], while higher-level discussions of the algorithms can be found, e.g., in [76–78]. A number of powerful ODE solvers have been developed over the past decades and integrated into readily available libraries, and we choose to rely on one of these here, namely ODEPACK (see www.netlib.org/odepack and [79–81]). The SciPy package provides convenient Python wrappers for the ODEPACK solvers.

The following source code shows the essential part of our Python implementation of the SRG flow for the pairing model. The full program with additional features can be downloaded from https://github.com/ManyBodyPhysics/LectureNotesPhysics/tree/master/Programs/Chapter10-programs/python/srg_pairing.

```
import numpy as np
from numpy import array, dot, diag, reshape
from scipy.linalg import eigvalsh
from scipy.integrate import odeint

# Hamiltonian for the pairing model
def Hamiltonian(delta,g):

    H = array(
      [[2*delta-g, -0.5*g, -0.5*g,  -0.5*g, -0.5*g,      0.],
       [  -0.5*g, 4*delta-g, -0.5*g, -0.5*g,    0.,   -0.5*g ],
       [  -0.5*g, -0.5*g, 6*delta-g,    0.,  -0.5*g,  -0.5*g ],
       [  -0.5*g, -0.5*g,       0., 6*delta-g, -0.5*g, -0.5*g ],
```

```
      [  -0.5*g,      0.,    -0.5*g,   -0.5*g, 8*delta-g, -0.5*g ],
      [      0.,  -0.5*g,    -0.5*g,   -0.5*g,   -0.5*g, 10*delta-g
          ]]
    )

  return H

# commutator of matrices
def commutator(a,b):

   return dot(a,b) - dot(b,a)

# right-hand side of the flow equation
def derivative(y, t, dim):

   # reshape the solution vector into a dim x dim matrix
   H = reshape(y, (dim, dim))

   # extract diagonal Hamiltonian...
   Hd = diag(diag(H))

   # ... and construct off-diagonal the Hamiltonian
   Hod = H-Hd

   # calculate the generator
   eta = commutator(Hd, Hod)

   # dHdt is the derivative in matrix form
   dH = commutator(eta, H)

   # convert dH into a linear array for the ODE solver
   dy = reshape(dH, -1)

   return dy


#---------------------------------------------------------------
# Main program
#---------------------------------------------------------------
def main():
   g    = 0.5
   delta = 1

   H0   = Hamiltonian(delta, g)
   dim  = H0.shape[0]

   # calculate exact eigenvalues
   eigenvalues = eigvalsh(H0)

   # turn initial Hamiltonian into a linear array
   y0 = reshape(H0, -1)
```

```
# flow parameters for snapshot images
flowparams = array([0.,0.001,0.01,0.05,0.1, 1., 5., 10.])

# integrate flow equations - odeint returns an array of

# solutions, which are 1d arrays themselves
ys = odeint(derivative, y0, flowparams, args=(dim,))

# reshape individual solution vectors into dim x dim

# Hamiltonian matrices
Hs = reshape(ys, (-1, dim,dim))
```

The routine `Hamiltonian` sets up the Hamiltonian matrix (10.42) for general values of δ and g. The right-hand side of the flow equation is implemented in the routine `derivative`, which splits $H(s)$ in diagonal and off-diagonal parts, and calculates both the generator and the commutator $[\hat{\eta}, \hat{H}]$ using matrix products. Since the interface of essentially all ODE libraries require the ODE system and derivative to be passed as a one-dimensional array, NumPy's `reshape` functionality used to rearrange these arrays into 6×6 matrices and back again.

The main routine calls the ODEPACK wrapper `odeint`, passing the initial Hamiltonian as a one-dimensional array `y0` as well as a list of flow parameters s for which a solution is requested. The routine `odeint` returns these solutions as a two-dimensional array.

10.2.3.4 A Numerical Example

As a numerical example, we solve the pairing Hamiltonian for $\delta = 1.0$ and $g = 0.5$. In Fig. 10.2, we show snapshots of the matrix $H(s)$ at different stages of the flow. We can nicely see how the SRG evolution drives the off-diagonal matrix elements to zero. The effect becomes noticeable on our logarithmic color scale around $s = 0.05$, where the outermost off-diagonal matrix elements start to lighten. At $s = 0.1$, $H_{05}(s)$ has been reduced by four to five orders of magnitude, and at $s = 1.0$, essentially all of the off-diagonal matrix elements have been affected to some extent. Note that the strength of the suppression depends on the distance from the diagonal, aside from H_{05} itself, which has a slightly larger absolute value than H_{04} and H_{15}. The overall behavior is as expected from our approximate solution (10.50), and the specific deviations can be explained by the approximate nature of that result. Once we have reached $s = 10$, the matrix is essentially diagonal, with off-diagonal matrix elements reduced to 10^{-10} or less. Only the 2×2 block spanned by the states labeled 2 and 3 has slightly larger off-diagonal matrix elements remaining, which is due to the degeneracy of the corresponding eigenvalues.

In Fig. 10.3, we compare the flowing diagonal matrix elements $H_{ii}(s)$ to the eigenvalues of the pairing Hamiltonian. As we have just mentioned, the pairing Hamiltonian has a doubly degenerate eigenvalue $E_2 = E_3 = 6\delta - g$, which is why we see only five curves in these plots. For our choice of parameters, the diagonal

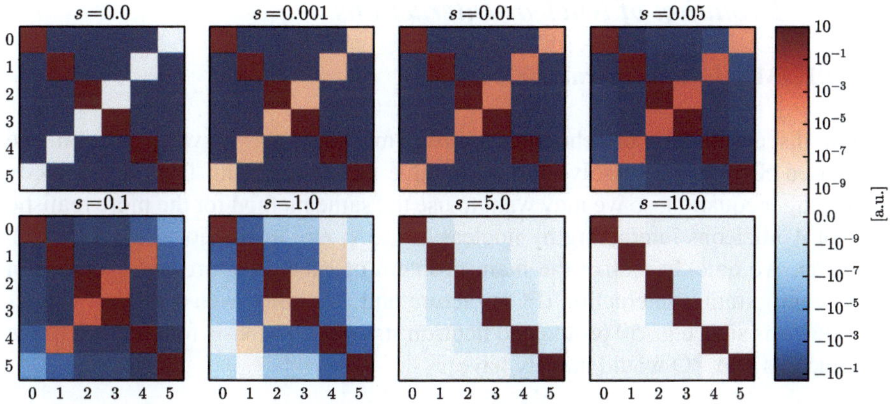

Fig. 10.2 SRG evolution of the pairing Hamiltonian with $\delta = 1, g = 0.5$. The figures show snapshots of the Hamiltonian's matrix representation at various stages of the flow, indicated by the flow parameters s. Note the essentially logarithmic scales of the positive and negative matrix elements, which are bridged by a linear scale in the vicinity of 0

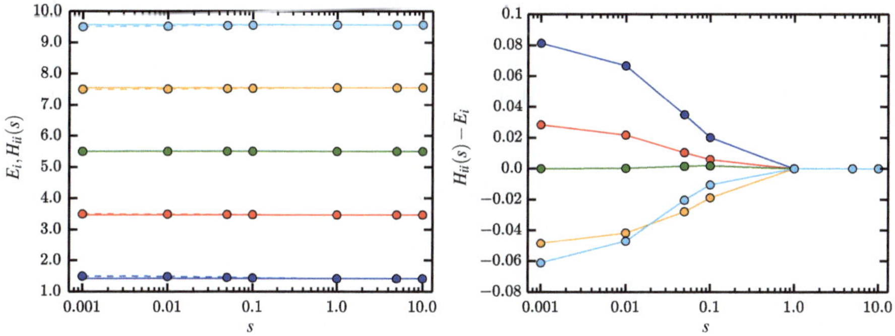

Fig. 10.3 SRG evolution of the pairing Hamiltonian with $\delta = 1, g = 0.5$. The *left panel* shows the diagonal matrix elements $H_{ii}(s)$ as a function of the flow parameter s (*dashed lines*) and the corresponding eigenvalues (*solid lines*), the *right panel* the difference of the two numbers. The color coding is the same in both panels

matrix elements are already in fairly good agreement with the eigenvalues to begin with. Focusing on the right-hand panel of the figure in particular, we see that H_{00} (blue) and H_{11} (red) approach their eigenvalues from above, while H_{44} (orange) and H_{55} (light blue) approach from below as we evolve to large s. It is interesting that the diagonal matrix elements are already practically identical to the eigenvalues once we have integrated up to $s = 1.0$, despite the non-vanishing off-diagonal matrix elements that are visible in the $s = 1.0$ snapshot shown in Fig. 10.2.

10.2.4 Evolution of Nuclear Interactions

10.2.4.1 Matrix and Operator Flows

In our discussion of the schematic pairing model in the previous section, we have used SRG flows to solve the eigenvalue problem arising from a four-body Schrödinger equation, so we may want to use the same method for the more realistic case of A nucleons interacting by nuclear NN, $3N$, etc. interactions (see Chap. 8). However, we quickly realize the main problem of such an approach: Working in a full configuration interaction (FCI) picture and assuming even a modest single-particle basis size, e.g., 50 proton and neutron states each, a basis for the description of a nucleus like ^{16}O would naively have

$$d(^{16}\mathrm{O}) = \binom{50}{8} \times \binom{50}{8} \approx 2.88 \times 10^{17} \tag{10.51}$$

configurations, i.e., we would need about 2 exabytes (EB) of memory to store all the coefficients of just one eigenvector (assuming double precision floating-point numbers), and 7×10^{17} EB to construct the complete Hamiltonian matrix! State-of-the-art methods for large-scale diagonalization are able to reduce the memory requirements and computational effort significantly by exploiting matrix sparseness, and using modern versions of Lanczos-Arnoldi [82, 83] or Davidson algorithms [84], but nuclei in the vicinity of the oxygen isotopic chain are among the heaviest accessible with today's computational resources (see, e.g., [44, 85] and references therein). A key feature of Lanczos-Arnoldi and Davidson methods is that the Hamiltonian matrix only appears in the calculation of matrix-vector products. In this way, an explicit construction of the Hamiltonian matrix in the CI basis is avoided, because the matrix-vector product can be calculated from the input NN and $3N$ interactions that only require $\mathcal{O}(n^4)$ and $\mathcal{O}(n^6)$ storage, respectively, where n is the size of the single-particle basis (see Sect. 10.3). However, the SRG flow of the previous section clearly forces us to construct and store the Hamiltonian matrix in its entirety—at best, we could save some storage by resizing the matrix once its off-diagonal elements have been sufficiently suppressed.

Instead of trying to evolve the many-body Hamiltonian matrix, we therefore focus on the Hamiltonian operator itself instead. Let us consider a nuclear Hamiltonian with a two-nucleon interaction for simplicity:

$$\hat{H}_{\mathrm{int}} = \hat{T}_{\mathrm{int}} + \hat{V}^{[2]} . \tag{10.52}$$

Since nuclei are self-bound objects, we have to consider the intrinsic form of the kinetic energy,

$$\hat{T}_{\mathrm{int}} \equiv \hat{T} - \hat{T}_{\mathrm{cm}} . \tag{10.53}$$

It is straightforward to show that $\hat{T}_{\text{int}}$ can be written either as a sum of one- and two-body operators,

$$\hat{T}_{\text{int}} = \left(1 - \frac{1}{\hat{A}}\right)\sum_i \frac{\hat{\mathbf{p}}_i^2}{2m} - \frac{1}{\hat{A}}\sum_{i<j} \frac{\hat{\mathbf{p}}_i \cdot \hat{\mathbf{p}}_j}{m} \tag{10.54}$$

or as a pure two-body operator

$$\hat{T}_{\text{int}} = \frac{2}{\hat{A}}\sum_{i<j} \frac{\hat{\mathbf{q}}_{ij}^2}{2\mu}, \quad \hat{\mathbf{q}}_{ij} \equiv \hat{\mathbf{p}}_i - \hat{\mathbf{p}}_j \,. \tag{10.55}$$

Here, $\hat{A}$ should be treated as a particle-number *operator* (see [86]), and $\mu = m/2$ is the reduced nucleon mass (neglecting the proton-neutron mass difference). Using Eq. (10.55) for the present discussion, we can write the intrinsic Hamiltonian as

$$\hat{H}_{\text{int}} = \frac{2}{\hat{A}}\sum_{i<j} \frac{\hat{\mathbf{q}}_{ij}^2}{2\mu} + \sum_{i<j} \hat{v}_{ij}^{[2]} \,, \tag{10.56}$$

and directly consider the evolution of this operator via the flow equation (10.8). It is customary to absorb the flow-parameter dependence completely into the interaction part of the Hamiltonian, and leave the kinetic energy invariant—in our previous examples, this simply amounts to moving the s-dependent part of $\hat{H}_d(s)$ into $\hat{H}_{od}(s)$. We end up with a flow equation for the two-body interaction:

$$\frac{d}{ds}\hat{v}_{ij}^{[2]} = [\hat{\eta}, \hat{v}_{ij}^{[2]}] \,. \tag{10.57}$$

In cases where we can expand the two-body interaction in terms of a finite algebra of "basis" operators, Eq. (10.57) becomes a system of ODEs for the expansion coefficients, the so-called running couplings of the Hamiltonian, as explained in earlier chapters of this book. An example is the toy problem discussed in Sect. 10.2.2: We actually expanded our 2×2 in terms of the algebra $\{\hat{I}, \hat{\sigma}_1, \hat{\sigma}_2, \hat{\sigma}_3\}$, and related the matrix elements to the coefficients in this expansion. While the representation of the basis operators of our algebra would force us to use extremely large matrices when we deal with an A-body system, we may be able to capture the SRG flow completely with a small set of ODEs for the couplings of the Hamiltonian!

If we cannot identify a set of basis operators for the two-body interaction, we can still resort to representing it as a matrix between two-body states. For a given choice of single-particle basis with size n, $\hat{v}^{[2]}$ is then represented by $\mathscr{O}(n^4)$ matrix elements, as mentioned above. In general, we will then have to face the issue of induced many-body forces, as discussed in Sect. 10.2.4.5.

10.2.4.2 SRG in the Two-Nucleon System

Let us now consider the operator flow of the *NN* interaction in the two-nucleon system, Eq. (10.57). Since the nuclear Hamiltonian is invariant under translations and rotations, it is most convenient to work in momentum and angular momentum eigenstates of the form

$$|q(LS)JMTM_T\rangle\,. \tag{10.58}$$

Because of the rotational symmetry, the *NN* interaction conserves the total angular momentum quantum number J, and it is easy to show that the total spin S of the nucleon pair is a conserved quantity as well. The orbital angular momentum is indicated by the quantum number L, and we remind our readers that L is *not* conserved, because the nuclear tensor operator

$$S_{ij}(\hat{\mathbf{r}},\hat{\mathbf{r}}) = \frac{3}{\hat{\mathbf{r}}^2}(\hat{\boldsymbol{\sigma}}_i\cdot\hat{\mathbf{r}})(\hat{\boldsymbol{\sigma}}_j\cdot\hat{\mathbf{r}}) - \hat{\boldsymbol{\sigma}}_i\cdot\hat{\boldsymbol{\sigma}}_j \tag{10.59}$$

can couple states with $\Delta L = \pm 2$. We assume that the interaction is charge-dependent in the isospin channel $T = 1$, i.e., matrix elements will depend on the projection $M_T = -1, 0, 1$, which indicates the neutron-neutron, neutron-proton, and proton-proton components of the nuclear Hamiltonian.

In Fig. 10.4 we show features of the central and tensor forces of the Argonne V18 (AV18) interaction [87] in the $(S,T) = (1,0)$ channel, which has the quantum numbers of the deuteron. This interaction belongs to a group of so-called realistic interactions that describe nucleon-nucleon scattering data with high accuracy, but precede the modern chiral forces (see Chap. 8, [6, 7]). AV18 is designed to be maximally local in order to be a suitable input for nuclear Quantum Monte Carlo calculations [10, 11, 88]. Because of the required locality, AV18 has a strong repulsive core in the central part of the interaction. Like all *NN* interactions, it also has a strong tensor force that results from pion exchange. The radial dependencies of these interaction components are shown in the left panel of Fig. 10.4.

When we switch to the momentum representation, we see that the $^3S_1$ partial wave[2] which gives the dominant contribution to the deuteron wave function has strong off-diagonal matrix elements, with tails extending over the entire shown range and as high as $|\hat{\mathbf{q}}| \sim 20\,\text{fm}^{-1}$. The matrix elements of the $^3S_1 - {}^3D_1$ mixed partial wave, which are generated exclusively by the tensor force, are sizable as well. The strong coupling between states with low and high relative momenta forces us to use large Hilbert spaces in few- and many-body calculations, even if we are only interested in the lowest eigenstates. Methods like the Lanczos algorithm (see Chap. 8 and [82]) extract eigenvalues and eigenvectors by repeatedly acting

[2]We use the conventional partial wave notation $^{2S+1}L_J$, where $L = 0, 1, 2, \ldots$ is indicated by the letters $S, P, D, \ldots$. The isospin channel is fixed by requiring the antisymmetry of the *NN* wavefunction, leading to the condition $(-1)^{L+S+T} = -1$.

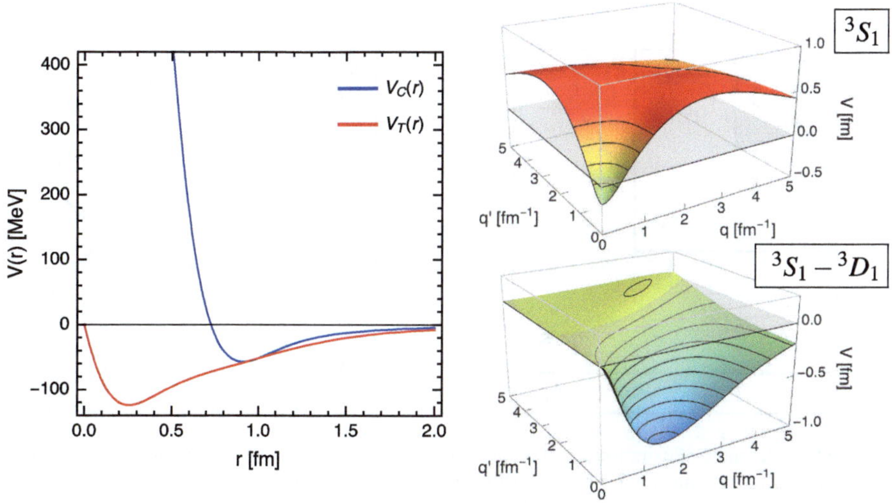

Fig. 10.4 Repulsive core and tensor force of the Argonne V18 *NN* interaction [87] in the $(S, T) = (1, 0)$ channel. In the *left panel*, the radial dependencies of the central ($V_C(r)$) and tensor components ($V_T(r)$) of Argonne V18 are shown, while the *right panel* shows its momentum space matrix elements in the deuteron partial waves

with the Hamiltonian on an arbitrary starting vector in the many-body space, i.e., by repeated matrix-vector products. Even if that vector only has low-momentum or low-energy components in the beginning, an interaction like AV18 will mix in high-momentum components even after a single matrix-vector multiplication, let alone tens or hundreds as in typical many-body calculations. Consequently, the eigenvalues and eigenstates of the nuclear Hamiltonian converge very slowly with respect to the basis size of the Hilbert space (see, e.g., [44]). To solve this problem, we perform an RG evolution of the *NN* interaction.

In Fig. 10.5, we show examples for two types of RG evolution that decouple the low- and high-momentum pieces of *NN* interactions. The first example, Fig. 10.5a, is a so-called *RG decimation*, in which the interaction is evolved to decreasing cutoff scales $\Lambda_0 > \Lambda_1 > \Lambda_2$, and high-momentum modes are "integrated out". This is the so-called $V_{\text{low-k}}$ approach, which was first used in nuclear physics in the early 2000s [21, 23]. Note that the resulting low-momentum interaction is entirely confined to states with relative momentum $q \leq \Lambda$. In contrast, Fig. 10.5b shows the SRG evolution of the *NN* interaction to a band-diagonal shape via the flow equation (10.57), using a generator built from the relative kinetic energy in the two-nucleon system [22, 23]:

$$\eta(\lambda) \equiv [\frac{\hat{\mathbf{q}}^2}{2\mu}, v(\lambda)]\,. \tag{10.60}$$

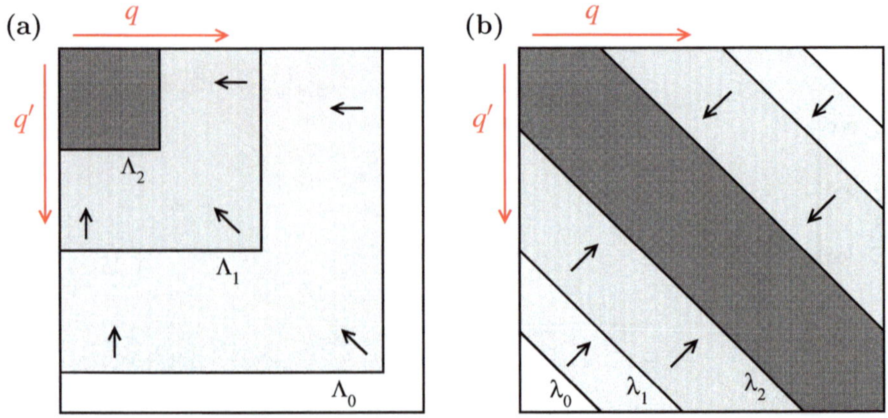

Fig. 10.5 Schematic illustration of two types of RG evolution for *NN* potentials in momentum space: (**a**) $V_{\text{low-k}}$ running in Λ, and (**b**) SRG running in λ (see main text). Here, q and q' denote the relative momenta of the initial and final state, respectively. At each Λ_i or λ_i, the matrix elements outside of the corresponding blocks or bands are negligible, implying that high- and low-momentum states are decoupled

Instead of the flow parameter s, we have parameterized the evolution by $\lambda = s^{-1/4}$, which has the dimensions of a momentum in natural units. Note that the generator (10.60) would vanish if the interaction were diagonal in momentum space. As suggested by Fig. 10.5b, λ is a measure for the "width" of the band in momentum space. Thus, momentum transfers between nucleons are limited according to

$$Q \equiv |\hat{\mathbf{q}}' - \hat{\mathbf{q}}| \lesssim \lambda\,, \tag{10.61}$$

and low- and high-lying momenta are decoupled as λ is decreased.

Equation (10.61) implies that the spatial resolution scale of an SRG-evolved interaction (or a $V_{\text{low-k}}$ if we determine the maximum momentum transfer in the low-momentum block) is $\sim 1/Q \geq 1/\lambda$, i.e., only long-ranged components of the *NN* interaction are resolved explicitly and short-range components of the interaction can just as well be replaced by contact interactions [20, 21, 23, 89]. This is the reason why the realistic *NN* interactions that accurately describe *NN* scattering data collapse to a universal long-range interaction when RG-evolved, namely one-pion exchange (OPE). This universal behavior emerges in the range $1.5\,\text{fm}^{-1} \leq \lambda \leq 2.5\,\text{fm}^{-1}$. Any further evolution to lower λ starts to remove pieces of OPE, and eventually generates a pion-less theory that is essentially parameterized in terms of contact interactions. While it is possible to implement such an evolution in the two-body system without introducing pathological behavior [90], such an interaction must be complemented by strong induced many-nucleon forces once it is applied in finite nuclei, as discussed in Sect. 10.2.4.5. For this reason, *NN* interactions are only ever evolved to the aforementioned range of λ values.

Nowadays, SRG evolutions are preferred over $V_{\text{low-k}}$ style decimations in nuclear many-body theory, because they can be readily extended to $3N, \ldots$ interactions and to general observables [23, 91–96]. Moreover, we could easily achieve a block decoupling as in Fig. 10.5a by using a generator like [97]

$$\eta(\lambda) \equiv [\underbrace{P_\Lambda H(\lambda) P_\Lambda + Q_\Lambda H(\lambda) Q_\Lambda}_{\equiv H_d(\lambda)}, H(\lambda)] \,. \tag{10.62}$$

where the projection operators P_Λ and Q_Λ partition the relative momentum basis in states with $|\hat{\mathbf{q}}| \leq \Lambda$ and $|\hat{\mathbf{q}}| > \Lambda$, respectively. In this case, λ is an auxiliary parameter that is eliminated by evolving $\lambda \to 0$, just like we evolved $s \to \infty$ in Sects. 10.2.2 and 10.2.3.

10.2.4.3 Implementation of the Flow Equations

We are now ready to implement the flow equations for the *NN* interaction in the momentum-space partial-wave representation. Using basis states that satisfy the orthogonality and completeness relations

$$\langle qLSJMTM_T | q'L'S'J'M'T'M_T' \rangle = \frac{\delta(q-q')}{qq'} \delta_{LL'} \delta_{SS'} \delta_{JJ'} \delta_{MM'} \delta_{TT'} \delta_{M_T M_T'} \tag{10.63}$$

and

$$\hat{I} = \sum_{LSJMTM_T} \int_0^\infty dq\, q^2 \, |qLSJMTM_T\rangle \, \langle qLSJMTM_T| \;, \tag{10.64}$$

respectively, we obtain [22, 23]

$$\begin{aligned}\left(-\frac{\lambda^5}{4}\right) \frac{d\langle qL|\, \hat{v}\, |q'L'\rangle}{d\lambda} =& -(q^2 - q'^2)^2 \langle qL|\, \hat{v}\, |q'L'\rangle \\ &+ \sum_{\bar{L}} \int_0^\infty dp p^2 \,(q^2 + q'^2 - 2p^2) \langle qL|\, \hat{v}\, |p\bar{L}\rangle \langle p\bar{L}|\, \hat{v}\, |q'L'\rangle \,,\end{aligned} \tag{10.65}$$

where we have used scattering units ($\hbar^2/m = 1$) and suppressed the λ-dependence of $\hat{v}$ as well as the conserved quantum numbers for brevity. Note that a prefactor $-\lambda^5/4$ appears due to our change of variables from s to λ.

We can turn this integro-differential equation back into a matrix flow equation by discretizing the relative momentum variable, e.g., on uniform or Gaussian quadrature meshes. The matrix elements of the relative kinetic energy operator are then simply given by

$$\langle q_i L | \hat{t} | q_j L' \rangle = q_i^2 \delta_{q_i q_j} \delta_{LL'} \tag{10.66}$$

(with $\hbar^2/m = 1$). The discretization turns the integration into a simple summation,

$$\int_0^\infty dq\, q^2 \rightarrow \sum_i w_i q_i^2 , \tag{10.67}$$

where the weights w_i depend on our choice of mesh. For a uniform mesh, all weights are identical and correspond to the mesh spacing, while for Gaussian quadrature rules the mesh points and weights have to be determined numerically [75]. For convenience, we absorb the weights and q^2 factors from the integral measure into the interaction matrix element,

$$\langle q_i L | \overline{v} | q_j L' \rangle \equiv \sqrt{w_i w_j} q_i q_j \langle q_i L | \hat{v} | q_j L' \rangle . \tag{10.68}$$

The discretized flow equation can then be written as

$$\frac{d}{d\lambda} \langle q_i L | \overline{v} | q_j L' \rangle = -\frac{4}{\lambda^5} \langle q_i L | [[\hat{t}, \overline{v}], \hat{t} + \overline{v}] | q_j L' \rangle . \tag{10.69}$$

We can solve Eq. (10.69) using a modified version of our Python code for the pairing model, discussed in Sect. 10.2.3. The Python code and sample inputs can be downloaded from https://github.com/ManyBodyPhysics/LectureNotesPhysics/tree/master/Programs/Chapter10-programs/python/srg_nn. Let us briefly discuss the most important modifications.

First, we have a set of functions that read the momentum mesh and the input matrix elements from a file:

```
def uniform_weights(momenta):
    weights = np.ones_like(momenta)
    weights *= abs(momenta[1]-momenta[0])
    return weights

def read_mesh(filename):
    data = np.loadtxt(filename, comments="#")
    dim = data.shape[1]

    momenta = data[0,:dim]

    return momenta

def read_interaction(filename):
    data = np.loadtxt(filename, comments="#")
    dim = data.shape[1]
    V = data[1:,:dim]
    return V
```

The matrix element files have the following format:

```
# momentum space matrix elements
# partial wave J=1, L=0/0, S=1, T=0, MT=0
#
# momentum grid [fm^-1]
0.000000 0.050000 0.100000 0.150000 0.200000 0.250000 0.300000
     0.350000 0.400000 0.450000
0.500000 0.550000 0.600000 0.650000 0.700000 0.750000 0.800000
     0.850000 0.900000 0.950000
...
6.300000 6.350000 6.400000 6.450000 6.500000 6.550000 6.600000
     6.650000 6.700000 6.750000
6.800000 6.850000 6.900000 6.950000 7.000000
#
# matrix elements [MeV fm^3]
-36.94918 -36.83554 -36.49896 -35.95649 -35.23306 -34.35875
     -33.36536 -32.28337 -31.13985
-29.95722 -28.75281 -27.53904 -26.32400 -25.11217 -23.90513
     -22.70231 -21.50158 -20.29973
...
0.00000 0.00000 0.00000 0.00000 0.00000 0.00000 0.00000 0.00000
     0.00000
0.00000 0.00000 0.00000 0.00000 0.00000 0.00000 0.00000 0.00000
     0.00000
```

Comments, indicated by the # character, are ignored. The first set of data is a row containing the mesh points. Here, we have 141 points in total, ranging from 0 to 7 fm^{-1} with a spacing of 0.05 fm^{-1}. The range of momenta is sufficient for the chiral *NN* interaction we use in our example, the N^3LO potential by Entem and Machleidt with cutoff $\Lambda = 500\,\text{MeV}$ [7, 98], which is considerably softer than the AV18 interaction discussed above. This is followed by a simple 141×141 array of matrix elements. It is straightforward to adapt the format and I/O routines to Gaussian quadrature meshes by including mesh points (i.e., the abscissas) and weights in the data file.

The derivative routine is almost unchanged, save for the prefactor due to the use of λ instead of s to parameterize the flow, and the treatment of the kinetic energy operator as explicitly constant:

```
def derivative(lam, y, T):
   dim = T.shape[0]

   # reshape the solution vector into a dim x dim matrix
   V = reshape(y, (dim, dim))

   # calculate the generator
   eta = commutator(T, V)

   # dV is the derivative in matrix form
   dV = -4.0/(lam**5) * commutator(eta, T+V)

   # convert dH into a linear array for the ODE solver
```

```
    dy = reshape(dV, -1)

    return dy
```

In the main routine of the program, we first set up the mesh and then proceed to read the interaction matrix elements for the different partial waves. We are dealing with a coupled-channel problem because the tensor forces connects partial waves with $\Delta L = 2$ in all $S = 1$ channels. In our example, we restrict ourselves to the partial waves that contribute to the deuteron bound state, namely ${}^3S_1$, ${}^3D_1$, and ${}^3S_1 - {}^3D_1$. Indicating the orbital angular momenta of these partial waves by indices, we have

$$\mathbf{T} = \begin{pmatrix} t & \\ & t \end{pmatrix}, \quad \mathbf{V} = \begin{pmatrix} \overline{v}_{00} & \overline{v}_{02} \\ \overline{v}^{\dagger}_{02} & \overline{v}_{22} \end{pmatrix}, \tag{10.70}$$

where

$$t = \text{diag}\left(q_0^2, \ldots, q_{\max}^2\right), \tag{10.71}$$

since the kinetic energy is independent of L. As soon as we pass from the S- into the D-wave in either the rows or the columns, the momentum mesh simply starts from the lowest mesh point again. We use NumPy's `hstack` and `vstack` functions to assemble the interaction matrix from the partial-wave blocks:

```
def main():
...

    # read individual partial waves
    partial_waves=[]
    for filename in ["n3lo500_3s1.meq", "n3lo500_3d1.meq", "
        n3lo500_3sd1.meq"]:
      partial_waves.append(read_interaction(filename))
      # print partial_waves[-1].shape

    # assemble coupled channel matrix
    V = np.vstack((np.hstack((partial_waves[0], partial_waves
        [2])),
                   np.hstack((np.transpose(partial_waves[2]),
                       partial_waves[1]))
                  ))

    # switch to scattering units
    V = V/hbarm

...
```

As discussed earlier, we work in scattering units with $\hbar^2/m = 1$. Thus, we have to divide the input matrix elements by this factor. We also need to absorb the weights and explicit momentum factors into the interaction matrix. It is convenient to define a conversion matrix for this purpose, which can be multiplied element-wise with the entries of **V** using the regular $\star$ operator (recall that the matrix product is implemented by the NumPy function dot).

Since we changed variables from s to λ, we now start the integration at $\lambda = \infty$, or $\lambda \gg 1\,\mathrm{fm}^{-1}$ in practice. As discussed above, we do not evolve all the way to $\lambda = 0\,\mathrm{fm}^{-1}$, but typically stop before we start integrating out explicit pion physics, e.g., at $\lambda = 1.5\,\mathrm{fm}^{-1}$. For typical *NN* interactions, especially those with a hard core like AV18, the flow equations tend to become stiff because they essentially depend on cubic products of the kinetic energy and interaction. For this reason, we use SciPy's ode class, which provides access to a variety of solvers and greater control over the parameters of the integration process. Specifically, we choose the VODE solver package and its fifth-order Backward Differentiation method [99], which is efficient and works robustly for a large variety of input interactions.

```
...

    lam_initial = 20.0
    lam_final = 1.5

    # integrate using scipy.ode instead of scipy.odeint - this

    # gives us more control over the solver
    solver = ode(derivative,jac=None)

    # equations may get stiff, so we use VODE and Backward
        Differentiation
    solver.set_integrator('vode', method='bdf', order=5, nsteps
        =1000)
    solver.set_f_params(T)
    solver.set_initial_value(y0, lam_initial)

...
```

Finally, we reach the loop that integrates the ODE system. We request output from the solver in regular intervals, reducing these intervals as we approach the region of greatest practical interest, $1.5\,\mathrm{fm}^{-1} \leq \lambda \leq 2.5\,\mathrm{fm}^{-1}$:

```
...

    while solver.successful() and solver.t > lam_final:
       # adjust the step size in different regions of the flow
           parameter
       if solver.t >= 6.0:
         ys = solver.integrate(solver.t-1.0)
       elif solver.t < 6.0 and solver.t >= 2.5:
         ys = solver.integrate(solver.t-0.5)
       elif solver.t < 2.5 and solver.t >= lam_final:
         ys = solver.integrate(solver.t-0.1)

       # add evolved interactions to the list
       flowparams.append(solver.t)
       Vtmp = reshape(ys,(dim,dim))
```

```
        Vs.append(Vtmp)

        print("%8.5f %14.8f"%(solver.t, eigvalsh((T + Vtmp)*hbarm
            )[0]))

...
```

Of course, the ODE solver will typically take several hundred adaptive steps to propagate the solution with the desired accuracy between successive requested values of λ. At the end of each external step, we diagonalize the evolved Hamiltonian and check whether the lowest eigenvalue, i.e., the deuteron binding energy, remains invariant within the numerical tolerances we use for the ODE solver. To illustrate the evolution of the *NN* interaction, the code will also generate a sequence of matrix plots at the desired values of λ, similar to Fig. 10.2 for the pairing Hamiltonian.

10.2.4.4 Example: Evolution of a Chiral *NN* Interaction

As an example of a realistic application, we discuss the SRG evolution of the chiral N^3LO nucleon-nucleon interaction by Entem and Machleidt with initial cutoff $\Lambda = 500\,\text{MeV}$ [7, 98]. The momentum-space matrix elements of this interaction in the deuteron partial waves are distributed with the Python code discussed in the previous section.

In the top row of Fig. 10.6, we show the matrix elements of the initial interaction in the $^3S_1$ partial wave; the $^3S_1 - {}^3D_1$ and $^3D_1$ are not shown to avoid clutter. Comparing the matrix elements to those of the AV18 interaction we discussed in Sect. 10.2.4.2, shown in Fig. 10.4, we note that the chiral interaction has much weaker off-diagonal matrix elements to begin with. While the AV18 matrix elements extend as high as $|\hat{\mathbf{q}}| \sim 20\,\text{fm}^{-1}$, the chiral interaction has no appreciable strength in states with $|\hat{\mathbf{q}}| \sim 4.5\,\text{fm}^{-1}$. In nuclear physics jargon, AV18 is a much harder interaction than the chiral interaction because of the former's strongly repulsive core. By evolving the initial interaction to $3\,\text{fm}^{-1}$ and then to $2\,\text{fm}^{-1}$, the offdiagonal matrix elements are suppressed, and the interaction is almost entirely contained in a block of states with $|\hat{\mathbf{q}}| \sim 2\,\text{fm}^{-1}$, except for a weak diagonal ridge.

Next to the matrix elements, we also show the deuteron wave functions that we obtain by solving the Schrödinger equation with the initial and SRG-evolved chiral interactions. For the unevolved *NN* interaction, the *S*-wave ($L = 0$) component of the wave function is suppressed at small relative distances, which reflects short-range correlations between the nucleons. (For AV18, the *S*-wave component of the deuteron wave function vanishes at $r = 0\,\text{fm}$ due to the hard core.) There is also a significant *D*-wave ($L = 2$) admixture due to the tensor interaction. As we lower the resolution scale, the "correlation hole" in the wave function is filled in, and all but eliminated once we reach $\lambda = 2.0\,\text{fm}^{-1}$. The *D*-wave admixture is reduced significantly, as well, because the evolution suppresses the matrix elements in the $^3S_1 - {}^3D_1$ wave, which are responsible for this mixing [23]. Focusing just on the *S*-wave, the wave function is extremely simple and matches what we would

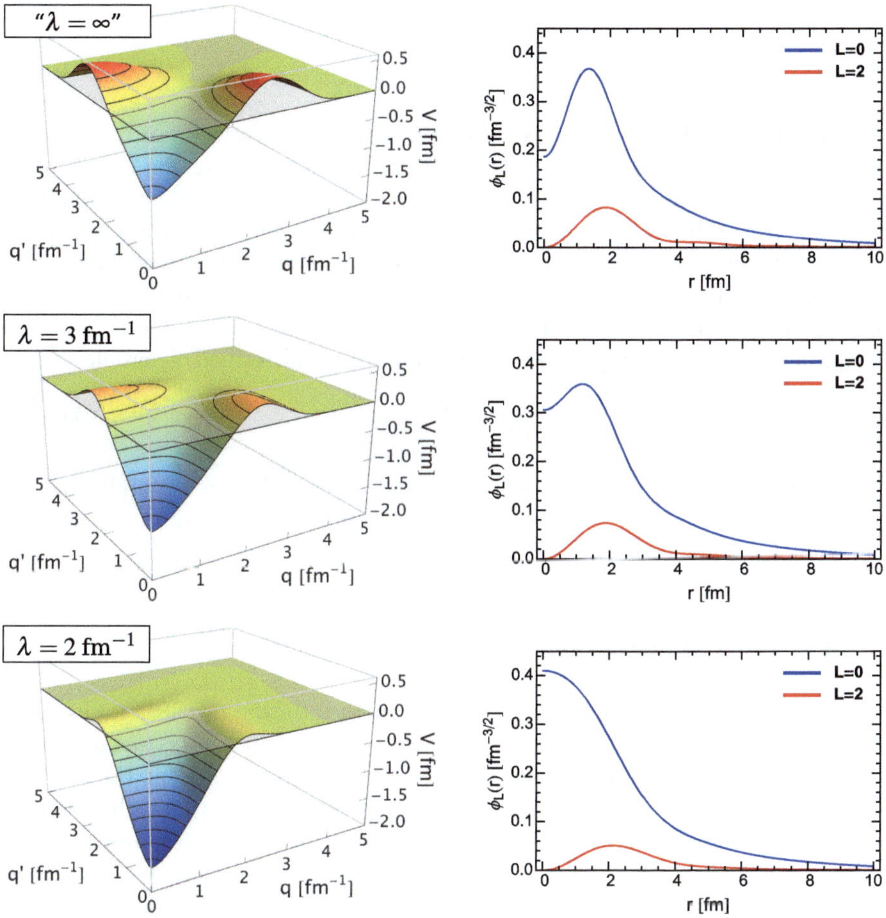

Fig. 10.6 SRG evolution of the chiral N^3LO nucleon-nucleon interaction by Entem and Machleidt, with initial cutoff $\Lambda = 500\,\text{MeV}$ [7, 98]. In the *left column*, we show the momentum-space matrix elements of the interaction in the $^3S_1$ partial wave for different values of the SRG resolution scale λ. The *top-most row* shows the initial interaction at $s = 0\,\text{fm}^4$, i.e., "$\lambda = \infty$". In the *right column*, we show the *S*- and *D*-wave components of the deuteron wave function that is obtained by solving the Schrödinger equation with the corresponding SRG-evolved interaction

expect for two almost independent, uncorrelated nucleons. The Pauli principle does not affect the coordinate-space part of the wave function here because the overall antisymmetry of the deuteron wave function is ensured by its spin and isospin parts.

Let us dwell on the removal of short-range correlations from the wave function for another moment, and consider the exact eigenstates of the initial *NN* Hamiltonian,

$$\hat{H}(0)\,|\psi_n\rangle = E_n\,|\psi_n\rangle\,. \tag{10.72}$$

The eigenvalues are invariant under a unitary transformation, e.g., an SRG evolution,

$$\hat{H}(\lambda)\hat{U}(\lambda)|\psi_n\rangle \equiv \hat{U}(\lambda)\hat{H}(0)\hat{U}^\dagger(\lambda)\hat{U}(\lambda)|\psi_n\rangle = E_n\hat{U}(\lambda)|\psi_n\rangle\,. \tag{10.73}$$

We can interpret this equation as a shift of correlations from the wave function into the effective, *RG-improved* Hamiltonian. When we solve the Schrödinger equation numerically, we can usually only obtain an approximation $|\phi_n\rangle$ of the exact eigenstate. In the ideal case, this is merely due to finite-precision arithmetic on a computer, but more often, we also have systematic approximations, e.g., mesh discretizations, finite basis sizes, many-body truncations (think of the cluster operator in Coupled Cluster, for instance, cf. Chap. 8), etc. If we use the evolved Hamiltonian $\hat{H}(\lambda)$, we only need to approximate the transformed eigenstate,

$$|\phi_n\rangle \approx \hat{U}(\lambda)|\psi_n\rangle \tag{10.74}$$

instead of $|\psi_n\rangle$, which is often a less demanding task. This is certainly true for our deuteron example at $\lambda = 2.0\,\mathrm{fm}^{-1}$, where we no longer have to worry about short-range correlations.

10.2.4.5 Induced Interactions

As discussed earlier in this section, our motivation for using the SRG to decouple the low- and high-lying momentum components of *NN* interactions is to improve the convergence of many-body calculations. The decoupling prevents the Hamiltonian from scattering nucleon pairs from low to high momenta or energies, which in turn allows configuration-space based methods to achieve convergence in much smaller Hilbert spaces than for a "bare", unevolved interaction. This makes it possible to extend the reach of these methods to heavier nuclei [44–48, 52, 53, 100–108].

In practical applications, we pay a price for the improved convergence. To illustrate the issue, we consider the Hamiltonian in a second-quantized form, assuming only a two-nucleon interaction for simplicity [cf. Eq. (10.56)]:

$$\hat{H}_{\text{int}} = \frac{1}{4}\sum_{pqrs}\langle pq|\,\frac{\hat{\mathbf{q}}^2}{2\mu} + \hat{v}\,|rs\rangle a^\dagger_p a^\dagger_q a_s a_r\,. \tag{10.75}$$

If we plug the kinetic energy and interaction into the commutators in Eqs. (10.60) and (10.8), we obtain

$$[a^\dagger_i a^\dagger_j a_l a_k, a^\dagger_p a^\dagger_q a_s a_r] = \delta_{lp} a^\dagger_i a^\dagger_j a^\dagger_q a_s a_r a_k + a^\dagger a^\dagger a^\dagger aaa - \delta_{lp}\delta_{kq} a^\dagger_i a^\dagger_j a_s a_r + a^\dagger a^\dagger aa\,, \tag{10.76}$$

where the terms with suppressed indices schematically stand for additional two- and three-body operators. Even if we start from a pure *NN* interaction, the SRG flow will induce operators of higher rank, i.e., 3*N*, 4*N*, and in general up to *A*-nucleon

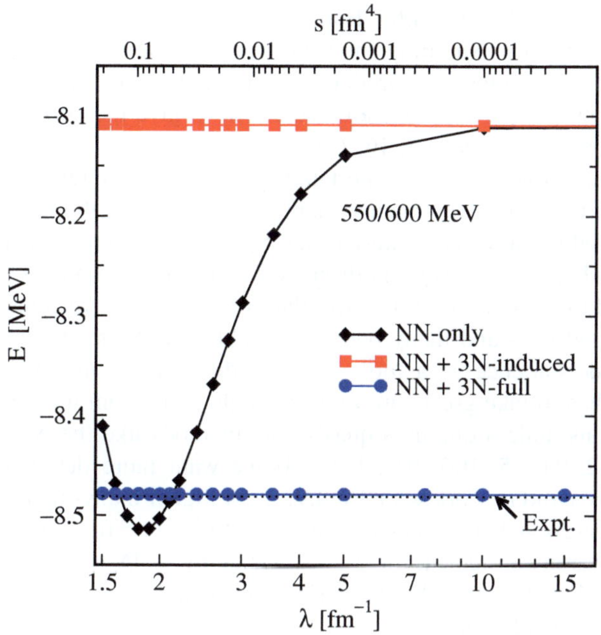

Fig. 10.7 Ground state energy of ^{3}H as a function of the flow parameter λ for chiral NNLO *NN* and *NN*+3*N* interactions (see [95] for details). *NN*-only means initial and induced 3*N* interactions are discarded, *NN* + 3*N*-induced takes only induced 3*N* interactions into account, and 3*N*-full contains initial 3*N* interactions as well. The *black dotted line* shows the experimental binding energy [111]. Data for the figure courtesy of K. Hebeler

interactions. Of course, these induced interactions are only probed if we study an *A*-nucleon system. If we truncate the SRG flow equations at the two-body level, the properties of the two-nucleon system are preserved, in particular the *NN* scattering phase shifts and the deuteron binding energy. A truncation at the three-body level ensures the invariance of observables in $A = 3$ nuclei, e.g. ^{3}H and ^{3}He ground-state energies, and so on.

Nowadays, state-of-the-art SRG evolutions of nuclear interactions are performed in the three-body system [45, 94, 95, 109, 110]. In Fig. 10.7, we show ^{3}H ground-state energies that have been calculated with a family of SRG-evolved interactions that is generated from a chiral NNLO *NN* interaction by Epelbaum, Glöckle, and Meißner [112, 113], and a matching 3*N* interaction (see [95] for full details). As mentioned above, the SRG evolution is not unitary in the three-body system if we truncate the evolved interaction and the SRG generator at the two-body level (*NN-only*). The depends strongly on λ, varying by 5–6% over the typical range that we consider here (cf. Sect. 10.2.4.2). If we truncate the operators at the three-body level instead, induced 3*N* interactions are properly included and the unitarity of the transformation is restored (*NN* + 3*N-induced*): The energy does not change as λ is varied. Finally, the curve *NN* + 3*N-full* shows results of calculations in which a

$3N$ force was included in the initial Hamiltonian and evolved consistently to lower resolution scale as well. Naturally, the triton ground-state energy is invariant under the SRG flow, and it closely reproduces the experimental value because the $3N$ interaction's low-energy constants are usually fit to give the correct experimental ^{3}H ground-state energy (see, e.g., [6, 7, 114]).

Our example shows that it is important to track induced interactions, especially when we want to use evolved nuclear Hamiltonians beyond the few-body systems we have focused on here. The nature of the SRG as a continuous evolution works at least somewhat in our favor: As discussed above, truncations of the SRG flow equations lead to a violation of unitarity that manifests as a (residual) dependence of our calculated few- and many-body observables on the resolution scale λ. We can use this dependence as a tool to assess the size of missing contributions, although one has to take great care to disentangle them from the effects of many-body truncations, unless one uses quasi-exact methods like the NCSM (see, e.g., [23, 46, 48, 52, 94, 95, 100, 108, 115]). If we want more detailed information, then we cannot avoid to work with $3N, 4N, \ldots$ or higher many-nucleon forces. The empirical observation that SRG evolutions down to $\lambda \sim 1.5\,\text{fm}^{-1}$ appear to preserve the natural hierarchy of nuclear forces, i.e., $NN > 3N > 4N > \ldots$, suggests that we can truncate induced forces whose contributions would be smaller than the desired accuracy of our calculations.

While we may not have to go all the way to the treatment of induced A-nucleon operators, which would be as expensive as implementing the matrix flow in the A-body system (cf. Sect. 10.2.4.2), dealing with induced $3N$ operators is already computationally expensive enough. Treating induced $4N$ forces explicitly is out of the question, except in schematic cases. However, there is a way of accounting for effects of induced $3N, \ldots$ forces in an implicit manner, by performing SRG evolutions in the nuclear medium.

10.3 The In-Medium SRG

As discussed in the previous section, we now want to carry out the operator evolution (10.8) in the nuclear medium. The idea is to decompose a given N-body operator into in-medium contributions of lower rank and residual components that can be truncated safely. To this end, we first have to lay the groundwork by reviewing the essential elements of normal ordering, as well as Wick's theorem.

10.3.1 Normal Ordering and Wick's Theorem

10.3.1.1 Normal-Ordered Operators

To construct normal-ordered operators, we start from the usual Fermionic creation and annihilation operators, $a^{\dagger}_{i}$ and a_i, which satisfy the canonical anticommutation relations

$$\{a^{\dagger}_{i}, a^{\dagger}_{j}\} = \{a_i, a_j\} = 0\,, \quad \{a^{\dagger}_{i}, a_j\} = \delta_{ij}\,. \tag{10.77}$$

The indices are collective labels for the quantum numbers of our single-particle states. Using the creators and annihilators, we can express any given A-body operator in second quantization. Moreover, we can construct a complete basis for a many-body Hilbert space by acting with products of $a^{\dagger}_{i}$ on the particle vacuum,

$$|\Phi\{i_1 \dots i_A\}\rangle = \prod_{k=1}^{A} a^{\dagger}_{i_k} |\text{vac}\rangle\,, \tag{10.78}$$

and letting the indices $i_1, \dots, i_A$ run over all single-particle states. The states $|\Phi\{i_1 \dots i_A\}\rangle$ are, of course, nothing but antisymmetrized product states, i.e., Slater determinants.

Of course, not all of the Slater determinants in our basis are created equal. We can usually find a Slater determinant that is a fair approximation to the nuclear ground state, and use it as a *reference state* for the construction and organization of our many-body basis. By simple energetics, the ground state and low-lying excitation spectrum of an A-body nucleus are usually dominated by excitations of particles in the vicinity of the reference state's Fermi energy. This is especially true for *NN* interactions that have been evolved to a low resolution scale λ (see Sect. 10.2.4.2). For such forces, the coupling between basis states whose energy expectation values differ by much more than the characteristic energy $\hbar^2\lambda^2/m$ is suppressed.

Slater determinants that are variationally optimized through a Hartree-Fock (HF) calculation have proven to be reasonable reference states for interactions with $\lambda \approx 2.0\,\text{fm}^{-1}$ (see, e.g., [23, 43, 44, 49, 53, 116] and references therein), allowing post-HF methods like MBPT, CC, or the IMSRG discussed below to converge rapidly to the exact result. Starting from such a HF reference state $|\Phi\rangle$, we can obtain a basis consisting of the state itself and up to A-particle, A-hole ($ApAh$) excitations:

$$|\Phi\rangle,\ a^{\dagger}_{p_1} a_{h_1} |\Phi\rangle,\ \dots\ ,\ a^{\dagger}_{p_1} \dots a^{\dagger}_{p_A} a_{h_A} \dots a_{h_1} |\Phi\rangle\,. \tag{10.79}$$

Here, indices p_i and h_i run over all one-body basis states with energies above (*particle* states) and below the Fermi level (*hole* states), respectively. Such bases work best for systems with large gaps in the single-particle spectrum, e.g., closed-shell nuclei. If the gap is small, excited basis states can be nearly degenerate with the reference state, which usually results in spontaneous symmetry breaking and strong configuration mixing.

We can now introduce a one-body operator that is normal-ordered with respect to the reference state $|\Phi\rangle$ by defining

$$a_i^\dagger a_j \equiv \{a_i^\dagger a_j\} + \overbracket{a_i^\dagger a_j}\,, \tag{10.80}$$

where the brackets $\{\ldots\}$ indicate normal ordering, and the brace over a pair of creation and annihilation operators means that they have been *contracted*. The contraction itself is merely the expectation value of the operator in the reference state $|\Phi\rangle$:

$$\overbracket{a_i^\dagger a_j} \equiv \langle\Phi|\, a_i^\dagger a_j\, |\Phi\rangle \equiv \rho_{ji}\,. \tag{10.81}$$

By definition, the contractions are identical to the elements of the one-body density matrix of $|\Phi\rangle$ [117]. Starting from the one-body case, we can define normal-ordered A-body operators recursively by evaluating all contractions between creation and annihilation operators, e.g.,

$$\begin{aligned}
&a_{i_1}^\dagger \ldots a_{i_A}^\dagger a_{j_A} \ldots a_{j_1}\\
&\equiv \{a_{i_1}^\dagger \ldots a_{i_A}^\dagger a_{j_A} \ldots a_{j_1}\}\\
&\quad + \overbracket{a_{i_1}^\dagger a_{j_1}}\{a_{i_2}^\dagger \ldots a_{i_A}^\dagger a_{j_A} \ldots a_{j_2}\} - \overbracket{a_{i_1}^\dagger a_{j_2}}\{a_{i_2}^\dagger \ldots a_{i_A}^\dagger a_{j_A} \ldots a_{j_3} a_{j_1}\} + \text{singles}\\
&\quad + \left(\overbracket{a_{i_1}^\dagger a_{j_1}}\overbracket{a_{i_2}^\dagger a_{j_2}} - \overbracket{a_{i_1}^\dagger a_{j_2}}\overbracket{a_{i_2}^\dagger a_{j_1}}\right)\{a_{i_3}^\dagger \ldots a_{i_A}^\dagger a_{j_A} \ldots a_{j_3}\} + \text{doubles}\\
&\quad + \ldots + \text{full contractions}\,.
\end{aligned} \tag{10.82}$$

Here, we have followed established quantum chemistry jargon (singles, doubles, etc.) for the number of contractions in a term (cf. Chap. 8). Note that the double contraction shown in the next-to-last line is identical to the factorization formula for the two-body density matrix of a Slater determinant,

$$\rho_{j_1 j_2 i_1 i_2} \equiv \langle\Phi|\, a_{i_1}^\dagger a_{i_2}^\dagger a_{j_2} a_{j_1}\, |\Phi\rangle = \rho_{i_1 j_1}\rho_{i_2 j_2} - \rho_{i_1 j_2}\rho_{i_2 j_1}\,. \tag{10.83}$$

From Eq. (10.80), it is evident that $\langle\Phi|\, \{a_i^\dagger a_j\}\, |\Phi\rangle$ must vanish, and this is readily generalized to expectation values of arbitrary normal-ordered operators in the reference state $|\Phi\rangle$,

$$\langle\Phi|\, \{a_{i_1}^\dagger \ldots a_{i_1}\}\, |\Phi\rangle = 0\,. \tag{10.84}$$

This property of normal-ordered operators greatly facilitates calculations that require the evaluation of matrix elements in a space spanned by excitations of

$|\Phi\rangle$. Another important property is that we can freely anticommute creation and annihilation operators within a normal-ordered string (see Problem 10.2):

$$\{\dots a_i^\dagger a_j \dots\} = -\{\dots a_j a_i^\dagger \dots\}. \tag{10.85}$$

As an example, we consider an intrinsic nuclear A-body Hamiltonian containing both NN and $3N$ interactions,

$$H = \left(1 - \frac{1}{\hat{A}}\right)\hat{T}^{[1]} + \frac{1}{\hat{A}}\hat{T}^{[2]} + \hat{V}^{[2]} + \hat{V}^{[3]}, \tag{10.86}$$

where the one- and two-body kinetic energy terms are

$$\hat{T}^{[1]} \equiv \sum_i \frac{\hat{\mathbf{p}}_i^2}{2m}, \tag{10.87}$$

$$\hat{T}^{[2]} \equiv -\frac{1}{m}\sum_{i<j} \hat{\mathbf{p}}_i \cdot \hat{\mathbf{p}}_j \tag{10.88}$$

(see Sect. 10.2.4.2 and [86]). Choosing a single Slater determinant $|\Phi\rangle$ as the reference state, we can rewrite the Hamiltonian *exactly* in terms of normal-ordered operators,

$$\hat{H} = E + \sum_{ij} f_{ij}\{a_i^\dagger a_j\} + \frac{1}{4}\sum_{ijkl}\Gamma_{ijkl}\{a_i^\dagger a_j^\dagger a_l a_k\} + \frac{1}{36}\sum_{ijklmn} W_{ijklmn}\{a_i^\dagger a_j^\dagger a_k^\dagger a_n a_m a_l\}, \tag{10.89}$$

where the labels for the individual contributions have been chosen for historical reasons. For convenience, we will work in the eigenbasis of the one-body density matrix in the following, so that

$$\rho_{ab} = n_a \delta_{ab}, \quad n_a \in \{0, 1\}. \tag{10.90}$$

The individual normal-ordered contributions in Eq. (10.89) are then given by

$$\begin{aligned} E = &\left(1 - \frac{1}{A}\right)\sum_a \langle a|\,\hat{t}^{[1]}\,|a\rangle n_a + \frac{1}{2}\sum_{ab}\langle ab|\,\frac{1}{A}\hat{t}^{[2]} + \hat{v}^{[2]}\,|ab\rangle n_a n_b \\ &+ \frac{1}{6}\sum_{abc}\langle abc|\,\hat{v}^{[3]}\,|abc\rangle n_a n_b n_c, \end{aligned} \tag{10.91}$$

$$f_{ij} = \left(1 - \frac{1}{A}\right)\langle i|\,\hat{t}^{[1]}\,|j\rangle + \sum_a \langle ia|\,\frac{1}{A}\hat{t}^{[2]} + \hat{v}^{[2]}\,|ja\rangle n_a + \frac{1}{2}\sum_{ab}\langle iab|\,\hat{v}^{[3]}\,|jab\rangle n_a n_b, \tag{10.92}$$

$$\Gamma_{ijkl} = \langle ij| \frac{1}{A}\hat{t}^{(2)} + \hat{v}^{[2]} |kl\rangle + \sum_a \langle ija| \hat{v}^{[3]} |kla\rangle n_a \,, \tag{10.93}$$

$$W_{ijklmn} = \langle ijk| \hat{v}^{[3]} |lmn\rangle \,. \tag{10.94}$$

Due to the occupation number factors in Eqs. (10.91)–(10.93), the sums run only over states that are occupied in the reference state. This means that the zero-, one-, and two-body parts of the Hamiltonian all contain in-medium contributions from the free-space 3N interaction.

For low-momentum interactions, it has been shown empirically that the omission of the normal-ordered three-body piece of the Hamiltonian causes a deviation of merely 1–2% in ground-state and (absolute) excited state energies of light and medium-mass nuclei [100, 103, 118–120]. This *normal-ordered two-body approximation* (NO2B) to the Hamiltonian is useful for practical calculations, because it provides an efficient means to account for $3N$ force effects in nuclear many-body calculations without incurring the computational expense of explicitly treating three-body operators. In Sect. 10.3.2, we will see that the NO2B approximation also meshes in a natural way with the framework of the IMSRG, which makes it especially appealing for our purposes.

10.3.1.2 Wick's Theorem

The normal-ordering formalism has additional benefits for the evaluation of products of normal-ordered operators. Wick's theorem (see, e.g., [60]), which is a direct consequence of Eq. (10.82), allows us to expand such products in the following way:

$$\begin{aligned}
&\{a^\dagger_{i_1} \ldots a^\dagger_{i_N} a_{j_N} \ldots a_{j_1}\}\{a^\dagger_{k_1} \ldots a^\dagger_{k_M} a_{l_M} \ldots a_{l_1}\} \\
&\quad= (-1)^{M\cdot N}\{a^\dagger_{i_1} \ldots a^\dagger_{i_N} a^\dagger_{k_1} \ldots a^\dagger_{k_M} a_{j_N} \ldots a_{j_1} a_{l_M} \ldots a_{l_1}\} \\
&\quad+ (-1)^{M\cdot N}\, \contraction{}{a^\dagger_{i_1}}{}{a_{l_1}} a^\dagger_{i_1} a_{l_1}\, \{a^\dagger_{i_2} \ldots a^\dagger_{k_M} a_{j_N} \ldots a_{l_2}\} \\
&\quad+ (-1)^{(M-1)(N-1)}\, \contraction{}{a_{j_N}}{}{a^\dagger_{k_1}} a_{j_N} a^\dagger_{k_1}\, \{a^\dagger_{i_1} \ldots a^\dagger_{k_M} a_{j_N} \ldots a_{j_2}\} \\
&\quad+ \text{singles} + \text{doubles} + \ldots \,.
\end{aligned} \tag{10.95}$$

The phase factors appear because we anti-commute the creators and annihilators until they are grouped in the canonical order, i.e., all $a^\dagger$ appear to the left of the a. In the process, we also encounter a new type of contraction,

$$\contraction{}{a_i}{}{a^\dagger_j} a_i a^\dagger_j \equiv \langle \Phi| a_i a^\dagger_j |\Phi\rangle = \delta_{ij} - \rho_{ij} \equiv \overline{\rho}_{ij} \,, \tag{10.96}$$

as expected from the canonical anti-commutator algebra. $\overline{\rho}$ is the so-called *hole density matrix*.

The defining feature of Eq. (10.95) is that only contractions between one index from each of the two strings of creation and annihilation operators appear in the expansion, because contractions between indices within a single operator string have already been subtracted when we normal ordered it initially. In practical calculations, this leads to a substantial reduction of terms. An immediate consequence of Eq. (10.95) is that a product of normal-ordered M and N-body operators has the general form

$$\hat{A}^{[M]}\hat{B}^{[N]} = \sum_{k=|M-N|}^{M+N} \hat{C}^{[k]} . \tag{10.97}$$

Note that zero-body contributions, i.e., plain numbers, can only be generated if both operators have the same particle rank.

10.3.2 In-Medium SRG Flow Equations

10.3.2.1 Induced Forces Revisited

In Sect. 10.2.4.5, we discussed how SRG evolutions naturally induce $3N$ and higher many-nucleon forces, because every evaluation of the commutator on the right-hand side of the operator flow equation (10.8) increases the particle rank of $\hat{H}(s)$, e.g.,

$$\begin{aligned}\sum_{ijklpqrs} &\eta_{ijkl} H_{pqrs} [a^{\dagger}_i a^{\dagger}_j a_l a_k, a^{\dagger}_p a^{\dagger}_q a_s a_r] \\ &= -\sum_{ijkqrs} \eta_{ijkl} H_{kqrs} a^{\dagger}_i a^{\dagger}_j a^{\dagger}_q a_s a_r a_l + \text{3N terms} + \text{2N terms} .\end{aligned} \tag{10.98}$$

Note that there are no induced $4N$ interactions, and that commutators involving at least one one-body operator do not change the particle rank (see Problem 10.2). In the free-space evolution, we found that the truncation of $3N$ forces in the flowing Hamiltonian caused a significant flow-parameter dependence of observables in $A \geq 3$ systems.

Working in the medium and using normal-ordered operators, we can expand the induced $3N$ operators:

$$\begin{aligned}\sum_{ijkqrs} &\eta_{ijkl} H_{kqrs} a^{\dagger}_i a^{\dagger}_j a^{\dagger}_q a_s a_r a_l \\ &= \sum_{ijkqrs} \eta_{ijkl} H_{kqrs} \Big(\{a^{\dagger}_i a^{\dagger}_j a^{\dagger}_q a_s a_r a_l\} + n_q \delta_{qs} \{a^{\dagger}_i a^{\dagger}_j a_r a_l\} + n_j n_q \delta_{jr} \delta_{qs} \{a^{\dagger}_i a_l\} \\ &\quad + n_i n_j n_q \delta_{il} \delta_{jr} \delta_{qs} + \text{permutations} \Big) .\end{aligned} \tag{10.99}$$

If we now truncate operators to the normal-ordered two-body level, we keep all the in-medium contributions of the induced $3N$ terms, and retain information that we would have lost in the free-space evolution. These in-medium contributions continuously feed into the 0B, 1B, and 2B matrix elements of the flowing Hamiltonian as we integrate Eq. (10.8).

10.3.2.2 The IMSRG(2) Scheme

The evolution of the Hamiltonian or any other observable by means of the flow equation (10.8) is a continuous unitary transformation in A-nucleon space only if we keep up to induced A-nucleon forces. Because an explicit treatment of induced contributions up to the A-body level is simply not feasible, we have to introduce a truncation to close the system of flow equations.

As explained in the previous subsection, we can make such truncations more robust if we normal order all operators with respect to a reference state that is a fair approximation to the ground state of our system (or another exact eigenstate we might want to target). Here, we choose to truncate operators at the two-body level, to avoid the computational expense of treating explicit three-body operators. For low-momentum $NN + 3N$ Hamiltonians, the empirical success of the NO2B approximation mentioned at the end of Sect. 10.3.1.1 seems to support this truncation: The omission of the normal-ordered $3N$ term in exact calculations causes deviations of only $\sim 1\%$ in the oxygen, calcium, and nickel isotopes [48, 103, 119].

Following this line of reasoning, we demand that for all values of the flow parameter s

$$\hat{\eta}(s) \approx \hat{\eta}^{(1)}(s) + \hat{\eta}^{(2)}(s)\,, \tag{10.100}$$

$$\hat{H}(s) \approx E(s) + f(s) + \Gamma(s)\,, \tag{10.101}$$

$$\frac{d}{ds}\hat{H}(s) \approx \frac{d}{ds}E(s) + \frac{d}{ds}f(s) + \frac{d}{ds}\Gamma(s)\,. \tag{10.102}$$

This is the so-called IMSRG(2) truncation, which has been our primary workhorse in past applications [46, 51–53, 72, 101, 121]. It is the basis for all results that we will discuss in the remainder of this chapter. The IMSRG(2) is a cousin to Coupled Cluster with Singles and Doubles (CCSD) and the ADC(3) scheme in Self-Consistent Green's Function Theory (see Chaps. 8 and 11). Since all three methods (roughly) aim to describe the same type and level of many-body correlations, we expect to obtain similar results for observables.

Let us introduce the permutation symbol P_{ij} to interchange the indices of any expression, i.e.,

$$P_{ij}g(\ldots,i,\ldots,j) \equiv g(\ldots,j,\ldots,i)\,, \tag{10.103}$$

Plugging Eqs. (10.100)–(10.102) into the operator flow equation (10.8) and evaluating the commutators with the expressions from Appendix, we obtain the following system of IMSRG(2) flow equations:

$$\frac{dE}{ds} = \sum_{ab}(n_a - n_b)\eta_{ab}f_{ba} + \frac{1}{2}\sum_{abcd}\eta_{abcd}\Gamma_{cdab}n_a n_b \bar{n}_c \bar{n}_d\,, \tag{10.104}$$

$$\begin{aligned}\frac{df_{ij}}{ds} &= \sum_{a}(1+P_{ij})\eta_{ia}f_{aj} + \sum_{ab}(n_a - n_b)(\eta_{ab}\Gamma_{biaj} - f_{ab}\eta_{biaj})\\ &\quad + \frac{1}{2}\sum_{abc}(n_a n_b \bar{n}_c + \bar{n}_a \bar{n}_b n_c)(1+P_{ij})\eta_{ciab}\Gamma_{abcj}\,,\end{aligned} \tag{10.105}$$

$$\begin{aligned}\frac{d\Gamma_{ijkl}}{ds} &= \sum_{a}\left\{(1-P_{ij})(\eta_{ia}\Gamma_{ajkl} - f_{ia}\eta_{ajkl}) - (1-P_{kl})(\eta_{ak}\Gamma_{ijal} - f_{ak}\eta_{ijal})\right\}\\ &\quad + \frac{1}{2}\sum_{ab}(1 - n_a - n_b)(\eta_{ijab}\Gamma_{abkl} - \Gamma_{ijab}\eta_{abkl})\\ &\quad + \sum_{ab}(n_a - n_b)(1-P_{ij})(1-P_{kl})\eta_{aibk}\Gamma_{bjal}\,.\end{aligned} \tag{10.106}$$

Here, $\bar{n}_i = 1 - n_i$, and the s-dependence has been suppressed for brevity. To obtain ground-state energies, we integrate Eqs. (10.104)–(10.106) from $s = 0$ to $s \to \infty$, starting from the initial components of the normal-ordered Hamiltonian [Eqs. (10.91)–(10.93)] (see Sects. 10.3.6 and 10.3.7 for numerical examples).

By integrating the flow equations, we absorb many-body correlations into the flowing normal-ordered Hamiltonian, summing certain classes of terms in the many-body expansion to all orders [53]. We can identify specific structures by looking at the occupation-number dependence of the terms in Eqs. (10.104)–(10.106): for instance, $\bar{n}_i$ and n_i restrict summations to particle and hole states, respectively [cf. Eq. (10.90)]. Typical IMSRG generators (see Sect. 10.3.4) are proportional to the (offdiagonal) Hamiltonian, which means that the two terms in the zero-body flow equation essentially have the structure of second-order energy corrections, but evaluated for the *flowing* Hamiltonian $\hat{H}(s)$. Thus, we can express the equation in terms of Hugenholtz diagrams as

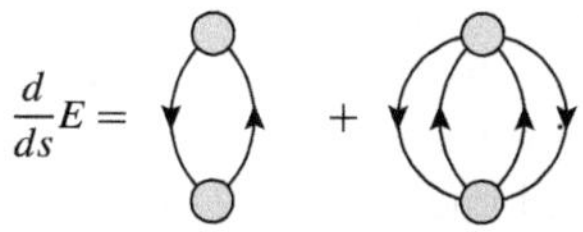

(10.107)

Note that the energy denominators associated with the propagation of the intermediate state are consistently calculated with $\hat{H}(s)$ here.

In the flow equation for the two-body vertex Γ, terms that are proportional to

$$1 - n_a - n_b = \bar{n}_a \bar{n}_b - n_a n_b \tag{10.108}$$

will build up a summation of particle-particle and hole-hole *ladder diagrams* as we integrate the flow equations $s \to \infty$. Similarly, the terms proportional to

$$n_a - n_b = n_a \bar{n}_b - \bar{n}_a n_b \,, \tag{10.109}$$

will give rise to a summation of *chain diagrams* representing particle-hole terms at all orders. We can illustrate this by expanding the vertex we obtain after two integration steps, $\Gamma(2\delta s)$, in terms of the prior vertices $\Gamma(\delta s)$ and $\Gamma(0)$. Indicating these vertices by light gray, dark gray, and black circles, we schematically have

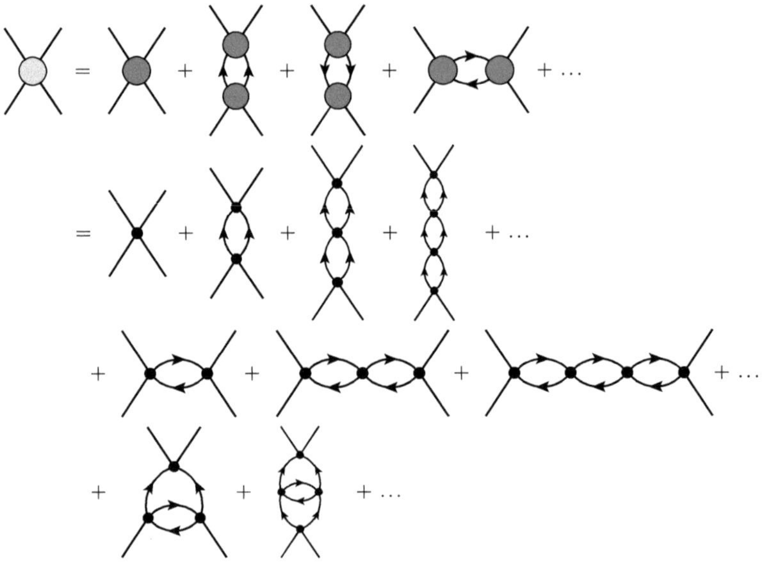

(10.110)

In the first line, we see that $\Gamma(2\delta s)$ is given by the vertex of the previous step, $\Gamma(\delta s)$, plus second-order corrections. As in the energy flow equation, it is assumed that the energy denominators associated with the propagation of the intermediate states are calculated with $\hat{H}(\delta s)$. For brevity, we have suppressed additional permutations of the shown diagrams, as well as the diagrams that result from contracting one- and two-body operators in Eq. (10.106).

In the next step, we expand each of the $\Gamma(\delta s)$ vertices in terms of $\Gamma(0)$, and assume that energy denominators are now expressed in terms of $\hat{H}(0)$. In the second line of Eq. (10.110), we explicitly show the ladder-type diagrams with intermediate particle-particle states that are generated by expanding the first two diagrams for $\Gamma(\delta s)$, many additional diagrams are suppressed. Likewise, the third line illustrates the emergence of the chain summation via the particle-hole diagrams that are generated by expanding the fourth diagram for $\Gamma(\delta s)$. In addition to the ladder and chain summations, the IMSRG(2) will also sum interference diagrams like the ones shown in the last row of Eq. (10.110). Such terms are not included in traditional summation methods, like the G-matrix approach for ladders, or the Random Phase Approximation (RPA) for chains [31, 32, 122]. We conclude our discussion at this point, and refer interested readers to the much more detailed analysis in [53].

10.3.2.3 Computational Scaling

Let us briefly consider the computational scaling of the IMSRG(2) scheme, ahead of the discussion of an actual implementation in Sect. 10.3.5. When performing a single integration step, the computational effort is dominated by the two-body flow equation (10.106), which naively requires $\mathcal{O}(N^6)$ operations, where N denotes the size of the single-particle basis. This puts the IMSRG(2) in the same category as CCSD and ADC(3) (see Chaps. 8 and 11). Fortunately, large portions of the flow equations can be expressed in terms of matrix products, allowing us to use optimized linear algebra libraries provided by high-performance computing vendors.

Moreover, we can further reduce the computational cost by distinguishing particle and hole states, because the number of hole states N_h is typically much smaller than the number of particle states $N_p \sim N$. The best scaling we can achieve in the IMSRG(2) depends on the choice of generator (see Sect. 10.3.4). If the one- and two-body parts of the generator only consist of *ph* and *pphh* type matrix elements and their Hermitian conjugates, the scaling is reduced to $\mathcal{O}(N_h^2 N_p^4)$, which matches the cost of solving the CCSD amplitude equations.

10.3.3 Decoupling

10.3.3.1 The Off-Diagonal Hamiltonian

Having set up the IMSRG flow equations, we now need to specify our decoupling strategy, i.e., how we split the Hamiltonian into diagonal parts we want to keep, and off-diagonal parts we want to suppress (cf. Sect. 10.2). To this end, we refer to the matrix representation of the Hamiltonian in a basis of A-body Slater determinants, but let us stress that we never actually construct the Hamiltonian matrix in this representation.

Our Slater determinant basis consists of a reference determinant and all its possible particle-hole excitations (cf. Sect. 10.3.1):

$$|\Phi\rangle,\ \left\{a^\dagger_p a_h\right\}|\Phi\rangle,\ \left\{a^\dagger_p a^\dagger_{p'} a_{h'} a_h\right\}|\Phi\rangle,\ldots. \tag{10.111}$$

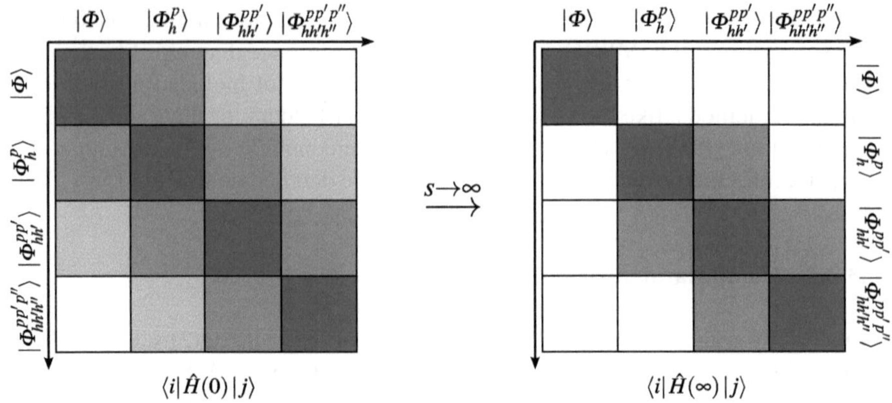

Fig. 10.8 Schematic view of single-reference IMSRG decoupling in a many-body Hilbert space spanned by a Slater determinant reference $|\Phi\rangle$ and its particle-hole excitations $|\Phi_{h\ldots}^{p\ldots}\rangle$

Note that

$$\{a_{p_1}^{\dagger}\ldots a_{p_i}^{\dagger}a_{h_i}\ldots a_{h_1}\} = a_{p_1}^{\dagger}\ldots a_{p_i}^{\dagger}a_{h_i}\ldots a_{h_1} \tag{10.112}$$

because contractions of particle and hole indices vanish by construction. Using Wick's theorem, one can show that the particle-hole excited Slater determinants are orthogonal to the reference state as well as each other (see Problem 10.3). In the Hilbert space spanned by this basis, the matrix representation of our initial Hamiltonian in the NO2B approximation (cf. Sect. 10.3.1.1) has the structure shown in the left panel of Fig. 10.8, i.e., it is band-diagonal, and can at most couple *npnh* and $(n \pm 2)p(n \pm 2)h$ excitations.

We now have to split the Hamiltonian into appropriate diagonal and off-diagonal parts on the operator level [123–125]. Using a broad definition of diagonality is ill-advised because we must avoid inducing strong in-medium $3N, \ldots$ interactions to maintain the validity of the IMSRG(2) truncation. For this reason, we choose a so-called *minimal decoupling scheme* that only aims to decouple the one-dimensional block spanned by the reference state from all particle-hole excitations, as shown in the right panel of Fig. 10.8.

If we could implement the minimal decoupling without approximations, we would extract a single eigenvalue and eigenstate of the many-body Hamiltonian for the nucleus of interest in the limit $s \to \infty$. The eigenvalue would simply be given by the zero-body piece of $H(\infty)$, while the eigenstate is obtained by applying the unitary IMSRG transformation to the reference state, $\hat{U}^{\dagger}(\infty)|\Phi\rangle$. In practice, truncations cannot be avoided, of course, and we only obtain an approximate eigenvalue and mapping. We will explicitly demonstrate in Sect. 10.3.6 that the chosen reference state plays an important role in determining which eigenvalue and eigenstate of the Hamiltonian we end up extracting in our minimal decoupling scheme. An empirical rule of thumb is that the IMSRG flow will connect the reference state to the eigenstate with which it has the highest overlap. If we are interested in the exact ground state, this is typically the case for a HF Slater determinant, because it minimizes both the absolute energy and the correlation energy.

Analyzing the matrix elements between the reference state and its excitations with the help of Wick's theorem, we first see that the Hamiltonian couples the $0p0h$ block to $1p1h$ excitations through the matrix elements

$$\begin{aligned} \langle\Phi|\hat{H}\{a^\dagger_p a_h\}|\Phi\rangle &= E\langle\Phi|\{a^\dagger_p a_h\}|\Phi\rangle + \sum_{ij} f_{ij}\langle\Phi|\{a^\dagger_i a_j\}\{a^\dagger_p a_h\}|\Phi\rangle \\ &\quad + \frac{1}{4}\sum_{ijkl}\Gamma_{ijkl}\langle\Phi|\{a^\dagger_i a^\dagger_j a_l a_k\}\{a^\dagger_p a_h\}|\Phi\rangle \\ &= \sum_{ij} f_{ij}\delta_{ih}\delta_{pj}n_i\bar{n}_j = f_{hp} \end{aligned} \tag{10.113}$$

and their Hermitian conjugates. The contributions from the zero-body and two-body pieces of the Hamiltonian vanish because they are expectation values of normal-ordered operators in the reference state [cf. Eq. (10.84)]. Likewise, the $0p0h$ and $2p2h$ blocks are coupled by the matrix elements

$$\langle\Phi|\hat{H}\{a^\dagger_p a^\dagger_{p'} a_{h'} a_h\}|\Phi\rangle = \Gamma_{hh'pp'} \tag{10.114}$$

and their conjugates. It is precisely these two-body matrix elements that couple $npnh$ and $(n \pm 2)p(n \pm 2)h$ states and generate the outermost side diagonals of the Hamiltonian matrix. This suggests that we can transform the Hamiltonian to the shape shown in the top right panel of Fig. 10.8 by defining its offdiagonal part as

$$\hat{H}_{od} \equiv \sum_{ph} f_{ph}\{a^\dagger_p a_h\} + \frac{1}{4}\sum_{pp'hh'}\Gamma_{pp'hh'}\{a^\dagger_p a^\dagger_{p'} a_{h'} a_h\} + \text{H.c.}. \tag{10.115}$$

In Sect. 10.3.6, we will show that the IMSRG flow does indeed exponentially suppress the matrix elements of $\hat{H}_{od}$ and achieve the desired decoupling in the limit $s \to \infty$.

10.3.3.2 Variational Derivation of Minimal Decoupling

Our minimal decoupling scheme is very reminiscent of the strategy followed in Coupled Cluster approaches [49, 60], except that we specifically use a unitary transformation instead of a general similarity transformation. It is also appealing for a different reason: As we will discuss now, it can be derived from a variational approach, tying the seemingly unrelated ideas of energy minimization and renormalization in the many-body system together.

Consider the energy expectation value of the final IMSRG evolved Hamiltonian,

$$\overline{H} \equiv \hat{H}(\infty)\,, \tag{10.116}$$

in the reference state (which is assumed to be normalized):

$$E = \langle \Phi | \, \overline{H} \, | \Phi \rangle\,. \tag{10.117}$$

We can introduce a *unitary* variation, which we are free to apply either to the reference state ,

$$|\Phi\rangle \to e^{\hat{Z}} \, |\Phi\rangle\,, \quad \hat{Z}^{\dagger} = -\hat{Z}\,, \tag{10.118}$$

or, equivalently, to the Hamiltonian:

$$e^{\hat{Z}^{\dagger}} \overline{H} e^{\hat{Z}} = e^{-\hat{Z}} \overline{H} e^{\hat{Z}}\,. \tag{10.119}$$

The variation of the energy is

$$\delta E = \langle \Phi | \, e^{-\hat{Z}} (\overline{H} - E) e^{\hat{Z}} \, | \Phi \rangle = \langle \Phi | \, \overline{H} - E \, | \Phi \rangle + \langle \Phi | \, [\overline{H} - E, \hat{Z}] \, | \Phi \rangle + O(||\hat{Z}||^2)\,, \tag{10.120}$$

where $|| \cdot ||$ is an appropriate operator norm. The first term obviously vanishes, as does the commutator of $\hat{Z}$ with the energy, because the latter is a mere number. Thus, the energy is stationary if

$$\delta E = \langle \Phi | \, [\overline{H}, \hat{Z}] \, | \Phi \rangle = 0\,. \tag{10.121}$$

Expanding

$$\hat{Z} = \sum_{ph} Z_{ph} \{a^{\dagger}_p a_h\} + \frac{1}{4} \sum_{pp'hh'} Z_{pp'hh'} \{a^{\dagger}_p a^{\dagger}_{p'} a_{h'} a_h\} + \text{H.c.} + \ldots\,, \tag{10.122}$$

and using the independence of the expansion coefficients (save for the unitarity conditions), we obtain the system of equations

$$\langle \Phi | \, [\overline{H}, \{a^{\dagger}_p a_h\}] \, | \Phi \rangle = 0\,, \tag{10.123}$$

$$\langle \Phi | \, [\overline{H}, \{a^{\dagger}_h a_p\}] \, | \Phi \rangle = 0\,, \tag{10.124}$$

$$\langle \Phi | \, [\overline{H}, \{a^{\dagger}_p a^{\dagger}_{p'} a_{h'} a_h\}] \, | \Phi \rangle = 0\,, \tag{10.125}$$

$$\langle \Phi | \, [\overline{H}, \{a^{\dagger}_h a^{\dagger}_{h'} a_{p'} a_p\}] \, | \Phi \rangle = 0\,, \tag{10.126}$$

$$\ldots$$

which are special cases of the so-called *irreducible Brillouin conditions (IBCs)* [64, 126–128]. Writing out the commutator in the first equation, we obtain

$$\begin{aligned}\langle\Phi|\,[\overline{H},\{a^\dagger_p a_h\}]\,|\Phi\rangle &= \langle\Phi|\,\overline{H}\{a^\dagger_p a_h\}\,|\Phi\rangle - \langle\Phi|\,\{a^\dagger_p a_h\}\overline{H}\,|\Phi\rangle\\ &= \langle\Phi|\,\overline{H}\{a^\dagger_p a_h\}\,|\Phi\rangle = 0\,,\end{aligned} \tag{10.127}$$

where the second term vanishes because it is proportional to $n_p\bar{n}_h = 0$. The remaining equations can be evaluated analogously, and we find that the energy is stationary if the IMSRG evolved Hamiltonian $\overline{H}$ no longer couples the reference state and its particle-hole excitations, as discussed above. However, we need to stress that the IMSRG is *not* variational, because any truncation of the flow equation breaks the unitary equivalence of the initial and evolved Hamiltonians. Thus, the final IMSRG(2) energy cannot be understood as an upper bound for the true eigenvalue in a strict sense, although the qualitative behavior might suggest so in numerical applications.

10.3.4 Choice of Generator

In the previous section, we have identified the matrix elements of the Hamiltonian that couple the ground state to excitations, and collected them into a definition of the off-diagonal Hamiltonian that we want to suppress with an IMSRG evolution. While we have decided on a decoupling "pattern" in this way, we have a tremendous amount of freedom in implementing this decoupling. As long as we use the same off-diagonal Hamiltonian, many different types of generators will drive the Hamiltonian to the desired shape in the limit $s \to \infty$, and some of these generators stand out when it comes to numerical efficiency [53].

10.3.4.1 Construction of Generators for Single-Reference Applications

A wide range of suitable generators for the single-reference case is covered by the ansatz

$$\eta = \sum_{ph} \eta_{ph}\{a^\dagger_p a_h\} + \frac{1}{4}\sum_{pp'hh'} \eta_{pp'hh'}\{a^\dagger_p a^\dagger_{p'} a_{h'} a_h\} - \text{H.c.}\,, \tag{10.128}$$

constructing the one- and two-body matrix elements directly from those of the offdiagonal Hamiltonian and a tensor G that ensures the anti-Hermiticity of η:

$$\eta_{ph} \equiv G_{ph} f_{ph}\,, \tag{10.129}$$

$$\eta_{pp'hh'} \equiv G_{pp'hh'}\Gamma_{pp'hh'}\,. \tag{10.130}$$

To identify possible options for G, we consider the flow equations in perturbation theory (see [53] for a detailed discussion). We assume a Hartree-Fock reference state, and partition the Hamiltonian as

$$\hat{H} = \hat{H}_0 + \hat{H}_I \,, \tag{10.131}$$

with

$$\hat{H}_0 \equiv E + \sum_i f_{ii}\{a_i^\dagger a_i\} + \frac{1}{4}\sum_{ij} \Gamma_{ijij}\{a_i^\dagger a_j^\dagger a_j a_i\} \,, \tag{10.132}$$

$$\hat{H}_I \equiv \sum_{ij}^{i\neq j} f_{ij}\{a_i^\dagger a_j\} + \frac{1}{4}\sum_{ijkl}^{(ij)\neq(kl)} \Gamma_{ijkl}\{a_i^\dagger a_j^\dagger a_l a_k\} \,. \tag{10.133}$$

We introduce a power counting in terms of the auxiliary parameter g, and count the diagonal Hamiltonian $\hat{H}_0$ as unperturbed ($\mathcal{O}(1)$), while the perturbation $\hat{H}_I$ is counted as $\mathcal{O}(g)$. In the space of up to $2p2h$ excitations, our partitioning is a second-quantized form of the one used by Epstein and Nesbet [129, 130].

We now note that the one-body piece of the initial Hamiltonian is diagonal in the HF orbitals, which implies

$$f_{ph} = 0, \quad \eta_{ph} = 0 \,. \tag{10.134}$$

Inspecting Eq. (10.105), we see that corrections to f that are induced by the flow are at least of order $\mathcal{O}(g^2)$, because no diagonal matrix elements of Γ appear:

$$\frac{d}{ds} f_{ij}\bigg|_{s=0} = \frac{1}{2}\sum_{abc}\left(\eta_{iabc}\Gamma_{bcja} - \Gamma_{iabc}\eta_{bcja}\right)\left(n_a\bar{n}_b\bar{n}_c + \bar{n}_a n_b n_c\right) = \mathcal{O}(g^2) \,. \tag{10.135}$$

Using this knowledge, the two-body flow equation for the $pphh$ matrix elements of the off-diagonal Hamiltonian reads

$$\begin{aligned}
\frac{d}{ds}\Gamma_{pp'hh'} &= -\left(f_{pp} + f_{p'p'} - f_{hh} - f_{h'h'}\right)\eta_{pp'hh'} - \left(\Gamma_{hh'hh'} + \Gamma_{pp'pp'}\right)\eta_{pp'hh'} \\
&\quad + \left(\Gamma_{p'h'p'h'} + \Gamma_{phph} + \Gamma_{ph'ph'} + \Gamma_{p'hp'h}\right)\eta_{pp'hh'} + \mathcal{O}(g^2) \\
&= -\Delta_{pp'hh'}\eta_{pp'hh'} + \mathcal{O}(g^2) \,.
\end{aligned} \tag{10.136}$$

Note that $\hat{\eta}$, which is of order $\mathcal{O}(g)$, is multiplied by unperturbed, diagonal matrix elements of the Hamiltonian in the leading term. Because of this restriction, the sums in the particle-particle and hole-hole ladder terms [line 2 of Eq. (10.106)] collapse, and the pre-factors $\frac{1}{2}$ are canceled by factors 2 from the unrestricted summation over indices, e.g.,

$$\frac{1}{2}\sum_{h_1h_2}\eta_{pp'h_1h_2}\Gamma_{h_1h_2hh'}(1-n_{h_1}-n_{h_2}) = -\frac{1}{2}\eta_{pp'hh'}\Gamma_{hh'hh'} - \frac{1}{2}\eta_{pp'h'h}\Gamma_{h'hhh'}$$
$$= -\eta_{pp'hh'}\Gamma_{hh'hh'}\,. \tag{10.137}$$

In Eq. (10.136), we have introduced the quantity

$$\begin{aligned}\Delta_{pp'hh'} &\equiv f_{pp}+f_{p'p'}-f_{hh}-f_{h'h'}+\Gamma_{hh'hh'}+\Gamma_{pp'pp'}\\ &\quad-\Gamma_{phph}-\Gamma_{p'h'p'h'}-\Gamma_{ph'ph'}-\Gamma_{p'hp'h}\\ &=\langle\Phi|\,\{a^\dagger_h a^\dagger_{h'}a_{p'}a_p\}\hat{H}\{a^\dagger_p a^\dagger_{p'}a_{h'}a_h\}\,|\Phi\rangle-\langle\Phi|\,\hat{H}\,|\Phi\rangle\\ &=\langle\Phi|\,\{a^\dagger_h a^\dagger_{h'}a_{p'}a_p\}\hat{H}_0\{a^\dagger_p a^\dagger_{p'}a_{h'}a_h\}\,|\Phi\rangle-\langle\Phi|\,\hat{H}_0\,|\Phi\rangle\,,\end{aligned} \tag{10.138}$$

i.e., the unperturbed energy difference between the two states that are coupled by the matrix element $\Gamma_{pp'hh'}$, namely the reference state $|\Phi\rangle$ and the excited state $\{a^\dagger_p a^\dagger_{p'}a_{h'}a_h\}\,|\Phi\rangle$. Since it is expressed in terms of diagonal matrix elements, $\Delta_{pp'hh'}$ would appear in precisely this form in appropriate energy denominators of Epstein-Nesbet perturbation theory.

Plugging our ansatz for η into Eq. (10.136), we obtain

$$\frac{d}{ds}\Gamma_{pp'hh'} = -\Delta_{pp'hh'}G_{pp'hh'}\Gamma_{pp'hh'} + \mathcal{O}(g^2)\,, \tag{10.139}$$

Neglecting $\mathcal{O}(g^2)$ terms in the flow equations, the one-body part of $\hat{H}$ remains unchanged, and assuming that G itself is independent of s at order $\mathcal{O}(g)$, we can integrate Eq. (10.136):

$$\Gamma_{pp'hh'}(s) = \Gamma_{pp'hh'}(0)e^{-\Delta_{pp'hh'}G_{pp'hh'}s}\,. \tag{10.140}$$

Clearly, the offdiagonal matrix elements of the Hamiltonian will be suppressed for $s\to\infty$ if the product $\Delta_{pp'hh'}G_{pp'hh'}$ is positive. $G_{pp'hh'}$ also allows us to control the details of this suppression, e.g., the decay scales. To avoid misconceptions, we stress that we do not impose perturbative truncations in practical applications, and treat all matrix elements and derived quantities, including the $\Delta_{pp'hh'}$, as s-dependent.

10.3.4.2 White's Generators

A generator that is particularly powerful in numerical applications is inspired by the work of White on canonical transformation theory in quantum chemistry [51, 53, 61]. In the language we have set up above, it uses $G_{pp'hh'}$ to *remove* the scale dependence of the IMSRG flow. This so-called White generator is defined as

$$\hat{\eta}^{\mathrm{W}}(s) \equiv \sum_{ph} \frac{f_{ph}(s)}{\Delta_{ph}(s)}\{a^{\dagger}_{p}a_{h}\} + \frac{1}{4}\sum_{pp'hh'} \frac{\Gamma_{pp'hh'}(s)}{\Delta_{pp'hh'}(s)}\{a^{\dagger}_{p}a^{\dagger}_{p'}a_{h'}a_{h}\} - \text{H.c.}\,, \tag{10.141}$$

where the Epstein-Nesbet denominators use the energy differences defined in Eqs. (10.138) and (10.145).

For the White generator, we find

$$\Gamma_{pp'hh'}(s) = \Gamma_{pp'hh'}(0)e^{-s}\,, \tag{10.142}$$

i.e., *all* off-diagonal matrix elements are suppressed *simultaneously* with a decay scale identical (or close to) 1 [53]. While this means that η^{W} does *not* generate a proper RG flow, this is inconsequential if we are only interested in the final Hamiltonian $\hat{H}(\infty)$, because all unitary transformations which suppress $\hat{H}_{od}$ must be equivalent up to truncation effects [53].

A benefit of the White generator is that its matrix elements are defined as ratios of energies, and therefore the Hamiltonian only contributes linearly to the magnitude of the right-hand side of the flow equations (10.104)–(10.106). This leads to a significant reduction of the ODE system's stiffness compared to the other generators discussed here or in [53], and greatly reduces the numerical effort for the ODE solver. However, the dependence of $\hat{\eta}^{\mathrm{W}}$ on energy denominators can also be a drawback if Δ_{ph} and/or $\Delta_{pp'hh'}$ become small, which would cause the generator's matrix elements to diverge. This can be mitigated by using an alternative ansatz that is also inspired by White's work [61]:

$$\hat{\eta}^{\mathrm{W'}}(s) \equiv \frac{1}{2}\sum_{ph} \arctan\frac{2f_{ph}(s)}{\Delta_{ph}(s)}\{a^{\dagger}_{p}a_{h}\} + \frac{1}{8}\sum_{pp'hh'} \arctan\frac{2\Gamma_{pp'hh'}(s)}{\Delta_{pp'hh'}(s)}\{a^{\dagger}_{p}a^{\dagger}_{p'}a_{h'}a_{h}\} - \text{H.c.}\,. \tag{10.143}$$

This form emphasizes that the unitary transformation can be thought of as an abstract rotation of the Hamiltonian. The matrix elements of $\eta^{\mathrm{W'}}$ are regularized by the arctan function, and explicitly limited to the interval $]-\frac{\pi}{4},\frac{\pi}{4}[$. Expanding the function for small arguments, we recover our initial ansatz for the White generator, Eq. (10.141).

10.3.4.3 The Imaginary-Time Generator

Using $G_{pp'hh'}$ to ensure that the energy denominator is always positive, we obtain the so-called *imaginary-time generator* [53, 72, 101], which is inspired by imaginary-time evolution techniques that are frequently used in Quantum Monte Carlo methods, for instance (see Chap. 9, [88] and references therein). Explicitly indicating the flow parameter dependence of all quantities, we define

$$\hat{\eta}^{\mathrm{IT}}(s) \equiv \sum_{ph} \operatorname{sgn}\left(\Delta_{ph}(s)\right) f_{ph}(s)\{a_p^\dagger a_h\} + \frac{1}{4}\sum_{pp'hh'} \operatorname{sgn}\left(\Delta_{pp'hh'}(s)\right) \Gamma_{pp'hh'}(s)\{a_p^\dagger a_{p'}^\dagger a_{h'} a_h\} - \mathrm{H.c.}\,, \tag{10.144}$$

where

$$\Delta_{ph} \equiv f_{pp} - f_{hh} + \Gamma_{phph} = \langle\Phi|\,\{a_h^\dagger a_p\}\hat{H}\{a_p^\dagger a_h\}\,|\Phi\rangle - \langle\Phi|\,\hat{H}\,|\Phi\rangle\,. \tag{10.145}$$

For this generator, the perturbative analysis of the offdiagonal two-body matrix elements yields

$$\Gamma_{pp'hh'}(s) = \Gamma_{pp'hh'}(0)e^{-|\Delta_{pp'hh'}|s}\,, \tag{10.146}$$

ensuring that they are driven to zero by the evolution. We also note that the energy difference $\Delta_{pp'hh'}$ controls the scales of the decay. Matrix elements between states with large energy differences are suppressed more rapidly than those which couple states that are close in energy. This means that η^{IT} generates a proper renormalization group flow [53, 54].

10.3.4.4 Wegner's Generator

Last but not least, we want to discuss Wegner's original ansatz [40], which we have used in the free-space SRG applications in Sect. 10.2:

$$\hat{\eta}^{\mathrm{WE}}(s) \equiv [\hat{H}_d(s), \hat{H}_{od}(s)]\,. \tag{10.147}$$

Truncating $\hat{H}_d(s)$ and $\hat{H}_{od}(s)$ at the two-body level and using the commutators from Appendix, it is straightforward to derive the one- and two-body matrix elements of $\hat{\eta}(s)$; the operator has no zero-body component because of its anti-Hermiticity. We obtain

$$\eta_{ij} = \sum_a (1-P_{ij}) f^d_{ia} f^{od}_{aj} + \sum_{ab}(n_a - n_b)(f^d_{ab}\Gamma^{od}_{biaj} - f^{od}_{ab}\Gamma^d_{biaj}) + \frac{1}{2}\sum_{abc}(n_a n_b \bar{n}_c + \bar{n}_a \bar{n}_b n_c)(1-P_{ij})\Gamma^d_{ciab}\Gamma^{od}_{abcj}\,, \tag{10.148}$$

$$\eta_{ijkl} = \sum_a \{(1-P_{ij})(f^d_{ia}\Gamma^{od}_{ajkl} - f^{od}_{ia}\Gamma^d_{ajkl}) - (1-P_{kl})(f^d_{ak}\Gamma^{od}_{ijal} - f^{od}_{ak}\Gamma^d_{ijal})\}$$
$$+ \frac{1}{2}\sum_{ab}(1-n_a-n_b)(\Gamma^d_{ijab}\Gamma^{od}_{abkl} - \Gamma^{od}_{ijab}\Gamma^d_{abkl})$$
$$+ \sum_{ab}(n_a-n_b)(1-P_{ij})(1-P_{kl})\Gamma^d_{aibk}\Gamma^{od}_{bjal}\,. \tag{10.149}$$

Structurally, Eqs. (10.148) and (10.149) are identical to the flow IMSRG(2) equations except for signs stemming from the anti-Hermiticity of the generator.

Superficially, the Wegner generator is quite different from the imaginary-time and White generators, but we can uncover commonalities by carrying out a perturbative analysis along the lines of the previous sections. For a HF Slater determinant, the one-body part of the off-diagonal Hamiltonian vanishes at $s = 0$, and corrections that are induced by the flow start at $\mathcal{O}(g^2)$ [see Eq. (10.135)]. This means that the one-body part of the Wegner generator has the form

$$\eta_{ij} = \frac{1}{2}\sum_{abc}(n_a n_b \bar{n}_c + \bar{n}_a \bar{n}_b n_c)(1-P_{ij})\Gamma^d_{ciab}\Gamma^{od}_{abcj} + \mathcal{O}(g^2)\,. \tag{10.150}$$

In the minimal decoupling scheme, the matrix elements appearing here are counted as follows:

$$\Gamma^d_{ijij} = -\Gamma^d_{jiij} = \mathcal{O}(1), \quad \Gamma^d_{ijkl} = \mathcal{O}(g) \text{ for } (ij) \neq (kl), \quad \Gamma^{od}_{ijkl} = \mathcal{O}(g)\,. \tag{10.151}$$

To obtain a $\mathcal{O}(g)$ contribution to the one-body generator, we need either $a = c$ and $b = i$, or $a = i$ and $b = c$, but then the occupation number factor becomes

$$n_i n_c \bar{n}_c + \bar{n}_i \bar{n}_c n_c = 0\,. \tag{10.152}$$

This implies that the leading contributions to η_{ij} are of order $\mathcal{O}(g^2)$.

A similar analysis for the two-body part of $\hat{\eta}^{\mathrm{WE}}$ (see Problem 10.4) shows that

$$\eta_{ijkl} = \big(f^d_{ii} + f^d_{jj} - f^d_{kk} + f^d_{ll} + (1-n_i-n_k)\Gamma^d_{ijij} - (1-n_k-n_l)\Gamma^d_{klkl}$$
$$+ (n_i-n_k)\Gamma^d_{ikik} + (n_j-n_k)\Gamma^d_{jkjk} + (n_i-n_l)\Gamma^d_{ilil} + (n_j-n_l)\Gamma^d_{jljl}\big)\,\Gamma^{od}_{ijkl} + \mathcal{O}(g^2)\,. \tag{10.153}$$

Since Γ^{od}_{ijkl} is restricted to *pphh* matrix elements, we immediately obtain

$$\eta_{pp'hh'} = \Delta_{pp'hh'}\Gamma_{pp'hh'}\,, \tag{10.154}$$

and plugging this into Eq. (10.136), we have

$$\frac{d}{ds}\Gamma_{pp'hh'} = -\left(\Delta_{pp'hh'}\right)^2 G_{pp'hh'}\Gamma_{pp'hh'} + \mathcal{O}(g^2)\,. \tag{10.155}$$

Neglecting the $\mathcal{O}(g^2)$ terms, we can integrate the flow equation and find that the Wegner generator suppresses off-diagonal matrix elements with a Gaussian exponential function,

$$\Gamma_{pp'hh'}(s) = \Gamma_{pp'hh'}(0)e^{-(\Delta_{pp'hh'})^2 s}\,, \tag{10.156}$$

and therefore generates a proper RG flow. Note that this result matches our findings for the SRG flows of the 2 × 2 matrix toy model (Sect. 10.2.2) and the pairing Hamiltonian (Sect. 10.2.3), which were using matrix versions of the Wegner generator.

In numerical applications, Wegner generators are less efficient than our other choices. The cost for constructing $\hat{\eta}^{\mathrm{WE}}$ is of order $\mathcal{O}(N_p^4 N_h^2)$, compared to $\mathcal{O}(N_p^2 N_h^2)$ for the White and imaginary-time generators. More importantly, $\hat{\eta}^{\mathrm{WE}}$ generates very stiff flow equations because the RHS terms are cubic in the Hamiltonian. This forces us to use ODE solvers that are appropriate for stiff systems, which have higher storage requirements and need more computing time than solvers for non-stiff systems, which can be used for imaginary-time and White IMSRG flows.

10.3.5 Implementation

Now that we have all the necessary ingredients, it is time to discuss the numerical implementation of IMSRG flows. As an example, we use a Python code that is designed for solving the pairing Hamiltonian (see Sect. 10.3.6) but easily adaptable to other problems. The IMSRG solver and tools for visualizing the flow can be found at https://github.com/ManyBodyPhysics/LectureNotesPhysics/tree/master/Programs/Chapter10-programs/python/imsrg_pairing.

10.3.5.1 Basis and Matrix Element Handling

The code `imsrg_pairing.py` uses `NumPy` arrays to store the one- and two-body matrix elements of the normal-ordered operators. The underlying one- and two-nucleon states are indexed as integers and pairs of integers, respectively. The complete lists of states are stored in the variables `bas1B` and `bas2B`, respectively. We also define an additional set of two-nucleon states in the list `basph2B`, for reasons that will be explained shortly. The lists are all initialized at the beginning of the program's main routine:

```
def main():
...
    particles = 4

    # setup shared data
    dim1B    = 8

    # this defines the reference state
    holes    = [0,1,2,3]
    particles = [4,5,6,7]

    # basis definitions
    bas1B    = range(dim1B)
    bas2B    = construct_basis_2B(holes, particles)
    basph2B = construct_basis_ph2B(holes, particles)

    idx2B    = construct_index_2B(bas2B)
    idxph2B = construct_index_2B(basph2B)

  ...
```

Aside from a distinction between occupied (hole) and unoccupied (particle) single-particle states, we do not impose any constraints on the one- and two-nucleon bases. All permutations of two-body matrix elements Γ_{ijkl} are stored explicitly, and so are the vanishing matrix elements like Γ_{iikl} or Γ_{ijkk} that are forbidden by the Pauli principle. The benefit of using this "naive" basis construction is that we do not need to worry whether certain combinations of quantum numbers are allowed or forbidden by symmetries. This is relevant for flow equation terms that cannot be expressed as matrix products, i.e., the contractions of one- and two-body operators that appear in Eqs. (10.105) and (10.106). To implement such terms, we also need to translate pairs of single-particle indices into collective two-nucleon state indices and back, which is achieved with the help of the lookup arrays `idx2B` and `idxph2B`.

The state and lookup lists `basph2B` and `idxph2B` are used to work with matrices in the so-called *particle-hole representation*. This representation allows us to write the particle-hole contributions in the third line of Eq. (10.106) as matrix products. Superficially, these terms look like they require explicit loop summations:

$$\frac{d}{ds}\Gamma^{(ph)}_{ijkl} = (1 - P_{ij})(1 - P_{kl}) \sum_{ab} (n_a - n_b)\eta_{aibk}\Gamma_{bjal} \,. \tag{10.157}$$

Since we are working with a Slater determinant reference state and therefore able to distinguish particle and hole states in the single-particle basis, we can also define hole creation and annihilation operators (see, e.g., [131]):

$$h_i^\dagger \equiv a_i \,, \quad h_i \equiv a_i^\dagger \,, \tag{10.158}$$

that satisfy the same anticommutation relations as the regular creators and annihilators. In addition, we also have

$$\{h_i^\dagger, a_j\} = \{h_i, a_j^\dagger\} = 0\,,$$
$$\{h_i^\dagger, a_j^\dagger\} = \{h_i, a_j\} = \delta_{ij}\,. \tag{10.159}$$

Using Eq. (10.85) for the hole operators, we can rewrite a generic normal-ordered two-body operator in the following way:

$$\hat{A} \equiv \frac{1}{4}\sum_{ijkl} A_{ijkl}\{a_i^\dagger a_j^\dagger a_l a_k\} = \frac{1}{4}\sum_{ijkl} A_{ijkl}\{a_i^\dagger h_j h_l^\dagger a_k\} = -\frac{1}{4}\sum_{ijkl} A_{ijkl}\{a_i^\dagger h_l^\dagger h_j a_k\}\,. \tag{10.160}$$

We can also define $\hat{A}$ directly in the particle-hole representation,

$$\hat{A} \equiv \frac{1}{4}\sum_{ijkl} A_{i\bar{j}k\bar{l}}\{a_i^\dagger h_j^\dagger h_l a_k\}\,, \tag{10.161}$$

where we have indicated the hole states by lines over the indices. Thus, the operator's matrix elements in the regular particle representation and the particle-hole representation are related by the following expression:

$$A_{i\bar{j}k\bar{l}} = -A_{ilkj}\,. \tag{10.162}$$

We see that the switch to particle-hole representation is achieved by a simple rearrangement of matrix elements. The situation is more complicated if one works with angular-momentum coupled states, because then the angular momenta must be recoupled in a different order, giving rise to a so-called *Pandya transformation* [131–133].

Using particle-hole matrix elements, the right-hand side of Eq. (10.157) can be written as

$$\sum_{ab}(n_a - n_b)\eta_{iabk}\Gamma_{bjla} = \sum_{ab}(n_a - n_b)\eta_{i\bar{k}b\bar{a}}\Gamma_{b\bar{a}l\bar{j}} \equiv M_{i\bar{k}l\bar{j}}\,, \tag{10.163}$$

which makes it possible to evaluate the term using matrix product routines, treating $(n_a - n_b)$ as a diagonal matrix. The resulting product matrix M can then be transformed back into the particle representation, where it will be completely antisymmetrized by the permutation symbols in Eq. (10.157). The necessary transformations between the particle and particle-hole representations are implemented in the routines `ph_transform_2B` and `inverse_ph_transform_2B`, respectively.

10.3.5.2 Reference States

After the basis initialization, we need to define the reference Slater determinant $|\Phi\rangle$ for the subsequent normal ordering. To this end, we simply create lists of the hole (occupied) and particle (unoccupied) single-particle states:

```
def main():
...

    particles = 4

    # setup shared data
    dim1B    = 8

    # this defines the reference state
    holes    = [0,1,2,3]
    particles = [4,5,6,7]

    # basis definitions
...

    # occupation number matrices
    occ1B    = construct_occupation_1B(bas1B, holes, particles)
    occA_2B  = construct_occupationA_2B(bas2B, occ1B)
    occB_2B  = construct_occupationB_2B(bas2B, occ1B)
    occC_2B  = construct_occupationC_2B(bas2B, occ1B)

    occphA_2B = construct_occupationA_2B(basph2B, occ1B)

...
```

In addition to the elementary lists, we set up diagonal matrices for the various occupation number factors that appear in the IMSRG(2) flow equations (10.104)–(10.106), i.e., $(n_a - n_b)$, which is required both in particle and particle-hole representation (`construct_occupationA_2B`), $(1 - n_a - n_b)$ (`construct_occupationB_2B`), and $n_a n_b$ (`construct_occupationC_2B`). The latter appears when we rewrite the last occupation factor in the one-body flow equation (10.105):

$$n_a n_b \bar{n}_c + \bar{n}_a \bar{n}_b n_c = n_a n_b - n_a n_b n_c + (1 - n_a - n_b + n_a n_b) n_c = n_a n_b + (1 - n_a - n_b) n_c \,. \tag{10.164}$$

10.3.5.3 Sharing Data and Settings

Since the routines for normal ordering the Hamiltonian and calculating the generator and derivatives need to access the bases, index lookups, and occupation number matrices, we store this shared data in a Python dictionary:

```
...
   user_data = {
      "dim1B":   dim1B,
      "holes":   holes,
      "particles": particles,
      "bas1B":   bas1B,
      "bas2B":   bas2B,
      "basph2B": basph2B,
      "idx2B":   idx2B,
      "idxph2B": idxph2B,
      "occ1B":   occ1B,
      "occA_2B": occA_2B,
      "occB_2B": occB_2B,
      "occC_2B": occC_2B,
      "occphA_2B": occphA_2B,

      "eta_norm": 0.0,             # variables for sharing data
                                   # between ODE solver
      "dE":      0.0,              # and main routine

      "calc_eta": eta_white,       # specify the generator
                                   # (function object)
      "calc_rhs": flow_imsrg2      # specify the right-hand side
                                   # and truncation
   }
...
```

Rather than passing all of the data structures as separate parameters, we can then pass `user_data` as a parameter. By passing the data as a dictionary, we can also avoid the creation of global variables, and make it easier to reuse individual routines in other projects.

We also want to direct our readers' attention to the last two entries of the dictionary. These are function objects that are used to define which generator and flow equation routines the ODE solver will call (see below). Through this abstraction, users can easily add additional generators, or implement different truncations of the flow equations. The current version of `imsrg_pairing.py` implements all of the generators discussed in Sect. 10.3.4, and the standard IMSRG(2) truncation.

10.3.5.4 Normal Ordering

The next task of the main routine is the normal ordering of the initial Hamiltonian:

```
def main():
...
   # set up initial Hamiltonian
   H1B, H2B = pairing_hamiltonian(delta, g, user_data)

   E, f, Gamma = normal_order(H1B, H2B, user_data)
...
```

In order to facilitate the reuse of our code, we proceed in two steps: First, we set up the Hamiltonian in the vacuum, in this case the pairing Hamiltonian (10.38) with single-particle spacing δ and pairing strength g. The one- and two-body matrix elements are then passed to a generic routine that performs the normal ordering:

```
def normal_order(H1B, H2B, user_data):
   bas1B     = user_data["bas1B"]
   bas2B     = user_data["bas2B"]
   idx2B     = user_data["idx2B"]
   particles = user_data["particles"]
   holes     = user_data["holes"]

   # 0B part
   E = 0.0
   for i in holes:
     E += H1B[i,i]

   for i in holes:
     for j in holes:
       E += 0.5*H2B[idx2B[(i,j)],idx2B[(i,j)]]

   # 1B part
   f = H1B
   for i in bas1B:
     for j in bas1B:
       for h in holes:
         f[i,j] += H2B[idx2B[(i,h)],idx2B[(j,h)]]

   # 2B part
   Gamma = H2B

   return E, f, Gamma
```

10.3.5.5 Integration

Once the initial Hamiltonian is set up, we use the `SciPy` `ode` class to integrate the flow equations, so that we can switch between non-stiff and stiff solvers, and give the user as much control over the solver as possible.

The ODE solver calls the following derivative wrapper function:

```
def derivative_wrapper(t, y, user_data):
...
   calc_eta = user_data["calc_eta"]
   calc_rhs = user_data["calc_rhs"]

   # extract operator pieces from solution vector
   E, f, Gamma = get_operator_from_y(y, dim1B, dim2B)


   # calculate the generator
   eta1B, eta2B = calc_eta(f, Gamma, user_data)

   # calculate the right-hand side
   dE, df, dGamma = calc_rhs(eta1B, eta2B, f, Gamma, user_data)

   # convert derivatives into linear array
   dy  = np.append([dE], np.append(reshape(df, -1), reshape(dGamma,
       -1)))

   # share data
   user_data["dE"]  = dE
   user_data["eta_norm"] = np.linalg.norm(eta1B,ord='fro')+np.linalg.
       norm(eta2B,ord='fro')

   return dy
```

This routine is very similar to the ones we used in the SRG codes in Sects. 10.2.3.3 and 10.2.4.3. It extracts $E(s)$,$f(s)$, and $\Gamma(s)$ from the solution vector, and calls appropriate routines to construct the generator and the derivatives. This is where the function object entries of the `user_data` dictionary come into play: We use them as an abstract interface to call the routines we assigned to the dictionary in our `main` routine. This is much more elegant than selecting the generator and flow equation truncation scheme via `if...elif` clauses. Most importantly, we do not need to modify the derivative routine at all if we want to add new generators and truncation schemes, but only need to assign the new functions to `calc_eta` and `calc_rhs`.

In the ODE loop, we do not require output at an externally chosen value of s, but check and process the intermediate solution after each accepted (not attempted!) internal step of the solver. This is achieved by setting the option `step=True`. We extract $E(s)$,$f(s)$, and $\Gamma(s)$, and use it to calculate diagnostic quantities like the second- and third-order energy corrections, $\Delta E^{(2)}(s)$ and $\Delta E^{(3)}(s)$, as well as the norms of $f_{od}(s)$ and $\Gamma_{od}(s)$:

```
def main():
...
   while solver.successful() and solver.t < sfinal:
      ys = solver.integrate(sfinal, step=True)

      dim2B = dim1B*dim1B
      E, f, Gamma = get_operator_from_y(ys, dim1B, dim2B)
```

```
DE2 = calc_mbpt2(f, Gamma, user_data)
DE3 = calc_mbpt3(f, Gamma, user_data)

norm_fod  = calc_fod_norm(f, user_data)
norm_Gammaod = calc_Gammaod_norm(Gamma, user_data)

print("%8.5f %14.8f %14.8f %14.8f %14.8f %14.8f %14.8f
    %14.8f %14.8f"%(
  solver.t, E , DE2, DE3, E+DE2+DE3, user_data["dE"],
     user_data["eta_norm"], norm_fod, norm_Gammaod))

if abs(DE2/E) < 10e-8: break
```

As discussed in earlier sections, the off-diagonal matrix elements create $1p1h$ and $2p2h$ admixtures to the reference state wave function which give rise to the energy corrections. These admixtures are suppressed as we decouple, hence the size of the energy corrections must decrease. In the limit $s \to \infty$, we expect them to be completely absorbed into the RG-improved Hamiltonian (see Sect. 10.3.6). It is by this reasoning that we use the relative size of the second-order correction to the flowing energy, $\Delta E^2(s)/E(s)$, as the stopping criterion for the flow. Once this quantity falls below 10^{-8}, we terminate the evolution.

10.3.5.6 Optimizations

The Python implementation of the IMSRG that we describe here can solve problems with small single-particle basis sizes in reasonable time. To tackle large-scale calculations, we implement the IMSRG in languages like C/C++ or Fortran that are closer to the hardware.

As mentioned in the course of the discussion, we have favored simplicity in the design of the Python code, which leaves significant room for optimization. For instance, we can exploit that nucleons are Fermions, and antisymmetrize the two-nucleon basis states. This reduces the storage for matrix elements involving identical particles by a factor of four, because we only need to store one of the antisymmetrized matrix elements

$$\Gamma_{ijkl} = -\Gamma_{jikl} = -\Gamma_{ijlk} = \Gamma_{jilk} \,. \tag{10.165}$$

The storage can be reduced even further if we also exploit the Hermiticity and anti-Hermiticity of $\hat{H}$ and $\hat{\eta}$, respectively.

Another important tool for optimization are symmetries of the Hamiltonian. Nuclear Hamiltonians conserve the total angular momentum, parity, and the isospin projection of quantum states, which implies that f and Γ are block-diagonal in the corresponding quantum numbers. Since the generator and the derivatives are constructed from the Hamiltonian, they have the same symmetries, and are block-diagonal as well. Exploiting this block structure, we can reduce the storage requirements, which are naively $\mathcal{O}(N^4)$ for each operator, by about one to two orders of magnitude. We can also explicitly work on the blocks instead of the full matrices when we evaluate the right-hand sides of the IMSRG flow equations. The complex couplings between blocks prevents us from evolving individual blocks or small groups separately, in contrast to the free-space SRG case, where this was possible (see Sect. 10.2.4.3).

10.3.6 IMSRG Solution of the Pairing Hamiltonian

Let us now use the code from the previous section to solve the Schrödinger equation for four particles that interact via the pairing Hamiltonian (10.38).

10.3.6.1 Ground-State Calculations

As a first application, we calculate the ground-state energy for the pairing Hamiltonian with $\delta = 1.0$ and $g = 0.5$, which we studied using SRG matrix flows in Sect. 10.2.3.4. In the left panel of Fig. 10.9, we show the flowing ground-state energy $E(s)$. Starting from the energy of the uncorrelated reference state, which is $E(0) = 2\delta - g = 1.5$, we obtain a final energy $E(\infty) = 1.4133$, which is slightly below the exact result 1.4168.

The mechanism by which the flowing ground-state energy is absorbing correlation energy can be understood by considering the zero-body flow equation (10.104)

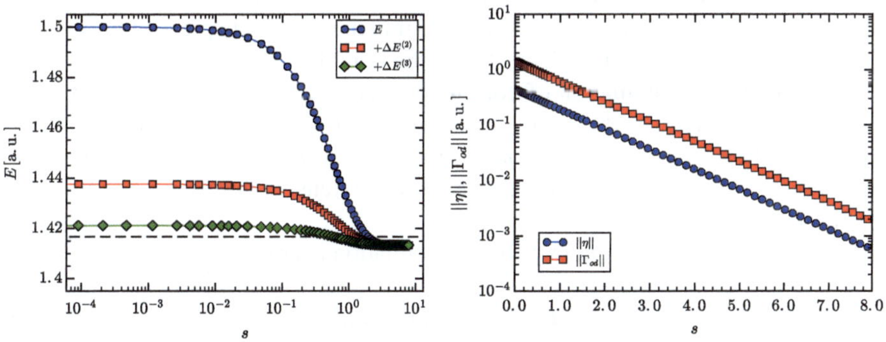

Fig. 10.9 IMSRG(2) flow for the ground state of the pairing Hamiltonian with $\delta = 1.0, g = 0.5$ (cf. Sect. 10.2.3.4). Calculations were performed with the White generator, Eq. (10.141). *Left panel:* Flowing ground-state energy $E(s)$ plus perturbative second and third-order energy corrections for $\hat{H}(s)$. The exact ground-state energy is indicated by the *dashed line*. *Right panel:* Norm of the White generator and the off-diagonal Hamiltonian $\hat{H}_{od}(s)$ (note that $||f_{od}(s)|| = 0$)

in the perturbative approach we introduced in Sect. 10.3.4. We note that

$$\frac{dE}{ds} = \underbrace{\sum_{ab}(n_a - n_b)\eta^a_b f^b_a}_{\mathcal{O}(g^4)} + \underbrace{\frac{1}{4}\sum_{abcd}\left(\eta^{ab}_{cd}\Gamma^{cd}_{ab} - \Gamma^{ab}_{cd}\eta^{cd}_{ab}\right) n_a n_b \bar{n}_c \bar{n}_d}_{\mathcal{O}(g^2)} . \tag{10.166}$$

For the White generator (10.141),

$$\Gamma^{pp'}_{hh'}(s) = \Gamma^{pp'}_{hh'}(0)e^{-s} , \quad \Gamma^{hh'}_{pp'}(s) = \Gamma^{hh'}_{pp'}(0)e^{-s} . \tag{10.167}$$

As we can see in the left panel of Fig. 10.9, the off-diagonal matrix elements and the generator indeed decay exponentially with a single, state-independent scale. Plugging the matrix elements into the energy flow equations to $\mathcal{O}(g^2)$, we have

$$\frac{dE}{ds} = \frac{1}{2}\sum_{pp'hh'}\frac{|\Gamma^{pp'}_{hh'}(0)|^2}{\Delta^{pp'}_{hh'}(0)}e^{-2s} . \tag{10.168}$$

Integrating over the flow parameter, we obtain

$$E(s) = E(0) - \frac{1}{4}\sum_{pp'hh'}\frac{|\Gamma^{pp'}_{hh'}(0)|^2}{|\Delta^{pp'}_{hh'}(0)|}\left(1 - e^{-2s}\right) . \tag{10.169}$$

We recognize the second-order energy correction, evaluated with the initial Hamiltonian, and see that $E(s)$ will decrease with s (i.e., the binding energy increases). In the limit $s \to \infty$, the entire correction is shuffled into the zero-body piece of the evolved Hamiltonian. As discussed in Sect. 10.3.2, the complete IMSRG(2) flow performs a more complex re-summation of correlations, but we can see from Fig. 10.9 that it certainly encompasses the complete second order. In fact, we see that the third-order correction is completely absorbed into the final $E(\infty)$ as well. Readers who are interested in more details are referred to the extensive discussion in [53] (also see [72]).

In Fig. 10.10, we show the IMSRG(2) correlation energy

$$E_{\text{corr}} = E(\infty) - E_{\text{HF}} \tag{10.170}$$

as a function of the pairing strength g, holding the single-particle level spacing constant at $\delta = 1.0$. As we see in the left panel, the IMSRG(2) results for the White [Eq. (10.141)], imaginary time [Eq. (10.144)] and Wegner generators [Eq. (10.147)] are practically identical as we evolve to $s \to \infty$. This behavior is expected, because all three generators are based on the same off-diagonal Hamiltonian, and we have explained in the previous sections that the specific choice of the generator then only affects the numerical aspects of the flow, with the potential exception of accumulated truncation errors. Such errors clearly do not matter for the present case.

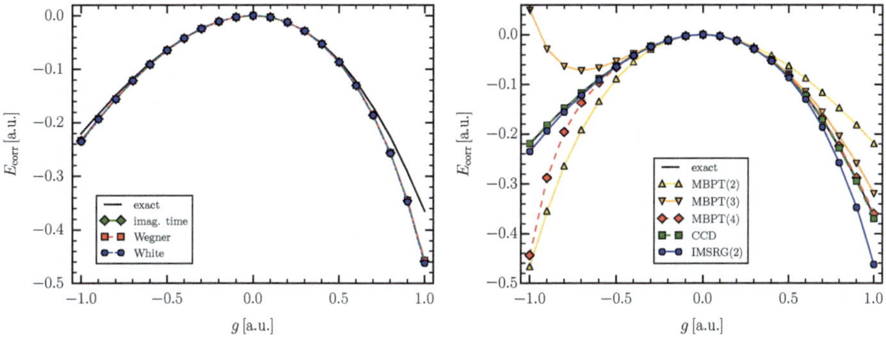

Fig. 10.10 Ground-state correlation energies as a function of the pairing strength, for $\delta = 1.0$. *Left panel:* IMSRG(2) correlation energies for different flow generators. *Right panel:* Comparison of IMSRG(2) and other many-body methods

In the range $-0.5 \lesssim g \lesssim 0.5$, the IMSRG(2) is in excellent agreement with the exact diagonalization. This range corresponds to a region of weak correlations between the four particles that we consider in our system, because the ratio between the characteristic excitation scale of the uncorrelated many-body states [diagonal matrix elements of Eq. (10.42)] and the pairing strength (off-diagonal matrix elements) is small,

$$\left|\frac{g}{2\Delta E}\right| = \left|\frac{g}{4\delta}\right| \leq 0.125\,. \tag{10.171}$$

Going beyond $g \geq 0.5$, the correlations grow stronger, and the IMSRG(2) starts to overestimate the size of the correlation energy. The absolute deviation is about 0.1 at $g = 1.0$, where the exact eigenvalue of the pairing Hamiltonian is 0.6355.

Interestingly, the deviations are much smaller in the opposite case, $g = -1.0$. The right-hand panel of Fig. 10.10 sheds further light on this matter. There, we compare the IMSRG(2) correlation energy to results from finite-order many-body perturbation theory (MBPT) and Coupled Cluster with Doubles Excitations (CCD)—see Chap. 8 for details. We see that the correlation energies from finite-order MBPT alternate in sign for a repulsive interaction, which suggests that the agreement between the exact solution, IMSRG(2), and CCD is due to cancellations at all orders that these methods take into account. Looking back at the attractive pairing force for $g \sim 1.0$, we observe that the MBPT results have converged to the exact solution once fourth-order corrections are taken into account, and CCD gives the same result. The IMSRG(2) overestimates the correlation energy in this region because it undercounts a set of four repulsive fourth-order diagrams by a factor $1/2$, while CCD takes them into account completely (see [53, 134]).

10.3.6.2 Targeting Excited States

In Sect. 10.3.3.1, we mentioned that the choice of reference state will affect which eigenstate of the Hamiltonian the IMSRG evolution is targeting. To illustrate this, Fig. 10.11 shows the IMSRG(2) flows starting from uncorrelated states with energies $4 - g$ (with nucleons occupying the $p = 1$ and $p = 2$ single-particle states, see Table 10.1) and $6 - g$ (with $p = 0$ and $p = 2$ occupied), respectively. For $g = 0.5$ and $\delta = 1.0$, the exact eigenvalues are 3.4707 and 5.5.

As we can see in the left panel, the IMSRG(2) flow towards the correct energy. Between $s = 1$ and $s = 3$, it overshoots the exact energy. As discussed in Sect. 10.3.3.2, the IMSRG(2) is not a variational method, so this is unproblematic. In the limit $s \to \infty$, the flow converges to 3.4708, matching the energy of the first excited state. The flow shown in the right panel of Fig. 10.11 is particularly interesting. Recall from Sect. 10.2.3.4 that the state with energy 5.5 is already a degenerate eigenstate of the Hamiltonian, and therefore supposed to be invariant under a unitary flow. This is indeed the case, at least to the eight digits recorded in the flow data file `imsrg-white_d1.0_g+0.5_N4_ev3.flow`. It also means that all perturbative corrections in the IMSRG(2) summation must cancel out. While the second- and third-order energy corrections are of opposite sign, they are not of the same size. Thus, contributions from fourth and higher orders are involved in the cancellation, and the invariance of $E(s)$ demonstrates that they are indeed generated by the IMSRG(2) flow. We conclude our discussion of the IMSRG treatment of the pairing Hamiltonian here.

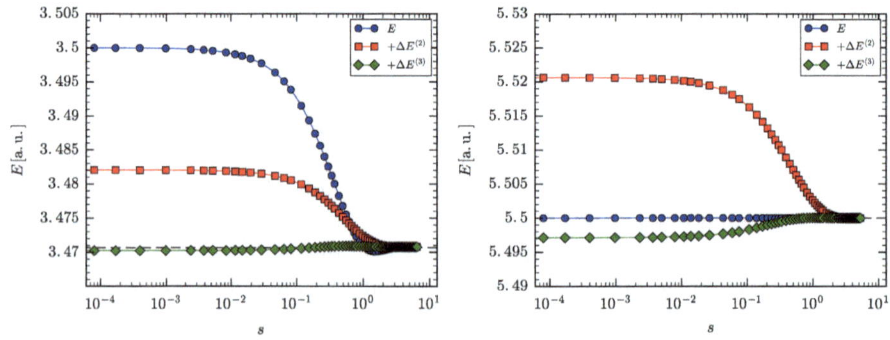

Fig. 10.11 IMSRG(2) flow for different reference states, using the pairing Hamiltonian with $\delta = 1.0$, $g = 0.5$. Calculations were performed with the White generator, Eq. (10.141). For Slater determinants with uncorrelated energies $4 - g = 3.5$ (*left panel*) and $6 - g = 5.5$ (*right panel*), the IMSRG(2) flow targets the 2nd and 3rd excited states, see Sect. 10.2.3.4

10.3.7 Infinite Neutron Matter

After discussing the pairing Hamiltonian, we now want to apply the IMSRG(2) to a large-scale problem, namely the calculation of the equation of state for pure neutron matter. Python scripts and data are available at https://github.com/ManyBodyPhysics/LectureNotesPhysics/tree/master/Programs/Chapter10-programs/python/imsrg_pnm. The C++ program is available from https://github.com/ManyBodyPhysics/LectureNotesPhysics/tree/master/Programs/Chapter10-programs/cpp/imsrg_pnm.

We work in a basis of plane wave states, which is set up just like in the CC case discussed in Sect. 8.7. We work in a spherical periodic cell $L = (N/\rho)^{1/3}$, where N is the number of neutrons in the cell, and ρ the neutron matter density. Because of the periodic boundary conditions, the momenta p_x, p_y, p_z are discretized, and we can write the single-particle states as $|n_x, n_y, n_z, s_z\rangle$ (s_z is the spin projection of the neutron). We impose the truncation

$$n_x^2 + n_y^2 + n_z^2 \leq N_{\max} . \tag{10.172}$$

In Fig. 10.12, we show the IMSRG(2) results for neutron matter for the semi-realistic Minnesota *NN* potential [135], using $N = 66$ neutrons in an $N_{\max} = 36$ basis. The left panel shows the IMSRG(2) equation of state (EOS) which is essentially the same as that of methods with comparable correlation content, namely CCD, the Configuration Interaction Monte Carlo based on CCD wave functions (CIMC, see Sect. 9.6.1) and the Self-Consistent Green's Functions in the ADC(3) scheme (Sect. 11.3). We also include the MBPT(2) EOS, which is in very good agreement with the more sophisticated methods. This shows that pure neutron matter is only weakly correlated, and the many-body expansion is clearly converging rapidly. The IMSRG(2) (and other methods) gain an additional 10% additional

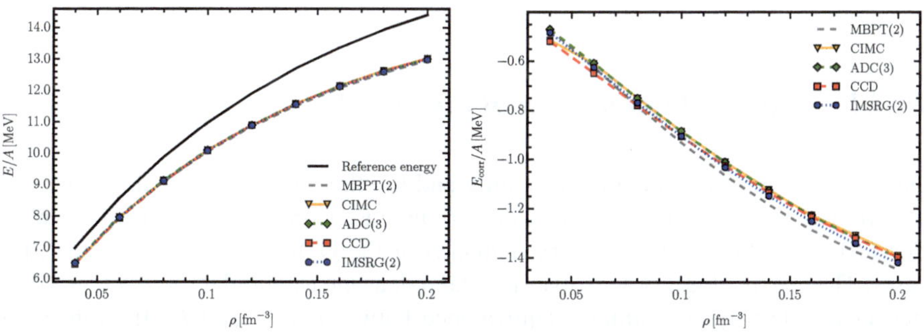

Fig. 10.12 Equation of state for pure neutron matter from IMSRG(2) and other many-body methods, based on the Minnesota *NN* potential [135]. The *left panel* shows the energy per particle obtained with IMSRG(2), CCD (Chap. 8), CIMC (Chap. 9), and the ADC(3) Self-Consistent Green's Function scheme (Chap. 11), the *right panel* the correlation energy per particle from the same methods

energy compared to the uncorrelated HF EOS, all the way from dilute matter at $\rho = 0.05\,\mathrm{fm}^{-3}$ to $\rho = 0.2\,\mathrm{fm}^{3}$, which is more than twice the neutron density of typical stable nuclei.

In the right panel of Fig. 10.12, we show the correlation energy per particle, which reveals some differences between the various methods. As the neutron matter density comes and correlations are expected to become increasingly important, MBPT(2) gains the highest amount of correlation energy, just as in our results for the pairing Hamiltonian (cf. Fig. 10.10). Curiously, CCD gives the most binding of all methods in dilute neutron matter, but eventually, the CCD correlation energy is very similar to that of CIMC and ADC(3), which should be superior approximations to the exact ground state (see Chaps. 9 and 11). As the density increases the IMSRG(2) starts to gain more binding from correlations than CCD, its closest cousin among the considered methods, but not as much additional binding energy as MBPT(2). This reflects our findings for the pairing Hamiltonian, where we observed the same phenomenon (see Fig. 10.10). As explained in Sect. 10.3.6, the reason for this energy gain compared to CCD is the under-counting of certain repulsive fourth-order diagrams in the IMSRG(2), see [53].

10.4 Current Developments

After covering the essential concepts of the SRG and IMSRG, and discussing both the formal and technical aspects of their applications, we want to introduce our readers to the three major directions of current IMSRG research: These are the use of the so-called Magnus expansion to explicitly construct the IMSRG transformation (Sect. 10.4.1), the Multireference IMSRG for generalizing the method to correlated reference states (Sect. 10.4.2), and the construction of effective Hamiltonians for use in configuration interaction and Equation-of-Motion methods, which allows us to tackle excited states (Sect. 10.4.3).

10.4.1 Magnus Formulation of the IMSRG

Despite its modest computational scaling and the flexibility to tailor the generator to different applications, IMSRG calculations based on the direct integration of Eqs. (10.8) are limited by memory demands of the ODE solver in many realistic cases. The use of a high-order solver is essential, as the accumulation of integration-step errors destroys the unitary equivalence between $H(s)$ and $H(0)$ even if no truncations are made in the flow equations. State-of-the-art solvers can require the storage of 15–20 copies of the solution vector in memory, which is the main computational bottleneck of the method (see, e.g., [79, 81, 136]).

Matters are complicated further if we also want to calculate expectation values for observables besides the Hamiltonian. General operators have to be evolved consistently using the flow equation

$$\frac{d}{ds}\hat{O}(s) = [\hat{\eta}(s), \hat{O}(s)] , \tag{10.173}$$

but since storage of $\hat{\eta}(s)$ at every point of the flow trajectory is prohibitively expensive, we are forced to solve Eq. (10.173) simultaneously with the flow equation for the Hamiltonian. The evaluation of N observables besides the Hamiltonian implies that the dimension of the ODE system (10.104)–(10.106) grows by a factor $N + 1$. In addition, generic operators can evolve with rather different characteristic scales than the Hamiltonian, increasing the likelihood of the ODEs becoming stiff.

We can now overcome these limitations by re-formulating the IMSRG using the Magnus expansion from the theory of matrix differential equations [137, 138]. Magnus proved that the path-ordered series defining the IMSRG transformation, Eq. (10.12), can be summed into a true exponential expression if the generator $\hat{\eta}$ meets certain conditions (see [137]):

$$\hat{U}(s) \equiv e^{\hat{\Omega}(s)} . \tag{10.174}$$

This allows us to derive a flow equation for the anti-Hermitian Magnus operator $\hat{\Omega}(s)$:

$$\frac{d\hat{\Omega}}{ds} = \sum_{k=0}^{\infty} \frac{B_k}{k!} \mathrm{ad}^k_{\hat{\Omega}}(\hat{\eta}) , \tag{10.175}$$

where B_k are the Bernoulli numbers, and

$$\mathrm{ad}^0_{\hat{\Omega}}(\hat{\eta}) = \hat{\eta} \tag{10.176}$$

$$\mathrm{ad}^k_{\hat{\Omega}}(\hat{\eta}) = [\hat{\Omega}, \mathrm{ad}^{k-1}_{\hat{\Omega}}(\hat{\eta})] . \tag{10.177}$$

As in the standard IMSRG(2), we truncate $\hat{\eta}$ and $\hat{\Omega}$ as well as their commutator at the two-body level. We refer to the resulting calculation scheme as the Magnus(2) formulation of the IMSRG. The series of nested commutators generated by $\mathrm{ad}^k_{\hat{\Omega}}$ is evaluated recursively, until satisfactory convergence of the right-hand side of Eq. (10.175) is reached [72]. At each integration step, we use $\hat{U}(s)$ to construct the Hamiltonian $\hat{H}(s)$ via the Baker-Campbell-Hausdorff (BCH) formula

$$\hat{H}(s) \equiv e^{\hat{\Omega}(s)} \hat{H}(0) e^{-\hat{\Omega}(s)} = \sum_{k=0}^{\infty} \frac{1}{k!} \mathrm{ad}^k_{\hat{\Omega}(s)} \left(\hat{H}(0)\right) , \tag{10.178}$$

(the flow parameter dependence is stated explicitly here for clarity). Like $\hat{\eta}$ and $\hat{\Omega}$, the Hamiltonian is truncated at the two-body level.

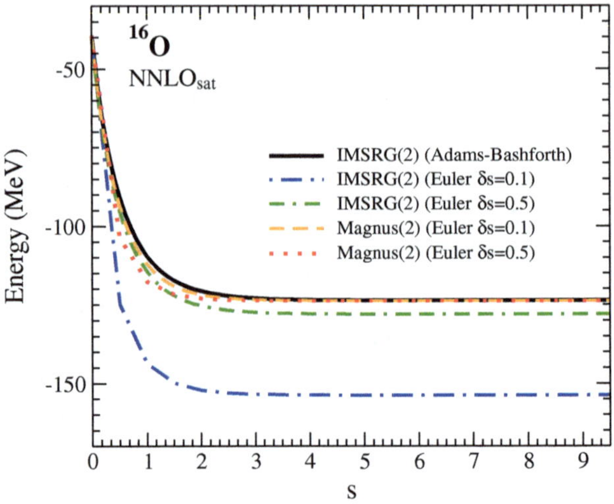

Fig. 10.13 IMSRG(2) and Magnus(2) ground-state energy of ^{16}O, starting from the NNLO$_{\text{sat}}$ $NN + 3N$ interaction [17]. The *solid black line* is the IMSRG(2) flow obtained with an adaptive solver based on the Adams-Bashforth method [80], while the *other lines* show Magnus(2) and IMSRG(2) flows obtained with a simple Forward Euler method and different fixed step sizes. All calculations were done in an $e_{\max} = 8$ model space, with $\hbar\omega = 24\,\text{MeV}$ for the underlying harmonic-oscillator basis

A major advantage of the Magnus formulation stems from the fact that the flow equations for $\hat{\Omega}(s)$ can be solved using a simple first-order Euler step method without any loss of accuracy, resulting in substantial memory savings and a modest reduction in CPU time. While sizable integration-step errors accumulate in $\Omega(s)$ with a first-order method, upon exponentiation the transformation is still unitary, and the transformed $H(s) = \hat{U}(s)H\hat{U}^{\dagger}(s)$ is unitarily equivalent to the initial Hamiltonian aside from the truncations made while evaluating the BCH formula. For further details on the implementation of the Magnus formulation, see [72].

The insensitivity of the Magnus scheme to integration-step errors is illustrated in Fig. 10.13, which shows flowing ground-state energy for ^{16}O, calculated with the chiral NNLO$_{\text{sat}}$ $NN + 3N$ interaction by Ekström et al. [17]. The black solid line denotes the results of a standard IMSRG(2) calculation using a high-order predictor-corrector solver [79, 136], while the other curves denote IMSRG(2) and Magnus(2) calculations using a first-order Euler method with different step sizes δs. Unsurprisingly, the IMSRG(2) calculations using a first-order Euler method are very poor, with different step sizes converging to different large-s limits. The Magnus(2) calculations, on the other hand, converge to the same large-s limit in excellent agreement with the standard IMSRG(2) results.

A second major advantage of the Magnus formulation of the IMSRG is that we can evaluate arbitrary observables by simply using the final Magnus operator $\hat{\Omega}(\infty)$ to calculate

$$\hat{O}(\infty) \equiv e^{\hat{\Omega}(\infty)} \hat{O}(0) e^{-\hat{\Omega}(\infty)} . \tag{10.179}$$

This is obviously much more convenient than dealing with observables in the direct IMSRG(2) approach, because we do not have to deal with the doubling, tripling, ... of the already large system of flow equations. In contrast to the prohibitive space required to store $\hat{\eta}(s)$ for all values of s, we can also easily archive the $\Omega(\infty)$ for long-term use, e.g., when we want to look at new observables in future studies.

10.4.2 The Multi-Reference IMSRG

10.4.2.1 Correlated Reference States

Many-body bases built from a single Slater determinant and its particle-hole excitations work best for systems with large gaps in the single-particle spectrum, e.g., closed-shell nuclei. If the gap is small, particle-hole excited basis states can be near-degenerate with the reference determinant, which results in strong configuration mixing. When the mixing involves configurations in which many nucleons are excited simultaneously, many-body physicists speak of *static* or *collective correlations* in the wave function, as opposed to *dynamic correlations* that are caused by the excitations of a small number of nucleons only.

Important examples are the emergence of nuclear superfluidity [139] or diverse rotational and vibrational bands in open-shell nuclei (see, e.g., [140]). These phenomena are conveniently described by using the concept of intrinsic wave functions that explicitly break appropriate symmetries of the Hamiltonian. For instance, nuclear superfluidity can be treated in to leading-order in the self-consistent Hartree-Fock-Bogoliubov (HFB) approach, which is formulated in terms of anti-symmetrized product states of fermionic quasi-particles that are superpositions of particles and holes. Because of this, the intrinsic HFB wave functions are superpositions of states with different particle numbers. The broken symmetries must eventually be restored by means of projection methods, which have a long history in nuclear many-body theory [117, 141–151]).

The standard IMSRG framework as described in Sect. 10.3 works with an uncorrelated reference state, and therefore puts the entire onus of describing correlations on the transformation $\hat{U}(s)$. The computational cost limits us to the IMSRG(2) scheme, or an eventual approximate IMSRG(3) that is roughly analogous to completely renormalized Coupled Cluster schemes with approximate triples [119]. An IMSRG(4), let alone the complete IMSRG(A) scheme, are not feasible. Thus, the IMSRG, like CC, is best suited to the description of dynamic correlations. We can mitigate this shortcoming by extending the IMSRG to work with correlated reference states, and building static correlation that would otherwise require an IMSRG(4),... scheme directly into the reference state. This leads us to the Multi-Reference IMSRG (MR-IMSRG) [46, 73, 101], which is constructed using the generalized normal ordering and Wick's theorem developed by Kutzelnigg and Mukherjee [152, 153].

10.4.2.2 Generalized Normal Ordering

In [152], Kutzelnigg and Mukherjee developed a generalized normal ordering for arbitrary reference states. In the brief discussion that follows, we use the slightly different notation of Kong et al. [154].

First, we introduce a pseudo-tensorial notation for strings of creation and annihilation operators, to facilitate book-keeping and make the formalism more compact. A product of k creators and annihilators each is written as

$$\hat{A}^{i_1\ldots i_k}_{j_1\ldots j_k} \equiv a^\dagger_{i_1}\ldots a^\dagger_{i_k} a_{j_k}\ldots a_{j_1}\,. \tag{10.180}$$

We do not consider particle-number changing operators in the present work, because they cause ambiguities in the contraction and sign rules for the A operators that are defined in the following. The anticommutation relations imply

$$\hat{A}^{\mathscr{P}(i_1\ldots i_k)}_{\mathscr{P}'(j_1\ldots j_k)} = (-1)^{\pi(\mathscr{P})+\pi(\mathscr{P}')}\hat{A}^{i_1\ldots i_k}_{j_1\ldots j_k}\,, \tag{10.181}$$

where $\pi(\mathscr{P}) = \pm 1$ indicates the parity (or signature) of a permutation $\mathscr{P}$. A general k-body operator can now be written as

$$O^{(k)} = \frac{1}{(k!)^2}\sum_{\substack{i_1\ldots i_k\\ j_1\ldots j_k}} o^{i_1\ldots i_k}_{j_1\ldots j_k}\hat{A}^{i_1\ldots i_k}_{j_1\ldots j_k}\,, \tag{10.182}$$

where we assume that the coefficients $o^{i_1\ldots i_k}_{j_1\ldots j_k}$ are antisymmetrized, and therefore also obey Eq. (10.181) under index permutations.

Next, we introduce *irreducible k-body density matrices* $\lambda^{(k)}$. In the one-body case, we have the usual density matrix

$$\lambda^i_j \equiv \langle\Phi|\,\hat{A}^i_j\,|\Phi\rangle\,, \tag{10.183}$$

and for future use, we also define

$$\xi^i_j \equiv \lambda^i_j - \delta^i_j\,. \tag{10.184}$$

Up to a factor (-1) that unifies the sign rules for one-body contractions presented below, $\xi^{(1)}$ is simply the generalization of the hole density matrix for a correlated state [cf. Eq. (10.96)]. In the natural orbital basis, both one-body density matrices are diagonal, with fractional occupation numbers $0 \leq n_i, \bar{n}_i \leq 1$ as eigenvalues.

For $k \geq 2$, we denote full density matrices by

$$\rho^{i_1...i_k}_{j_1..j_k} = \langle \Phi | \hat{A}^{i_1...i_k}_{j_1..j_k} | \Phi \rangle \,, \tag{10.185}$$

and define

$$\lambda^{ij}_{kl} \equiv \rho^{ij}_{kl} - \mathscr{A}\{\lambda^i_k \lambda^j_l\}\,, \tag{10.186}$$

$$\lambda^{ijk}_{lmn} \equiv \rho^{ijk}_{lmn} - \mathscr{A}\{\lambda^i_l \lambda^{jk}_{mn}\} - \mathscr{A}\{\lambda^i_l \lambda^j_m \lambda^k_n\}\,, \tag{10.187}$$

etc., where $\mathscr{A}\{\ldots\}$ fully antisymmetrizes the indices of the expression within the brackets, e.g.,

$$\mathscr{A}\{\lambda^i_k \lambda^j_l\} = \lambda^i_k \lambda^j_l - \lambda^i_l \lambda^j_k\,. \tag{10.188}$$

From Eq. (10.186), it is easy to see that $\lambda^{(2)}$ encodes the two-nucleon correlation content of the reference state $|\Phi\rangle$. If the reference state is a Slater determinant, i.e., an independent-particle state, the full two-body density matrix factorizes, and $\lambda^{(2)}$ vanishes:

$$\lambda^{ij}_{kl} = \rho^{ij}_{kl} - \mathscr{A}\{\lambda^i_k \lambda^j_l\} = \lambda^i_k \lambda^j_l - \lambda^i_k \lambda^j_l - \left(\lambda^i_k \lambda^j_l - \lambda^i_k \lambda^j_l\right) = 0\,. \tag{10.189}$$

Equation (10.187) shows that $\lambda^{(3)}$ is constructed by subtracting contributions from three independent particles as well as two correlated nucleons in the presence of an independent spectator particle from the full three-body density matrix, and therefore encodes the genuine three-nucleon correlations. This construction and interpretation generalizes to irreducible density matrices of rank k.

Normal-ordered one-body operators are constructed in the same manner as in the standard normal ordering of Sect. 10.3.1:

$$\{A^a_b\} \equiv A^a_b - \langle \Phi | A^a_b | \Phi \rangle = A^a_b - \lambda^a_b\,. \tag{10.190}$$

For a two-body operator, we have the expansion

$$A^{ab}_{cd} = \{A^{ab}_{cd}\} + \lambda^a_c\{A^b_d\} - \lambda^a_d\{A^b_c\} + \lambda^b_d\{A^a_c\} - \lambda^b_c\{A^a_d\} + \lambda^a_c\lambda^b_d - \lambda^a_d\lambda^b_c + \lambda^{ab}_{cd}\,. \tag{10.191}$$

As a consequence of Eq. (10.181), the sign of each term is determined by the product of the parities of the permutations that map upper and lower indices to their ordering in the initial operator. Except for the last term, this expression looks like the result for the regular normal ordering, with pairwise contractions of indices giving rise to one-body density matrices. The last term, a contraction of four indices, appears because we are dealing with an arbitrary, correlated reference state here. For a three-body operator, we obtain schematically

$$\begin{aligned} A^{abc}_{def} &= \{A^{abc}_{def}\} + \mathscr{A}\{\lambda^{a}_{d}\{A^{bc}_{ef}\}\} + \mathscr{A}\{\lambda^{a}_{d}\lambda^{b}_{e}\{A^{c}_{f}\}\} + \mathscr{A}\{\lambda^{ab}_{de}\{A^{c}_{f}\}\} \\ &\quad + \lambda^{abc}_{def} + \mathscr{A}\{\lambda^{a}_{d}\lambda^{bc}_{ef}\} + \mathscr{A}\{\lambda^{a}_{d}\lambda^{b}_{e}\lambda^{c}_{f}\}\,, \end{aligned} \tag{10.192}$$

and the procedure can be extended to higher particle rank in an analogous fashion.

When we work with arbitrary reference states, the regular Wick's theorem of Sect. 10.3.1 is extended with additional contractions:

$$\{A^{a\boxed{b}}_{cd}\}\{A^{ij}_{\boxed{k}l}\} = -\lambda^{\boxed{b}}_{\boxed{k}}\{A^{aij}_{cdl}\}\,, \tag{10.193}$$

$$\{A^{ab}_{\boxed{c}d}\}\{A^{i\boxed{j}}_{kl}\} = -\xi^{\boxed{j}}_{\boxed{c}}\{A^{bia}_{dkl}\}\,, \tag{10.194}$$

$$\{A^{\boxed{ab}}_{cd}\}\{A^{ij}_{\boxed{kl}}\} = +\lambda^{\boxed{ab}}_{\boxed{kl}}\{A^{ij}_{cd}\}\,, \tag{10.195}$$

$$\{A^{a\boxed{b}}_{cd}\}\{A^{\boxed{i}j}_{\boxed{kl}}\} = -\lambda^{\boxed{ib}}_{\boxed{kl}}\{A^{aj}_{cd}\}\,, \tag{10.196}$$

$$\{A^{ab}_{\boxed{c}d}\}\{A^{\boxed{ij}}_{\boxed{k}l}\} = -\lambda^{\boxed{ij}}_{\boxed{ck}}\{A^{ab}_{dl}\}\,, \tag{10.197}$$

$$\{A^{\boxed{ab}}_{c\boxed{d}}\}\{A^{\boxed{i}j}_{\boxed{kl}}\} = -\lambda^{\boxed{abi}}_{\boxed{dkl}}\{A^{j}_{c}\}\,, \tag{10.198}$$

$$\{A^{\boxed{ab}}_{\boxed{cd}}\}\{A^{\boxed{ij}}_{\boxed{kl}}\} = +\lambda^{\boxed{abij}}_{\boxed{cdkl}}\,. \tag{10.199}$$

The new contractions (10.195)–(10.199) increase the number of terms when we expand operator products. Fortunately, the overall increase in complexity is manageable.

Applying the generalized normal ordering to the intrinsic nuclear A-body Hamiltonian (10.86) we obtain

$$\hat{H} = E + \sum_{ij} f^{i}_{j}\{A^{i}_{j}\} + \frac{1}{4}\sum_{ijkl}\Gamma^{ij}_{kl}\{A^{ij}_{kl}\} + \frac{1}{36}\sum_{ijklmn} W^{ijk}_{lmn}\{A^{ijk}_{lmn}\}\,, \tag{10.200}$$

with the individual contributions

$$E \equiv \left(1 - \frac{1}{A}\right)\sum_{ab} t^{a}_{b}\lambda^{a}_{b} + \frac{1}{4}\sum_{abcd}\left(\frac{1}{A}t^{ab}_{cd} + v^{ab}_{cd}\right)\rho^{ab}_{cd} + \frac{1}{36}\sum_{abcdef} v^{abc}_{def}\rho^{abc}_{def}\,, \tag{10.201}$$

$$f^{i}_{j} \equiv \left(1 - \frac{1}{A}\right)t^{i}_{j} + \sum_{ab}\left(\frac{1}{A}t^{ia}_{jb} + v^{ia}_{jb}\right)\lambda^{a}_{b} + \frac{1}{4}\sum_{abcd} v^{iab}_{jcd}\rho^{ab}_{cd}\,, \tag{10.202}$$

$$\Gamma^{ij}_{kl} \equiv \frac{1}{A}t^{ij}_{kl} + v^{ij}_{kl} + \sum_{ab} v^{ija}_{klb}\lambda^{a}_{b}\,, \tag{10.203}$$

$$W^{ijk}_{lmn} \equiv v^{ijk}_{lmn}\,. \tag{10.204}$$

Here, we use the full density matrices for compactness, but it is easy to express Eqs. (10.201)–(10.203) completely in terms of irreducible density matrices by using Eqs. (10.186) and (10.187).

10.4.2.3 MR-IMSRG Flow Equations

We evaluate the operator flow equation (10.8) using the generalized Wick's theorem, truncating all operators the two-body level, and obtain the MR-IMSRG(2) flow equations [46, 73, 101]:

$$\begin{aligned}\frac{dE}{ds} &= \sum_{ab}(n_a - n_b)\eta^{a}_{b}f^{b}_{a} + \frac{1}{4}\sum_{abcd}\left(\eta^{ab}_{cd}\Gamma^{cd}_{ab} - \Gamma^{ab}_{cd}\eta^{cd}_{ab}\right) n_a n_b \bar{n}_c \bar{n}_d \\ &\quad + \frac{1}{4}\sum_{abcd}\left(\frac{d}{ds}\Gamma^{ab}_{cd}\right)\lambda^{ab}_{cd} + \frac{1}{4}\sum_{abcdklm}\left(\eta^{ab}_{cd}\Gamma^{kl}_{am} - \Gamma^{ab}_{cd}\eta^{kl}_{am}\right)\lambda^{bkl}_{cdm}\,, \end{aligned} \tag{10.205}$$

$$\begin{aligned}\frac{d}{ds}f^{i}_{j} &= \sum_{a}\left(\eta^{i}_{a}f^{a}_{j} - f^{i}_{a}\eta^{a}_{j}\right) + \sum_{ab}\left(\eta^{a}_{b}\Gamma^{bi}_{aj} - f^{a}_{b}\eta^{bi}_{aj}\right)(n_a - n_b) \\ &\quad + \frac{1}{2}\sum_{abc}\left(\eta^{ia}_{bc}\Gamma^{bc}_{ja} - \Gamma^{ia}_{bc}\eta^{bc}_{ja}\right)(n_a\bar{n}_b\bar{n}_c + \bar{n}_a n_b n_c) \\ &\quad + \frac{1}{4}\sum_{abcde}\left(\eta^{ia}_{bc}\Gamma^{de}_{ja} - \Gamma^{ia}_{bc}\eta^{de}_{ja}\right)\lambda^{de}_{bc} + \sum_{abcde}\left(\eta^{ia}_{bc}\Gamma^{be}_{jd} - \Gamma^{ia}_{bc}\eta^{be}_{jd}\right)\lambda^{ae}_{cd} \\ &\quad - \frac{1}{2}\sum_{abcde}\left(\eta^{ia}_{jb}\Gamma^{cd}_{ae} - \Gamma^{ia}_{jb}\eta^{cd}_{ae}\right)\lambda^{cd}_{be} + \frac{1}{2}\sum_{abcde}\left(\eta^{ia}_{jb}\Gamma^{bc}_{de} - \Gamma^{ia}_{jb}\eta^{bc}_{de}\right)\lambda^{ac}_{de}\,, \end{aligned} \tag{10.206}$$

$$\begin{aligned}\frac{d}{ds}\Gamma^{ij}_{kl} &= \sum_{a}\left(\eta^{i}_{a}\Gamma^{aj}_{kl} + \eta^{j}_{a}\Gamma^{ia}_{kl} - \eta^{a}_{k}\Gamma^{ij}_{al} - \eta^{a}_{l}\Gamma^{ij}_{ka} - f^{i}_{a}\eta^{aj}_{kl} - f^{j}_{a}\eta^{ia}_{kl} + f^{a}_{k}\eta^{ij}_{al} + f^{a}_{l}\eta^{ij}_{ka}\right) \\ &\quad + \frac{1}{2}\sum_{ab}\left(\eta^{ij}_{ab}\Gamma^{ab}_{kl} - \Gamma^{ij}_{ab}\eta^{ab}_{kl}\right)(1 - n_a - n_b) \\ &\quad + \sum_{ab}(n_a - n_b)\left(\left(\eta^{ia}_{kb}\Gamma^{jb}_{la} - \Gamma^{ia}_{kb}\eta^{jb}_{la}\right) - \left(\eta^{ja}_{kb}\Gamma^{ib}_{la} - \Gamma^{ja}_{kb}\eta^{ib}_{la}\right)\right)\,. \end{aligned} \tag{10.207}$$

All single-particle indices and occupation numbers (cf. Sect. 10.3.1) refer to natural orbitals, and the s-dependence has been suppressed for brevity. Because we use general reference states, the MR-IMSRG flow equations also include couplings to correlated pairs and triples of nucleons in that state through the irreducible density matrices $\lambda^{(2)}$ and $\lambda^{(3)}$. The single-reference limit [Eqs. (10.104)–(10.106)] can be obtained by setting the irreducible density matrices $\lambda^{(2)}$ and $\lambda^{(3)}$ to zero in the previous expressions.

Superficially, the computational cost for the evaluation of the MR-IMSRG(2) flow equations is dominated by the final term of Eq. (10.205), which is of $\mathcal{O}(N^7)$. However, since storage of the complete $\lambda^{(3)}$ is prohibitive in large-scale calculations, we impose certain constraints on the reference state, which in turn restrict the non-zero matrix elements to small subsets of the entire matrix. For example, for particle-number projected HFB reference states, $\lambda^{(3)}$ is almost diagonal, which reduces the effort for the zero-body flow equation to $\mathcal{O}(N^4)$. We have also explored reference states from No-Core Shell Model calculations in a small model space [155], which limit the indices of $\lambda^{(3)}$ to 5–10 single-particle states out of a complete single-particle basis that is one to two orders of magnitude larger. In a similar scenario, we have used reference states consisting of a valence space (or active space, in chemistry parlance) on top of an inert core, as in the traditional nuclear Shell model (cf. Sect. 10.4.3). In that case, the correlations are restricted to this valence space, and $\lambda^{(3)}$ is only non-zero if all indices refer to valence space (active space) single-particle states. Thus, the main driver of the computational effort is still the two-body flow equation, at $\mathcal{O}(N^6)$, just like in the regular IMSRG(2). Equation (10.207) actually has exactly the same for as its single-reference counterpart, Eq. (10.106), except that the occupation numbers can now have arbitrary values between 0 and 1.

10.4.2.4 Decoupling and Generators

In the multireference case, we choose a suitable correlated reference state, and construct its excitations by applying all possible one- and two-body operators:

$$|\Phi\rangle,\ \{\hat{A}^{i}_{j}\}\,|\Phi\rangle,\ \{\hat{A}^{ij}_{kl}\}\,|\Phi\rangle, \ldots\,. \tag{10.208}$$

The properties of the normal ordering ensure that the excited states are orthogonal to the reference state, but they are in general not orthogonal to each other: for instance,

$$\langle\Phi|\,\{A^{i}_{j}\}\{A^{k}_{l}\}\,|\Phi\rangle = -\lambda^{i}_{l}\xi^{k}_{j} + \lambda^{ij}_{kl} = n_i\bar{n}_j\delta^{i}_{l}\delta^{k}_{j} + \lambda^{ij}_{kl}\,, \tag{10.209}$$

where $0 \leq n_i, \bar{n}_i \leq 1$. Moreover, there can be linear dependencies between the excitations of the correlated reference state, so the matrix representations of the Hamiltonian and other operators in this basis can be rank deficient. While the rank deficiency poses a major challenge for multireference CC methods (see, e.g., [156]), for us it only means that we are implementing the MR-IMSRG flow on a matrix that has spurious zero eigenvalues that are typically far removed from the low-lying part of the spectrum in which we are most interested.

To identify the off-diagonal Hamiltonian, we can proceed like in the single-reference IMSRG, and try to satisfy the decoupling conditions

$$\langle\Phi|\,\hat{H}(\infty)\{\hat{A}^{i}_{j}\}\,|\Phi\rangle = 0\,, \tag{10.210}$$

$$\langle\Phi|\,\hat{H}(\infty)\{\hat{A}^{ij}_{kl}\}\,|\Phi\rangle = 0\,, \tag{10.211}$$

$$\dots$$

and corresponding conditions for the conjugate matrix elements. The matrix elements can be evaluated with the generalized Wick's theorem, e.g.,

$$\begin{aligned}\langle\Phi|\,H\{\hat{A}^{i}_{j}\}\,|\Phi\rangle &= \bar{n}_i n_j f^{j}_{i} + \sum_{ab} f^{a}_{b}\lambda^{ai}_{bj} + \frac{1}{2}\sum_{abc}\left(\bar{n}_i\lambda^{bc}_{ja}\Gamma^{bc}_{ia} - n_j\Gamma^{ja}_{bc}\lambda^{ia}_{bc}\right)\\ &+ \frac{1}{4}\sum_{abcd}\Gamma^{ab}_{cd}\lambda^{iab}_{jcd}\,.\end{aligned} \tag{10.212}$$

The first term is merely the generalization of the one-body particle-hole matrix element from Sect. 10.3.3: In the single-reference limit, the occupation number prefactor is nonzero if i and j are particle and hole indices, respectively. In addition, the matrix element depends on the irreducible densities $\lambda^{(2)}$ and $\lambda^{(3)}$ due to the coupling of the Hamiltonian to correlated pairs and triples of nucleons in the reference state. The coupling condition to two-nucleon excitations, Eq. (10.211), not only has a much more complicated structure than its single-reference counterpart, but even depends on $\lambda^{(4)}$ (see [73] for details). Constructing and storing $\lambda^{(4)}$ is essentially out of the question in general MR-IMSRG applications, hence we are forced to introduce truncations to evaluate Eq. (10.212) and similar matrix elements. This implies that we can only achieve approximate decoupling in general.

In recent applications, we have found the variational perspective introduced in Sect. 10.3.3.2 to be useful. We can write the decoupling conditions as

$$\langle\Phi|\,H\{\hat{A}^{i}_{j}\}\,|\Phi\rangle = \frac{1}{2}\langle\Phi|\,\{H,\{\hat{A}^{i}_{j}\}\}\,|\Phi\rangle + \frac{1}{2}\langle\Phi|\,[H,\{\hat{A}^{i}_{j}\}]\,|\Phi\rangle\,, \tag{10.213}$$

$$\langle\Phi|\,H\{\hat{A}^{ij}_{kl}\}\,|\Phi\rangle = \frac{1}{2}\langle\Phi|\,\{H,\{\hat{A}^{ij}_{kl}\}\}\,|\Phi\rangle + \frac{1}{2}\langle\Phi|\,[H,\{\hat{A}^{ij}_{kl}\}]\,|\Phi\rangle\,, \tag{10.214}$$

and at least suppress the second terms in both equations through what amounts to a minimization of the ground-state energy under unitary variation. This means that we aim to satisfy the IBCs introduced in Sect. 10.3.3.2, (also see [64, 126–128]).

Evaluating the commutators, we obtain

$$\langle\Phi|\,[H,\{\hat{A}^i_j\}]\,|\Phi\rangle = (n_j - n_i)f^j_i - \frac{1}{2}\sum_{abc}\left(\Gamma^{ja}_{bc}\lambda^{ia}_{bc} - \Gamma^{ab}_{ic}\lambda^{ab}_{jc}\right), \tag{10.215}$$

$$\begin{aligned}\langle\Phi|\,[H,\{\hat{A}^{ij}_{kl}\}]\,|\Phi\rangle &= \Gamma^{kl}_{ij}(\bar{n}_i\bar{n}_j n_k n_l - n_i n_j \bar{n}_k\bar{n}_l) + \sum_a\left((1-P_{ij})f^a_i\lambda^{aj}_{kl} - (1-P_{kl})f^k_a\lambda^{ij}_{al}\right)\\ &+ \frac{1}{2}\left((\lambda\Gamma)^{kl}_{ij}\,(1-n_i-n_j) - (\Gamma\lambda)^{kl}_{ij}\,(1-n_k-n_l)\right)\\ &+ (1-P_{ij})(1-P_{kl})\sum_{ac}\left(n_j - n_k\right)\Gamma^{ak}_{cj}\lambda^{ai}_{cl}\\ &+ \frac{1}{2}\sum_{abc}\left((1-P_{kl})\Gamma^{ka}_{bc}\lambda^{aij}_{bcl} - (1-P_{ij})\Gamma^{ab}_{ic}\lambda^{abj}_{ckl}\right).\end{aligned} \tag{10.216}$$

Like the MR-IMSRG(2) flow equations (10.205)–(10.207), these expressions only depend linearly on $\lambda^{(2)}$ and $\lambda^{(3)}$, which makes untruncated implementations feasible.

We use the IBCs to define the so-called Brillouin generator as

$$\eta^i_j \equiv \langle\Phi|\,[H,:\hat{A}^i_j:]\,|\Phi\rangle \tag{10.217}$$

$$\eta^{ij}_{kl} \equiv \langle\Phi|\,[H,:\hat{A}^{ij}_{kl}:]\,|\Phi\rangle \tag{10.218}$$

Because the matrix elements of $\hat{\eta}$ are directly given by the residuals of the IBCs, it can be interpreted as the *gradient* of the energy with respect to the parameters of the unitary transformation at each step of the flow. At the fixed point of the flow, $\hat{\eta} = 0$, and the flowing zero-body part of the Hamiltonian, $E(\infty)$, will be an extremum of the energy. Indeed, $\hat{\eta}$ has behaved in this manner in all numerical applications to date, generating a monotonic flow of the energy towards the converged results [73].

10.4.2.5 Example: The Oxygen Isotopic Chain

As a sample application of the MR-IMSRG(2), we use spherical, particle-number projected HFB vacua (see, e.g., [117, 157] to compute the ground-state energies and radii of the even oxygen isotopes (odd isotopes have irreducible densities that are non-scalar under rotation, which requires a future extension of our framework) [46, 158]. Our results are shown in Fig. 10.14. We use various chiral $NN + 3N$ interactions.

The $NN + 3N(400)$ Hamiltonian consists of the N^{3}LO interaction by Entem and Machleidt [7, 98], with cutoff $\Lambda_{NN} = 500\,\mathrm{MeV}$, and a local NNLO interaction with a reduced cutoff $\Lambda_{NN} = 400\,\mathrm{MeV}$ [100, 114]. The low-energy constants (LECs),

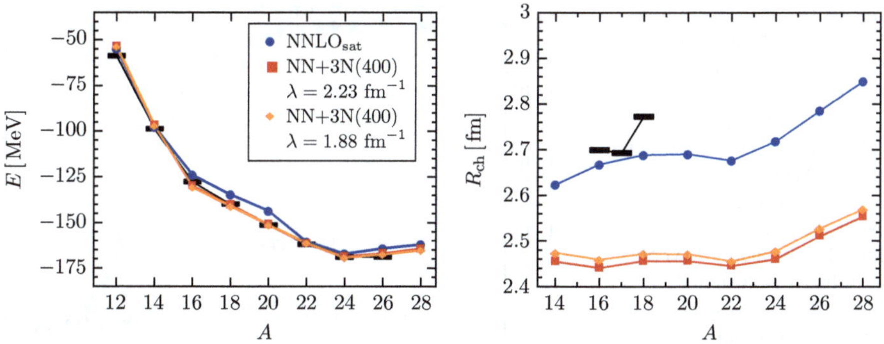

Fig. 10.14 MR-IMSRG(2) ground-state energies and charge radii of the oxygen isotopes for NNLO$_{sat}$ and $NN+3N(400)$ at $\lambda = 1.88, \ldots, 2.24\,\text{fm}^{-1}$ ($e_{max} = 14, E_{3max} = 14$, and optimal $\hbar\omega$). *Black bars* indicate experimental data [111, 159]

i.e., the parameters of the chiral Hamiltonian, are entirely fixed by fitting data in the $A = 2, 3, 4$ systems, and it is evolved to lower resolution scales λ via free-space SRG, as discussed in Sect. 10.2.4.2. In contrast, the LECs of NNLO$_{sat}$ are also optimized with respect to selected many-body data [17], and it is sufficiently soft that we use it as is.

While $NN + 3N(400)$ gives a good reproduction of the oxygen ground-state energies, an issue with the Hamiltonian's saturation properties is revealed by inspecting the oxygen charge radii (see Fig. 10.14). The theoretical charge radii are about 10% smaller than the experimental charge radius of ^{16}O, $R_{ch} = 2.70$ fm [159], and the sharp increase for ^{18}O is missing entirely. The variation of λ produces only a 0.2% change in the ground-state energies, but this is the result of a fine-tuned cancellation between induced 4N forces that are generated by the NN and 3N pieces of the Hamiltonian, and should not be seen as representative for chiral interactions in general. The charge radii grow larger as λ decreases, which is consistent with a study for light nuclei by Schuster et al. [92]. The authors found that two- and three-body terms that are induced by consistently evolving the charge radius operator to lower λ have the opposite effect and *reduce* its expectation value. These terms have not been included here, but need to be considered for complete consistency in the future.

The MR-IMSRG(2) ground-state energies obtained with NNLO$_{sat}$ are slightly *lower* than those for $NN+3N(400)$ in the proton-rich isotopes ^{12,14}O, and above the $NN+3N(400)$ energies in $^{16-28}$O. From $^{16-22}$O, the NNLO$_{sat}$ ground-state energies exhibit a parabolic behavior as opposed to the essentially linear trend we find for $NN+3N(400)$. A possible cause is the inclusion of the ^{22,24}O ground-state energies in the optimization protocol, which constrains the deviation of the energies from experimental data in these nuclei. NNLO$_{sat}$ predicts the drip line at ^{24}O, and the trend for the ^{26,28}O resonance energies is similar to the $NN+3N(400)$ case. For NNLO$_{sat}$, the charge radii for the bound oxygen isotopes are about 10% larger than for $NN+3N(400)$, which is expected given the use of the ^{16}O charge radius in

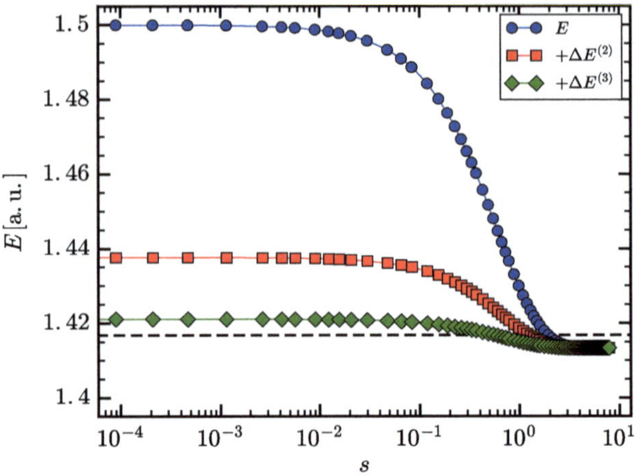

Fig. 10.15 IMSRG(2) flow for the ground state of the pairing Hamiltonian with $\delta = 1.0, g = 0.5$ (cf. Sect. 10.2.3.4). Calculations were performed with the White generator, Eq. (10.141). The figures shows the flowing ground-state energy $E(s)$ plus perturbative second and third-order energy corrections for $\hat{H}(s)$. The exact ground-state energy is indicated by the *dashed line*

the optimization of the LECs (also see [158]). For the resonant states, the increase is even larger, but continuum effects must be considered to make a meaningful comparison. We note that NNLO_{sat} also fails to describe the sharp jump in R_{ch} at ^{18}O.

10.4.3 Effective Hamiltonians

A recurring theme of this chapter has been the transformation of nuclear Hamiltonians to a shape that facilitates their subsequent application in many-body calculations. We have stressed this point in our discussion of the free-space SRG, in particular (see Sect. 10.2.4.2), but it applies to the IMSRG (or the MR-IMSRG) as well. Recall our application of the IMSRG to the pairing Hamiltonian in Sect. 10.3.6, where we primarily focused on how correlations that are usually probed by perturbative corrections are shuffled into the flowing ground-state energy $E(s)$ in the limit $s \to \infty$ (see Fig. 10.15). We can also interpret the results shown in this figure in a slightly different way: At any given value of s, $E(s)$ would result from a simple HF calculation with the Hamiltonian $\hat{H}(s)$, which has absorbed correlations because of an RG improvement. The result of the simple HF calculation approaches the exact result as we evolve, aside from truncation errors, of course. The same is true for the MBPT(2) and MBPT(3) calculations, summing $E(s)$ plus perturbative corrections through the indicated order. Of course, these approaches are already closer to the exact result in the first place.

This example illustrates the potential benefits of using Hamiltonians that have been improved through IMSRG evolution as input for other many-body methods. In this section, we will briefly discuss applications in the traditional nuclear Shell model, which will give us access to a wealth of spectroscopic observables like excitation energies and transition rates. We will also look at the use of IMSRG

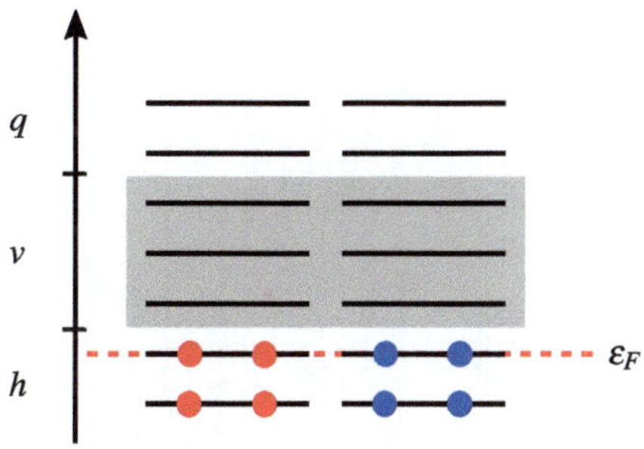

Fig. 10.16 Separation of the single-particle basis into hole (h), valence particle (v) and non-valence particle (q) states. The Fermi energy of the fully occupied core, ε_F, is indicated by the *red dashed line*

Hamiltonians in Equation-of-Motion methods, which are an alternative approach to the computation of excited-state properties.

10.4.3.1 Non-empirical Interactions for the Nuclear Shell Model

In IMSRG ground-state applications, we use the RG flows to decouple a suitable reference state from *npnh* excitations (see Sect. 10.3.3). From a more general perspective, we can view this as a decoupling of different *sectors* of the many-body Hilbert space by driving the couplings of these sectors to zero. We are not forced to restrict the decoupling to a single state, but could target multiple states at once [73, 121, 134]—all we need to do is tailor our definition of the off-diagonal Hamiltonian to the problem, as in all SRG and IMSRG applications!

In the nuclear Shell model, we split the single-particle basis in our calculation into core or hole (h), valence particle (v) and non-valence particle (q) orbitals (see Fig. 10.16).[3] The actual many-body calculation for a nucleus with A nucleons is an exact diagonalization of the Hamiltonian matrix in a subspace of the Hilbert space that is spanned by configurations of the form

$$|a^{\dagger}_{v_1} \dots a^{\dagger}_{v_{A_v}}\rangle \equiv a^{\dagger}_{v_1} \dots a^{\dagger}_{v_{A_v}} |\Phi\rangle \,, \tag{10.219}$$

where $|\Phi\rangle$ is the wave function for a suitable core with A_c nucleons, and the A_v valence nucleons are distributed over the valence orbitals v_i in all possible ways. Since the core is assumed to be inert, it can be viewed as a vacuum state for the valence configurations. The matrix representation of the Hamiltonian in the space spanned by these configurations is

$$\langle v'_1 \dots v'_{A_v}| \hat{H} |v_1 \dots v_{A_v}\rangle = \langle\Phi| a_{v'_{A_v}} \dots a_{v'_1} \hat{H} a^{\dagger}_{v_1} \dots a^{\dagger}_{v_{A_v}} |\Phi\rangle \,. \tag{10.220}$$

This expression suggests that we normal order the Hamiltonian and other operators with respect to the core wave function $|\Phi\rangle$, which can be obtained from a simple

[3] In quantum chemistry, what we call a valence space is usually refereed to as the *active space*.

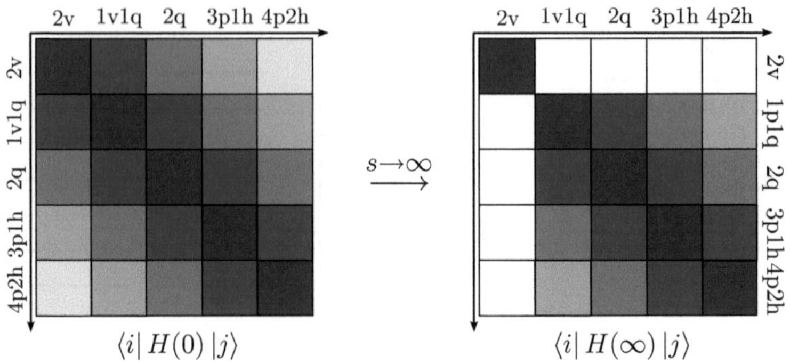

Fig. 10.17 Schematic view of IMSRG valence-space decoupling for two valence nucleons (p = v, q)

spherical HF calculation. The state $|\Phi\rangle$ takes on the role of the reference state for the IMSRG flow, but recent studies have shown that choosing either individual configurations for the target nucleus or ensembles of configurations as references will reduce truncation errors due to omitted induced terms, see [160, 161]. When the resulting valence-space interactions are used to calculate nuclear ground-state energies, we find excellent agreement with direct IMSRG ground-state calculations, which indicates that the introduction of the inert core is justified, at least for the used $NN + 3N$ forces with low resolution scales λ.

We want to use the IMSRG evolution to decouple the configurations (10.219) from states that involve excitations of the core, just as in the ground-state calculations. In addition, we need to decouple them from states containing nucleons in non-valence particle states (see Fig. 10.17). Working in IMSRG(2) truncation, i.e., assuming up to two-body terms in $\hat{H}(s)$, we can identify the matrix elements that couple pairs of valence-space particles to $1q1v$, $2q$, $3p1h$, and $4p2h$ excitations, respectively, where $p = v, q$. For each type of matrix element, we show the antisymmetrized Goldstone diagrams [60] that represent the excitation process. Additional diagrams due to permutations of the nucleons or Hermitian adjoints are suppressed for brevity.

Diagrams (I) and (II) are eliminated if matrix elements of f and Γ that contain at least one q index are chosen to be off-diagonal. Diagrams (III) and (V) are eliminated by decoupling the reference-state, i.e., the core, which requires f^p_h and $\Gamma^{pp'}_{hh'}$ to be off-diagonal (cf. Sect. 10.3.3). This only leaves diagram (IV), which vanishes if matrix elements of the type $\Gamma^{pp'}_{vh}$ vanish. Thus, we define [73, 121]

$$H_{od} \equiv \sum_{i\neq i'} f^i_{i'}\{\hat{A}^i_{i'}\} + \frac{1}{4}\left(\sum_{pp'hh'} \Gamma^{pp'}_{hh'}\{\hat{A}^{pp'}_{hh'}\} + \sum_{pp'vh} \Gamma^{pp'}_{vh}\{\hat{A}^{pp'}_{vh}\} + \sum_{pqvv'} \Gamma^{pq}_{vv'}\{\hat{A}^{pq}_{vv'}\}\right) + \text{H.c.}. \tag{10.221}$$

Table 10.2 Classification of matrix elements of the many-body Hamiltonian in the many-body Hilbert space spanned by $(n+2)pnh$ excitations of the reference state (cf. Fig. 10.17)

No.	Type	Diagram	Energy difference Δ
I	$\langle 2p\vert \hat{H} \vert 2p\rangle$	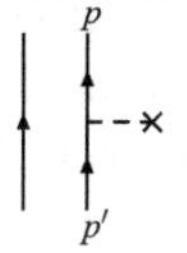	$f_p^p - f_{p'}^{p'}$
II	$\langle 2p\vert \hat{H} \vert 2p\rangle$	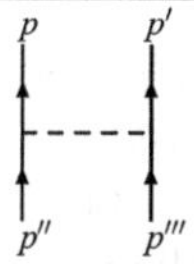	$f_p^p + f_{p'}^{p'} - f_{p''}^{p''} - f_{p'''}^{p'''} + \Gamma_{pp'}^{pp'} - \Gamma_{p''p'''}^{p''p'''}$
III	$\langle 3p1h\vert \hat{H} \vert 2p\rangle$	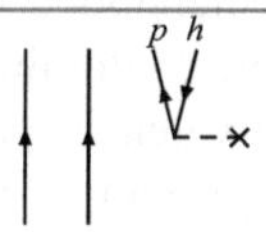	$f_p^p - f_h^h - \Gamma_{ph}^{ph}$
IV	$\langle 3p1h\vert \hat{H} \vert 2p\rangle$	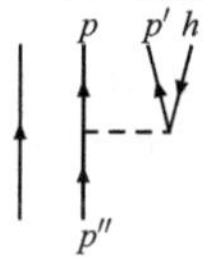	$f_p^p + f_{p'}^{p'} - f_{p''}^{p''} - f_h^h + \Gamma_{pp'}^{pp'} - \Gamma_{ph}^{ph} - \Gamma_{p'h}^{p'h}$
V	$\langle 4p2h\vert \hat{H} \vert 2p\rangle$	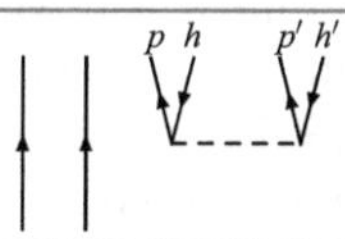	$f_p^p + f_{p'}^{p'} - f_h^h - f_{h'}^{h'} + \Gamma_{pp'}^{pp'} + \Gamma_{hh'}^{hh'} - \Gamma_{ph}^{ph} - \Gamma_{p'h'}^{p'h'} - \Gamma_{ph'}^{ph'} - \Gamma_{p'h}^{p'h}$

For each matrix element, we show the corresponding antisymmetrized Goldstone diagrams [60] involving the one- and two-body parts of $\hat{H}$ (permutations involving spectator particles which are required by antisymmetry are implied), as well as the energy differences appearing in the matrix elements for $\eta(s)$ in each case (see text)

This definition of the off-diagonal Hamiltonian holds for an arbitrary number of valence particles A_v. For $A_v = 1$, diagram (II) vanishes, while diagrams (I) and (III)-(V) have the same topology, but one less spectator nucleons. Analogously, diagrams (I)-(V) merely contain additional spectator nucleons for $A_v > 2$ (Table 10.2).

Using $\hat{H}_{od}$ in the construction of generators, we evolve the Hamiltonian by solving the IMSRG(2) flow equations (10.104)–(10.106). The evolved Hamiltonian is given by

$$\hat{H}(\infty) = E + \sum_v f_v^v \{\hat{A}_v^v\} + \frac{1}{4} \sum_{v_i,v_j,v_k,v_l} \Gamma_{v_k v_l}^{v_i v_j} \{\hat{A}_{v_k v_l}^{v_i v_j}\} + \dots, \tag{10.222}$$

where the explicitly shown terms are the core energy, single-particle energies, and two-body matrix elements that are used as input for a subsequent Shell model diagonalization.

A possible subtlety is associated with the treatment of the mass-number dependence of the intrinsic Hamiltonian (10.86). We interpret it as a dependence on the mass-number *operator* $\hat{A}$, which acts on the many-body states on which we are operating, i.e., configurations in the *target* nucleus, and therefore the mass number of the target should be used in the intrinsic Hamiltonian at all stages of a calculation [86, 160, 161]. This is appropriate because the combined IMSRG + Shell model calculation is supposed to approximate the results of an exact diagonalization for that particular nucleus.

The naive computational scaling for the valence-decoupling procedure described here is $\mathcal{O}(N^6)$, just like that of IMSRG(2) ground-state calculations. On 2015/16 computing hardware, typical evolutions require about 100–1000 core hours, assuming a single major shell as the valence space. The Shell model calculation is typically less expensive in that case. However, it will start to dominate the computational scaling as soon as we have to consider extended valence spaces consisting of two or more major shells, because of the factorial growth of the Shell model basis (cf. Sect. 10.2.3).

As an example, Fig. 10.18 shows the low-lying excitation spectra of $^{22-24}$O from Shell model calculations with IMSRG-derived valence-space interactions. These interactions were generated from an underlying chiral $NN + 3N$ Hamiltonian consisting of the N^{3}LO NN interaction by Entem and Machleidt with $\Lambda_{NN} =$ 500 MeV, and an NNLO $3N$ interaction with $\Lambda_{3N} = 400$ MeV, which has been evolved to $\lambda = 1.88\,\text{fm}^{-1}$ [7, 98, 100, 114] (also cf. Sects. 10.2.4.2, 10.4.2). We compare our results to the gold-standard phenomenological USDB interaction by Brown and Richter, which describes more than 600 ground-state and excitation energies in *sd*-shell ($8 \leq Z, N \leq 20$) nuclei with an rms deviation of merely ~ 130 keV [162], and to experimental data. In the nuclei shown, the agreement is quite satisfactory given that the chiral input Hamiltonian is entirely fixed by $A \leq 4$

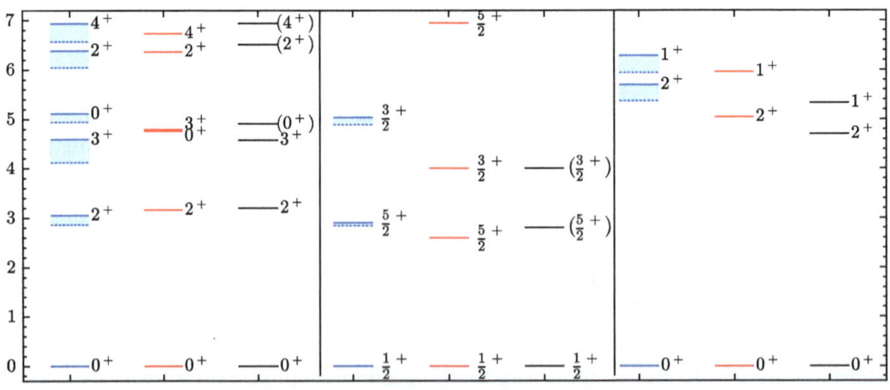

Fig. 10.18 Excited-state spectra of ^{22,23,24}O based on the chiral $NN + 3N(400)$ interaction with $\lambda = 1.88\,\text{fm}^{-1}$, compared to results with the phenomenological USDB interaction [162] and experimental data, see [73, 163] for full details. *Dotted and solid lines* are results for $\hbar\omega = 20$ MeV and $\hbar\omega = 24$ MeV, respectively, which are shown as an indicator of convergence

data. The IMSRG interactions turn out to perform quite well in the entire lower *sd*-shell, achieving an rms deviation of $\sim$ 580 keV in about 150 states [73, 160]. The chiral 3*N* interactions are found to be of crucial importance for the correct reproduction of level orderings and spacings.

10.4.3.2 Equation-of-Motion Methods

Equation-of-Motion (EOM) methods [164] are a useful alternative to the Shell model when it comes to the calculation of excited states, in particular when extended valence spaces lead to prohibitively large Shell model basis dimensions. In these approaches, the Schrödinger equation is rewritten in terms of ladder operators that create excited eigenstates from the exact ground state:

$$\hat{H}\left|\Psi_n\right\rangle = E_n\left|\Psi\right\rangle \qquad \longrightarrow \qquad \hat{H}\hat{X}_n^\dagger\left|\Psi_0\right\rangle = E_n\hat{X}_n^\dagger\left|\Psi_0\right\rangle . \tag{10.223}$$

Formally, $\hat{X}_n^\dagger$ is given by the dyadic product $\left|\Psi_n\right\rangle\left\langle\Psi_0\right|$, and by thinking of the exact eigenstates in a CI sense, it is easy to see that they can be expressed as a linear combination of up to A-body excitation and de-excitation operators acting on the ground state. We can further rewrite Eq. (10.223) as the Equation of Motion

$$[\hat{H}, \hat{X}_n^\dagger]\left|\Psi_0\right\rangle = (E_n - E_0)\hat{X}_n^\dagger\left|\Psi_0\right\rangle \equiv \omega_n\hat{X}_n^\dagger\left|\Psi_0\right\rangle , \tag{10.224}$$

and introduce systematic approximations to the $\hat{X}_n^\dagger$ and the ground-state $\left|\Psi_0\right\rangle$. For example, by replacing $\left|\Psi_0\right\rangle$ with a simple Slater determinant and using the ansatz

$$\hat{X}_n^\dagger = \sum_k X_{nk}a_k^\dagger , \tag{10.225}$$

we obtain Hartree-Fock theory, for

$$\hat{X}_n^\dagger = \sum_{ph} X_{ph}^{(n)}a_p^\dagger a_h , \tag{10.226}$$

we have the Tamm-Dancoff Approximation (TDA) for excited states, and

$$\hat{X}_n^\dagger = \sum_{ph} X_{ph}^{(n)}a_p^\dagger a_h - Y_{ph}^{(n)}a_h^\dagger a_p , \tag{10.227}$$

yields the Random Phase Approximation (RPA) in quasi-boson approximation [117, 131]. Plugging the Slater determinant reference state and the ansatz for the ladder operators into Eq. (10.224), we end up with a regular or a generalized eigenvalue problem, which we solve for the amplitudes appearing in the $\hat{X}_n^\dagger$ operators. Since the computed amplitudes can be used to improve the ground-state ansatz, it is usually possible to construct self-consistent solutions of the EoM (10.224) in a given truncation.

Since we have casually referred to the approximate ground state as a reference state already, it will not come as a surprise to our readers that we can quite naturally combine EOM methods with the IMSRG. Per construction, the reference state $|\Phi_0\rangle$ will be the ground state of the final IMSRG Hamiltonian

$$\overline{H} \equiv \hat{U}(\infty)\hat{H}(0)\hat{U}^{\dagger}(\infty). \tag{10.228}$$

Multiplying Eq. (10.224) by $\hat{U}(\infty)$ and recalling that

$$\hat{U}(\infty)\,|\Psi_0\rangle = |\Phi_0\rangle\,, \tag{10.229}$$

we obtain the unitarily transformed EOM

$$[\overline{H}, \overline{X}_n^{\dagger}]\,|\Phi_0\rangle = \omega_n \overline{X}_n^{\dagger}\,|\Phi_0\rangle\,. \tag{10.230}$$

The solutions $\overline{X}_n^{\dagger}$ can be used to obtain the eigenstates of the unevolved Hamiltonian via

$$|\Psi_n\rangle = \hat{U}^{\dagger}(\infty)\overline{X}_n^{\dagger}\,|\Phi_0\rangle\,. \tag{10.231}$$

In current applications, we include up to $2p2h$ excitations in the ladder operator [134]:

$$\overline{X}_n^{\dagger} = \sum_{ph} \bar{X}_{ph}^{(n)}\{a_p^{\dagger}a_h\} + \frac{1}{4}\sum_{pp'hh'} \bar{X}_{pp'hh'}^{(n)}\{a_p^{\dagger}a_{p'}^{\dagger}a_{h'}a_h\}\,. \tag{10.232}$$

Note that the operator only contains excitation operators because de-excitation operators annihilate the reference state $|\Phi_0\rangle$ and therefore do not contribute in the EOM:

$$[\{a_h^{\dagger}a_{h'}^{\dagger}a_{p'}a_p\}, \overline{H}]\,|\Phi_0\rangle = (\overline{H} - E)\{a_h^{\dagger}a_{h'}^{\dagger}a_{p'}a_p\}\,|\Phi_0\rangle = 0\,. \tag{10.233}$$

The ladder operator $\overline{X}^{\dagger}$ can be systematically improved by including higher particle-hole excitations, until we reach the $ApAh$ level which would amount to an exact diagonalization of $\overline{H}$. Denoting the operator rank of the ladder operator by m and the IMSRG truncation by n, we refer to a specific combined scheme as EOM-IMSRG(m,n).

Like TDA and RPA, the EOM (10.230) can be implemented as an eigenvalue problem that can be tackled with the Lanczos-Arnoldi or Davidson algorithms [82–84] that are also used in CI approaches. As mentioned throughout this chapter, these algorithms only require knowledge of matrix-vector products. In the EOM-IMSRG, the required product is identified once we realize that the commutator in

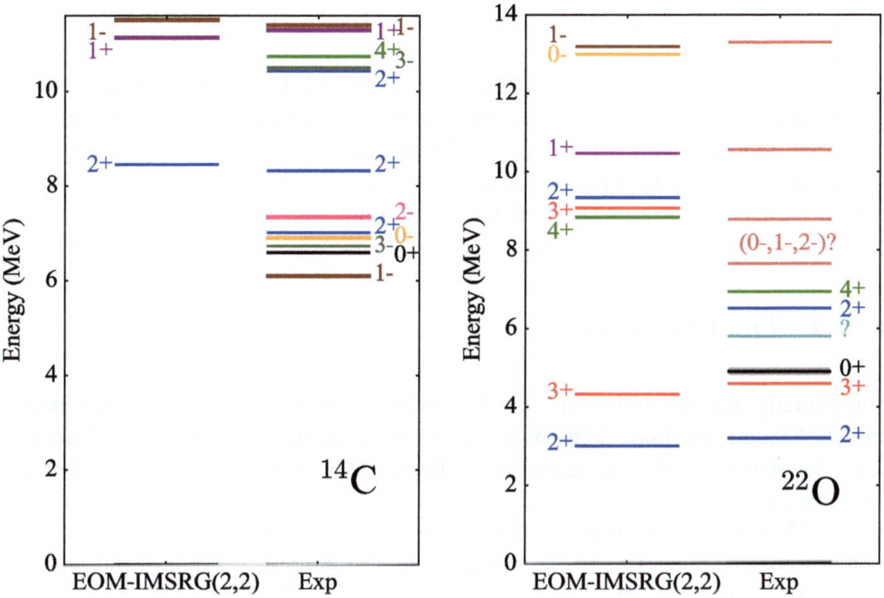

Fig. 10.19 EOM-IMSRG(2,2) excitation spectra of ^{14}C and ^{22}O, calculated with the chiral $NN+3N(400)$ Hamiltonian for $\lambda = 1.88\,\mathrm{fm}^{-1}$ (cf. Fig. 10.18), compared with experimental data

Eq. (10.230) can be rewritten as the connected product of $\overline{H}$ and $\overline{X}^{\dagger}$:

$$[\overline{H}, \overline{X}_n^{\dagger}] = \{\overline{H}\overline{X}_n^{\dagger}\}_C . \tag{10.234}$$

Thus, the matrix-vector product can be computed with the same commutator routines that are used in the evaluation of the IMSRG flow equations. Thus, the scaling of EOM-IMSRG(m,n) is the same as that of the IMSRG($\max m, n$), e.g., $\mathcal{O}(N^6)$ for the EOM-IMSRG(2,2) scheme that is the method we primarily use at this point [134].

In Fig. 10.19, we show sample spectra of ^{14}C and ^{22}O from EOM-IMSRG(2,2) calculations with the chiral $NN + 3N(400)$ Hamiltonian ($\lambda = 1.88\,\mathrm{fm}^{-1}$). We see that certain levels seem to be reproduced quite well, while experimentally observed states below 8 MeV are either missing in the calculation in the case of ^{14}C), or found at higher excitation theoretical excitation energy in the case of ^{22}O. In the latter case, we can compare the EOM-IMSRG(2,2) calculation to the IMSRG+SM spectrum shown in Fig. 10.18, in which the group of $2^+, 3^+, 4^+$ states is found much closer to their experimental counterparts. This suggests that the overestimation of the excitation energy in the EOM approach is caused by the truncation of the ladder operator at the $2p2h$ level, while the Shell model contains all allowed $npnh$ excitations in the valence space. Conversely, we can conclude that the states that the

EOM-IMSRG(2,2) is reproducing well are dominated by $1p1h$ and $2p2h$ excitations [134].

An advantage of the EOM-IMSRG(2,2) over the IMSRG+SM is that it allows us to compute negative parity states, to the level of accuracy that the truncation allows. In the Shell model, the description of such states requires a valence space consisting of two major shells, which usually makes the exact diagonalization prohibitively expensive, especially in *sd*-shell nuclei and beyond.

10.4.4 Final Remarks

Superficially, the discussion in Sect. 10.4 focused on three different subject areas, namely Magnus methods, MR-IMSRG, and the construction of effective Hamiltonians, but we expect that our readers can already tell that these directions are heavily intertwined.

The Magnus formulation of the IMSRG will greatly facilitate the evaluation of general observables not just in the regular single-reference version of the method, but also in the MR-IMSRG. Magnus methods can also be readily adapted to the construction of effective valence-space operators besides the nuclear interaction, e.g., radii and electroweak transitions. In addition, it allows us to construct systematic and computationally tractable approximations to the full IMSRG(3) [134, 165], similar to the non-iterative treatment of triples in Coupled Cluster methods [119, 166–168].

While the current version of our EOM technology is based on the single-reference IMSRG, we will formulate a multireference EOM scheme based on final Hamiltonians from the MR-IMSRG evolution next. Such a MR-EOM-IMSRG will hold great potential for the description of excitations in deformed or weakly bound nuclei, which would require excessively large valence spaces in traditional CI approaches, and at the same time require an explicit treatment of static correlations that the uncorrelated reference wave functions used in the IMSRG and EOM-IMSRG cannot provide.

The MR-IMSRG has also recently been used to pre-diagonalize Hamiltonians that serve as input for the No-Core Shell Model (NCSM), merging the two approaches into an iterative scheme that we call the In-Medium NCSM (IM-NCSM for short) [155]. In this combined approach, we can converge NCSM results in model spaces whose dimensions are orders of magnitude smaller than those of the regular NCSM or its importance-truncated variant [47, 169, 170], which should significantly expand its range of applicability. A similar idea might be used fruitfully in the traditional Shell model with a core as well, where the use of statically correlated reference states and an MR-IMSRG evolution might help to overcome shortcomings in the Shell model interactions for specific nuclei.

10.5 Conclusions

In this chapter, we have presented a pedagogical introduction to the SRG and IMSRG, and discussed their applications in the context of the nuclear many-body problem. The former has become maybe the most popular tool for pre-processing nuclear interactions and operators, leading to vast improvements in the rate of convergence of many-body calculations, and extending the range of nuclei that can be tackled in ab initio approaches [44, 48, 49, 53, 73, 108]. The IMSRG implements SRG concepts directly in the A-body system, relying on normal-ordering methods to control the size of induced operators and make systematic truncations feasible. As we have demonstrated through a variety of applications, the IMSRG is an extremely versatile and powerful addition to the canon of quantum many-body methods. In Sect. 10.4, we have given an overview of the main thrusts of current IMSRG research, and we hope that our readers will be inspired to contribute to these developments, or find ways in which the IMSRG framework can be useful to their own research programs.

The explicit RG aspect of the IMSRG framework is a unique feature that sets it apart from most other many-body methods on the market. When the comparison with those other methods and experimental data is our first and foremost concern, we are primarily interested in the $s \to \infty$ limit of the IMSRG or MR-IMSRG evolution, but the flow trajectory is an enormous source of additional insight. By studying the flows, not just the final fixed points, we can gain a new understanding of how many-body correlations are reshuffled between the wave function and the Hamiltonian, or different pieces of the Hamiltonian, making transparent what is only implicitly assumed in other methods. Like in the free-space SRG (or other RG methods), we have the freedom to work at intermediate values of s if this is more practical than working at $s = 0$ or in the limit $s \to \infty$, especially if either of these extremes would lead to the accrual of unacceptable numerical errors in our results (see, e.g., [155, 171, 172]). This is the inherent power of a framework that integrates many-body and renormalization group techniques, and the reason why we consider the IMSRG to be an extremely valuable tool for quantum many-body theory.

10.6 Exercises and Projects

10.1

(a) Prove that the two forms of the intrinsic kinetic energy operator given in Eqs. (10.54) and (10.55) are equivalent.

(b) Now consider the expectation values of the two forms of $\hat{T}_{\text{int}}$ in a state that does not have a fixed particle number, e.g., as in Bardeen-Cooper-Schrieffer (BCS) [173, 174] or Hartree-Fock-Bogoliubov (HFB) theory (see, e.g., [117]). Expand the $\frac{1}{\hat{A}}$ dependence of Eqs. (10.54) and (10.55) into series around $\langle \hat{A} \rangle$ by introducing $\Delta\hat{A} = \hat{A} - \langle\hat{A}\rangle$, and compare the series expansions order by order. (A thorough discussion of the issue can be found in [86].)

10.2

(a) Prove that the expectation value of a normal-ordered operator in the reference state vanishes [Eq. (10.84)]:

$$\langle\Phi|\left\{a^\dagger_{i_1}\ldots a_{j_1}\right\}|\Phi\rangle = 0\,. \tag{10.235}$$

Start by considering a one-body operator, and extend your result to the general case by induction.

(b) Show that $a^\dagger_i$ and a_j anticommute freely in a normal-ordered string [Eq. (10.85)].

$$\left\{\ldots a^\dagger_i a_j \ldots\right\} = -\left\{\ldots a_j a^\dagger_i \ldots\right\}. \tag{10.236}$$

Consider the one-body case first, as in Problem 10.2(a).

(c) Prove the following schematic expression for products of normal-ordered operators:

$$\hat{A}^{[M]}\hat{B}^{[N]} = \sum_{k=|M-N|}^{M+N} \hat{C}^{[k]}\,. \tag{10.237}$$

(d) Show that the following rule applies for commutators of normal-ordered operators:

$$[\hat{A}^{[M]}, \hat{B}^{[N]}] = \sum_{k=|M-N|}^{M+N-1} \hat{C}^{[k]}\,. \tag{10.238}$$

Thus, the largest particle rank appearing in the expansion of the commutator of normal-ordered M- and N-body operators is $M+N-1$.

(e) We can view the free-space operators as being normal-ordered with respect to the vacuum state. How are the expansion formulas for products and commutators modified in that case?

10.3 Use Wick's theorem to show that the basis consisting of a Slater determinant $|\Phi\rangle$ and its particle-hole excitations,

$$|\Phi\rangle,\ \left\{a^\dagger_p a_h\right\}|\Phi\rangle,\ \left\{a^\dagger_p a^\dagger_{p'} a_{h'} a_h\right\}|\Phi\rangle, \ldots\,, \tag{10.239}$$

is orthogonal if the underlying single-particle basis is orthonormal.

10.4

(a) Validate the leading-order perturbative expression for the Wegner generator, Eq. (10.153).

(b) Let us now assume that we have used a Slater determinant reference state that has *not* been optimized by performing a Hartree-Fock calculation. Using Epstein-Nesbet partitioning, the one-body part of the off-diagonal Hamiltonian is then counted as $f^{od}_{ij} = \mathscr{O}(g)$ instead of $\mathscr{O}(g^2)$ during the flow. Show that the one-body part of the Wegner generator has the following perturbative expansion in this case:

$$\eta_{ij} = \left(f^d_{ii} - f^d_{jj} - (n_i - n_j)\Gamma^d_{ijij}\right) f^{od}_{ij} + \mathscr{O}(g^2) . \tag{10.240}$$

Interpret the expression in the parentheses.

10.5 (Project: Optimization of the IMSRG(2) Code) In Sect. 10.3.5, we mention several ways of optimizing the performance of the Python code, like taking into account antisymmetry of two-body (and three-body) states, or exploiting symmetries and the resulting block structures.

(a) Optimize the storage requirements and speed of the Python code by taking the antisymmetry of states as well as the Hermiticity (anti-Hermiticity) of $\hat{H}$ (η) into account.

(b) Identify the symmetries of the pairing Hamiltonian, and construct a variant of `imsrg_pairing.py` that is explicitly block diagonal in the irreducible representations of the corresponding symmetry group.

(c) A significant portion of the code `imsrg_pairing.py` consists of infrastructure routines that are used to convert between one- and two-body bases. We could avoid this inversion if we treat Γ and the two-body part of the generator as *rank-four tensors* instead of matrices. `NumPy` offers tensor routines that can be used to evaluate tensor contractions and products, in particular `numpy.tensordot()`. Rewrite `imsrg_pairing.py` in terms of tensors, and compare the performance of your new code to the original version.

10.6 (Project: IMSRG(3) for the Pairing Hamiltonian) Throughout this chapter, we used a single-particle basis consisting of only 8 states in our discussions of the pairing Hamiltonian. For such a small basis size, it is possible to implement to include explicit 3N operators in the IMSRG evolution, that is, to work in the IMSRG(3) scheme.

(a) Derive the IMSRG(3) flow equations. (Note: compare your results with [53]).

(b) Implement the IMSRG(3). The Python code discussed in Sect. 10.3.5 provides a good foundation, but you may find it necessary to switch to a language like C/C++ or Fortran for performance reasons.

(c) "Interpolate" between IMSRG(2) and IMSRG(3) by selectively activating flow equation terms, and document the impact of these intermediate steps.

Acknowledgements The authors are indebted to a multitude of colleagues for many stimulating discussions of the SRG and IMSRG that are reflected in this work. We are particularly grateful to Angelo Calci, Thomas Duguet, Dick Furnstahl, Kai Hebeler, Morten Hjorth-Jensen, Jason Holt, Robert Roth, Achim Schwenk, Ragnar Stroberg, and Kyle Wendt.

The preparation of this chapter was supported in part by NSF Grant No. PHY-1404159 and the NUCLEI SciDAC Collaboration under the U.S. Department of Energy Grant No. DE-SC0008511. H. H. gratefully acknowledges the National Superconducting Cyclotron Laboratory (NSCL)/Facility for Rare Isotope Beams (FRIB) and Michigan State University (MSU) for startup support during the preparation of this work. Computing resources were provided by the MSU High-Performance Computing Center (HPCC)/Institute for Cyber-Enabled Research (iCER).

Appendix: Products and Commutators of Normal-Ordered Operators

In this appendix, we collect the basic expressions for products and commutators of normal-ordered one- and two-body operators. All single-particle indices refer to the natural orbital basis, where the one-body density matrix is diagonal

$$\rho_{kl} = \langle \Phi | \, a^{\dagger}_{l} a_{k} \, | \Phi \rangle = n_k \delta_{kl} \,, \quad n_k \in \{0, 1\} \,, \tag{10.241}$$

(notice the convention for the indices of ρ, cf. [117]). We also define the hole density matrix

$$\bar{\rho}_{kl} \equiv \langle \Phi | \, a_k a^{\dagger}_{l} \, | \Phi \rangle = \delta_{kl} - \rho_{kl} \equiv \bar{n}_k \delta_{kl} \tag{10.242}$$

whose eigenvalues are

$$\bar{n}_k = 1 - n_k \,, \tag{10.243}$$

i.e., 0 for occupied and 1 for unoccupied single-particle states. Finally, we will again use the permutation symbol $\hat{P}_{ij}$ to interchange the indices in any expression,

$$\hat{P}_{ij} g(\ldots, i, \ldots, j) \equiv g(\ldots, j, \ldots, i) \tag{10.244}$$

(see Sect. 10.3 and Chap. 8).

Operator Products

$$\{a^{\dagger}_{a} a_{b}\}\{a^{\dagger}_{k} a_{l}\} = \{a^{\dagger}_{a} a^{\dagger}_{k} a_{l} a_{b}\} - n_a \delta_{al} \{a^{\dagger}_{k} a_{b}\} + \bar{n}_b \delta_{bk} \{a^{\dagger}_{a} a_{l}\} + n_a \bar{n}_b \delta_{al} \delta_{bk} \tag{10.245}$$

$$\{a_a^\dagger a_b\}\{a_k^\dagger a_l^\dagger a_n a_m\} = \{a_a^\dagger a_k^\dagger a_l^\dagger a_n a_m a_b\} + \left(1-\hat{P}_{mn}\right) n_a \delta_{an} \{a_k^\dagger a_l^\dagger a_m a_b\}$$
$$- \left(1-\hat{P}_{kl}\right) \bar{n}_b \delta_{bl} \{a_a^\dagger a_k^\dagger a_n a_m\}$$
$$- \left(1-\hat{P}_{kl}\right)\left(1-\hat{P}_{mn}\right) n_a \bar{n}_b \delta_{am} \delta_{bl} \{a_k^\dagger a_n\} \tag{10.246}$$

$$\{a_a^\dagger a_b^\dagger a_d a_c\}\{a_k^\dagger a_l^\dagger a_n a_m\}$$
$$= \{a_a^\dagger a_b^\dagger a_k^\dagger a_l^\dagger a_n a_m a_d a_c\}$$
$$+ (1-\hat{P}_{ab})(1-\hat{P}_{mn}) n_a \delta_{am} \{a_b^\dagger a_k^\dagger a_l^\dagger a_n a_d a_c\}$$
$$- (1-\hat{P}_{cd})(1-\hat{P}_{kl}) \bar{n}_c \delta_{ck} \{a_a^\dagger a_b^\dagger a_l^\dagger a_n a_m a_d\}$$
$$+ (1-\hat{P}_{mn}) n_a n_b \delta_{am} \delta_{bn} \{a_k^\dagger a_l^\dagger a_d a_c\} + (1-\hat{P}_{cd}) \bar{n}_c \bar{n}_d \delta_{ck} \delta_{dl} \{a_a^\dagger a_b^\dagger a_n a_m\}$$
$$+ (1-\hat{P}_{ab})(1-\hat{P}_{cd})(1-\hat{P}_{kl})(1-\hat{P}_{mn}) n_a \bar{n}_c \delta_{am} \delta_{ck} \{a_b^\dagger a_l^\dagger a_n a_d\}$$
$$+ (1-\hat{P}_{ab})(1-\hat{P}_{kl})(1-\hat{P}_{mn}) n_b \bar{n}_c \bar{n}_d \delta_{bn} \delta_{ck} \delta_{dl} \{a_a^\dagger a_m\}$$
$$+ (1-\hat{P}_{cd})(1-\hat{P}_{kl})(1-\hat{P}_{mn}) \bar{n}_d n_a n_b \delta_{dl} \delta_{an} \delta_{bm} \{a_k^\dagger a_c\}$$
$$+ (1-\hat{P}_{kl})(1-\hat{P}_{mn}) n_a n_b \bar{n}_c \bar{n}_d \delta_{am} \delta_{bn} \delta_{ck} \delta_{dl} \tag{10.247}$$

Commutators

$$[\{a_a^\dagger a_b\}, \{a_k^\dagger a_l\}] = \delta_{bk}\{a_a^\dagger a_l\} - \delta_{al}\{a_k^\dagger a_b\} + (n_a - n_b)\delta_{al}\delta_{bk} \tag{10.248}$$

$$[\{a_a^\dagger a_b\}, \{a_k^\dagger a_l^\dagger a_n a_m\}] = \left(1-\hat{P}_{kl}\right)\delta_{bk}\{a_a^\dagger a_l^\dagger a_n a_m\} - \left(1-\hat{P}_{mn}\right)\delta_{am}\{a_k^\dagger a_l^\dagger a_n a_b\}$$
$$+ \left(1-\hat{P}_{kl}\right)\left(1-\hat{P}_{mn}\right)(n_a - n_b)\delta_{an}\delta_{bl}\{a_k^\dagger a_m\} \tag{10.249}$$

$$[\{a_a^\dagger a_b^\dagger a_d a_c\}, \{a_k^\dagger a_l^\dagger a_n a_m\}]$$
$$= \left(1-\hat{P}_{ab}\right)\left(1-\hat{P}_{mn}\right)\delta_{am}\{a_b^\dagger a_k^\dagger a_l^\dagger a_n a_d a_c\}$$
$$- \left(1-\hat{P}_{cd}\right)\left(1-\hat{P}_{kl}\right)\delta_{kc}\{a_a^\dagger a_b^\dagger a_l^\dagger a_n a_m a_d\}$$

$$
\begin{aligned}
&+ (1-\hat{P}_{cd})\,(\bar{n}_c\bar{n}_d - n_c n_d)\,\delta_{ck}\delta_{dl}\{a^\dagger_a a^\dagger_b a_n a_m\}\\
&+ (1-\hat{P}_{ab})\,(n_a n_b - \bar{n}_a\bar{n}_b)\,\delta_{am}\delta_{bn}\{a^\dagger_k a^\dagger_l a_d a_c\}\\
&+ (1-\hat{P}_{ab})(1-\hat{P}_{cd})(1-\hat{P}_{kl})(1-\hat{P}_{mn})\,(n_b - n_d)\,\delta_{bn}\delta_{dl}\{a^\dagger_a a^\dagger_k a_m a_c\}\\
&+ (1-\hat{P}_{ab})(1-\hat{P}_{mn})\,(n_b\bar{n}_c\bar{n}_d - \bar{n}_b n_c n_d)\,\delta_{bn}\delta_{ck}\delta_{dl}\{a^\dagger_a a_m\}\\
&- (1-\hat{P}_{cd})(1-\hat{P}_{kl})\,(n_d\bar{n}_a\bar{n}_b - \bar{n}_d n_a n_b)\,\delta_{dl}\delta_{am}\delta_{bn}\{a^\dagger_k a_c\}\\
&+ (1-\hat{P}_{ab})(1-\hat{P}_{cd})\,(n_a n_b\bar{n}_c\bar{n}_d - \bar{n}_a\bar{n}_b n_c n_d)\,\delta_{am}\delta_{bn}\delta_{ck}\delta_{dl}
\end{aligned}
\tag{10.250}
$$

References

1. J.D. Jackson, *Classical Electrodynamics*, 3rd edn. (Wiley, New York, 1999)
2. E. Fermi, Z. Phys. **88**, 161 (1934)
3. W. Detmold, *Nuclear Physics from Lattice QCD* (Springer International Publishing, Cham, 2015), p. 153
4. D.J. Gross, F. Wilczek, Phys. Rev. Lett. **30**, 1343 (1973)
5. H.D. Politzer, Phys. Rev. Lett. **30**, 1346 (1973)
6. E. Epelbaum, H.W. Hammer, U.G. Meißner, Rev. Mod. Phys. **81**, 1773 (2009)
7. R. Machleidt, D. Entem, Phys. Rep. **503**, 1 (2011)
8. E. Epelbaum, H. Krebs, U.G. Meißner, Phys. Rev. Lett. **115**, 122301 (2015)
9. D.R. Entem, N. Kaiser, R. Machleidt, Y. Nosyk, Phys. Rev. C **91**, 014002 (2015)
10. A. Gezerlis, I. Tews, E. Epelbaum, M. Freunek, S. Gandolfi, K. Hebeler, A. Nogga, A. Schwenk, Phys. Rev. C **90**, 054323 (2014)
11. J.E. Lynn, I. Tews, J. Carlson, S. Gandolfi, A. Gezerlis, K.E. Schmidt, A. Schwenk, Phys. Rev. Lett. **116**, 062501 (2016)
12. S. Pastore, L. Girlanda, R. Schiavilla, M. Viviani, R.B. Wiringa, Phys. Rev. C **80**, 034004 (2009)
13. S. Pastore, L. Girlanda, R. Schiavilla, M. Viviani, Phys. Rev. C **84**, 024001 (2011)
14. M. Piarulli, L. Girlanda, L.E. Marcucci, S. Pastore, R. Schiavilla, M. Viviani, Phys. Rev. C **87**, 014006 (2013)
15. S. Kölling, E. Epelbaum, H. Krebs, U.G. Meißner, Phys. Rev. C **80**, 045502 (2009)
16. S. Kölling, E. Epelbaum, H. Krebs, U.G. Meißner, Phys. Rev. C **84**, 054008 (2011)
17. A. Ekström, G.R. Jansen, K.A. Wendt, G. Hagen, T. Papenbrock, B.D. Carlsson, C. Forssén, M. Hjorth-Jensen, P. Navrátil, W. Nazarewicz, Phys. Rev. C **91**, 051301 (2015)
18. A.M. Shirokov, I.J. Shin, Y. Kim, M. Sosonkina, P. Maris, J.P. Vary, Phys. Lett. B **761**, 87 (2016)
19. G.P. Lepage (1989). arXiv:hep-ph/0506330
20. G.P. Lepage (1997). arXiv:nucl-th/9706029
21. S.K. Bogner, T.T.S. Kuo, A. Schwenk, Phys. Rep. **386**, 1 (2003)
22. S.K. Bogner, R.J. Furnstahl, R.J. Perry, Phys. Rev. C **75**, 061001(R) (2007)
23. S.K. Bogner, R.J. Furnstahl, A. Schwenk, Prog. Part. Nucl. Phys. **65**, 94 (2010)
24. R.J. Furnstahl, K. Hebeler, Rep. Prog. Phys. **76**, 126301 (2013)
25. H.A. Bethe, Ann. Rev. Nucl. Sci. **21**, 93 (1971)
26. W. Polyzou, W. Glöckle, Few-Body Syst. **9**, 97 (1990)
27. K.A. Brueckner, C.A. Levinson, H.M. Mahmoud, Phys. Rev. **95**, 217 (1954)
28. K.A. Brueckner, C.A. Levinson, Phys. Rev. **97**, 1344 (1955)
29. H.A. Bethe, J. Goldstone, Proc. R. Soc. Lond. A Math. Phys. Eng. Sci. **238**, 551 (1957)

30. J. Goldstone, Proc. R. Soc. Lond. A Math. Phys. Eng. Sci. **239**, 267 (1957)
31. B.D. Day, Rev. Mod. Phys. **39**, 719 (1967)
32. B.H. Brandow, Rev. Mod. Phys. **39**, 771 (1967)
33. B.R. Barrett, M.W. Kirson, Nucl. Phys. A **148**, 145 (1970)
34. M.W. Kirson, Ann. Phys. **66**, 624 (1971)
35. B.R. Barrett, Phys. Lett. B **38**, 371 (1972)
36. M.W. Kirson, Ann. Phys. **82**, 345 (1974)
37. P. Goode, M.W. Kirson, Phys. Lett. B **51**, 221 (1974)
38. J.P. Vary, P.U. Sauer, C.W. Wong, Phys. Rev. C **7**, 1776 (1973)
39. S.D. Glazek, K.G. Wilson, Phys. Rev. D **48**, 5863 (1993)
40. F. Wegner, Ann. Phys. (Leipzig) **3**, 77 (1994)
41. S.K. Bogner, R.J. Furnstahl, S. Ramanan, A. Schwenk, Nucl. Phys. A **773**, 203 (2006)
42. R. Roth, J. Langhammer, Phys. Lett. B **683**, 272 (2010)
43. A. Tichai, J. Langhammer, S. Binder, R. Roth, Phys. Lett. B **756**, 283 (2016)
44. B.R. Barrett, P. Navrátil, J.P. Vary, Prog. Part. Nucl. Phys. **69**, 131 (2013)
45. E.D. Jurgenson, P. Maris, R.J. Furnstahl, P. Navrátil, W.E. Ormand, J.P. Vary, Phys. Rev. C **87**, 054312 (2013)
46. H. Hergert, S. Binder, A. Calci, J. Langhammer, R. Roth, Phys. Rev. Lett. **110**, 242501 (2013)
47. R. Roth, A. Calci, J. Langhammer, S. Binder, Phys. Rev. C **90**, 024325 (2014)
48. S. Binder, J. Langhammer, A. Calci, R. Roth, Phys. Lett. B **736**, 119 (2014)
49. G. Hagen, T. Papenbrock, M. Hjorth-Jensen, D.J. Dean, Rep. Prog. Phys. **77**, 096302 (2014)
50. G. Hagen, M. Hjorth-Jensen, G.R. Jansen, T. Papenbrock, Phys. Scr. **91**, 063006 (2016)
51. K. Tsukiyama, S.K. Bogner, A. Schwenk, Phys. Rev. Lett. **106**, 222502 (2011)
52. H. Hergert, S.K. Bogner, S. Binder, A. Calci, J. Langhammer, R. Roth, A. Schwenk, Phys. Rev. C **87**, 034307 (2013)
53. H. Hergert, S.K. Bogner, T.D. Morris, A. Schwenk, K. Tsukiyama, Phys. Rep. **621**, 165 (2016)
54. S. Kehrein, *The Flow Equation Approach to Many-Particle Systems*. Springer Tracts in Modern Physics, vol. 237 (Springer, Berlin, Heidelberg, 2006)
55. C. Heidbrink, G. Uhrig, Eur. Phys. J. B **30**, 443 (2002). 10.1140/epjb/e2002-00401-9
56. N.A. Drescher, T. Fischer, G.S. Uhrig, Eur. Phys. J. B **79**, 225 (2011). 10.1140/epjb/e2010-10723-6
57. H. Krull, N.A. Drescher, G.S. Uhrig, Phys. Rev. B **86**, 125113 (2012)
58. B. Fauseweh, G.S. Uhrig, Phys. Rev. B **87**, 184406 (2013)
59. J. Krones, G.S. Uhrig, Phys. Rev. B **91**, 125102 (2015)
60. I. Shavitt, R.J. Bartlett, *Many-Body Methods in Chemistry and Physics: MBPT and Coupled-Cluster Theory* (Cambridge University Press, Cambridge, 2009)
61. S.R. White, J. Chem. Phys. **117**, 7472 (2002)
62. T. Yanai, G.K.L. Chan, J. Chem. Phys. **127**, 104107 (2007)
63. H. Nakatsuji, Phys. Rev. A **14**, 41 (1976)
64. D. Mukherjee, W. Kutzelnigg, J. Chem. Phys. **114**, 2047 (2001)
65. D.A. Mazziotti, Phys. Rev. Lett. **97**, 143002 (2006)
66. F.A. Evangelista, J. Chem. Phys. **141**, 054109 (2014)
67. V. Bach, J.B. Bru, J. Evol. Equ. **10**, 425 (2010)
68. B. Boutin, N.J. Raymond, Evol. Equ. (2016). doi:10.1007/s00028-016-0337-3
69. R. Brockett, Linear Algebra Appl. **146**, 79 (1991)
70. M.T. Chu, Fields Inst. Commun. **3**, 87 (1994)
71. M.T. Chu, Linear Algebra Appl. **215**, 261 (1995)
72. T.D. Morris, N.M. Parzuchowski, S.K. Bogner, Phys. Rev. C **92**, 034331 (2015)
73. H. Hergert, Phys. Scr. **92**, 023002 (2017)
74. G. Golub, C. Van Loan, *Matrix Computations*. Johns Hopkins Studies in the Mathematical Sciences (Johns Hopkins University Press, Baltimore, MD, 2013)
75. W.H. Press, S.A. Teukolsky, W.T. Vetterling, B.P. Flannery, *Numerical Recipes: The Art of Scientific Computing*, 3rd edn. (Cambridge University Press, Cambridge, 2007)

76. L.F. Shampine, M.K. Gordon, *Computer Solution of Ordinary Differential Equations: The Initial Value Problem* (W. H. Freeman, San Francisco, 1975)
77. R.H. Landau, M.J. Paez, C.C. Bordeianu, *A Survey of Computational Physics: Introductory Computational Science* (Princeton University Press, Princeton, NJ, 2012)
78. M. Hjorth-Jensen, *Computational Physics Lecture Notes* (2015). https://github.com/CompPhysics/ComputationalPhysics
79. A.C. Hindmarsh, *ODEPACK, A Systematized Collection of ODE Solvers*. IMACS Transactions on Scientific Computation, vol. 1 (North-Holland Publishing Company, Amsterdam, 1983), p. 54
80. K. Radhakrishnan, A.C. Hindmarsh, Description and use of LSODE, the Livermore solver for ordinary differential equations. Technical Report, Lawrence Livermore National Laboratory (1993)
81. P.N. Brown, A.C. Hindmarsh, Appl. Math. Comput. **31**, 40 (1989)
82. C. Lanczos, J. Res. Natl. Bur. Stand. **45**, 255 (1950)
83. W.E. Arnoldi, Q. Appl. Math. **9**, 17 (1951)
84. E.R. Davidson, Comput. Phys. Commun. **53**, 49 (1989)
85. C. Yang, H.M. Aktulga, P. Maris, E. Ng, J.P. Vary, in *Proceedings of International Conference 'Nuclear Theory in the Supercomputing Era — 2013' (NTSE-2013), Ames, IA, May 13–17, 2013*, ed. by A.M. Shirokov, A.I. Mazur (2013), p. 272
86. H. Hergert, R. Roth, Phys. Lett. B **682**, 27 (2009)
87. R.B. Wiringa, V.G.J. Stoks, R. Schiavilla, Phys. Rev. C **51**, 38 (1995)
88. J. Carlson, S. Gandolfi, F. Pederiva, S.C. Pieper, R. Schiavilla, K.E. Schmidt, R.B. Wiringa, Rev. Mod. Phys. **87**, 1067 (2015)
89. J.D. Holt, T.T.S. Kuo, G.E. Brown, S.K. Bogner, Nucl. Phys. A **733**, 153 (2004)
90. K.A. Wendt, R.J. Furnstahl, R.J. Perry, Phys. Rev. C **83**, 034005 (2011)
91. E.R. Anderson, S.K. Bogner, R.J. Furnstahl, R.J. Perry, Phys. Rev. C **82**, 054001 (2010)
92. M.D. Schuster, S. Quaglioni, C.W. Johnson, E.D. Jurgenson, P. Navrátil, Phys. Rev. C **90**, 011301 (2014)
93. S.N. More, S. König, R.J. Furnstahl, K. Hebeler, Phys. Rev. C **92**, 064002 (2015)
94. E.D. Jurgenson, P. Navrátil, R.J. Furnstahl, Phys. Rev. Lett. **103**, 082501 (2009)
95. K. Hebeler, Phys. Rev. C **85**, 021002 (2012)
96. K.A. Wendt, Phys. Rev. C **87**, 061001 (2013)
97. E. Anderson, S.K. Bogner, R.J. Furnstahl, E.D. Jurgenson, R.J. Perry, A. Schwenk, Phys. Rev. C **77**, 037001 (2008)
98. D.R. Entem, R. Machleidt, Phys. Rev. C **68**, 041001 (2003)
99. P. Brown, G. Byrne, A. Hindmarsh, SIAM J. Sci. Stat. Comput. **10**, 1038 (1989)
100. R. Roth, J. Langhammer, A. Calci, S. Binder, P. Navrátil, Phys. Rev. Lett. **107**, 072501 (2011)
101. H. Hergert, S.K. Bogner, T.D. Morris, S. Binder, A. Calci, J. Langhammer, R. Roth, Phys. Rev. C **90**, 041302 (2014)
102. G. Hagen, T. Papenbrock, D.J. Dean, M. Hjorth-Jensen, Phys. Rev. C **82**, 034330 (2010)
103. R. Roth, S. Binder, K. Vobig, A. Calci, J. Langhammer, P. Navrátil, Phys. Rev. Lett. **109**, 052501 (2012)
104. S. Binder, J. Langhammer, A. Calci, P. Navrátil, R. Roth, Phys. Rev. C **87**, 021303 (2013)
105. V. Somà, T. Duguet, C. Barbieri, Phys. Rev. C **84**, 064317 (2011)
106. V. Somà, C. Barbieri, T. Duguet, Phys. Rev. C **87**, 011303 (2013)
107. V. Somà, C. Barbieri, T. Duguet, Phys. Rev. C **89**, 024323 (2014)
108. V. Somà, A. Cipollone, C. Barbieri, P. Navrátil, T. Duguet, Phys. Rev. C **89**, 061301 (2014)
109. E.D. Jurgenson, P. Navrátil, R.J. Furnstahl, Phys. Rev. C **83**, 034301 (2011)
110. K.A. Wendt, Advances in the application of the similarity renormalization group to strongly interacting systems. Ph.D. thesis, The Ohio State University (2013)
111. M. Wang, G. Audi, A. Wapstra, F. Kondev, M. MacCormick, X. Xu, B. Pfeiffer, Chin. Phys. C **36**, 1603 (2012)
112. E. Epelbaum, A. Nogga, W. Glöckle, H. Kamada, U.G. Meißner, H. Witała, Phys. Rev. C **66**, 064001 (2002)

113. E. Epelbaum, Prog. Part. Nucl. Phys. **57**, 654 (2006)
114. D. Gazit, S. Quaglioni, P. Navrátil, Phys. Rev. Lett. **103**, 102502 (2009)
115. H.W. Griesshammer, in *Proceedings of the "8th International Workshop on Chiral Dynamics"*, vol. PoS(CD15) (2015), p. 104
116. R. Roth, T. Neff, H. Feldmeier, Prog. Part. Nucl. Phys. **65**, 50 (2010)
117. P. Ring, P. Schuck, *The Nuclear Many-Body Problem*, 1st edn. (Springer, Berlin, 1980)
118. G. Hagen, T. Papenbrock, D.J. Dean, A. Schwenk, A. Nogga, M. Włoch, P. Piecuch, Phys. Rev. C **76**, 034302 (2007)
119. S. Binder, P. Piecuch, A. Calci, J. Langhammer, P. Navrátil, R. Roth, Phys. Rev. C **88**, 054319 (2013)
120. E. Gebrerufael, A. Calci, R. Roth, Phys. Rev. C **93**, 031301 (2016)
121. K. Tsukiyama, S.K. Bogner, A. Schwenk, Phys. Rev. C **85**, 061304 (2012)
122. A. Fetter, J. Walecka, *Quantum Theory of Many-particle Systems*. Dover Books on Physics (Dover, New York, 2003)
123. W. Kutzelnigg, J. Chem. Phys. **77**, 3081 (1982)
124. W. Kutzelnigg, S. Koch, J. Chem. Phys. **79**, 4315 (1983)
125. W. Kutzelnigg, J. Chem. Phys. **80**, 822 (1984)
126. W. Kutzelnigg, D. Mukherjee, J. Chem. Phys. **116**, 4787 (2002)
127. W. Kutzelnigg, D. Mukherjee, J. Chem. Phys. **120**, 7340 (2004)
128. W. Kutzelnigg, D. Mukherjee, J. Chem. Phys. **120**, 7350 (2004)
129. P.S. Epstein, Phys. Rev. **28**, 695 (1926)
130. R.K. Nesbet, Proc. R. Soc. Lond. A Math. Phys. Eng. Sci. **230**, 312 (1955)
131. J. Suhonen, *From Nucleons to Nucleus. Concepts of Microscopic Nuclear Theory*, 1st edn. (Springer, Berlin, 2007)
132. S.P. Pandya, Phys. Rev. **103**, 956 (1956)
133. T.T.S. Kuo, J. Shurpin, K.C. Tam, E. Osnes, P.J. Ellis, Ann. Phys. **132**, 237 (1981)
134. N.M. Parzuchowski, T.D. Morris, S.K. Bogner, Phys. Rev. C (2017, in press). arXiv:1611.00661
135. D.R. Thompson, M. Lemere, Y.C. Tang, Nucl. Phys. A **286**(1), 53 (1977)
136. A.C. Hindmarsh, P.N. Brown, K.E. Grant, S.L. Lee, R. Serban, D.E. Shumaker, C.S. Woodward, ACM Trans. Math. Softw. **31**, 363 (2005)
137. W. Magnus, Commun. Pure Appl. Math. **7**, 649 (1954)
138. S. Blanes, F. Casas, J. Oteo, J. Ros, Phys. Rep. **470**, 151 (2009)
139. D.J. Dean, M. Hjorth-Jensen, Rev. Mod. Phys. **75**, 607 (2003)
140. A. Bohr, B.R. Mottelson, *Nuclear Structure, Vol. II: Nuclear Deformations* (World Scientific, Singapore, 1999)
141. R.E. Peierls, Proc. R. Soc. Lond. A **333**, 157 (1973)
142. J.L. Egido, P. Ring, Nucl. Phys. A **383**, 189 (1982)
143. L.M. Robledo, Phys. Rev. C **50**, 2874 (1994)
144. H. Flocard, N. Onishi, Ann. Phys. **254**, 275 (1997)
145. J.A. Sheikh, P. Ring, Nucl. Phys. A **665**, 71 (2000)
146. J. Dobaczewski, M.V. Stoitsov, W. Nazarewicz, P.G. Reinhard, Phys. Rev. C **76**, 054315 (2007)
147. M. Bender, T. Duguet, D. Lacroix, Phys. Rev. C **79**, 044319 (2009)
148. T. Duguet, M. Bender, K. Bennaceur, D. Lacroix, T. Lesinski, Phys. Rev. C **79**, 044320 (2009)
149. D. Lacroix, T. Duguet, M. Bender, Phys. Rev. C **79**, 044318 (2009)
150. D. Lacroix, D. Gambacurta, Phys. Rev. C **86**, 014306 (2012)
151. T. Duguet, J. Phys. G **42**, 025107 (2015)
152. W. Kutzelnigg, D. Mukherjee, J. Chem. Phys. **107**, 432 (1997)
153. D. Mukherjee, Chem. Phys. Lett. **274**, 561 (1997)
154. L. Kong, M. Nooijen, D. Mukherjee, J. Chem. Phys. **132**, 234107 (2010)
155. E. Gebrerufael, K. Vobig, H. Hergert, R. Roth, Phys. Rev. Lett. (2017, in press). arXiv:1610.05254
156. D.I. Lyakh, M. Musiał, V.F. Lotrich, R.J. Bartlett, Chem. Rev. **112**, 182 (2012)

157. H. Hergert, R. Roth, Phys. Rev. C **80**, 024312 (2009)
158. V. Lapoux, V. Somà, C. Barbieri, H. Hergert, J.D. Holt, S.R. Stroberg, Phys. Rev. Lett. **117**, 052501 (2016)
159. I. Angeli, K.P. Marinova, At. Data Nucl. Data Tables **99**, 69 (2013)
160. S.R. Stroberg, H. Hergert, J.D. Holt, S.K. Bogner, A. Schwenk, Phys. Rev. C **93**, 051301 (2016)
161. S.R. Stroberg, A. Calci, H. Hergert, J.D. Holt, S.K. Bogner, R. Roth, A. Schwenk, Phys Rev. Lett. 118, 032502 (2017)
162. B.A. Brown, W.A. Richter, Phys. Rev. C **74**, 034315 (2006)
163. S.K. Bogner, H. Hergert, J.D. Holt, A. Schwenk, S. Binder, A. Calci, J. Langhammer, R. Roth, Phys. Rev. Lett. **113**, 142501 (2014)
164. D.J. Rowe, Rev. Mod. Phys. **40**, 153 (1968)
165. T.D. Morris, Systematic improvements of ab initio in-medium similarity renormalization group calculations. Ph.D. thesis, Michigan State University (2016)
166. A.G. Taube, R.J. Bartlett, J. Chem. Phys. **128**, 044110 (2008)
167. A.G. Taube, R.J. Bartlett, J. Chem. Phys. **128**, 044111 (2008)
168. P. Piecuch, M. Włoch, J. Chem. Phys. **123**, 224105 (2005)
169. R. Roth, P. Navrátil, Phys. Rev. Lett. **99**, 092501 (2007)
170. R. Roth, Phys. Rev. C **79**, 064324 (2009)
171. C. Li, F.A. Evangelista, J. Chem. Theory Comput. **11**, 2097 (2015)
172. C. Li, F.A. Evangelista, J. Chem. Phys. **144**, 164114 (2016)
173. J. Bardeen, L.N. Cooper, J.R. Schrieffer, Phys. Rev. **108**, 1175 (1957)
174. J. Bardeen, L.N. Cooper, J.R. Schrieffer, Phys. Rev. **106**, 162 (1957)

Chapter 11
Self-Consistent Green's Function Approaches

Carlo Barbieri and Arianna Carbone

11.1 Introduction

Ab initio methods that present polynomial scaling with the number of particles have proven highly useful in reaching finite systems of rather large sizes up to medium mass nuclei [1–3] and even infinite matter [4–6]. Most approaches of this type that are discussed in previous chapters aim at the direct evaluation of the ground state energy of the system, where several other quantities of interest can be addressed in a second stage through the equation of motion and particle removal or attachment techniques. Here, we will follow a different route and focus on gaining from the start a global view of the spectral structure of a system of fermions. Our approach will be that of calculating directly the self-energy (also know as *mass operator*), which describes the complete response of a particle embedded in the true ground state of the system. This not only provides an effective in medium interaction for the nucleon, but it is also the optical potential for elastic scattering and it yields the spectral information relative to the attachment and removal of a particle. Once the one body Green's function has been obtained, the total energy of the system is calculated, as the final step, by means of appropriate sum rules [7, 8].

Two main approaches have become standard choices for calculations of Green's functions in nuclear many-body theory. The Algebraic Diagrammatic Construction

C. Barbieri (✉)
Department of Physics, University of Surrey, Guildford GU2 7XH, UK
e-mail: C.Barbieri@surrey.ac.uk

A. Carbone
Institut für Kernphysik, Technische Universität Darmstadt, 64289 Darmstadt, Germany

ExtreMe Matter Institute EMMI, GSI Helmholtzzentrum für Schwerionenforschung GmbH, 64291 Darmstadt, Germany
e-mail: arianna@theorie.ikp.physik.tu-darmstadt.de

M. Hjorth-Jensen et al. (eds.), *An Advanced Course in Computational Nuclear Physics*, Lecture Notes in Physics 936, DOI 10.1007/978-3-319-53336-0_11

(ADC) method, that was originally devised for quantum chemistry applications [9, 10], has proven to be optimal for discrete bases, as it is normally necessary to exploit for finite nuclei. However, this can also be applied to fermion gases in a box with periodic boundary conditions, which simplifies the analysis even more thanks to translational invariance. We will focus on the case of infinite nucleonic matter and provide an example of a working numerical code. ADC(n) methods are part of a larger class of approaches based on intermediate-state representations (ISRs) to which also the equation-of-motion coupled cluster belongs [11, 12]. The other method consists in solving directly the nucleon-nucleon ladder scattering matrix for dressed particles in the medium, which can be done effectively in a finite temperature formalism [13, 14]. Hence, this makes possible to study thermodynamic properties of the infinite and liquid matter. For these studies to be reliable, it is mandatory to ensure the satisfaction of fundamental conservation laws and to maintain thermodynamic consistency in the infinite size limit. We show here how to achieve this by preforming fully *self-consistent* calculations of the Green's function. In this context, 'self-consistency' means that the input information about the ground state and excitations of the systems no longer depend on any reference state but instead it is taken directly from the computed correlated wave function (or propagator, in our case). To achieve this, the computed spectral function is fed back into the working equations and calculations are repeated until a consistency between input and output is obtained. This approach is referred to as self-consistent Green's function (SCGF) method and it is always implemented, partially or in full, for nuclear structure applications.

Very recent advances in computational applications concern the extension of SCGF theory to the Gorkov-Nambu formalism for the breaking of particle number symmetry [15–17]. This allows to treat pairing systematically in systems with degenerate reference states and, therefore, to calculate open-shell nuclei directly. As a result, these developments have opened the possibility of studying large set of semi-magic nuclei that were previously beyond the reach of ab initio theory. We will not discuss the Gorkov-SCGF method here, but we will focus on the fundamental features of the standard approaches instead. The interested reader is referred to recent literature on the topic [16, 18–20].

In the process of discussing the relevant working equations of SCGF theory, we will also deal with applications to the same pairing model and the neutron matter with the Minnesota potential already discussed in the previous chapters. Together with presenting the most important steps for their numerical implementations, this book provides two examples of working codes in FORTRAN and C++ that can solve these models. Results for the self-energy and spectral functions should serve to gain a deeper understanding of the many-body physics that is embedded in the SCGF method. In discussing this, we will also benchmark the SCGF results with those obtained in other chapters of this book: coupled cluster (Chap. 8), Monte Carlo (Chap. 9) and in-medium similarity renormalization group (Chap. 10).

11.2 Many-Body Green's Function Theory

This chapter will focus on many-body *Green's functions*, which are also referred to as *propagators*. These are defined in the second quantization formalism by assuming the knowledge of the true ground state $|\Psi_0^A\rangle$ of a target system of A nucleons, which is taken to be a vacuum of excitations. The one-body Green's function (or propagator) is then defined as [21, 22]:

$$i\hbar\, g_{\alpha\beta}(t-t') = \langle\Psi_0^A|\mathcal{T}[a_\alpha(t)a_\beta^\dagger(t')]|\Psi_0^A\rangle\,, \tag{11.1}$$

where $\mathcal{T}$ is the time ordering operator, $a_\alpha^\dagger(t)$ ($a_\alpha(t)$) are the creation (annihilation) operators in Heisenberg picture, and Greek indices α, β, ... label a complete single particle basis that defines our model space. These can be the continuum spaces of momentum or coordinate, or any other discrete set of single particle states. Note that $g(t-t')$ depends only on the time difference $t-t'$ due to time translation invariance. For $t > t'$, Eq. (11.1) gives the probability amplitude to add a particle to $|\Psi_0^A\rangle$ in state β at time t' and then to let it propagate to reach state α at a later time t. Vice versa, for $t < t'$ a particle is removed from state α at t and added to β at t'.

In spite of being the simplest type of propagator, the one-body Green's function does contain a wealth of information regarding single particle behavior inside the many-body system, one-body observables, the total binding energy, and even elastic nucleon-nucleus scattering. The propagation of a particle or a hole excitation corresponds to the time evolution of an intermediate many-body system with $A+1$ or $A-1$ particles. One can better understand the physics information included in Eq. (11.1) from considering the eigenstates $|\Psi_n^{A+1}\rangle$, $|\Psi_k^{A-1}\rangle$ and eigenvalues E_n^{A+1}, E_k^{A-1} of these intermediate systems. By expanding on these eigenstates and Fourier transforming from time to frequency, one arrives at the spectral representation of the one-body Green's function:

$$\begin{aligned} g_{\alpha\beta}(\omega) &= \int \mathrm{d}\tau\, e^{i\omega\tau} g_{\alpha\beta}(\tau) \\ &= \sum_n \frac{\langle\Psi_0^A|a_\alpha|\Psi_n^{A+1}\rangle\langle\Psi_n^{A+1}|a_\beta^\dagger|\Psi_0^A\rangle}{\hbar\omega-(E_n^{A+1}-E_0^A)+\mathrm{i}\eta} + \sum_k \frac{\langle\Psi_0^A|a_\beta^\dagger|\Psi_k^{A-1}\rangle\langle\Psi_k^{A-1}|a_\alpha|\Psi_0^A\rangle}{\hbar\omega-(E_0^A-E_k^{A-1})-\mathrm{i}\eta}\,, \\ &\equiv \sum_n \frac{(\mathcal{X}_\alpha^n)^*\mathcal{X}_\beta^n}{\hbar\omega-\varepsilon_n^+ + i\eta} + \sum_k \frac{\mathcal{Y}_\alpha^k(\mathcal{Y}_\beta^k)^*}{\hbar\omega-\varepsilon_k^- - i\eta}\,, \end{aligned} \tag{11.2}$$

where the operators $a_\alpha^\dagger$ and a_α are now in Schördinger picture. Equation (11.2) was derived by a number of authors in the 1950s but is usually referred to as the 'Lehmann' representation in many-body physics [23–25]. For the rest of this chapter (with the only exception of Appendix 1) we will work in dimensionless units $\hbar = c = 1$ to avoid carrying over unnecessary $\hbar$ terms. From Eq. (11.2), we see that the poles of the Green's function, $\varepsilon_n^+ \equiv (E_n^{A+1} - E_0^A)$ and $\varepsilon_k^- \equiv (E_0^A - E_k^{A-1})$, are one-nucleon addition and removal energies, respectively. Note that these are generically referred to in the literature as "separation" or "quasiparticle" energies although the first naming should normally refer to transitions involving only $(A \pm 1)$-nucleon ground states. We will use the second convention in the following, unless the two naming are strictly equivalent. In the last line of Eq. (11.2) we have also introduced short notations for the spectroscopic amplitudes associated with the addition ($\mathcal{X}_\alpha^n \equiv \langle \Psi_n^{A+1} | a_\alpha^\dagger | \Psi_0^A \rangle$) and the removal ($\mathcal{Y}_\alpha^k \equiv \langle \Psi_k^{A-1} | a_\alpha | \Psi_0^A \rangle$) of a particle to and from the initial ground state $|\Psi_0^A\rangle$. We will use the Latin letter n to label one-particle excitations and to distinguish them from one-hole states that are indicated by k instead. This compact form will simplify deriving the working formalism in the following sections.

The one-body Green's function (11.2) is completely determined by solving the Dyson equation:

$$g_{\alpha\beta}(\omega) = g_{\alpha\beta}^{(0)}(\omega) + \sum_{\gamma\delta} g_{\alpha\gamma}^{(0)}(\omega)\, \Sigma_{\gamma\delta}^{\star}(\omega)\, g_{\delta\beta}(\omega) \tag{11.3a}$$

$$= g_{\alpha\beta}^{(0)}(\omega) + \sum_{\gamma\delta} g_{\alpha\gamma}(\omega)\, \Sigma_{\gamma\delta}^{\star}(\omega)\, g_{\delta\beta}^{(0)}(\omega)\,, \tag{11.3b}$$

where we have put in evidence that there exists two different conjugate forms of this equation, corresponding to the first and second lines. In Eqs. (11.3), the unperturbed propagator $g_{\alpha\beta}^{(0)}(\omega)$ is the initial reference state (usually a mean-field or Hartree-Fock state), while $g_{\alpha\beta}(\omega)$ is called the *correlated* or *dressed* propagator. The quantity $\Sigma_{\gamma\delta}^{\star}(\omega)$ is the *irreducible self-energy* and it is often referred to as the *mass operator*. This operator plays a central role in the GF formalism and can be interpreted as the non-local and energy-dependent potential that each fermion feels due to the interactions with the medium. For frequencies $\omega > 0$, the solution of Eqs. (11.3) yields a continuum spectrum with $E_n^{A+1} > E_0^A$ and the state $|\Psi_n^{A+1}\rangle$ describes the elastic scattering of the additional nucleon off the $|\Psi_0^A\rangle$ target. It can be shown that $\Sigma^{\star}(\omega)$ is an exact optical potential for scattering of a particle from the many-body target [26–28]. The Dyson equation is nonlinear in its solution, $g(\omega)$, and thus it corresponds to an all-orders resummation of diagrams involving the self-energy. The Feynman diagrams representations of both forms of the Dyson equation are shown in Fig. 11.1. In both cases, by recursively substituting the exact Green's function (indicated by double lines) that appears on the right hand side with the whole equation, one finds a unique expansion in terms of the unperturbed $g^{(0)}(\omega)$ and the irreducible self-energy. The solution of Eqs. (11.3) is referred to as *dressed* propagators since it formally results by 'dressing' the free particle (shown as a single line) by repeated interactions with the system $(\Sigma^{\star}(\omega))$.

A full knowledge of the self-energy $\Sigma^{\star}(\omega)$ [see Eqs. (11.3)] would yield the exact solution for $g(\omega)$ but in practice this has to be approximated somehow. Standard perturbation theory, expands $\Sigma^{\star}(\omega)$ in a series of terms that depend on the

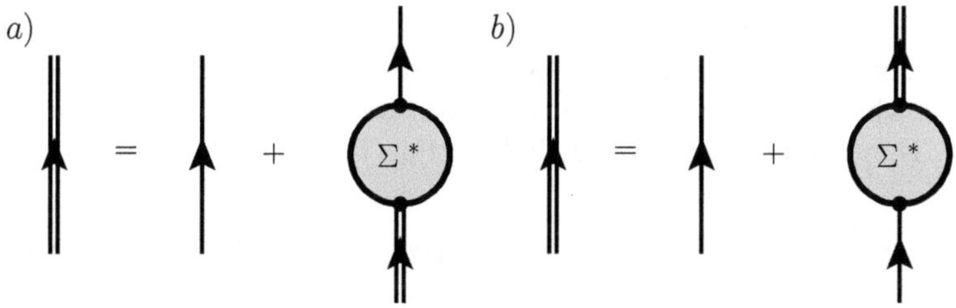

Fig. 11.1 Diagrammatic representations of the Dyson equation. The diagram on the *left* represents Eq. (11.3a), while its conjugate equation (11.3b) is shown to the *right*. *Single lines with an arrow* represent the unperturbed propagator $g^{(0)}(\omega)$ and *double lines* are the fully dressed propagator $g(\omega)$ of Eq. (11.2). Both equations, when expanded in terms of $g^{(0)}(\omega)$, give rise to the same series of diagrams for the correlated propagator

interactions and on the unperturbed propagator $g^{(0)}(\omega)$. However, it is also possible to rearrange the perturbative expansion in diagrams that depend only on the exact dressed propagator itself (that is, $\Sigma^{\star} = \Sigma^{\star}[g(\omega)]$). Since any propagator in this diagrammatic expansion is already dressed, one only needs to consider a smaller set of contributions—the so-called *skeleton* diagrams. These are diagrams that do not explicitly include any self-energy insertion, as these are already generated by Eqs. (11.3). We will discuss these aspects in more detail in Sect. 11.2.2. For the present discussion, we only need to be aware that the functional dependence of $\Sigma^{\star}[g(\omega)]$ requires an iterative procedure in which $\Sigma^{\star}(\omega)$ and Eqs. (11.3) are calculated several times until they converge to a unique solution. This approach defines the SCGF method and it is particularly important since it can be shown that full self-consistency allows to exactly satisfy fundamental symmetries and conservation laws [29, 30]. In practical applications, and especially in finite systems, this scheme may not be achievable exactly and self-consistency is implemented only partially for the most important contributions. Normally this is still sufficient to obtain highly accurate results. We will present suitable approximation schemes to calculate the self-energy in the following sections. In particular, we will focus on the ADC(n) method that can be applied with discretized bases in finite and infinite systems in Sects. 11.3 and 11.4. The case of extended systems at finite temperature is discussed in Sect. 11.5. Before going into the actual approximation schemes, we need to see how experimental quantities can be calculated once the one-body propagator is known, as well as to discuss the basic results of perturbation theory.

11.2.1 Spectral Function and Relation to Experimental Observations

Once the one-body Green's function is known, it can be used to calculate the total binding energy and the expectation values of all one-body observables. The attractive feature of the SCGF approach is that $g(\omega)$ describes the one-body dynamics completely. This information can be recast in the particle and hole spectral functions, which contain the separate responses for the attachment and removal of a nucleon. They can be obtained directly from Eq. (11.2), as follows:

$$S^{p}_{\alpha\beta}(\omega) = -\frac{1}{\pi}\text{Im}\, g_{\alpha\beta}(\omega) = \sum_n \left(\mathcal{X}^n_\alpha\right)^* \mathcal{X}^n_\beta\, \delta\Big(\omega - (E^{A+1}_n - E^A_0)\Big)\,, \qquad \text{for } \omega \geq \varepsilon^+_0\,,$$

$$S^{h}_{\alpha\beta}(\omega) = \frac{1}{\pi}\text{Im}\, g_{\alpha\beta}(\omega) = \sum_k \mathcal{Y}^k_\alpha (\mathcal{Y}^k_\beta)^*\, \delta\Big(\omega - (E^A_0 - E^{A-1}_k)\Big)\,, \qquad \text{for } \omega \leq \varepsilon^-_0\,. \tag{11.4}$$

The diagonal parts of Eqs. (11.4), have a straightforward physical interpretation [21, 22]. The particle part, $S^p_{\alpha\alpha}(\omega)$, is the joint probability of adding a nucleon with quantum numbers α to the A-body ground state, $|\Psi^A_0\rangle$, and then to find the system in a final state with energy $E^{A+1} = E^A_0 + \omega$. Likewise, $S^h_{\alpha\alpha}(\omega)$ gives the probability of removing a particle from state α while leaving the nucleus in an eigenstate of energy $E^{A-1} = E^A_0 - \omega$. These are demonstrated in coordinate space in Fig. 11.2 for neutrons around ^{56}Ni. Below the Fermi energy, $E_F \equiv \frac{1}{2}(\varepsilon^+_0 + \varepsilon^-_0)$, one can see a single dominant quasihole peak corresponding to the $f_{7/2}$ orbit. The states from the *sd* shell are at lower energies and are instead very fragmented. Just above E_F, there are sharp quasiparticles corresponding to the attachment of a neutron to the remaining *pf* orbits. Finally, for $\omega > 0$, one has neutron-^{56}Ni elastic scattering states. Remarkably, one can see that dominant quasiparticle peaks persist around the Fermi surface, which confirms the underlying shell structure outside the ^{40}Ca core for this nucleus.

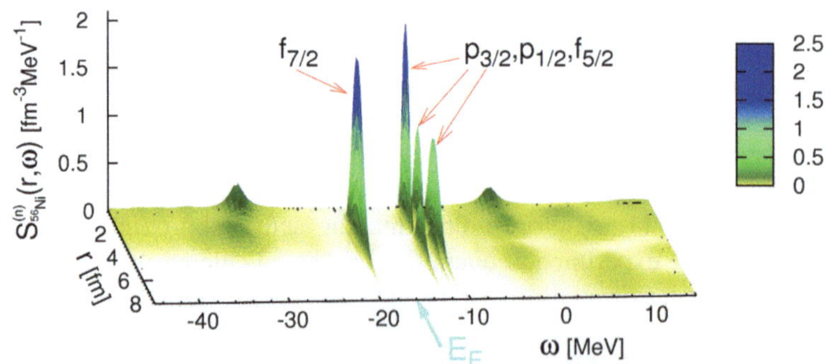

Fig. 11.2 Calculated single-particle spectral function for the addition and removal of a neutron to and from ^{56}Ni, from [31]. The diagonal part, $S_{r,r}(\omega)$, is shown in coordinate space. Energies below the Fermi level E_F correspond to the one-hole spectral function $S^h_{r,r}(\omega)$ which describes the distribution of nucleons in energy and coordinate space. Integrating over all the quasihole energies yields the matter density distribution, Eq. (11.9). Energies above E_F are for the one-particle spectral function $S^p_{r,r}(\omega)$

The existence of isolated dominant peaks as those shown in Fig. 11.2 indicates that the eigenstates $|\Psi_n^{A+1}\rangle$ and $|\Psi_k^{A-1}\rangle$ are to a very good approximation constructed of a nucleon or a hole independently orbiting the ground state $|\Psi_0^A\rangle$. This is the basic hypothesis at the origin of the nuclear shell-model. How much a real nucleus deviates from this assumption can be gauged by the deviations in the values of their *spectroscopic factors*. These are defined as the normalization overlap of the spectroscopic amplitudes for the attachment or removal of a particle:

$$SF_n^+ = \sum_\alpha |\mathcal{X}_\alpha^n|^2 , \qquad\qquad SF_k^- = \sum_\alpha |\mathcal{Y}_\alpha^k|^2 . \tag{11.5}$$

The energy distribution of spectroscopic factors is given by

$$\begin{aligned} S(\omega) &= \sum_\alpha S^p_{\alpha\alpha}(\omega) \quad + \quad \sum_\alpha S^h_{\alpha\alpha}(\omega) \\ &= \sum_n SF_n^+\, \delta(\omega - E_n^{A+1} + E_0^A) \;+\; \sum_k SF_k^-\, \delta(\omega - E_0^A + E_k^{A-1}) , \end{aligned} \tag{11.6}$$

where each δ-peak corresponds to eigenstates of a neighboring isotope with $A \pm 1$ particles. These quasiparticle energies are directly observed in nucleon addition and removal experiments. Note that the total strength seen in similar experiments results from a convolution of the spectroscopic amplitudes with the dynamics of the reaction mechanisms. Hence, while the quasiparticle energies appearing in the poles of Eq. (11.2) are strictly observed, the magnitude of the spectral strength $S(\omega)$ only gives a semi-quantitative description of the strength of the observed cross sections.

Any one-body observable can be calculated via the one-body density matrix $\rho_{\alpha\beta}$, which is obtained from $g_{\alpha\beta}(\omega)$ as follows:

$$\rho_{\alpha\beta} \;\equiv\; \langle\Psi_0^A|a_\beta^\dagger a_\alpha|\Psi_0^A\rangle \;=\; \int_{-\infty}^{\varepsilon_0^-} S^h_{\alpha\beta}(\omega)\, d\omega = \sum_k (\mathcal{Y}_\beta^k)^* \mathcal{Y}_\alpha^k . \tag{11.7}$$

The expectation value of a one-body operator, $\widehat{O}^{1B}$, can then be written in terms of the $\mathcal{Y}$ amplitudes as:

$$\langle\widehat{O}^{1B}\rangle = \sum_{\alpha\beta} O^{1B}_{\alpha\beta}\, \rho_{\beta\alpha} = \sum_k \sum_{\alpha\beta} (\mathcal{Y}_\alpha^k)^*\, O^{1B}_{\alpha\beta}\, \mathcal{Y}_\beta^k . \tag{11.8}$$

However, evaluating two- and many-nucleon observables requires the knowledge of many-body propagators. Equation (11.7) also implies that the density profile of the system can be obtained by integrating over the hole spectral function in coordinate space (cf. Fig. 11.2):

$$\rho(\mathbf{r}) = \int_{-\infty}^{\varepsilon_0^-} S^h_{\mathbf{r},\mathbf{r}}(\omega)\, d\omega . \tag{11.9}$$

Likewise, a second sum (or integration) over the coordinate space yields the total number of particles,

$$\int d\mathbf{r} \int_{-\infty}^{\varepsilon_0^-} S^h_{\mathbf{r},\mathbf{r}}(\omega)\, d\omega \;=\; \sum_{\alpha} \int_{-\infty}^{\varepsilon_0^-} S^h_{\alpha\alpha}(\omega)\, d\omega \;=\; A\,. \tag{11.10}$$

A very special case is the Koltun sum-rule that allows calculating the total energy of the system by means of the exact one-body propagator alone, $g(\omega)$ [7, 32]. This relation is exact for any Hamiltonian containing at most one- and two body interactions. When many-particle interactions are present, it is necessary to correct for the over countings that arise from these additional terms [33]. For the specific case in which a three-body interaction $\widehat{W}$ is included, the exact relation for the ground state energy is given by the following modified Koltun rule:

$$E_0^A = \sum_{\alpha\beta} \frac{1}{2} \int_{-\infty}^{\varepsilon_0^-} [\, T_{\alpha\beta} + \omega\, \delta_{\alpha\beta}\,]\, S^h_{\beta\alpha}(\omega)\, d\omega \;-\; \frac{1}{2} \langle \widehat{W} \rangle\,. \tag{11.11}$$

This still relies on the use of a one-body propagator but it requires the additional evaluation of the expectation value of the three-body interaction, $\langle \widehat{W} \rangle$ (which in principle requires the knowledge of more complex Green's functions). Thankfully, in most cases the total strength of $\widehat{W}$ is much smaller than other terms in the Hamiltonian. Thus, one can safely approximate its expectation value at lowest order, in terms of three correlated density matrices, as

$$\langle \widehat{W} \rangle \simeq \frac{1}{6} \sum_{\alpha\beta\mu\gamma\delta\nu} W_{\alpha\beta\mu,\gamma\delta\nu}\; \rho_{\gamma\alpha}\; \rho_{\delta\beta}\; \rho_{\nu\mu}\,. \tag{11.12}$$

As a typical example in finite nuclei, the error from this approximation has been estimated not to exceed 250 keV for the total binding energies for ^{16}O and ^{24}O [34]. However, the accuracy of Eq. (11.12) is not guaranteed and needs to be verified case by case.

11.2.2 Perturbation Expansion of the Green's Function

In order to understand the following sections and to devise appropriate approximations to the self-energy $\Sigma^\star(\omega)$ it is necessary to understand the basic elements of perturbation theory. These will be also fundamental to derive all-order summation schemes leading to non-perturbative solutions and to discuss the concept of self-consistency. We summarize here the material needed to understand the following sections, while the full set of Feynman rules is reviewed in Appendix 1.

We work with a system of A non-relativistic fermions interacting by means of two-body and three-body interactions. We divide the Hamiltonian into two parts, $\widehat{H} = \widehat{H}_0 + \widehat{H}_1$. The unperturbed term, $\widehat{H}_0 = \widehat{T} + \widehat{U}$, is given by the sum of the kinetic term and an auxiliary one-body operator $\widehat{U}$. Its choice defines the reference state, $|\Phi_0^A\rangle$, and the corresponding unperturbed propagator $g^{(0)}(\omega)$ that are the starting point for the perturbative expansion.[1] The perturbative term is then $\widehat{H}_1 = -\widehat{U} + \widehat{V} + \widehat{W}$, where $\widehat{V}$ denotes the two-body interaction operator and $\widehat{W}$ is the three-body interaction. In a second-quantized framework, the full Hamiltonian reads:

$$\widehat{H} = \sum_{\alpha} \varepsilon_\alpha^0 \, a_\alpha^\dagger a_\alpha - \sum_{\alpha\beta} U_{\alpha\beta} \, a_\alpha^\dagger a_\beta + \frac{1}{4} \sum_{\substack{\alpha\gamma \\ \beta\delta}} V_{\alpha\gamma,\beta\delta} \, a_\alpha^\dagger a_\gamma^\dagger a_\delta a_\beta$$
$$+ \frac{1}{36} \sum_{\substack{\alpha\gamma\epsilon \\ \beta\delta\eta}} W_{\alpha\gamma\epsilon,\beta\delta\eta} \, a_\alpha^\dagger a_\gamma^\dagger a_\epsilon^\dagger a_\eta a_\delta a_\beta \,. \tag{11.13}$$

In Eq. (11.13) we continue to use Greek indices $\alpha,\beta,\gamma,\ldots$ to label the single particle basis that defines the model space. But we make the additional assumption that these are the same states which diagonalize the unperturbed Hamiltonian, $\widehat{H}_0$, with eigenvalues ε_α^0. This choice is made in most applications of perturbation theory but it is not strictly necessary here and it will not affect our discussion in the following sections. The matrix elements of the one-body operator $\widehat{U}$ are given by $U_{\alpha\beta}$. And we work with properly antisymmetrized matrix elements of the two-body and three-body forces, $V_{\alpha\gamma,\beta\delta}$ and $W_{\alpha\gamma\epsilon,\beta\delta\eta}$.

In time representation, the many-body Green's functions are defined as the expectation value of time-ordered products of annihilation and creation operators in the Heisenberg picture. This is shown by Eq. (11.1) for the single particle propagator. Every Green's function can be expanded in a perturbation series in powers of $\widehat{H}_1$. For the one-body propagator this reads [22, 35]:

$$g_{\alpha\beta}(t_\alpha - t_\beta) = (-i) \sum_{n=0}^{\infty} (-i)^n \frac{1}{n!} \int \mathrm{d}t_1 \ldots \int \mathrm{d}t_n$$
$$\times \langle \Phi_0^A | \mathcal{T}[\widehat{H}_1^I(t_1) \ldots \widehat{H}_1^I(t_n) a_\alpha^I(t_\alpha) a_\beta^{I\,\dagger}(t_\beta)] | \Phi_0^A \rangle_{\mathrm{conn}} \,, \tag{11.14}$$

where $\widehat{H}_1^I(t)$, $a_\alpha^I(t)$ and $a_\beta^{I\,\dagger}(t)$ are now intended as operators in the interaction picture with respect to $\widehat{H}_0$. The subscript "conn" implies that only *connected* diagrams have to be considered when performing the Wick contractions of the time-ordered product $\mathcal{T}$. Each Wick contraction generates an uncorrelated single particle propagator, $g^{(0)}(\omega)$, which is associated with the system governed by the

[1] A typical choice in nuclear physics would be a Slater determinant such as the solution of the Hartree-Fock problem or a set of single-particle harmonic oscillator wave functions.

Hamiltonian H_0. At order $n = 0$, the expansion of Eq. (11.14) simply gives $g^{(0)}(\omega)$. H_1 contains contributions from one-body, two-body and three-body interactions that come from the last three terms on the right hand side of Eq. (11.13). Thus, for $n \geq 1$ the expansion involves terms with individual contributions of each force, or combinations of them, that are linked by uncorrelated propagators. To each term in the expansion there corresponds a Feynman diagram that gives an intuitive picture of the physical process accounted by its contribution. The full set of Feynman diagrammatic rules that stems out of Eq. (11.14) in the presence of three-body interactions is detailed in Appendix 1.

A first reorganization of the contributions generated by Eq. (11.14) is obtained by considering *one-particle reducible* diagrams, that is diagrams that can be disconnected by cutting a single fermionic line. In general, the reducible diagrams generated by expansion (11.14) will always have separate structures that are linked together by only one $g^{(0)}(\omega)$ line. These are the same class of diagrams that are created implicitly in the all-orders resummation of the Dyson equation (11.3). Thus, the irreducible self-energy $\Sigma^{\star}(\omega)$ is defined as the kernel that collects all the *one-particle irreducible* (1PI) diagrams (with the external legs stripped off). As already discussed above, $\Sigma^{\star}(\omega)$ plays the role of an effective potential that is seen by a nucleon inside the system. It splits in static and frequency dependent terms:

$$\Sigma^{\star}_{\alpha\beta}(\omega) \;=\; -\,U_{\alpha\beta} \;+\; \Sigma^{(\infty)}_{\alpha\beta} \;+\; \widetilde{\Sigma}_{\alpha\beta}(\omega)\,, \tag{11.15}$$

where we have separated $\widehat{U}$ since this is auxiliary defined and it eventually cancels out when solving the Dyson equation. The term $\Sigma^{(\infty)}$ plays the role of the static mean-field that a nucleon feels due to the average interactions will all other particles in the system. The frequency-dependent part, $\widetilde{\Sigma}(\omega)$, describes the effects of dynamical excitations of the many-body state that are induced by the nucleon itself. In general, this means the propagation of (complex) intermediate excitations and therefore it must have a Lehmann representation analogous to that of Eq. (11.2). For very large energies ($\omega \to \pm\infty$) the poles of such Lehmann representation become vanishingly small and one is left with just $\Sigma^{(\infty)}$ and the auxiliary potential $\widehat{U}$.

A further level of simplification in the self-energy expansion can be obtained if unperturbed propagators, $g^{(0)}(\omega)$, in the internal fermionic lines are replaced by dressed Green's functions, $g(\omega)$. This choice further restricts the set of diagrams to the so-called *skeleton* diagrams [22], which are defined as 1PI diagrams that do not contain any portion that can be disconnected by cutting any two fermion lines at different points. These portions would correspond to self-energy insertions, which are already re-summed into the dressed propagator $g(\omega)$ by Eq. (11.3). The SCGF approach is precisely based on expressing the irreducible self-energy in terms of such skeleton diagrams with dressed propagators. The SCGF framework offers great advantages. First, it is intrinsically non-perturbative and completely independent from any choice of the reference state and auxiliary one-body potential. This is so because $\Sigma^{\star}(\omega)$ no longer depends on $g^{(0)}(\omega)$ and $\widehat{U}$ always drops out of the Dyson equation (see Eq. (11.45) below). Second, many-body correlations are expanded directly in terms of single particle excitations of the true propagator, which are generally closer to the exact solution than those associated with the unperturbed state, $|\Phi_0^A\rangle$. Third, given an appropriate truncation of self-energy, if a full SCGF calculation is possible then it automatically satisfies the basic conservation laws of particle number, angular momentum, etc… [22, 29, 30]. Finally, the number of diagrams to be considered is vastly reduced to 1PI skeletons one. However, this is not always a simplification since a dressed propagator contains a very large number of poles, which can be much more difficult to deal with than for the corresponding uncorrelated $g^{(0)}(\omega)$.

If three- or many-body forces are included in the Hamiltonian, the number of Feynman diagrams that need to be considered at a given order increases very rapidly. In this case it becomes very useful and instructive to restrict the attention to an even smaller class of diagrams that are *interaction-irreducible* [33]. An interaction vertex is said to be reducible if the whole diagram can be disconnected in two parts by cutting the vertex itself. In general, this happens for an m-body interaction when there is a smaller number of n lines ($n < m$) that leave the interaction, may interact only among themselves, and eventually all return to it. The net outcome is that one is left with a $(m-n)$-body operator that results from the average interactions with other n-spectator nucleons. This plays the role of a system dependent effective force that is irreducible. Figure 11.3 shows diagrammatically how $\widehat{V}$ and $\widehat{W}$ can be reduced to one- and two-body *effective interactions* in this way.

Hence, for a system with up to three-body forces, we define an effective Hamiltonian

$$\widetilde{H}_1 = \widetilde{U} + \widetilde{V} + \widehat{W} , \tag{11.16}$$

where $\widetilde{U}$ and $\widetilde{V}$ are the effective interaction operators. The diagrammatic expansion arising from Eq. (11.14) with the effective Hamiltonian $\widetilde{H}_1$ is formed only of (1PI, skeleton) interaction-irreducible diagrams. Note that the three-body interaction, $\widehat{W}$, remains the same as in Eq. (11.13) but enters only diagrams as an interaction-irreducible three-body force. The explicit expressions for the one-body and two-

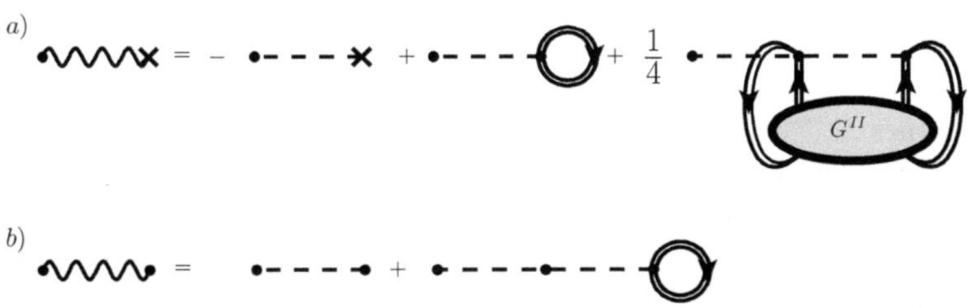

Fig. 11.3 Graphical representation of the effective one-body interaction of Eq. (11.17a), *top row* (**a**), and the effective two-body interaction (11.17b), *bottom row* (**b**). *Dashed lines* represent the one-, two,- and three-body interactions entering Eq. (11.13) and *wavy lines* are the effective operators $\widetilde{U}$ and $\widetilde{V}$

body effective interaction operators can be obtained form the Feynman diagrams of Fig. 11.3 and they are given by:

$$\widetilde{U}_{\alpha\beta} = -U_{\alpha\beta} + \sum_{\delta\gamma} V_{\alpha\gamma,\beta\delta}\, \rho_{\delta\gamma} + \frac{1}{4}\sum_{\mu\nu\gamma\delta} W_{\alpha\mu\nu,\beta\gamma\delta}\, \Gamma_{\gamma\delta,\mu\nu} \,, \tag{11.17a}$$

$$\widetilde{V}_{\alpha\beta,\gamma\delta} = V_{\alpha\beta,\gamma\delta} + \sum_{\mu\nu} W_{\alpha\beta\mu,\gamma\delta\nu}\, \rho_{\nu\mu} \,. \tag{11.17b}$$

where we used the reduced two-body density matrix Γ, which can be computed from the exact two-body Green's function:

$$\Gamma_{\gamma\delta,\mu\nu} = \lim_{\tau\to 0^-} -i\, G^{II}_{\gamma\delta,\mu\nu}(\tau) = \langle \Psi_0^A | a_\nu^\dagger a_\mu^\dagger a_\gamma a_\delta | \Psi_0^A \rangle \,. \tag{11.18}$$

The effective Hamiltonian of Eq. (11.16) not only regroups Feynman diagrams in a more efficient way, but also defines the effective one-body and two-body terms from higher order interactions. As long as interaction-irreducible diagrams are used together with the effective Hamiltonian, $\widetilde{H}_1$, this approach provides a systematic way to incorporate many-body forces in the calculations and to generate effective in-medium interactions. More importantly, the formalism is such that all symmetry factors are guaranteed to be correct and no diagram is over-counted [33]. Equations (11.17) can be seen as a generalization of the normal ordering of the Hamiltonian with respect to the reference state $|\Phi_0^A\rangle$ discussed in Chap. 8. However, these contractions go beyond normal ordering because they are performed with respect to the exact correlated density matrices. To some extent, one can intuitively think of the effective Hamiltonian $\widetilde{H}_1$ as being ordered with respect to the interacting many-body ground-state $|\Psi_0^A\rangle$, rather than the non-interacting $|\Phi_0^A\rangle$.

Since the static self-energy does not propagate any intermediate excitations, it can only receive a contribution when the incoming and outgoing lines of a Feynman diagram are attached to the same interaction vertex. Thus, by definition, $\Sigma^{(\infty)}$ must include the one body term in $\widehat{H}_1$ plus any higher order interaction that are reduced to effective one-body interactions, hence:

$$\widetilde{U}_{\alpha\beta} = -U_{\alpha\beta} + \Sigma^{(\infty)}_{\alpha\beta} \,, \tag{11.19}$$

which defines $\Sigma^{(\infty)}$ by comparison with Eq. (11.17a). The two terms that contribute to $\Sigma^{(\infty)}$ represent extensions of the Hartree-Fock (HF) potentials to correlated ground states. The correlated Hartree-Fock potential from $\widehat{V}$ is the only effective operator when just two-body forces are present. In this case there is very little gain in using the concept of the effective Hamiltonian (11.16). However, with three-body interactions, additional effective interaction terms appear in both $\widetilde{U}$ and $\widetilde{V}$. From Eq. (11.19) we see that the perturbative SCGF expansion with the $\widetilde{H}_1$ Hamiltonian has only one (1PI, skeleton and interaction-irreducible) term at first order. The first contributions to $\widetilde{\Sigma}(\omega)$ appear at second order with the two diagrams in Fig. 11.4.

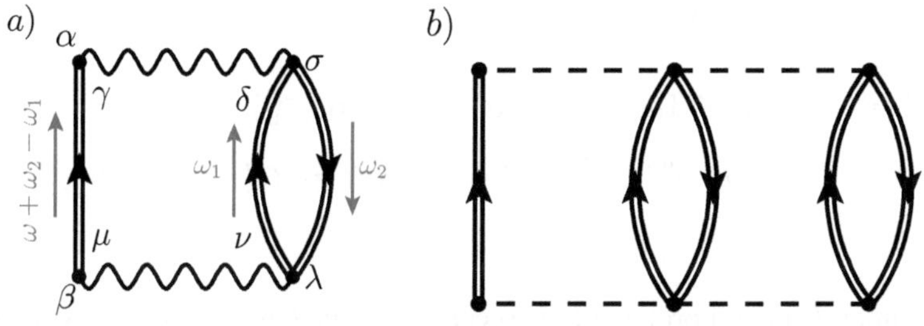

Fig. 11.4 Second-order interaction-irreducible contributions to the self-energy arising from both two-nucleon [diagram (**a**)] and three-nucleon forces [diagram (**b**)]. The diagram (**a**) that depends on the effective two-body interactions also shows the indices and labels that are used for calculating its contribution in Example 11.2

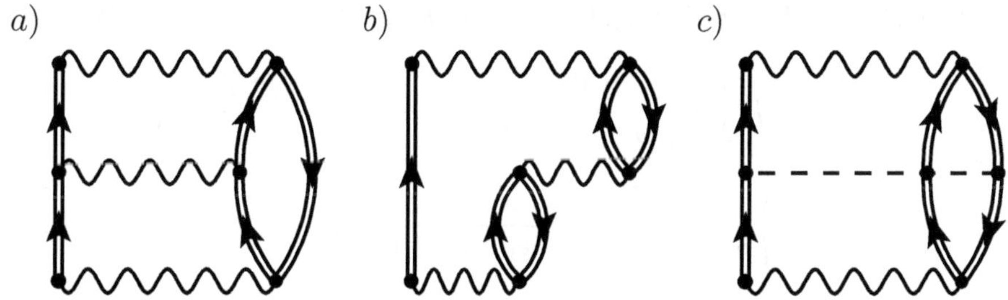

Fig. 11.5 The three simplest skeleton and interaction-irreducible diagrams contributing to the self-energy at third order. All these terms involve intermediate state configurations of at most 2p1h and 2h1p. The first two contain only two-nucleon interactions and are the first terms in the resummation of ladders [diagram (**a**)] and rings [diagram (**b**)]. The diagram (**c**) is the first contribution containing an irreducible three-nucleon interaction. All the remaining 14 diagrams at third order require explicit three-body interactions and ISCs with 3p2h and 3h2p excitations [33, 36]

Expanding with respect to $\widehat{H}_1$, there would have been five diagrams instead of only the two interaction-irreducible ones shown in Fig. 11.4. These diagrams indeed have a proper Lehmann representation (see Example 11.2 below and Exercise 11.2) and propagate *intermediate state configurations* (ISCs) of type 2-particle 1-hole (2p1h), 2h1p, 3p2h, etc... At third order, $\widetilde{H}_1$ generates 17 SCGF diagrams two of which contain only two-body interactions. The simplest of these, that involve at most 2p1h and 2h1p ISCs, are shown in Fig. 11.5. All interaction-irreducible contributions to the proper self-energy up to third order in perturbation theory are discussed in details in [33].

Example 11.1 Calculate the Feynman-Galitskii propagator, $G^{II,f}(\tau)$, that corresponds to the propagation of two particles or two holes that do not interact with each other.

This is the lowest order approximation to the two-times and two-body propagator which evolves two particle from states α and β to states γ and δ after a time $\tau > 0$, or two holes from γ and δ to α and β when $\tau < 0$. By applying the perturbative expansion equivalent to Eq. (11.14) at order $n = 0$, we find:

$$G^{II\,(0)}_{\alpha\beta,\gamma\delta}(\tau) = -i\langle\Phi_0^A|\mathcal{T}[a^I_\beta(\tau)\,a^I_\alpha(\tau)\,a^{I\,\dagger}_\gamma(0)\,a^{I\,\dagger}_\delta(0)]|\Phi_0^A\rangle$$
$$= ig^{(0)}_{\alpha\gamma}(\tau)\;g^{(0)}_{\beta\delta}(\tau) - ig^{(0)}_{\alpha\delta}(\tau)\;g^{(0)}_{\beta\gamma}(\tau) \equiv G^{II\,(0)f}_{\alpha\beta,\gamma\delta}(\tau) - G^{II\,(0)f}_{\alpha\beta,\delta\gamma}(\tau)\,. \tag{11.20}$$

The Feynman-Galitskii propagator is precisely defined as the non antisymmetrized part of Eq. (11.20). We now transform this to frequency space and apply the Feynman rules of Appendix 1 to calculate the $G^{II f}$ for the more general case of two *dressed* propagator lines:

$$G^{II f}_{\alpha\beta,\gamma\delta}(\omega) = \int \mathrm{d}\tau\; e^{i\omega\tau}\, G^{II f}_{\alpha\beta,\gamma\delta}(\tau) = (-i)\int \frac{\mathrm{d}\omega_1}{2\pi}\; i\,g_{\alpha\gamma}(\omega-\omega_1)\; i\,g_{\beta\delta}(\omega_1)$$
$$= -\int \frac{\mathrm{d}\omega_1}{2\pi i}\left\{\frac{(\mathcal{X}^{n_1}_\alpha)^*\mathcal{X}^{n_1}_\gamma}{\omega-\omega_1-\varepsilon^+_{n_1}+i\eta} + \frac{\mathcal{Y}^{k_1}_\alpha(\mathcal{Y}^{k_1}_\gamma)^*}{\omega-\omega_1-\varepsilon^-_{k_1}-i\eta}\right\}$$
$$\times\left\{\frac{(\mathcal{X}^{n_2}_\beta)^*\mathcal{X}^{n_2}_\delta}{\omega_1-\varepsilon^+_{n_2}+i\eta} + \frac{\mathcal{Y}^{k_2}_\beta(\mathcal{Y}^{k_2}_\delta)^*}{\omega_1-\varepsilon^-_{k_2}-i\eta}\right\}, \tag{11.21}$$

where we have used the convention that repeated indices are summed over. The integral in the above equation can be performed with the Cauchy theorem by closing an arch on either the positive or the negative imaginary half planes. Hence, contributions where all the poles are on the same side of the real axis cancel out. Extracting the residues of the other contributions leads to the following result:

$$G^{II f}_{\alpha\beta,\gamma\delta}(\omega) = \sum_{n_1,n_2}\frac{(\mathcal{X}^{n_1}_\alpha\mathcal{X}^{n_2}_\beta)^*\mathcal{X}^{n_1}_\gamma\mathcal{X}^{n_2}_\delta}{\omega-(\varepsilon^+_{n_1}+\varepsilon^+_{n_2})+i\eta} - \sum_{k_1,k_2}\frac{\mathcal{Y}^{k_1}_\alpha\mathcal{Y}^{k_2}_\beta(\mathcal{Y}^{k_1}_\gamma\mathcal{Y}^{k_2}_\delta)^*}{\omega-(\varepsilon^-_{k_1}+\varepsilon^-_{k_2})-i\eta}\,. \tag{11.22}$$

11.1 *E*xercise: Calculate the contribution of the three-body force $\widehat{W}$ to the effective one body potential, in the approximation of two dressed but non interacting spectator nucleons.

Solution

This is the last term in Fig. 11.3a and Eq. (11.17a) but with $G^{II}(\tau)$ approximated by two independent fermion lines, as for the dressed Feynman-Galitskii propagator. Using Eq. (11.18) and re-expressing the second line of (11.20) in terms of $g(\tau)$, we arrive at:

$$\widetilde{U}^{(W)}_{\alpha\beta} = \frac{1}{2}\sum_{\mu\nu\gamma\delta} W_{\alpha\mu\nu,\beta\gamma\delta}\;\rho_{\gamma\mu}\;\rho_{\delta\nu}\,. \tag{11.23}$$

Example 11.2 Calculate the expression for the second-order contribution to $\Sigma^{\star}(\omega)$ from two-nucleon interactions only.

This is the diagram of Fig. 11.4a. By applying the Feynman rules of Appendix 1 we have:

$$\begin{aligned}
\Sigma^{(2,2N)}_{\alpha\beta}(\omega) &= -\frac{(i)^2}{2}\int\frac{d\omega_1}{2\pi}\frac{d\omega_2}{2\pi}V_{\alpha\sigma,\gamma\delta}\,g_{\gamma\mu}(\omega+\omega_2-\omega_1)\,g_{\delta\nu}(\omega_1)\,g_{\lambda\sigma}(\omega_2)\,V_{\mu\nu,\beta\lambda}\\
&= +\frac{1}{2}\int\frac{d\omega_2}{2\pi i}V_{\alpha\sigma,\gamma\delta}\,G^{II,f}_{\gamma\delta,\mu\nu}(\omega+\omega_2)\,g_{\lambda\sigma}(\omega_2)\,V_{\mu\nu,\beta\lambda}\\
&= \frac{1}{2}\int\frac{d\omega_2}{2\pi i}V_{\alpha\sigma,\gamma\delta}\left\{\frac{(\mathcal{X}^{n_1}_{\gamma}\mathcal{X}^{n_2}_{\delta})^{*}\mathcal{X}^{n_1}_{\mu}\mathcal{X}^{n_2}_{\nu}}{\omega+\omega_2-(\varepsilon^{+}_{n_1}+\varepsilon^{+}_{n_2})+i\eta}\right.\\
&\qquad\left.-\frac{\mathcal{Y}^{k_1}_{\gamma}\mathcal{Y}^{k_2}_{\delta}(\mathcal{Y}^{k_1}_{\mu}\mathcal{Y}^{k_2}_{\nu})^{*}}{\omega+\omega_2-(\varepsilon^{-}_{k_1}+\varepsilon^{-}_{k_2})-i\eta}\right\}\\
&\qquad\times\left\{\frac{(\mathcal{X}^{n_3}_{\lambda})^{*}\mathcal{X}^{n_3}_{\sigma}}{\omega_2-\varepsilon^{+}_{n_3}+i\eta}+\frac{\mathcal{Y}^{k_3}_{\lambda}(\mathcal{Y}^{k_3}_{\sigma})^{*}}{\omega_2-\varepsilon^{-}_{k_3}-i\eta}\right\}V_{\mu\nu,\beta\lambda}\,,
\end{aligned}\tag{11.24}$$

where we have used the two-body interaction $\widehat{V}$, but it could have been equally calculated with the effective interaction $\widetilde{V}$. Note that the integration over ω_1 is exactly the same as in Eq. (11.21). Thus, we can directly substitute the expression for the Feynman-Galitskii propagator (11.22) in the last two lines above. By performing the last Cauchy integral we find that only two out of four possible terms survive. The final result for the second-order irreducible self-energy is:

$$\begin{aligned}
\Sigma^{(2,2N)}_{\alpha\beta}(\omega) = \frac{1}{2}V_{\alpha\sigma,\gamma\delta}&\left\{\sum_{\substack{n_1,n_2\\k_3}}\frac{(\mathcal{X}^{n_1}_{\gamma}\mathcal{X}^{n_2}_{\delta}\mathcal{Y}^{k_3}_{\sigma})^{*}\mathcal{X}^{n_1}_{\mu}\mathcal{X}^{n_2}_{\nu}\mathcal{Y}^{k_3}_{\lambda}}{\omega-(\varepsilon^{+}_{n_1}+\varepsilon^{+}_{n_2}-\varepsilon^{-}_{k_3})+i\eta}\right.\\
&\left.+\sum_{\substack{k_1,k_2\\n_3}}\frac{\mathcal{Y}^{k_1}_{\gamma}\mathcal{Y}^{k_2}_{\delta}\mathcal{X}^{n_3}_{\sigma}(\mathcal{Y}^{k_1}_{\mu}\mathcal{Y}^{k_2}_{\nu}\mathcal{X}^{n_3}_{\lambda})^{*}}{\omega-(\varepsilon^{-}_{k_1}+\varepsilon^{-}_{k_2}-\varepsilon^{+}_{n_3})-i\eta}\right\}V_{\mu\nu,\beta\lambda}\,,
\end{aligned}\tag{11.25}$$

where repeated Greek indices are summed over implicitly but we show the explicit summation over the poles corresponding to 2p1h and 2h1p ISCs.

11.2 *E*xercise: Calculate the expression for the other second-order contribution to $\Sigma^{\star}(\omega)$ arising from three-nucleon interactions (diagram of Fig. 11.4b). Show that this contains ISCs of 3p2h and 3h2p.

Solution

Upon performing the four frequency integrals, one obtains:

$$\Sigma^{(2,3N)}_{\alpha\beta}(\omega) = \frac{1}{12} W_{\alpha\gamma\delta,\mu\nu\lambda} \left\{ \sum_{\substack{n_1,n_2,n_3 \\ k_4,k_5}} \frac{(\mathcal{X}^{n_1}_{\mu} \mathcal{X}^{n_2}_{\nu} \mathcal{X}^{n_3}_{\lambda} \mathcal{Y}^{k_4}_{\gamma} \mathcal{Y}^{k_5}_{\delta})^* \mathcal{X}^{n_1}_{\mu'} \mathcal{X}^{n_2}_{\nu'} \mathcal{X}^{n_3}_{\lambda'} \mathcal{Y}^{k_4}_{\gamma'} \mathcal{Y}^{k_5}_{\delta'}}{\omega - (\varepsilon^+_{n_1} + \varepsilon^+_{n_2} + \varepsilon^+_{n_3} - \varepsilon^-_{k_4} - \varepsilon^-_{k_5}) + i\eta} \right.$$
$$\left. + \sum_{\substack{k_1,k_2,k_3 \\ n_4,n_5}} \frac{\mathcal{Y}^{k_1}_{\mu} \mathcal{Y}^{k_2}_{\nu} \mathcal{Y}^{k_3}_{\lambda} \mathcal{X}^{n_4}_{\gamma} \mathcal{X}^{n_5}_{\delta} (\mathcal{Y}^{k_1}_{\mu'} \mathcal{Y}^{k_2}_{\nu'} \mathcal{Y}^{k_3}_{\lambda'} \mathcal{X}^{n_4}_{\gamma'} \mathcal{X}^{n_5}_{\delta'})^*}{\omega - (\varepsilon^-_{k_1} + \varepsilon^-_{k_2} + \varepsilon^-_{k_3} - \varepsilon^+_{n_4} - \varepsilon^+_{n_5}) - i\eta} \right\} W_{\mu'\nu'\lambda',\beta\gamma'\delta'} \, . \tag{11.26}$$

11.3 The Algebraic Diagrammatic Construction Method

The most general form of the irreducible self-energy is given by Eq. (11.15). The $\Sigma^{(\infty)}$ is defined by the mean-field diagrams of Fig. 11.3a and Eq. (11.17a), while $\widetilde{\Sigma}(\omega)$ has a Lehmann representation as seen in the examples of Eqs. (11.25) and (11.26). Similarly to the case of a propagator, the pole structure of the energy-dependent part is dictated by the principle of causality with the correct boundary conditions coded by the $\pm i\eta$ terms in the denominators. This implies a dispersion relation that can link the real and imaginary parts of the self-energy [22, 26]. Correspondingly, the direct coupling of single particle orbits to ISCs (of 2p1h and 2h1p character or more complex) imposes the separable structure of the residues. In this section we consider the case of a finite system, for which it is useful to use a discretized single particle basis $\{\alpha\}$ as the model space. From now on we will use the Einstein convention that repeated indices (n, k, $\alpha \ldots$) are summed over even if not explicitly stated. Thus, the above constraints impose the following analytical form for the self-energy operator:

$$\Sigma^{(\star)}_{\alpha\beta}(\omega) = -U_{\alpha\beta} + \Sigma^{(\infty)}_{\alpha\beta} + M^{\dagger}_{\alpha,r} \left[\frac{1}{\omega - [E^{>} + C]_{r,r'} + i\eta} \right]_{r,r'} M_{r',\beta}$$
$$+ N_{\alpha,s} \left[\frac{1}{\omega - (E^{<} + D) - i\eta} \right]_{s,s'} N^{\dagger}_{s',\beta} \, , \tag{11.27}$$

where, here and in the following, ω and $\pm i\eta$ are to be intended as multiplication operators (that is, with matrix elements $[\omega + i\eta]_{s,s'} = (\omega + i\eta)\delta_{s,s'}$) and the fraction means a matrix inversion. In Eq. (11.27), the $E^{>}$ and $E^{<}$ are the unperturbed energies for the forward and backward ISCs and r and s are collective indices that label sets of configurations beyond single particle structure. Specifically, r is for particle addition and will label 2p1h, 3p2h, 4p3h, ... states, in the general case. Likewise, s is for particle removal and we will use it to label 2h1p states (or higher configurations).

However, for the approximations presented in this chapter and for our discussion below we will only be limited to 2p1h and 2h1p ISCs.

The expansion of the self-energy at second order in perturbation theory trivially satisfies Eq. (11.27). In the results of Eq. (11.25), the sums over r and s can be taken to run over ordered configurations $r \equiv \{n_1 < n_2, k_3\}$ and $s \equiv \{k_1 < k_2, n_3\}$. Because of the Pauli principle, the half residues of each pole are antisymmetric with respect to exchanging two quasiparticle or two quasihole indices. Therefore the constraints $n_1 < n_2$ and $k_1 < k_2$ can be imposed to avoid counting the same configurations twice. Thus, we can identify the expressions for the residues and poles as follows:

$$M_{r,\alpha} = \mathcal{X}^{n_1}_{\mu} \mathcal{X}^{n_2}_{\nu} \mathcal{Y}^{k_3}_{\lambda} \, V_{\mu\nu,\alpha\lambda} \tag{11.28a}$$

$$E^{>}_{r,r'} = \mathrm{diag}\left(\varepsilon^{+}_{n_1} + \varepsilon^{+}_{n_2} - \varepsilon^{-}_{k_3}\right) \tag{11.28b}$$

$$C_{r,r'} = 0 \tag{11.28c}$$

and

$$N_{\alpha,s} = V_{\alpha\lambda,\mu\nu} \, \mathcal{Y}^{k_1}_{\mu} \mathcal{Y}^{k_2}_{\nu} \mathcal{X}^{n_3}_{\lambda} \tag{11.29a}$$

$$E^{<}_{s,s'} = \mathrm{diag}\left(\varepsilon^{-}_{k_1} + \varepsilon^{-}_{k_2} - \varepsilon^{+}_{n_3}\right) \tag{11.29b}$$

$$D_{s,s'} = 0 \,, \tag{11.29c}$$

where the factor $1/2$ from Eq. (11.25) disappears because we restricted the sums to triplets of indices where $n_1 < n_2$ and $k_1 < k_2$. As we will discuss in the next section, Eqs. (11.28) and (11.29) define the algebraic diagrammatic method at second order [ADC(2)].

Unfortunately, $\Sigma^{\star}(\omega)$ loses its analytical form of Eq. (11.27) as soon as one moves to higher orders in perturbation theory. To demonstrate this, let us calculate the contribution of the third order 'ladder' diagram of Fig. 11.5a. By exploiting the Feynman rules and Eq. (11.21) we obtain:

$$\begin{aligned}
\Sigma^{(3,ld)}_{\alpha\beta}(\omega) &= -\frac{i^3}{4}\int \frac{d\omega_1}{2\pi}\int \frac{d\omega_2}{2\pi}\int \frac{d\omega_3}{2\pi} V_{\alpha\sigma,\gamma\delta}\, g_{\gamma\gamma'}(\omega+\omega_3-\omega_1)\, g_{\delta\delta'}(\omega_1)\, V_{\gamma'\delta',\mu'\nu'} \\
&\quad \times g_{\mu'\mu}(\omega+\omega_3-\omega_2)\, g_{\nu'\nu}(\omega_2)\, V_{\mu\nu,\beta\lambda}\, g_{\lambda\sigma}(\omega_3) \\
&= \frac{1}{4}\int \frac{d\omega_3}{2\pi i} V_{\alpha\sigma,\gamma\delta}\, G^{II,f}_{\gamma\delta,\gamma'\delta'}(\omega+\omega_3)\, V_{\gamma'\delta',\mu'\nu'}\, G^{II,f}_{\mu'\nu',\mu\nu}(\omega+\omega_3)\, V_{\mu\nu,\beta\lambda}\, g_{\lambda\sigma}(\omega_3) \\
&= \frac{1}{4}\int \frac{d\omega_3}{2\pi i} V_{\alpha\sigma,\gamma\delta} \left\{ \frac{(\mathcal{X}^{n_1}_{\gamma}\mathcal{X}^{n_2}_{\delta})^{*}\mathcal{X}^{n_1}_{\gamma'}\mathcal{X}^{n_2}_{\delta'}}{\omega+\omega_3-(\varepsilon^{+}_{n_1}+\varepsilon^{+}_{n_2})+i\eta} - \frac{\mathcal{Y}^{k_1}_{\gamma}\mathcal{Y}^{k_2}_{\delta}(\mathcal{Y}^{k_1}_{\gamma'}\mathcal{Y}^{k_2}_{\delta'})^{*}}{\omega+\omega_3-(\varepsilon^{-}_{k_1}+\varepsilon^{-}_{k_2})-i\eta} \right\} \\
&\quad \times V_{\gamma'\delta',\mu'\nu'} \left\{ \frac{(\mathcal{X}^{n_4}_{\mu'}\mathcal{X}^{n_5}_{\nu'})^{*}\mathcal{X}^{n_4}_{\mu}\mathcal{X}^{n_5}_{\nu}}{\omega+\omega_3-(\varepsilon^{+}_{n_4}+\varepsilon^{+}_{n_5})+i\eta} - \frac{\mathcal{Y}^{k_4}_{\mu'}\mathcal{Y}^{k_5}_{\nu'}(\mathcal{Y}^{k_4}_{\mu}\mathcal{Y}^{k_5}_{\nu})^{*}}{\omega+\omega_3-(\varepsilon^{-}_{k_4}+\varepsilon^{-}_{k_5})-i\eta} \right\} \\
&\quad \times V_{\mu\nu,\beta\lambda} \left\{ \frac{(\mathcal{X}^{n_3}_{\lambda})^{*}\mathcal{X}^{n_3}_{\sigma}}{\omega_3-\varepsilon^{+}_{n_3}+i\eta} + \frac{\mathcal{Y}^{k_3}_{\lambda}(\mathcal{Y}^{k_3}_{\sigma})^{*}}{\omega_3-\varepsilon^{-}_{k_3}-i\eta} \right\} \,.
\end{aligned} \tag{11.30}$$

Performing the Cauchy integrals, only six terms out of the eight combinations of poles survive. To simplify the discussion we will focus on the three integrals that contribute to the forward propagation of the self-energy [third term on the r.h.s. of (11.27)]. This is done by retaining only the poles $(\omega_3 - \varepsilon^-_{k_3} - i\eta)^{-1}$ in the last propagator of Eq. (11.30), which lie above the real axis with respect to the integrand ω_3. Thus, we have:

$$
\begin{aligned}
\Sigma^{(ld,>)}_{\alpha\beta}(\omega) = &\frac{1}{4}\int \frac{d\omega_3}{2\pi i} V_{\alpha\sigma,\gamma\delta} \left\{ -\frac{\mathcal{Y}^{k_1}_{\gamma}\mathcal{Y}^{k_2}_{\delta}\,(\mathcal{Y}^{k_1}_{\gamma'}\mathcal{Y}^{k_2}_{\delta'})^*}{\omega + \omega_3 - (\varepsilon^-_{k_1} + \varepsilon^-_{k_2}) - i\eta} \right\} \\
&\times V_{\gamma'\delta',\mu'\nu'} \left\{ \frac{(\mathcal{X}^{n_4}_{\mu'}\mathcal{X}^{n_5}_{\nu'})^*\,\mathcal{X}^{n_4}_{\mu}\mathcal{X}^{n_5}_{\nu}}{\omega + \omega_3 - (\varepsilon^+_{n_4} + \varepsilon^+_{n_5}) + i\eta} \right\} V_{\mu\nu,\beta\lambda} \left\{ \frac{\mathcal{Y}^{k_3}_{\lambda}(\mathcal{Y}^{k_3}_{\sigma})^*}{\omega_3 - \varepsilon^-_{k_3} - i\eta} \right\} \\
&+ \frac{1}{4}\int \frac{d\omega_3}{2\pi i} V_{\alpha\sigma,\gamma\delta} \left\{ \frac{(\mathcal{X}^{n_1}_{\gamma}\mathcal{X}^{n_2}_{\delta})^*\,\mathcal{X}^{n_1}_{\gamma'}\mathcal{X}^{n_2}_{\delta'}}{\omega + \omega_3 - (\varepsilon^+_{n_1} + \varepsilon^+_{n_2}) + i\eta} \right\} \\
&\times V_{\gamma'\delta',\mu'\nu'} \left\{ -\frac{\mathcal{Y}^{k_4}_{\mu'}\mathcal{Y}^{k_5}_{\nu'}\,(\mathcal{Y}^{k_4}_{\mu}\mathcal{Y}^{k_5}_{\nu})^*}{\omega + \omega_3 - (\varepsilon^-_{k_4} + \varepsilon^-_{k_5}) - i\eta} \right\} V_{\mu\nu,\beta\lambda} \left\{ \frac{\mathcal{Y}^{k_3}_{\lambda}(\mathcal{Y}^{k_3}_{\sigma})^*}{\omega_3 - \varepsilon^-_{k_3} - i\eta} \right\} \\
&+ \frac{1}{4}\int \frac{d\omega_3}{2\pi i} V_{\alpha\sigma,\gamma\delta} \left\{ \frac{(\mathcal{X}^{n_1}_{\gamma}\mathcal{X}^{n_2}_{\delta})^*\,\mathcal{X}^{n_1}_{\gamma'}\mathcal{X}^{n_2}_{\delta'}}{\omega + \omega_3 - (\varepsilon^+_{n_1} + \varepsilon^+_{n_2}) + i\eta} \right\} \\
&\times V_{\gamma'\delta',\mu'\nu'} \left\{ \frac{(\mathcal{X}^{n_4}_{\mu'}\mathcal{X}^{n_5}_{\nu'})^*\,\mathcal{X}^{n_4}_{\mu}\mathcal{X}^{n_5}_{\nu}}{\omega + \omega_3 - (\varepsilon^+_{n_4} + \varepsilon^+_{n_5}) + i\eta} \right\} V_{\mu\nu,\beta\lambda} \left\{ \frac{\mathcal{Y}^{k_3}_{\lambda}(\mathcal{Y}^{k_3}_{\sigma})^*}{\omega_3 - \varepsilon^-_{k_3} - i\eta} \right\} \\
= &\frac{\frac{1}{2}V_{\alpha\sigma,\gamma\delta}\,\mathcal{Y}^{k_1}_{\gamma}\mathcal{Y}^{k_2}_{\delta}\,(\mathcal{Y}^{k_1}_{\gamma'}\mathcal{Y}^{k_2}_{\delta'})^*\,V_{\gamma'\delta',\mu'\nu'}\,(\mathcal{X}^{n_4}_{\mu'}\mathcal{X}^{n_5}_{\nu'}\mathcal{Y}^{k_3}_{\sigma})^*}{[\varepsilon^-_{k_1} + \varepsilon^-_{k_2} - \varepsilon^+_{n_4} - \varepsilon^+_{n_5}]} \\
&\times \frac{1}{2}\frac{\mathcal{X}^{n_4}_{\mu}\mathcal{X}^{n_5}_{\nu}\mathcal{Y}^{k_3}_{\lambda}}{\omega - (\varepsilon^+_{n_4} + \varepsilon^+_{n_5} - \varepsilon^-_{k_3}) + i\eta} V_{\mu\nu,\beta\lambda} + V_{\alpha\sigma,\gamma\delta}\frac{(\mathcal{X}^{n_1}_{\gamma}\mathcal{X}^{n_2}_{\delta}\mathcal{Y}^{k_3}_{\sigma})^*}{\omega - (\varepsilon^+_{n_1} + \varepsilon^+_{n_2} - \varepsilon^-_{k_3}) + i\eta} \\
&\times \frac{1}{2}\frac{\mathcal{Y}^{k_3}_{\lambda}\mathcal{X}^{n_1}_{\gamma'}\mathcal{X}^{n_2}_{\delta'}\,V_{\gamma'\delta',\mu'\nu'}\,\mathcal{Y}^{k_4}_{\mu'}\mathcal{Y}^{k_5}_{\nu'}\,(\mathcal{Y}^{k_4}_{\mu}\mathcal{Y}^{k_5}_{\nu})^*\frac{1}{2}V_{\mu\nu,\beta\lambda}}{[\varepsilon^-_{k_4} + \varepsilon^-_{k_5} - \varepsilon^+_{n_1} - \varepsilon^+_{n_2}]} \\
&+ \frac{V_{\alpha\sigma,\gamma\delta}\,(\mathcal{X}^{n_1}_{\gamma}\mathcal{X}^{n_2}_{\delta}\mathcal{Y}^{k_3}_{\sigma})^*}{\omega - (\varepsilon^+_{n_1} + \varepsilon^+_{n_2} - \varepsilon^-_{k_3}) + i\eta}\,\frac{1}{2}\mathcal{X}^{n_1}_{\gamma'}\mathcal{X}^{n_2}_{\delta'}\,V_{\gamma'\delta',\mu'\nu'}\,(\mathcal{X}^{n_4}_{\mu'}\mathcal{X}^{n_5}_{\nu'})^* \\
&\times \frac{1}{2}\frac{\mathcal{X}^{n_4}_{\mu}\mathcal{X}^{n_5}_{\nu}\mathcal{Y}^{k_3}_{\lambda}\,V_{\mu\nu,\beta\lambda}}{\omega - (\varepsilon^+_{n_4} + \varepsilon^+_{n_5} - \varepsilon^-_{k_3}) + i\eta}
\end{aligned}
$$

$$
\begin{aligned}
&\equiv \quad M^{(2,ld)\,\dagger}\frac{1}{\omega - E^{>} + i\eta}M^{(1)} \\
&+ M^{(1)\,\dagger}\frac{1}{\omega - E^{>} + i\eta}M^{(2,ld)} \\
&+ M^{(1)\,\dagger}\frac{1}{\omega - E^{>} + i\eta}C^{(ld)}\frac{1}{\omega - E^{>} + i\eta}M^{(1)}\,,
\end{aligned}
\tag{11.31}
$$

where $M^{(1)}$ and $E^{>}$ are the same as in Eqs. (11.28) and the factors 1/2 are again absorbed by summing over the ordered configurations for r and r'. The 2p1h ladder interaction $C^{(ld)}$ is at first order in V, while the coupling matrix $M^{(2,ld)}$ is at second order. These can be read from the previous lines of Eq. (11.31) and turn out to be (showing all summations explicitly):

$$
\begin{aligned}
M^{(2,ld)}_{r,\alpha} &= \sum_{k_4,k_5} \sum_{\substack{\sigma,\zeta,\gamma,\delta\\ \mu,\nu,\lambda}} \frac{\mathcal{X}^{n_1}_{\gamma}\,\mathcal{X}^{n_2}_{\delta}\;V_{\gamma\delta,\sigma\zeta}\;\mathcal{Y}^{k_4}_{\sigma}\mathcal{Y}^{k_5}_{\zeta}\;(\mathcal{Y}^{k_4}_{\mu}\mathcal{Y}^{k_5}_{\nu})^{*}\,\mathcal{Y}^{k_3}_{\lambda}}{[\varepsilon^{-}_{k_4} + \varepsilon^{-}_{k_5} - \varepsilon^{+}_{n_1} - \varepsilon^{+}_{n_2}]}\,\frac{1}{2}V_{\mu\nu,\alpha\lambda} \\
C^{(ld)}_{r,r'} &= \sum_{\alpha,\beta,\gamma,\delta} \mathcal{X}^{n_1}_{\alpha}\,\mathcal{X}^{n_2}_{\beta}\;V_{\alpha\beta,\gamma\delta}\;(\mathcal{X}^{n'_1}_{\gamma}\,\mathcal{X}^{n'_2}_{\delta})^{*}\;\delta_{k_3,k'_3}\,.
\end{aligned}
\tag{11.32}
$$

Equation (11.31) clearly breaks the known Lehmann representation for the self-energy and would even lead to inconsistent results unless its contribution is very small compared to the second-order contribution of Eq. (11.25). That is, Eq. (11.31) would invalidate the perturbative expansion unless V is small. Therefore, we need to identify proper corrections that allow to retain these third order contributions but at the same time let us recover the correct analytical form (11.27). For the first two terms on the right hand side of Eq. (11.31), this issue can be easily solved by remembering that the corresponding diagram from $\Sigma^{(2)}(\omega)$ [see Eq. (11.25)] is to be included. If then one adds an extra term that is quadratic in $M^{(2,ld)}$, this leads to:

$$
\begin{aligned}
&\Sigma^{(2)}(\omega) + \Sigma^{(3,ld)}(\omega) + M^{(2,ld)\,\dagger}\frac{1}{\omega - E^{>} + i\eta}M^{(2,ld)} \\
&\quad\longrightarrow \left[M^{(1)} + M^{(2,ld)}\right]^{\dagger}\frac{1}{\omega - E^{>} + i\eta}\left[M^{(1)} + M^{(2,ld)}\right],
\end{aligned}
\tag{11.33}
$$

which resolves the issue of obtaining the residues in separable form. Note that this new correction is just one specific Goldstone diagram among the many that contribute to the self-energy at *fourth order*. On the other hand, adding all of the fourth-order diagrams would lead to new terms that break the Lehmann representation themselves and that in turn would call for the inclusions of selected Goldstone terms at even higher orders. In other words, we have achieved to recover the structure of Eq. (11.27) but at the price of giving up a systematic perturbative expansion that is complete at each order in $\widetilde{V}$. Given that the Lehmann representation is dictated by physical properties, this is a more satisfactory rearrangement of the perturbation series.

The last term in Eq. (11.31) is more tricky to correct since it contains second-order poles as $(\omega - E - i\eta)^{-2}$, which cannot be canceled by single contributions at higher order. Instead, we are forced to perform a non-perturbative resummation of Goldstone diagrams to all orders that results in a geometric series. This is done by considering the relation

$$\frac{1}{A-B} = \frac{1}{A} + \frac{1}{A}B\frac{1}{A-B} = \frac{1}{A} + \frac{1}{A}B\frac{1}{A} + \frac{1}{A}B\frac{1}{A}B\frac{1}{A} + \frac{1}{A}B\frac{1}{A}B\frac{1}{A}B\frac{1}{A} + \dots \tag{11.34}$$

for two operators A and B. If we chose $A \equiv \omega - E^{>} + i\eta$ and $B \equiv C^{(ld)}$, the first and second term on the right hand side can then be identified respectively with the contribution from $\Sigma^{(2)}(\omega)$ and the last term of Eq. (11.31). Also in this case, all perturbative terms up to third order have been kept unchanged but we are forced to select a series of Goldstone diagrams up to infinite order.

If then one adds an extra term that is quadratic in $M^{(2,ld)}$, this lead to:

$$\Sigma^{(2)}(\omega) + \Sigma^{(3,ld)}(\omega) + \begin{matrix}\text{terms beyond}\\ 3^{rd}\text{ order}\end{matrix}$$

$$\longrightarrow \left[M^{(1)} + M^{(2,ld)}\right]^{\dagger} \frac{1}{\omega - E^{>} - C^{(ld)} + i\eta} \left[M^{(1)} + M^{(2,ld)}\right], \tag{11.35}$$

which now contains *all* the perturbation theory terms at second (11.25) and third order (11.31) while preserving the expected analytical form for $\widetilde{\Sigma}(\omega)$.

It can be shown that the summation implicit in Eq. (11.35) is equivalent to a full resummation of two-particle ladder diagrams in the Tamn-Dancoff approximation (TDA) [37]. In this sum the remaining quasi hole state appearing in the 2p1h ISC remains uncoupled from the ladder series, as it can be seen in Fig. 11.5a, which is the first term in the series. Likewise, one would find that the remaining backward-going terms in Eq. (11.30) would lead to resumming the two-hole TDA ladders within the 2h1p configurations. Instead, diagram in Fig. 11.5b involves a resummation of ph ring diagrams. Extensions of these series to random-phase approximation (RPA) is also possible, this would introduce a larger set of high-order Goldstone diagrams but it would not be required to enforce consistency with perturbation theory at third order.

11.3 *E*xercise: Complete the calculation of Eq. (11.30) and derive the remaining corrections to the 2h1p interaction $D^{(ld)}$ and the 1h-2h1p coupling term $N^{(2,ld)}$.

11.3.1 The ADC(n) Approach and Working Equations at Third Order

The procedure discussed above to devise reliable approximations for the self-energy is at the heart of the ADC method, originally introduced by J. Schirmer and collaborators [9, 10]. This approach generates a hierarchy of approximations of increasing accuracy such that, at a given order n, the ADC(n) equations will maintain the analytic form of Eq. (11.27) and will be consistent with perturbation theory up to order n. Note that this does not mean that ADC(n) is a perturbative truncation but that it must contain at least all the Feynman diagrams for $\Sigma^{\star}(\omega)$ up to order n, among higher terms. In fact, we will see below that for $n > 2$ it always involve an infinite resummation of diagrams [see also Eqs. (11.34) and (11.35)]. To implement this scheme for the dynamic self-energy, $\widetilde{\Sigma}(\omega)$, we expand its Lehmann representation in powers of the perturbation interaction $\widehat{H}_1$ (or, equivalently, $\widetilde{H}_1$). The interaction matrices C and D appearing in the denominators of Eq. (11.27) can only be of first order in either $\widehat{U}$, $\widehat{V}$ or $\widehat{W}$. However, the coupling matrices can contain terms of any order:

$$
\begin{aligned}
M &= M^{(1)} + M^{(2)} + M^{(3)} + \ldots \\
N &= N^{(1)} + N^{(2)} + N^{(3)} + \ldots
\end{aligned}
\tag{11.36}
$$

Using Eqs. (11.34) and (11.36) one finds the following expansion for Eq. (11.27):

$$
\begin{aligned}
\Sigma^{\star}(\omega) = & -\widehat{U} \; + \; \Sigma^{(\infty)} \\
& + M^{(1)\dagger}\frac{1}{\omega - E^{>} + i\eta}M^{(1)} + N^{(1)}\frac{1}{\omega - E^{<} - i\eta}N^{(1)\dagger} \\
& + M^{(2)\dagger}\frac{1}{\omega - E^{>} + i\eta}M^{(1)} + M^{(1)\dagger}\frac{1}{\omega - E^{>} + i\eta}M^{(2)} \\
& + M^{(1)\dagger}\frac{1}{\omega - E^{>} + i\eta}C\frac{1}{\omega - E^{>} + i\eta}M^{(1)} \\
& + N^{(2)}\frac{1}{\omega - E^{<} - i\eta}N^{(1)\dagger} \; + N^{(1)}\frac{1}{\omega - E^{<} - i\eta}N^{(2)\dagger} \\
& + N^{(1)}\frac{1}{\omega - E^{<} - i\eta}D\frac{1}{\omega - E^{<} - i\eta}N^{(1)\dagger} \\
& + \mathcal{O}(\widehat{H}_1{}^4) \, ,
\end{aligned}
\tag{11.37}
$$

where all terms up to third order in $\widehat{H}_1$ are shown explicitly. The ADC procedure is then to simply calculate all possible diagrams up to order n. By comparing them to Eq. (11.37), one then reads the minimum expressions for the coupling and interaction matrices, M, N, C and D that are needed to retain all the n-order diagrams for $\widetilde{\Sigma}(\omega)$. Correspondingly, the energy-independent self-energy $\Sigma^{(\infty)}$ needs to be computed at least up to order n as well. Note that the dynamic part of the self-energy, which propagates ISCs, appears only starting from second order. This is so because any such diagram needs at least one perturbing interaction V to generate an ISC and a second one to annihilate it back to a single particle state. In general, if the Hamiltonian contains up to m-body forces and i is an integer, then the ADC($2i$) and ADC($2i+1$) will require ISCs up to (k+1)-particle–k-hole and ($k+1$)-hole–k-particle, where $k = (m-1) * i$. Thus, with two-nucleon forces ADC(2) and ADC(3) include 2p1h and 2h1p states, ADC(4) and ADC(5) need up to 3p2h and 2h3p states, and so on. However, the full ADC(2/3) sets with three-nucleon forces already includes 3p2h and 3h2p configurations [36].

At first order, ADC(1) requires to only calculate diagram(s) that contribute to $\widetilde{U} = -\widehat{U} + \Sigma^{(\infty)}$, see Fig. 11.3a, and thus the scheme reduces to Hartree-Fock theory. At second order and with at most two-body interactions, there is only one diagram contributing to $\widetilde{\Sigma}(\omega)$ which is already in the proper Lehmann form. Hence, Eqs. (11.25), (11.28) and (11.29) fully define the ADC(2) approximation. In this case, $\Sigma^{(\infty)}$ also requires a second-order non-skeleton term.

Higher order cases are more complicated. For a two-body Hamiltonian, the only skeleton diagrams at third order are the ladder and ring diagrams shown in Fig. 11.5a, b. As long as one works with a Hartree-Fock reference state or a fully self-consistent (dressed) propagators, no other diagram is needed because the additional non-skeleton terms either vanish or must not be included (see Exercise 11.5). In these cases, one obtains the following working expressions the for the ADC(3) approximation:

$$M_{r,\alpha} = \mathcal{X}^{n_1}_{\mu}\mathcal{X}^{n_2}_{\nu}\mathcal{Y}^{k_3}_{\lambda}\,V_{\mu\nu,\alpha\lambda} + \frac{\mathcal{X}^{n_1}_{\gamma}\mathcal{X}^{n_2}_{\delta}\,V_{\gamma\delta,\sigma\zeta}\,\mathcal{Y}^{k_4}_{\sigma}\mathcal{Y}^{k_5}_{\zeta}}{2\,[\varepsilon^-_{k_4}+\varepsilon^-_{k_5}-\varepsilon^+_{n_1}-\varepsilon^+_{n_2}]}\,(\mathcal{Y}^{k_4}_{\mu}\mathcal{Y}^{k_5}_{\nu})^*\,\mathcal{Y}^{k_3}_{\lambda}\,V_{\mu\nu,\alpha\lambda} \tag{11.38a}$$

$$+ \frac{\mathcal{X}^{n_2}_{\rho}\mathcal{Y}^{k_3}_{\sigma}V_{\rho\delta,\sigma\gamma}\mathcal{Y}^{k_5}_{\gamma}\mathcal{X}^{n_6}_{\delta}}{[\varepsilon^-_{k_3}-\varepsilon^+_{n_2}+\varepsilon^-_{k_5}-\varepsilon^+_{n_6}]}(\mathcal{Y}^{k_5}_{\nu}\mathcal{X}^{n_6}_{\lambda})^*\,\mathcal{X}^{n_1}_{\mu}\,V_{\mu\nu,\alpha\lambda}$$

$$- \frac{\mathcal{X}^{n_1}_{\rho}\mathcal{Y}^{k_3}_{\sigma}V_{\rho\delta,\sigma\gamma}\mathcal{Y}^{k_5}_{\gamma}\mathcal{X}^{n_6}_{\delta}}{[\varepsilon^-_{k_3}-\varepsilon^+_{n_1}+\varepsilon^-_{k_5}-\varepsilon^+_{n_6}]}(\mathcal{Y}^{k_5}_{\nu}\mathcal{X}^{n_6}_{\lambda})^*\,\mathcal{X}^{n_2}_{\mu}\,V_{\mu\nu,\alpha\lambda}$$

$$E^{>}_{r,r'} = \mathrm{diag}\left(\varepsilon^+_{n_1}+\varepsilon^+_{n_2}-\varepsilon^-_{k_3}\right) \tag{11.38b}$$

$$C_{r,r'} = \mathcal{X}^{n_1}_{\alpha}\mathcal{X}^{n_2}_{\beta}\,V_{\alpha\beta,\gamma\delta}\,(\mathcal{X}^{n'_1}_{\gamma}\mathcal{X}^{n'_2}_{\delta})^*\,\delta_{k_3,k'_3}$$

$$+ \mathcal{X}^{n_1}_{\alpha}\mathcal{Y}^{k_3}_{\beta}\,V_{\alpha\delta,\beta\gamma}\,(\mathcal{X}^{n'_1}_{\gamma}\mathcal{Y}^{k'_3}_{\delta})^*\,\delta_{n_2,n'_2} - \mathcal{X}^{n_2}_{\alpha}\mathcal{Y}^{k_3}_{\beta}\,V_{\alpha\delta,\beta\gamma}\,(\mathcal{X}^{n'_1}_{\gamma}\mathcal{Y}^{k'_3}_{\delta})^*\,\delta_{n_1,n'_2} \tag{11.38c}$$

$$- \mathcal{X}^{n_1}_{\alpha}\mathcal{Y}^{k_3}_{\beta}\,V_{\alpha\delta,\beta\gamma}\,(\mathcal{X}^{n'_2}_{\gamma}\mathcal{Y}^{k'_3}_{\delta})^*\,\delta_{n_2,n'_1} + \mathcal{X}^{n_2}_{\alpha}\mathcal{Y}^{k_3}_{\beta}\,V_{\alpha\delta,\beta\gamma}\,(\mathcal{X}^{n'_2}_{\gamma}\mathcal{Y}^{k'_3}_{\delta})^*\,\delta_{n_1,n'_1}$$

and

$$
\begin{aligned}
N_{\alpha,s} = {} & V_{\alpha\lambda,\mu\nu}\, \mathcal{Y}^{k_1}_{\mu} \mathcal{Y}^{k_2}_{\nu} \mathcal{X}^{n_3}_{\lambda} \; + \; V_{\alpha\lambda,\mu\nu}\, \mathcal{X}^{n_3}_{\lambda} \, (\mathcal{X}^{n_7}_{\mu} \mathcal{X}^{n_8}_{\nu})^* \, \frac{\mathcal{X}^{n_7}_{\gamma} \mathcal{X}^{n_8}_{\delta} V_{\gamma\delta,\sigma\rho} \mathcal{Y}^{k_1}_{\sigma} \mathcal{Y}^{k_2}_{\rho}}{2\,[\varepsilon^-_{k_1} + \varepsilon^-_{k_2} - \varepsilon^+_{n_7} - \varepsilon^+_{n_8}]} \\
& + V_{\alpha\lambda,\mu\nu}\, \mathcal{Y}^{k_1}_{\mu} \, (\mathcal{X}^{n_5}_{\nu} \mathcal{Y}^{k_6}_{\lambda})^* \, \frac{\mathcal{X}^{n_5}_{\gamma} \mathcal{Y}^{k_6}_{\delta} V_{\gamma\rho,\delta\sigma} \mathcal{Y}^{k_2}_{\sigma} \mathcal{X}^{n_3}_{\rho}}{[\varepsilon^-_{k_2} - \varepsilon^+_{n_3} + \varepsilon^-_{k_6} - \varepsilon^+_{n_5}]} \\
& - V_{\alpha\lambda,\mu\nu}\, \mathcal{Y}^{k_2}_{\mu} \, (\mathcal{X}^{n_5}_{\nu} \mathcal{Y}^{k_6}_{\lambda})^* \, \frac{\mathcal{X}^{n_5}_{\gamma} \mathcal{Y}^{k_6}_{\delta} V_{\gamma\rho,\delta\sigma} \mathcal{Y}^{k_1}_{\sigma} \mathcal{X}^{n_3}_{\rho}}{[\varepsilon^-_{k_1} - \varepsilon^+_{n_3} + \varepsilon^-_{k_6} - \varepsilon^+_{n_5}]}
\end{aligned}
\tag{11.39a}
$$

$$
E^{<}_{s,s'} = \mathrm{diag}\left(\varepsilon^-_{k_1} + \varepsilon^-_{k_2} - \varepsilon^+_{n_3} \right) \tag{11.39b}
$$

$$
\begin{aligned}
\mathcal{D}_{s,s'} = {} & -(\mathcal{Y}^{k_1}_{\alpha} \mathcal{Y}^{k_2}_{\beta})^* \, V_{\alpha\beta,\gamma\delta} \, \mathcal{Y}^{k'_1}_{\gamma} \mathcal{Y}^{k'_2}_{\delta} \; \delta_{n_3,n'_3} \\
& - (\mathcal{Y}^{k_1}_{\alpha} \mathcal{X}^{n_3}_{\beta})^* V_{\alpha\delta,\beta\gamma} \, \mathcal{Y}^{k'_1}_{\gamma} \mathcal{X}^{n'_3}_{\delta} \; \delta_{k_2,k'_2} + (\mathcal{Y}^{k_2}_{\alpha} \mathcal{X}^{n_3}_{\beta})^* V_{\alpha\delta,\beta\gamma} \, \mathcal{Y}^{k'_1}_{\gamma} \mathcal{X}^{n'_3}_{\delta} \; \delta_{k_1,k'_2} \\
& + (\mathcal{Y}^{k_1}_{\alpha} \mathcal{X}^{n_3}_{\beta})^* V_{\alpha\delta,\beta\gamma} \, \mathcal{Y}^{k'_2}_{\gamma} \mathcal{X}^{n'_3}_{\delta} \; \delta_{k_2,k'_1} - (\mathcal{Y}^{k_2}_{\alpha} \mathcal{X}^{n_3}_{\beta})^* V_{\alpha\delta,\beta\gamma} \, \mathcal{Y}^{k'_2}_{\gamma} \mathcal{X}^{n'_3}_{\delta} \; \delta_{k_1,k'_1} \, ,
\end{aligned}
\tag{11.39c}
$$

where only ordered configurations $r = \{n_1 < n_2, k_3\}$ and $s = \{k_1 < k_2, n_3\}$ need to be considered, in accordance with the Pauli principle. Note that these equations apply to the case of two-body interactions but they remain unchanged for an effective operator $\widetilde{V}$ that is derived from three-body forces. However the full inclusion of $\widehat{W}$ would require the inclusion of the diagram of Fig. 11.4b at the ADC(2) level and several other interaction-irreducible diagrams for ADC(3). The non-skeleton contributions to $\widetilde{\Sigma}(\omega)$ that arise at third order when the reference propagator is not dressed are shown in Fig. 11.6. The case of three-nucleon forces is discussed in full detail in [36].

To remain consistent with the ADC(n) formulation, the static self-energy $\Sigma^{(\infty)}$ must also be computed at least to the same order n. However, this involves a large number of non-skeleton diagrams when self-consistency is not implemented. In practice, it is relatively inexpensive to compute it directly from dressed propagators, as given by (11.17a) and therefore it can be iterated to self-consistency. This prescription, in which $\widetilde{\Sigma}(\omega)$ is calculated from an unperturbed reference state $g^{0)}(\omega)$ but $\Sigma^{(\infty)}$ is obtained self-consistently, is often used in nuclear physics applications and we refer to it as the *sc0* approximation [19]. When dealing with the Coulomb force in molecular systems, the dynamic self-energy can be simply calculated in terms of a Hartree-Fock reference state. In nuclear physics, a Hartree-Fock reference state is adequate only if the chosen Hamiltonian is particularly soft. Otherwise, it is necessary to optimize the reference state by choosing a $\widehat{H}_0$ and

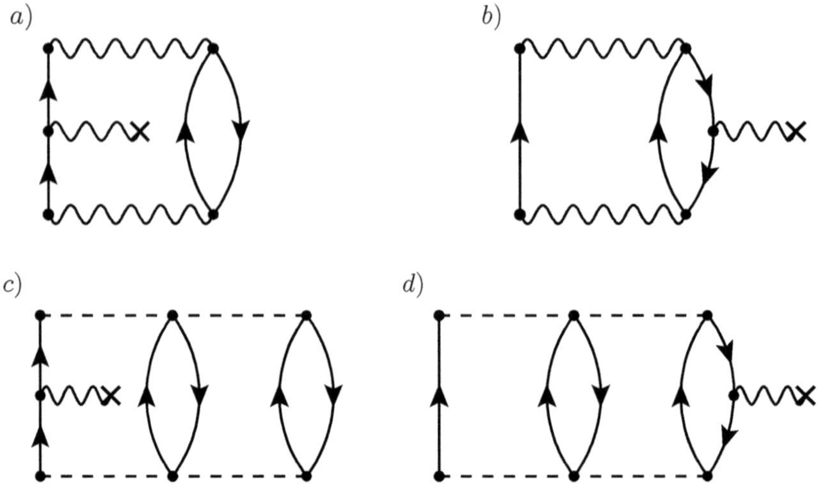

Fig. 11.6 Self-energy insertion diagrams that appear, at third order, in the perturbative expansion for $\widetilde{\Sigma}(\omega)$ with two- and three-nucleon interactions. These non-skeleton diagrams need to be considered when the reference propagators are not self-consistent. Diagrams (**a**) and (**b**) involve only one- and two-body interactions and results from self-energy insertions into the diagram of Fig. 11.4a. With the inclusion of three-nucleon interactions, the diagrams (**c**) and (**d**) arise from the one of Fig. 11.4b. When a Hartree-Fock reference state is used all these contributions cancel out (see Exercise 11.5)

$g^{0)}(\omega)$ that better represent the correlated single particle energies in the dressed propagator. In all cases, at least the *sc0* approach to $\Sigma^{(\infty)}$ is always required when computing finite nuclei and infinite nucleonic matter.

The standard ADC(n) prescription is to identify the *minimal* matrices M, N, C and D that make the self-energy consistent with perturbation theory up to order n. However, other intermediate approximations are also possible and have been exploited in the past. The so-called *2p1h-TDA* method is an extension of the second order scheme of Eqs. (11.28) and (11.29) where the matrices C and D are calculated at first order instead, as given by Eqs. (11.38c) and (11.39c). As a rule of thumb, the ADC(2) approximation yields roughly 90% of the total correlation energy in most applications, while the ADC(3) can account for about 99% of it—hence, with a 1% error in binding energies. The 2p1h-TDA contains the ADC(2) in full but it further resumes the full set of two-particle (pp), two-holes (hh) and particle-hole (ph) diagrams. This can result in a sensible improvement in the accuracy of binding energies but without the price of computing corrections to the M and N coupling matrices. Nevertheless, the 2p1h-TDA misses the second order terms from Eqs. (11.36) that are known to contribute strongly to quasiparticle energies. As a consequence the one nucleon addition and separation energies (or, equivalently, ionization potentials and electron affinities in molecules) would be predicted poorly in 2p1h-TDA and in general they require full ADC(3) calculations [38]. In nuclear physics applications, the description of collective excitations often requires that particle-hole configurations are diagonalized at least in the RPA scheme. While this is similar to the TDA all-order summations included in 2p1h-TDA and in ADC(3), extra ground state correlations effects from the RPA series are deemed important to reproduce collective modes typical of nuclear systems [37]. To account for these effects on needs to separate the partial summations in the pp, hh and ph channels, substitute them with equivalent RPA series and recouple these through a Faddeev-like expansion, in order to eventually reconstruct the self-energy [39–42]. The Faddeev-RPA (FRPA) method contains the ADC(3) in full but it also generates additional ground state correlation terms that are induced by the RPA summation and are at fourth and higher order in the perturbative expansion of the self-energy. The working implementation of the FRPA approach has been formulated in [40, 43, 44].

Another important extension of the ADC(3) framework comes from the realization that Eqs. (11.36) still imply a perturbative truncation for M and N. This causes the energy denominators in Eqs. (11.38a) and (11.39a) to become unstable if the system is close to being degenerate. The way out from this situation is again to perform an all-orders summation. Since the coupling matrices correspond to specific energy-independent parts of Goldstone diagrams, they can be resummed in the same way as for the coupled cluster (CC) technique (C. Barbieri, 2016, unpublished). We show this for the second term on the right hand side of (11.38a), which can be rewritten as follows:

$$\frac{\mathcal{X}^{n_1}_{\gamma}\mathcal{X}^{n_2}_{\delta}\, V_{\gamma\delta,\sigma\zeta}\, \mathcal{Y}^{k_4}_{\sigma}\mathcal{Y}^{k_5}_{\zeta}\, (\mathcal{Y}^{k_4}_{\mu}\mathcal{Y}^{k_5}_{\nu})^* \mathcal{Y}^{k_3}_{\lambda}}{2\,[\varepsilon^-_{k_4} + \varepsilon^-_{k_5} - \varepsilon^+_{n_1} - \varepsilon^+_{n_2}]}\, V_{\mu\nu,\alpha\lambda} \quad\longrightarrow\quad \frac{1}{2}\, t^{(0)\, n_1\, n_2}_{\;\;k_4\, k_5}\, (\mathcal{Y}^{k_4}_{\mu}\mathcal{Y}^{k_5}_{\nu})^* \mathcal{Y}^{k_3}_{\lambda}\, V_{\mu\nu,\alpha\lambda}\,, \tag{11.40}$$

where the amplitude

$$t^{(0)\, n_1\, n_2}_{\;\;k_4\, k_5} \equiv \frac{\mathcal{X}^{n_1}_{\gamma}\mathcal{X}^{n_2}_{\delta}\, V_{\gamma\delta,\sigma\zeta}\, \mathcal{Y}^{k_4}_{\sigma}\mathcal{Y}^{k_5}_{\zeta}}{\varepsilon^-_{k_4} + \varepsilon^-_{k_5} - \varepsilon^+_{n_1} - \varepsilon^+_{n_2}} \tag{11.41}$$

generalizes the zeroth approximation to the CC operator $\hat{T}_2$ (see Sect. 8.7.4). In case of a dressed propagator, the spectroscopic amplitudes $\mathcal{X}$ ($\mathcal{Y}$) account for the fragmentation of single particle strength. However, for a standard mean-field reference, they simply select the particle (hole) reference orbits and $t^{(0)}$ is exactly the same as for the CC approach. In order to mitigate effects of the perturbative truncation in Eqs. (11.38a) and (11.39a) (and to resum the 2p2h ISCs) one simply substitutes $t^{(0)}$ with the corresponding CC solution. In general, when t is computed using the CC doubles (CCD) approach we refer to the whole self-energy as being in the ADC(3)-D approximation, when t is obtained by resumming both singles and doubles (CCSD) it will give the ADC(3)-SD approximation, and so on. In Sect. 11.3.3, we will see a case when these corrections are important.

The working equations for the self-energy at the ADC(4) level and beyond are discussed in [10].

11.4 *E*xercise: Calculate the ladder and ring diagrams in Fig. 11.5 and prove Eqs. (11.38) and (11.39) in full. [Hint: for the ring diagrams it is simpler to first perform integrations for the free polarization propagator, $\Pi^{f}_{\alpha\beta,\gamma\delta}(\omega) = \int \frac{d\omega_1}{2\pi i} g_{\alpha\gamma}(\omega + \omega_1) g_{\delta\beta}(\omega_1)$, which describes non interacting particle-hole states.]

11.5 *E*xercise: In case of a reference propagator that is not fully self-consistent, it is necessary to also include non-skeleton diagrams. For $\widetilde{\Sigma}(\omega)$ these first appear at third order with the diagrams shown in Fig. 11.6. Calculate the expressions for diagrams in (a) and (b), then:

- Deduct the corresponding corrections to Eqs. (11.38) and (11.39). These will be the complete ADC(3) working equations.
- Show that they cancel out exactly if the reference propagator is of Hartree-Fock type. Hence these corrections do not need to be taken into account even tough the Hartree-Fock reference state is *not* a dressed—and fully self-consistent—input in this case.

[Hint: In Hartee-Fock theory, the static self-energy $\Sigma^{(\infty)}$ reduces to the Hartree-Fock potential. The reference state in this case is given by $\widehat{H}_0 = \widehat{T} + \widehat{U}^{HF} \equiv \widehat{H}^{HF}$, which is also the Hartree-Fock Hamiltonian. Additionally, in the notation of Eqs. (11.42) below, the (orthogonal) single particle wave functions are the solutions of $\{T + \Sigma^{HF}\}\mathcal{Z}^i = \varepsilon^i \mathcal{Z}^i$.]

11.3.2 Solving the Dyson Equation

Once we have a suitable approximation to the self-energy, it is necessary to solve the Dyson equation (11.3) to obtain the single particle propagator, the associated observables and the spectral function. The latter will also yield spectroscopic amplitudes and their spectroscopic factor for the addition and removal of a nucleon form the correlated state $|\Psi_0^A\rangle$. In doing this, Eqs. (11.3) take the form of a one-body Schrödinger equation for the scattering of a particle or a hole inside the medium. Given that all the Cauchy integrals associated with Feynman diagrams have been carried out, we can safely take the limit $\pm i\eta \to 0$ in all denominators for simplicity. The same equation applies to states both above and below the Fermi surface. Thus, it is convenient to take a general index i and using ε_i and $\mathcal{Z}^i$ to label energies and spectroscopic amplitudes for all quasiparticle and quasihole states. Specifically,

$$\varepsilon_i \longrightarrow \begin{cases} \varepsilon_n^+ & \text{for } i = n, \text{ particle,} \\ \varepsilon_k^- & \text{for } i = k, \text{ hole,} \end{cases} \quad \text{and} \quad \mathcal{Z}_\alpha^i \longrightarrow \begin{cases} (\mathcal{X}_\alpha^n)^* & \text{for } i = n, \text{ particle,} \\ \mathcal{Y}_\alpha^k & \text{for } i = k, \text{ hole.} \end{cases} \tag{11.42}$$

In order to extract the solution for the pole i in the Lehmann representation, we extract the corresponding residue on both the left and right hand side of Eq. (11.3a):

$$\lim_{\omega\to\varepsilon_i} (\omega - \varepsilon_i) \left\{ g_{\alpha\beta}(\omega) = g^{(0)}_{\alpha\beta}(\omega) + g^{(0)}_{\alpha\gamma}(\omega)\Sigma^{\star}_{\gamma\delta}(\omega) g_{\delta\beta}(\omega) \right\} , \tag{11.43}$$

which gives

$$\mathcal{Z}^i_\alpha (\mathcal{Z}^i_\beta)^* = g^{(0)}_{\alpha\gamma}(\omega)\Sigma^{\star}_{\gamma\delta}(\omega)\mathcal{Z}^i_\delta(\mathcal{Z}^i_\beta)^* \Big|_{\omega=\varepsilon_i} . \tag{11.44}$$

By dividing out $(\mathcal{Z}^i_\beta)^*$ and using the fact that $[g^{(0)}(\omega)]^{-1} = \omega - \widehat{H}_0$ we finally obtain the eigenvalue equation

$$\begin{aligned}
\varepsilon_i \mathcal{Z}^i_\alpha &= \left\{ \widehat{T} + \widehat{U} + \Sigma^{\star}(\omega) \right\}_{\alpha\,\delta} \mathcal{Z}^i_\delta \Big|_{\omega=\varepsilon_i} \\
&= \left\{ \widehat{T} + \Sigma^{(\infty)} + M^\dagger \frac{1}{\omega - E^{>} - C + i\eta} M + N \frac{1}{\omega - E^{<} - D - i\eta} N^\dagger \right\}_{\alpha\,\delta} \mathcal{Z}^i_\delta \Big|_{\omega=\varepsilon_i} ,
\end{aligned} \tag{11.45}$$

where the potential $\widehat{U}$ defining the unperturbed state completely cancels out. From here we see that the true irreducible self-energy $\Sigma^{(\infty)} + \widetilde{\Sigma}(\omega)$ acts as a non-local and energy dependent potential that accounts for the motion of both particles and holes inside the system and for their coupling to intermediate excitations. At positive energies ($\omega > 0$) this equation describes the elastic scattering of a nucleon off the $|\Psi^A_0\rangle$ ground state and the self-energy can be identified with a fully microscopic optical potential [27, 28, 45]. In this case the spectroscopic amplitudes $\mathcal{Z}^i$ correspond to scattering wave functions with the usual asymptotic normalization. Instead, at $\omega < 0$, Eq. (11.45) describes the transition to states of $|\Psi^{A+1}_i\rangle$ with bound amplitudes. The norm of each $\mathcal{Z}^i$ gives the corresponding spectroscopic factor and it is obtained as

$$SF_i = \sum_\alpha |\mathcal{Z}^i_\alpha|^2 = \frac{1}{1 - (\overline{\mathcal{Z}}^i_\beta)^* \left.\frac{d\Sigma^{\star}_{\beta\gamma}(\omega)}{d\omega}\right|_{\omega=\varepsilon_i} \overline{\mathcal{Z}}^i_\gamma} , \tag{11.46}$$

where $\overline{\mathcal{Z}}^i \equiv \mathcal{Z}^i/\sqrt{SF_i}$ is the spectroscopic amplitude normalized to 1.

Equations (11.45) and (11.46) are the central equations of the Green's function formalism and show how the single-particle propagator is the solution of an effective one-body Schrödinger equation for a nucleon or a hole propagating inside the correlated system. The energy dependence of $\Sigma^{\star}(\omega)$ and its non-locality are a consequence of the underlying many-body dynamics. Equation (11.46) also shows that the reduction of spectral strength commonly observed in correlated systems arises from the dispersion properties of the self-energy.

In spite of its beauty, Eq. (11.45) is also the worst starting point to solve the Dyson equation in a discretized finite basis. Unless one is interested in just a few solutions near the Fermi surface or the model space is extremely small, this approach will require high computational times due to the large amounts of diagonalizations required to extract the correct eigenvalues. The reason is that root-finding algorithms are needed to match the eigenvalues ε_i with the argument of $\Sigma^\star(\varepsilon_i)$, but simple searching algorithms may miss a large amount of solutions. The consequences of missing a large portion of spectral strength are that wrong results would be obtained for the ground state observables computed as in Sect. 11.2.1. This can also deteriorate the self-consistency already at the level of the static self-energy, $\Sigma^{(\infty)} = \widetilde{U}$. If Eq. (11.45) must be used, it is possible to gather all the necessary solutions by starting from extremely fine energy meshes to be sure that all eigenvalues are bracketed first. However, this easily becomes suicidal in terms of the increase of computing time. We discuss here a different approach that is not affected by these problems and that will also give some further insight into the physics content of the Dyson equation.

First, for each solution of the Dyson equation we define two new vectors $\mathcal{W}^i$ and $\mathcal{V}^i$ which live in the ISCs space as follows:

$$
\begin{aligned}
[\omega - E^> - C]_{r,r'}\, \mathcal{W}^i_{r'} &\equiv M_{r,\delta} Z^i_\delta \,, \\
[\omega - E^< - D]_{s,s'}\, \mathcal{V}^i_{s'} &\equiv N^\dagger_{s,\delta} Z^i_\delta \,,
\end{aligned}
\tag{11.47}
$$

where we have let $i\eta \to 0$ as this is no longer needed in a finite and discretized basis. With these definitions, Eq. (11.45) is easily rearranged into a single eigenvalue problem of larger dimensions but where the corresponding matrix is energy independent:

$$
\begin{pmatrix}
\widehat{T} + \Sigma^{(\infty)} & M^\dagger & N \\
M & E^> + C & \\
N^\dagger & & E^< + D
\end{pmatrix}
\begin{pmatrix} \mathcal{Z}^i \\ \mathcal{W}^i \\ \mathcal{V}^i \end{pmatrix}
=
\begin{pmatrix} \mathcal{Z}^i \\ \mathcal{W}^i \\ \mathcal{V}^i \end{pmatrix} \varepsilon_i
\tag{11.48}
$$

and the normalization condition (11.46) becomes

$$
\sum_\alpha |\mathcal{Z}^i_\alpha|^2 + \sum_r |\mathcal{W}^i_r|^2 + \sum_s |\mathcal{V}^i_s|^2 = 1 \,. \tag{11.49}
$$

The advantage of this approach is that it linearizes the Dyson equation and yields all solutions in one single diagonalization. Although the dimension of the Dyson matrix in Eq. (11.48) is much larger than a one-body Schrödinger problem and that it requires a substantial amount of memory storage, it typically provides the full spectral strength 100 times faster than using Eq. (11.45) directly. Furthermore, it is possible to reduce the dimensionality of the eigenvalue problem by projecting matrices $[E^{>} + C]$ and $[E^{<} + D]$ (separately!) onto smaller Lanczos/Krylov subspaces [19, 46]. In this way one reduces the number of poles of $g(\omega)$ far away from the Fermi surface—where only their average is physically meaningful—but conserves the overall strength needed to compute ground state observables.

Equation (11.48) also puts in evidence how the Dyson equation is very closely related to a configuration interaction (CI) approach. For solutions $(\varepsilon_n^+, \mathcal{X}^n)$ in the single particle spectrum, the eigenstates of $|\Psi_n^{A+1}\rangle$ are expanded in terms of 1p configurations (from the $\widehat{T} + \Sigma^{(\infty)}$ sector) and 2p1h or larger configurations, which is evident from the matrix C, in Eq. (11.38c). However, additional 2h1p configurations are included through matrix D. This is in spirit very similar to how ground state correlations are included in the random phase approximation approach [37]. Furthermore, the matrices that couples these subspaces are the same as in CI only at first order ($M^{(1)}$ and $N^{(1)}$). The eigenstates of Eq. (11.48) will approach the exact solution as the approximation of the self-energy is systematically improved. Similarly, the propagation of hole states that correspond to the eigenstates of $|\Psi_k^{A-1}\rangle$ are obtained in a CI fashion. Equation (11.49) is then the natural normalization condition for the CI expansion and shows that the spectroscopic amplitudes are the projection of more complex many-body wave functions onto a single-particle space.

11.6 *E*xercise: Perform a Taylor expansion of the propagator $g(\omega)$ at zeroth order around a given pole $\varepsilon_i^{\pm}$. Then, use this and the conjugate Dyson equation (11.3b) to obtain the normalization condition for spectroscopic factors given in Eq. (11.46).

11.7 *E*xercise: Based on the definitions of vectors $\mathcal{W}^i$ and $\mathcal{V}^i$, Eqs. (11.47), show that (11.46) and (11.49) are equivalent.

11.3.3 A Simple Pairing Model

As a first demonstration of the ADC formalism, we consider the pairing Hamiltonian already discussed in Chap. 8. This is a system of four spin-1/2 fermions in a 4-level model space that interact through a pairing force:

$$\widehat{H} = \widehat{H}_0 + \widehat{V} = \xi \sum_{p=1}^{4} \sum_{\sigma=+,-} (p-1) a_{p\sigma}^{\dagger} a_{p\sigma} \;-\; \frac{g}{2} \sum_{p,q=1}^{4} a_{p+}^{\dagger} a_{p-}^{\dagger} a_{q-} a_{q+} \,. \tag{11.50}$$

In spite of its simplicity, this model poses a particularly difficult test for many-body approximations based on ISRs because the Hamiltonian (11.50) does not allow for admixtures of leading order excitations, that is of the particle-hole type. The ground state contains only 2p2h and higher excitations. Correspondingly, the pairing interactions $\widehat{V}$ cannot couple particle states to 2p1h configurations, neither hole states with 2h1p ones. This is obvious looking at the leading terms,

Eqs. (11.28a) and (11.29a), that would involve interactions between a particle and a hole (which cannot be connected by pairing) but it applies to the full ADC(3) couplings (11.38a) and (11.39a) as well. It follows that the spectra for particle attachment and removal are dominated by 3p2h and 3h2p ISCs. These are partially included in the Dyson equation by couplings between particles and backward going, 2h1p, terms in the self-energy (or between holes and the forward 2p1h terms). However, a complete account of them would require many-body truncations at the ADC(4) level and higher. Remarkably, it is still possible to reach rather accurate results as demonstrated by Figs. 11.7 and 11.8.

The unperturbed propagator, associated with the $\widehat{H}_0$ term of Eq. (11.50), is given by

$$g^{(0)}_{p\sigma_p\,,\,q\sigma_q}(\omega) = \delta_{pq}\delta_{\sigma_p\,\sigma_q}\left\{\sum_{n=3,4}\frac{\delta_{np}}{\omega-\varepsilon^{(0)}_n+i\eta}+\sum_{k=1,2}\frac{\delta_{kp}}{\omega-\varepsilon^{(0)}_k-i\eta}\right\} \tag{11.51}$$

where $\varepsilon^{(0)}_p = \xi(p-1)$ are the unperturbed single particle energies and the gap at the Fermi surface is $E^{(0)}_{ph} = \varepsilon^{(0)}_3 - \varepsilon^{(0)}_2 = \xi$. For this particular model, the unperturbed state is also the same state that solves the HF equations. Thus, the HF propagator is written exactly in the same way but with only a shift in the single particle energies

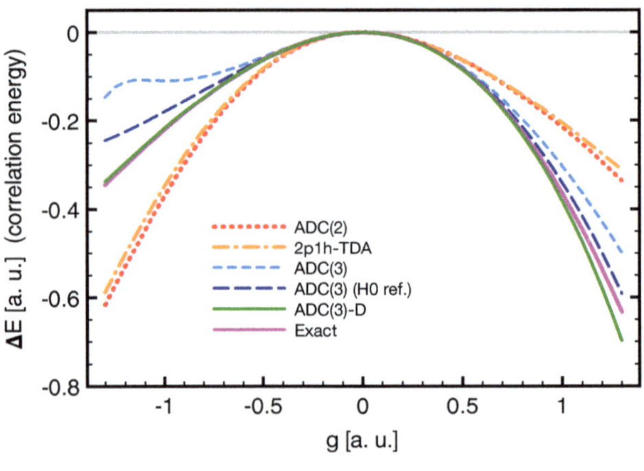

Fig. 11.7 Correlation energy for the pairing Hamiltonian of Eq. (11.50) as a function of the coupling g, obtained for different ADC(n) approximations to the self-energy and in the *sc0* scheme. The *dotted, dot-dashed, short dashed and full lines* are all obtained from the HF reference of Eq. (11.52) and show successive approximations of the ADC(n) hierarchy [respectively: ADC(2), 2p1h-TDA, ADC(3) and ADC(3)-D]. The *long dashed line* is the same ADC(3) truncation but based on the unperturbed reference propagator of Eq. (11.51). The *purple line* shows the exact result calculated from a full configuration interaction diagonalization

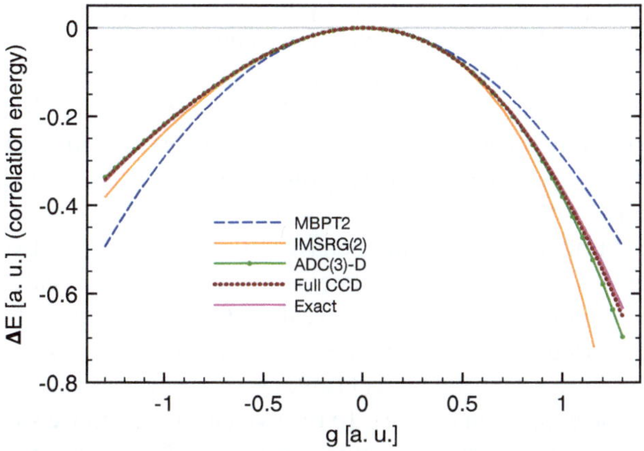

Fig. 11.8 Correlation energy for the pairing Hamiltonian of Eq. (11.50) as a function of the coupling g, for different many-body methods discussed in this book. The *purple line* is the exact results from configuration interaction theory. The results for second-order perturbation theory (MBPT2), for IMSRG(2), for the CC-corrected ADC(3)-D and for the standard CC with doubles (CCD) are compared. See also Sect. 10.3.6 for higher truncations of MBPT

of the hole states (see also Sect. 8.7.4 and Table 8.11):

$$g^{HF}_{p\sigma_p\,,q\sigma_q}(\omega) = \delta_{pq}\delta_{\sigma_p\,\sigma_q}\left\{\sum_{n=3,4}\frac{\delta_{np}}{\omega-\varepsilon^{HF}_n+i\eta}+\sum_{k=1,2}\frac{\delta_{kp}}{\omega-\varepsilon^{HF}_k-i\eta}\right\} \tag{11.52}$$

where

$$\varepsilon^{HF}_p = \begin{cases}\xi(p-1)\,, & \text{for } p=3,4\\[2ex] \xi(p-1)-g/2\,, & \text{for } p=1,2\end{cases} \tag{11.53}$$

and the particle-hole gap now depends on the coupling constant, $E^{HF}_{ph} = \varepsilon^{HF}_3 - \varepsilon^{HF}_2 = \xi + g/2$. One may chose either of these propagators as the reference state for calculating the ADC(n) self-energy. However, $g^{(0)}(\omega)$ will also require additional corrections terms for the interactions matrices C and D, as seen in Exercise 11.5. In practice, these corrections are already included in the shifts of Eq. (11.53) and the HF reference is normally a better starting point for calculating the self-energy.

We now set $\xi = 1$ and perform calculations at different levels of approximations in the ADC(n) approach, by using the $g^{HF}(\omega)$ as reference (except when indicated) and by calculating $\Sigma^{(\infty)}$ self-consistently in the *sc0* scheme. After solving the Dyson equation, we extract the ground state energy from the Koltun sum rule (11.11) and calculate the correlation energy $\Delta E = E_{g.s.} - (2\xi - g)$. The result of the ADC(2) equations (11.28) and (11.29) is shown by the dotted line in Fig. 11.7.

The 2p1h-TDA approximation improves upon this by using the interaction matrices from Eqs. (11.38c) and (11.39c), which resumes infinite ladders of 2p and 2h states. However, this brings only a very small improvement to this system. The ADC(3) approximation gives better results and it is shown for both the $g^{(0)}(\omega)$ and $g^{HF}(\omega)$ choices of the reference state with long dashed and short dashed lines, respectively. Remarkably, these results depend strongly on the reference state and are much closer to the exact solution for the $g^{(0)}(\omega)$ case, which would have been expected to be a poorer choice. Furthermore, $g^{HF}(\omega)$ behaves erratically for negative values of g, corresponding to a repulsive pairing interaction $\widehat{V}$. These two calculations differ only in the single particle energies used to calculate the coupling matrices M and N. Such behavior is simply explained by the dependence of E_{ph}^{HF} on g, which can make the denominator in Eqs. (11.38c) and (11.39c) very small and causes the breakdown of the perturbation expansion (11.36). To resolved his problem we substitute the $t^{(0)}$ of Eq. (11.41) with the converged solution from the CCD equations. The resulting ADC(3)-D is now completely independent of the choice between the two reference states and it also reproduces the exact result closely. This is shown by the two solid lines in Fig. 11.7.

Figure 11.8 compares the ADC approach with CC, in-medium similarity renormalization group (IMSRG) and second-order perturbation theory. The ADC(3)-D, the two-body truncation of IMSRG (IMSRG(2), see Chap. 10) and the CC methods perform similarly at $g < 0$, where they are all close to the exact solution all the way to $g \approx -1.3$. For smaller values of the coupling the CCD iterations stop converging. At large positive values of g (corresponding to attractive pairing) the various approaches eventually deviate from the exact result but with CCD being slightly better. Clearly the full ADC(3)-D is a more complex calculation than CCD but leads to similar results for the binding energy. On the other hand, this does not only yield the ground state energy but also the whole spectral function for the addition and removal of a particle is generated when solving the Dyson equation (11.45) or (11.48). The next section will demonstrate examples of the self-energy and the spectral distribution obtained when calculating the single particle propagator.

The FORTRAN code that generated these results is available online at https://github.com/ManyBodyPhysics/LectureNotesPhysics/tree/master/Programs/Chapter11-programs/Pair_Model. We do not examine this code here but we will give a detailed discussion of how to structure a complex ADC(n) code for infinite matter computations in the next section.

11.4 Numerical Solutions for Infinite Matter

In this section we discuss how to implement the calculation of the self-energy and the single particle propagator in the ADC(n) formalism. We will demonstrate this for the case of infinite nucleonic matter and use our results to discuss general features of the spectral function. A general code that can solve for both symmetric and pure neutron matter up to ADC(3) is provided with this chapter at the URL https://github.com/ManyBodyPhysics/LectureNotesPhysics/tree/master/Programs/Chapter11-programs/Inf_Matter. We will use the C++ programming language and will refer to this code for describing the technical details of the implementation. We then show results based on the Minnesota nuclear potential from [47]. This is a very simplified model of the nuclear interaction that allows for an easy implementation. On the other hand, it still retains some physical properties of the nuclear Hamiltonian that will allow us to discuss the basic features of the spectral function of nucleonic matter (and of infinite fermionic systems in general). The reader interested in these physics aspects could refer directly to Sect. 11.4.2.

11.4.1 Computational Details for ADC(n)

The first fundamental step to set up a SCGF computation is the choice of the model space. For infinite matter, translational invariance imposes that the Dyson equation is diagonal in momentum and therefore it becomes much easier to solve the problem in momentum space. However, there remain two possible choices for how to encode single particle degrees of freedom. The first one is to subdivide the infinite space in boxes of finite size and to impose periodic boundary conditions (see also Chap. 8). In this way, the number of fermions included in each box is finite and determined by the particle density of the system. The resulting model space is naturally expressed by a set of discretized single particle states and one solves the working equations in the form of Eqs. (11.38), (11.39) and (11.48). This path requires the same technical steps needed to calculate finite systems in a box. Numerical results then need to be converged with respect to the truncation of the k-space (and, for an infinite system, with respect to the number of nucleons inside each periodic box). We will follow this approach for the present computational project. The other approach is to retain the full momentum space and write the SCGF equations already in the full thermodynamic limit. This choice is best suited to solve the Dyson equation at finite temperatures and in a full SCGF fashion and will be discussed further in Sect. 11.5.

Construction of the Model Space For simplicity, we assume a total number A of nucleons in each (cubic) periodic box. For boxes of length L, the density and the Fermi momentum are expressed, respectively as ($\hbar$=1):

$$\rho = \frac{A}{L} \qquad \text{and} \qquad p_F = \sqrt[3]{\frac{6\pi^2\rho}{\nu_d}}\,, \tag{11.54}$$

where the degeneracy ν_d is twice the number of different spin-$\frac{1}{2}$ fermions and the basis states are defined by the cartesian quantum numbers n_x, n_y, n_z=0, 1, 2... with momentum

$$\mathbf{p} = \frac{2\pi}{L}\begin{pmatrix} n_x \\ n_y \\ n_z \end{pmatrix}. \tag{11.55}$$

The kinetic energies, and hence the unperturbed single particle energies, will depend on $|\,\mathbf{p}|^2$ and hence the values of $N_{sq} = n_x^2 + n_y^2 + n_z^2$ define a set of separate shells. Since we need closed shell reference states, only certain values for the number of nucleons in each box, A, are possible. The size of the model space is given by $N_{sq}^{\max} = \max\{n_x^2 + n_y^2 + n_z^2\}$. The construction of the single particle model space is then straightforward. We will do it constructing a specific class with pointers to arrays for each relevant quantum number and additional arrays for the kinetic energies or any other useful quantity associated with each state.

```
class SpBasisK {

  public:
    int SpNmax, SpNAlloc; // total number of s.p. states and
        allocated space
    int *nx, *ny, *nz, *spin; // quantum numbers
    double *e_kin; // kinetic energy

    double Lbox; // side length of the periodic box
    int N_holes; // number of nucleons in a box (# of occupied
        states)

    // grouping s.p. states of equal symmetry
    int N_grps; // number of different groups
    int *gr_mlt, *gr_rep;

    // functions
  public:
    void Build_sp_basis(int, double, int);

    int Build_groups_table(void );

  };
```

The constructor for the model space will be necessary to order the basis with increasing values of N_{sq}, so that the orbits corresponding to the A hole states come first. This will become useful to construct ISCs later on. We first count the total number of possible (n_x, n_y, n_z) configurations. Once it is known how many single particle $\mathbf{k}$ states there are, we can allocate arrays in memory to store the relevant quantum numbers of each of them:

```
const double PI    = 3.141592653589793;
const double hbarc = 197.326968; // [MeV*fm]
const double NUCLEONmass = 939.565; // [MeV]

void SpBasisK::Build_sp_basis(int Nsq_max, double Lbox, int A) {

  int imax = int( sqrt(double Nsq_max) + 1 ); // max value of |
      n_x|, |n_y| or |n_z|

  int i_count = 0; // counts the number of basis states:
  for (int ix=-imax; ix<=imax; ++ix)
    for (int iy=-imax; iy<=imax; ++iy)
```

```
    for (int iz=-imax; iz<=imax; ++iz)
     if (ix*ix + iy*iy + iz*iz <= Nsq_max) ++i_count;


SpNAlloc = 2 * i_count; // 2 is the spin-1/2 degeneracy; we
    assume PNM here

cout << "\n Allocating space for "<< SpNAlloc << " sp states
    ... \n";

nx   = new int[SpNAlloc]; // Allocate basis' arrays
ny   = new int[SpNAlloc];
nz   = new int[SpNAlloc];
spin = new int[SpNAlloc];
e_kin = new double[SpNAlloc];

double xek;

cout << "\n Single particle basis:\n ----------------------";
cout << "\n orbit n_x n_y n_z  Nsq   E_kin\n";

i_count = 0;
for (int isq=0; isq<=Nsq_max; ++isq) {
  for (int ix=-imax; ix<=imax; ++ix)
    for (int iy=-imax; iy<=imax; ++iy)
      for (int iz=-imax; iz<=imax; ++iz) {
        if ((ix*ix + iy*iy + iz*iz) != isq) continue;

        xek = double(isq) * pow((hbarc * 2.0 * PI / Lbox), 2.0)
            / 2.0 / NUCLEONmass;
        cout <<i_count <<" " <<ix <<" " <<iy <<" " <<iz <<" " <<
            isq <<" " <<xek <<endl;

        for (int is=-1; is<2; is+=2) {
          nx[i_count] = ix;
          ny[i_count] = iy;
          nz[i_count] = iz;
          spin[i_count] = is;
          e_kin[i_count] = xek;
          ++i_count;
        }

      } // end of ix, iy, iz loop
} // end of isq loop
SpNmax = i_count;

this->N_holes = A; // very important! Must set the # of
    occupied states

return;}
```

Construction of the ISCs Due to translational invariance the Dyson equation (11.3) separates in a set of uncoupled equations for each values of $\{\mathbf{p}_i, s_z^i\}$ in the model space (where s_z is the spin projection and i labels the basis states):

$$g(\mathbf{p}_i, s_z^i; \omega) = g^{(0)}(\mathbf{p}_i, s_z^i; \omega) + g^{(0)}(\mathbf{p}_i, s_z^i; \omega)\, \Sigma^\star(\mathbf{p}_i, s_z^i; \omega)\, g(\mathbf{p}_i, s_z^i; \omega)\,. \quad (11.56)$$

This diagonal equation can be formally inverted as shown in Eqs. (11.83) and (11.86) below. However, we will solve for all of its eigenstates instead and this is better done by diagonalizing Eq. (11.48). For each state i, we need to generate tables for the relevant 2p1h and 2h1p ISCs and then calculate the elements of the Dyson matrix. One can build a class whose objects are associated to a particular orbit of the given model space and then construct the ISCs in accordance with the conservation of momentum and other symmetries of the Hamiltonian, which are implicit in the matrix elements for the coupling (M and N) and interaction (C and D) matrices. Schematically, looking only at the 2p1h configurations for simplicity, this will be:

```
class ADC3BasisK {

  public:
    int *Bas_2p1h, *Bas_2h1p; // pointers to 2p1h/2h1p bases
    int Nbas_2p1h, Nbas_2h1p; // dimensions of the 2p1h/2h1p
        bases

    int iSpLoc; // {p,s_z} state in the s.p. basis associated
        with the 2p1h/2h1p
    SpBasisK *SpBasLoc;

    // functions
  public:
    void Build_2p1h_basis(SpBasisK*, int );
  };

void ADC3BasisK::Build_2p1h_basis(SpBasisK *InBasis, int isp) {

  this->SpBasLoc = InBasis; // keep track of the basis and the s
      .p. states associated
  this->iSpLoc = isp; // to this 2p1h ICSs, for use by other
      functions

  Nbas_2p1h = ... ; // must compute the number of expected 2p1h
      configurations

  if (NULL != Bas_2p1h) delete [] Bas_2p1h;
  this->Bas_2p1h = new int[3*(Nbas_2p1h)]; // need 3 indices for
       each config (n1, n2, k3)

  int  k3_x, k3_y, k3_z, k3_sp;

  i_count = 0;
  for (int n1=SpBasLoc->N_holes; n1<SpBasLoc->SpNmax; ++n1) {
```

```
    for (int n2=n1+1; n2<SpBasLoc->SpNmax; ++n2) { // n1 < n2 due
         to Pauli

      // expected q.#s for 3rd index (k3), imposed by the
          Hamiltonian's symmetries:
      k3_x = SpBasLoc->nx[n1] + SpBasLoc->nx[n2] - SpBasLoc->nx[
          isp];
      k3_y = SpBasLoc->ny[n1] + SpBasLoc->ny[n2] - SpBasLoc->ny[
          isp];
      k3_z = SpBasLoc->nz[n1] + SpBasLoc->nz[n2] - SpBasLoc->nz[
          isp];
      k3_sp = SpBasLoc->spin[n1] + SpBasLoc->spin[n2] - SpBasLoc
          ->spin[isp];

      for (int k3=0; k3<SpBasLoc->N_holes; ++k3) {
        if ( (k3_x != SpBasLoc->nx[k3]) || (k3_y != SpBasLoc->ny[
            k3] ) ||
             (k3_z != SpBasLoc->nz[k3]) || (k3_sp != SpBasLoc->
                 spin[k3]) ) continue;

        this->Bas_2p1h[3*i_count ] = n1;
        this->Bas_2p1h[3*i_count + 1] = n2;
        this->Bas_2p1h[3*i_count + 2] = k3;
        ++i_count;

      } // end k3 loop
    } // end n2 loop
  } // end n1 loop
  if (i_count > Nbas_2p1h) {/* This is a trouble */} else {
      Nbas_2p1h = i_count;}

  return;}
```

Spectral Representation Both the propagator and the self-energy have spectral representations in terms of poles, with residues in separable form. Hence, we can devise a general class that could store both objects. Specifically, by using the conservation of spin and the fact that the propagator is diagonal in momentum space, one can write the Lehmann representation (11.2) as

$$g(\mathbf{p}_i, s_z^i; \omega) = \sum_n \frac{S^p(\mathbf{p}_i, s_z^i; \varepsilon_n^{\mathbf{p}_i +})}{\omega - \varepsilon_n^{\mathbf{p}_i +} + i\Gamma} + \sum_k \frac{S^h(\mathbf{p}_i, s_z^i; \varepsilon_k^{\mathbf{p}_i -})}{\omega - \varepsilon_k^{\mathbf{p}_i -} - i\Gamma}, \tag{11.57}$$

where $S^{p(h)}(\mathbf{p}_i, s_z^i; \omega)$ are the particle and hole parts of the spectral function [see Eqs. (11.4)]. Hence, it is simpler and more efficient to store the full residues rather than separate spectroscopic amplitudes. The self-energy can be casted in the same simple pole structure by diagonalizing the interactions matrices. Assuming that $U_C\,(E^> + C)\,U_C^\dagger = \text{diag}(\lambda_C^r)$ and $U_D\,(E^< + D)\,U_D^\dagger = \text{diag}(\lambda_D^q)$, with $\lambda_{C,D}$ being the eigenvalues, we rewrite Eq. (11.27) as follows:

$$\Sigma^{\star}(\mathbf{p}_i, s_z^i; \omega) = \Sigma^{(\infty)}(\mathbf{p}_i, s_z^i) + \sum_r \frac{|\widetilde{M}_{r;\mathbf{p}_i, s_z^i}|^2}{\omega - \lambda_C^r + i\Gamma} + \sum_q \frac{|\widetilde{N}_{\mathbf{p}_i, s_z^i; q}|^2}{\omega - \lambda_D^q - i\Gamma}, \tag{11.58}$$

where $\widetilde{M} = U_C M$ and $\widetilde{N} = N U_D^{\dagger}$. A full pre-diagonalization of the interaction matrices C and D is not needed to construct the Dyson matrix. Thus, storing the self-energy in the form of Eq. (11.58) is worth only if self-energy is to be calculated for specific values of its arguments (for example to plot it). However, in most cases, a reduction of these matrices through a Lanczos algorithm is still necessary to reduce the dimensionality of the problem, as discussed below here. The resulting tridiagonal matrices can be accommodated in the same structure as for the propagator by simply adding an extra array for the sub-diagonal elements. Thus, the class for the Lehmann representation has the following structure:

```
class SpctDist {

  public:
    SpBasisK *SpBasLoc; // associated s.p. basis

    int N_LEH_ALLOC; // number of Lehmann representations to
        store

    int    *N_fw_pls, *N_bk_pls, *N_PLS_ALLOC;
    double **ek_fw,  **ek_bk; // - poles of the propagator/self-
        energy
    double **eb_fw,  **eb_bk; // - eb_xx Lanczos subdiagonal for
        storing self-energy
    double **Sk_fw,  **Sk_bk; // - this is the FULL residue (not
        the amplitude X,Y)
    double *Sig_inf; // static self-energy

  // functions
  public:
    SpctDist(SpBasisK* ); // constructor
    int add_k_channel(int, int, double*, double*, int, double*,
        double*,
                        double in_Sig_inf=0.0, double *B_fw_in=NULL
                              , double *B_bk_in=NULL);
  };

void SpctDist::SpctDist(SpBasisK *InBasis ) {
  //
  // Use constructor to initialize the object with a table
  // of pointers for all basis states

  this->SpBasLoc = InBasis; // keeps track of the associated
      model space
```

```
  N_LEH_ALLOC = this->SpBasLoc->SpNmax;


  Sig_inf = new double[N_LEH_ALLOC];

  N_fw_pls = new int[N_LEH_ALLOC]; N_bk_pls = new int[
      N_LEH_ALLOC];
  N_PLS_ALLOC = new int[N_LEH_ALLOC];
  ek_fw    = new double*[N_LEH_ALLOC]; ek_bk = new double*[
      N_LEH_ALLOC];
  Sk_fw    = new double*[N_LEH_ALLOC]; Sk_bk = new double*[
      N_LEH_ALLOC];
  eb_fw    = new double*[N_LEH_ALLOC]; eb_bk = new double*[
      N_LEH_ALLOC];

  for (int isp=0; isp<N_LEH_ALLOC; ++isp) {
   Sig_inf[isp] = 0.0;

   N_fw_pls [isp] = -100; N_bk_pls [isp] = -100;
   N_PLS_ALLOC[isp] = -100;
   ek_fw[isp]   = NULL; ek_bk[isp]   = NULL;
   Sk_fw[isp]   = NULL; Sk_bk[isp]   = NULL;
   eb_fw[isp]   = NULL; eb_bk[isp]   = NULL;
  }

  return;}

void SpctDist::add_k_channel(int i_Leh, int N_fw_in, double *
    A_fw_in, double *E_fw_in,
                                    int N_bk_in, double *A_bk_in,
                                        double *E_bk_in,
                                    double in_Sig_inf /*=0.0*/,
                                    double *B_fw_in/*=NULL*/, double *
                                        B_bk_in/*=NULL*/){
  //
  // This function is to load and store the spectral
      representation of a s.p. propagator
  // or a self-energy, if the additional array for the
      subdiagonal elements the self-energy
  // are not provided, they are set automatically to zero.

  // Allocate memory for the basis' state i_Leh; only one array
      is allocate for both hole
  // and particle poles, the xx_fw[] arrays will just point to
      where the particles begin
  N_PLS_ALLOC[i_Leh] = N_bk_in + N_fw_in;
  ek_bk[i_Leh] = new double[N_PLS_ALLOC[i_Leh]]; ek_fw[i_Leh] =
      ek_bk[i_Leh] + N_bk_in;
  eb_bk[i_Leh] = new double[N_PLS_ALLOC[i_Leh]]; eb_fw[i_Leh] =
      eb_bk[i_Leh] + N_bk_in;
  Sk_bk[i_Leh] = new double[N_PLS_ALLOC[i_Leh]]; Sk_fw[i_Leh] =
      Sk_bk[i_Leh] + N_bk_in;

  // store hole poles
```

```
  N_bk_pls[i_Leh] = N_bk_in;
  for (int ibk=0; ibk<N_bk_in; ++ibk) {
    ek_bk[i_Leh][ibk] = E_bk_in[ibk];
    Sk_bk[i_Leh][ibk] = A_bk_in[ibk];
    eb_bk[i_Leh][ibk] = 0.0;
    if (NULL != B_bk_in) eb_bk[i_Leh][ibk] = B_bk_in[ibk];
  }

  // store particle pole
  N_fw_pls[i_Leh] = N_fw_in;
  for (int ifw=0; ifw<N_fw_in; ++ifw) {
    ek_fw[i_Leh][ifw] = E_fw_in[ifw];
    Sk_fw[i_Leh][ifw] = A_fw_in[ifw];
    eb_fw[i_Leh][ifw] = 0.0;
    if (NULL != B_fw_in) eb_fw[i_Leh][ifw] = B_fw_in[ifw];
  }

  Sig_inf[i_Leh] = in_Sig_inf; // stores the static self-energy;
       == 0.0 if default

  return;}
```

The above classes simplify the calculation of quantities related to SCGF. For example, let us assume a function, `Vpotential(ia,ib,ic,id)`, that returns the matrix elements of the two-body interaction. The ADC(2) coupling matrix (11.28a) could be calculated using the following code:

```
// Configurations for s.p. state iL:
ADC3BasisK ISC2p1h(); ISC2p1h.Build_2p1h_basis(SpBasis, iL);

// Array to store the coupling matrix M:
double M_rp = new double[ISC2p1h.Nbas_2p1h];

for (int ir = 0; ir<ISC2p1h.Nbas_2p1h; ++ir) {
  // no need to loop over s.p. states since we are diagonal in
      the channel ia

  // Single particle states for the ir-th 2p1h configuration:
  im = Bas_2p1h[3*ir ];
  iv = Bas_2p1h[3*ir + 1 ];
  iL = Bas_2p1h[3*ir + 2 ];

  // Apply Eq. (11.28a) [a HF ref. state is assumed here... X=Y
      =1]
  M_rp[ir] = V_potential(im,iv,ia,iL);
}
```

Likewise, the correlated HF diagram that contributes to $\Sigma^{(\infty)}$ [second term on the right hand side of Eq. (11.17a)] could be obtained as follows:

```
// To calculate the HF potential (V_HF) between states ia and ib
     we do:
```

```
double Sh, Vhf_ab;
int nHoles;
SpBasisK *Bas = ; // point to some object containing the model
    space
SpctDist SpProp(Bas); // sp propagator, contains spectral
    distribution of every (p_i,s_z)

Vhf_ab = 0.0;
for (ic = 0; ic<Bas->SpNmax; ++ic) {
 nHoles = SpProp.N_bk_poles[ic];
 Sh = 0.0;
 for (int k=0; k<nHoles; ++k) Sh += SpProp[ic].Sh[k];
  Vhf_ab += V_potential(ia,ic,ib,ic) * Sh;
  }
```

Reducing the Computational Load Practical applications often require rather large model spaces to achieve convergence. This poses a major hindrance since the number of ISCs can grow very fast with the size of the space. The strongest constraint comes from 2p1h configurations (that is, the dimension of the C matrix), which increases quadratically with the number of unoccupied states and linearly with the number of occupied ones. As a consequence, it is almost never possible to attempt a fully self-consistent calculations of the dynamic self-energy because these would be based on the huge number of poles in Eqs. (11.2) or (11.57). In fact, the dimensionality wall not only prohibits going beyond a *sc0* calculation but the dimensions of the Dyson matrix can become prohibitive even for a mean-field reference state and models spaces of moderate size.

As already mentioned in Sect. 11.3.2, the way out from this situation is to substitute the denominators in the Lehmann representation of the self-energy (11.58) with a much smaller numbers of effective poles. This is done by projecting the sub-matrices $E^{>} + C$ and $E^{<} + D$ onto Krylov spaces of much smaller dimensions by using a Lanczos algorithm (or Block Lanczos, in the general case when the self-energy is not diagonal in $\mathbf{p}_i$) [48]. This approach is usually more efficient if the vectors corresponding to the columns of M and $N^\dagger$ are taken as the pivots. For example, if $\mathcal{L}$ is the $N_{\text{red}} \times N_{\text{2p1h}}$ matrix that projects from the full space of 2p1h configurations to the Krylov space of dimension N_{red} ($<< N_{\text{2p1h}}$), then the third term on the right hand side of Eq. (11.27) is modified as follows:

$$M^\dagger \frac{1}{\omega - [E^{>} + C] + i\eta} M \quad \longrightarrow \quad M^\dagger \mathcal{L}^\dagger \frac{1}{\omega - \mathcal{L}[E^{>} + C]\mathcal{L}^\dagger + i\eta} \mathcal{L} M \tag{11.59}$$

and similarly for the 2h1p sector. In most cases, a number of Lanczos vectors between $N_{\text{red}} = 50$ and 300 is sufficient, depending on model space size and the accuracy required. The reason for choosing a Krylov type of projection to reduce the dimensionality of the Dyson eigenvalue problem is that this allows to preserve two crucial properties of the spectral distribution of $\Sigma^\star(\omega)$. First, the lowest $2N_{\text{red}}$ moments of the spectral distribution are conserved, which guarantees to reproduce well the average spectral function at medium and large energies. Second, the eigenvectors at the extremes of the (2p1h or the 2h1p) spectrum converge first in the Lanczos algorithm. This implies that the self-energy and the particle attachment or removal distributions converge fast to the exact one near the Fermi energy. For this reason it is crucial that both the $E^{>} + C$ and $E^{<} + D$ matrices are projected and that they are handled separately. See [19] for details of the implementation in the SCGF approach.

In addition to the dimensions problem, one also needs to diagonalize Eq. (11.48) for each separate channel ($\mathbf{p}_i$, s_z^i) in the basis. On the other hand, some single particle states are equivalent. For example, the momentum states with n_x=3, n_y=2 and n_z=1 is the same as n_x=2, n_y=-3 and n_z=1 except for a rotation around the z-axis. Likewise, n_x=3, n_y=2 and n_z=-1 differs only by a parity inversion. The diagonalization of each of these channel would yield exactly the same results and needs to be performed only once. The obvious procedure is that of grouping the model space states according to the same symmetries of the Hamiltonian. In this way, Eq. (11.48) is typically solved a few tens of times even when the model space is two orders of magnitude larger. For an Hamiltonian that is invariant under rotation, parity inversion and spin flipping, the algorithm to separate the basis in groups of the same symmetry is as follows:

```
int SpBasisK::Build_groups_table(void ) {

  int AbsN_mx = ... // Maximum absolute value of n_x, n_y or n_z

  int N_ALLOC_GRPS = ... //Max number of different groups
      expected

  gr_rep = new int[N_ALLOC_GRPS]; // for each group, keep track
      of a representative state
  gr_mlt = new int[N_ALLOC_GRPS]; // number of basis states
      belonging to a group

  int i_mult, i_rep, n1, n2, n3, itmp;


  int count=0;
  for (int i1=0; i1<=AbsN_mx; ++i1)
    for (int i2=i1; i2<=AbsN_mx; ++i2)
      for (int i3=i2; i3<=AbsN_mx; ++i3) {

        i_mult = 0;
        i_rep = -100;
        for (int isp=0; isp<this->SpNmax; ++isp) {

          n1 = abs(nx[isp]); n2 = abs(ny[isp]); n3 = abs(nz[isp]);

          if (n1 > n2) {itmp=n1; n1=n2; n2=itmp;} // order the q.#
              s of the orbit isp in
          if (n1 > n3) {itmp=n1; n1=n3; n3=itmp;} // increasing
              values, according
```

```
        if (n2 > n3) {itmp=n2; n2=n3; n3=itmp;} // to i1 < i2 <
            i3

        if ((n1==i1) && (n2==i2) && (n3==i3)) {
          ++i_mult;
          if (i_rep < 0) i_rep = isp;
        }

      } // end loop over isp

      if (i_rep >= 0) {
        gr_mlt[count] = i_mult;
        gr_rep[count] = i_rep;
        ++count;
      }

    }

  this->N_grps = count;

  cout << "\n\n A total of " << N_grps << " independent groups
      of single particle basis \n";
  cout <<     "states has been found. All states within one group
       are equivalent \n";
  cout <<     "under rotation, spin and/or parity inversion.\n";

  return N_grps;}
```

11.4.2 Spectral Function in Pure Neutron and Symmetric Nuclear Matter

We test the ADC approach for pure neutron matter (PNM) and symmetric nuclear matter (SNM) using the Minnesota nuclear force [47]. This is a simple semi-realistic potential that contains only central terms, for different spin and isospin, but no tensor force. It has often been used in structure studies of light neutron-rich nuclei, although it fails to predict any saturation of infinite nuclear matter up to very high densities. Nevertheless, it is a good toy model for describing certain salient features of nucleonic matter and of quantum liquids in general. In pure neutron matter, we computed A=N=66 neutrons in a model space truncated at $N_{sq}^{\max}$=36, which is enough to converge the total energy per particle. For symmetric nuclear matter, we fill the same unperturbed orbits with Z=66 protons and N=66 neutron. Thus, we have a total of A=132 nucleons and truncate the model space at $N_{sq}^{\max}$=26. This requires up to 30 Gb of memory but it is still small enough to be computed on a high-end desktop. In both cases, the Dyson equation is solved for each value of the momentum $\mathbf{p}_i$ as discussed above. We retained N_{red}=300 Lanczos vectors in every channel, which is even more than necessary for converging the binding energies and spectral functions with respect to the Krylov projection.

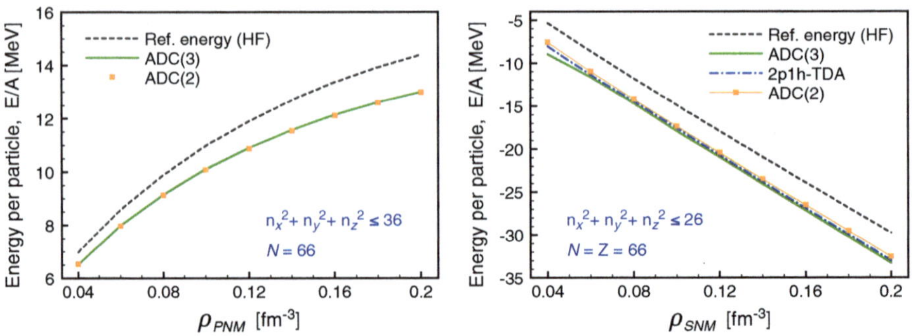

Fig. 11.9 Equation of state for PNM (*left*) and SNM (*right*) as predicted by the Minnesota two-nucleon interaction. Different curves show results for different ADC approximations. The ADC(2) (*filled squares*), 2p1h-TDA (*dot-dashed line*) and full ADC(3) (*full lines*) are calculated using a Hartree-Fock reference state and unperturbed single particle energies

Total energies per particle are shown in Fig. 11.9, for the reference state (which is HF) and for different approximations that show the convergence with respect to the many-body truncation: in order ADC(2), 2p1h-TDA and ADC(3). These plots already demonstrate one general feature of infinite nucleonic matter: PNM is relatively weakly correlated and may allow for solutions in MBPT, while SNM is more correlated and requires more sophisticated all-orders methods. The correlations energy with respect to the HF reference, $E_{\text{corr.}} = Eg.s. - E_{HF}$, varies between 0.5 and 2 MeV for neutrons but it is twice as much ($\approx$4 MeV) for symmetric matter and independent of the density (note the different scales in the two panels). Furthermore, the ADC(2) energies for PNM are already very close to the full ADC(3) results, showing that the calculation is extremely well converged. In SNM, the situation is different and truncations beyond the second order contribute to the calculated correlation energy. The difference between 2p1h-TDA and the ADC(3) is always about 300 keV/A and the trend shows convergence with respect to the many-body truncation.

The resulting spectral functions from ADC(3) are shown in Fig. 11.10 and compared to the unperturbed (HF) reference state. Since we are working in a discrete basis, the results are given for the cartesian momenta $\mathbf{p}_i$ and only discrete quasiparticle energies are obtained from Eq. (11.48) [also compare Eqs. (11.4) and (11.57)]. In order to give a clearer visualization of the spectral distribution, we fold each state along the energy axis with Lorentzians of width Γ=1.2 MeV near the Fermi energy and Γ=7 MeV otherwise. The corresponding expression of the spectral function in the HF approximation has no fragmentation and displays only isolated δ-peaks for each momenta:

$$S^{HF}(\mathbf{p}, s_z; \omega) = S^{h,HF}(\mathbf{p}, s_z; \omega) + S^{p,HF}(\mathbf{p}, s_z; \omega) = \delta\Big(\omega - \varepsilon^{HF}(\mathbf{p})\Big) , \qquad (11.60)$$

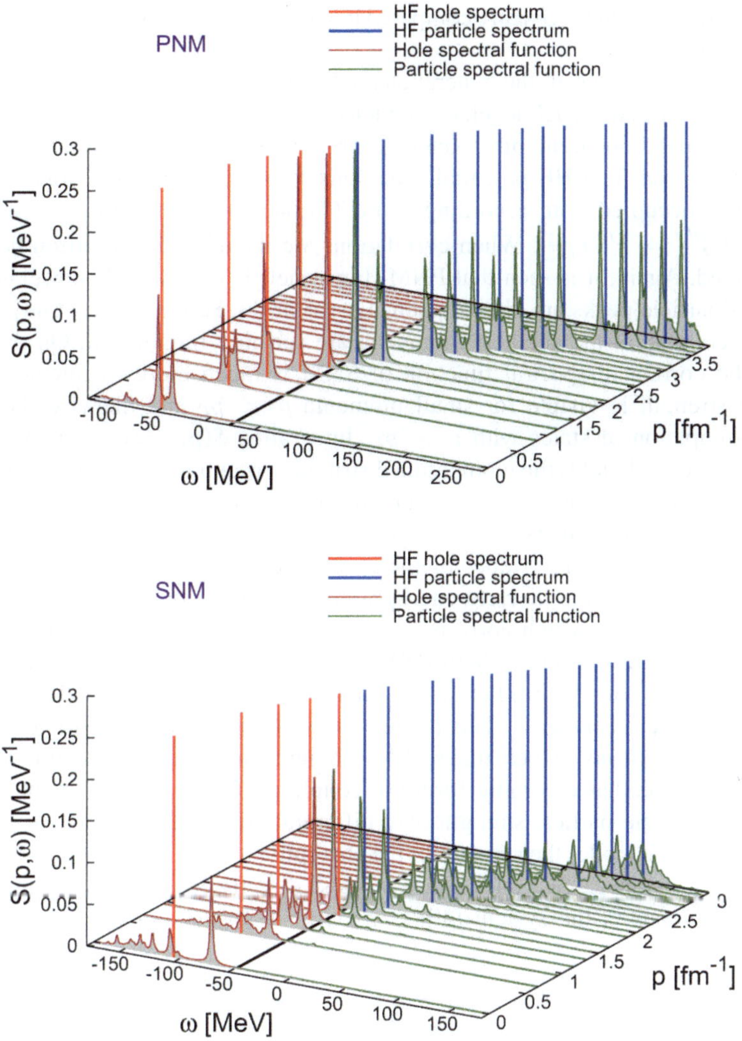

Fig. 11.10 Spectral function of PNM (*top*) and SNM (*bottom*) at nominal saturation density ($\rho = 0.16\,\text{fm}^{-3}$) from ADC(3). The correlated strength distribution is folded with Lorentzians along the energy axis. The isolated *vertical lines* mark the unperturbed HF spectrum and are normalized to the same height assumed for the Lorentzians, so that a visual comparison with the correlated distribution is meaningful. The *thick line* at constant ω marks the Fermi energy, E_F, for the correlated ADC(3) results, which separates the quasihole from the quasiparticle spectrum

where $\varepsilon^{HF}(\mathbf{p}) = \frac{p^2}{2m} + v_{HF}(\mathbf{p})$ are the HF single particle energies. Equation (11.60) is plotted as separate spikes in Fig. 11.10, with their height taken to be the same as for the (normalized) Lorentzians near the Fermi surface. Thus, the unperturbed spectral function can be visually compared to the fragmented distribution plotted for the ADC(3).

Figure 11.10 shows all the general characteristics of the spectral distribution for infinite systems. At the HF level, each nucleon has an energy spectrum $\varepsilon^{HF}(\mathbf{p})$ that follows the parabolic trend of its kinetic energy but it is otherwise shifted in energy due to the mean-field HF potential. The density ρ determines the momentum p_F of the last occupied state according to Eq. (11.54), which in turn sets the Fermi energy, $E_F^{(HF)} = \varepsilon^{HF}(p_F)$. When correlations are included the spectrum becomes fragmented. Again, it is seen that PNM (top panel) is only weakly correlated and the quasiparticle peaks are almost unchanged near the Fermi surface. Only deeply bound neutrons, at the smallest momenta, are sensibly fragmented. On the other hand, the correlated spectral function of SNM is much more fragmented, some particle strength is visible for small momenta $p < p_F$ and likewise there is a small occupation of states with $p > p_F$. Integrating $S(\mathbf{p}, s_z; \omega)$ over the energy interval $]-\infty, E_F]$ yields the momentum distribution (per unit volume), while further integrating over momenta gives the total nucleon density ρ [see Eq. (11.10)].

The real and imaginary parts of the self-energy, $\Sigma^{\star}(\mathbf{p}, s_z; \omega)$, are shown in Figs. 11.11 and 11.12 for values of the momentum $\mathbf{p}_i$ both below and above p_F. Also in these plots, the discrete energy poles are folded by taking a finite value of Γ in Eq. (11.58), which corresponds to using finite width Lorentzians for the imaginary part. In Fig. 11.11, both PNM and SNM have a similar dependence on momentum that comes form the kinetic energy term in $\Sigma^{(\infty)}(\mathbf{p})$ but there is more attraction in the second case. This is due to the additional attractive force between protons and neutrons, which makes SNM bound. Superimposed to this trend is the energy dependence coming form the coupling to ISCs, which fragments and spreads the single particle strength over different energies. The imaginary part of the self-energy encodes the strength of the absorption effects that mix single particle degrees of freedom to ISCs ones. Thus, it is also directly connected to the mean free path of nucleons in the system [49]. This term is always positive (negative) for

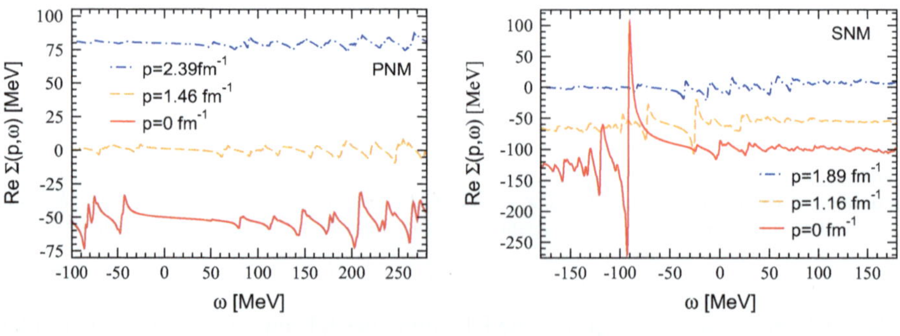

Fig. 11.11 Real part of the nuclear self-energy, Re $\Sigma(p, \omega)$, of PNM (*left*) and SNM (*right*) at nominal saturation density ($\rho = 0.16\,\text{fm}^{-3}$), obtained from ADC(3). The Fermi momentum is k_F=1.68 fm^{-1} for PNM and k_F=1.33 fm^{-1} for SNM. The plots are shown for fixed momenta at $p = 0\,\text{fm}^{-1}$, at $p \approx 0.87 p_F$ (just *below* p_F) and at $p \approx 1.42 p_F$ (*above* p_F)

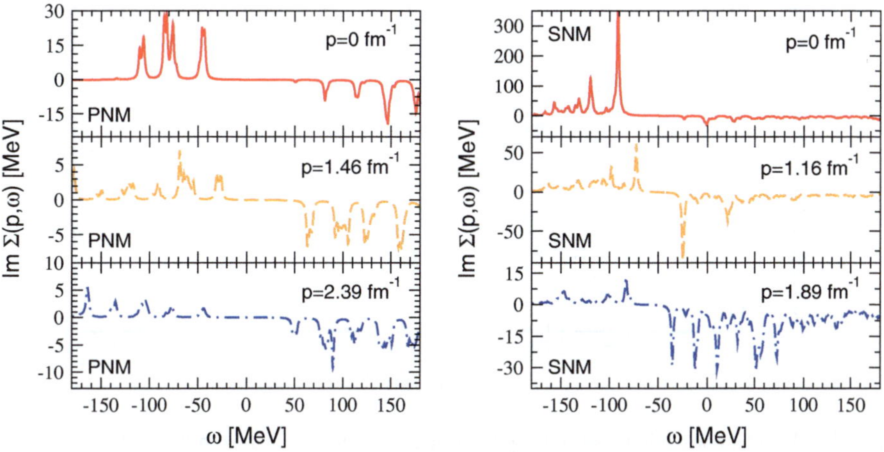

Fig. 11.12 Imaginary part of the nuclear self-energy, Im $\Sigma(p, \omega)$, of PNM (*left*) and SNM (*right*) at nominal saturation density ($\rho = 0.16\,\text{fm}^{-3}$), as calculated from ADC(3). The Fermi momentum is k_F=1.68 fm^{-1} for PNM and k_F=1.33 fm^{-1} for SNM. Fixed momenta of $p = 0\,\text{fm}^{-1}$, at $p \approx 0.87 p_F$ and $p \approx 1.42 p_F$ are shown

energies below (above) the fermi surface. For $\mathbf{p}_i \approx 0$ the absorption is strongest at low energies. As one increases p, this becomes weak in the energy region of hole states and much more stronger correlations are seen for quasiparticle energies and momenta outside the Fermi sea. Once again the PNM panel shows weak and more isolated peaks, while SNM is characterized by stronger fragmentation and absorption (hence, a more collective behavior).

Most of the qualitative features of these self-energies and of the spectral functions just shown are general for extended correlated fermion systems and are also seen, for example, in the electron gas or liquid ^{3}He. It is interesting to compare the plots of Fig. 11.10 to the analogous distribution of a finite system, like the one shown in Fig. 11.2. In the latter case, the spectral function displays orbits form the shell structure rather than peaks distributed according to kinetic energy. In all cases, correlations alter the simple mean-field view. However, the strength near the Fermi energy tends to remain dominated by single particle structures because of the low density of ISCs (2p1h, 2h1p and beyond) in that region.

Figure 11.13 compares the results for PNM with the coupled cluster and Monte Carlo methods introduced in previous chapters. Note that we show correlation energies, rather than the total energy per particle, to amplify differences among many-body methods. The largest discrepancy is at the lowest density and amounts to ≈50 keV/A. This is 10% of the correlation energy but less than 0.5% of the total energy. At larger densities, all methods agree to higher accuracy. It is interesting to see that ADC(3) initially follows configuration interaction Monte Carlo (CIMC) and then shifts to being closer to CCD as the density increases.

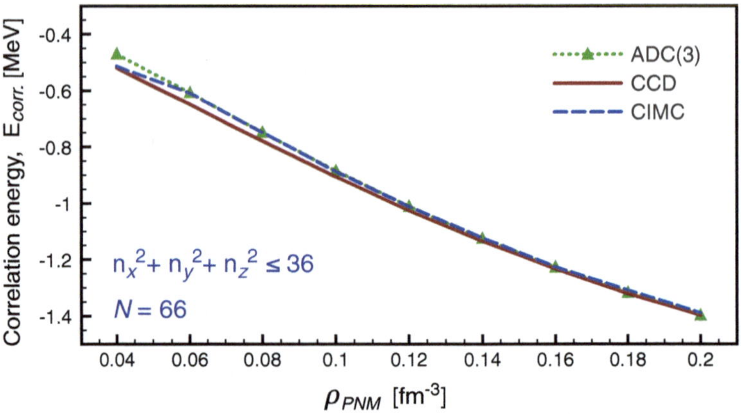

Fig. 11.13 Total correlation energy for pure neutron matter obtained from the CCD, the configuration interaction Monte Carlo (CIMC) and the ADC(3)-*sc0* methods that are presented in this book. See also Sect. 10.3.7 for results based on the IMSRG(2) approach

11.5 Self-Consistent Green's Functions at Finite Temperature in the Thermodynamic Limit

We now concentrate on the study of infinite systems at finite temperature and will set ourselves in the thermodynamic limit, that is number of particles N and volume V going to infinity with density $\rho = N/V$ kept constant. The many-body SCGF approach at finite temperature is particularly suited for this kind of study because, for appropriate approximations of the self-energy, it is thermodynamically consistent: a quantity calculated from the microscopic point of view yields the same result as the thermodynamical macroscopic quantity [30]. This consistency is strictly related to the fact that a fully dressed propagator, obtained via iterative solution of the Dyson equation, Eqs. (11.3), is used in the calculation of the partition function in the Luttinger-Ward formalism [50], from which one extracts the thermodynamical properties of the system. Furthermore, it can be demonstrated that this method fulfills the Hugenholtz van-Hove theorem [51], and this once again relates to the fact that the conservation laws of particle number, momentum and energy are preserved in this kind of approximation [29, 30].

We will show in this section how to calculate the self-consistent propagator in the *ladder approximation*, a specific approximation for the self-energy $\Sigma^{\star}(\omega)$ where particle-particle and hole-hole intermediate scattering states are resummed to all orders in the so called in-medium T-matrix. We will be working with the effective Hamiltonian of Eq. (11.16), considering the two-body averaged three-body force that enters $\widetilde{U}$ as given in Eq. (11.23), and disregarding all irreducible three-body terms. The Koltun sum rule of Eq. (11.11) is then used to obtain the total energy of the many-body system. The great advantage of working at finite temperature is that the appearance of pairing when considering hole-hole intermediate states is washed out by thermal effects [52]. Note that a different possibility is to account for pairing by implementing analogous calculations but in a formalism with both normal and anomalous propagators (as done in Gorkov theory) [53, 54]. Recently, an improved treatment of pairing in the SCGF method when going to zero temperature has been presented in [55]. Within the Luttinger-Ward formalism at finite temperatures, the entropy can then be calculated via the knowledge of the self-consistent propagator, and from the entropy all other thermodynamical quantities are accessible. We will not treat here the calculation of the entropy, for a detailed description we refer the reader to Chap. 3 of [14].

In the next section, we will give a few hints on the theoretical formalism and then sketch in the following section the working equations necessary to perform the numerical implementation. The full self-consistent numerical calculation considering the complete off-shell properties of the system and considering fully microscopic potentials was performed by the Gent [56], the Tübingen and Barcelona [4, 13, 14, 57–60] and the Cracow groups [61–64].

11.5.1 Finite-Temperature Green's Function Formalism

In a similar way to Sect. 11.2, we start by defining the one-body Green's function, however this time as a statistical average in the grand-canonical ensemble:

$$ig(\mathbf{x}t, \mathbf{x}'t') = \mathrm{Tr}\{\widehat{\rho}\mathcal{T}[\widehat{\psi}(\mathbf{x}t)\widehat{\psi}^{\dagger}(\mathbf{x}'t')]\}\,; \tag{11.61}$$

here $\mathcal{T}$ describes the Wick time-ordered product of the quantum field operators for the creation, $\widehat{\psi}^{\dagger}(\mathbf{x}'t')$, and destruction, $\widehat{\psi}(\mathbf{x}t)$, of a single-particle state in the Heisenberg picture. The field operators are related to the operators of creation and destruction, i.e. $a^{\dagger}_{\alpha}$ and a_{α}, via $\widehat{\psi}^{\dagger}(\mathbf{x}') = \sum_{\alpha} \psi^{\dagger}_{\alpha}(\mathbf{x})a^{\dagger}_{\alpha}$ and $\widehat{\psi}(\mathbf{x}) = \sum_{\alpha} \psi_{\alpha}(\mathbf{x})a_{\alpha}$, where the coefficients are the single-particle wave functions of state α and the sum is over the complete basis set of single-particle quantum numbers. The statistical factor $\widehat{\rho}$ is defined by:

$$\widehat{\rho} = \frac{1}{Z} e^{-\beta(\widehat{H}-\mu\widehat{N})}\,, \tag{11.62}$$

where β=1/T is the inverse temperature, μ is the chemical potential and Z is the grand-partition function

$$Z = \mathrm{Tr}\, e^{-\beta(\widehat{H}-\mu\widehat{N})}\,, \tag{11.63}$$

with $\widehat{H}$ the Hamiltonian given in Eq. (11.13), and $\widehat{N}$ the particle number operator. The trace in Eq. (11.63) is to be taken over a full set of energy and particle number eigenstates of the system. The two possible time-ordering products in Eq. (11.61) are given by:

$$\mathcal{T}[\widehat{\psi}(\mathbf{x}t)\widehat{\psi}^\dagger(\mathbf{x}'t')] = \begin{cases} \widehat{\psi}(\mathbf{x}t)\widehat{\psi}^\dagger(\mathbf{x}'t'), & t > t' \\ -\widehat{\psi}^\dagger(\mathbf{x}'t')\widehat{\psi}(\mathbf{x}t), & t' > t\,. \end{cases} \tag{11.64}$$

The first time-ordered product in Eq. (11.64) describes the creation of a particle state at time t' with position $\mathbf{x}'$, and the destruction of the propagated particle state at time t with position $\mathbf{x}$. Analogously, the second time-ordered product describes the destruction of a particle state, or creation of a hole state, at time t with position $\mathbf{x}$, and the destruction of the propagated hole state at time t' with position $\mathbf{x}'$. Using Eq. (11.64) one can define the correlation functions:

$$ig^>(\mathbf{x}t, \mathbf{x}'t') = \mathrm{Tr}\{\widehat{\rho}[\widehat{\psi}(\mathbf{x}t)\widehat{\psi}^\dagger(\mathbf{x}'t')]\} \tag{11.65}$$

$$ig^<(\mathbf{x}t, \mathbf{x}'t') = -\mathrm{Tr}\{\widehat{\rho}[\widehat{\psi}^\dagger(\mathbf{x}'t')\widehat{\psi}(\mathbf{x}t)]\}\,. \tag{11.66}$$

Depending on the specific time ordering, the Green's function defined in Eq. (11.61) corresponds to one correlation function or the other, i.e. either to Eq. (11.65) or to Eq. (11.66). It is also useful to define the retarded propagator; this is that part of the one-body Green's function which is related only to the causal propagation of events, i.e. forward in time:

$$g^R(\mathbf{x}t, \mathbf{x}'t') = \theta(t - t')[g^>(\mathbf{x}t, \mathbf{x}'t') - g^<(\mathbf{x}t, \mathbf{x}'t')]\,. \tag{11.67}$$

In the following we will be dealing with the imaginary time domain, also known as Matsubara formalism to solve for the Green's function. One could equivalently well work in the real-time domain and reach the same result [64]. The quantum field operators of creation and destruction in Heisenberg picture

$$\widehat{\psi}^{(\dagger)}(\mathbf{x}t) = e^{i\widehat{H}t}\widehat{\psi}^{(\dagger)}(\mathbf{x}0)e^{-i\widehat{H}t} \tag{11.68}$$

carry a resemblance between the thermal weight factor $e^{\beta\widehat{H}}$ and the time evolution operator $e^{i\widehat{H}t}$ when considering the imaginary time domain $t = -i\beta$. If one includes the expression (11.68) in the definition of the correlation functions, Eqs. (11.65) and (11.66), it can be proved that for a certain imaginary time domain there is absolute convergence of the two expressions, specifically in the intervals $-i\beta < t - t' < 0$ for $g^>$ and $0 < t - t' < i\beta$ for $g^<$. Furthermore, it can be shown that the two correlation functions are related to one another at one of their imaginary time boundaries, providing the important relation:

$$g^<(\mathbf{x}, t = 0; \mathbf{x}', t') = e^{\beta\mu}g^>(\mathbf{x}, t = -i\beta; \mathbf{x}', t')\,. \tag{11.69}$$

Thanks to the invariance under space translation of an infinite system and to time translational invariance, the Green's function only depends on the differences

$\mathbf{r} = \mathbf{x} - \mathbf{x}'$ and $\tau = t - t'$. Consequently, by exploiting the quasi-periodicity relation of the Green's function along the imaginary time axis given in Eq. (11.69), one can write a discrete Fourier representation for the one-body Green's function in the frequency domain:

$$g(\mathbf{r}, \tau) = \int \frac{d^3p}{(2\pi)^3} e^{i\mathbf{pr}} \frac{1}{-i\beta} \sum_{\nu} e^{-iz_\nu \tau} g(\mathbf{p}, z_\nu) \,, \tag{11.70}$$

where $z_\nu = \frac{\pi\nu}{-i\beta} + \mu$ are the Matsubara frequencies for odd integers $\nu = \pm 1, \pm 3, \pm 5, \ldots$ The Fourier coefficients are then given by the inverse transformation:

$$g(\mathbf{p}, z_\nu) = \int d^3r \int_0^{-i\beta} d\tau \, e^{-i\mathbf{pr} + iz_\nu \tau} g(\mathbf{r}, \tau) \,. \tag{11.71}$$

These coefficients are evaluated for an infinite set of complex frequencies z_ν, corresponding to the imaginary time domain, however one would like to understand the properties of the physical propagator, i.e. in the real time and frequencies domain. To do so let's go back to the expressions of the correlation functions, Eqs. (11.65) and (11.66), and write down their Fourier transform:

$$g^>(\mathbf{p}, \omega) = i \int d^3r \int_{-\infty}^{+\infty} d\tau \, e^{-i\mathbf{pr} + i\omega\tau} g^>(\mathbf{r}, \tau) \,, \tag{11.72}$$

$$g^<(\mathbf{p}, \omega) = -i \int d^3r \int_{-\infty}^{+\infty} d\tau \, e^{-i\mathbf{pr} + i\omega\tau} g^<(\mathbf{r}, \tau) \,. \tag{11.73}$$

These two quantities now define the spectral probability to attach or remove a particle with an energy ω and momentum $\mathbf{p}$ to or from the many-body system (we omit for simplicity the spin and isospin quantum numbers). The sum of these two functions is a positive quantity and yields the spectral function at finite temperatures:

$$A(\mathbf{p}, \omega) = g^>(\mathbf{p}, \omega) + g^<(\mathbf{p}, \omega) \,. \tag{11.74}$$

An important feature of the spectral function is that it fulfills the sum rule

$$\int_{-\infty}^{+\infty} \frac{d\omega}{2\pi} A(\mathbf{p}, \omega) = 1 \,, \tag{11.75}$$

which is consistent with the interpretation of $A(\mathbf{p}, \omega)$ as a probability of leaving the system in a state of energy ω by either adding or removing a nucleon of momentum $\mathbf{p}$. Below, we show how $A(\mathbf{p}, \omega)$ relates to its zero temperature counterpart, Eqs. (11.4).

Using Eq. (11.69) in Eqs. (11.72) and (11.73), we can write the Fourier transform of the periodicity condition

$$g^{>}(\mathbf{p},\omega) = e^{\beta(\omega-\mu)} g^{<}(\mathbf{p},\omega)\,, \tag{11.76}$$

and considering the definition of the spectral function, we can write the correlation functions in momentum and frequency as:

$$g^{<}(\mathbf{p},\omega) = f(\omega) A(\mathbf{p},\omega)\,, \tag{11.77}$$

$$g^{>}(\mathbf{p},\omega) = [1 - f(\omega)] A(\mathbf{p},\omega)\,, \tag{11.78}$$

where $f(\omega) = \frac{1}{e^{\beta(\omega-\mu)}+1}$ is the Fermi-Dirac distribution function. These expressions show that, once the spectral function is known, it is easy to access the correlation functions. A similar relation can be found between the spectral function and the Fourier coefficients of Eq. (11.71):

$$g(\mathbf{p}, z_\nu) = \int_{-\infty}^{+\infty} \frac{d\omega'}{2\pi} \frac{A(\mathbf{p},\omega')}{z_\nu - \omega'}\,. \tag{11.79}$$

The previous expression is performed for a given infinite set of Matsubara frequencies in the complex plane. However we would like to extend this to the entire complex plane, especially close to the real axis, which corresponds to physical frequencies. It can be demonstrated that this analytical continuation is possible and one can safely replace $z_\nu \rightarrow z$, where z is a continuous energy variable in the complex plane [29]. Equation (11.79) then relates the Green's function $g(\mathbf{p}, z)$ in the complex plane to the spectral function $A(\mathbf{p},\omega)$ and is referred to as the *spectral decomposition* of the single-particle propagator. Similarly, one could write the real-time Fourier transform for the retarded propagator defined in Eq. (11.67):

$$g^R(\mathbf{p},\omega) = \int_{-\infty}^{+\infty} \frac{d\omega'}{2\pi} \frac{A(\mathbf{p},\omega')}{\omega_+ - \omega'}\,, \tag{11.80}$$

with $\omega_+ = \omega + i\eta$. This quantity is equal to evaluating the Green's function slightly above the real axis, i.e. $g^R(\mathbf{p},\omega) = g(\mathbf{p},\omega_+)$. This equality is of fundamental importance. In fact, it tells us that, by knowledge of the spectral function, there exists a Green's function $g(\mathbf{p}, z)$ which corresponds both to the Green's function at the Matsubara frequencies, $z = z_\nu$, and also to the retarded propagator for frequencies slightly above the real axis, $z = \omega + i\eta$. This means that the information carried by the coefficients in Eq. (11.79) can be analytically continued to the real axis, and so to a physical propagator. Furthermore, exploiting the Plemelj identity,

$$\frac{1}{\omega \pm i\eta} = \frac{\mathcal{P}}{\omega} \mp i\pi\delta(\omega)\,, \tag{11.81}$$

one can separate the real and imaginary part of the propagator in Eq. (11.80), and it can be checked that the imaginary part of the retarded propagator is proportional to the spectral function:

$$A(\mathbf{p},\omega) = -2\,\mathrm{Im}\,g(\mathbf{p},\omega_+)\,. \tag{11.82}$$

Furthermore, the Dyson equation given in Eq. (11.3) can be rewritten in an algebraic form as follows:

$$g(\mathbf{p},\omega_+) = \frac{1}{[g^{(0)}(\mathbf{p},\omega_+)]^{-1} - \Sigma^\star(\mathbf{p},\omega_+)}\,, \tag{11.83}$$

and combining Eq. (11.83) with Eq. (11.82), one can express the spectral function as:

$$A(\mathbf{p},\omega) = \frac{-2\,\mathrm{Im}\,\Sigma^\star(\mathbf{p},\omega_+)}{[\omega - \frac{p^2}{2m} - \mathrm{Re}\Sigma^\star(\mathbf{p},\omega)]^2 + [\mathrm{Im}\,\Sigma^\star(\mathbf{p},\omega_+)]^2}\,. \tag{11.84}$$

The numerical calculation that one has to perform requires self-consistency between Eq. (11.84) and an appropriate approximation for $\Sigma^\star(\mathbf{p},\omega)$. Self-consistency is achieved once the spectral function inserted in the calculation of the irreducible self-energy is equal to the one obtained by solving Eq. (11.84).

Before going on, it is interesting to point out that in the limit of zero temperature, the spectral decomposition of the one-body propagator given in Eq. (11.79) can be separated into two pieces:

$$g(\mathbf{p},\omega) = \int_{\varepsilon_F}^{\infty} d\omega' \frac{S^p(\mathbf{p},\omega')}{\omega - \omega' + i\eta} + \int_{-\infty}^{\varepsilon_F} d\omega' \frac{S^h(\mathbf{p},\omega')}{\omega - \omega' - i\eta}\,, \tag{11.85}$$

The $S^p(\mathbf{p},\omega)$ and $S^h(\mathbf{p},\omega)$ correspond to the particle and hole spectral functions, which were already introduced in Eqs. (11.4). Notice however that, unlike in Eqs. (11.4), we have one single Fermi energy ε_F ($\varepsilon_F = \varepsilon_0^+ = \varepsilon_0^-$) in the integrals domain because the gap disappears in an infinite gas or a normal Fermi liquid. In an uncorrelated system, this energy defines the last filled level and hence corresponds to the energy needed to remove a particle from the many-body ground state. In the case of an interacting system, not in the superfluid nor in the superconducting phase, ε_F equals the chemical potential μ, and corresponds to the minimum energy necessary to add or remove a particle to/from the many-body system. Consequently, the expression for the spectral function given in Eq. (11.84) can be divided into two parts:

$$S^p(\mathbf{p},\omega) = -\frac{1}{\pi}\frac{\mathrm{Im}\,\Sigma^\star(\mathbf{p},\omega)}{(\omega - \frac{p^2}{2m} - \mathrm{Re}\,\Sigma^\star(\mathbf{p},\omega))^2 + (\mathrm{Im}\,\Sigma^\star(\mathbf{p},\omega))^2} \qquad \omega > \varepsilon_F\,, \tag{11.86a}$$

$$S^h(\mathbf{p},\omega) = \frac{1}{\pi}\frac{\mathrm{Im}\,\Sigma^\star(\mathbf{p},\omega)}{(\omega - \frac{p^2}{2m} - \mathrm{Re}\,\Sigma^\star(\mathbf{p},\omega))^2 + (\mathrm{Im}\,\Sigma^\star(\mathbf{p},\omega))^2} \qquad \omega < \varepsilon_F\,, \tag{11.86b}$$

resembling the structure of Eqs. (11.4).

With all the basic formalism in place, we still need to devise a proper conserving approximation to the self-energy $\Sigma^{\star}(\mathbf{p}, \omega)$. For applications to infinite nucleonic matter this is done by summing infinite ladders of two-particle and two-hole configurations inside the medium. Hence, the first two diagrams of this expansion are those of Figs. 11.4a and 11.5a. This approximation is analogous to the series generated in Eq. (11.35) and Exercise 11.3 except that it is resummed in the RPA way.[2] In the next subsection we will sketch the main steps that have to be taken to perform the numerical implementation of SCGF calculations at finite temperature and introduce the working equations of the ladder expansion of the self-energy along the way.

11.5.2 Numerical Implementation of the Ladder Approximation

Figure 11.14 shows the iterative scheme that needs to be implemented numerically for self-consistent calculations. This is for the case of both two-body and three-body forces, when working with the Hamiltonian $\widetilde{H}_1$ of Eq. (11.16) and disregarding irreducible three-body terms. The fundamental quantities that one has to compute are the non-interacting two-body Green's function, the in-medium T-matrix and the irreducible self-energy, which are depicted in the three light blue boxes with their respective Feynman diagrams. The diagrams are a direct way to write down the complicated mathematical expressions that one has to solve numerically (see Appendix 1 for the T=0 case). The one-body and two-body effective nuclear potentials are depicted in the central orange boxes. These are similar to the contributions given in Fig. 11.3, except for the one-body effective potential in which the first term is zero for infinite matter and the last term where we approximate the contribution of the three-nucleon force by averaging only with two independent correlated density matrices. Figure 11.14 shows the correct multiplying factor, as in Eq. (11.23) [6, 65]. As can be seen from this scheme, all quantities in blue or orange boxes are fed with the spectral function, the left red box, which is then computed iteratively by solving the Dyson equation, in the form of Eq. (11.84), until convergence. The criteria for reaching self-consistency is usually to compare the chemical potentials of two consecutive iterations, which is computed using $A(\mathbf{p}, \omega)$.

[2] Note that particle-hole summations, corresponding to the *ring* diagram of Fig. 11.5b, represent a formidable task in nucleonic matter and have been almost always disregarded in SCGF studies of infinite matter. This is in contrast to Green's function studies of the electron gas and solid state materials, where rings are necessary to screen the Coulomb interaction while ladders can often be neglected.

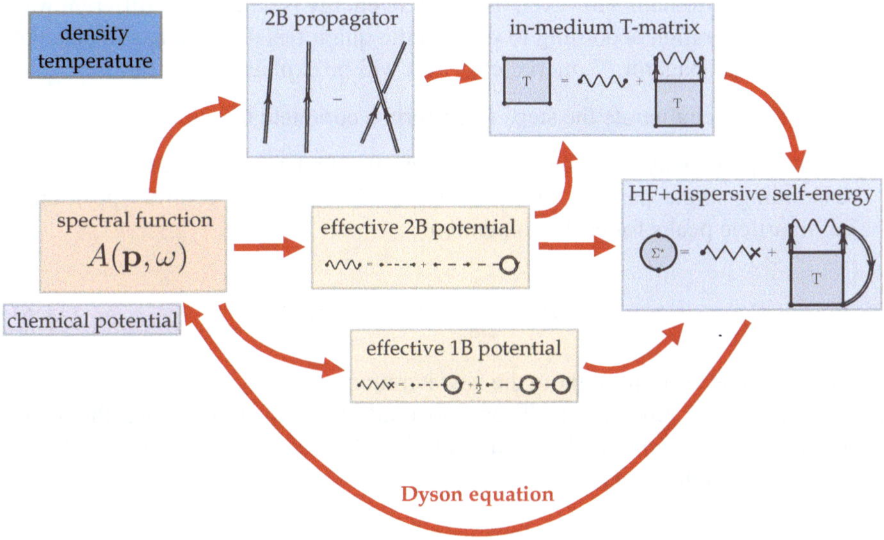

Fig. 11.14 The structure of a ladder SCGF calculation including both two-body and three-body forces through the definition of effective interactions (see text for details). Each quantity is also represented via the corresponding Feynman diagram

For clarity, we will distinguish between the wording *calculation* and *iteration*: we will refer to calculation as the whole set of several iterations necessary to get to a converged result for the spectral function, in this sense Fig. 11.14 depicts exactly one iteration. For a more in depth explanation of the numerical details the reader can refer to [13, 14].

Each calculation is performed at a specific density ρ and temperature T of the system. One starts the first iteration with a guess of the spectral function, which is given in terms of the imaginary, Im $\Sigma^\star(\mathbf{p},\omega)$, and real, Re $\Sigma^\star(\mathbf{p},\omega)$, parts of the irreducible self-energy. When possible, it is convenient to start with a converged solution for these quantities at different values of ρ and T.

- **Numerical tips for the $(\mathbf{p},\omega)$ meshes:** The mesh of the single-particle momentum $\mathbf{p}$ for the self-energy is adjusted during the first iteration to be more dense around the Fermi momentum p_F corresponding to the specific density considered: $N_\mathbf{p} = 70$ mesh points are enough, considering linear meshes at low momentum and around the Fermi momentum, and a logarithmic mesh for the tail all the way up to a value $\sim 10p_\mathrm{F}$. The mesh in the single-particle energy ω has to be very dense because of the complicated features of the spectral function, especially near the quasiparticle peak. Storing a dense mesh at each iteration is demanding in terms of memory; for this reason one saves separately the imaginary and real part in a dilute linear mesh, typically of $N_\omega \approx 6000$ points in the interval [−2000:15000] MeV. This is interpolated during the iterations to denser meshes of $N_\omega \approx 30000$ points, in order to have a good description of the spectral function in the energy domain. However, the mesh in energy is adjusted in different ways during the iteration according to the specific quantities that one has to calculate (two-body propagator, T-matrix, etc.) as it will be explained later on.

We will now enumerate the steps to perform a complete iteration.

1. Given a previously computed self-energy, the first step is to extract the corresponding single particle spectrum, which describes the centroid position of quasiparticle peaks for each momentum:

$$\varepsilon(\mathbf{p}) = \frac{p^2}{2m} + \mathrm{Re}\,\Sigma(\mathbf{p}, \varepsilon(\mathbf{p}))\,, \tag{11.87}$$

which will be used throughout the new iteration.

2. The density ρ, temperature T and spectral function $A(\mathbf{p}, \omega)$ are the inputs to calculate the next fundamental quantity: the chemical potential μ. This is obtained from the sum rule for the density:

$$\rho = \nu_d \int \frac{d\mathbf{p}}{2\pi^3} \int_{-\infty}^{+\infty} \frac{d\omega}{2\pi} A(\,\mathbf{p}, \omega) f(\omega, \mu)\,, \tag{11.88}$$

where ν_d is the degeneracy of the system (ν_d=2 for pure neutron and ν_d=4 for isospin symmetric matter), the temperature enters through the Fermi-Dirac function, $f(\omega, \mu)$, and we have also made explicit its dependence on μ.

- **Numerical tips for the μ mesh:** One chooses a sample mesh of chemical potentials μ to insert in $f(\omega, \mu)$ and then solves Eq. (11.88). For each point μ one gets a value of density ρ. Parametrizing ρ as a function of μ, one can then find the value of μ which corresponds to the correct density of the system. The mesh of μ can be initially distributed around the value of the single-particle spectrum calculated at p_F (in the case of a zero temperature calculation the relation $\varepsilon(p_F) = \mu$ holds), and then adjust the mesh testing if the limits include the value of the external density.

It must be noted that both the single-particle spectrum $\varepsilon(\mathbf{p})$ and the spectral function $A(\mathbf{p}, \omega)$ that enter Eq. (11.88) come from a previous iteration that was based on a different value of μ. In this sense, the old value of the chemical potential is implicitly carried over throughout the new iteration. However, these will end up coinciding when self-consistency is reached at the end of the calculation.

3. At this point the imaginary part of the non-interacting two-body Green's function can be computed. The lowest order approximation of the two-body propagator corresponds to the independent propagation of two fully dressed particles and was discussed in Example 11.1 for the case of zero temperature. This includes two terms, a direct and an exchange one, as depicted diagrammatically in Fig. 11.14. Since we are working with dressed propagators we should refer to Eqs. (11.21) and (11.22), however both the direct and exchange terms must be

included in our G^{IIf}_{pphh}. The imaginary part of this quantity extended to finite temperatures reads:

$$\operatorname{Im} G^{IIf}_{pphh}(\Omega_+;\mathbf{p},\mathbf{p}') = -\frac{1}{2}\int_{-\infty}^{+\infty}\frac{d\omega}{2\pi}A(\mathbf{p},\omega)A(\mathbf{p}',\Omega-\omega)[1-f(\omega)-f(\Omega-\omega)]\,. \tag{11.89}$$

where Ω_+ is the sum of the energies of the two particles close to the real axis. This expression is derived from a sum over Matsubara frequencies of a function with a double pole on the real-energy axis via use of the Cauchy theorem [14].

- **Numerical tips for the ω mesh:** The integrand of Eq. (11.89) will be particularly hard to resolve in regions where the two spectral functions are peaked, at energies where $\omega \sim \varepsilon(\mathbf{p})$ and $\omega \sim \Omega - \varepsilon(\mathbf{p}')$. It can be shown that a convenient variable change makes these energies independent of the momenta $\mathbf{p}$ and $\mathbf{p}'$, so that one is safe with defining an energy mesh accurately distributed around only two specific points (see [14] for details). To obtain the spectral function in this specific mesh one interpolates the imaginary and real self-energies to this mesh and then uses Eq. (11.84).

4. From the imaginary part it is then possible to obtain the real-part of the non-interacting two-body Green's function via a dispersion relation:

$$\operatorname{Re} G^{IIf}_{pphh}(\Omega;\mathbf{p},\mathbf{p}') = -\mathcal{P}\int_{-\infty}^{+\infty}\frac{d\Omega'}{\pi}\frac{\operatorname{Im} G^{IIf}_{pphh}(\Omega'_+;\mathbf{p},\mathbf{p}')}{\Omega-\Omega'}\,. \tag{11.90}$$

5. In practice, G^{IIf}_{pphh} has to be averaged over angles. This is necessary to circumvent the coupling of partial waves with different total angular momentum J which appear in G^{IIf}_{pphh}. The average is performed over the angle formed by the center of mass momentum $\mathbf{P} = \mathbf{p} + \mathbf{p}'$ and the relative momentum of the two nucleons $\mathbf{k} = (\mathbf{p} - \mathbf{p}')/2$. This strategy will facilitate solving the in-medium T-matrix equations to evaluate the effective interaction in the medium. The average reads:

$$\overline{G^{IIf}_{pp,hh}}(\Omega_+;P,k) = \frac{1}{2}\int_{-1}^{+1} d\cos\theta\; G^{IIf}_{pp,hh}(\Omega_+;|\mathbf{P}/2+\mathbf{k}|,|\mathbf{P}/2-\mathbf{k}|)\,. \tag{11.91}$$

6. The two-body propagator together with the nuclear potential are then used to obtain the in-medium T-matrix. The exact equation for this is of Lippmann-Schwinger type:

$$\langle\mathbf{k}'|T(\Omega_+,\mathbf{P})|\mathbf{k}\rangle = \langle\mathbf{k}'|\widetilde{V}|\,\mathbf{k}\rangle + \int d\mathbf{k}_1\langle\mathbf{k}'|\widetilde{V}|\mathbf{k}_1\rangle G^{IIf}_{pp,hh}(\Omega_+;\mathbf{P},\mathbf{k}_1)\langle\mathbf{k}_1|T(\Omega_+;\mathbf{P})|\mathbf{k}\rangle\,. \tag{11.92}$$

As explained previously, this is a ladder resummation of particle-particle and hole-hole diagrams, this differs with respect to the Brueckner G-matrix presented in Chap. 8 because it includes hole-hole diagrams and considers the full off-shell description of the spectral function (that is, the dressed propagator). As seen from Fig. 11.14, the potential to be included is the sum of a bare two-body potential and an averaged three-body one. Details on the numerical solution for the averaged three-body force are given in the next section, while working equations for three-nucleon chiral forces are reported in Appendix 2.

Here, we make an approximation and substitute the two-body propagator with its angle-averaged version (11.91). Since the latter depends only on the magnitudes of momenta P and k, our Eq. (11.92) reduces to a one dimensional integral and decouples in total angular momentum, spin and isospin:

$$
\begin{aligned}
&\langle k'|T^{JST}(\Omega_+, P)|k\rangle \\
&\quad = \langle k'|\widetilde{V}^{JST}|k\rangle + \int_0^\infty \mathrm{d}k_1\, {k_1}^2\, \langle k'|\widetilde{V}^{JST}|k_1\rangle\, \overline{G^{II,f}_{pp,hh}}(\Omega_+; P, k_1) \langle k_1|T^{JST}(\Omega_+; P)|k\rangle\, .
\end{aligned}
\tag{11.93}
$$

Going beyond Eq. (11.91) with fully dressed (off shell) propagators is extremely difficult and, to our knowledge, there is no available implementation of SCGF that can treat Eq. (11.92) exactly. However, estimates in Brueckner-type calculations suggests that the error introduced by the angle averaging is small [66, 67]. Equation (11.93) is a one dimensional integral equation for each allowed combination of J, S, T and has at most two coupled values of L, due to the tensor component of the nuclear interaction. It must be noted that the nuclear interaction $\widetilde{V}$ considered in Eq. (11.92) is the effective two-body operator given in Eq. (11.17b). By means of a discretization procedure, the equation for the T-matrix is converted into a complex matrix equation which can be solved via standard numerical techniques [14]. A matrix inversion has to be performed to solve this equation. This can be quite demanding if the dimension of the matrix is large.

- **Numerical tips for the $\mathbf{k_1}$ and Ω mesh:** It is important to sample in a correct manner the number of integration mesh points without loosing physical information. This is achieved by sampling conveniently the region where $G^{II,f}_{pp,hh}$ is maximum in the relative momentum $\mathbf{k_1}$ (for $\Omega > 0$ this is close to the pole $k_1 = \sqrt{m\Omega}$) and the high relative momentum region, where $G^{II,f}_{pp,hh}$ might not be negligible due to correlations. Furthermore the T-matrix has a node for $\Omega = 2\mu$, so an accurate mesh for the bosonic energies around this value is needed for the computation of the self-energy in the next steps.

At low temperatures, the appearance of bound states signals the onset of the pairing instability. This would directly appear as a pole in the matrix which has to be inverted to solve the Lippmann-Schwinger equation, for $\mathbf{P} = 0$ and $\Omega = 2\mu$ [68]. However, this should be seen only below a critical temperature which is around $T_c \sim 4$ MeV. For this reason, calculations should not go below this border line in temperature. Especially in the case of symmetric nuclear matter, convergence starts to become slow and difficult to control when approaching this temperature and for increasing densities. This is due to the neutron-proton pairing in the coupled $^3S_1 -^3 D_1$ channel. In pure neutron matter, where this channel is not available, convergence is good for higher densities, and even for lower temperatures.

7. The remaining step in the SCGF method is the computation of the self-energy from the T-matrix. The first quantity to be obtained is the imaginary part of the self-energy:

$$\mathrm{Im}\,\Sigma^\star(\mathbf{p},\omega_+) = \int \frac{\mathrm{d}\mathbf{p}'}{(2\pi)^3} \int_{-\infty}^{+\infty} \frac{\mathrm{d}\omega'}{2\pi} \langle \mathbf{p}\mathbf{p}'|\mathrm{Im}\,T(\omega_+ + \omega', \mathbf{P})|\mathbf{p}\mathbf{p}'\rangle A(\mathbf{p}',\omega')[f(\omega')+b(\omega+\omega')], \tag{11.94}$$

where $b(\Omega) = \frac{1}{e^{\beta(\Omega-2\mu)}-1}$ is the Bose function. We recall that the expression (11.94) is also obtained from a summation over Matsubara frequencies of a function with two poles on the real energy axis [14].

- **Numerical tips for the p′ and ω' meshes:** A momenta and energy integrals have to be performed, taking special care for the pole in energy of the Bose function $b(\Omega)$. This pole is canceled by the node we had previously mentioned in the T-matrix, for this reason it comes in hand that we had already defined a convenient mesh for Ω around the node.

8. The real part of the self-energy is then obtained from its imaginary part by means of the dispersion relation:

$$\mathrm{Re}\,\Sigma^\star(\mathbf{p},\omega) = \Sigma^{(\infty)}(\mathbf{p}) - \mathcal{P}\int_{-\infty}^{+\infty} \frac{\mathrm{d}\omega'}{\pi} \frac{\mathrm{Im}\,\Sigma^\star(\mathbf{p},\omega'_+)}{\omega-\omega'}. \tag{11.95}$$

The $\Sigma^{(\infty)}$ is the correlated Hartree-Fock part of the single-particle self-energy that is defined by Eqs. (11.17a) and (11.19). We now approximate this according to Eq. (11.23), where the three-body interaction is averaged over two non interacting particles. This can be explicitly written as:

$$\Sigma^{(\infty)}(\mathbf{p}) = \int \frac{\mathrm{d}\mathbf{p}'}{(2\pi)^3} n(\mathbf{p}') \Big[\langle \mathbf{p}\mathbf{p}'|V^{2\mathrm{NF}}|\mathbf{p}\mathbf{p}'\rangle + \frac{1}{2}\langle \mathbf{p}\mathbf{p}'|\widetilde{V}^{3\mathrm{NF}}|\mathbf{p}\mathbf{p}'\rangle \Big], \tag{11.96}$$

where $V^{2\mathrm{NF}}$ and $\widetilde{V}^{3\mathrm{NF}}$ correspond respectively to the first and second terms in Eq. (11.17b). $\widetilde{V}^{3\mathrm{NF}}$ is a one-body averaged three-nucleon force, detailed description on how to calculate this quantity and the momentum distribution $n(\mathbf{p})$, together with an additional numerical sample code, are given in Sect. 11.5.3.

Finally, the spectral function can be obtained via Eq. (11.84) and the procedure starts again from step **1.** until a consistent result is achieved for the chemical potential. It must be kept in mind that, according to the mesh points in which the spectral function is needed, the interpolation is done on the imaginary and real part of the self-energy, and not directly on the spectral function. This is done in order to avoid incorrect samplings of the structure of the spectral function which could induce numerical inaccuracies. We must point out that the energy mesh for the evaluation of the spectral function must be accurate enough to reproduce not only the quasiparticle peak region but also the low and high-energy tails that characterize the spectral function (especially for large momenta in the case of hard interactions).

To calculate the total energy of the system, we make use of the modified Koltun sum rule given in Eq. (11.11). Consequently we need to evaluate the expectation value of the three-body operator $\langle \widehat{W} \rangle$. As already stated in Sect. 11.2.1, we approximate this expectation value to its first-order term, which in infinite matter corresponds to the integral over three independent but fully correlated momentum distributions $n(\mathbf{p})$. The integral to be evaluated is given by the expression:

$$\langle \widehat{W} \rangle \simeq \frac{\nu_d}{\rho} \frac{1}{6} \int \frac{d\,\mathbf{p}}{(2\pi)^3} \int \frac{d\mathbf{p}'}{(2\pi)^3} n(\mathbf{p}) n(\mathbf{p}') \langle \mathbf{p}\mathbf{p}' | \widetilde{V}^{3NF} | \mathbf{p}\mathbf{p}' \rangle \,, \tag{11.97}$$

with ν_d the degeneracy of the system and the averaged three-body force, $\widetilde{V}^{3NF}$, is discussed in the next section. Once $\widetilde{V}^{3NF}$ is known, the total energy per nucleon of the system can be calculated via the modified Koltun sum rule:

$$\frac{E}{A} = \frac{\nu_d}{\rho} \int \frac{d\mathbf{p}}{(2\pi)^3} \int \frac{d\omega}{2\pi} \frac{1}{2} \left\{ \frac{p^2}{2m} + \omega \right\} A(\mathbf{p}, \omega) f(\omega) - \frac{1}{2} \langle \widehat{W} \rangle \,, \tag{11.98}$$

which is equivalent to Eq. (11.11).

11.5.3 Averaged Three-Body Forces: Numerical Details

The inclusion of one-body averaged three-nucleon forces $\widetilde{V}^{3NF}$ enters the calculations presented in the previous section through Eqs. (11.92), (11.96) and (11.97). Its computation requires traces over the spin and isospin quantum numbers of the averaged particle, in this case the third particle, and an integration over its momentum $\mathbf{p}_3$:

$$\langle \mathbf{p}_1' \mathbf{p}_2' | \widetilde{V}^{3NF} | \mathbf{p}_1 \mathbf{p}_2 \rangle_A = \mathrm{Tr}_{\sigma_3} \mathrm{Tr}_{\tau_3} \int \frac{d\mathbf{p}_3}{(2\pi)^3} n(\mathbf{p}_3) \langle \mathbf{p}_1' \mathbf{p}_2' \mathbf{p}_3 | W^{3NF} (1 - P_{13} - P_{23}) | \mathbf{p}_1 \mathbf{p}_2 \mathbf{p}_3 \rangle_{A_{12}} \,, \tag{11.99}$$

where $\mathbf{p}_i$ are single-particle momenta of particles $i = 1, 2, 3$ and W^{3NF} is the third term on the right hand side of Eq. (11.13); we have omitted the spin/isospin indices in the potential matrix elements for simplicity. The ket on the right hand side of Eq. (11.99) is antisymmetrized only with respect to particles 1 and 2, i.e. $A_{12} = (1 - P_{12})/2$; this part is not affected by the averaging procedure over the third particle. $P_{ij} = (1 + \boldsymbol{\sigma}_i \cdot \boldsymbol{\sigma}_j)(1 + \boldsymbol{\tau}_i \cdot \boldsymbol{\tau}_j)/4$ is the permutation operator of momentum and spin/isospin quantum numbers of particles i and j. The momentum distribution that appears in Eq. (11.99) can be obtained directly from the spectral function, via the relation:

$$n(\mathbf{p}) = \int_{-\infty}^{+\infty} \frac{d\omega}{2\pi} A(\mathbf{p}, \omega) f(\omega) \tag{11.100}$$

Let us give some details on the numerical implementation of Eq. (11.99) with regards to the mesh for the internal momentum $\mathbf{p}_3$ and the calculation of the distribution $n(\mathbf{p}_3)$ via Eq. (11.100):

- We start with the definition of the mesh necessary to calculate the integral over the internal momenta $\mathbf{p}_3$. Considering that in the integral we deal with a dressed distribution function $n(\mathbf{p}_3)$, which may have populated states at high momentum, we need to cover momenta up to a certain high value in which it is sure that the $n(\mathbf{p}_3)$ has reached zero. One may choose to compose this of an arbitrary number `imesh-1` of Gauss-Legendre meshes (in the example shown below, `imesh=4`), with each mesh spanning a region of width $2/3p_F$. This width is chosen to cover accurately the behavior of the distribution function below, across and above the Fermi momentum p_F. Finally, high-momentum points are included through an additional tangential mesh. We have 100 points in the Gauss-Legendre meshes, and 50 in the tangential one.
- One then needs to calculate the momentum distribution function by means of Eq. (11.100). To do so, we extract the spectral function on the fly, from the self-energy of the previous iterative step. The values of the imaginary and real part of the self-energy are stored at each iteration for different points in the momentum and energy space: for the momentum we typically have $N_{\mathbf{p}} = 70$ mesh points with values ranging from 0 to 3000 MeV/c; for the energy, it is sufficient to cover a smaller range of values than the one actually stored, but a much finer mesh is useful to simplify the integrations. We perform a spline interpolation of the stored energy values of the imaginary and real parts of the self-energy to a fine linear energy mesh of $N_{\omega,\text{spline}} = 30000$ in the interval of $\approx[-2000{:}5000]$ MeV. These values are used to calculate the spectral function [see Eq. (11.84)] necessary to evaluate Eq. (11.100) correctly: with the fine energy mesh this integration is easily performed across the quasiparticle peak via the trapezoidal rule. Finally, one can linearly interpolate the values obtained for $n(\mathbf{p})$ to the mesh of $\mathbf{p}_3$ in order to perform the integration of the averaged force, Eq. (11.99). In doing this, the values of $n(\mathbf{p}_3)$ outside the range of the original $N_{\mathbf{p}} = 70$ mesh are set to zero.

Here we show a simple Fortran code to perform the previous two steps (`gauss()` is a standard routine to generate a Gauss-Legendre mesh; `splin()` and `splin2()` are used to perform spline interpolations; `linint()` performs linear interpolations):

```
! ... MOMENTA MESH FOR INTEGRALS OVER MOMENTUM DISTRIBUTION

write(*,*) "Correlated distribution function for averaged 3BF
    integration"

! choose number of mesh regions for momenta p3, (imesh-1)
    gauss set + 1 tangent set for farther points
imesh = 4

! choose number of points for gauss and tangent sets
Np1=100 ! gauss
Np2=50   !tangent
Np3=(imesh-1)*Np1+Np2 ! total number of mesh points

itmp = MAX(Np1,Np2) ! for the auxiliary arrays always allocate
     the largest between Np1 and Np2
ALLOCATE(xp3(Np3),wp3(Np3))
ALLOCATE(xaux( itmp ),waux( itmp ))

! initialize variables
xp3=0d0
wp3=0d0

! first mesh point
pin = 0d0

do im = 1, imesh-1 ! loop over linear regions

   ! reset auxiliary variables at each region
   xaux=0d0
   waux=0d0

   pfin = im*(2d0/3d0)*pF ! set final point of mesh region
       according to Fermi momentum pF

   ! ... gaussian set of points for momenta p3 from pin to
       pfin
   call gauss(pin,pfin,Np1,xaux,waux)

   ! copy points to final vector for mesh p3
   do ip3=1,Np1
      xp3(ip3+(im-1)*Np1)= xaux(ip3)
      wp3(ip3+(im-1)*Np1)= waux(ip3)
   enddo

   pin = pfin ! set last point of previous region to first
       point of next region
enddo

! ... create the tangent set for higher momenta
call gauss(0d0,1d0,Np2,xaux,waux) ! gauss set [0,1] to be
    mapped to the interval [pin,+infinity]
```

```
c=10d0*pF/tan(pi/2.d0*xaux(Np2))
do ip3=1,Np2
   xp3(ip3+(im-1)*Np1)=c*tan(pi/2.d0*xaux(ip3))+pin
   xxw=cos(pi/2.d0*xaux(ip3))
   xxw=xxw*xxw
   wp3(ip3+(im-1)*Np1)=pi/2.d0*c/xxw*waux(ip3)
enddo


! ... obtaining correlated momentum distribution

! ... FINE ENERGY MESH WHERE CALCULATIONS ARE DONE
! ... allocate energy mesh for calculation of momentum
    distribution
N_fine=30000
ALLOCATE(xmom(N_fine))
ALLOCATE(xmp(Np))

wi=-2000.d0 !MeV initial energy for spectral function
wf=5000.d0 !MeV final energy for spectral function
dw=(wf-wi)/dble(N_fine-1)

! ... LOOP OVER PMESH
do ip=1,Np ! this is the mesh of stored momenta (usually Np ~
    70)

   edp=xpmesh(ip)**2/(2.d0*xmass) ! kinetic spectrum

   do iw=1,Nwac
      auxre(iw)=xreal_sigma(ip,iw) ! real part of self-energy
      auximi(iw)=ximag_sigma(ip,iw) !imaginary part of self-
          energy
   enddo

   ! obtain derivatives of the self-energy for later splines
   call spline(w_actual,auxim,Nwac,yspl,yspl,d2im)
   call spline(w_actual,auxre,Nwac,yspl,yspl,d2re)

   ! ... LOOP OVER WFINE
   do iif=1,N_fine
      w_fine = wi + dble(iif-1)*dw
      wfine(iif)=w_fine
      fdfine=fermi(t,xmu,w_fine) !Fermi-Dirac distribution

      ! .. Spline interpolation in fine energy mesh
      call splin2(w_actual,auxim,d2im,Nwac,w_fine,ximsig)
      call splin2(w_actual,auxre,d2re,Nwac,w_fine,xresig)

      ! ... Spectral function
      sf=-ximsig/( (w_fine - edp - xresig)**2 + ximsig**2 )/pi

      ! ... momentum distribution
      xmom(iif)=sf*fdfine
```

```
    enddo ! END LOOP OVER WFINE

    ! performs the integration over energy (trapezoidal rule)
    call trapz(w_fine,xmom,N_fine,mom)
    xmp(ip)=mom

enddo ! END LOOP OVER MOMENTA

! ... interpolation of momentum distribution to mesh xp3 for
     integrals
call linint(xpmesh,xmp,Np,xp3,xnp3,Np3)

! ... set the extrapolated values of n(p) to zero, mesh points
      xp3 beyond initial mesh xpmesh
do ip3=1,Np3
   xnp0=xp3(ip3)
   if(xnp0.gt.xpmesh(Np)) xnp3(ip3)=0d0
   if(xnp0.lt.0d0) xnp3(ip3)=0d0
enddo

DEALLOCATE(xaux,waux,xmom,xmp)
```

Note that the chemical potential μ enters the calculation of the averaged three-body force, via the Fermi-Dirac function in the expression for momentum distribution, Eq. (11.100). For this reason it is best to compute Eq. (11.99) after step **2** of the iterative procedure presented in the previous section. For further details on including three-body forces in a SCGF infinite matter calculation we refer the reader to [69].

11.6 Concluding Remarks

This chapter concludes an overview of the major methods based on Fock space, which are covered in Chaps. 8, 10 and 11 of this book. All these approaches have the common feature that their computing requirements scale only polynomially with the increase of particle number. This feature has permitted to push ab initio studies of atomic nuclei up to medium-mass isotopes: a progress that would have seemed unthinkable until just a decade ago.

Here, we have focused on many-body Green's function theory, which is arguably the most complex of these formalisms but it has the advantage of providing a unique and global view of the many-particle structure and dynamics. The spectral function is extracted directly from the physics information contained in the one-body Green's function and gives an intuitive understanding of correlations (that is, features that go beyond a simple mean-field description) of the system. Besides, expectation values of observables can be calculated easily, including binding energies. The formalism of SCGF is so vast that even a dedicated monograph would not be able to cover it in full. In this chapter, we have focused on presenting the two most important techniques that are currently used in modern ab initio nuclear theory. In the first case, the Algebraic Diagrammatic Construction method proves to be particularly suited for the study of finite nuclei, but can as well be applied to infinite matter, as was demonstrated in this chapter. In the second case, we looked at how one can solve the Dyson equation directly in momentum space for extended systems. The latter is an important aspect since the formalism allows to construct a fully-dressed propagator at finite temperature, which grants the method to be thermodynamically consistent, preserving all the fundamental laws of conservation. For these cases we also discussed the most relevant steps and knowhow necessary for implementing SCGF calculations. Furthermore, we provided working numerical codes that can solve the same toy models used as examples throughout this book: a four-level pairing Hamiltonian and neutron matter with a Minnesota force. While these applications are simple, the codes we provide already contain the most crucial elements and could be easily extended to real applications (in nuclear physics and other fields too!). We hope this chapter can be the starting point for readers interested in working with many-body Green's functions, starting from the sample codes presented and making use of numerous tips provided for the numerical solutions.

What we did not touch upon, due to lack of space, are the most advanced techniques that have been introduced in recent years or that are still under development. Improving accuracy in calculating open shell isotopes, describing excited spectra, accessing deformed nuclei and describing pairing and superfluidity at finite temperatures are some among the compelling challenges that are to be addressed and that will be crucial to the study of exotic nuclei at future radioactive beam facilities. Likewise, the methods described in this chapter can be extended to novel applications in nuclear physics, besides the structure and reactions with unstable nuclei. Examples are: understanding the response to electroweak probes and the interaction of high energy neutrinos with matter; the spectral function (and hence the individual behavior) of hyperons in finite nuclei and neutron star matter; how thermodynamic properties of nuclear matter impact stellar evolution. With still much room for further development, Fock space methods, and the SCGF approach in particular, are possibly the most promising frontier for advancing first principle computations on large and complex nuclei. All in all, this is an exciting time not only for computational nuclear physics itself but also for the quest of an accurate understanding of nuclear structure and related topics.

Acknowledgements We thank O. Benhar, A. Cipollone, W.H. Dickhoff, C. Drischler, T. Duguet, K. Hebeler, M. Hjorth-Jensen, J.W. Holt, A. Lovato, G. Martínez-Pinedo, H. Müther, P. Navrátil, A. Polls, A. Rios, J. Schirmer, A. Schwenk, V. Somà and D. Van Neck for several fruitful collaborations and enlightening discussions over the years. This work was supported by the UK Science and Technology Facilities Council (STFC) under Grant No. ST/L005743/1, the Deutsche Forschungsgemeinschaft through Grant SFB 1245 and by the Alexander von Humboldt Foundation through a Humboldt Research Fellowship for Postdoctoral Researchers.

Appendix 1: Feynman Rules for the One-Body Propagator and the Self-Energy

We present the Feynman rules associated with the diagrams arising in the perturbative expansion of Eq. (11.14) at zero temperature. The rules are given both in time and energy formulation and a specific example is given at the end. We provide the general rules for p-body propagators. These arise from a trivial generalization of the perturbative expansion of the one-body propagator in Eq. (11.14) [33]. At kth order in perturbation theory, any contribution from the time-ordered product in Eq. (11.14), or its generalization, is represented by a diagram with $2p$ external lines and k interaction lines (called *vertices*), all connected by means of oriented fermion lines. These fermion lines arise from contractions between annihilation and creation operators. In the following we will explicitly include the $\hbar$ factors. Applying the Wick theorem to the terms at each order of the above expansion results in the following Feynman rules. At order k in the perturbation series:

Rule 1: Draw all, topologically distinct and connected diagrams with k vertices, p incoming and p outgoing external lines, using directed arrows. Each vertex representing a n-body interaction must have n incoming and n outgoing lines. For diagrams describing interaction kernels the external lines are not present.

Rule 2: Each oriented fermion line represents a Wick contraction, leading to the unperturbed propagator $\mathrm{i}\hbar g^{(0)}_{\alpha\beta}(t_\alpha - t_\beta)$ [or $\mathrm{i}\hbar g^{(0)}_{\alpha\beta}(\omega_i)$]. In time formulation, the t_α and t_β label the times of the vertices respectively at the end and at the beginning of the line. In energy formulation, ω_i denotes the energy carried by the propagator along its oriented line.

Rule 3: Each fermion line starting from and ending at the *same* vertex is an equal-time propagator and contributes: $-\mathrm{i}\hbar g^{(0)}_{\alpha\beta}(0^-) = \rho^{(0)}_{\alpha\beta}$.

Rule 4: For each one-body, two-body or three-body vertex, write down a factor $\frac{\mathrm{i}}{\hbar}U_{\alpha\beta}$, $-\frac{\mathrm{i}}{\hbar}V_{\alpha\gamma,\beta\delta}$ or $-\frac{\mathrm{i}}{\hbar}W_{\alpha\gamma\xi,\beta\delta\theta}$, respectively. For effective interactions, the factors are $-\frac{\mathrm{i}}{\hbar}\widetilde{U}_{\alpha\beta}$, $-\frac{\mathrm{i}}{\hbar}\widetilde{V}_{\alpha\gamma,\beta\delta}$.

When propagator renormalization is considered, only skeleton diagrams are used in the expansion. In that case, the previous rules apply with the substitution $\mathrm{i}\hbar g^{(0)}_{\alpha\beta} \to \mathrm{i}\hbar g_{\alpha\beta}$. Furthermore, note that Rule 3 generates interaction-reducible diagrams and therefore it is not encountered when working with the effective Hamiltonian (11.16). However, the correlated density matrix $\rho_{\alpha\beta}$ enters the calculations of $\widetilde{U}$ and $\widetilde{V}$ through Eqs. (11.17).

Rule 5: Include a factor $(-1)^L$ where L is the number of closed fermion loops. This sign comes from the odd permutation of operators needed to create a loop. The loops of a single propagator are already accounted for by Rule 3 and must *not* be included in the count for L.

Rule 6: For a diagram representing a $2p$-point Green's function, add a factor $(-\mathrm{i}/\hbar)$, whereas for a $2p$-point interaction kernel without external lines (such as $\Sigma^\star(\omega)$) add a factor $\mathrm{i}\hbar$.

The next two rules require a distinction between the time and the energy representations. In the time representation:

Rule 7: Assign a time to each interaction vertex. All the fermion lines connected to the same vertex i share the same time, t_i.

Rule 8: Sum over all the internal quantum numbers and integrate over all internal times from $-\infty$ to $+\infty$.

Alternatively, in energy representation:

Rule 7': Label each fermion line with an energy ω_i, under the *constraint* that the total incoming energy equals the total outgoing energy at each interaction vertex, $\sum_i \omega_i^{in} = \sum_i \omega_i^{out}$.

Rule 8': Sum over all the internal quantum numbers and integrate over each independent internal energy, with an extra factor $\frac{1}{2\pi}$, i.e. $\int_{-\infty}^{+\infty} \frac{d\omega_i}{2\pi}$.

Each diagram is then multiplied by a combinatorial factor S that originates from the number of equivalent Wick contractions that lead to it. This symmetry factor represents the order of the symmetry group for one specific diagram or, in other words, the order of the permutation group of both open and closed lines, once the vertices are fixed. Its structure, assuming only 2BFs and 3BFs, is the following:

$$S = \frac{1}{k!} \frac{1}{[(2!)^2]^q [(3!)^2]^{k-q}} \binom{k}{q} C = \prod_i S_i \, . \tag{11.101}$$

Here, k represents the order of expansion. q $(k-q)$ denotes the number of two-body (three-body) vertices in the diagram. The binomial factor counts the number of terms in the expansion $(\widehat{V} + \widehat{W})^k$ that have q factors of $\widehat{V}$ and $k-q$ factors of $\widehat{W}$. Finally, C is the overall number of *distinct* contractions and reflects the symmetries of the diagram. Stating general rules to find C is not simple. For example, explicit simple rules valid for the well-known $\lambda\phi^4$ scalar theory are still an object of debate [70]. An explicit calculation for C has to be carried out diagram by diagram [70]. Equation (11.101) can normally be factorized in a product factors S_i, each due to a particular symmetry of the diagram. In the following, we list a series of specific examples which is, by all means, not exhaustive.

Rule 9: For each group of n symmetric lines, or symmetric groups-of-lines as defined below, multiply by a symmetry factor $S_i = \frac{1}{n!}$. The overall symmetry factor of the diagram will be $S = \prod_i S_i$. Possible cases include:

(i) *Equivalent lines.* n equally-oriented fermion lines are said to be equivalent if they start from the same initial vertex and end on the same final vertex.

(ii) *Symmetric and interacting lines.* n equally-oriented fermion lines that start from the same initial vertex and end on the same final vertex, but are linked via an interaction vertex to one or more close fermion line blocks. The factor arises as long as the diagram is *invariant* under the permutation of the two blocks.

(iii) *Equivalent groups of lines*. These are blocks of interacting lines (e.g. series of bubbles) that are equal to each other: they all start from the same initial vertex and end on the same final vertex.

Rule 9(i) is the most well-known case and applies, for instance, to the two second-order diagrams of Fig. 11.4. Diagram (a) in Fig. 11.4 has 2 upward-going equivalent lines and requires a symmetry factor $S_e = \frac{1}{2!}$. In contrast, diagram (b) in Fig. 11.4 has 3 upward-going equivalent lines and 2 downward-going equivalent lines, that give a factor $S_e = \frac{1}{2!3!} = \frac{1}{12}$. For an extended explanation on how to calculate this combinatorial factor and examples for rules 9(ii) and 9(iii) we refer to [33].

As an example of the application of the above Feynman rules, we give here the formulae for diagram (c) in Fig. 11.5. There are two sets of upward-going equivalent lines, which contribute to a symmetry factor $S_e = \frac{1}{2^2}$. Considering the overall factor of Eq. (11.101) and the presence of one closed fermion loop, one finds:

$$\Sigma^{(c)}_{\alpha\beta}(\omega) = -\frac{(i\hbar)^4}{4} \int \frac{d\omega_1}{2\pi} \cdots \int \frac{d\omega_4}{2\pi} \sum_{\substack{\gamma\delta\nu\mu\epsilon\lambda \\ \xi\eta\theta\sigma\tau\chi}} \widetilde{V}_{\alpha\gamma,\delta\nu}\, g_{\delta\mu}(\omega_1)\, g_{\nu\epsilon}(\omega_2)\, W_{\mu\epsilon\lambda,\xi\eta\theta}\, g_{\xi\sigma}(\omega_3)\, g_{\eta\tau}(\omega_4)$$

$$\times\, g_{\theta\gamma}(\omega_1 + \omega_2 - \omega)\, \widetilde{V}_{\sigma\tau,\beta\chi}\, g_{\chi\lambda}(\omega_3 + \omega_4 - \omega)\,. \tag{11.102}$$

Appendix 2: Chiral Next-to-Next-to-Leading Order Three-Nucleon Forces

We report the working equations that result from performing analytically the average of Eq. (11.99) in the specific case of leading order three-nucleon forces, i.e. next-to-next-to-leading order (NNLO), in the chiral effective filed theory expansion [71, 72]. At NNLO we have a two-pion exchange (TPE), one-pion exchange (OPE) and a contact three-nucleon forces (3NF), given respectively by the following expressions:

$$W^{3NF}_{TPE} = \sum_{i\neq j\neq k} \frac{g_A^2}{8F_\pi^4} \frac{(\boldsymbol{\sigma}_i\cdot\mathbf{q}_i)(\boldsymbol{\sigma}_j\cdot\mathbf{q}_j)}{(\mathbf{q}_i^2+M_\pi^2)(\mathbf{q}_j^2+M_\pi^2)} F^{\alpha\beta}_{ijk}\, \tau_i^\alpha \tau_j^\beta\,, \tag{11.103}$$

$$W^{3NF}_{OPE} = -\sum_{i\neq j\neq k} \frac{c_D g_A}{8F_\pi^4 \Lambda_\chi} \frac{\boldsymbol{\sigma}_j\cdot\mathbf{q}_j}{\mathbf{q}_j^2+M_\pi^2} (\boldsymbol{\tau}_i\cdot\boldsymbol{\tau}_j)(\boldsymbol{\sigma}_i\cdot\mathbf{q}_j)\,, \tag{11.104}$$

$$W^{3NF}_{cont} = \sum_{j\neq k} \frac{c_E}{2F_\pi^4\Lambda_\chi} \boldsymbol{\tau}_j\cdot\boldsymbol{\tau}_k\,, \tag{11.105}$$

where the $\mathbf{p}_i$ are the initial and $\mathbf{p}'_i$ are the final single-particle momenta of the ith nucleon ($i = 1, 2, 3$), the $\mathbf{q}_i = \mathbf{p}'_i - \mathbf{p}_i$ are the transferred momenta and $\boldsymbol{\sigma}_i$ and $\boldsymbol{\tau}_i$ are the spin and isospin matrices. The physical constants appearing in these expressions are the axial-vector coupling constant g_A, the average pion mass M_π, the weak pion decay constant F_π and the chiral symmetry breaking constant $\Lambda_\chi \sim$700 MeV. The quantity $F_{ijk}^{\alpha\beta}$ in the TPE contribution (11.103) is

$$F_{ijk}^{\alpha\beta} = \delta^{\alpha\beta}[-4M_\pi^2 c_1 + 2c_3 \mathbf{q}_i \cdot \mathbf{q}_j] + \sum_\gamma c_4 \epsilon^{\alpha\beta\gamma} \tau_k^\gamma \boldsymbol{\sigma}_k \cdot [\mathbf{q}_i \times \mathbf{q}_j]\,. \tag{11.106}$$

The force is regularized with a function that in Jacobi momenta reads:

$$f(\mathbf{p}_1, \mathbf{p}_2, \mathbf{p}_3) = f(p,q) = \exp\left[-\frac{(p^2 + 3q^2/4)}{\Lambda_{3\mathrm{NF}}^2}\right]^n, \tag{11.107}$$

where $\mathbf{p} = (\mathbf{p}_1 - \mathbf{p}_2)/2$ and $\mathbf{q} = 2/3(\mathbf{p}_3 - (\mathbf{p}_1 + \mathbf{p}_2)/2)$ are identified only in this expression as the Jacobi momenta. $\Lambda_{3\mathrm{NF}}$ defines the cutoff value applied to the 3NF in order to obtain a three-body contribution which dies down similarly to the two-body part one. The regulator function is applied both on incoming $(\mathbf{p}, \mathbf{q})$ and outgoing $(\mathbf{p}', \mathbf{q}')$ Jacobi momenta. In present numerical calculations, the approximation of $\mathbf{P} \equiv \mathbf{p}_1 + \mathbf{p}_2 = 0$ is used to facilitate the solution of equations. The averaged terms presented in the following are calculated only for equal relative incoming and outgoing momentum, i.e. $\mathbf{k} = \mathbf{k}'$ with $\mathbf{k} = |\mathbf{p}_1 - \mathbf{p}_2|/2$ and $\mathbf{k}' = |\mathbf{p}_1' - \mathbf{p}_2'|/2$; an extrapolation is then applied to obtain the off-diagonal potential matrix elements [6]. Given these conditions, the regulator on incoming and outgoing momenta can be defined as a function of $f(k, p_3)$.

Symmetric Nuclear Matter Let's start with the isospin-symmetric case of nuclear matter. Evaluating Eq. (11.99) for the TPE term of Eq. (11.103) leads to three contracted in-medium two-body interactions.

TPE-1 The first term is an isovector tensor term, this corresponds to a 1π exchange contribution with an in-medium pion propagator:

$$\widetilde{V}_{\mathrm{TPE}-1}^{3\mathrm{NF}} = \frac{g_A \rho_f}{2F_\pi^4} \frac{(\boldsymbol{\sigma}_1 \cdot \mathbf{q})(\boldsymbol{\sigma}_2 \cdot \mathbf{q})}{[q^2 + M_\pi^2]^2} \boldsymbol{\tau}_1 \cdot \boldsymbol{\tau}_2 [2c_1 M_\pi^2 + c_3 q^2]\,. \tag{11.108}$$

ρ_f defines the integral of the correlated momentum distribution function weighed by the regulator function $f(k, p_3)$

$$\frac{\rho_f}{\nu_d} = \int \frac{\mathrm{d}\mathbf{p}_3}{(2\pi)^3} n(\mathbf{p}_3) f(k, p_3)\,, \tag{11.109}$$

where ν_d is the degeneracy of the system, $\nu_d = 2$ for pure neutron matter and $\nu_d = 4$ in the isospin symmetric case. If the regulator function included in Eq. (11.109) were not dependent on the internal integrated momentum p_3, the integral would reduce to the value of the total density of the system, ρ, divided by the degeneracy and multiplied by an external regulator function.

TPE-2 The second term is also a tensor contribution to the in-medium nucleon-nucleon interaction. It adds up to the previous term. This term includes vertex corrections to the 1π exchange due to the presence of the nuclear medium:

$$\widetilde{V}^{3NF}_{TPE-2} = \frac{g_A^2}{8\pi^2 F_\pi^4} \frac{(\boldsymbol{\sigma}_1 \cdot \mathbf{q})(\boldsymbol{\sigma}_2 \cdot \mathbf{q})}{q^2 + M_\pi^2} \boldsymbol{\tau}_1 \cdot \boldsymbol{\tau}_2$$
$$\times \Big\{ - 4c_1 M_\pi^2 \left[\Gamma_1(k) + \Gamma_0(k)\right] - (c_3 + c_4) \big[q^2(\Gamma_0(k) + 2\Gamma_1(k) + \Gamma_3(k))$$
$$+4\Gamma_2(k)\big] + 4c_4 \mathcal{I}(k)\Big\} . \qquad (11.110)$$

We have introduced the functions $\Gamma_i(k)$ $(i = 0 - 3)$ and $\mathcal{I}(k)$, which are integrals over a single pion propagator:

$$\Gamma_0(k) = \int \frac{d\mathbf{p}_3}{2\pi} n(\mathbf{p}_3) \frac{1}{[\mathbf{k} + \mathbf{p}_3]^2 + M_\pi^2} f(k, p_3) , \qquad (11.111)$$

$$\Gamma_1(k) = \frac{1}{k^2} \int \frac{d\mathbf{p}_3}{2\pi} n(\mathbf{p}_3) \frac{\mathbf{k} \cdot \mathbf{p}_3}{[\mathbf{k} + \mathbf{p}_3]^2 + M_\pi^2} f(k, p_3) , \qquad (11.112)$$

$$\Gamma_2(k) = \frac{1}{2k^2} \int \frac{d\mathbf{p}_3}{2\pi} n(\mathbf{p}_3) \frac{p_3^2 k^2 - (\mathbf{k} \cdot \mathbf{p}_3)^2}{[\mathbf{k} + \mathbf{p}_3]^2 + M_\pi^2} f(k, p_3) , \qquad (11.113)$$

$$\Gamma_3(k) = \frac{1}{2k^4} \int \frac{d\mathbf{p}_3}{2\pi} n(\mathbf{p}_3) \frac{3(\mathbf{k} \cdot \mathbf{p}_3)^2 - p_3^2 k^2}{[\mathbf{k} + \mathbf{p}_3]^2 + M_\pi^2} f(k, p_3) , \qquad (11.114)$$

$$\mathcal{I}(k) = \int \frac{d\mathbf{p}_3}{2\pi} n(\mathbf{p}_3) \frac{[\mathbf{p}_3 \pm \mathbf{k}]^2}{[\mathbf{p}_3 + \mathbf{k}]^2 + M_\pi^2} f(k, p_3) . \qquad (11.115)$$

TPE-3 The last TPE contracted term includes in-medium effects for a 2π exchange two-body term:

$$\widetilde{V}^{3NF}_{TPE-3} = \frac{g_A^2}{16\pi^2 F_\pi^4} \Big\{ - 12c_1 M_\pi^2 \big[2\Gamma_0(k) - G_0(k, q)(2M_\pi^2 + q^2)\big]$$
$$- c_3\big[12\pi^2 \rho_f - 12(2M_\pi^2 + q^2)\Gamma_0(k) - 6q^2 \Gamma_1(k) + 3(2M_\pi^2 + q^2)^2 G_0(k, q)\big]$$
$$+ 4c_4 \boldsymbol{\tau}_1 \cdot \boldsymbol{\tau}_2 \big[(\boldsymbol{\sigma}_1 \cdot \boldsymbol{\sigma}_2)\, q^2 - (\boldsymbol{\sigma}_1 \cdot \mathbf{q})(\boldsymbol{\sigma}_2 \cdot \mathbf{q})\big] G_2(k, q)$$
$$- (3c_3 + c_4 \boldsymbol{\tau}_1 \cdot \boldsymbol{\tau}_2)\, i(\boldsymbol{\sigma}_1 + \boldsymbol{\sigma}_2) \cdot (\mathbf{q} \times \mathbf{k})$$
$$\times\big[2\Gamma_0(k) + 2\Gamma_1(k) - (2M_\pi^2 + q^2)G_0(k, q) + 2G_1(k, q)\big]$$
$$- 12c_1 M_\pi^2\, i(\boldsymbol{\sigma}_1 + \boldsymbol{\sigma}_2) \cdot (\mathbf{q} \times \mathbf{k})\big[G_0(k, q) + 2G_1(k, q)\big]$$
$$+ 4c_4 \boldsymbol{\tau}_1 \cdot \boldsymbol{\tau}_2 \boldsymbol{\sigma}_1 \cdot (\mathbf{q} \times \mathbf{k}) \boldsymbol{\sigma}_2 \cdot (\mathbf{q} \times \mathbf{k})\big[G_0(k, q) + 4G_1(k, q) + 4G_3(k, q)\big]\Big\} .$$
$$(11.116)$$

Here we have introduced the function $G_0(k,q)$, which is an integral over the product of two different pion propagators and defined as follows:

$$G_{0,\star,\star\star}(k,q) = \int \frac{d\mathbf{p}_3}{2\pi} n(\mathbf{p}_3) \frac{\{p_3^0, p_3^2, p_3^4\}}{[[\mathbf{k}+\mathbf{q}+\mathbf{p}_3]^2 + M_\pi^2][[\mathbf{p}_3 + \mathbf{k}]^2 + M_\pi^2]} f(k,p_3)\,, \tag{11.117}$$

where the subscripts 0, $\star$ and $\star\star$ refer respectively to the powers p_3^0, p_3^2 and p_3^4 in the numerator. The functions $G_\star(k,q)$ and $G_{\star\star}(k,q)$ have been introduced to define the remaining functions, $G_1(k,q)$, $G_2(k,q)$ and $G_3(k,q)$:

$$G_1(k,q) = \frac{\Gamma_0(k) - (M_\pi^2 + k^2)G_0(k,q) - G_\star(k,q)}{4k^2 - q^2}\,, \tag{11.118}$$

$$G_{1\star}(k,q) = \frac{3\Gamma_2(k) + k^2\Gamma_3(k) - (M_\pi^2 + k^2)G_\star(k,q) - G_{\star\star}(k,q)}{4k^2 - q^2}\,, \tag{11.119}$$

$$G_2(k,q) = (M_\pi^2 + k^2)G_1(k,q) + G_\star(k,q) + G_{1\star}(k,q)\,, \tag{11.120}$$

$$G_3(k,q) = \frac{\Gamma_1(k)/2 - 2(M_\pi^2 + k^2)G_1(k,q) - 2G_{1\star}(k,q) - G_\star(k,q)}{4k^2 - q^2}\,. \tag{11.121}$$

Note that $G_{1\star}(k,q)$ is needed only to define $G_2(k,q)$ and $G_3(k,q)$.

Integrating Eq. (11.99) for the OPE 3NF term, given in Eq. (11.104), leads to two contributions.

OPE-1 The first one is a tensor contribution which defines a vertex correction to a 1π exchange nucleon-nucleon term. It is proportional to the quantity ρ_f, similar to what was obtained for the TPE 3NF contracted term $\widetilde{V}^{3\text{NF}}_{\text{TPE}-1}$ [see Eq. (11.108)]:

$$\widetilde{V}^{3\text{NF}}_{\text{OPE}-1} = -\frac{c_D\, g_A\, \rho_f}{8\, F_\pi^4\, \Lambda_\chi} \frac{(\boldsymbol{\sigma}_1\cdot\mathbf{q})(\boldsymbol{\sigma}_2\cdot\mathbf{q})}{q^2 + M_\pi^2} (\boldsymbol{\tau}_1\cdot\boldsymbol{\tau}_2)\,. \tag{11.122}$$

As for the $\widetilde{V}^{3\text{NF}}_{\text{TPE}-1}$ term, $\widetilde{V}^{3\text{NF}}_{\text{OPE}-1}$ is an isovector tensor term.

OPE-2 The second term derived from the 3NF OPE defines a vertex correction to the short-range contact nucleon-nucleon interaction. It reads:

$$\begin{aligned}\widetilde{V}^{3\text{NF}}_{\text{OPE}-2} = \frac{c_D g_A}{16\pi^2 F_\pi^4 \Lambda_\chi}\Big\{&\big(\Gamma_0(k) + 2\Gamma_1(k) + \Gamma_3(k)\big)\Big[\boldsymbol{\sigma}_1\cdot\boldsymbol{\sigma}_2\Big(2k^2 - \frac{q^2}{2}\Big)\\ &+ (\boldsymbol{\sigma}_1\cdot\mathbf{q}\,\boldsymbol{\sigma}_2\cdot\mathbf{q})\Big(1 - \frac{2k^2}{q^2}\Big) - \frac{2}{q^2}\boldsymbol{\sigma}_1\cdot(\mathbf{q}\times\mathbf{k})\boldsymbol{\sigma}_2\cdot(\mathbf{q}\times\mathbf{k})\frac{1}{q^2}\Big](\boldsymbol{\tau}_1\cdot\boldsymbol{\tau}_2)\\ &+ 2\Gamma_2(k)(\boldsymbol{\sigma}_1\cdot\boldsymbol{\sigma}_2)(\boldsymbol{\tau}_1\cdot\boldsymbol{\tau}_2) + 6\mathcal{I}(k)\Big\}\,.\end{aligned} \tag{11.123}$$

11.8 *E*xercise: Compute Eq. (11.99) for the contact term given in Eq. (11.105). Demonstrate that it yields a scalar central contribution to the in-medium nucleon-nucleon interaction proportional to ρ_f with formal expression:

$$\widetilde{V}^{3NF}_{\rm cont} = -\frac{3c_E \rho_f}{2F_\pi^4 \Lambda_\chi} \,. \tag{11.124}$$

Pure Neutron Matter In the case of pure neutron matter, the evaluation of Eq. (11.99) is simplified. In fact, the trace over isospin is trivial because pairs of neutrons can only be in total isospin $T = 1$, thus $\boldsymbol{\tau}_1 \cdot \boldsymbol{\tau}_2 = 1$. Consequently the exchange operators reduces only to the momentum and spin part. In operator form it reads:

$$P_{ij} = \frac{1 + \boldsymbol{\sigma}_i \cdot \boldsymbol{\sigma}_j}{2} \,. \tag{11.125}$$

Furthermore it can also be proved that for a non-local regulator, such as Eq. (11.107), the 3NF terms proportional to c_4, c_D and c_E vanish [73, 74]. Therefore the only non zero density-dependent contributions in neutron matter are those containing the low-energy constants c_1 and c_3 in Eq. (11.103). All of their expressions seen from above remain valid except for the change in the trace over isospin indices. It follows that the density-dependent interacting terms obtained in neutron matter will only differ with respect to the symmetric case ones by different prefactors.

In order to obtain the correct degeneracy for neutron matter, i.e. $\nu_d = 2$, we need to replace $\rho_f \to 2\rho_f$ in the $\widetilde{V}^{3NF}_{\rm TPE-1}$ contribution of Eq. (11.108) and the $\widetilde{V}^{3NF}_{\rm TPE-3}$ contribution of Eq. (11.116), [see also Eq. (11.109)]. The isovector tensor terms $\widetilde{V}^{3NF}_{\rm TPE-1}$ and $\widetilde{V}^{3NF}_{\rm TPE-2}$, given in Eqs. (11.108) and (11.110) must then change prefactor according to:

$$\widetilde{V}^{3NF}_{\rm TPE-1} : \boldsymbol{\tau}_1 \cdot \boldsymbol{\tau}_2 \to \frac{1}{2} \boldsymbol{\tau}_1 \cdot \boldsymbol{\tau}_2 \,, \tag{11.126}$$

$$\widetilde{V}^{3NF}_{\rm TPE-2} : \boldsymbol{\tau}_1 \cdot \boldsymbol{\tau}_2 \to \frac{1}{4} (\boldsymbol{\tau}_1 \cdot \boldsymbol{\tau}_2 - 2) \,. \tag{11.127}$$

The isoscalar part of the density-dependent potential appearing in $\widetilde{V}^{3NF}_{\rm TPE-3}$, which contributes to both a central and spin-orbit terms, must change prefactor according to:

$$\widetilde{V}^{3NF}_{\rm TPE-3} : 1 \to \frac{1}{3} \,. \tag{11.128}$$

References

1. V. Somà, A. Cipollone, C. Barbieri, P. Navrátil, T. Duguet, Phys. Rev. C **89**, 061301 (2014)
2. C. Barbieri, J. Phys. Confer. Ser. **529**, 012005 (2014)
3. G. Hagen, T. Papenbrock, M. Hjorth-Jensen, D.J. Dean, Rep. Prog. Phys. **77**, 096302 (2014)
4. T. Frick, H. Müther, Phys. Rev. C **68**, 034310 (2003)
5. G. Baardsen, A. Ekström, G. Hagen, M. Hjorth-Jensen, Phys. Rev. C **88**, 054312 (2013)
6. A. Carbone, A. Rios, A. Polls, Phys. Rev. C **90**, 054322 (2014)
7. D.S. Koltun, Phys. Rev. C **9**, 484 (1974)
8. W. Dickhoff, C. Barbieri, Prog. Part. Nucl. Phys. **52**, 377 (2004)
9. J. Schirmer, Phys. Rev. A **26**, 2395 (1982)
10. J. Schirmer, L.S. Cederbaum, O. Walter, Phys. Rev. A **28**, 1237 (1983)
11. F. Mertins, J. Schirmer, Phys. Rev. A **53**, 2140 (1996)
12. F. Mertins, J. Schirmer, A. Tarantelli, Phys. Rev. A **53**, 2153 (1996)
13. T. Frick, Self-consistent green's functions in nuclear matter at finite temperature. Ph.D. thesis, University of Tübingen (2004)
14. A. Rios, Thermodynamical properties of nuclear matter from a self-consistent green's function approach. Ph.D. thesis, Universitat de Barcelona (2007)
15. V. Van der Sluys, D. Van Neck, M. Waroquier, J. Ryckebusch, Nucl. Phys. A **551**, 210 (1993)
16. V. Somà, T. Duguet, C. Barbieri, Phys. Rev. C **84**, 064317 (2011)
17. A. Idini, F. Barranco, E. Vigezzi, Phys. Rev. C **85**, 014331 (2012)
18. V. Somà, C. Barbieri, T. Duguet, Phys. Rev. C **87**, 011303 (2013)
19. V. Somà, C. Barbieri, T. Duguet, Phys. Rev. C **89**, 024323 (2014)
20. A. Cipollone, C. Barbieri, P. Navrátil, Phys. Rev. C **92**, 014306 (2015)
21. A.L. Fetter, J. Walecka, *Quantum Theory of Many-Particle Systems* (Dover, New York, 2003)
22. W. Dickhoff, D. Van Neck, *Many-Body Theory Exposed!: Propagator Description of Quantum Mechanics in Many-Body Systems*, 2nd edn. (World Scientific, Hackensack, 2008)
23. H. Umezawa, S. Kamefuchi, Prog. Theor. Phys. **6**, 543 (1951)
24. G. Källén, Helv. Phys. Acta **25**, 417 (1952)
25. H. Lehmann, Il Nuovo Cimento (1943–1954) **11**, 342 (1954)
26. C. Mahaux, R. Sartor, Adv. Nucl. Phys. **20**, 1 (1991)
27. F. Capuzzi, C. Mahaux, Ann. Phys. **245**, 147 (1996)
28. L.S. Cederbaum, Ann. Phys. **291**, 169 (2001)
29. G. Baym, L.P. Kadanoff, Phys. Rev. **124**, 287 (1961)
30. G. Baym, Phys. Rev. **127**, 1391 (1962)
31. C. Barbieri, M. Hjorth-Jensen, Phys. Rev. C **79**, 064313 (2009)
32. V.M. Galitskii, A.B. Migdal, Sov. Phys. JETP **7**, 96 (1958)
33. A. Carbone, A. Cipollone, C. Barbieri, A. Rios, A. Polls, Phys. Rev. C **88**, 054326 (2013)
34. A. Cipollone, C. Barbieri, P. Navrátil, Phys. Rev. Lett. **111**, 062501 (2013)
35. R.D. Mattuck, *A Guide to Feynman Diagrams in the Many-Body Problem*, 2nd edn. (Dover, New York, 1992)
36. F. Raimondi, C. Barbieri, (2016). In preparation, arXiv:1701.08127
37. P. Ring, P. Schuck, *The Nuclear Many-Body Problem* (Springer, New York, 1980)
38. W. von Niessen, J. Schirmer, L. Cederbaum, Comput. Phys. Rep. **1**, 57 (1984)
39. P. Danielewicz, P. Schuck, Nucl. Phys. A **567**, 78 (1994)
40. C. Barbieri, W.H. Dickhoff, Phys. Rev. C **63**, 034313 (2001)
41. C. Barbieri, W.H. Dickhoff, Phys. Rev. C **68**, 014311 (2003)
42. C. Barbieri, Phys. Lett. B **643**, 268 (2006)
43. C. Barbieri, D. Van Neck, W.H. Dickhoff, Phys. Rev. A **76**, 052503 (2007)
44. M. Degroote, D. Van Neck, C. Barbieri, Phys. Rev. A **83**, 042517 (2011)
45. C. Barbieri, B.K. Jennings, Phys. Rev. C **72**, 014613 (2005)
46. J. Schirmer, G. Angonoa, J. Chem. Phys. **91**, 1754 (1989)
47. D.R. Thompson, M. Lemere, Y.C. Tang, Nucl. Phys. A **286**, 53 (1977)

48. G.H. Golub, C.F.V. Loan, *Matrix Computations*, 3rd edn. (Johns-Hopkins University Press, Baltimore/London, 1996)
49. A. Rios, V. Somà, Phys. Rev. Lett. **108**, 012501 (2012)
50. J.M. Luttinger, J.C. Ward, Phys. Rev. **118**, 1417 (1960)
51. N.M. Hugenholtz, L. van Hove, Physica **24**, 363 (1958)
52. T. Alm, G. Röpke, A. Schnell, N.H. Kwong, H.S. Köhler, Phys. Rev. C **53**, 2181 (1996)
53. P. Bożek, Phys. Rev. C **65**, 034327 (2002)
54. H. Müther, W.H. Dickhoff, Phys. Rev. C **72**, 054313 (2005)
55. D. Ding, A. Rios, H. Dussan, W.H. Dickhoff, S.J. Witte, A. Carbone, A. Polls, Phys. Rev. C **94**, 025802 (2016)
56. Y. Dewulf, D. Van Neck, M. Waroquier, Phys. Rev. C **65**, 054316 (2002)
57. T. Frick, H. Müther, A. Rios, A. Polls, A. Ramos, Phys. Rev. C **71**, 014313 (2005)
58. A. Rios, A. Polls, A. Ramos, H. Muther, Phys. Rev. C **74**, 054317 (2006)
59. A. Rios, A. Polls, A. Ramos, H. Muther, Phys. Rev. C **78**, 044314 (2008)
60. A. Rios, A. Polls, I. Vidana, Phys. Rev. C **79**, 025802 (2009)
61. V. Somà, P. Bozek, Phys. Rev. C **74**, 045809 (2006)
62. V. Somà, P. Bożek, Phys. Rev. C **78**, 054003 (2008)
63. V. Somà, P. Bożek, Phys. Rev. C **80**, 025803 (2009)
64. V. Somà, Short range correlations in nuclear matter. Ph.D. thesis, Institute of Nuclear Physics, Krakow (2009)
65. A. Carbone, A. Polls, A. Rios, Phys. Rev. C **88**, 044302 (2013)
66. R. Sartor, Phys. Rev. C **54**, 809 (1996)
67. K. Suzuki, R. Okamoto, M. Kohno, S. Nagata, Nucl. Phys. A **665**, 92 (2000)
68. D.J. Thouless, Ann. Phys. **10**, 553 (1960)
69. A. Carbone, Self-consistent green's functions with three-body forces. Ph.D. thesis, University of Barcelona (2014)
70. L. Hue, H. Hung, H. Long, Rep. Math. Phys. **69**, 331 (2012)
71. U. van Kolck, Phys. Rev. C **49**, 2932 (1994)
72. E. Epelbaum, A. Nogga, W. Glöckle, H. Kamada, U.G. Meißner, H. Witała, Phys. Rev. C **66**, 064001 (2002)
73. L. Tolos, B. Friman, A. Schwenk, Nucl. Phys. A **806**, 105 (2008)
74. K. Hebeler, A. Schwenk, Phys. Rev. C **82**, 014314 (2010)

MIX
Papier aus verantwortungsvollen Quellen
Paper from responsible sources
FSC® C105338

If you have any concerns about our products,
you can contact us on
ProductSafety@springernature.com

In case Publisher is established outside the EU,
the EU authorized representative is:
Springer Nature Customer Service Center GmbH
Europaplatz 3, 69115 Heidelberg, Germany

Printed by Libri Plureos GmbH
in Hamburg, Germany